T0189563

Texts in Computer Science

Editors
David Gries
Fred B. Schneider

For further volumes:
www.springer.com/series/3191

Rodney G. Downey · Michael R. Fellows

Fundamentals of Parameterized Complexity

 Springer

Rodney G. Downey
School of Mathematics, Statistics and
    Operations Research
Victoria University
Wellington, New Zealand

Michael R. Fellows
School of Engineering and Information
    Technology
Charles Darwin University
Darwin, Northern Territory, Australia

Series Editors
David Gries
Department of Computer Science
Cornell University
Ithaca, NY, USA

Fred B. Schneider
Department of Computer Science
Cornell University
Ithaca, NY, USA

ISSN 1868-0941   ISSN 1868-095X (electronic)
Texts in Computer Science
ISBN 978-1-4471-7164-5   ISBN 978-1-4471-5559-1 (eBook)
DOI 10.1007/978-1-4471-5559-1
Springer London Heidelberg New York Dordrecht

Downey dedicates this book to his wife Kristin, and Fellows to his wife Frances.

Preface

Parameterized complexity/multivariate complexity algorithmics is an exciting field of modern algorithm design and analysis, with a broad range of theoretical and practical aspects that answers the vital need for efficient algorithms by almost every facet of modern society. The last decade and a half has seen remarkable progress. There are now whole conferences devoted even to subareas of the field. There has been an exhilarating development of techniques where we see an extended discourse with a combinatorial problem via parameters as articulated in the original monograph, Downey and Fellows [247]. We have seen the development of deep techniques for systematically attacking the tractability of problems, as well as techniques for showing that the techniques are nearly optimal. This "yin and yang" of parameterized complexity is a stunning endorsement of the methodology, although many excellent questions remain, not the least of which is why things like SAT-solvers work so well, although the parameterized framework is providing answers through work by Gaspers and Szeider [340] and others.

In the preface to our original book, we described how the project began with some concrete puzzlements that seemed elemental. In particular, the graph minors project of Robertson and Seymour had suggested the notion, central to this field, of fixed-parameter tractability.

At that time, we knew that what is seemingly the most important problem in combinatorial optimization, the VERTEX COVER problem, is fixed-parameter tractable, parameterized in the natural way, by solution size. We knew that the similarly defined graph optimization problems INDEPENDENT SET and DOMINATING SET seemed to resist the power of well-quasiordering. And we knew a reduction carrying the issue of fixed-parameter tractability from the INDEPENDENT SET problem to the DOMINATING SET problem, but not *vice versa*. It seemed like an interesting theory should be possible! We had no clear idea if it would be useful. Fired by a common love of surfing and wine, we embarked!

After legendary adventures, numerous papers, and the sacrificing of many graduate students to what was long considered a cult, especially in North America (a reception to our ideas that we did not anticipate), we amassed enough interesting results to offer the first book, conceived in 1990 and finally published in 1999. The first book was full of bugs (we hope this one has fewer), but that does not seem to have mattered so much, as it had fresh ideas and conceptual approaches, and a nice list of challenges at the end. So does this one.

The field of parameterized/multivariate algorithms and complexity is now firmly, and permanently, a vibrant part of theoretical computer science. There is a thriving conference series, hundreds of papers, two further books,[1] and any number of new fields of applications. The subject is evolving very rapidly as brilliant young authors remake it in their own image. In writing this book we were especially keen to showcase the wealth of recent techniques for proving parameterized tractability and to showcase the powerful new lower bound techniques.

The multivariate perspective has proved useful, even arguably essential, to modern science, starting with bioinformatics as a charter area for applications. Two volumes of the *Computer Journal* [250] essay just some of the areas of applications of multivariate algorithms. As we mention in this book, this work remains hand-in-hand with algorithm engineering.

It has become clear that, far from being an accidental incitement to the central notion of fixed-parameter tractability, deep mathematical structure theory, such as represented by the graph minors project, is a fundamental companion of the central complexity notion.

When this book was being prepared, the first author gave a series of tutorials at *a Coruna* for the conference LATA 2012. Someone in the audience asked whether mastery of this (upcoming) book would suffice to understand complexity issues in their own area of research. The answer is definitely "Yes". Mastering a reasonable fragment of this book will enable a researcher to use the positive and negative toolkits in their own research. The multivariate framework allows an extended dialog with a problem, which we describe in this book.

This book is targeted at the beginning graduate student, and accessible to an advanced senior (i.e. final year undergraduate student). It is also aimed at the general computer scientist, and the mathematically aware scientist seeking tools for their research. We have tried to make the material as self-contained as possible.

For many problems we have given a large number of methods for algorithmic solutions for the *same* problem. Similarly for the same problem we have often given a series of lower bounds based on ever stronger complexity hypotheses, showing how the methods and ideas have evolved over the past 25 years.

We have enjoyed teaching many young researchers to surf, and trotting the ideas all over the globe. We welcome you to enjoy the far-reaching ideas of multivariate algorithmics, the heartbeat of algorithms and complexity.

Wellington, New Zealand   Rodney G. Downey
Darwin, Australia   Michael R. Fellows

---

[1]Namely Flum and Grohe [312] and Niedermeier [546].

Acknowledgements

First thanks to Springer for being enthusiastic about this project, and to Neeldhara Misra for being so persistent with her requests for a new book.

As always the staff at Springer were professional, helpful, and tolerant. Special thanks to Ronan Nugent, Simon Rees, and Rajiv Monsurate.

This project would not have happened without the exceptional support provided by the Marsden Fund of New Zealand, and to the wonderful *Turing Centenary Programme* of 2012 at the Newton Institute where Downey spent the first six months of 2012. Fellows thanks the Australian Research Council, and the Alexander von Humboldt Foundation for providing a marvelous year in Germany as a Humboldt Research Awardee.

Special thanks go to those workers in the parameterized world who provided insight into their work, provided corrections and offered corrections, suggestions, and exercises. There are many, but special thanks go to Hans Bodlaender, Jianer Chen, Andrew Drucker, Martin Grohe, Mike Langston, Daniel Lokshtanov, Stephan Kreutzer, Dániel Marx, Neeldhara Misra, Frances Rosamond, Saket Saurabh, and Dimitrios Thilikos.

We drew on many sources while writing this book, most especially of course on the insights and research of the parameterized/multivariate community. Helpful tracking of recent developments in parameterized complexity is provided by Bart Jansen who maintains a list of *FPT Papers in Conferences*, and a list of *FPT papers on arXiv* on the parameterized complexity wiki www.fpt.wikidot.com. Conferences and seminars, new results, awards, and prizes are announced in the *Parameterized Complexity Newsletter* edited by Frances Rosamond, and archived at the wiki.

We thank Cathy Young and Gabrielle Schubert for their wonderful editorial support. Thanks to Judith Egan, Ralph Bottesch, and Wolfgang Merkle for being master spotters of errors, and thanks to Frances Rosamond for expert help with the galleys.

Introduction

The world is a highly structured place and the data upon which we wish to run algorithms is usually highly structured as a consequence. Virtually all arenas of applications of algorithmics are replete with parameters. These parameters are aspects of size, topology, shape, logical depth and the like and reflect this structure. Multivariate algorithmics/parameterized complexity systematically seeks to understand the contribution of such parameters to the overall complexity of the problem. As this process of deconstruction runs its course, each problem generates its own extended dialog with multivariate algorithmics. The intellectual framework that is multivariate algorithmics has delivered a remarkably rich set of distinctive tools for algorithm design and analysis. It delivers tools for *how to attack and analyze a problem.* More techniques are being developed each day. Amazing progress has been made in the last 20 years of multivariate research. We have even reached the point where we have almost alignment of upper and lower bounds of algorithmics.

The goal of the present book is to give, as best we can, an accessible account of these developments. This gives a snapshot of the current state of the art of multivariate algorithmics and complexity. It is our belief that algorithmics are developed for real applications. Because of this, our fundamental belief is the following.

> **The future of algorithmics is multivariate.**

The reader who is new to this thought, and to the big picture ideas of parameterized complexity, should begin by reading the concrete and hopefully entertaining introduction to the first book [247] from which this book has grown. That introduction, and many surveys, two other books (Niedermeier [546], Flum-Grohe [312]) and two issues of *The Computer Journal* [250] explore the ideas underpinning the area, and for our purposes, we will not deal with motivating the area nor the ranges of parameters or applications in this introduction.

Since this book is a follow up to [247], in this introduction, we turn to the question:

What have we learned since then?

We have learned a lot about the parameterized/multivariate perspective on algorithms and complexity since 1999!

> The multivariate revolution in algorithmics is as much about intellectual workflow as it is about mathematics.

Being a new science, Computer Science is largely misunderstood. High school students often have the impression that it is a nerdy thing, having to do with learning to use various software packages. Why major in something so boring?

Chemists are amazed that Ph.D.'s are awarded in "programming" (think Fortran, even until now). Biologists think that if we have worked out a combinatorial model for the computational problem of deducing gene regulatory networks in the presence of some noise in the primary data, the next step is to "program it", which they regard as "routine".

Pretty well everybody outside of the area of computer science thinks that if your program is running too slowly, what you need is a *faster machine*.

But at the beating heart of Computer Science is *algorithms and complexity*. This book is about that beating heart: how mathematical ideas are deployed to design useful algorithms—for everyone! Google, for example, is essentially an algorithm to find and rank webpages, based on surprisingly sophisticated mathematical ideas.

This book is fundamentally about a multivariate mathematical approach to the beating heart of Computer Science: algorithms and complexity.

Physicists are perhaps our most closely related family of scientists. If you told a physicist that until recently,[1] the beating mathematical heart of Computer Science, the subject of algorithms and complexity, was dominated by a one-dimensional theoretical framework (the one dimension being the number of bits in a valid input file), the most likely response would be, "You must be joking. That cannot possibly be true!"

Yet, it has been so—until the subject of this book, and its predecessor. So, how did that happen?

In Physics, there is a generally healthy dynamic (taking the long view, as one can) between Theoretical Physics and Experimental Physics. In Computer Science there has been nothing analogous.

The multivariate point of view of the beating heart of the subject, algorithms and complexity, has its compass firmly pointed in the direction of an interaction between the Theoretical and Applied branches of Computer Science, much more like our elder intellectual family in Physics, and that direction is emphasized in this book.

So where did the one-dimensional framework in Theoretical Computer Science come from? It is largely an historical accident. In the 1930's (before there were computers) there began a branch of pure mathematics concerned with the study of computable and noncomputable functions called *recursion theory*.

A memorable quote of the first author (Downey) at an important meeting in Leiden in 2010 is:

---

[1] When the first book was published in 1999, there is no question, it *was* dominated by the one-dimensional outlook of P vs. NP.

"Most people wouldn't know a noncomputable function if it hit them on the head."

One of the founders of recursion theory, Anil Nerode, a mentor to the authors since the beginnings of parameterized complexity,[2] can be quoted as follows.

"My perspective is as one who has participated in the development of theoretical computer science for over sixty years. Though of mathematical interest, almost the whole of classical complexity theory ... was of no use whatsoever for the development of new fast-running algorithms and therefore of little interest to code writers of any sort."

In contrast to Physics, Theoretical Computer Science essentially diverged from Practical Computer Science, enforced by a disastrously designed elite conference system, in stark contrast to the research culture in Mathematics, from which content-wise it differs by ε. In this work, we do not have as our goal separating P from NP, but *trying to deliver quality tools to practitioners working in applications*. This is a paradigmatic shift. We are naturally led to fine-grained analysis of problems via their parameters. In contrast to Physics, the young field of Theoretical Computer Science has never had a fundamental intellectual paradigm shift. This book is about such a shift: from *univariate* to *multivariate* complexity analysis and algorithmics.

We have learned that this shift involves more than just the mathematics—generalizing a univariate mathematical framework to a multivariate analog. That is part of the picture, but more importantly, the shift is about the *intellectual workflow*. This turns on the questions:

What are you trying to accomplish? Why are you asking the mathematical questions you are asking? How will this work eventually lead to something useful?

In the classical era, the story (or "business model") was essentially as follows:
- Combinatorially model the computational problem of interest.
- Analyze the computational complexity of the general abstract problem (usual outcome: NP-hardness, non-approximability, etc.).
- If you happen to find a P-time algorithm, publish. Someone else will implement it.

In the emerging multivariate era, the intellectual workflow is envisaged as much richer, largely because of the aspiration and opportunity to really engage mathematical science in algorithm design for real datasets with useful outcomes. There will be more on this further in this Introduction, after some of the most central ideas are introduced. And certainly in the book!

> The multivariate program is amazingly doable mathematically on a routine basis.

---

[2]It was Anil who first suggested, at a meeting in Siena in 2007, that the subject would be more effectively named *multivariate algorithmics* rather than *parameterized complexity* to communicate the central idea—a suggestion we have adopted.

When the authors give introductory survey lectures on the subject of this book, we like to describe how "in the beginning" we knew two things about three basic parameterized decision problems about graphs.

In classical complexity, a decision problem is specified by two items of information:

- The input to the problem.
- The question to be answered.

In parameterized complexity there are three parts of a parameterized problem specification:

- The input to the problem.
- The aspects of the input that constitute the parameter.
- The question to be answered.

It is important to observe that in a parameterized problem specification, the parameter can be an aggregate of information, such as a vector (k, r, t, ε), where k is the solution size, r measures *solution robustness* of some sort, t measures some *structural property* of the input, and ε specifies the goodness of approximation required (just to make up an illustrative example).

It is also important to observe that specifying a parameterized decision problem also specifies a decision problem that can be considered classically—just forget the parameter declaration! But not *vice versa*. Accordingly, a classical decision problem can lead to an unlimited number of different parameterized decision problems, depending on what is specified to be the parameter.

These two observations turn out to be highly important in how the field of multivariate algorithmics eventually delivers useful results.

But in the beginnings of this research area, the authors only knew two things about three basic parameterized problems about graphs, with quite simple parameters: in each case, the parameter is just the *size of the solution*, a quite natural initial parameterization for many NP-hard problems.

The three basic problems refereed to are: VERTEX COVER, INDEPENDENT SET, and DOMINATING SET. We define them collectively as follows.

Input: A graph $G = (V, E)$ and a positive integer k.
Parameter: A positive integer k.
Question: Is there a set of vertices $V' \subseteq V$ of size k that is (respectively), a *vertex cover*, an *independent set*, or a *dominating set*?

A *vertex cover* is a set of vertices V' such that every edge in G has at least one endpoint in V'. An *independent set* is a set of vertices V' such that no two vertices in V' are adjacent in G. A *dominating set* is a set V' of vertices such that every vertex in G has at least one neighbor in V'. These three properties are illustrated in Fig. 1.

All three problems are obviously solvable by the "try all k subsets" brute-force algorithm that requires time $O(n^{O(k)})$. XP is the class of parameterized problems that are solvable in time $O(n^{g(k)})$ for some function g.

Fig. 1 Three vertex set problems

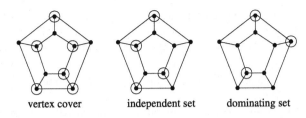

vertex cover   independent set   dominating set

Table 1 The ratio $\frac{n^{k+1}}{2^k n}$ for various values of n and k

	$n = 50$	$n = 100$	$n = 150$
$k = 2$	625	2,500	5,625
$k = 3$	15,625	125,000	421,875
$k = 5$	30,625 [247]	6,250,000	31,640,625
$k = 10$	1.0×10^{12} [247]	0.8×10^{14} [247]	3.7×10^{16}
$k = 20$	1.8×10^{26}	0.5×10^{31} [247]	2.1×10^{35}

The two things we knew about these three simple graph problems:

- We knew an algorithm for VERTEX COVER with running time $O(2^k n)$, but no algorithms for INDEPENDENT SET or DOMINATING SET with running times of the form $O(f(k)n^c)$ for some function f, and constant c. Are such algorithms possible? We did not know, but this was the issue we were focused on. Parameterized problems having algorithms with such running times are termed *fixed-parameter tractable* and constitute the class of parameterized problems FPT.

- We knew a somewhat complicated polynomial-time combinatorial reduction (just as in the theory of NP-completeness), taking (G, k) to $(G', 2k + 1)$ such that G' has a dominating set of size $2k + 1$ if and only if G has an independent set of size k. So we knew a reduction from INDEPENDENT SET to DOMINATING SET, showing that if DOMINATING SET is in FPT, then so is INDEPENDENT SET. We did not know a relevant reduction in the opposite direction.

In the beginning, it seemed to us that from this small seed, a theory might grow. It did. It is not hard to go from these concrete examples to the relevant definitions to frame the theory. The complexity issue under investigation (FPT *versus* XP) is compelling, as shown in Table 1.

So we had the definitions, and the motivation, and we embarked! But what we did not know was whether we would find any more interesting reductions, or if our theory might wither by being too technically difficult to carry out on a routine, daily basis, by ordinary mathematical workers in Computer Science.

What we have learned is that the multivariate algorithmics program is *amazingly doable*—*far* beyond anything we ever expected or *hoped*!

Not only are there now thousands of relevant parameterized reductions—rivaling the theory of NP-completeness in terms of do-ability, but there are *kinds of information* about the parameterized complexity of concrete parameterized problems that turn out to be routinely accessible, far beyond any dreaming when we began.

Some concrete examples:

(1) We know that the "general shape" of our FPT algorithm mentioned above for VERTEX COVER is "qualitatively optimal"—there can be no $O(2^{o(k)}n^c)$ (... for example, no $O(2^{\sqrt{k}}n^c)$ ...) algorithm for VERTEX COVER unless the ETH complexity hypothesis fails. This has become a program known as *FPT Optimality*.

(2) We know that the "general shape" of our XP algorithm mentioned above for INDEPENDENT SET and DOMINATING SET is similarly "qualitatively optimal"— there can be no $O(n^{o(k)})$ algorithms for these problems, unless ETH fails. This has become a program known as *XP Optimality*.

(3) We know a polynomial-time kernelization algorithm for the FPT FEEDBACK VERTEX SET problem where the reduced instances have $O(n^2)$ vertices, and we know that there can be no polynomial-time kernelization algorithm for the FEEDBACK VERTEX SET problem where the reduced instances have total size $O(n^{2-\varepsilon})$, for any $\varepsilon > 0$, unless the classical Polynomial Hierarchy collapses to the third level.

Similar results are now pouring in for hundreds of concrete parameterized problems. And the mathematical technology appears to be routinely doable.

The authors are frankly staggered that such fine-grained information about the parameterized complexity of concrete problems is routinely mathematically accessible!

> FPT is about the possibility of deep and useful mathematical structure theory.

Many of the motivations and rough ideas for the development of parameterized complexity originated in early work on algorithmic aspects and consequences of graph minor theory, a story well-told in a recent survey by Langston [479].

An important thing we have learned since 1999, is that this motivational origin was no accident whatsoever.

Lovasz's Thesis In the early days of Theoretical Computer Science, L. Lovasz made the observation that "nice combinatorial structure theory" (for example, concerning a property of graphs) seemed to go hand-in-hand with (classical) polynomial-time computational complexity.

Independently, Jack Edmonds made similar observations, offering something slightly more general:

Edmond's Thesis The possibility of elegant combinatorial structure theory goes hand-in-hand with membership in NP intersect coNP.

The thing we have learned, we call:

Langston's Thesis The possibility of deep and algorithmically useful combinatorial structure theory goes hand-in-hand with FPT.

It is no longer the case that there is just the one example of graph minor theory in support of Langston's Thesis. With new compasses, the boats have left the shore!

There seems to be more FPT than we expected.

One thing we have learned that we continue to find surprising, is that there seems to be a lot more FPT "out there" than we imagined in 1999, or 2007, or even two years ago! This may be due to a psychological mirage where technical difficulties become interpreted as likely impossibilities. FPT seems to be nearly as common (though sometimes as mathematically hard to get to) as earth-like planets!

But there is hardness out there as well. We seem to need better mathematically intuitive means for differentiating between "mathematical hardness" and "current frustration".

The importance of a mind meld with algorithms engineers and collaboration with researchers who have real datasets.

An important lesson we have learned since 1999 is that we can and should be much more aggressive about parameterization, in the quest for real applications. In the "Era of the First Book", problem parameterization tended to be quite modest and simple (much as one might pick out a single natural parameter, by perusing the Garey and Johnson catalog [337]).

What we have learned is that the multivariate mathematical perspective allows for, and will serve the world better, if it more aggressively and bravely exploits its ability to set up a sustained dialog with NP-hard problems.

There are two key complementary principles that codify this lesson:

(1) *If your problem is* FPT, *enrich the model, in the interests of a better grip on real applications.*

(2) *If your problem is* $M[1]$- *or* $W[1]$-*hard, enrich the parameterization.*

For an example illustrating (1), GRAPH COLORING (which famously models *scheduling*) is FPT when the parameter is the treewidth of the instance graph. But for applications of graph coloring to scheduling problems, GRAPH LIST COLORING is a much more realistic model (because Bob cannot go to any meetings on Thursday, etc.). GRAPH LIST COLORING, parameterized by the treewidth of the instance graph, turns out to be hard for $W[1]$.

For an example illustrating (2), and continuing the story ... the $W[1]$-hardness proof exploits that the lists can be arbitrary, and makes a construction where most people are available for a meeting only two different hours (two different colors on their lists) in the next year. In terms of the proposed application, this is ridiculous. If we escape the silliness of the hardness proof by aggregately parameterizing on the treewidth of the instance graph, together with the average *deficiency* of the lists—the average number of colors (time blocks) *not* on a vertex's list, then we may go back to FPT—having then perhaps an FPT algorithm with a much better case to serve real applications.

In order to pursue such projects, we need to engage with algorithms engineers and researchers with real datasets, who can tell us: (1) what further enrichment of the model is relevant, and (2) why the hardness proofs make no sense for real datasets.

The importance of Scientific Sociology.

The classical complexity concept of NP-completeness has thoroughly penetrated modern science and scholarship. As a rough measure, the rightly famous book of Garey and Johnson [337] has been cited more than 50,000 times (according to Google Scholar).

Everywhere that NP-completeness has gone, multivariate algorithmics should follow, usefully! The first book [247] has been cited about 3,000 times—so there is still a long way to go.

But how do you get there, to deliver these useful new tools for algorithm design, relevant to just about everybody?

We have learned that this is primarily a people-problem, not a publication-problem, nor a curriculum-problem.

We have learned that it is *not enough* to send a nice paper exploring the parameterized complexity of fundamental (NP-hard) problems of specialized area X, to the annual specialized conference or workshop series concerning X. There are hundreds of such X, where algorithms and complexity are a relevant, even central issue, that have not yet been turned on to the multivariate point of view.

To move the word along, what really is necessary is to:

- Seriously collaborate and publish with a leader of the targeted application area.
- Make a commitment to the area, and to moving the *interdisciplinary* collaboration forward, on a big scale.

As Langston pointed out at a memorable talk at APAC 2012 (easily located on the Web), this means you will be publishing at meetings you have *never heard about before*, but you will have fun, learn new things, profit with research funding, and do good.

This we have learned: the outward-bound potential of multivariate algorithmics is largely a person-to-person and community formation problem.

Connections with classical complexity are possible, but not important.

In the first book, we made much of the quest to find parameterized analogs of classical complexity results (such as Ladner's Density Theorem), and to somehow find connections to classical complexity theory.

What we have learned is that this is just not important! There are so many forward opportunities! Connections with the old and original one-dimensional complexity theory have turned up, as might happen. It's not important.

The productivity of multiple views of FPT.

It was several years after our first paper on the subject that we realized that FPT parameterized problem complexity, could be completely equivalently viewed in terms of polynomial-time kernelization [247]. The characterization is as follows:

Lemma 1 *A parameterized problem Π is FPT if and only if there is a polynomial transformation from Π to itself, taking (x, k) to (x', k') such that:*
1. *(x, k) is a* YES-*instance if and only if (x', k') is a* YES-*instance.*
2. *$k' \leq k$*
3. *$|x'| \leq g(k)$ for some function g*

This lemma can be found in Chap. 5 of the first book [247], where the matter is treated as a structural complexity theory abstraction. Apparently, we both forgot about this, and rediscovered the result with the help of Neal Koblitz on a pleasant drive to Sombrio. The resulting, stripped-down simple proof was published in [247], in the context of more concrete parameterized problem research intentions. This all came to a focus around 2006 (see Fellows [285]) when we began programmatically to ask the question:

What parameterized problems in FPT admit P-time many : 1 kernelization algorithms to reduced instances of size bounded by a polynomial function of the parameter?

We have learned that this is a profoundly productive question, admitting a deep mathematical theory, and with strong connections to pre-processing, a very general heuristic algorithmic strategy for NP-hard combinatorial optimization problems. A key idea of [247] is the following:

For the first time in the history of computing, we have a mathematically disciplined theory of pre-processing.

That seems a big deal. Certainly the associated multivariate complexity theory aspects are flourishing.

Are there other equivalent ways of viewing FPT? One that has recently appeared on the horizon establishes that FPT is equivalent to "nice well-quasi-ordering with nice (FPT) order testing"—which recalls us to the origins of the entire project.

The lesson remains: FPT can be viewed in a variety of ways.

The W-hierarchy is not of essential importance.

The reader will discover in Chap. 21 that parameterized complexity offers a profound analog of Cook's Theorem, centered on the parameterized problem complexity class $W[1]$. This has some mathematical interest. A lot about the W-hierarchy has been carried along from the first book to this one, but our perceived lesson since 1999, is that the W-hierarchy is a much less important mathematical construct than we thought then.

The fine-grained lower bounds methods that the field of multivariate algorithmics has now, are largely based on variations on the ETH, and we find this reasonable and modern. If there is an even better plausibly fundamental intractability hypothesis with respect to which observed concrete problem complexities can be explained and unified even better—bring it on! Parameterized complexity is not centrally about the W-hierarchy, machine models, circuit models, or what was fashionable to think about in what was called Theoretical Computer Science in the 1980s—it is about FPT or the lack of it.

> The multivariate toolkit really *is* useful.

As we remarked at the end of the Introduction to the first book, there were then a number of mathematically interesting FPT methods, "... for which practicality is still not yet clearly established ... " (such as color-coding and bounded treewidth).

What we have learned is that, with development, and with algorithms engineering, most FPT methods *really are useful*. Increasingly, multivariate algorithmics is having significant practical impact in many application domains. More such developments are on the horizon, as, for example, via the interaction with the design of heuristics.

This book is divided into three main sections. The first gives the reader access to many of the rich variety of standard algorithmic techniques available to establish parametric tractability. The second section is devoted to the classical hardness classes of the area. The last section is devoted to exploring various limitations and relaxations of the technology. Questions explored are those like *What is the limitation of this technique? What is the best running time I might ever get for this problem? What do I get if I use approximation or randomization?* The material often treats the same problem many times, with a view towards understanding how each technique yields new insight into how to attack the problem.

The key missing ingredient from this already long book, is how to actually implement the techniques for large datasets (load balancing, data structures, incrementalization, etc.). This is an area of significant current research, but beyond the expertise of the authors.

The field has developed enormously and our perspective has changed a lot since 1999. But the last line of the first Introduction remains true:

Much remains to be explored!

Contents

Part III Techniques Based on Graph Structure

Part I
Parameterized Tractability

Preliminaries

1.1 Basic Classical Complexity

In this chapter, we will give a basic primer in the concepts of classical complexity. The treatment is brief and we expect this chapter to be more of a reminder.

We will fix an alphabet, which will usually be taken as $\Sigma = \{0, 1\}$, unless we specifically talk about *unary* alphabets such as $\Sigma = \{1\}$. The strings obtainable from Σ will be denoted by Σ^*. A subset $L \subseteq \Sigma^*$ is called a *language*.

We will fix on some notion of a computational device, such as a Turing machine, and will use the ideas of classical computability theory. A function is called *computable* if it can be implemented on the relevant computation device. By counting the steps in the computation or the amount of memory used, we can define notions of the time or space requirements of a computation. The reader should recall that a Turing machine is called *deterministic* if and only if for all configurations, there is at most one next configuration. If we allow a Turing machine to potentially have more than one possible next move, then it is called *non-deterministic*.

Complexity theory is concerned with the amount of time and space resources required for computations. If the resource is *time*, we simply count the number of steps in the performance of a computation. In the case of *space* we will consider the work space needed for the computation. In the case of space, with the Turing machine model, this can be taken as the number of squares used in an additional work tape. For the purposes of this book, we will assume that all reasonable mathematical models of a computing device are polynomially related, in the sense that for devices M_1, M_2, if M_1 requires time (or space) $f(n)$ for an input of *length* n, then M_2 will require time (resp. space) bounded by $f(n)^c$ for some fixed constant c.

For simplicity, we will assume that if a computation takes as input a string of symbols of length at least one, then the computation makes at least one step. The following are classical definitions.

Definition 1.1.1 Let $f : \mathbb{N} \to \mathbb{N}$. We say that a language L is in $\mathrm{DTIME}(f)$ if and only if there is some deterministic Turing machine M such that for all $\sigma \in \Sigma^*$,

$$\sigma \in L \quad \text{if and only if} \quad M(\sigma) = 1,$$

R.G. Downey, M.R. Fellows, *Fundamentals of Parameterized Complexity*,
Texts in Computer Science, DOI 10.1007/978-1-4471-5559-1_1,
© Springer-Verlag London 2013

and for all x, $M(x)$ runs in time $O(f(|x|))$.

Similarly, we can define $\text{NTIME}(f)$ using a non-deterministic Turing machine accepting L, and $\text{DSPACE}(f)$, etc. Two famous complexity classes of interest are

$$P = \bigcup_{c \in \mathbb{N}} \text{DTIME}(n^c) \quad \text{and}$$

$$NP = \bigcup_{c \in \mathbb{N}} \text{NTIME}(n^c).$$

The most famous question in computer science is whether the two classes coincide.

Other solution complexity classes that will be of relevance to us, are LOGSPACE and NLOGSPACE, and time classes $E = \bigcup_{c \in \mathbb{N}} \text{DTIME}(2^{cn})$ and $\text{EXP} = \bigcup_{c \in \mathbb{N}} (2^{n^c})$. LOGSPACE and NLOGSPACE allow us to have counters. For example, the problem of determining whether two points are path connected in a graph is in NLOGSPACE.

We also need a sense of *completeness* with respect to complexity classes. This notion came to computer science from computability theory, through the early work of Church, Turing, Post, and others. In order to measure the relative complexity of problems coded by languages, we investigate the relation between languages denoted $L \le \hat{L}$, defined: if we had some process to solve $\hat{L}$ (that is, if $\hat{L}$ is computable), then we could use this to solve L. In classical computational complexity, the two key ideas are polynomial-time many-one (or Karp) reducibility $\le_m^P$ and polynomial-time Turing (or Cook) reducibility $\le_T^P$. The reader should recall that $A \le_m^P B$ if and only if there is a polynomial-time computable function g such that for all x, $x \in A$ if and only if $g(x) \in B$. Similarly $A \le_T^P B$ if and only if there is a polynomial-time Turing procedure M such that $x \in A$ if and only if $M^B(x) = 1$.

For a complexity class $\mathcal{C}$, a language L is *hard* for $\mathcal{C}$ if and only if every language $\hat{L} \in \mathcal{C}$ reduces to L. In common usage, "reduces" here typically means $\le_m^P$, and we will adopt this convention unless otherwise stated. In space, sometimes we will use logspace many-one reductions. We will say that L is *complete* for $\mathcal{C}$ if it is hard for $\mathcal{C}$ and $L \in \mathcal{C}$.

The foundational result for computational complexity theory is that satisfaction for propositional calculus (or in Levin's case tiling) can be used to efficiently simulate a non-deterministic Turing machine computation.

Theorem 1.1.1 (Cook [163], Levin [488]) CNFSAT, *the problem of determining if a formula of propositional calculus consisting of clauses has a satisfying truth assignment, is NP-complete.*

Proof (sketch) We sketch the proof, and we will later refer back to it.

Now, A is accepted by some NTM M. We will describe a function σ that will generate a formula $\varphi = \sigma(x)$ such that $x \in A$ iff φ is satisfiable.

The possible executions of M on input x form a branching tree of configurations, where each configuration gives information on the current instantaneous state of the computation. It includes all relevant information, such as: tape contents, read head position, current state, etc. M is polynomially time bounded, so we can assume the depth of this tree is at most $N = |x|^k$ for some fixed k (not dependent on x).

A valid computation sequence of length N can use no more than N tape cells, since at the very worst m moves right one tape cell in each step. Thus there are at most N time units and N tape cells to consider.

We will encode computations of M on input x as truth assignments to various arrays of Boolean variables, which describe things like where the read head is at time i, which symbol is occupying cell j at time i, etc. We will write down clauses involving these variables that will describe legal moves of the machine and unique legal starting and accepting configurations of M on x.

A truth assignment will simultaneously satisfy all these clauses if and only if it describes a valid computation sequence of M on input x. So we take $\varphi = \sigma(x)$ to be the conjunction of all these clauses. Then the satisfying truth assignments of φ correspond one-to-one to the accepting computations of M on x.

We generate the propositional variables:

- $Q[i, k]$, $0 \leq i \leq N$ "at time i, M is in state q_k";
- $H[i, j]$, $0 \leq i, j \leq N$ "at time i, M is scanning tape cell j";
- $S[i, j, k]$, $0 \leq i, j \leq N$ "at time i, tape cell j contains symbol s_k".

We have six restrictions on M, from which we construct six clauses groups.

- G_1: at each time i, M is in exactly one state

$$\bigwedge_{0 \leq i \leq N} \left[(Q[i, 0] \vee \cdots \vee Q[i, r]) \wedge \bigwedge_{j_1 \neq j_2} (\neg Q[i, j] \vee \neg Q[i, j_2]) \right].$$

The others are similar and left to the reader:

- G_2: at each time i, M is scanning exactly one square.
- G_3: at each time i, each square contains exactly one symbol.
- G_4: at time 0, computation is in the initial configuration of reading input $x = x_0 x_1 \ldots x_{n-1}$.
- G_5: by time N, M has entered state q_y and accepted x $Q[N, y]$.
- G_6: for each i, configuration of M at time $i + 1$ follows by a single application of transition function from the configuration at time i, i.e., the action is faithful.

Then the final formula φ is simply the conjunction of the G_i. It is clear that M has a satisfying assignment if and only if φ is satisfiable. □

We also know that 3SAT is NP-complete by a reduction: given a clause c of size m, like $(x_1 + \overline{x}_3 + x_5 + \overline{x}_7 + x_8)$, we replace it by introducing new variables z_i^c, and then we use this conjunction of clauses $(x_1 + \overline{x}_3 + z_1^c)(\overline{z}_1^c + x_5 + z_2^c)(\overline{z}_2^c + \overline{x}_7 + x_8)$. The new formula φ' so obtained is satisfiable if and only if the original one φ was.

Theorem 1.1.2 (Karp [431]) 3SAT *is NP-complete*.

A basic course in complexity theory will identify a number of NP-complete problems such as CLIQUE, INDEPENDENT SET, HAMILTON CYCLE, LONG PATH, VERTEX COVER, DOMINATING SET etc. Soon, we will be studying parameterized versions of these problems, so we will not define them here.

We will find it is possible to extend ideas of NP and P to higher levels. The guess-and-check paradigm for NP views it as a search for a certificate y of length $\leq |x|^c$ such that $Q(x, y)$ holds, where Q is a polynomial-time predicate. An example would be that x is an instance of SATISFIABILITY and y is a satisfying assignment, but more generally $Q(x, y)$ simply encodes the computation path y in the action of the polynomial-time non-deterministic Turing machine Q. This means that we can represent any language $L \in$ NP in Σ_1^p-form.

Proposition 1.1.1 $L \in$ NP *if and only if* $L \in \Sigma_1^p$ *where this means there is a polynomial-time relation Q and a constant $c \in \mathbb{B}$ with $x \in L \Leftrightarrow \exists y[|y| \leq |x|^c \wedge Q(x, y)]$.*

Notice that we can have any polynomially bounded number of $\exists^p x_1 \exists^p x_2 \ldots$ and still stay in Σ_1^p. The complement of an NP language is called co-NP, and that similarly corresponds to what is called a Π_1^p language, where now the quantifier is a universal polynomial-time bounded search $x \in L$ if and only if $\forall y[|y| \leq |x|^c \rightarrow Q(x, y)]$. However, allowing alternations of the quantifiers cannot (apparently) be coalesced, and this leads to the *polynomial-time hierarchy*, PH. In this, Σ_n^p has n-alternations of polynomially bounded quantifiers beginning with a $\exists^p$, and similarly Π_n^p. Then we would define PH$= \bigcup_n \Sigma_n^p = \bigcup_n \Pi_n^p$. It is not hard to use an analogue of the proof of Cook's Theorem to establish that there is an analogue of SATISFIABILITY for Σ_n^p. For example, the language which consists of propositional formulas $\varphi(y_1, \ldots, y_n, z_1, \ldots, z_m)$, with variables partitioned into two sets given by the y_i and z_j, will be Σ_2^p complete when we ask if there is an assignment for the literals y_i which satisfies every assignment for the z_j. We would write this as $\exists^p \overline{y} \forall^p \overline{z} \varphi(\overline{y}, \overline{z})$. When the fact we were using tuples was obvious or irrelevant, we would likely write y for $\overline{y}$. Since polynomial space is big enough to simulate all of the relevant searches, we have the following.

Proposition 1.1.2 (Folklore) PH $\subseteq$ PSPACE.

Less obvious, but also true, is that the "diagonal" of PH is complete for PSPACE. That is, we can define QBFSAT as the collection of codes for true quantified Boolean formulas. That is $\exists^p x_1 \forall^p x_2 \ldots \varphi(x_1, x_2, \ldots, x_n)$. The difference here is that PH is only concerned with $\bigcup_n \Sigma_n^p$, the union of *fixed levels*, whereas in the case of QBFSAT we are concerned with all the levels at once. The following result is not altogether obvious.

Theorem 1.1.3 (Stockmeyer and Meyer [631]) QBFSAT *is* PSPACE-*complete*.

Proof (sketch) Fix a PSPACE bounded machine M with bound n^c. The proof uses the formula access$_j(\alpha, \beta)$ where α and β are free variables representing M-configurations, and the formula asserts that we can produce β from α in $\leq 2^j$ steps.

The key idea is that the formula is re-usable: for $j = 1$, $access_j$ can be read off from the transition diagram, and for $j > 1$, you can express $access_j$ as

$$\exists \gamma \left(\text{config}_M(\gamma) \wedge \left(\forall \delta \forall \lambda \left[(\delta = \alpha \wedge \lambda = \gamma) \vee (\delta = \gamma \wedge \lambda = \beta) \right] \right. \right.$$
$$\left. \left. \rightarrow access_{j-1}(\delta, \lambda) \right) \right).$$

Here $\text{config}_M(\gamma)$ asserts that γ is a configuration of M. The formula asserts that there is an accepting β such that on input α $access_{n^c}(\alpha, \beta)$. □

It is possible to identify a number of natural problems complete for PSPACE. Typically they are inspired by the above characterization of PSPACE. Quantifier alternations naturally correspond to strategies in games. Player A has a winning strategy, by definition, if there is first move that A can make, such that for every possible second move by Player B, there exists a third move in the game that A can make, such that for every fourth move that B can make ... A wins. For example, for the natural analog of the game GO, determining who has a winning strategy is PSPACE complete. We will return to this and to combinatorial games when we look at the parameterized version of space. We leave further discussion of concrete examples till that section.

A good general reference to classical complexity theory is Garey and Johnson's classic [337] (*so good*, in our opinion, that it should be on everyone's bookshelf—we will assume it is on *yours*—it is of historical importance to the story we wish to tell here). A more recent reference for the basics of classical computational complexity theory is Kozen [465].

1.2   Advice Classes

Further along in this book, when we come to considering the astonishing new lower bound methods in the study of kernelization for FPT parameterized problems, we will need to make reference to some heretofore relatively obscure classical complexity classes, which the reader might likely not have met.

Definition 1.2.1 (Karp and Lipton [432])
1. An *advice* function $f : \Sigma^* \rightarrow \Sigma^*$ has the property that if $|x| = |y|$ then $f(x) = f(y)$, and hence for a length we can write unambiguously $f(n)$ for $f(x)$ if $|x| = n$. The advice is *polynomial* if there is a constant c such that $|f(n)| < n^c$. That is, the advice has polynomially bounded length.
2. For a language L and complexity class $\mathcal{C}$ we say that $L \in \mathcal{C}/\text{Poly}$ if there is a polynomial advice function, f, and a language $\hat{L} \in \mathcal{C}$ such that for all x,

$$x \in L \quad \text{if and only if} \quad \langle x, f(|x|) \rangle \in \hat{L}.$$

In particular, P/Poly is the collection of languages which are in P with nonuniform advice. This may seem a strange concept at first glance, since for any language L, the language $S_L = \{1^n : n \in L\}$ is in P/Poly, as the only thing of length n

that could *possibly* be in S_L is a string of the form 1^n; and that is the advice. Hence there will be uncountably many languages in P/Poly.

One of the great results of classical structural complexity theory is that P/Poly is exactly the languages L that can be decided by a (nonuniform) family of circuits C_n, one for each input size n, where $|C_n| \le p(n)$ for some polynomial p (Pippenger [561]).

In this section we will define an *advice* hierarchy which is an advice analog of the polynomial hierarchy. We will see that a collapse of an advice hierarchy implies a related collapse in the polynomial-time hierarchy.

To begin this analysis, we define the advice hierarchy by appending "/Poly" to the end of everything. We show that a collapse of some level of PH/Poly propagates upwards. For ease of notation, we will drop the superscript "p" from, for instance, Σ_i^p, and write Σ_i, when the context is clear that we are concerning ourselves with the *polynomial-time hierarchy* or its analogues.

Lemma 1.2.1 (Yap [670]) *For all $i \ge 1$, $\Pi_i \subseteq \Sigma_i/\text{Poly}$ implies $\Sigma_{i+1}/\text{Poly} = \Sigma_i/\text{Poly}$.*

Proof First it is easy to see that for any complexity class $\mathcal{C}$, (co-$\mathcal{C}$)/Poly is the same as co-($\mathcal{C}$/Poly). It is enough to prove this for $i = 1$ and the rest works entirely analogously. Suppose that $\Pi_1 \subseteq \Sigma_1/\text{Poly}$ and let $L \in \Sigma_2/\text{Poly}$. Since $L \in \Sigma_2/\text{Poly}$, we have some advice function f, and $Q \in \Sigma_2$ with $x \in L$ iff $\langle x, f(x) \rangle \in Q$, and hence for some polynomial-time computable R, $x \in L$ if and only if $\exists y [\forall z R(\langle x, f(x) \rangle, y, z)]$. Because $\Pi_1 \subseteq \Sigma_1/\text{Poly}$, we can replace the Π_1 relation in the brackets [,] by a Σ_1/Poly one. We can foresee that for each y, there are polynomial-time relations R_y' and advice function g such that $x \in L$ if and only if $\exists y \exists y' (R_y'(\langle x, f(x) \rangle, y, \langle y, g(x, f(x)), y \rangle))$. To finish we need to amalgamate all the advice so that the advice will depend only upon x. However, all the advice depends only on x and y and the range of these is $\le (|x| + |y|)^c = |x|^{c'}$ for some c, c'. As it is only the lengths, we can concatenate the advice together in a self-delimiting way, and only have a polynomial amount of advice that depends on $|x|^{c'}$. Thus we can construct a polynomial-time relation R'' and an advice function h, such that $x \in L$ if and only if $\exists y R''(\langle x, h(x) \rangle, y)$, as required. $\square$

Corollary 1.2.1 (Yap [670]) *The following are equivalent.*
1. $\Pi_i \subseteq \Sigma_i/\text{Poly}$.
2. $\Sigma_i \subseteq \Pi_i/\text{Poly}$.
3. $\text{PH} = \Pi_i/\text{Poly} = \Sigma_i/\text{Poly}$.

The result we will need in Chap. 30 is the following. It is a translation of an advice collapse to a PH collapse.

Theorem 1.2.1 (Yap [670]) *For $i \geq 1$,*

$$\Sigma_i \subseteq \Pi/\text{Poly} \quad implies \quad \Sigma_{i+2} = \text{PH}.$$

Proof We show that $\Sigma_3 \subseteq \Pi_3$, assuming $\text{NP} \subseteq \text{CO-NP/Poly}$. The general case is analogous. Let $L \in \Sigma_3$ so that $x \in L$ if and only if $\exists y \forall z R(x, y, z) \in \text{SAT}$ for some poly relation R, since SAT is NP-complete. Now the assumption is that $\text{NP} \subseteq \text{CO-NP/Poly}$. This gives us an advice function α and a poly relation C such that

$$\psi \in \text{SAT} \quad \text{if and only if} \quad \forall v C\big(\psi, \alpha(|\psi|), v\big).$$

We see $x \in L$ if and only if $\exists y \forall z \forall v C(R(x, y, z), \alpha(|R(x, y, z)|), v)$.

Using the fact that all strings, etc., are poly bounded in $|x|$, say by n^c, we can build an advice function β which concatenates all advice strings of length $\leq n^c$. So we would have (without loss of generality; w.l.o.g.), $x \in L$ if and only if $\exists y \forall z \forall v C(R(x, y, z), \beta(|R(x, y, z)|), v)$. Notice that this shows that $L \in \Sigma_2/\text{Poly}$. The idea of the proof is to now "guess" the advice and arrange matters so that we can use a self-reduction to verify we have the correct advice. That is, we will modify the relation C to get a relation H so that $x \in L$ if and only if $\exists w \exists y \forall z \forall z \forall v H(R(x, y, z), w, v)$. This self-reduction process is achieved as follows. Let $R(x, y, z) = R$ have variables $x_1, \ldots, x_m$. We describe the procedure Red(R) as follows.

1. Ask $\forall v C(R, w, v)$? If yes, continue; or else no, reject.
2. Set $x_1 = 0$, and ask if $\forall v C(R(x_1 = 0), w, v)$? If yes, continue, else set $x_1 = 1$.
3. Repeat this for $x_2, \ldots, x_m$.
4. If we get to the end then we have an assignment for R, and we can check if that assignment satisfies R. If yes, we return a yes, if not reject.

We claim that $x \in L$ if and only if $\exists w \exists y \forall z \forall v C[(R(x, y, z), w, v) \wedge \text{Red}(R(x, y, z))]$.

For the forward implication, we can set $w = \beta(x)$. For the reverse implication, if some w actually passes Red($R(x, y, z)$) then $R(x, y, z) \in \text{SAT}$ for all z and some y, and hence $x \in L$. Moreover, if $x \in L$ then some w will work. The claim now follows and hence $L \in \Sigma_2$, giving the result. $\square$

1.3 Valiant–Vazirani and BPP

In this section we will meet another complexity class of interest which might not be known to someone who has not had a course in computational complexity. It concerns *randomized reductions*.

Definition 1.3.1 We say that a language $L \in$ RP if there is a non-deterministic polynomial-time Turing machine M such that for all x,
- $x \in L$ implies $M(x)$ accepts for at least $\frac{3}{4}$ many computation paths on input x.
- $x \notin L$ implies that $M(x)$ has no accepting paths.

We define the class BPP similarly, except we replace the second item by
- $x \notin L$ implies that $M(x)$ has at most $\frac{1}{4}$ paths accepting.

In the above, the numbers $\frac{3}{4}$ and $\frac{1}{4}$ are arbitrary. For a randomly chosen computation path y, the probability that M accepts for certificate y, is $\Pr M(x, y) \geq \frac{3}{4}$, and likewise that M rejects. Using repetition, we can amplify the computational path as we like. In particular, for any $L \in$ RP or BPP, and any polynomial n^d, if you build a new Turing machine M which replicates the original one n^{c+d} many times, you can make sure that
- $x \in L$ implies that the probability that M accepts is $\geq 1 - 2^{-n^d}$ and
- $x \notin L$ implies M does not accept (RP) or the probability that M accepts is $\leq 2^{-n^d}$ (BPP).

This process is referred to as (probability) *amplification*.

It is easily seen that P $\subseteq$ RP $\subseteq$ NP, and that RP $\subseteq$ BPP. There is one relationship known.

Theorem 1.3.1 (Sipser [620]) BPP $\subseteq \Sigma_2 \cap \Pi_2$.

Proof Let $L \in$ BPP. Assume that we have a machine M amplified as above, with acceptance probability $\geq 1 - 2^{-n}$ and rejection $\leq 2^{-n}$. Let

$$A_x = \{ y \in 2^m \mid M(x, y) \text{ accepts} \},$$
$$R_x = \{ y \in 2^m \mid M(x, y) \text{ rejects} \}.$$

Here $m = n^c$ is the length of a computation path of $M(x)$. Notice that for $x \in L$, $|A_x| \geq 2^m - 2^{m-n}$ and $|R_x| \leq 2^{m-n}$. Conversely, if $x \notin L$, $|R_x| \geq 2^m - 2^{m-n}$ and $|A_x| \leq 2^{m-n}$.

We need the following technical lemma. Let $\oplus$ denote the *exclusive or* operation on binary vectors, that is, bitwise sum modulo 2.

Lemma 1.3.1 $x \in L$ *if and only if* $\exists z_1, \ldots, z_m [|z_m| = m \wedge \{ y \oplus z_j \mid 1 \leq j \leq m \wedge y \in A_x \} = 2^m]$.

Proof of the lemma First suppose $x \notin L$. Then $|A_x| \leq 2^{m-n}$. For any $z_1, \ldots, z_m$, $\{ y \oplus z_j \mid 1 \leq j \leq m \wedge y \in A_x \} = \bigcup_{j=1}^m \{ y \oplus z_j \mid y \in A_x \}$. Note that for sufficiently large n, $\sum_{j=1}^m |\{ y \oplus z_j \mid y \in A_x \}| = \sum_{j=1}^m |A_x| \leq m 2^{m-n} < 2^m$. Therefore, $\{ y \oplus z_j \mid 1 \leq j \leq m \wedge y \in A_x \} \neq 2^m$. Conversely, if $x \in L$, $|R_x| \leq 2^{m-n}$. We need $z_1, \ldots, z_m$

such that for all d, $\{d \oplus z_j \mid 1 \leq j \leq m\} \not\subseteq R_x$. Each $z_1, \ldots, z_m$ for which there is a d with $\{d \oplus z_j \mid 1 \leq j \leq m\} \subseteq R_x$ is determined by a subset of R_x of size m, and a string d of length m. There are $\leq (2^{m-n})^m$. of the former, and $\leq 2^m$ of the latter, giving a total of $\leq (2^{m-n})^m 2^m = 2^{m(m-n+1)} < 2^{m^2}$. But there are 2^{m^2} many choices for $z_1, \ldots, z_m$ of length m, and hence there is some choice $z_1, \ldots, z_m$, with $\{d \oplus z_j \mid 1 \leq j \leq m\} \not\subseteq R_x$. □

Now we can complete the proof of Sipser's Theorem. Since BPP is closed under complementation, it is enough to show $L \in \Sigma_2$. To decide if $x \in L$,

- Guess $z_1, \ldots, z_m$.
- Generate all potential d of length m.
- See, in polynomial time, if $d \in \{y \oplus z_j \mid 1 \leq j \leq m \wedge y \in A_x\}$, which is the same as $\{d \oplus z_j \mid 1 \leq j \leq m\} \cap A_x \neq \emptyset$; by checking $M(x, d \oplus z_j)$ for each j.

This matrix defines a Σ_2 predicate. Hence $L \in \Sigma_2$ by the lemma. □

We next explore means for showing that problems are likely to be intractable, using randomized polynomial-time reductions. In some sense, this result and the one above were the key to many results in modern complexity like Toda's Theorem (Toda [647])[1] and the PCP Theorem.[2] We will see later that the Valiant–Vazirani Theorem admits a parameterized analogue. The problem of interest is whether there is unique satisfiability for a CNF formula.

UNIQUE SATISFIABILITY (USAT)

Instance: A CNF formula φ.
Question: Does φ have exactly one satisfying assignment?

Theorem 1.3.2 (Valiant and Vazirani [650]) NP $\subseteq$ RP^{USAT}.

Proof The method is based around a certain algebraic construction. Let V be an n dimensional vector space over GF[2] the two element field. For this field we remind the reader that for any subspace Q of V, $Q^{\perp} = \{z \mid z \cdot w = 0 \text{ for all } w \in Q\}$, is the *orthogonal complement* of Q.

Lemma 1.3.2 (Valiant and Vazirani [650]) *Let S be a nonempty subset of V. Let $\{b_1, \ldots, b_n\}$ be a randomly chosen basis of V and let $V_j = \{b_1, \ldots, b_{n-j}\}^{\perp}$, so that $\{0\} = V_0 \subset V_1 \subset \ldots \subset V_n = V$ is a randomly chosen tower of subspaces of V, and V_j has dimension j. Then the probability*

$$\Pr\bigl(\exists j \bigl(|V_j \cap S| = 1\bigr)\bigr) \geq \frac{3}{4}.$$

---

[1] This theorem states that if we consider polynomial-time algorithms that have access to an oracle giving the number of accepting paths on a non-deterministic polynomial-time Turing machine (P$^{\#P}$) then we can compute any level of the polynomial-time hierarchy in polynomial time; that is, PH $\subseteq$ P$^{\#P}$.

[2] This is a characterization of NP in terms of "polynomially checkable proofs," which is beyond the scope of this book.

Proof of lemma First note that if either $|S| = 1$ or $0 \in S$ then we are done. Hence we assume $|S| \geq 2$ and $S \cap V_0 = \emptyset$. It is easy to see that there is some least j with $|S \cap V_j| \geq 2$. Since we are working in GF[2], this means that the dimension of the span sp$(V \cap V_j)$ is also at least 2.

Consider the hyperplane $H = \{b_{n-k+1}\}^{\perp}$ and I a maximal (linearly) independent subset of $S \cap V_k$. Notice that $V_{k-1} = V_k \cap H$, and $I \cap H \subseteq S \cap V_k \cap H = S \cap V_{k-1}$.

Now to establish the result, it is enough to show that $\Pr(S \cap V_{k-1} = \emptyset) \leq \frac{1}{4}$. Hence it is enough to show that $\Pr(I \cap H = \emptyset) \leq \frac{1}{4}$. Notice that $\Pr[(I \cap H) = \emptyset] \leq \max_{m \geq 2}\{\Pr(I \cap H = \emptyset \mid \dim(\text{sp}(I)) = m)\}$. To estimate $\Pr(I \cap H = \emptyset \mid \dim(\text{sp}(I)) = m)$, notice that since H is a hyperplane in a vector space, if $\dim(\text{sp}(I)) = m$, then $\dim(\text{sp}(I \cap H)) \in \{m, m-1\}$. However, since if $\dim(\text{sp}(I)) = m$ and $\dim(\text{sp}(H)) = m$ then $I \subseteq H$ and hence $I \cap H \neq \emptyset$. This means that $\Pr(I \cap H = \emptyset \mid \dim(\text{sp}(I)) = m \wedge \dim(\text{sp}(I)) = m)$ will be bounded by $\Pr(I \cap H = \emptyset \mid \dim(\text{sp}(I)) = m \wedge \dim(\text{sp}(I \cap H)) = m - 1)$. Since we are using random bases, the probability we are after, therefore, is the probability that we have a set of m independent vectors $\hat{I}$ and a random hyperplane $\hat{H} = \{x\}^{\perp}$ and $\hat{I} \cap \hat{H} = \emptyset$. That is, for all $y \in I$, $y \cdot x = 0$. Now we calculate using inclusion/exclusion: $\Pr(y \cdot x = 0$ for all $y \in \hat{I}) = 1 - \Pr(\exists y \in \hat{I} \mid y \cdot x = 0) = 1 - \sum_{i=1}^{|\hat{I}|}(-1)^{i+1}\sum_{J \subseteq \hat{I} \wedge |J| = i} \Pr(x \in J^{\perp})$. This quantity equals $1 - \sum_{i=1}^{m}(-1)^{i+1}\binom{m}{i}2^{-i} = \sum_{i=0}^{m}\binom{m}{i}(-1)^i 2^{-i} = (1 - \frac{1}{2})^m$. This quantity is maximized for $m = 2$, giving the bound $\frac{1}{4}$. $\square$

Now we turn to the proof of the Valiant–Vazirani Theorem. Let φ be an instance of SATISFIABILITY. Let n be the number of variables $\{x_1, \ldots, x_n\}$ in φ. We use the random basis $\{b_i : 1 \leq i \leq n\}$ to construct the tower $V_0 \subset \cdots \subset V_n = V$ of the lemma. This requires $O(n^2)$ random bits. The set S will be the set of *satisfying* assignment of $\{x_1, \ldots, x_n\}$. The key idea is that we can regard $(x_1, \ldots, x_n)$ as n-bit vectors in GF$[2]^n$. All we need to do is to construct a formula ξ_i stating that a particular assignment lies in V_i. Then the relevant formula is $\varphi \wedge \xi_i$. The reduction, regarded as a Turing reduction, is essentially: for each $i = 0, \ldots, n$, see if $\varphi \wedge \xi_i$ is in USAT. By the lemma above, this succeeds with probability $\frac{3}{4}$, if there is some satisfying assignment for φ (i.e. $S \neq \emptyset$), and never returns "yes" if φ is not satisfiable. This can also be regarded as a randomized polynomial-time m-reduction by guessing i, and making the error as small as desired using amplification. $\square$

1.4   Historical Notes

The origins of complexity theory are discussed in many places, and we will not do this here. The key papers were Edmonds [267], Hartmanis and Stearns [381], and early Russian work on *perebor* algorithms, or perebor for short, a "brute-force" or "exhaustive" search method. Hartmanis [380] also gives an interesting account anticipating the importance of problems perhaps in NP − P, in his discussion of a letter from von Neumann to Gödel. The Cook–Levin theorem to determine NP-completeness, which we tagged CNFSAT, can be found in Cook [163] and Levin [488]. The PSPACE completeness of QBFSAT is from Stockmeyer and Meyer [631]. Likely the most important early paper in the area is that of Karp [431] who showed that the NP-completeness phenomenon was *ubiquitous* and *widespread*. The idea of advice is from Karp and Lipton [432]. Yap's Theorem, on an advice collapse to a PH collapse, will be important for us in Chap. 30, and is from Yap [670]. The Valiant–Vazirani Theorem concerning randomized reductions is from [650]. For more on these issues, one fine older (1987) text is Balcázar, Díaz, and Gabarró [52], with a modern (2011) text being Homer and Selman [399].

1.5 Summary

Basic results and definitions in classical complexity theory are given including the Valiant–Vazirani Theorem and Yap's Theorem on advice classes.

The Basic Definitions

2

2.1   The Basic Definition

Our concerns are (decision) problems with two or more inputs. Thus we will be considering languages $L \subseteq \Sigma^* \times \Sigma^*$. We refer to such languages as *parameterized languages*. If $\langle x, k \rangle$ is in a parameterized language L, we call k the parameter.[1] Usually the parameter will be a positive integer, but it might be a graph or algebraic structure. However, in the interest of readability and with no loss of generality, we will usually identify the domain of the parameter as the natural numbers (in *unary*) $\mathbb{N}$ and hence consider languages $L \subseteq \Sigma^* \times \mathbb{N}$. For a fixed k, we call $L_k = \{\langle x, k \rangle : \langle x, k \rangle \in L\}$ the kth *slice* of L.

As we have seen in the introduction, our main idea is to study languages that are tractable "by the slice." As the reader will recall, being tractable by the slice meant that there is a constant c, independent of k, such that for all k, membership of L_k can be determined in time $O(|x|^c)$. There are various levels of non-uniformity and of non-computability available for such a definition, but for almost all of this book, the reader can take the following as the *definition* of fixed-parameter tractability.

> **Definition 2.1.1** (The basic definition) We say that a parameterized language L is (strongly uniformly) *fixed-parameter tractable* (FTP) iff there exists an algorithm Φ and a constant c and a computable function f such that, for all x, k, $\Phi(\langle x, k \rangle)$ runs in time at most $f(k)|x|^c$ and
>
> $$\langle x, k \rangle \in L \quad \text{iff} \quad \Phi(\langle x, k \rangle) = 1.$$

---

[1] There is another tradition here suggested by Flum and Grohe [312] that the parameter be a *function* $\kappa : \Sigma \times \Sigma \to \Sigma^*$. We believe that the original definition is explicit enough and certainly appropriate in practical applications. We only mention this fact for the reader who looks at material in the literature using this notation.

R.G. Downey, M.R. Fellows, *Fundamentals of Parameterized Complexity*, Texts in Computer Science, DOI 10.1007/978-1-4471-5559-1_2, © Springer-Verlag London 2013

The reader may be alerted by the presence of "computable function f" in Definition 2.1.1. Isn't this supposed to be a theory designed to address complexity in practical computation? Definition 2.1.1 allows for functions like $f(k) = 2^{2^{\cdot^{\cdot^{2}}}} \quad k \text{ times}$, or even worse! Of course, it could well be argued that for polynomial time similarly horrible polynomial-time running times could happen, and hence this is also an issue in the definition of P.

As it turns out, generally if we use the elementary toolkit described in the first chapters, c is usually small (1, 2 or maybe 3) and the constant $f(k)$ is usually a manageable functions of k, like 2^k.

For example, consider VERTEX COVER parameterized by k, defined as follows.

VERTEX COVER

Instance: A graph $G = (V, E)$.
Parameter: A positive integer k.
Question: Does G have a vertex cover of size $\leq k$? (A vertex cover of a graph
              G is a collection of vertices V' of G such that for all edges $v_1 v_2$ of G
              either $v_1 \in V'$ or $v_2 \in V'$.)

Then, as we soon see, VERTEX COVER, parameterized by k, can be solved in time $2^k |G|$. This running time can be written in a more convenient form.

Definition 2.1.2 (The O^* notation) If a parameterized algorithm has running time $f(k)|x|^c$, we will write that the algorithm has a running time of $O^*(f(k))$, i.e. ignoring the polynomial part and concentrating on the exponential part.

In the case of VERTEX COVER as above, we would say we can solve it in time $O^*(2^k)$. Some authors (such as Flum and Grohe [312]) choose to write p-VERTEX COVER to emphasize that we are dealing with the problem as a parameterized one.

2.2   The Other Flavors of Parameterized Tractability

Now, even though at this stage it might be somewhat mysterious, we feel that it is important to point out that there really are other flavors of parameterized tractability. These other definitions involve the use of non-computability, first in terms of the constant, and secondly as a non-uniformity in the algorithm itself. The use of non-computability in complexity is not new, with classes like P/Poly being uncountable. However, we will try to motivate these classes with examples. Before we give some precise definitions, we invite the reader to consider the following examples to add to the one we have seen so far, VERTEX COVER.

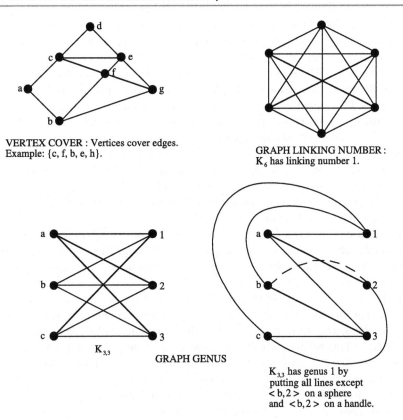

VERTEX COVER : Vertices cover edges.
Example: {c, f, b, e, h}.

GRAPH LINKING NUMBER :
K_6 has linking number 1.

$K_{3,3}$

GRAPH GENUS

$K_{3,3}$ has genus 1 by
putting all lines except
< b, 2 > on a sphere
and < b, 2 > on a handle.

Fig. 2.1 Examples of FPT problems

Example 2.2.1 GRAPH GENUS

Instance: A graph $G = (V, E)$.
Parameter: A positive integer k.
Question: Does G have genus k? (That is, can G be embedded with no edges crossing on a surface with k handles?)

Example 2.2.2 GRAPH LINKING NUMBER

Instance: A graph $G = (V, E)$.
Parameter: A positive integer k.
Question: Can G be embedded into 3-space such that the maximum size of a collection of topologically linked disjoint cycles is bounded by k?

In Fig. 2.1 we give some examples to illustrate the problems above. The fact that K_6 has linking number 1 is due to Sachs [597] and Conway and Gordon [162]. If we consider the classical versions of the problems above where k is not fixed, then they are all NP-hard. Each of the above problems exhibits some form of parameterized tractability. As we mentioned above, in the next section, we will look at a simple technique called the method of bounded search trees that can solve VERTEX COVER using a single algorithm Φ running in time $2^k|G|$ for each k. Fellows and Langston [297, 299] introduced a method which allowed them to use the deep results of Robertson and Seymour [588, 589] to construct a single algorithm Φ which accepts $\langle G, k \rangle$ as input instances of GRAPH GENUS such that Φ determines if G has genus k in time $O(|G|^3)$. This running time was improved by Mohar [533] to $O(|G|)$. Finally, Fellows and Langston [297, 299] used the Robertson–Seymour methods to prove that for each k there is an algorithm Ψ_k which runs in time $O(|G|^3)$ and which determines if the graph G has linking number k. (We will look at the proofs of these results and the techniques used in Chap. 17.)

Notice the differences between the three varieties of fixed-parameter tractability. They are all $O(|G|^c)$ for some c slicewise but:

- In the case of VERTEX COVER, we have a single *known* algorithm which works for all k, and, moreover, we can compute the constant and hence the exact running time for each k. This is the behavior we gave in Definition 2.1.1 and called it there *strongly uniform fixed-parameter tractability*.
- In Example 2.2.1, for GRAPH GENUS, we still have a single algorithm Φ for all k, but this time we have *no* way of computing the constant in the running time. We merely know that for each k, the running time of Φ on input $\langle G, k \rangle$ is $O(|G|^3)$. This behavior is called *uniform fixed-parameter tractability*.
- In Example 2.2.2, for GRAPH LINKING NUMBER, all we know is the exponent of the running time for the algorithms. For each k, we have a different (and unknown) algorithm running in $O(|G|^3)$ with unknown constants. This behavior is called *non-uniform fixed-parameter tractability*.

Considerations such as the examples above naturally lead us to the definitions below.

Definition 2.2.1 (Uniform and non-uniform FPT) Let A be a parameterized problem.
  (i) We say that A is *uniformly fixed-parameter tractable* if there is an algorithm Φ, a constant c, and an arbitrary function $f : \mathbb{N} \mapsto \mathbb{N}$ such that: the running time of $\Phi(\langle x, k \rangle)$ is at most $f(k)|x|^c$,
 (ii) We say that A is *non-uniformly fixed-parameter tractable* if there is a constant c, a function $f : \mathbb{N} \mapsto \mathbb{N}$, and a collection of procedures $\{\Phi_k : k \in \mathbb{N}\}$ such that for each $k \in \mathbb{N}$, and the running time of $\Phi_k(\langle x, k \rangle)$ is $f(k)|x|^c$ and $\langle x, k \rangle \in A$ iff $\Phi_k(\langle x, k \rangle) = 1$.

In [242, 247], Downey and Fellows give proofs constructing languages showing that the definitions above generate three distinct classes of computable parameterized languages. The diagonalization arguments used construct artificial languages to provably separate the classes. At present, we are not aware of any natural problem that is provably (say) in the class of problems that are uniformly but not strong uniformly fixed-parameter tractable. We remark, however, that there are examples of natural problems such as GRAPH LINKING NUMBER which can at present only be classified as non-uniformly tractable. Nevertheless, in practice, it seems that the most important types of parameterized tractability are the two varieties of uniform tractability.

The reader may well ask why we bothered to introduce these apparently exotic classes. One of the reasons concerns lower bounds for algorithms. That is, later in Chaps. 29 and 30, we will prove results about lower bounds on the power of certain kinds of parameterized algorithms. These lower bounds require some complexity assumptions[2] which need the more general notions of FPT to be used. Additionally, the uniform and non-uniform versions come to center stage in the chapter where we look at applications of the Robertson–Seymour methodology, Chap. 17. For the present the readers should file the notions above in the backs of their minds.

Therefore until Chaps. 17 and 30, the reader can assume that unless otherwise specified, fixed-parameter tractability will mean *strongly uniform fixed-parameter tractability.*

We remark that most natural FPT problems which have been studied so far seem to exhibit a general migration toward strongly uniform tractability; that is, once a problem has been identified as FPT, then with more precise combinatorics, we eventually show the problem to be strongly uniformly FPT.

One reason for this general process of improvement is that often the initial classification of the problem as FPT came from the application of some general result pertaining to a class of problems containing the one at hand. Later, more uniformity, together with better constants and algorithms, can be obtained after studying the particular combinatorics of the problem considered alone. It is an open question whether there are natural examples of languages which are FPT but provably not strongly uniformly FPT.

Parameterized complexity is orthogonal to classical complexity; that is, the parameterized complexity of a problem can bear no relationship to the classical complexity of the problem. For instance, let L be *any* computable language. Perhaps L is not even elementary recursive and the time complexity for L might dwarf Ackermann's function. But consider the parameterized language $L' = \{\langle x, x \rangle : x \in L\}$. Classically, L' has the same complexity as L, but as a parameterized problem, L' is in FPT, with a constant time algorithm. Later, we will see examples where the parameterized versions are hard but classical versions are easy.

2.3 Exercises

Exercise 2.3.1 List 20 aspects of a graph you might regard as a parameter.

Exercise 2.3.2 Recall that GRAPH 3 COLORING is NP-complete. Use this to prove that GRAPH k COLORING (i.e. $\langle G, k \rangle \in L$ iff G is k-colorable) is FPT iff $P = NP$.

---

[2]Like $W[1] \neq$ FPT, as we will later define.

Exercise 2.3.3 The notion of parameterization is not restricted to P. Give a definition of a language $L \subseteq \Sigma^* \times \mathbb{N}$ being in parameterized LOGSPACE.

Exercise 2.3.4 (Cai, Chen, Downey, and Fellows [122])

1. We say that a parameterized language L is (uniformly) *eventually* in DTIME(n^d) iff there is an algorithm Φ such that, for each k, there exists $m = m(k) \in \mathbb{N}$ such that, for all x with $|x| > n$, $\langle x, k \rangle \in L$ iff $\Phi(\langle x, k \rangle) = 1$ and Φ runs in time $|x|^d$ on input $\langle x, k \rangle$. Prove that L is in uniform FPT iff there is a d such that L is uniformly eventually in DTIME(n^d). Moreover, show that we can replace "uniform" by "strongly uniform" if the function $k \mapsto m(k)$ is computable.

2. (This is called the "advice view".) $L \in$ FPT (non-uniform) iff there is a polynomial-time oracle Turing machine Φ and a function $f : \mathbb{N} \to \Sigma^*$ such that for all x and k

$$\langle x, k \rangle \in L \quad \text{iff} \quad \Phi^{f(k)}(\langle x, k \rangle) = 1.$$

3. Furthermore, $L \in$ FPT (*strongly uniform*) iff there is a polynomial-time oracle Turing machine Φ and a *computable* function $f : \mathbb{N} \to \Sigma^*$ such that for all x and k

$$\langle x, k \rangle \in L \quad \text{iff} \quad \Phi^{f(k)}(\langle x, k \rangle) = 1.$$

Exercise 2.3.5 (Challenging) Prove that the following problems are in FPT.
 (i) VERTEX COVER.
(ii) PLANAR INDEPENDENT SET.

2.4   Historical Notes

Ever since the discovery of the fact that many natural problems are seemingly intractable, authors have looked for feasible partial solutions. We refer the reader to Garey and Johnson [337], particularly Chap. 4. In that chapter, Garey and Johnson look at what they call "Analyzing Subproblems". The reason for Garey and Johnson's interest in subproblems is encapsulated in the following quote:

> "If a general problem is NP-complete, we know that an exponential time algorithm will be needed (unless P = NP), but there are a variety of ways which the time complexity of an algorithm can be "exponential," some of which might be preferable to others." (Garey and Johnson [337, p. 106].)

In retrospect, it is obvious that many authors have devised algorithms which demonstrate the fact that the parameterized problem is FPT. For instance, as we will see in Chap. 9, all pseudo-polynomial-time algorithms actually demonstrate the fact that the relevant problem is FPT. To our knowledge, the first author to explicitly note the fact that as k varied, problems such as DOMINATING SET seemed to take time $\Omega(n^{f(k)})$ with $f(k) \to \infty$ was Ken Regan in some comments in [577]. There was also a reference by Moshe Vardi [655], who suggested that classical complexity was the wrong notion for databases, since the queries were very small compared to the database. In particular, Vardi pointed out that the input for database-query evaluation consists of two components, query and database. For first-order queries, query evaluation is PSPACE-complete, and for fixpoint query it is EXPTIME-complete, but, if you fix the query, the complexity goes down to LOGSPACE and PTIME correspondingly. In particular, the size of the database was not the right complexity for database-query complexity and the size of parameter counted. Also in the 1980s were the papers Vardi and Wolper [656] and Lichtenstein and A. Pnueli [491] who pointed out that the input for LTL model checking consists of two components, formula and transition system. LTL model checking is PSPACE-complete, but if you fix the formula, the complexity goes down to LOGSPACE.

So people in the database community were very aware that fixing a parameter makes an intractable problem tractable. In retrospect the key if that they did miss the big difference between query evaluation and model checking. In query evaluation the dependence on the formula is exponential, while in model checking it is multiplicative. Indeed, as was shown later, model checking is FPT and query evaluation is likely not FPT.

The first paper to address the "asymptotic" fixed-parameter complexity of parameterized problems (i.e. the behavior as $k \to \infty$), was Abrahamson, Ellis, Fellows and Mata [6]. Roughly speaking, those authors looked at the comparison between languages that were non-uniformly FPT and those that were "P-complete" or "dual P-complete" by the slice. There are a number of severe limitations and other more technical problems with the Abrahamson et al. approach. Those authors certainly could not address comparisons of the parameterized complexities of, say, VERTEX COVER and DOMINATING SET, nor parameterized versions of problems outside of NP. The basic definitions of this chapter were given in Downey and Fellows [241, 243, 244].

We will give more detailed historical remarks concerning the issue of fixed-parameter tractability versus intractability in Part II, where they will be more in context. We will also give historical comments concerning the various techniques examined in this book in the relevant chapters and sections.

Finally, recent histories of the early years of parameterized complexity, and the preceding work on Fellows and Langston can be found in Downey [237] and Langston [479]. Other related materials about the early years of parameterized complexity can be found in the festscrift volume Bodlaender, Downey, Fomin and Marx [85].

2.5 Summary

This chapter introduced the main definition of FPT, and introduces two less uniform variations which are important in some later chapters; particularly those describing methods based on graph minor techniques (Chap. 19) and for lower bound arguments (Chap. 30).

Part II
Elementary Positive Techniques

Bounded Search Trees

3

3.1 The Basic Method

The method of *bounded search trees* is fundamental to FPT algorithmic results in a variety of ways.

- First, we compute some search space which is often a search tree of size bounded by a function of the parameter (typically exponential).
- Thereafter, we run some relatively efficient algorithm on each branch of the tree.

The exponential worst-case complexity of such algorithms comes from problem instances where complete exploration of the search space is required. Probabilistic or experimental analysis of the performance of such algorithms may hinge on how often significant exploration of the search space is required. Thus, worst-case analysis may diverge significantly from the behavior of such algorithms on real datasets.

Real algorithmics must be involved with practical implementations. In practical implementations, a key strategy is to *eliminate branchings whenever possible*. To this end, it is often effective to combine the bounded search tree approach with other strategies, such as kernelization—re-kernelizing on each node of the search tree.

In this chapter, we detail three exemplary applications of the FPT algorithmic strategy of bounded search trees.

3.2 VERTEX COVER

3.2.1 A Basic Search Tree

Theorem 3.2.1 (Monien *via* Mehlhorn [527], Downey and Fellows [241]) VERTEX COVER *is solvable in time* $O(2^k|V(G)|)$, *where the hidden constant is independent of k and $|V(G)|$.*[1]

---

[1] Henceforth, we will drop the "where the hidden constant is independent of k and $|V(G)|$" as this should always be understood.

R.G. Downey, M.R. Fellows, *Fundamentals of Parameterized Complexity*,
Texts in Computer Science, DOI 10.1007/978-1-4471-5559-1_3,
© Springer-Verlag London 2013

25

Proof We construct a binary tree of height k as follows. Label the root of the tree with the empty set and the graph G. Choose an edge $uv \in E$. In any vertex cover V' of G, we must have either $u \in V'$ or $v \in V'$, so we create children of the root node corresponding to these two possibilities, one labeled u and one with v. The set of vertices labeling a node represents a potential vertex cover, and the graph labeling the node represents what remains to be covered in G. In general, for a node labeled with the set of vertices S, we choose an edge $wq \in E(G)$ with neither w nor q connected to any vertex in S, and create the two child nodes labeled, respectively, $S \cup \{w\}$ and $S \cup \{q\}$. If we create a node at height at most k in the tree that covers G, then a vertex cover of cardinality at most k has been found. There is no need to explore the tree beyond height k.   □

Corollary 3.2.1 *The problem* LOG VERTEX COVER *which asks if a given graph G has a vertex cover of size $\leq \log(|G|)$ is in P and is solvable in time $O(|G|^2)$.*

Corollary 3.2.1 improves the work of Papadimitriou and Yannakakis [553].

3.2.2 Shrinking the Search Tree

In some cases, it is possible to significantly improve the constants in the algorithms. We only look at the example of VERTEX COVER, which, despite its apparent simplicity, has a rich structure and admits many parameterized algorithms. Fix k and consider a graph G. Now if G does not have a vertex of degree 3 or more, then it is a fairly trivial graph consisting of a collection of paths and cycles and isolated vertices. If such a G has more than $2k$ edges then it cannot have a size k vertex cover. Thus, without loss of generality (at the expense of an additive constant factor), we will only concern ourselves with graphs with many vertices of degree 3 or greater. Now choose a vertex of degree 3 or greater, say v_0. Now, either v_0 is in a vertex cover or *all of its neighbors are*. Thus, we can begin a search tree with one branch labeled v_0 and the other labeled $\{w_1, w_2, \ldots, w_p\}$, where $p \geq 3$ and $\{w_1, w_2, \ldots, w_p\}$ are the neighbors of v_0. Again, consider the subgraphs of G not covered by the appropriate sets $\{v_0\}$ and $\{w_1, w_2, \ldots, w_p\}$, respectively. In one, we need a size $k - 1$ vertex cover, but in the other, we only need a size $k - p$ vertex cover. Clearly, again we only consider graphs where we always have a degree 3 or greater vertex in the relevant subgraph. Thus, the complexity of the search is now determined not by a tree of size $O(2^k)$ but by a smaller one. The recurrence relation which generates the number of nodes in this new search tree is

$$a_{k+3} = a_{k+2} + a_k + 1, \qquad a_0 = 0, \qquad a_1 = a_2 = 1.$$

Here, we always generate one node as k increases by 1 and we get a splitting on the nodes "3 back".

To estimate the solution for this recurrence, we will try $a_k = c^k$ and we want the best c. To take care of the 1 in the right-hand side, let us try $a_k = c^k - 1$, so that now we have to solve

$$c^{k+3} = c^{k+2} + c^k$$

or

$$c^3 = c^2 + 1.$$

When $c = 5^{1/4}$, it is true that $1 + \sqrt{(5)} \leq 5^{3/4}$. Hence, $a_k \leq 5^{k/4} - 1$ can be verified by induction. This reasoning yields the following result.

Theorem 3.2.2 (Balasubramanian, Fellows, and Raman) VERTEX COVER *can be solved in time* $O([5^{1/4}]^k |G|)$.

Note that $5^{1/4}$ is bounded above by 1.466, and hence this solution is feasible for $k \leq 70$.

The current champion of VERTEX COVER algorithms is Chen, Kanj, and Xia [148], improving Chen, Kanj, and Jia [146]. The earlier algorithm in [146] gives an $O(1.286^k + k|G|)$ running time. The [146] algorithm's methodology is along the lines of that of Theorem 3.2.2, but further shrinks the search tree using more intricate combinatorics based on analysis of possible graph structure. The Chen et al. [146] algorithm additionally needs the graph to be pre-processed first. We will give a simpler algorithm along similar lines in the next chapter (which deals with pre-processing) in Sect. 4.1. Also, in the next chapter, the reader will see how the multiplicative part of the algorithm becomes an additive factor when we look at pre-processing/kernelization. (Basically, we do the search tree method on a small pre-processed graph.) The algorithm of [148] runs in time $O(1.2738^k + k|G|)$. It uses an array of additional techniques, some introduced in [148], and some simply reflecting a decade's development in parameterized algorithm design. These methods will be discussed in Sect. 4.2.

Such algorithms combined with the methods of the next section *and some practical tricks* (heuristics) are used in many applications for very large datasets. We refer, for example, to Langston, Perkins, Saxton, Scharff and Voy [480] where the datasets come from computational biology and are astronomical in size. These algorithms often perform better than we expect because they are based on simple local rules, are parallelizable, scalable, and have simple data structures.

A natural question to ask is: can we do even better? Can we have a $O^*((1+\epsilon)^k)$ algorithm for VERTEX COVER for any $\epsilon > 0$ (i.e. assuming something like P $\neq$ NP). The answer is *no* and we return to this important point in Chap. 29 where it is shown in Theorem 29.5.9 that assuming a reasonable complexity hypothesis called the EXPONENTIAL TIME HYPOTHESIS (namely that n-variable 3SAT cannot be solved in DTIME($2^{o(n)}$)) there is an $\epsilon > 0$ such that k-VERTEX COVER cannot be solved in $O^*((1+\epsilon)^k)$.

3.3   PLANAR INDEPENDENT SET

Later we will see that (general) INDEPENDENT SET is likely not FPT as it is complete for one of our basic hardness classes, $W[1]$. However, parameterizing further by genus allows us to extract tractability.

PLANAR INDEPENDENT SET

Input: A planar graph G.
Parameter: A positive integer k.
Question: Does G have an independent set of size k? Here, a set S of vertices
               of G is called an independent set iff for all $x, y \in S$, xy is not an edge
               of G.

One slightly subtle point is that we have not argued that planarity is FPT. Planarity is not only FPT, but there are linear time algorithms for it. However, we would also like to take this opportunity to make two salient points. First we would mention that the argument below would work without even knowing that planarity is FPT since if it returns no then the graph either does not have a size k independent set or it is not planar. Second, the reader should note that when we apply further parameterizations to a problem there will be at least three kinds of problems arising. The first is to simply solve the problem with the two parameters fixed, as we are doing here for INDEPENDENT SET. The second kind of problem comes from *presenting* the problem with the extra parameter (such as a graph width metric) *displayed.* Here this would be asking that the input be *a plane presentation of the graph.* The certificate is visible. The third possibility is the *promise* version where we simply *promise* that the problem obeys the second parameter, but no certificate is given. In the present example, this would correspond to inputting G and promising it is planar. Notice that in this last case we would likely only ask that the algorithm work for graphs obeying the promise. In the case of PLANAR INDEPENDENT SET, all of these are more or less equivalent (except up to polynomial running time), but in the case where one parameter is not known to be FPT, these really do diverge. We will return to these points many times in the book.

We also point out another aspect of this easy example. We have begun an extended discourse with the problem at hand to explore the frontiers of tractability via new parameters. In this case the parameter is genus, but it could be many others, such a width metrics, degree bounds, colorings etc, as we will see in this book. The tendency is that each such class of parameterizations has it own set of tools.

We give a simple algorithm to prove that PLANAR INDEPENDENT SET is FPT. This algorithm uses a calculation as to how large a graph can be and not have a size k independent set.

Proposition 3.3.1 PLANAR INDEPENDENT SET *is in FPT by an algorithm running in time* $O^*((6k + 1)^k)$.

Proof First note that if S is an independent set and $v \in G - S$, then $S \cup \{v\}$ is an independent set unless $N[v] \cap S \neq \emptyset$. If S has fewer than $6k + 1$ many vertices, then use complete search. This takes at worst $(6k)^k$ many steps. Now, assume that graph has at least $6k + 1$ many vertices. Choose $v_1 \in G$ of degree ≤ 5. Put v_1 into S. Remove v_1 and all of its at most five neighbors. Repeat to get v_2 and add v_2, etc. Each time we remove at most six vertices. $\qquad \square$

Another similar proof can be based around the 4-color theorem. By Robertson, Sanders, Seymour, and Thomas [580], there is a polynomial-time algorithm which 4-colors a planar graph. Therefore, if $|G| > 4k$, 4-color the graph and any of the color classes will be an independent set. If $|G| \leq 4k$, use complete search. We remark that the polynomial-time coloring algorithm for 4-coloring planar graphs is horrible. The method here would be most easily implemented via the 5-coloring theorem which has a straightforward polynomial-time interpretation.

We can use bounded search trees to improve the constants above, via very similar reasoning. The following was noted early on by several authors, notably Rolf Niedermeier.

Proposition 3.3.2 PLANAR INDEPENDENT SET *is solvable in time* $O(6^k|G|)$.

Proof The argument is nearly the same as the first one we gave. We begin at a vertex of degree ≤ 5, v_1. What we do is to begin a search tree with six nodes at the first level, one for v_1 and one for each of the members of $N[v_1]$. Then at each node we delete all of these vertices and repeat with the new graph. The crucial observation is that for any independent set I in G, if I does not contain any of these at most 6 vertices, at least one could be added. Again, the depth of the tree is at most k. $\qquad \square$

The current best algorithm for PLANAR INDEPENDENT SET uses combinatorics and bounded branching and kernelization from the next section as well as some other methods like dynamic programming. It runs in time $O^*(2^{11.98\sqrt{k}})$ (Dorn [229], Chen, Fernau, and Kanj [140]).

3.4 PLANAR DOMINATING SET and Annotated Reduction Rules

3.4.1 The Basic Analysis

In this section, we mention a technique called *annotation*, which sometimes makes the kernelization more easily manipulated.

The problem we consider is PLANAR DOMINATING SET. As we know, a planar graph G we have a degree 5 vertex v. Evidently, either v or one of its five neighbors can be chosen for an optimal dominating set. But if we remove v from the graph, we still keep the five elements of $N[v]$ being already dominated. If you consider this, then it is reasonable to consider a tree of possibilities partitioning into two possibilities, in this case *black* and *white* vertices. Such considerations lead to the following problem.

Annotated Dominating Set

Instance: A graph $G = (V = B \sqcup W, E)$.
Parameter: A positive integer k.
Question: Is there a choice of k vertices $V' \subseteq V$ such that for every vertex $u \in B$, there is a vertex $u' \in N[u] \cap V'$?

That is, is there a set of k vertices dominating all the black ones? In the algorithm below, we would like a search tree based on branching on low degree black vertices. Note that, by planarity, we can guarantee that there is always a vertex $v \in B \sqcup W$ with degree $d \leq 5$. However, not all such vertices need to be black, but this is why the algorithm becomes $8^k n$.

Theorem 3.4.1 (Alber, Fan, Fellows, Fernau, Niedermeier, Rosamond, and Stege [24]) Planar Dominating Set *is solvable by an FPT algorithm running in time* $O(8^k |G|)$.

Proof The algorithm is based on easily described reduction rules. Verification that they are sound is the hard part—we do not cover all the details here. In the following, we always assume that we are branching from a reduced instance. When a vertex u is placed in a dominating set S the target size is reduced by 1. Moreover the neighbors of u are *whitened*. The algorithm begins by coloring all the vertices black.

1. Delete an edge between white vertices.
2. Delete a pendant white vertex.
3. If there is a pendant black vertex v with neighbor u (either black or white), delete v, put u into S and reduce k by 1.
4. If there is a white vertex u of degree 2, with two black neighbors u_1, u_2 with $u_1 u_2 \in E$, delete u.
5. If there is a white vertex u of degree 2, with black neighbors u_1, u_3 such that u_1 has at most seven neighbors of degree at least 4 and there is a black u_2 with $u_1 u_2, u_2 u_3 \in E$, delete u.
6. If there is a white u of degree 2 with black neighbors u_1, u_3 such that u_1 has at most seven neighbors of degree at least 4, and there is a white u_2 with $u_1 u_2, u_2 u_3 \in E$, delete u.
7. If there is a white u of degree 3 with black neighbors u_1, u_2, u_3 with $u_1 u_2, u_2 u_3 \in E$, delete u.

Thus the algorithm at each stage deletes a white vertex or edge, the exception being Rule 3 which adds u to S, and we call this *branching* (on a black vertex of degree 1). In the discussion of Hagerup's algorithm below, we will say that in this case that the *depth of the recursion increases by* 1. For the present algorithm, the decisive fact, proven in [24], is that given $G = (B \sqcup W, E)$, in linear time we can produce a reduced graph G' to which none of the reduction rules above can be applied. We argue that in G' there is always a *black* vertex of degree ≤ 7. Using this it is possible to build a search tree with branching factor at most 8, allowing us to solve the Annotated Planar Dominating Set problem in time $8^k n$ where n is the number of vertices in G. The [24] verification of this fact is quite technical.

We will generalize the above result when we look at d-degenerate graphs later in Sect. 16.7, beginning with work by Alon and Gutner [34]. There we will give another proof of parameterized tractability of PLANAR DOMINATING SET, but with slower running time, as it will be derived as a special case of the more general result.

□

3.4.2 An Improved Analysis

Quite recently, the [24] analysis of PLANAR DOMINATING SET above was simplified and improved by Torben Hagerup [374]. Hagerup noted that if the algorithm branches at a vertex of degree d, then there will be $d + 1$ recursive invocations. We conclude that it is important to find a black vertex of small degree when it is ready to branch. As mentioned above Alber et al. [24] show that there is always one of degree ≤ 7. Hagerup shows that for an n vertex input graph, so long as the current depth of recursion is bounded by $\frac{n}{19}$, there is always a black vertex of degree ≤ 6.

Hagerup phrases his analysis in terms of multigraphs, where he calls a plane embedding *stout* if no face of the embedding is bordered by only two parallel edges. The following lemma can be established by Euler's Formula, and its proof is left to the reader (Exercise 3.8.7)

Lemma 3.4.1 *Let $G = (V, E)$ be an n-vertex multigraph and $U \subseteq V$ a set of vertices of degree ≤ 2 in G such that $G \setminus U$ is stout. Then G has at most $3n - |U| - 3$ edges.*

Lemma 3.4.2 (Hagerup [374]) *Let G be an n-vertex planar graph and $S \subseteq V$. Suppose that each vertex in $N(S) \setminus S$ has at least two neighbors in $B = V \setminus (S \cup N(S))$ and each vertex in B has degree at least 7. Then $|S| > \max\{\frac{n}{19}, \frac{|B|}{6}\}$.*

Proof Call the vertices of S red, those of $N(S) \setminus S$ white, and those of B black. We modify the graph. First delete every edge between two black or two white vertices. For each white vertex w delete all edges that joint w to a red neighbor except one. Note that since w is white it has at least one red neighbor. Then ensure that each white vertex has exactly two adjacent black neighbors. This last step is done as flows. Let $b_1, \dots, b_d$ list the black neighbors of w in the cyclic order in which they occur around w for a fixed plane embedding of G, numbered so that the red neighbor u of w occurs between b_d and v_1. Split w into $d - 1$ new white vertices $w_1, \dots, w_{d-1}$, and for $i \in [d - 1]$, connect w_i to u, b_i, b_{i+1}, and if b_i and b_{i+1} are not adjacent, add and edge $b_i b_{i+1}$.

Next, for each white vertex w whose red neighbor u has degree ≥ 3, insert edges from w to the predecessor and the successor of w in the cyclic order of the (white) neighbors of u around u in the plane embedding (if such edges are not already present). If u has degree 2, if necessary insert an edge between w and the neighbor of u.

Define the *black degree* of a vertex to be the number of black neighbors. No black vertex is isolated. The last step of the construction is to ensure that the black degree of each black vertex is at least three. We allow this step to make G into a multigraph, but make sure that it is stout.

Consider a black vertex v of black degree 1. It must have at least six white neighbors, and all of these have the black neighbor of u of v as their other black neighbor. Therefore it is possible to insert two new edges between u and v whilst keeping G stout. Let $w_1, \ldots, w_p$ (for $p \geq 6$) be the white neighbors of v, in the cyclic order they occur around v numbered so that u occurs between w_p and w_1. For simplicity let $p = 6$. If a new edge is embedded in a face whose boundary includes two edges between w_i and its black neighbors for $i \in \{2, 4\}$, the resulting multigraph is stout, since between any two parallel edges will be one of w_1, w_3, w_5. We can repeat this as we run through the black vertices and repeat it for all black vertices that are still of black degree 1. Similarly, for a black vertex v of black degree 2, some black neighbor u of v shares at least three white neighbors with v, so that a single new edge can be added between u and v preserving stoutness. After the processing of the black vertices of black degree 1, we can then run through the vertices again for each vertex of black degree 2. Thus we can ensure that all the black vertices have black degree at least 3.

Let G' be the multigraph obtained after the process above is applied. Let W be the set of white vertices, S_i the set of red vertices of degree i for $i \in \{1, 2\}$. Let $U = S_1 \cup S_2$. Let $n' = |V(G')| = |S| + |W| + |B| \geq n$. A white vertex whose red neighbor is of degree d has two black neighbors in addition to its red neighbor; as well as $\min\{2, d+1\}$ white neighbors. This gives a total of $\min\{5, d+2\}$ neighbors. Vertices of S_1 and S_2 have $|S_1|$ and $2|S_2|$ neighbors, respectively. The sum of the degrees of the white vertices is $5|W| - 2|S_1| - 2|S_2| = 5|W| - 2|U|$. The sum of the degrees of the red vertices is $|W|$. The sum of the degrees of the black vertices is at least $\max\{7|B|, 2|W| + 3|B|\}$. Thus the sum of the degrees of the n' vertices of G' is

$$5|W| - 2|U| + |W| + \max\{7|B|, 2|W| + 3|B|\}$$
$$= 6(n' - |S| - |B|) + \max\{7|B|, 2(n' - |S| - |B|) + 3|B|\} - 2|U|$$
$$= \max\{6n' - 6|S| + |B|, 8n' - 8|S| - 5|B|\} - 2|U|.$$

Since G' is stout, by Lemma 3.4.1, the degree sum in G' is $\leq 6n' - 2|U| - 6$. Now $6n' - 6|S| + |B| - 2|U| \leq 6n' - 2|U| - 6$ implies $|B| \leq 6|S|$. Therefore the inequality $8n' - 8|S| - 5|B| - 2|U| \leq 6n' - 2|U| - 6$ implies $2n \leq 2n' < 8|S| + 5|B| < 38|S|$. $\qquad\square$

Hagerup's new idea is to apply Lemma 3.4.2 when the Alber et al. algorithm is ready to branch. Let G be the original planar graph and assume that S is the set of vertices already chosen for the dominating set on the current branch, and that the current depth of the recursion is bounded by $\frac{n}{19}$ so that $|S| \neq \frac{n}{19}$. If a vertex $v \in N(S) \setminus S$ is still present in the current graph, H, then it must be white. Since we are ready to branch, Rules 1 and 2 do not apply, and hence v has at least two neighbors in H and these neighbors do not belong to $S \cup N(S)$. Lemma 3.4.2 implies that there is a vertex of G outside of $S \cup N(S)$ of degree at most 6. Since this vertex has never been colored white, it is present in H and is black. Thus, the algorithm will branch on a black vertex of degree at most 6. Since, for $k \leq \frac{n}{19}$ no node in the recursion tree has more than seven children, we conclude the following.

Theorem 3.4.2 (Hagerup [374]) *For an n-vertex graph G and parameter $k \leq \frac{n}{19}$, the Alber et al. algorithm computes a dominating set of size at most k, if one exists, in time $O(7^k n)$.*

3.5 FEEDBACK VERTEX SET

As another example of the technique of bounded search trees, we consider the parameterized version of a classical problem in Garey and Johnson called FEEDBACK VERTEX SET [337].

FEEDBACK VERTEX SET
Instance: A graph $G = (V, E)$.
Parameter: A positive integer k.
Question: Is there a set of vertices $V' \subseteq V$ of cardinality at most k such that $G - V'$ is acyclic?

FEEDBACK VERTEX SET arises in many settings. One important example in computer science is in resource allocation in operating systems. For example, FEEDBACK VERTEX SET is used for *deadlock recovery* where this applied to what is called the *system resource allocation graph*. This is discussed in, for instance, Silberschatz and Galvin [618].

In this chapter we will have a preliminary foray in our battle with FEEDBACK VERTEX SET. Later in Sect. 6.5, we will give the state of the art algorithm. The Sect. 6.5 algorithm's development is a benchmark for combining methods like bounded search trees, and new methods we will meet like measure and conquer, iterative compression and kernelization. For the simplified algorithm for this section, we only use elementary combinatorics.

Theorem 3.5.1 FEEDBACK VERTEX SET *is solvable in time $O((2k + 1)^k \cdot n^2)$.*

We will need the following lemma.

Lemma 3.5.1 (Itai and Rodeh [413]) *There is an $O(|G|^2)$ algorithm for the problem ALMOST MINIMUM below.*

Fig. 3.1 The cases of termination for ALMOST(x)

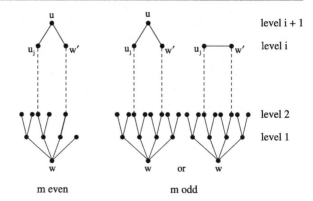

ALMOST MINIMUM

Input: *A graph G.*
Output: *Output ∞ if there is no cycle in G; otherwise output a cycle of length $\leq m + 1$, where m is the length of the shortest cycle in G.*

Proof The algorithm runs as follows. Let $G = \langle V, E \rangle$. For each $x \in V$, let AL-MOST(x) be the algorithm below.

Algorithm 3.5.1 (ALMOST(x))

Step 1. Declare that x has level(x) = 0. Set $k(x) := \infty$.

Step $i + 1$. Using, say, a queue, examine the vertices at level i in order. (This is breadth first.) For each such vertex x' at level i, consider the set of neighbors of x' not at level $i - 1$. Suppose that we are considering v, a neighbor of a vertex x' at level i. It will be the case that v has no level and we declare that v has level $i + 1$.

Now see if there is any vertex $q \neq x'$, either at level i or already at level $i + 1$ by the time we consider v, and additionally v is adjacent to q. If such a q exists, then terminate the algorithm and return $k(x) := i$.

If we consider all of level $i + 1$ and find no pair v, q, then move to Step $i + 2$, until there are no vertices without a level.

Now for each $x \in V$, ALMOST(x) takes $O(|V|)$ many steps. Let $k_{\min} = \min\{k(x) : x \in V\}$. The claim is that if $k(x) = k_{\min}$, then x belongs to a minimal circuit of G. To establish the claim, suppose that w is a vertex that belongs to a minimal circuit. If m is even, then ALMOST(w) stops when a vertex w' is found at level $i + 1$, completing a cycle at level i, and $k(w) = m/2 - 1$. (See Fig. 3.1, even case.)

If m is odd, then we can either find w' completing a cycle at level i or completing a cycle at level $i + 1$, as we see in the two odd cases of Fig. 3.1. In the odd case, $k(v) \leq (m - 1)/2$. Thus, ALMOST(v) finds a cycle that is no longer than $2k_{\min} + 2$ and hence no longer than $m + 1$. It is clear that the cycle contains w.

Thus, for each $x \in V$, we compute $k(x)$. From the collection of $k(x)$, we compute $k_{\min}$ and then we are able to compute an almost minimal cycle for G. The cost is the cost of calling ALMOST(x) $|V|$ many times. This is $O(|V|^2)$. $\square$

As Itai and Rodeh [413] observe, if we have some guarantee that the minimal cycle is of even length, then the algorithm above will return a minimal cycle. This observation applies to the case that G is bipartite.

Proof of Theorem 3.5.1 First, note that a graph G has a feedback vertex set of size k if and only if the *reduced* graph G' has one, where G' is obtained from G by replacing each maximal path in G having internal vertices all of degree 2 with a single edge. Note that the reduced graph G' may have loops and multiple edges, but that if G' is simple, then it has minimum degree 3. The reduced graph G' can be computed from G in linear time. Also, in linear time, a k-element feedback vertex set that has been identified in G' can be lifted to a k-element feedback vertex set in G.

As in the proofs of the theorems above, we build a search tree where each node is labeled with a set of vertices S representing a possible partial solution. The cardinality of a label corresponds to the height of the node in the tree, and we will, therefore, explore the tree to a height of no more than k. In linear time, we can check whether a set S is a solution. If the label set S of a node in the search tree is not a solution and the node has height less than k, then we can generate the children of the node, as follows.

Let H denote the graph $G - S$, and let H' be the reduction of H (as described above). If a vertex v of the H' has a self-loop, then v must belong to every feedback vertex set of H'. Corresponding to this observation, we create a single child node with label $S \cup \{v\}$.

If the reduced graph H' of the graph $H = G - S$ has multiple edges between a pair of vertices $u, v \in V(H)$, then either u or v must belong to every feedback vertex set of H', and we correspondingly create two child nodes with labels, respectively, $S \cup \{u\}$ and $S \cup \{v\}$.

If the reduced graph H' has no loops or multiple edges, then we can make use of the following.

Claim If a simple graph J of minimum degree 3 has a k-element feedback vertex set, then the *girth* of J (the length of a shortest cycle) is bounded above by $2k$.

Proof We prove the claim by induction on k. If J is simple, then by a standard result, J must contain a subdivision of K_4 (Lovász [499]), and this implies that a feedback vertex set must contain at least two elements.

For the induction step, suppose U' is a feedback vertex set consisting of $k + 1$ vertices of J. Suppose that $u, v \in U'$ with the distance from u to v, $d(u, v) \leq 2$ in J. Contracting the edges of a shortest path from u to v yields a graph J' of minimum degree 3 that has a feedback vertex set of k elements. By the induction hypothesis, there is a cycle C in J' of length at most $2k$. This implies that there is a cycle in J of length at most $2k + 2$. Otherwise, suppose no two vertices u and v of U' have $d(u, v) \leq 2$ in J. Then every vertex of $J - U'$ has degree at least 2, and so there is a cycle in J not containing any vertex of U', a contradiction. This establishes our claim.   □

By the above claim, we know that for the node of the search tree that we are processing, either H' contains a cycle of length at most $2l$ where $l = k - |S|$ or S cannot be extended to a k-element feedback vertex set. The algorithm of Itai and Rodeh [413] from Lemma 3.5.1 can be employed to find in H', a cycle of length $2l$ or $2l + 1$ in time $O(n^2)$. Thus, in time $O(n^2)$, we can either decide that the node should be a leaf of the search tree (because there is no cycle in H' of length at most $2l + 1$) or we can find a short cycle and create at most $2l + 1$ children, observing that at least one vertex of the short cycle that we discover in H' must belong to any feedback vertex set.   □

As we will see in Chap. 15, FEEDBACK VERTEX SET can be shown to be solvable in time $O(|G|)$ (Downey and Fellows [241], Bodlaender [72]). The running time is $O(17k^4|G|)$. After a series of improvements, the current best running time is an $O^*(3.83^k)$ algorithm due to Cao, Chen, and Liu [130] we meet in Sect. 6.5. As mentioned earlier, the [130] algorithm combines a number of techniques, notably *iterative compression*, and *measure and conquer*. We will look at these techniques in Chap. 6.

A major recent result is that DIRECTED FEEDBACK VERTEX SET, which is also called FEEDBACK ARC SET, is in FPT (Chen et al. [150]). The argument is beyond the scope of this book.

3.6   CLOSEST STRING

In this section we focus on a problem which arises in computational biology.

CLOSEST STRING
Input: A set of k length ℓ strings $s_1, \ldots, s_k$ over an alphabet Σ and an integer
                $d \geq 0$.
Parameter: Positive integers d and k.
Question: Find a center string, if any. That is find $s = s_j$ such that $d_H(s, s_i) \leq d$
                for all $i = i, \ldots, k$.

Here $d_H(s, s_i)$ denotes the (Hamming) distance between s and s_i. As discussed in Gramm, Niedermeier, and Rossmanith [357], like many problems motivating parameterized approaches to complexity, one setting for this problem is computational biology. The alphabet Σ should be considered the usual $\{A, C, T, G\}$, and the sequences strands of DNA. The goal is to find what is called a *primer* which is a short DNA sequence that binds to a set of longer ones as a starting point for replication. The biology tells us that the strength of the binding is determined by the number of positions in the sequence that "hybridize" it. Stojanovic et al. [632] suggested the approach of finding a well-binding sequence of length ℓ. First align the sequences. This means match as many sequences in columns as best we can. This itself is a very interesting computational problem and was one of the first investigated in parameterized complexity, since it has many possible relevant parameterizations. For example, it is discussed in Bodlaender et al. [82, 84], and Downey and Fellows [247].

All of this brings us to the nice clean combinatorial problem CLOSEST STRING above. The solution we give here is due to Gramm et al. [357], and remains the best algorithm at present.

Theorem 3.6.1 (Gramm, Niedermeier, and Rossmanith [357]) CLOSEST STRING *is solvable in time* $O((d+1)^d|G|)$.

Proof Since d_H is a metric, if we have $d_H(s_i, s_j) > 2d$, then there is no solution. The idea of the algorithm below is to keep track of a potential solution on the paths of the search tree. We can begin with $s = s_1$. If this is a solution then there is nothing more to do. If not then there is some s_i which differs from s in at least d and at most $2d$ positions. The children come from choosing $d + 1$ of these positions, and modifying s to move the distance to s_i closer. Now we repeat, but declaring this modified position permanently modified. Note that s_i could well be considered again. The depth of the tree is bounded by d. $\square$

3.7 JUMP NUMBER and Scheduling

So far, we have seen examples of parametric tractability in the areas of computational biology/stringology, and graph theory. We will examine another example of a branching algorithm, from a completely different area. This example involves combinatorics which are somewhat more involved than the ones met so far, but is a case study in the combinatorics needed to enable branching. This nice case study is in the theory of partially ordered sets which corresponds to the theory of scheduling.

We remark in passing that one of the earliest uses of the parameterized *hardness theory* was to give an indirect approach to proving likely intractability of problems which are not known to be NP-complete. A classic example of this is the following problem.

PRECEDENCE CONSTRAINED k-PROCESSOR SCHEDULING
Instance: A set T of unit-length jobs and a partial ordering $\preceq$ on T, a positive deadline D and a number of processors k.
Parameter: A positive integer k.
Question: Is there a mapping $f : T \to \{1, \ldots, D\}$ such that for all $t \prec t'$, $f(t) \prec f(t')$, and for all i $1 \le i \le D$, $|f^{-1}(i)| \le k$?

PRECEDENCE CONSTRAINED k-PROCESSOR SCHEDULING is NP-complete and is known to be in P for two processors. The question is what happens for $k > 2$ processors. For $k \ge 3$ the behavior of the problem remains one of the open questions from Garey and Johnson's famous book [337] (Open Problem 8), but we have the following *parametric analysis*: Later in this book, we will meet a complexity class called $W[2]$. At present it is immaterial exactly what it is, but suffice to say that it is a hardness class akin to NP-complete, and we do not think any problem hard for that class can be fixed-parameter tractable. What was shown was the following.

Theorem 3.7.1 (Bodlaender, Fellows, and Hallett [87]) PRECEDENCE CON-STRAINED k-PROCESSOR SCHEDULING *is* $W[2]$-*hard.*

The point here is that *even if* PRECEDENCE CONSTRAINED k-PROCESSOR SCHEDULING is DTIME$(n^{f(k)})$ for each k, there seems no way that it will be feasible for large k. Researchers in the area of parameterized complexity have long wondered whether this approach might be applied to other problems like COMPOSITE NUMBER or GRAPH ISOMORPHISM. For example, Luks [502] has shown that GRAPH ISOMORPHISM can be solved in $O(n^k)$ for graphs of maximum degree k, but any proof that the problem was hard for a parameterized class hard would clearly imply that the general problem was not feasible. We know that GRAPH ISOMORPHISM is almost certainly not NP-complete, since proving that would collapse the polynomial hierarchy to 2 or fewer levels (see the work of Schöning in [605]).

We return to our scheduling problem. Suppose a single machine performs a set of jobs, one at a time, with a set of precedence constraints prohibiting the start of certain jobs until some others are already completed. Think about "socks then shoes". We can think of such a schedule as a partial ordering. As is well known every partial ordering has a linear extension. That is, if $(P, \leq)$ is a poset (i.e. a partially ordered set) then there is a linear order $(P, \preceq)$ on the same set P such that if $a \leq b$ in the partial ordering, then $a \preceq b$ in the linear ordering. All that happens is that incomparables in $(P, \leq)$ are made comparable in $(L, \preceq)$.

However, there are often many ways to do this (leading to a dimension theory for posets). Whenever we have two successive elements in the linear ordering $a \prec b$ such that a and b were incomparable in $(P, \leq)$ then we have a *jump*. In an assembly line this corresponds to beginning a new task. For example, socks, shoes, hat, has fewer jumps than socks, hat, shoes. The computational goal we aim at is to construct a schedule to minimize the number of jumps. This is known as the "jump number problem".

Formally, the linear extension $L = (P, \preceq)$ of a finite ordered set $(P, \leq)$ is a total order that is the linear sum $L = C_1 \oplus C_2 \oplus \cdots \oplus C_m$ of disjoint chains $C_1, C_2, \ldots, C_m$ in P, whose union is all of P, such that $x \in C_i$, $y \in C_j$ and $x < y$ in P implies that $i \leq j$. If the maximum element of C_i is incomparable with the minimum element of C_{i+1} then $(\max C_i, \min C_{i+1})$ is a *jump* in L. The number of jumps in L is denoted by $s_L(P)$ and the *jump number* of P is

$$s(P) = \min\{s_L(P) : L \text{ is a linear extension of } P\}$$

By Dilworth's decomposition theorem, the minimum number of chains which form a partition of P is equal to the *width* $w(P)$ of P, that is, size of a maximum antichain. Thus $s(P) \geq w(P) - 1$. Pulleyblank [567], and also Bouchitte and Habib [109], have proved that the problem of determining the jump number $s(P)$ of any given ordered set P is NP-complete.

We will explore the following problem.

JUMP NUMBER

Instance: A finite ordered set P, and a positive integer k.

Parameter: k.

Question: Is $s(P) \le k$? If so, output $L(P)$ with $s_L(P) \le k$.

El-Zahar and Schmerl [270] have shown that the decision problem whether $s(P) \le k$, is in P but with a running time of $O(n^{k!})$.

We will use the method of bounded search trees to prove the following.

Theorem 3.7.2 (McCartin [521]) *There is an $O(k^2 k! |P|)$ algorithm that outputs $L(P)$ with $s_L(P) \le k$ if the answer is yes, and outputs no otherwise.*

Proof [2] This proof largely follows McCartin [521]. If $s(P) \le k$ then no element of P covers more than $k+1$ elements, or is covered by more than $k+1$ elements, as this would imply a set of incomparable elements of size $> k+1$. Thus, if we represent P as a Hasse diagram, then no element of P has indegree or outdegree $> k+1$. This observation allows us to input P as an adjacency list representation where we associate with each element an inlist and an outlist, each of length $\le k+1$. The inlist of an element a contains pointers to elements covered by a. The outlist of a contains pointers to elements which cover a. It is most efficient to use double pointers when implementing these lists, so each list entry consists of a pointer to an element, and a pointer from that element back to the list entry.

The key to the algorithm will be an analysis of the structure of posets with restricted jump number. This is typical of parameterized complexity, and a long article discussing how this might be systematized is Estivill-Castro, Fellows, Langston, and Rosamond [278].

If $k = 1$, then we are asking "is $s(P) \le 1$?". Then a *yes* answer means the structure of P is very restricted. If the answer is *yes*, then P must be able to be decomposed into at most two disjoint chains, C_1 and C_2, and any *links* between these chains must create only a *skewed ladder*, i.e. any element $a \in C_1$ comparable to any element $b \in C_2$ must obey the property $a < b$.

Now suppose we have a fixed k and find that $w(P) = k+1$ and $s(P) \le k$. Then P must be able to be decomposed into a width $k+1$ skewed ladder, i.e. $k+1$ disjoint chains such that these chains $C_1, C_2, \ldots, C_{k+1}$ can be ordered so that:

1. any element $a \in C_i$ comparable to any element $b \in C_j$ where $i < j$ must obey the property $a < b$.
2. $w(P_i) = i$, where $P_i = \bigcup \{C_1, \ldots, C_i\}$.

The FPT algorithm is now described using the structural result above. For any $a \in P$ let $D(a) = \{x \in P : x \le a\}$. We say that a is *accessible* in P if $D(a)$ is a chain. An element a is *maximally accessible* if $a < b$ implies b is not accessible.

---

[2]The algorithm in the proof of this result has been implemented in C code. It can be obtained by contacting McCartin. An interesting anecdote about this result was when the first algorithm and proof were implemented they did not work. There was a flaw in the original proof, and conceivably a lesson for theoreticians with implementations.

Fig. 3.2 A skewed ladder

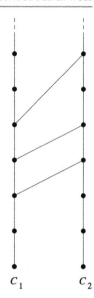

$$C_1 \qquad C_2$$

In any linear extension, $L = C_1 \oplus C_2 \oplus \cdots \oplus C_m$ of P, each $a \in C_i$ is accessible in $\bigcup_{j \geq i} C_j$. In particular, C_1 contains only elements accessible in P. If C_1 does not contain a maximally accessible element we can transform L to $L' = C_1' \oplus C_2' \oplus \cdots \oplus C_m'$ where C_1' does contain a maximally accessible element, without increasing the number of jumps [633] (Exercise 3.8.6). For this we are using the following lemma.

Lemma 3.7.1 *We use the fact that $s(P) \leq k$ iff $s(P - D(a)) \leq k - 1$ for some maximally accessible element $a \in P$.*

Proof Suppose $s(P) \leq k$. Then we have $L(P) = C_1 \oplus C_2 \oplus \cdots \oplus C_m$ where $s_L(P) \leq k$ and C_1 contains a maximally accessible element, say a. $L' = C_2 \oplus \cdots \oplus C_m$ is a linear extension of $P - D(a)$ with $s_{L'}(P - D(a)) \leq k - 1$. Suppose $s(P - D(a)) \leq k - 1$ for some maximally accessible element $a \in P$, then we have $L(P - D(a)) = C_1 \oplus \cdots \oplus C_m$ where $s_L(P - D(a)) \leq k - 1$. $L' = D(a) \oplus C_1 \oplus \cdots \oplus C_m$ is a linear extension of P with $s_{L'}(P) \leq k$. $\qquad \square$

Algorithm 3.7.1 We build a search tree of height at most k.
- Label the root of the tree with P.
- Find all maximally accessible elements in P. If there are more than $k + 1$ of these, then $w(P) > k + 1$ and the answer is *no*, so let us assume that there are $\leq k + 1$ maximally accessible elements.
- If there are exactly $k + 1$ maximally accessible elements, then P has width at least $k + 1$ so check whether P has a width $k + 1$ skewed ladder decomposition. If so, answer *yes* and output $L(P)$ as shown in the text following. If not, answer *no*.

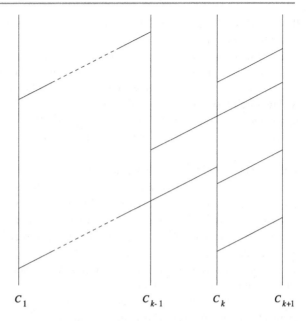

Fig. 3.3 A width $k+1$ skewed ladder

C_1 $\qquad\qquad\qquad$ $C_{k\text{-}1}$ $\quad$ C_k $\qquad$ C_{k+1}

- If there are $< k + 1$ maximally accessible elements, then for each of these elements $a_1, a_2, \ldots, a_m$ produce a child node labeled with $(D(a_i), P_i = P - D(a_i))$ and ask "Does P_i have jump number $k - 1$?" If so, we output $L(P) = D(a_i) \oplus L(P_i)$.

The search tree built by the algorithm will have height at most k, since at each level we reduce the parameter value by 1. The branching at each node is bounded by the current parameter value, since we only branch when we have strictly less than (parameter + 1) maximally accessible elements to choose from. Thus, the number of paths in the search tree is bounded by $k!$, and each path may contain at most k steps.

At each step we must find all maximally accessible elements in P, and either branch and produce a child node for each, or recognize and output the skewed ladder decomposition of the current correct width.

Lemma 3.7.2

(i) *Finding all maximally accessible elements can be done with $O((k+1)n)$ computations.*

(ii) *To produce a child node corresponding to some maximally accessible element a_i, labeled $(D(a_i), P_i = P - D(a_i))$, requires $O((k+1)n)$ computations.*

(iii) *To recognize and output the skewed ladder decomposition of width say m, requires at most $O((m(m+1)/2)n)$ computations.*

Assuming Lemma 3.7.2, we verify the running time of the algorithm. On each path, we will, at some point, need to recognize a skewed ladder. Suppose this occurs after q iterations of the "find all maximally accessible elements" phase, where the parameter decreases by 1 for each of these iterations. The skewed ladder to be recognized will have width $(k+1) - (q-1)$.

The first q steps will require $O((k+1)n)$, $O(((k+1)-1)n)$, ..., $O(((k+1)-(q-1))n)$ computations respectively. To recognize the skewed ladder will require $O((m(m+1)/2)n)$ computations, where $m = (k+1) - (q-1)$. Therefore, this path will require $O(((k+1)-(q-1))n + ((k+1)(k+2)/2)n)$ computations. Therefore, no path requires more than $O((k+1)n + ((k+1)(k+2)/2)n)$ computations.

Thus, the running time of the algorithm is $O(k^2 k! n)$.

Proof of Lemma 3.7.2 (i) Now we describe how we find all the *maximally accessible elements*. We input P to the algorithm as a list of elements, where we associate with each element an inlist and an outlist. The inlist of a contains pointers to elements covered by a. The outlist of a contains pointers to elements which cover a.

As mentioned earlier, if we take the current parameter to be k, then no element of P can have indegree or outdegree $> k + 1$, since otherwise we can immediately answer *no*. Therefore, we will access at most $k + 1$ elements in any inlist or outlist, and output *no* if further elements are present.

We begin by finding all minimal elements of P by running through the list of elements to find all those whose inlist is empty.

For each of these elements we create a chain structure, which to begin with is just the element itself, say a. We copy this chain structure for each element in the outlist of a that is accessible (i.e. whose inlist contains only a), and append the new element to its copy of a to create a new chain. We recursively apply this process to each of the new elements to build up at most $k + 1$ chains, each ending in a maximally accessible element.

We access each outlist at most once, each inlist at most twice, and copy any element at most $k + 1$ times. Therefore, for parameter k, we can find all maximally accessible elements in time $O((k+1)n)$. This proves (i).

(ii) If the current parameter is k and we find at most k maximally accessible elements, we branch and produce a child node for each maximally accessible element a_i, labeled with the chain $D(a_i)$, and a new ordered set $P_i = P - D(a_i)$.

To produce $P_i = P - D(a_i)$, we take a copy of P and alter it by removing $D(a_i)$ and all links to it. If we implement the inlists and outlists using double pointers, so that each list entry consists of a pointer to an element, and a pointer from that element back to the list entry, then we can do this in time $O((k+1)n)$. Each element of $D(a_i)$ has its inlist and outlist traversed to remove links that occur in the lists of other elements, then the element itself is removed. This proves (ii).

(iii) (Recognizing skewed ladders) We consider the $k + 1$ maximally accessible elements which force the skewed ladder recognition.

If P can indeed be decomposed into a width $k + 1$ skewed ladder, then at least one of these elements must be associated with a *top* chain, i.e. a maximal chain in which no element is below an element in any other chain, and whose removal leaves a skewed ladder of width k. Thus, for at least one of the maximally accessible elements it must be the case that all the elements above it have outdegree ≤ 1.

We build a chain structure recursively above each of the $k + 1$ maximally accessible elements, as described above, discarding it if we find an element with outdegree > 1, and marking it as a *possible top* chain if we reach a maximal element (having outdegree 0).

We then check, in turn, below each maximally accessible element that is the start of a *possible top* chain.

We add to the chain recursively downwards, by looking at any element in the inlist of the current element (there will be at most 1), and adding it to the chain if it has outdegree 1. If we reach an element that has outdegree > 1 we do not include it. Instead we need to ensure that this element lies below one of the remaining maximally accessible elements, since this element and all its predecessors will have to be incorporated into a width k skewed ladder containing all the remaining k maximally accessible elements (we obtain the necessary information by keeping track of the number of copies made of each element during the "find all maximally accessible elements" phase).

If we reach a minimal element, or an acceptable element of outdegree > 1, we remove the *top* chain found and all links to it, and then check that the remaining elements form a skewed ladder of width k.

We do $O((k + 1)n)$ computations to find a *top* chain of a width $k + 1$ skewed ladder, since we look at each element and its inlist and outlist at most once. However, each time we remove a *top* chain, the width parameter is reduced by 1, so the maximum number of elements accessed in any inlist or outlist is also reduced by 1.

Thus, we can recognize a skewed ladder of width $k + 1$ in time $O(((k + 1) \cdot (k + 2)/2)n)$, completing the proof of Lemma 3.7.2 and hence of the Theorem. $\square$

3.8 Exercises

Exercise 3.8.1 Prove that HITTING SET, which seeks to cover each hyperedge in a hypergraph, parameterized by k and hyperedge size d is FPT. Here we are given a hypergraph where the maximum number of vertices in a hyperedge is bounded by $d \geq 2$. Note that for $d = 2$ this is simply VERTEX COVER.

Exercise 3.8.2 (Damaschke [195]) For a fixed universe X of elements, a *bundle* is a subset of X.

The problem HITTING SET OF BUNDLES has as input a collection of sets B_i of bundles, $i \in [c]$. The question is whether it is possible to choose one bundle from each B_i so that the union of the c selected bundles has size at most k. Prove that if we parameterize by (k, b), where b is the maximum size of a bundle, then the problem is FPT.

Exercise 3.8.3 (Reidl, Rossmanith, Sánchez, Sidkar) The k-MAX SAT problem takes a CNF formula with m clauses and n variables and asks if at least k clauses can be satisfied. Prove that this is FPT.

(Hint: Branch on the possible assignments.)

Exercise 3.8.4 (Reidl, Rossmanith, Sánchez, Sidkar)   Construct an FPT algorithm for the following running in time $O^*(3^k)$.

TRIANGLE VERTEX DELETION
Input: A graph $G = (G, E)$.
Parameter: A positive integer k.
Question: Are there k vertices whose deletion results in a graph with no cycles of length 3?

Exercise 3.8.5 (Reidl, Rossmanith, Sánchez, Sidkar)   Construct an FPT algorithm for the following running in time $O^*(3^k)$. This problem models a number of questions in databases and bioinformatics.

EDGE EDIT TO CLIQUE
Input: A graph $G = (G, E)$.
Parameter: A positive integer k.
Question: Can I edit the graph by adding or deleting a total of at most k edges and obtain a clique?

Exercise 3.8.6   Prove the claim in the proof of the JUMP NUMBER Theorem. That is *If C_1 does not contain a maximally accessible element we can transform L to $L' = C_1' \oplus C_2' \oplus \cdots \oplus C_m'$ where C_1' does contain a maximally accessible element, without increasing the number of jumps.*

Exercise 3.8.7 (Hagerup [374])   Prove Lemma 3.4.1.

Exercise 3.8.8   Prove that DOMINATING SET restricted to graphs of maximum degree t is solvable in time $O((t+1)^k |G|)$.

Exercise 3.8.9 (Downey and Fellows [241, 245])   Let q be fixed. Prove that the following problems are solvable in time $O(q^k |E|)$.

WEIGHTED MONOTONE q-CNF SATISFIABILITY
Input: Boolean formula E in conjunctive normal form, with no negated variables and mum clause size q.
Parameter: A positive integer k.
Question: Does E have a weight k satisfying assignment? [The (Hamming) weight of an assignment is the number of variables made true in the assignment.]

WEIGHTED $\leq q$-CNF SATISFIABILITY
Input: Boolean formula E in conjunctive normal form, with maximum clause size q.
Parameter: A positive integer k.
Question: Does E have a weight k' satisfying assignment for some $k' \leq k$?

(Hint: For the first one, simply modify the method used for VERTEX COVER. For the other, first see if there is a clause with only unnegated variables. If not, then we are done. If there is one, fork on the clause variables.)

Exercise 3.8.10 (Gramm, Guo, Hüffner and Niedermeier [354]) Show that the following problem is FPT via a search tree of size $3^k n$.

CLUSTER EDITING

Instance: A graph $G = (V, E)$.
Parameter: A positive integer k.
Question: Can I add or delete at most k edges and turn the graph into a union of k cliques?

(Hint: Show that a graph G consists of disjoint cliques iff there are no three distinct vertices u, v, w with $uv, vw \in E$ and $uw \notin E$.)

Exercise 3.8.11 (Bodlaender, Jansen, and Kratsch [92]) Prove that the following is FPT in time $O(2^\ell |G|^c)$.

LONG CYCLE PARAMETERIZED BY A MAX LEAF NUMBER

Instance: A graph G, and two integer k and ℓ.
Parameter: A positive integer $\ell = |X|$.
Promise: The maximum number of leaves in any spanning tree is ℓ.
Question: Is there a cycle in G of length at least k?

In this promise problem if the max leaf conditions are not satisfied, then the output can be arbitrary.

Exercise 3.8.12 (Gramm, Guo, Hüffner, and Niedermeier [354]) Improve the previous question by modifying the search tree to get an $O^*(2.27^k)$ algorithm.

Exercise 3.8.13 (Kaplan, Shamir, and Tarjan [430]) A graph G is called *chordal* or *triangulated* if every cycle of length ≥ 4 contains a chord. If $G = \langle V, E \rangle$ is a graph, then a *fill-in* of G is a graph $G' = \langle V, E \cup F \rangle$ such that G' is chordal. The MINIMUM FILL-IN problem, which is also called CHORDAL GRAPH COMPLETION is the following:

MINIMUM FILL-IN

Input: A graph G.
Parameter: A positive integer k.
Question: Does G have a fill-in G' with $|F| \leq k$?

If k is allowed to vary, Yannakakis [668] has proven the problem to be NP-complete. Use the method of search trees to show that the parameterized version above is strongly uniformly FPT.[3]

---

[3]Kaplan, Shamir, and Tarjan [430] have used (essentially) the Problem Kernel Method of the next section to give a $O(k^5|V||E| + f(k))$ for suitably chosen f. Finally, Leizhen Cai [117] also used the problem kernel method another way to give a $O(4^k(k + 1)^{-3/2}[|E(G)||V(G)| + |V(G)|^2])$

(Hint: Let c_n denote $\binom{2n}{n} \cdot 1/1 + n$, the nth Catalan number. First, prove that the number of minimal triangulations of a cycle with n vertices is c_{n-2}. Build a search tree as follows. The root of the tree $\mathcal{T}$ is the graph G. To generate the children of a node G' of $\mathcal{T}$, first find a chordless cycle C of G' (i.e., of length ≥ 4) and let the children of the node labeled with G' correspond to the minimal triangulations of C. This only adds at most $c_{|C|-2}$ many children. Note that if we construct the complete tree, then each minimal triangulation of G will correspond to at least one leaf. Restrict the search to traverse paths in $\mathcal{T}$ that involve only k additional edges. One needs to prove that a chordless cycle can be found in a graph in time $O(|E|)$ (see Tarjan and Yannakakis [638]), to prove that all the minimal triangulations of a cycle C can be generated in $O(|C|)$ time (see Sleator, Tarjan, and Thurston [622]) and that the total number of nodes of $\mathcal{T}$ we need to visit is bounded by $2 \cdot 4^{2k}$. These observations give the running time of the algorithm to be $O(2^{4k}|E|)$.)

3.9   Historical Notes

Most of the results and exercises of this section are taken from Downey and Fellows [241, 247] and [245]. More detailed historical notes about VERTEX COVER can be found in the next section. PLANAR INDEPENDENT SET as known by several authors to be FPT, and probably the first explicit proof is found in Niedermeier's Ph.D. thesis. Alber, Bodlaender, Fernau, Kloks, and Niedermeier [21, 22], who gave an FPT algorithm for PLANAR DOMINATING SET. Those authors use annotation to correct the claimed $O(11^k|G|)$ algorithm from [247] and [245] which contained a flaw. The first use of annotation is in Alber, Fellows, and Niedermeier [25]. This algorithm had only two rules and gave a "polynomial kernel" as per the next chapter, and a running time of $O^*(335^k)$. The more intricate reduction rules given here are from Alber, Fan, Fellows, Fernau, Niedermeier, Rosamond, and Stege [24] which improve the running time to $O(8^k|G|)$. This was where matters stood until the analysis of Hagerup [374] yielding the $O(7^k|V(G)|)$ algorithm.

The archetypical application of branching in the solution of the JUMP NUMBER problem is from McCartin [521]. Kratsch and Kratsch [466] remark that the McCartin result is interesting since for the exact problem of finding the least jump number in a poset of size n, McCartin's $O(k^2k!n)$ algorithm does not imply a faster exact exponential algorithm than the trivial one based on listing all linear extensions of a poset P, running in time $O(n!)$. In [466], Kratsch and Kratsch give an exact exponential algorithm running in time $O^*(2^n)$ based on dynamic programming, a technique we meet in Chap. 10. They also demonstrate that there are faster algorithms on other parameterizations such as the pathwidth of the poset. The method for the important CLUSTER EDITING problem is due to Gramm, Guo, Hüffner and Niedermeier [354]. An extensive and highly instructive treatment of this problem can be found in Niedermeier's book [546]. The problem on MINIMUM FILL-IN is from Kaplan, Shamir, and Tarjan [430].

---

time algorithm. The current champion for this problem is Fomin and Villanger [326] and runs in time $0^*(2^{\sqrt{k}\log k})$.

3.10 Summary

We introduce the basic method of bounded search trees which limits the combinatorial explosion by making the depth of the search tree depend on the parameter alone, and the rest of the algorithm being a search along the paths. This method is applied to VERTEX COVER, PLANAR INDEPENDENT SET, CLOSEST STRING and JUMP NUMBER.

Kernelization

4

4.1 The Basic Method

The main idea of the method of reduction to a problem kernel is to reduce a problem instance I to an "equivalent" instance I', where the size of I' is bounded by some function of the parameter k. The instance I' is then analyzed (perhaps by an exponential subroutine), and a solution for I' can be lifted to a solution for I, in the case where a solution exists. Often this technique will lead to an *additive* rather than a *multiplicative* $f(k)$ exponential factor. We begin by re-examining the VERTEX COVER.

Theorem 4.1.1 (Buss and Goldsmith in [115]) VERTEX COVER *is solvable in time* $O(n + k^{4k})$.

Proof Phase 1. Observe that for a simple graph H any vertex of degree greater than k must belong to every k-element vertex cover of H. Also if a vertex has degree 1, then the vertex to which it is connected surely should be used without loss of generality. This all leads to the following *reduction rules*. We implement these at most k times and exhaustively. Initially we let $k' = k$.

Reduction Rule 1: If v has degree 1, delete v and its neighbor u and reduce the parameter to $k' := k' - 1$.

Reduction Rule 2: Locate all vertices in H of degree greater than k'; let p equal the number of such vertices. If $p > k'$, there is no k-vertex cover. Otherwise, find such a vertex v, delete v and let $k' := k' - 1$.

Phase 2. If the resulting graph H' has more than $k'(k' + 1)$ vertices, reject. The bound of $k'(k' + 1)$ in step 2 is justified by the fact that a simple graph with a k'-vertex cover, and degrees by bounded by k', has no more than $k'(k' + 1)$ vertices. Now to finish the algorithm, we simply search all size k' subsets of H' to see if they have formed a vertex cover. If yes, then the deleted vertices in the reduction and the found cover will constitute a vertex cover of G. If no, there is no such vertex cover. The worst case for this algorithm is that there are no reductions in the first phase.

R.G. Downey, M.R. Fellows, *Fundamentals of Parameterized Complexity*,
Texts in Computer Science, DOI 10.1007/978-1-4471-5559-1_4,
© Springer-Verlag London 2013

Then the graph H' will have size at most $k(k+1)$. The complete search would then take $\binom{k(k+1)}{k}$ many steps, which is bounded by $O(k^{4k})$.   □

First kernelize, and then apply the bounded search tree method to give additive bounds to the results. For example, we can improve Buss' kernelization by this two-phase technique, yielding the following.

Corollary 4.1.1 VERTEX COVER *is solvable in time* $O(n + f(k) \cdot k^2)$ *where* $f(k)$ *is any of the constants found by other methods provided that they work on a size* k^2 *kernel.*

Methods based in and around bounded search trees may need to assume that we begin with a well-reduced graph. We now give another algorithm for VERTEX COVER which achieves a better constant, is similar in structure to the Balasubramanian, Fellows, and Raman algorithm of Theorem 3.2.2 of Chap. 3, but is not as complex as the Chen, Kanj, and Jia [146] algorithm.

Theorem 4.1.2 (Downey, Fellow, and Stege [257]) VERTEX COVER *can be solved in time* $O(kn + 1.31951^k k^2)$. *Indeed, this works in time* $O(kn + 1.31951^k g(k))$ *where* $g(k)$ *is the size of any kernel.*

Proof Again the algorithm works in two phases.

Algorithm 4.1.1
Phase 1 (Reduction to a problem kernel): Starting with (G, k) we apply the following reduction rules until no further applications are possible. (Justifications for the reductions are given below.)
(0) If G has a vertex v of degree greater than k, then replace (G, k) with $(G - v, k - 1)$.
(1) If G has two nonadjacent vertices u, v such that $|N(u) \cup N(v)| > k$, then replace (G, k) with $(G + uv, k)$.
(2) If G has adjacent vertices u and v such that $N(v) \subseteq N[u]$, then replace (G, k) with $(G - u, k - 1)$.
(3) If G has a pendant edge uv with u having degree 1, then replace (G, k) with $(G - \{u, v\}, k - 1)$.
(4) If G has a vertex x of degree 2, with neighbors a and b, and none of the above cases applies (and thus a and b are not adjacent), then replace (G, k) with (G', k) where G' is obtained from G by:
   • Deleting the vertex x.
   • Adding the edge ab.
   • Adding all possible edges between $\{a, b\}$ and $N(a) \cup N(b)$.[1]

---

[1]This useful technique is often used as a heuristic anyway, and it is called *folding* or *struction* in the literature.

(5) If G has a vertex x of degree 3, with neighbors a, b, c, and none of the above cases applies, then replace (G, k) with (G', k) according to one of the following cases depending on the number of edges between a, b, and c.

(5.1) There are no edges between the vertices a, b, c. In this case G' is obtained from G by:
- Deleting vertex x from G.
- Adding edges from c to all the vertices in $N(a)$.
- Adding edges from a to all the vertices in $N(b)$.
- Adding edges from b to all the vertices in $N(c)$.
- Adding edges ab and bc.

(5.2) There is exactly one edge in G between the vertices a, b, c, which we assume to be the edge ab. In this case G' is obtained from G by
- Deleting vertex x from G.
- Adding edges from c to all the vertices in $N(b) \cup N(a)$.
- Adding edges from a to all the vertices in $N(c)$.
- Adding edges from b to all the vertices in $N(c)$.
- Adding edge bc.
- Adding edge ac.

The reduction rules described above are justified as follows:

(0) Any k-element vertex cover in G must contain v, since otherwise it would be forced to contain $N(v)$, which is impossible.

(1) It is impossible for a k-element vertex cover of G not to contain at least one of u, v, since otherwise it would be forced to contain all of the vertices of $N(u) \cup N(v)$.

(2) If a vertex cover C did not contain u then it would be forced to contain $N[v]$. But then there would be no harm in exchanging v for u.

(3) If G has a k-element vertex cover C that does not contain v, then C must contain u. But then $C - u + v$ is also a k-element vertex cover. Thus G has a k-element vertex cover if and only if it has one that contains v.

(4) We first argue that if G has a k-element vertex cover C, then C must have either the following form 1 or form 2:

1. C contains a and b, or
2. C contains x but neither of a, b, and therefore also contains $N(a) \cup N(b)$.

If C did not have either of these forms, then it must contain exactly one of a, b, and therefore also x. But this C can be modified to form 1. If G has a k-element vertex cover of the form 1, then this also forms a k-element vertex cover in G'. If G has a k-element vertex cover of the form 2, then this same set of vertices (with x replaced by either a or b) is a k-element vertex cover in G'. Conversely, suppose G' has a k-element vertex cover C. If C contains both a and b, then it is also a k-element vertex cover in G. Otherwise, it must contain at least one of a, b, suppose a. But then the edges from b to all of the vertices of $N(a) \cup N(b)$ in G' force C to contain $N(a) \cup N(b)$. So $C - a + x$ is a k-element vertex cover in G. Now consider the following case for reductions (5.1 and 5.2) to a problem kernel. Let C denote a k-element vertex cover in

G. If C does not contain x, then necessarily C contains $\{a, b, c\}$. In this case C is also a k-element vertex cover in G'. Assume that C contains x.

(5.1) We can assume that at most one of the vertices of $\{a, b, c\}$ belongs to C. If none of the vertices of $\{a, b, c\}$ belongs to C, then either b or c belongs to the vertex cover of G'. If $a \in C$ then $C - x + c$ is a vertex cover for G'; if $b \in C$ then $C - x + a$ is a vertex cover for G'; and if $c \in C$ then $C - x + a$ is a vertex cover for G'.

(5.2) We can assume (easy to check) that exactly one of the vertices of $\{a, b\}$ belongs to C. W.l.o.g. we assume $a \in C$. If $b \in C$ then $C - x + c$ is a vertex cover for G'.

It is easy to see that at the end of Phase 1 we have reduced (G, k) to (G', k') where G' has minimum degree 4, if we have not already answered the question. Furthermore, simply because of the reduction rules (1) and (2) we can conclude that the answer is "no" if the number of vertices in G' is more than k^2. Phase 1 is a good example of what is meant by the FPT technique of *reduction to a problem kernel*. If we still have no answer about the original input (G, k), then we are left with considering (G', k') where $|G'| \leq k^2$ and $k' \leq k$. A demonstration that the problem is in FPT is now trivial, since (in the absence of further ideas) we can just *exhaustively* answer the question for the kernel (G', k'). However, we can do a little better by analyzing the kernel instance by means of a search tree (Phase 2).

Phase 2 (Search tree): In this phase of the algorithm, we build a search tree of height at most k. The root of the tree is labeled with the output (G', k') of Phase 1. We will describe various rules for deriving the children of a node in the search tree. For example, we can note that for a vertex v in G', and for any vertex cover C, either $v \in C$ or $N(v) \subseteq C$. Consequently we could create two children, one labeled with $(G' - v, k' - 1)$, and the other labeled with $(G' - N[v], k' - \deg(v))$. In our algorithm, we perform this branching if there is a vertex of degree at least 6. By repeating this branching procedure, at each step reapplying the reductions of Phase 1, we can assume that at each leaf of the resulting search tree, we are left with considering a graph where every vertex has degree 4 or 5. If there is a vertex x of degree 4, then the following branching rules are applied. Suppose that the neighbors of a vertex x are $\{a, b, c, d\}$. We consider various cases according to the number of edges present between the vertices a, b, c, d.

Note that if not all of $\{a, b, c, d\}$ are in a vertex cover, then we can assume that at most two of them are.

Case 1. The subgraph induced by the vertices a, b, c, d has an edge, say ab.

Then c and d together cannot be in a vertex cover unless all four of a, b, c, d are there. We can conclude that one of the following is necessarily a subset of the vertex cover C, and branch accordingly:

1. $\{a, b, c, d\} \subseteq C$
2. $N(c) \subseteq C$
3. $\{c\} \cup N(d) \subseteq C$.

Case 2. The subgraph induced by the vertices a, b, c, d is empty. We consider three subcases.

Subcase 2.1. Three of the vertices (say a, b, c) have a common neighbor y other than x.

When not all of a, b, c, d are in a vertex cover, x and y must be. We can conclude that one of the following is a subset of the vertex cover C, and branch accordingly:
1. $\{a, b, c, d\} \subseteq C$
2. $\{x, y\} \subseteq C$.

Subcase 2.2. If Subcase 2.1 does not hold, then there may be a pair of vertices who have a total of six neighbors other than x, suppose a and b. If all of a, b, c, d are not in the vertex cover C then $c \notin C$, or $c \in C$ and $d \notin C$, or both $c \in C$ and $d \in C$ (in which case $a \notin C$ and $b \notin C$). We can conclude that one of the following is a subset of the vertex cover C and branch accordingly:
1. $\{a, b, c, d\} \subseteq C$
2. $N(c) \subseteq C$
3. $\{c\} \cup N(d) \subseteq C$
4. $\{c, d\} \cup N(\{a, b\}) \subseteq C$.

Subcase 2.3. If Subcases 2.1 and 2.2 do not hold, then the graph must have the following structure in the vicinity of x: (1) x has four neighbors a, b, c, d and each of these has degree four. (2) There is a set E of six vertices such that each vertex in E is adjacent to exactly two vertices in $\{a, b, c, d\}$, and the subgraph induced by $E \cup \{a, b, c, d\}$ is a subdivided K_4 with each edge subdivided once. In this case, we can branch according to:
1. $\{a, b, c, d\} \subseteq C$
2. $(E \cup \{x\}) \subseteq C$.

If the graph G is regular of degree 5 (there are no vertices of degree 4 to which one of the above branching rules apply) and none of the reduction rules of Phase 1 can be applied, then we choose a vertex x of degree 5 and do the following. First, we branch from (G, k) to $(G - x, k - 1)$ and $(G - N[x], k - 5)$. Then we choose a vertex u of degree 4 in $G - x$ and branch according to one of the above cases. The net result of these two combined steps is that from (G, k) we have created a subtree where one of the following cases holds:
1. There are four children with parameter values $k - 5$, from Case 1.
2. There are three children with parameter values $k_1 = k - 5$, $k_2 = k - 5$ and $k_3 = k - 3$, from Subcase 2.1.
3. There are five children with parameter values $k_1 = k - 5$, $k_2 = k - 5$, $k_3 = k - 5$, $k_4 = k - 6$ and $k_5 = k - 9$, from Subcase 2.2.

Note that if reduction rule (2) of Phase 1 cannot be applied to $G - x$, then at least one of the neighbors of u has degree 5, and so Subcase 2.3 is impossible.

The bottleneck recurrence comes from the degree 5 situation which produces four children with parameter values $k - 5$. The total running time of the kernelization algorithm is therefore $O(r^k k^2 + kn)$, where $r = 4^{1/5}$, or $r = 1.31951$ approximately. $\qquad\square$

The above algorithm offers a slight improvement on the $r = 1.32472$ algorithm of Balasubramanian, Fellows, and Raman [51].

4.2   Heuristics and Applications I

The parameter function of our analysis of the running time of Algorithm 4.1.1 indicates that this "exact" algorithm is useful for input graphs of any size, so long as the parameter k is no more than about 60. There are a number of uses of VERTEX COVER in biological "Data Analysis and Retrieval With Indexed Nucleotide-/peptide-sequences" (DARWIN), a programming language. For this reason, essentially the above algorithm was implemented by Hallett, Gonnet, and Stege in computational projects by the biosciences research group at the ETH-Zurich [376]. It turns out that the useful parameter range of $k \leq 60$ indicated by the parameter function above is overly pessimistic. In practice, the algorithm seems to be useful for k up to around 200. Note that if you run this algorithm on (G, k) where k is apparently too large, the worst that can happen is that the algorithm fails to finish in a reasonable amount of time. If it terminates, then it does give the correct answer. The reader should also note the rapid development here since 1998, where now Langston's group treat problems *much* larger quite routinely.

The two phases of the above algorithm are independent. If one intended to solve the general VERTEX COVER problem by simulated annealing (or any other method), it would still make sense to apply Phase 1 *before* doing the simulated annealing, since the "simplifications" accomplished by the kernelization require only polynomial time (that is, they in no way depend on k being small). Consequently, Phase 1 is a reasonable first step for *any* general algorithmic attack on the NP-complete VERTEX COVER problem. We can codify this discussion by describing the following heuristic algorithm for the *general* VERTEX COVER problem. That is, the following algorithm, although it is based on FPT methods, has nothing to do with small parameter ranges, and it runs in polynomial time.

Algorithm 4.2.1 (A general heuristic algorithm for VERTEX COVER based on kernelization.) The algorithm reduces the input graph G to nothing, by repeating the following two steps:
1. Apply Phase 1 of Algorithm 4.1.1.
2. Choose a vertex of maximum degree.

The reduction path gives an approximate minimum vertex cover for G. To see that this works correctly, it is necessary to observe that for each of the rules of Phase 1 reducing (G, k) to (G', k'), and given any vertex cover C' in G', we can reverse the reduction to obtain a vertex cover C in G.

In Algorithm 4.2.1, we have simply replaced Phase 2 of Algorithm 4.1.1, which is exponential in k, with the well-known "greedy" approximation heuristic for this problem. However, we could also *adapt* Phase 2 by *simply not exploring all of the branches of the search tree*, either by making a random selection, or by branch selection heuristics, etc. The following is a natural adaptation of Algorithm 4.1.1 that exploits this idea for designing heuristics based on the FPT search tree technique.

Algorithm 4.2.2 (A general heuristic algorithm for VERTEX COVER based on kernelization and a truncated search tree) This is the same as Algorithm 4.1.1, except
that in the second phase of building the search tree, we do two things. First of all,
we decide in advance how large of a search tree we can afford, and we build the tree
until we have this number of leaves. Then for each leaf of the search tree, we apply an approximation heuristic such as the greedy method used in Algorithm 4.2.1,
and take the best of all the solutions computed from the leaf instances of the search
tree.

The search tree is developed according to the following branching heuristic.
Given the instance (G, k) as the label on a node of the search tree to be expanded,
we calculate the most valuable branch in the following way:
1. Find a vertex a of maximum degree r in G.
2. Find a pair of vertices b, c such that $s = |S|$, where $S = N(b) \cap N(c)$ is maxi-
   mized.
3. Determine which of two equations ($x^r - x^{r-1} - 1 = 0$ or $x^s - x^{s-2} - 1 = 0$) has
   the smallest positive real root. If the first equation, then branch to $(G - a, k - 1)$
   and $(G - N[a], k - r)$. If the second, then branch to $(G - b - c, k - 2)$ and
   $(G - S, k - s)$. (This information can be precomputed and determined by table
   lookup.)

The branching heuristic above is justified because it essentially chooses the
branching rule to apply according to the heuristic criterion that if it were the only
rule applied in building a tree, then this would result in the smaller tree. (This could
obviously be generalized by considering 3-element subsets of the vertices, or by
employing a different criterion for choosing a branching rule.)

4.3 Shrink the Kernel: The Nemhauser–Trotter Theorem

Clearly the algorithms found so far will work better if we can run them on smaller
kernels. The following kernelization for VERTEX COVER is one of the most important kernels known, particularly for applications. We will assume that the reader is
familiar with (network flows and) matchings in bipartite graphs. While the previous
kernelizations have been based on rules exploiting local graph structure, this one
is based on the structure of the whole graph. It is based on matching in bipartite
graphs (see Chap. 34), and hence the performance depends on the running time of
the fastest algorithm for this problem (which works in $O(|G|)$ time).

Theorem 4.3.1 (Nemhauser and Trotter [544]) *Let $G = (V, E)$ have n vertices and
m edges. In time $O(\sqrt{n}m)$, we can compute two disjoint sets of vertices, C and H,
such that*
(1) *If $D \subseteq H$ is a vertex cover of $G[H]$, then $X = D \sqcup C$ is a vertex cover of G.*
(2) *There is a minimum sized vertex cover of G containing C.*
(3) *The minimum vertex cover for the induced subgraph $G[H]$ has size at least $\frac{|H|}{2}$.*

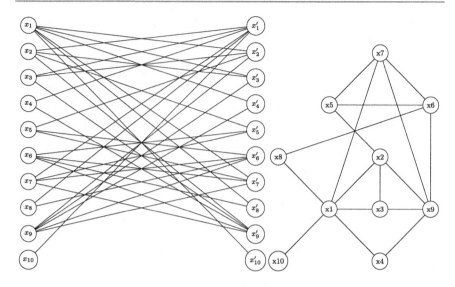

Fig. 4.1 Bipartite adjacency representation of the graph on the right

Here is the algorithm.

Algorithm 4.3.1 For $G = (V, E)$, construct a bipartite graph $B = (V, V', E_B)$ by defining $E_B = \{xy', x'y \mid xy \in E\}$, and $V' = \{v' \mid v \in V\}$.

An example is shown in Fig. 4.1.

Compute a maximum matching for B. By König's Minimax Theorem, this gives an optimal vertex cover C_B. Define

$$X = \{x \in V \mid x \in C_B \wedge x' \in C_B\},$$

$$H = \{x \in V \mid x \in C_B \vee x' \in C_B\} - X.$$

Evidently, $X \sqcup H$ is a vertex cover of G since no edge xy can be left out, lest we improve the matching via, for example, x, y' in B. We now verify Algorithm 4.3.1 and argue that it achieves the claims of Theorem 4.3.1.

Proof of Theorem 4.3.1 Define the independent set $I = \{x \in V \mid x \notin C_B \wedge x' \notin C_B\}$.

Claim 1: (1) is true. To see this, let D be a vertex cover of $G[H]$. Let $xy \in E(G)$. Note that if $x \in I$, it must be that y and y' are in C_B to have the edges xy' and $x'y$ covered in B. Thus $y \in X$. If $y \in I$ similarly. If x or y are in X, there is nothing to prove. Finally if $x, y \in H$, then $x \in D$ or $y \in D$, as D covers $G[H]$.

Claim 2: (2) is true, namely X is part of a minimum sized vertex cover of G. To prove this, we will take a minimum vertex cover C of G and then prove that $X \sqcup (C \cap H)$ is a vertex cover and has the same size. By (1), we know already that it is a vertex cover, since $C \cap H$ is a vertex cover of H. Let $C_1 = C \cap X$. Let $C_1' = \{x' : x \in C_1\}$.

First we show that $(V - \overline{(C \cap I)}) \cup C_1'$ is a vertex cover of B. Let $xy' \in E(C_B)$. If $x \notin \overline{(C \cap I)}$, then evidently $x \in C_1'$ by the first part of the union. If $x \in \overline{(C \cap I)}$, then $x \in I$, implying x is in neither X nor C. Since C is a vertex cover, $y \in C$. As the edges xy' and yx' are both covered *and* $x \in I$, it must be that these edges are both covered (in the covering C_B) by y or y'. Hence $y' \in X$. It follows that $y' \in C_1'$, and so, $F = (V - \overline{(C \cap I)}) \cup C_1'$ is a vertex cover of B.

We prove that $|X| \leq |C - H| = |C - (C \cap H)|$ as then $|X \cup (C \cap H)| \leq |C|$. Now $2|X| + |H| = |H \sqcup X \sqcup X'| = |C_B| \leq |F|$, by the minimality of C_B. Also, $|F| = |V - \overline{(C \cap I)}| + |X'| = |(H \sqcup X \sqcup I) - (I - (C \cap I))| + |X'|$. Thus $|F| = |H| + |X| + |C \cap I| + |\cap C \cap C|$. This all implies $|X| \leq |C \cap I| + |C \cap X| = |C \cap I| + |C \cap X| + |C \cap H| - |C \cap H| = |C| - |C \cap H|$.

This proves (2).

Claim 3: (3) is true, namely $G[H]$ has a minimum vertex cover of size $\leq \frac{|H|}{2}$. Let E be a minimum vertex cover of $G[H]$. By (1), $X \sqcup E$ is a vertex cover of G. Then we see that $X \sqcup X' \sqcup E \sqcup E'$ is a vertex cover of C_B. By definition, $|C_B| = |H| + 2|X|$ and by optimality of C_B, we see $|H| + 2|X| \leq |X \sqcup X' \sqcup E \sqcup E| = 2|X| + 2|E|$, and hence $|H| \leq 2|E|$. Thus, $|H| \leq 2|E|$, as required.   □

One corollary to the proof of Theorem 4.3.1, Claim 2, is the following.

Corollary 4.3.1 *Suppose that we have the decomposition of Theorem* 4.3.1. *Let C be any minimum vertex cover of G. Then* $X \sqcup (C \cap H)$ *is also a minimum vertex cover of G.*

Theorem 4.3.1 gives us our desired small kernel for VERTEX COVER.

Corollary 4.3.2 (Chen, Kanj, Jia [146]) VERTEX COVER *has a kernelization of size* $2k$. *We can compute this in time* $O(k|G| + kf(k))$ *where* $f(k)$ *is the size of any other kernelization whose size is a function of k and which can be found in time* $O(k|G|)$.

Proof Begin with some kernelization such as the Buss kernelization. Now apply Theorem 4.3.1 to get the set $\widehat{C}$ of vertices which must be in any vertex cover; and then the kernel is the subgraph $G[H]$ induced by H. If this has size $> 2k$ by Theorem 4.3.1 (3), then the answer is *no*. Else $G[H]$ is the desired kernel.   □

It has long been known that you can use linear programming to get a $2n$ approximation for (classical) VERTEX COVER. Indeed, it is a longstanding open question whether you can do better. The answer depends on something called the UNIQUE GAMES CONJECTURE. To sketch this rather deep area at a high level:

- To construct a kernel of size $(2 - \varepsilon)k$ will necessitate us solving a longstanding question in classical complexity theory called the UNIQUE GAMES CONJECTURE. This conjecture is by Khot and Regev [440]. One way to specify the UNIQUE GAMES CONJECTURE is to consider *colorings of a graph with unique constraints*. Here the possible colorings of the edges are restricted to a fixed set of ordered pairs, and for uniqueness we ensure that the allowable sets of pairs can color each vertex with only one color (for example, in the set of allowable colors, you could not have $(r, b), (g, b)$). The reader may see that this means when we color some vertex, the color of the component is determined. Deciding if we can satisfy all the constraints, is in P. If we cannot satisfy all the constraints, the fraction of the number of constraints which can be satisfied by some assignment is called the *value*. The UNIQUE GAMES CONJECTURE is to assert a gap phenomenon. The δ, ε-gap problem is the pair $L = \{G \mid$ some assignment satisfies at least $(1 - \delta)$ of the constraints$\}$ and $\widehat{L} = \{G \mid$ every assignment satisfies at most ε of the constraints$\}$. That is, for every sufficiently small δ, ε, there is a k such that the δ, ε-gap problem over an alphabet of size k is NP-complete. Khot and Regev applied this to get the $2 - \varepsilon$ bound for VERTEX COVER, and later others have used it similarly. For example, Feige and Goemans [283] used it for sharper bounds for MAX 2-SAT.

- Later, when we examine general methods for lower bounds, in Theorem 30.8.2, we will see Dell and van Melkebeek [206] prove that (assuming co-NP $\not\subseteq$ NP/poly) VERTEX COVER has no kernel of overall graph size $O(k^{2-\varepsilon})$ for any $\varepsilon > 0$.

- If we only assume that P $\neq$ NP, then the best we can prove using approximation lower bounds is a $1.36k$ lower bound using the "PCP" theorem by Dinur and Safra [223]. The celebrated PCP Theorem is due to Arora, Lund, Motwani, Sudan, and Szegedy in [47]. Its proof is beyond the scope of the present book but asserts that there is polynomial transformation which, given an instance of 3SAT, produces another such that the first is satisfiable iff the second is, but that if the first is not, then the second has only $1 - \delta$ of its clauses satisfiable. The reason that it could be used for lower bounds on approximation is clear. The proof uses sophisticated methods from algebraic coding theory.

The linear programming way to achieve the Nemhauser–Trotter kernel bound is described below. The running time depends on the implementation of linear programming, which is reasonably quick in practice if you use something like the SIMPLEX ALGORITHM but in that case there is no guarantee, and slower if you use polynomial-time algorithms which always work in polynomial time such as various ellipsoidal methods. Sometimes there are other methods which give somewhat bigger kernels but seem to run more quickly in practice. We will see this when we look at *crown reductions* in Sect. 5.2 where the kernel size is $3k$ rather than $2k$.

Theorem 4.3.2 (Bar-Yehuda and Even [54]) *The Nemhauser–Trotter kernel bound of $2k$ can be obtained using relaxation methods and linear programming.*

Proof We can formulate VERTEX COVER as an integer programming problem. Let $G = (V, E)$ be a graph. The problem is then formulated as follows. For each $v \in V$ we have a variable x_v with values $\{0, 1\}$. VERTEX COVER is then:

- Minimize $\sum_{v \in V} x_v$ subject to
- $x_u + v_v \geq 1$ for each $uv \in E$.

The classical relaxation of this makes it a linear programming problem by having x_v rational valued with $x_v \in [0, 1]$.

We can (more or less) define the sets of Theorem 4.3.1 by asking that the values be in the set $\{0, \frac{1}{2}, 1\}$.

$$S_1 = \left\{ v \in V \mid x_v > \frac{1}{2} \right\},$$

$$S_{\frac{1}{2}} = \left\{ v \in V \mid x_v = \frac{1}{2} \right\},$$

$$S_0 = \left\{ v \in V \mid x_v < \frac{1}{2} \right\}.$$

We claim that *there is a minimal sized (integral)* VERTEX COVER S *with* $S_1 \subseteq S \subseteq S_1 \sqcup S_{\frac{1}{2}}$, $S \cap S_0 = \emptyset$—*and there,* $(G[S_{\frac{1}{2}}], k - |S_1|)$ *is a kernel of size at most* $2k$. As in Theorem 4.3.1, let S be a minimal vertex cover. Now, as in the proof of Theorem 4.3.1, we consider $(S_1 - S) \cup (S - S_0)$. Since we need to make $x_u + x_v \geq 1$ for all $uv \in E$, we conclude that $N(S_0) \subseteq S_1$, and hence $|S_1 - S| \geq |S \cap S_0|$, since S is optimal. We claim that $|S_1 - S| \leq |S \cap S_0|$. Suppose otherwise. If we define $\delta = \min\{x_v - \frac{1}{2} \mid v \in S_1\}$, and for all $u \in S \cap S_0$ set $x_u := x_u + \delta$ and for $v \in S_1 - S$, set $x_v := x_v - \delta$, then $\sum_{v \in V(G)} x_v$ would be smaller if $|S_1 - S| > |S \cap S_0|$, which is impossible. Therefore $|S_1 - S| = |S \cap S_0|$ and as in the proof of Theorem 4.3.1, we substitute $S_1 - S$ for $S \cap S_0$ which is allowable since $N(S_0) \subseteq S_1$. Thus, we obtain a vertex cover of the same size.

To see that $|S_{\frac{1}{2}}| \leq 2k$, as with many relaxations, we note that the solution to the (objective function for the) linear programming problem is a lower bound for corresponding integer programming problem. Applying this to S_1, we see that any solution is bounded below by $\sum_{v \in S_{\frac{1}{2}}} x_v = |S_{\frac{1}{2}}|/2$. Therefore, if $|S_1| > 2(k - |S_{\frac{1}{2}}|)$, then this is a "no" instance. $\square$

4.4   Interleaving

There is another very important method of speeding up the process. The idea is that we should *re-kernelize* whenever possible.[2]

---

[2]There is another method commonly used in heuristics for speeding things up: *do not repeat computationally expensive tasks.* By this we mean: do not replicate things we have already done. Langston's group have used this in biology and a good example of this is the use of Brendan McKay's graph isomorphism package NAUTY in the computation of knot polynomials such as the study of Haggard, Pearce, and Royle [375]. The construction of what are called *Tutte* polynomials is known to be hard in general, as it is PSPACE complete. The usual method for the calculation of the Tutte polynomial (a kind of generalization of the chromatic polynomial) is a long exponential deletion/contraction algorithm, often with lots of replication of effort. There the idea is to test, reasonably efficiently, for isomorphism, and if we realize that we have already dealt with the current

The idea is to combine the two methods outlined *iteratively*. For example, for the k-VERTEX COVER problem, we can first reduce any instance to a problem kernel and then apply a search tree method to the kernel itself. Niedermeier and Rossmanith [547] developed the technique of *interleaving* depth-bounded search trees and kernelization. They show that applying kernelization repeatedly during the course of a search tree algorithm can significantly improve the overall time complexity in many cases.

Take any fixed-parameter algorithm that satisfies the following conditions. The algorithm works by first reducing an instance to a problem kernel, and then applying a depth-bounded search tree method to the kernel. Reducing any given instance to a problem kernel takes at most $P(|I|)$ steps and results in a kernel of size at most $q(k)$, where both P and q are polynomially bounded. The expansion of a node in the search tree takes $R(|I|)$ steps, where R is also bounded by some polynomial. The size of the search tree is bounded by $O(\alpha^k)$. The overall time complexity of such an algorithm running on instance (I, k) is $O(P(|I|) + R(q(k))\alpha^k)$.

The interleaving strategy developed in [547] is basically to apply re-kernelization to the problem instances at every node of the search tree algorithm. To expand a node in the search tree labeled by instance (I, k), we first check whether or not $|I| > c \cdot q(k)$, where $c \geq 1$ is a constant whose optimal value will depend on the implementation details of the algorithm. If $|I| > c \cdot q(k)$ then we apply the kernelization procedure to obtain a new instance (I', k'), with $|I'| \leq q(k)$, which is then expanded in place of (I, k). A careful analysis of this approach shows that the overall time complexity is reduced to $O(P(|I|) + \alpha^k)$.

This really does make a difference. The 3-HITTING SET problem is given as an example in [546]. An instance (I, k) of this problem can be reduced to a kernel of size k^3 in time $O(|I|)$, and the problem can be solved by employing a search tree of size 2.27^k. Compare a running time of $O(2.27^k \cdot k^3 + |I|)$ (without interleaving) with a running time of $O(2.27^k + |I|)$ (with interleaving).

Although the techniques of *kernelization* and *depth-bounded search tree* are simple algorithmic strategies, they are not part of the classical toolkit of polynomial-time algorithm design, since they both involve costs that are *exponential in the parameter.*

We will not dwell further on this technique here, because in Chap. 6 we will look at a technique called *measure and conquer* which uses these ideas and extends them further by a control structure for the branching in the tree.

4.5   Heuristics and Applications II

VERTEX COVER kernelizations are used extensively in practice. An excellent case study is provided by the work of Langston and his group in this area. In this section we will discuss the issues that arise in such practical deployments.

---

graph in the recursive calculation, then we can call what we have already done and terminate that computational branch.

One example is reported in Langston, Perkins, Saxton, Scharff, and Voy [480]. This paper describes techniques used in a groundbreaking study applying graph theoretical methods to understand the effects of low-level radiation on genes (Voy, Scharff, Perkins, Saxton, Borate, Chesler, Branstetter, and Langston [659]). These methods have also been used for studying genetic markers for chronic fatigue.

The study generated genome-scale gene expression data obtained via in vitro micro-array samples, collected after low dose exposure to radiation.[3] What comes out of all of this is a vast amount of data, within which one seeks to understand correlations between gene expression and cell behavior.

One popular method is to use what are called relevance networks, which are huge matrices of correlation coefficients between *all* pairwise combinations of genes. Now we are in the realm of graph theory. Genes are vertices and correlations are (weighted) edges. We allow an edge depending on the weight threshold determined by the biology (and engineering constraints). The project of evaluating relevance networks needs the cooperation of biologists, and computer scientists with goals ranging from very theoretical to completely practice-driven ones.

What we look for in this data, are *cliques*. From Langston et al. [480] we quote:

> The importance of cliques lies in the fact that each and every pair of vertices is joined by an edge, from which we infer some form of co-regulation among the corresponding edges.

Strictly speaking, cliques are too much to expect, but we look for something like that. As we will later see, CLIQUE is thought not to be FPT, so this looks bad also.

First, we notice that if the graph is sparse, then the techniques will be parallelizable, and we can work in small subregions of the graph. Moreover, with any luck, cliques will be small. In that case, using the complementary graph G' we seek vertex covers of size $n - k$ in subgraphs using one of the parameterized vertex cover algorithms, or even the heuristical versions.

For this to work, other techniques such as *load balancing* are needed. The graphs analyzed have size around 12,500 and the covers are around 12,000. The run time without load balancing on a supercomputer were around 6 days and with the balancing was around 2–4 hours.

---

[3]For those of us, like the authors, unfamiliar with the material here, we hope to give the flavor of the scientific work rather than the details. Almost every cell of the body contains identical genes. However, only a fraction of these genes are active in each cell. This gene "expression" acts both as an "on/off" switch as well as a "volume control", enabling a cell to respond dynamically to environmental stimuli and to its own changing needs. Gene expression is studied using micro-arrays. Small glass slides are set up as grids of tiny dots: samples of DNA sequences from cells to be studied and also from control cells with known properties. A micro-array may contain samples of DNA sequences from thousands of different genes. The micro-array is "washed" with four (fluorescently colored) nucleic molecules, which quickly pair with complementary gene sequences. A single experiment will soon show, in these four colors, the nucleotide bonding levels of hundreds or thousands of gene sequences within a cell. With the aid of a computer, the amount of bonding is precisely measured, giving a colorful profile of gene expression in the cell. Image analysis using sophisticated statistical methods then gains relatively clean data about how the micro-array reacted, and exactly which genes were expressing. Recently, this bonding method has been replaced with light sensitive chips on the assays, exploiting the reactivity of DNA to light. It all involves significant computer analysis. The numbers are very large.

As Langston remarked from a question of the first author:

In transcriptomic, proteomic, and other sort of -omics data we receive, the cliques are rarely exceedingly large. More importantly, these "real" graphs are typically sparse and nonregular. So the VC-FPT machine works quite well, at least as long as we do a lot of high performance computing and dynamic load balancing, etc.

But contrived graphs can still be quite difficult for us. Troublesome instances tend to be highly regular, so that kernelization is ineffective. A good example is the Keller graph, which we were recently able to solve only by using a great deal of symmetry and the Kraken supercomputer.

Other algorithms are extensively used. One outcome of the approach was the discovery of new links between low dose irradiation exposure and genes not previously linked to radiation, such as the wonderfully named tubby-like protein 4 (*Tulp4*, see Santagata, Boggon, Baird, Gomez, Zhao, Shan, Mysczka, and Shapiro [599]).

We believe that these methods will have many further applications.

4.6   Heuristics and Applications III

Kernelization is based on an old idea, that of pre-processing, or *reducing*, the input data of a computational problem. It often makes sense to try to eliminate those parts of the input data that are relatively easy to cope with, shrinking the given instance to some "hard core" that must be dealt with using a computationally expensive algorithm. In fact, this is the basis of many heuristic algorithms for NP-hard problems, in a variety of areas, that seem to work reasonably well in practice.

Kernelization is also applied to problems we do not think are FPT but still perform well in practice.

A compelling example of the effectiveness of data reduction, for a classically posed NP-hard problem, is given by Carsten Weihe [661]. He considered the following problem in the context of the European railroad network: given a set of trains, select a set of stations such that every train passes through at least one of those stations, and minimize the number of selected stations. Weihe modeled this problem as a path cover by vertices in an undirected graph. Here, we formulate the problem as domination of one set of vertices by another in a bipartite graph.

TRAIN COVERING BY STATIONS

Instance: A bipartite graph $G = (V_S \cup V_T, E)$, where the set of vertices V_S represents railway stations and the set of vertices V_T represents trains.
E contains an edge $(s, t), s \in V_s, t \in V_T$, iff the train t stops at the station s.

Problem: Find a minimum set $V' \subseteq V_S$ such that V' covers V_T.
That is, for every vertex $t \in V_T$, there is some $s \in V'$ such that $(s, t) \in E$.

Weihe employed two simple data reduction rules for this problem. For our problem formulation, these translate to the following TCS rules.

Reduction Rule TCS1 Let $N(t)$ denote the neighbors of t in V_S. If $N(t) \subseteq N(t')$ then remove t' and all adjacent edges of t' from G. If there is a station that covers t, then this station also covers t'.

Reduction Rule TCS2 Let $N(s)$ denote the neighbors of s in V_T. If $N(s) \subseteq N(s')$ then remove s and all adjacent edges of s from G. If there is a train covered by s, then this train is also covered by s'.

In practice, exhaustive application of these two simple data reduction rules allowed for the problem to be solved in minutes, for a graph modeling the whole European train schedule, consisting of around $1.6 \cdot 10^5$ vertices and $1.6 \cdot 10^6$ edges.

Weihe's problem is a essentially BIPARTITE DOMINATING SET, a problem known to be complete for the parameterized intractability class $W[2]$. This points to the possible importance of pre-processing as a heuristic, regardless of whether the problem is proved mathematically difficult (in either the classical or parameterized complexity frameworks). For other recent connections between parameterized complexity and heuristics we refer the reader to Abu-Khzam and Markarian [10] and Abu-Khzam and Mouawad [11] where issues such as load balancing and data structures are discused.

We next turn to the formal definition.

4.7 Definition of a Kernel

The method of the previous sections allows us to give a general definition of what we mean by a kernel.

The simplest version of kernelization is the following, although there are variations of the idea below.

Definition 4.7.1 (Kernelization) Let $L \subseteq \Sigma^* \times \Sigma^*$ be a parameterized language. A *reduction to a problem kernel,* or *kernelization,* replaces an instance (I, k) by a reduced instance (I', k'), called a *problem kernel,* such that

(i) $k' \leq k$,

(ii) $|I'| \leq g(k)$, for some function g depending only on k, and

(iii) $(I, k) \in L$ if and only if $(I', k') \in L$.

The reduction from (I, k) to (I', k') must be computable in time polynomial in $|I| + |k|$.

The definition above is good for *most* practical applications. The map $(I, k) \mapsto (I', k')$ is a *polynomial-time parameterized reduction* and will be used later in Chap. 30. Strictly speaking it could be a *parameterized reduction* as defined when we consider the hardness theory, in Chap. 20. Moreover, strictly speaking, it would be enough that we satisfied (ii) and (iii) without (i), so that it was not a *parameter decreasing* kernelization. Moreover, L could even be a different parameterized problem.

That is, we could have $(I, k) \in L$ iff $(I', k') \in L'$. For example, L' might be an annotated version of L. In the literature when L' is different from L kernelization is sometimes called *bi-kernelization*. This is sometimes also called *polynomial compression*. The term bi-kernel is from Alon, Gutin, Kim, Szeider, and Yeo [33]. In this book, we will not distinguish between kernel and bi-kernel when the context is clear.

Finally, there is also a *parameterized Turing kernel*, where the problem reduces to *many* different kernels, but we will leave this until Sect. 5.1.3 where it will be met in context.

> Until further notice, all kernelizations will be the simple parameter-reducing ones found in the basic definition.

It might seem that kernelization is a very specialized method of constructing FPT algorithms. So we offer the following rather fundamental observation. By way of motivation, the following result has a proof similar to that used in Chap. 3 for PLANAR INDEPENDENT SET in the first proof of its tractability (Proposition 3.3.1). There, recall that if the graph was big enough, then it must have a size k independent set. If the graph is large, then answering the decision problem question is easy. Otherwise, the (small) remaining case can be solved by (possibly non-uniform) table lookup. This situation applies all FPT problems.[4]

Proposition 4.7.1 *A language L is FPT if and only if L is kernelizable. (This only makes sense for the uniform and strongly uniform case.)*

Proof We do the uniform case; the other one is similar but more technical. For the nontrivial direction, let L be FPT in time $O(f(k)|x|^c)$ on input $\langle x, k \rangle$, via algorithm Φ. For the strongly uniform case, we suppose that f is computable.

The key observation is that from some length n onwards, $f(k)n^c < n^{c+1}$. Thus we can do the following. On input $\langle x, k \rangle$, for $|x|$ many steps, we will first try to compute $f(k)$, and see if $f(k)n^c < n^{c+1}$. If either we cannot yet compute $f(k)$ or the inequality is not satisfied, then we run Φ and put the answer in a look-up table.

On the other hand if the answer is yes, we run Φ on the input and the result will halt in $\leq |x|^{c+1}$ many steps. The kernel is the look-up table.   □

We remark that the theorem above is really only of academic interest. The kernels generated by the proof are far too large to be of use for practical computation, or for use in the elusive search for the best FPT algorithm, which sometimes is *not* a kernelization algorithm.

4.8   *k*-MAXIMUM LEAF SPANNING TREE

We can similarly solve the following problem.

k-LEAF SPANNING TREE
Instance: A graph $G = (V, E)$.

---

[4]The result should also be compared with the "advice view" of Chap. 2, Exercise 2.3.4.

Parameter: A positive integer k.
Question: Is there a spanning tree of G with at least k leaves?

The unparameterized version is NP-complete [485].

Theorem 4.8.1 (Downey, Doyle and Fellows)   *k*-LEAF SPANNING TREE *is solvable in time* $O(n + (2k)^{4k})$.[5]

Proof Note that any graph G that is a *yes* instance must be connected. A vertex v is called *useless* if (i) it has neighbors u and w of degree 2 and (ii) v has degree 2. We will argue that any sufficiently large graph without useless vertices of degree 2 is necessarily a *yes* instance. Note also that if G has a vertex of degree at least k, then G is a *yes* instance.

Say that a useless vertex v is *resolved* by deleting v from G and adding an edge between u and w. Let G' denote the (*resolved*) graph obtained from G (in linear time) by resolving all useless vertices.

Our algorithm for *k*-LEAF SPANNING TREE is very simply described:

Step 1.   Check whether G is a connected and whether there is a vertex of degree $\geq k$. If G has such a vertex, declare that G has no k-leaf spanning tree and stop.

Step 2.   If the answer is still undetermined, then compute G'. If G' has at least $4(k + 2)(k + 1)$ vertices, then the answer is *yes*.

Step 3.   Otherwise, exhaustively analyze G' and answer accordingly, since G' has a k-leaf spanning tree if and only if G does.

Our proof that the algorithm is correct employs the following fact.

Claim *If H is a connected simple resolved graph of order at least $4(k + 2)(k + 1)$, then H has a spanning tree with at least k leaves.*

To establish the claim, it can be shown by elementary induction that the following result holds (Exercise 4.11.3).

Proof of the Claim

> If a tree T has i internal (i.e., nonleaf) vertices of degree at least 3, then T has at least $i + 2$ leaves. $(\ast)$

Suppose H satisfies the hypotheses of the above Claim. If H has any vertex of degree k, we are done, and hence, without loss of generality, we may presuppose that H has no vertex of degree $\geq k$. Let T be a spanning tree of H having a maximum number of leaves l, and suppose $l \leq k - 1$. Now, as T has $4(k + 2)(k + 1)$ vertices but only l leaves, it must have at least $4(k + 2)(k + 1) - (k - 3) - (k - 1)$ vertices of degree 2 [by $(\ast)$]. Since the maximum degree in H of any vertex is $k - 1$ and

$$4(k + 1)(k + 2) - (k - 3) - (k - 1) - (k - 1)(k - 1) = 3k^2 + 12k + 11,$$

---

[5]The original proof of this theorem in [245] contained a combinatorial error.

Fig. 4.2 Increasing the
number of leaves

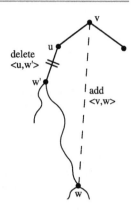

it follows that there are at least $3(k + 1)(k + 3) + 2$ vertices of degree 2 in T that
have the property that they are not connected to any leaf of T. So there are two
cases regarding the $3(k + 1)(k + 3) + 2$ internal vertices of degree 2 in T that are
not connected to any leaf of T.

Case 1. Of the $3(k + 1)(k + 3) + 2$ internal vertices of degree 2 in T that are not
connected to any leaf of T, at least $k - 2$ internal vertices have degree 3 or more in
the simple resolved graph H.

Let v be any such vertex not connected to any leaf of T having degree 3 or more
in H. For such vertices, it is convenient to regard T as rooted with root v. Let v be
such a degree 2 vertex of T. Regard T as a rooted tree with root v. Let $\{u_1, u_2\}$ be
the children of v. Now, as v has degree 2 in T, and yet degree 3 or more in H, it
follows that v is adjacent to some vertex w in H which is not a child of v (since H
is simple), and furthermore, w is not a leaf of T by choice of v. We claim that there
is such a choice of v such that one vertex of the path from v to w has degree 2 in T.
If not, then there are at least $k - 2$ internal vertices of degree 3 or more in T. But
then by $(*)$, there must be at least k leaves.

Thus, without loss of generality, suppose there is a node u of degree 2 in T on the
path from v to w. Let w' be the child of u. It is possible to change T into a new tree
T' with more leaves as follows. We can make u a leaf by deleting the edge (u, w')
and add an edge (v, w) where there is a path in T from w' to w. The reader should
refer to Fig. 4.2.

The point is that as w is not a leaf, this must increase the net number of leaves.

Case 2. Of the $3(k+1)(k+3)+2$ internal vertices of degree 2 in T not connected
to any leaf of T, at most $k - 3$ internal vertices have degree 3 or more in the simple
resolved graph H.

Then each of the remaining $(3k + 8)(k + 1)$ vertices of degree 2 in both T and
H are connected to at least one vertex of degree 3 or more in H (as H is resolved).
Since each vertex of H has maximum degree $k - 1$ and each of the $(3k + 8)(k + 1)$
vertices are connected to at most one other such vertex (again as the graph is re-
solved), there are at least $\lceil \frac{(3k+8)(k+1)}{k-1} \rceil \geq 3k + 1$ vertices of degree 3 or more in H
which are not leaves of T. These $3k + 1$ vertices of degree 3 or more in H must
therefore be internal vertices of T. At most $k - 3$ of them can have degree 2 in T

and therefore at least $2k + 4$ of them will have degree 3 or more in T. Therefore, by $(*)$, T has at least $2k + 6$ leaves.

This concludes the proof of the Claim. $\qquad\square$

It is now easy to see that the algorithm is correct since the derived graph, G', will satisfy the hypotheses of the Claim in step 2 of the algorithm. $\qquad\square$

Theorem 4.8.1 improves a result of Bodlaender, who showed that k-LEAF SPAN-NING TREE is linear-time fixed-parameter tractable with a multiplicative factor depending on k (Bodlaender [69]). We will look at sharper bounds in Chap. 7, specifically Sect. 7.1.2, where we find efficient reduction rules using the Method of Extremal Combinatorics.

In the next chapter, we will look at the version of k-LEAF SPANNING TREE for directed graphs. It has some unusual consequences for our studies, in that while it is FPT, it does not have a polynomial sized kernel unless the polynomial-time hierarchy collapses, but it does have what we call a *Turing Kernel*.

4.9 Hereditary Properties and Leizhen Cai's Theorem

Definition 4.9.1 (Hereditary Property) If Π is a property of graphs, we say a graph G is a Π graph iff G satisfies Π. We say that a property Π is a *hereditary property*, if, given any Π graph G, whenever H is an induced subgraph of G, then H is a Π graph.

There is a large literature concerning the parameterized complexity of hereditary graph properties. The definitional template for these problems is as follows.

Π-INDUCED SUBGRAPH
Instance: A graph $G = (V, E)$.
Parameter: An integer k.
Question: Does there exist an induced subgraph H of G on k vertices that belongs to Π?

If Π excludes both large cliques and independent sets, then by Ramsey's Theorem,[6] Π-INDUCED SUBGRAPH is FPT. The general problem otherwise is $W[1]$-hard[7] by Khot and Raman [439]. We will look further at general hereditary properties in Chap. 30.

Leizhen Cai's results are concerned with modifying graphs to get Π-graphs. The literature is filled with many such *graph modification* problems. As Leizhen Cai [117] remarks, in general these problems can be placed in the categories below.

(1) The *edge deletion problem*: Find a set of edges of minimum cardinality whose removal results in a Π graph.

---

[6]That is, for all k, we can compute $g(k)$ such that if G is a graph with at least $g(k)$ many vertices, it must have either a clique or an independent set of size k.

[7]That is, hard for the basic parameterized complexity class we meet in Chap. 21.

(2) The *vertex deletion problem*: Find a set of vertices of minimum cardinality whose removal results in a Π graph.

(3) The *edge/vertex deletion problem*: Categories (1) and (2) combined.

(4) The *edge addition problem*: Find a set of new edges of minimum cardinality whose addition results in a Π graph.

(1), (2) and (3) are usually NP-hard for any nontrivial hereditary property. Leizhen Cai [117] considered the following general parametric version of the modification problem.

$\Pi_{i,j,k}$ GRAPH MODIFICATION PROBLEM

Input: A graph $G = \langle V, E \rangle$.
Parameters: Non-negative integers i, j, and k.
Question: Can we delete at most i vertices, j edges, and add at most k edges and get a Π graph?

We will say that a property Π has a *forbidden set characterization* iff there exist a set $\mathcal{F}$ of graphs such that G is a Π graph iff G does not contain a member of $\mathcal{F}$ as an induced subgraph. If we can take $\mathcal{F}$ to be finite, then we say that Π has a finite forbidden set characterization.

Clearly, if $\mathcal{F}$ is a finite set characterizing Π, then there is a P-time recognition algorithm for the $\Pi_{i,j,k}$ GRAPH MODIFICATION PROBLEM for any fixed i, j and k as follows. Let N be the maximum number of vertices of any graph in $\mathcal{F}$. Given a graph Q, we can check whether Q is a Π graph in time $O(|Q|^N)$ by checking if Q' has an induced subgraph with $\leq N$ vertices in $\mathcal{F}$. Therefore, the $\Pi_{i,j,k}$ GRAPH MODIFICATION PROBLEM is solvable in time $O(|G|^{i+2j+2k+N})$. However, Cai used essentially the problem kernel method to prove the result below.

Theorem 4.9.1 (Leizhen Cai [117]) *Suppose that Π is any property with a finite forbidden set characterization. Then the $\Pi_{i,j,k}$ GRAPH MODIFICATION PROBLEM is FPT in time $O(N^{i+2j+2k}|G|^{N+1})$ where N denotes the maximum size of the vertex set of any graph in the forbidden set $\mathcal{F}$.*

We need the lemma below.

Lemma 4.9.1 (Leizhen Cai [117]) *For any hereditary property Π, if Π is recognizable in time $T(G)$ for any graph G, then for any graph G which is not a Π graph, a minimal forbidden induced subgraph for Π can be found in time $O(|V(G)|T(G))$.*

Proof Let $\mathcal{A}$ be the recognition algorithm for Π running in time $T(G)$, so that $\mathcal{A}(G) = 1$ iff G is a Π graph. So first use $\mathcal{A}(G)$ to see if G is a Π graph. Assuming that the answer is "no," we perform the following algorithm. Order the vertices of G as $\{v_0, \ldots, v_n\}$. The algorithm proceeds in stages. At stage 0, $G_0 = G$. At stage s, using $\mathcal{A}$ applied to $G_s - \{v_s\}$, see if $G_s - \{v_s\}$ is a Π graph. If the answer is yes, then v must belong to a minimal forbidden induced subgraph. We let $G_{s+1} = G_s$. If the answer is no, then we define $G_{s+1} = G_s - \{v_s\}$. In this case, $G_s - \{v_s\}$ must contain a minimal forbidden induced subgraph.

Clearly, if G is not a Π graph, then G_{n+1} is a minimal induced forbidden subgraph, as required. □

Proof of Theorem 4.9.1 Fix i, j and k. Assuming that G is not a Π graph, we will repeat the following steps until we either get a Π graph, or we have used up the deletion of $\leq i$ vertices, $\leq j$ edges, and the addition of $\leq k$ edges. In the latter case, we output "no".

Step 1. Find a minimal forbidden induced subgraph H of G.

Step 2. Modify G either by deleting an edge or a vertex, or adding an edge to H. (The crucial idea is to modify only minimal induced forbidden subgraphs.)

It is clear that the algorithm works. Now using the finite set characterization, we can recognize if G is a Π graph in time $O(|V(G)|^N)$. Therefore, by Lemma 4.9.1, we can compute a minimal induced forbidden subgraph H of G in time $O(|V(G)|^{N+1})$. Since N is the largest number of vertices of any member of the forbidden set $\mathcal{F}$, there are at most $\binom{N}{2}$ many ways to add or delete an edge of H, and at most N many ways to delete a vertex from H. The total number of graphs we can generate in the algorithm above is thus $N^i \cdot \binom{N}{2}^{j+k} = O(N^{i+2j+2k})$. The running time of the algorithm is thus $O(N^{i+2j+2k}|V(G)|^{N+1})$ as required. □

The reader should note that there are several interesting $\Pi_{i,j,k}$ graph modification problems that are *not* characterized by a finite forbidden set and yet are still FPT. One important example is MINIMUM FILL-IN (also called CHORDAL GRAPH COMPLETION), which asks if we can add k edges to G to make a chordal graph. (A chordal graph where any cycle of length greater than 3 contains a chord.) This problem is very important in the area of computational biology called perfect phylogeny or interval graph completion for DNA physical mapping. Interval graph completion for DNA physical mapping is a $\Pi_{0,0,k}$ graph modification problem, which is classically NP-hard (Yannakakis [668]). There is no finite forbidden set characterization (Exercise 4.11.2). Nevertheless, interval graph completion is FPT. Interval graph completion provides an interesting example of an FPT problem since there are several algorithms to demonstrate this fact. Kaplan, Shamir and Tarjan [430] used the method of bounded search trees to show that the parameterized version above is strongly uniformly FPT in time $O(2^{4k}|E(G)|)$ (Exercise 3.8.13). Kaplan, Shamir and Tarjan [430] used (essentially) the problem kernelization method to give a $O(k^5|V(G)||E(G)| + f(k))$ for suitably chosen recursive f. Leizhen Cai [117] also used the problem kernelization method to give a $O(4^k(k+1)^{-3/2}[|E(G)||V(G)| + |V(G)|^2])$ time algorithm.

4.10 MAXIMUM AGREEMENT FOREST

4.10.1 Phylogeny

We finish this chapter with another nice example of parametric reducing kernelization from computational biology. The problem is from *phylogeny*, which has been working with taxonomic data reduction analyses since its early years. Given modern phylogenetics molecular data, such as aligned DNA or amino acid sequences from each of $|L|$ species, there are now many parametric kernelization techniques for constructing a phylogenetic tree to represent the evolutionary history of a set of species. See, for example, Downey [236], analyzing several applications of parameterized kernelization to construct phylogenetic trees; and Gramm, Nickelsen and Tantau [356], which surveys (up to 2008) parameterized algorithms in phylogenetics. There are extensive applications in the book by Semple and Steel [613].

Definition 4.10.1 (Phylogenetic tree)   A *phylogenetic tree* $T = (V, E, L, \lambda)$ is an unrooted, binary tree where the leaves of T are labeled by a set of *species* L via a bijection λ. Each non-leaf vertex is unlabeled and has degree three.

The goal here to figure out the evolutionary history of the set of species in L. Note that this goal is also relevant to (e.g., European) languages evolution, as discussed in Downey [236], and especially in the work of Tandy Warnow and her group.

Varying the optimization criteria can lead to the construction of different trees. For example, two different phylogenetic tree construction algorithms on the same data might produce two trees T_1 and T_2 covering a common set of species L but such that T_1 is not isomorphic to T_2 as leaf-labeled trees. Similarly, two different data sets (e.g., two distinct gene families) considered to cover the same species, may induce different estimates of the gene trees, even when the same optimization criterion and technique is used. This disagreement motivates strategies for measuring similarity between two trees T_1 and T_2 and finding subtrees (subsets of the species) that do agree. Many techniques such as those in [28, 196] exist in the literature.

We consider the MAXIMUM AGREEMENT FOREST (MAF) problem. The intuition is as follows: given two phylogenetic trees T_1 and T_2 constructed from leaf label set L as input, the MAF problem asks to find the smallest number of trees such that each tree is a subtree of both T_1 and T_2, no two of these trees intersect, and these trees "cover" the entire leaf label set L.

To be more precise, let $T = (V, E, L, \lambda)$ be a phylogenetic tree. Given $U \subseteq V(T)$, let $T[U]$ denote the minimal subtree of T induced by all vertices that lie on paths between pairs of elements of U. Let $\sigma(T[U])$ denote the phylogenetic tree obtained from $T[U]$ by applying (recursively) *contractions* to remove all vertices of degree 2 and (recursively) removing any degree 1 vertex v when there does not exist an $l \in L$ s.t. $\lambda(l) = v$ (that is, v is not a leaf vertex in T).

Definition 4.10.2 (Agreement forest)   An *agreement forest* (AF) for two trees T_1 and T_2 is a collection $\mathcal{F} = \{t_1, \ldots, t_k\}$ of phylogenetic trees such that:

(i) $L(t_1), \ldots, L(t_k)$ partitions L,

(ii) $t_j = \sigma(T_1[L(t_j)]) = \sigma(T_2[L(t_j)])$ for all $j \in \{1, \ldots, k\}$, and

(iii) for $i = 1, 2$, the trees $\{T_i[L(t_j)] : 1 \le j \le k\}$ are vertex disjoint subtrees of T_i.

A *maximum agreement forest* (MAF) for T_1 and T_2 is an agreement forest $\mathcal{F}$ for T_1 and T_2 for which $|\mathcal{F}| = k$ is minimized.

Fig. 4.3 Two unrooted
binary phylogenetic trees T_1,
T_2 with the common leaf
label set
$L = \{a, b, c, d, e, f, g\}$, and
two different maximum
agreement forests, F and F',
for T_1 and T_2

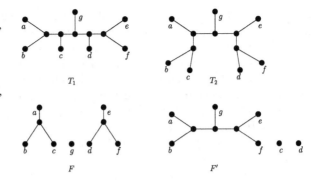

If $\mathcal{F}$ is a MAF for T_1 and T_2, then $k - 1$ is the number of edges that need to be
cut from each of T_1 and T_2 so that the resulting forests both agree with $\mathcal{F}$. We call
$k - 1$ the MAF distance between T_1 and T_2. Note that a MAF for T_1 and T_2 need
not be unique. See Fig. 4.3.

MAXIMUM AGREEMENT FOREST

Input: A pair of unrooted binary phylogenetic trees T_1 and T_2 over a common
leaf label set L.

Parameter: A positive integer k.

Output: An agreement forest $\mathcal{F} = \{t_1, \ldots t_{k'}\}$ where $k' \leq k$, or "no" if the MAF
distance between T_1 and T_2 is $> k$.

Classically, the MAXIMUM AGREEMENT FOREST problem is NP-complete
(Allen and Steel [28]). It can be approximated to within a factor of 3 (see Rodrigues,
Sagot and Wakabayashi [592]).

The MAF distance (introduced in Hein, Jiang, Wang, and Zhang [386]) between
two trees, is known to be equal to the *tree bisection and reconnection* (TBR) operation distance, and to be at least half of the *subtree prune and regraph* (SPR) operation distance [28]. The SPR distance (also known as the *subtree transfer distance*) is
bounded above by the *nearest neighbor interchange* (NNI) distance. Each of these
tree metrics is induced by operations that locally rearrange T_1 to produce T_2.

Allen and Steel [28] give a fixed-parameter tractable algorithm for the equivalent parameterized TBR problem where the number of TBR operations is bounded
by parameter k. The algorithm proceeds by the *kernelization* where the inputs are
phylogenetic trees T_1 and T_2. Specifically, they apply two rules below that are guaranteed to reduce the size of both T_1 and T_2, if they can be applied.

1. Replace any pendant subtree that occurs in both trees, by a single leaf with a new
   label.
2. Replace any chain of pendant subtrees that occur identically, by three new leaves
   with new labels correctly oriented to preserve the direction of the chain.

For both of these rules, the position of attachment of each pendant subtree must
be the same in the two trees. It is not difficult to show that the rules above can
be applied to reduce T_1 and T_2 until no further applications are possible in poly
time. After applying these rules recursively, the resultant kernelized trees have size

$k' \leq 4c(k-1)$, for constant $c < 7$, with the property that the TBR distance between them remains the same as the TBR distance between the two original trees. Note that k' is independent of n, the leaf set size of the original input trees. We prove this below, following the analysis of Allen and Steel. We define an *abc-tree* binary tree as one whose leaf set includes 3 leaves a, b, c with the property that is v_a, v_b, v_c are vertices of T adjacent to $a.b.c$ respectively then $v_a v_b$ and $v_b v_c$ are edges of T.

Lemma 4.10.1 *If T, T' are two abc trees with the same leaf sets, then there is a maximum agreement forest $\mathcal{F}$ for T_1, T_2 in which a, b, c are contained in the leaf set of one of the trees of F.*

Proof Let $\mathcal{F}$ be a MAF for T_1, T_2. Let L_a (L_c) be the set of leaves connected to a (c) once $v_a v_b$ ($v_b v_c$) is deleted from T, and $L'_a - \{a\}$; $L_c - \{c\}$. There are two cases.

Case 1. There is a tree $t \in \mathcal{F}$ with leaves from both L'_a and L'_c. Let $t_a = \sigma(t[L'_a])$ and similarly t_c. Let I denote the number of leaves in t amongst $\{a, b, c\}$. If $I = 0$, then a, b, and c must be isolated in $\mathcal{F}$, by the definition of an agreement forest. Define t_{abc} as the tree with three leaves a, b, c and let $\mathcal{F}' = (\mathcal{F} - \{a, b, c, t\} \cup \{t_a, t_b, t_{abc}\})$. Clearly $\mathcal{F}'$ is an agreement forest with fewer leaves than $\mathcal{F}$, a contradiction. Hence $I \neq 0$.

If $I = 1$, let x be the leaf in t and $\{a, b, c\}$, and y, z the other leaves. Again we see that y, z must be isolated so that $\mathcal{F}' = (\mathcal{F} - \{y, z, t\} \cup \{t_a, t_c, t_{abc}\})$ is also an agreement forest with the same number of leaves, and which has a, b, c in a single component.

If $I = 2$, one of the leaves $x \in \{a, b, c\}$ is an isolated vertex; by setting $t' = \sigma(T_1[\mathcal{L}(t) \cup \{x\}])$, we again contradict minimality via $\mathcal{F}' = (\mathcal{F} - \{x, t\} \cup \{t'\})$.

Finally if $I = 3$ we are done.

Case 2. No tree in $\mathcal{F}$ contains leaves from both L'_a and L'_c. First we can safely assume that no tree in $\mathcal{F}$ has leaves including all of $\{a, b, c\}$ else we are done. There are two subcases.

Subcase 1. At least one leaf $x \in \{a, b, c\}$ occurs as an isolated vertex in $\mathcal{F}$.

Then we delete a, b, c from any of the trees in $\mathcal{F}$ and replace the isolated leaf by t_{abc}.

Subcase 2. One component t_1 of $\mathcal{F}$ contains a, b and c is in another t_2.

Then we get a contradiction to the minimality, as we can let $t = \sigma(T_1[\mathcal{L}(t_1) \cup \mathcal{L}(t_2)])$, and $\mathcal{F}' = (\mathcal{F} - \{t_1, t_2\} \cup \{t\})$. $\qquad\square$

Lemma 4.10.2 *If T'_i denotes the outcome of applying the rules to T_i for $i \in \{1, 2\}$ then the TBR distance between the trees remains the same.*

Proof The proof for Rule 1 is left to the reader.

For Rule 2, let the subtrees in the chain shared by T_1 and T_2 be $t_1, \ldots, t_r$ with $r \geq 3$, and suppose they are replaced by new leaves a, b, c. Then T'_1 and T'_2 are a, b, c trees. Hence there is a MAF $\mathcal{F}$ satisfying Lemma 4.10.1. In these trees, re-insert $t_1, \ldots, t_r$ in this order in each of T'_1 and T'_2 to new vertices that subdivide the edge $v_a v_b$ where v_a is adjacent to a and similarly v_b. Call the result T_i''. Any MAF for T_1', T_2' which has a, b, c in the same component t can be modified to produce an agreement forest for T_1'' and T_2'' of the same size by attaching trees $t_1, \ldots, t_r$ along the edge $v_a v_b$ of t, or in the case $v_a = v_b$ in t, along the edge $a v_a$. Thus the TBR distance of T_1'', T_2'' is bounded by that of T_1', T_2'. It is easy to see that this implies the TBR distance of T_1, T_2 is bounded by that of T_1'', T_2''.

Conversely, suppose that we select leaves $a \in \mathcal{L}(t_1), b \in \mathcal{L}(t_2), c \in \mathcal{L}(t_3)$. Replacing the chain $t_1, \ldots, t_r$ in T_1, T_2 by leaves a, b, c correctly oriented, gives T_1', T_2'. The reader can verify that if U is the set of leaves of T_1 that do not form a chain plus a, b, c, then the TBR distance of $\sigma(T_1[U])$ and $\sigma(T_2[U])$ is bounded by that of T_1, T_2, and hence the TBR distance of T_1', T_2' is bounded by that of T_1, T_2 since $\sigma(T_i[U]) = T_i'$. $\qquad \square$

Now we suppose that we have exhaustively applied Rules 1 and 2, so that T_1 and T_2 are reduced. Suppose also that the TBR distance between them is k. We estimate the number of leaves in them. Let $t_1, \ldots, t_k$ be a MAF for T_1, T_2, and let $\deg^i(t_j)$ the number of edges of T_i incident with subtree t_j.

Lemma 4.10.3 $\sum_j \deg^i(t_j) \leq 2k - 2$.

Proof In T_i collapse each t_j for $j \in [k]$ to a single vertex of degree $\deg^i(t_j)$ to get a single tree $\widehat{T_i} = (V, E)$ consisting of these new vertices and $n_3^i \geq 0$ vertices of degree 3. Drop i, and we see $|V| = n_3 + k$, $\sum_j \deg^i(t_j) + 3n_3 = 2|E|$, and as we have a tree, $|V| = |E| + 1$. Thus, $\sum_j \deg^i(t_j) = 2k - 2 - n_3 \leq 2k - 2$. $\qquad \square$

Lemma 4.10.4 *Rules 1 and 2 result in a kernel size with at most* $c(\deg^1(t_j) + \deg^2(t_j))$ *leaves in* t_j, *for some constant* $c \leq 7$.

Proof Let I_j denote the set of edges of t_j incident with edges of either T_1 or T_2. Let t_j' denote the minimal subtree containing amongst its edges the set I_j. Let t_j'' denote the tree obtained from t_j'—by replacing each maximal path containing no edges from I_j, with a single edge—and let F_j denote this set of new edges. Let P_j denote the set of pendant edges of t_j''; and finally $i_j := |I_j|$, $f_j = |F_j|$, and $p_j = |P_j|$. Then $P_j \subseteq I_j$, $I_j \cup F_j$ is a partition of edges of t_j''. Each vertex of t_j'' of degree 2 is incident with at least one edge from I_j.

Having applied Rules 1 and 2, we see that the subtree t_j is reduced to one of size at most $s = p_j + 3f_j$ since: each subtree of t_j corresponding to an edge in P_j can be replaced by a single leaf (Rule 1); and each collection of subtrees corresponding to an edge in F_j can be replaced by 3 leaves (Rule 2).

Now we claim $s \leq 7i_j - 9$. To see this, let v_j^d denote the number of vertices of t_j'' of degree d. $v_j^1 = p_j$, $v_j^d = 0$ if $d > 3$. Since t_j'' is a tree, we calculate

$$v_j^1 + 2v_j^2 + 3v_j^3 = 2(i_j + f_j) = 2(v_j^1 + v_j^2 + v_j^3 - 1).$$

Since $v_j^1 = p_j$ this means $p_j - v_j^3 = 2$. Since f_j denotes the total number of edges of t_j'' minus i_j, we see

$$f_j = \left(v_j^1 + v_j^2 + v_j^3 - 1\right) - i_j.$$

Substituting, we obtain

$$s = p_j + 3\left(2p_j + v_j^2 - 3 - i_j\right) = 7p_j + 3\left(v_j^2 - i_j\right) - 9.$$

The observations above mean that: each edge in P_j generates at most one vertex of degree 2; and each edge of $I_j - P_j$ generates at most two vertices of degree 2. As this covers all vertices of degree 2, we see $v_j^2 \leq p_j + 2(i_j - p_j)$. Substituting, we get

$$s \leq 4p_j + 3i_j - 9 \leq 7i_j - 9,$$

and since $i_j \leq \deg^1(t_j) + \deg^2(t_j)$, the lemma follows. $\square$

Finally, observe that since $\sum_j (\deg^1(t_j) + \deg^2(t_j)) \leq 4c(k-1)$, the Rules 1 and 2 give the resultant kernelized trees of size $k' \leq 4c(k-1)$, for constant $c < 7$, with the property that the TBR distance between them remains the same as the TBR distance between the two original trees.

In total, there are $O(k^3)$ possible TBR operations on the resultant trees and an exhaustive search is used to determine if there exists a set of $\leq k$ such operations transforming T_1 into T_2. This requires a total time of $O(k^{3k}) + p(n)$ where $p(n)$ is a polynomial bounding the time required to apply the reduction rules.

In the next section, we examine the work of Hallett and McCartin [377], who used a simple bounded search tree approach to improve on the Allen-Steel parameterized TBR approach for determining MAF distance. The Hallett–McCartin algorithm runs in time $O(4^k \cdot k^5) + p(n)$, where k bounds the size of the agreement forest. In practice, the number of disagreements k between two trees T_1 and T_2 likely does grow as a slow function of n (for example, the MAF distance in some applications is proportional to the number of horizontal gene transfers). Nevertheless, for instances of MAF for pairs of trees that have been reconstructed from data for the same set of species, $k \sim 10$ would already represent a large amount of "disagreement" if the total number of species n is ≤ 50 (see Addario-Berry, Hallett, and Lagergren [14]).

4.10.2 The Algorithm and Correctness

The algorithm proceeds in two phases. In the first phase, we look for *minimal incompatible quartets*. These are subtrees that represent objects that must be eliminated in order to create a maximum agreement forest for the two given input trees, T_1 and T_2. We show that we can remove such a quartet by "depronging" the structure in exactly four ways, leading to four branches in our search tree, with a single edge removed from our current candidate agreement forest in each case. In the second phase, we look for *obstructions* that must be eliminated in order to create a maximum agreement forest. We show that each obstruction can be removed in two ways, leading to two branches in our search tree, with a single edge removed from our current candidate agreement forest in each case. Note that at most $k-1$ edges may be removed to produce any solution, so our search tree has depth bounded by k and size bounded by 4^k.

A *phylogenetic forest* $F = (V, E, L, \lambda)$ is the union $F^1 \cup \cdots \cup F^\varphi$ of unrooted, binary trees where the leaves of F are labeled by a leaf label set L via a bijection $\lambda : L \to L(V)$, for leaf set $L(V)$. When $\varphi = 1$, F a *phylogenetic tree*. We denote by $N_F(u)$ the set of vertices adjacent to u in F. For two vertices $u, v \in V(F^k)$, for some k, let $u = u_0 \cdot u_1 \cdots u_p = v$ represent the unique path between u and v in F^k. For $u, v \in V(F)$, let

$$P_F(u, v) = \begin{cases} u = u_0 \cdot u_1 \cdots u_p = v & \text{if } u_1, u_2 \in V(F^k) \text{ for some } k \\ \emptyset & \text{otherwise.} \end{cases}$$

For three distinct vertices $u_1, u_2, u_3 \in V(F)$, let

$$P_F(u_1, u_2, u_3)$$
$$= \begin{cases} x \in V(F) : x \in \bigcap_{1 \le i < j \le 3} P_F(u_i, u_j) & \text{if } u_1, u_2, u_3 \in V(F^k) \text{ for some } k, \\ \emptyset & \text{otherwise.} \end{cases}$$

For a set $L' \subseteq L$, we allow the notation $F[L']$ to denote that subforest

$$F\big[\big\{x \in V(F) : \text{there exists } l, l' \in L' \text{ s.t. } x \in P_F\big(\lambda(l), \lambda(l')\big)\big\}\big].$$

Definition 4.10.3 (Quartet) Let $F = (V, E, L, \lambda)$ be a phylogenetic forest and let $a, b, e, f \in L$. If $F[\{a, b, e, f\}]$ is connected, we call $Q = \sigma(F[\{a, b, e, f\}])$ a *quartet* in F. In Q, either (i) a, b, (ii) a, e or (iii) a, f are sibling leaves and we write $ab|ef$, $ae|bf$ or $af|be$ respectively as a shorthand to describe this graph.

Definition 4.10.4 (Quartet compatibility) Let F_1 and F_2 be a pair of phylogenetic forests with a common leaf label set L. Let $a, b, e, f \in L$. The quartet $Q = \sigma(F_1[\{a, b, e, f\}])$ is *compatible with* F_2 if and only if $\sigma(F_2[\{a, b, e, f\}])$ is isomorphic to Q as leaf labeled trees or $F_2[\{a, b, e, f\}]$ is disconnected. Otherwise, we say that Q *is incompatible with* F_2 (we also say that $\sigma(F_2[\{a, b, e, f\}])$ is incompatible with F_1). See Fig. 4.4.

Proposition 4.10.1 *Let F_1 and F_2 be phylogenetic forests with common leaf label set L and let $Q = ab|ef$ be a quartet in F_1. Q is compatible with F_2 iff $\sigma(F_2[\{a, b, e, f\}])$ is disconnected or $P_{F_2}(a, b) \cap P_{F_2}(e, f) = \emptyset$.*

Definition 4.10.5 (Forest compatibility) Let F_1 and F_2 be a pair of phylogenetic forests with a common leaf label set L. We say that F_1 and F_2 are *compatible* if and only if, for all quartets Q in F_1, Q is compatible with F_2. Otherwise, we say that F_1 and F_2 are *incompatible*.

Fig. 4.4 F_1 and F_2, a pair of phylogenetic forests (trees) over the same leaf set $L = \{a, b, c, d, e, f, g\}$, with $F_1[a, b, e, f]$ and $F_2[a, b, e, f]$ shown as unfilled nodes, and $Q = \sigma(F_1[a, b, e, f])$, a quartet in F_1. Q is incompatible with F_2

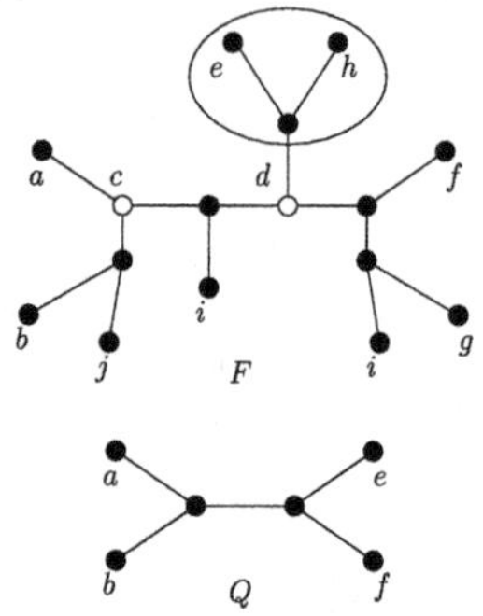

Fig. 4.5 F, a phylogenetic forest (tree) over the leaf set $L = \{a, b, e, f, g, h, i, j, k\}$, and $Q = ab|ef$ a quartet in F. The junction of a or b in F (labeled c) and the junction of e or f in F (labeled d) are shown as *unfilled nodes*. The *circled subtree* of F is $F^{(e, Q)}$

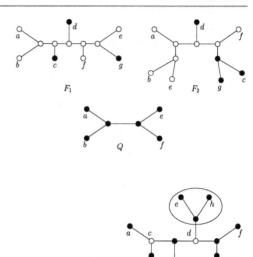

We emphasize here that a phylogenetic tree is uniquely characterized by the set of all compatible quartets over L. If F_1 and F_2 are both phylogenetic trees with a common leaf label set L, and the trees are compatible, then F_1 and F_2 are isomorphic as leaf labeled trees.

Definition 4.10.6 (Junctions) Let $F = (V, E, L, \lambda)$ be a phylogenetic forest and let $Q = ab|ef$ be a quartet in F. Let $c \in V(F)$ be the (unique) vertex s.t. a, b, and e are in different connected components in $F \setminus c$. We call c the junction of a or b in F. Let $d \in V(F)$ be the (unique) vertex s.t. e, f, and a are in different connected components in $F \setminus d$. We call d the junction of e or f in F. See Fig. 4.5.

Let F be a phylogenetic forest and let $Q = ab|ef$ be a quartet in F. Let $x \in \{a, b, e, f\}$ and let y be the junction of x in F; we denote by $F^{(x, Q)}$ the connected component of $F \setminus y$ that contains leaf x.

Definition 4.10.7 (Minimal incompatible quartets) Let F_1 be a phylogenetic forest and F_2 be a phylogenetic tree with common leaf label set L. We say that a quartet $Q = ab|ef$ in F_1 is a *minimal incompatible quartet in F_1* if and only if, (i) Q is incompatible with F_2, and (ii) for all $x \in \{a, b, e, f\}$ and for all $l \in L(F_1^{(x, Q)})$, the new quartet Q' formed by replacing the sibling leaf of x in Q with l is compatible with F_2.

Fig. 4.6 F, a phylogenetic forest (tree) over the leaf set $L = \{a, b, c, d, e, f, g, h, i\}$, and $Q = ab|ef$ a quartet in F. The fork R (induced by Q in F) is shown as unfilled nodes; the prongs of R are shown as *dotted lines*

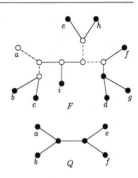

Definition 4.10.8 (Fork) Let $F = (V, E, L, \lambda)$ be a phylogenetic forest and let $Q = ab|ef$ be a quartet in F. Let c be the junction of a and d be the junction of e. The *fork R induced by Q in F* is equal to

$$F\big[\{x \in V(F) : x \in P_F(c, d) \text{ or } x \text{ is adjacent to } c \text{ or } d\}\big].$$

The *prongs* of the fork R are the edges incident to the leaves of R. For a prong (u, v) of R, we say that $\sigma(F \setminus \{(u, v)\})$ is the phylogenetic forest obtained by *depronging* F of (u, v). See Fig. 4.6.

Lemma 4.10.5 *Let F_1 be a phylogenetic forest and F_2 be a phylogenetic tree, with common leaf label set L. Let $Q = ab|ef$ be a minimal incompatible quartet in F_1. In any MAF $\mathcal{F} = \{t_1, \ldots, t_{k' \leq k}\}$ for F_1 and F_2,*

$$\{l_a, l_b, l_e, l_f\} \not\subseteq L(t_i),$$

for all $x \in \{a, b, e, f\}$, $l_x \in F_1^{(x, Q)}$, and i, $1 \leq i \leq k'$.

Proof Note that, since F_2 is a phylogenetic tree, $F_2[L(F_2)]$ is connected. Thus, all quartets compatible with F_2 are connected in F_2.

Suppose that there exists a t_i and a set $C = \{l_a, l_b, l_e, l_f\}$ s.t. $C \subseteq L(t_i)$. By Definition 4.10.2, it follows that $\sigma(t_i[C])$, $\sigma(F_1[C])$ and $\sigma(F_2[C])$ are all pairwise isomorphic as leaf labeled trees and $\sigma(t_i[C]) = l_a l_b | l_e l_f$. Let $Q_C = \sigma(t_i[C])$. Clearly, Q_C is compatible with F_2. Note that Q_C is not leaf label isomorphic to Q, since Q is incompatible with F_2. Thus, for at least one $x \in \{a, b, e, f\}$ and $l_x \in C$, $l_x \neq x$.

It must be the case that either (i) $ae|bf$ or (ii) $af|be$ is compatible with F_2. In either case, from the minimality of Q, it follows that the following quartets are also compatible with F_2. So we can establish that the following holds: either $a = l_a$ or $a l_a | ef$ is compatible with F_2; either $b = l_b$ or $b l_b | ef$ is compatible with F_2; either $e = l_e$ or $e l_e | ab$ is compatible with F_2; either $f = l_f$ or $f l_f | ab$ is compatible with F_2.

In case (i), it is easy to verify that $l_a l_e | l_b l_f$ is compatible with F_2. However, this is a contradiction since $l_a l_e | l_b l_f \neq Q_C = l_a l_b | l_e l_f$. A similar argument holds for case (ii). $\square$

Given two phylogenetic trees, F_1 and F_2, with a common leaf label set, the first phase of the algorithm proceeds by finding minimal incompatible quartets in F_1 (Definition 4.10.7). For each successive minimal incompatible quartet, we branch on the four possible ways to deprong the fork induced by the quartet (Definition 4.10.8), since, by Lemma 4.10.5, in any solution, at least one of the prongs must be removed. At the end of Phase I, we have a collection $\mathcal{U}$ of forests, where each forest $F \in \mathcal{U}$ is comprised of at most k components and s.t. F and F_2 are compatible. The number of elements in the collection $\mathcal{U}$ is at most 4^{k-1}. Let $\mathcal{U}'$ be a queue with initially one element F_1. Let $\mathcal{U}$ be initially empty.

Algorithm 4.10.1 (Algorithm Phase I for Maximum Agreement Forest)
Phase I—Depronging Forks
Input: Queue $\mathcal{U}'$ and phylogenetic forest F_2.
Step 0. $F \leftarrow Dequeue(\mathcal{U}')$.
Step 1. If there exists a minimal incompatible quartet Q in F and the number of connected components of F is $< k$, then proceed as follows:
Let t_i be the element of F that contains Q. Let R be the fork induced by Q in t_i. For each prong (u, v) of R, construct a new forest $F \setminus t_i \cup \{t_i', t_i''\}$ where t_i' and t_i'' are the two subforests obtained after depronging. Enqueue each of the resulting four forests in $\mathcal{U}'$.
If all quartets in F are compatible with F_2, then F and F_2 are compatible. In this case, add F to $\mathcal{U}$.
Step 2. If $\mathcal{U}'$ is empty, stop. Otherwise, goto Step 0.
Output: A collection $\mathcal{U}$ of forests, where each such forest F is the union of $\leq k$ phylogenetic trees, and subtrees (s.t.) F and F_2 are compatible.

Let $F = \{t_1, \ldots, t_{k' \leq k}\} \in \mathcal{U}$. F satisfies conditions (i) and (ii) of Definition 4.10.2. However, condition (iii) of Definition 4.10.2 may not be satisfied after Phase I. It will be the case that the trees $\{F_1[L(t_j)] : 1 \leq j \leq k\}$ are vertex disjoint subtrees of F_1, but it may not be the case that the trees $\{F_2[L(t_j)] : 1 \leq j \leq k\}$ are vertex disjoint subtrees of F_2. In Phase II of the algorithm (described after the following valuable definitions) we will correct this.

Phase II of the algorithm proceeds by finding *obstructions* that must be eliminated in order to produce an agreement forest for F_1 and F_2. For each obstruction found, we branch on two possible ways to remove it, in each case cutting a single edge.

For a phylogenetic forest $F = (V, E, L, \lambda) \in \mathcal{U}$ and a phylogenetic tree $F_2 = (V_2, E_2, L, \lambda_2)$, we use as a shorthand $\lambda^*(l)$ to represent the function $\lambda_2(\lambda^-(l))$, where $l \in L(F)$.

Definition 4.10.9 (Beta mapping) For $F = (V, E, L, \lambda)$, a phylogenetic forest, and $F_2 = (V_2, E_2, L, \lambda_2)$, a phylogenetic tree, let $\beta : V(F) \to V(F_2) : u \mapsto x$, where
(i) $x = \lambda^*(u)$, if $u \in L(F)$; otherwise,
(ii) $x = P_{F_2}(\lambda^*(l_1), \lambda^*(l_2), \lambda^*(l_3))$, where $l_i \in L(T_i)$, and T_1, T_2 and T_3 are the three subtrees obtained from the component of F containing u in $F \setminus u$.

Fig. 4.7 F, a phylogenetic
forest, and F_2, a phylogenetic
tree, over the same leaf set
$L = \{a, b, c, d, e, f, g\}$, with
beta mapping
$\beta : V(F) \rightarrow V(F_2)$ shown.
Edges (x, e) and (g, y) in F,
shown as *dotted lines*, are
obstructions for F and F_2

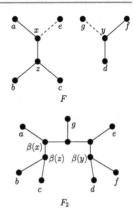

Proposition 4.10.2 *If F and F_2 are compatible, then $\beta(\cdot)$ is well-defined. That is, for F, F_2 and u, x as defined in Definition 4.10.9, for any $l_1, l_1' \in T_1$, $l_2, l_2' \in T_2$, $l_3, l_3' \in T_3$, it is the case that $P_{F_2}(\lambda^*(l_1), \lambda^*(l_2), \lambda^*(l_3)) = P_{F_2}(\lambda^*(l_1'), \lambda^*(l_2'), \lambda^*(l_3'))$.*

Proof Assume F and F_2 are compatible. W.l.o.g. let l_1 and l_1' be distinct leaves of T_1 and let

$$P_{F_2}\big(\lambda^*(l_1), \lambda^*(l_2), \lambda^*(l_3)\big) = a$$

and

$$P_{F_2}\big(\lambda^*(l_1'), \lambda^*(l_2), \lambda^*(l_3)\big) = b.$$

W.l.o.g. let the set $\{l_1, l_1', l_2, l_3\} = Q$ induce quartet $l_1 l_1' | l_2 l_3$ in F. Note that a is the unique vertex that lies on paths $P_{F_2}(\lambda^*(l_1), \lambda^*(l_2))$, $P_{F_2}(\lambda^*(l_1), \lambda^*(l_3))$ and $P_{F_2}(\lambda^*(l_2), \lambda^*(l_3))$, and that b also lies on path $P_{F_2}(\lambda^*(l_2), \lambda^*(l_3))$. If $a \neq b$, then it cannot be the case that $P_{F_2}(\lambda^*(l_1), \lambda^*(l_1')) \cap P_{F_2}(\lambda^*(l_2), \lambda^*(l_3)) = \emptyset$. Thus, Q is incompatible with F_2, contradicting the assumption that F and F_2 are compatible.

A similar argument holds if we consider $P_{F_2}(\lambda^*(l_1'), \lambda^*(l_2), \lambda^*(l_3))$ and replace l_2 with l_2'; and if we consider $P_{F_2}(\lambda^*(l_1'), \lambda^*(l_2'), \lambda^*(l_3))$ and replace l_3 with l_3'. $\square$

Definition 4.10.10 (Obstructions)   Let $F = \{t_1, \ldots, t_{k' \leq k}\} \in \mathcal{U}$ be a phylogenetic forest and F_2 be a phylogenetic tree as defined in Definition 4.10.9. Let (a, b) and (c, d) be distinct edges in $E(F)$. We say that (a, b) and (c, d) are *obstructions* for F and F_2 iff there exists an $x \in V(F_2)$ s.t. $x \in P_{F_2}(\beta(a), \beta(b)) \cap P_{F_2}(\beta(c), \beta(d))$. For an obstruction $(u, u') \in \{(a, b), (c, d)\}$, we say that $\sigma(F \setminus \{(u, u')\})$ is the phylogenetic forest obtained by clearing F of the obstruction (u, u'). See Fig. 4.7.

Lemma 4.10.6 *Let $F = \{t_1, \ldots, t_{k' \leq k}\} \in \mathcal{U}$ be a phylogenetic forest and F_2 be a phylogenetic tree. After Phase I of the algorithm, if (a, b), (c, d) are obstructions for F and F_2, then there does not exist an i **PLEASE DEFINE "i"; incompatible?** s.t. $(a, b) \in E(t_i)$ and $(c, d) \in E(t_i)$.*

Proof Let (a, b) and (c, d) be obstructions for F and F_2 in the same component t_i. W.l.o.g. assume that $b \in P_{t_i}(a, c)$ and $c \in P_{t_i}(b, d)$. Let T_a (resp. T_d) be the subtree of $t_i \setminus b$ (resp. $t_i \setminus c$) that contains a (resp. d). Let T_b be the subtree of $t_i \setminus b$ that contains the unique vertex $N_{t_i}(b) \setminus \{a, x\}$ where $x \in P_{t_i}(b, c)$. Let T_c be the subtree of $t_i \setminus c$ that contains the unique vertex $N_{t_i}(c) \setminus \{d, y\}$ where $y \in P_{t_i}(b, c)$.

It follows that there exist elements $l_x \in L(T_x)$, $x \in \{a, b, c, d\}$, from the definition of σ (that is, there must exist at least one leaf vertex in each such subtree). Any set $\{l_a, l_b, l_c, l_d\}$ where $l_x \in T_x$, $x \in \{a, b, c, d\}$ induces a quartet $l_a l_b | l_c l_d$ in F. From Definition 4.10.9, either $l_a l_c | l_b l_d$ or $l_a l_d | l_b l_c$ is a quartet in F_2. However, this contradicts the existence of t_i, since with the compatibility of F and F_2, all quartets in $t_i[\{L(T_x) : x \in \{a, b, c, d\}\}]$ are connected and compatible with F_2.   □

Lemma 4.10.7 *Let* $F = \{t_1, \ldots, t_{k' \le k}\} \in \mathcal{U}$ *be a phylogenetic forest and* F_2 *be a phylogenetic tree. Let* $(a, b), (c, d)$ *be obstructions for* F *and* F_2. *In any MAF* $\mathcal{F} = \{t_1^*, \ldots, t_{k'' \le k}^*\}$ *for* F_1 *and* F_2, *it is not the case that* $(a, b) \in t_i^*$ *and* $(c, d) \in t_j^*$, $1 \le i, j, \le k''$.

Proof Let (a, b) and (c, d) be obstructions for F and F_2 where $(a, b) \in V(t_i)$ and $(c, d) \in V(t_j)$. Let T_a be the subtree of T containing vertex a after removing (a, b), T_b be the subtree of T containing vertex b after removing (a, b). Define T_c and T_d similarly. Again, it follows that there exist elements $l_x \in L(T_x)$, $x \in \{a, b, c, d\}$ from the definition of σ (that is, there must exist at least one leaf vertex in each such subtree).

By Definition 4.10.9, $\beta(a) \in P_{F_2}(\beta(l_a), \beta(l_b))$, $\beta(b) \in P_{F_2}(\beta(l_a), \beta(l_b))$, $\beta(c) \in P_{F_2}(\beta(l_c), \beta(l_d))$, $\beta(d) \in P_{F_2}(\beta(l_c), \beta(l_d))$. Furthermore, there exists an $x \in V(F_2)$ s.t. $x \in P_{F_2}(\beta(a), \beta(b)) \cap P_{F_2}(\beta(c), \beta(d))$. It follows that $P_{F_2}(\beta(l_a), \beta(l_b)) \cap P_{F_2}(\beta(l_c), \beta(l_d)) \ne \emptyset$. Since $\mathcal{F}$ is a MAF, and therefore Condition (iii) must be satisfied, it follows that $\sigma(F_2[\{l_a, l_b, l_c, l_d\}])$ is connected and therefore $i = j$. However, this contradicts Lemma 4.10.6 $i \ne j$.   □

Note that if there are no obstructions for F and F_2, then F and F_2 satisfy all conditions of Definition 4.10.2. Let $\mathcal{U}_{\text{final}}$ be an empty queue. Let $\mathcal{U}'$ be the output queue $\mathcal{U}$ from Phase I.

Algorithm 4.10.2 (Algorithm Phase II for MAXIMUM AGREEMENT FOREST)
Phase II—Clearing Obstructions
Input: $\mathcal{U}'$ and F_2.
Step 0. $F \leftarrow Dequeue(\mathcal{U}')$.
Step 1. If there obstructions $(a, b), (c, d)$ for F and F_2, and the number of connected components of F is $< k$, then proceed as follows:
Let $F_{(a,b)}$ and $F_{(c,d)}$ be the forests obtained from F after clearing obstructions (a, b) and (c, d) respectively. Enqueue these two forests in $\mathcal{U}'$.
If no obstructions exist for F and F_2, enqueue F in $\mathcal{U}_{\text{final}}$.
Step 2. If $\mathcal{U}'$ is empty, stop. Otherwise, goto Step 0.
Output: A collection $\mathcal{U}_{\text{final}}$ of forests where each such forest F is the union of $\le k$ phylogenetic trees and is an agreement forest for F and F_2.

4.10.3 Analysis

Given phylogenetic trees T_1 and T_2 with common leaf label set L, $|L| = n$, we first apply the kernelization rules from Allen and Steele [28]. This produces trees F_1 and F_2 of size $k' \leq 4c(k-1)$, c a constant, in polynomial time $p(n)$. Phase I takes F_1 and F_2 as input. In this phase, the main effort in every node of the search tree is to find a minimal incompatible quartet in F. Since both $|F_1|$ and $|F_2|$ are $O(k)$, we can precompute a list of all quartets of F_1 that are incompatible with F_2 in time $O(k^5)$. This is accomplished by checking each quartet induced by a set of four connected leaves of F_1. There are $O(k^4)$ such quartets. The property stated in Observation 4.10.1 can be checked in $O(k) = |F_1|$.

It is easy to verify that the $O(k^4)$ list of incompatible quartets can be checked for minimality in total $O(k^5)$ time. Thus, the time to expand each node of the search tree in Phase I of the algorithm is bounded by $O(k^5)$. At most $k-1$ edges may be removed on any branch of the search tree, so the search tree has depth bounded by k, and size bounded by 4^k.

In Phase II of the algorithm, the main effort in each node of the search tree is to find obstructions (a, b) and (c, d) in F. In each branch of the search tree, at the start of Phase II, we can precompute a list X of all pairs of edges that are obstructions for F and F_2 in time $O(k^3)$. This requires only that we check each pair of edges, (a, b) and (c, d) that are in separate components of F. We first compute $\beta(x)$ for each $x \in V(F)$, in time $O(k^2)$. Whether (a, b) and (c, d) are obstructions can be tested in time $O(k)$ for each of the $O(k^2)$ pairs.

For each branch of the search tree in Phase II we remove an edge from F. We update X as follows. Suppose (a, b) is the edge removed from F. We remove from X any pair of edges that includes (a, b). This can be done in time $O(k^2)$. Thus, the time taken to expand each node of the search tree in Phase II of the algorithm is bounded by $O(k^3)$. At most $k-1$ edges may be removed on any branch of the search tree, so the search tree has depth bounded by k and size bounded by 2^k.

Through careful bookkeeping during the application of the reduction rules, and both Phase I/Phase II of the algorithm, we can reconstruct a MAF $\mathcal{F}$ for the original trees T_1 and T_2 from any solution $F \in \mathcal{U}_{final}$ within the same time bounds. This is sufficient to prove the following theorem.

Theorem 4.10.1 (Hallett and McCartin [377]) *The* MAXIMUM AGREEMENT FOREST *problem can be solved in time* $O(4^k \cdot k^5) + p(n)$, *where k bounds the size of the agreement forest.*

4.11 Exercises

Exercise 4.11.1 (Gramm, Niedermeier, and Rossmanith [357]) Use the method of kernelization to prove that CLOSEST STRING is solvable in time $O(k \cdot \ell + k \cdot (d+1)^d)$.

(Hint: The argument in the proof of Theorem 3.6.1 gives a running time of $O((d+1)^d|G|)$, and to reduce to a kernel, use the following fact. If the instance of CLOSEST STRING has more than $k \cdot \ell$ positions not identical for $s_1, \ldots, s_k$, then the answer is "no".)

Exercise 4.11.2 Consider the hereditary property $\Pi_{0,0,k}$ which states that we can add $\leq k$ edges to G and make a chordal graph. Prove that $\Pi_{0,0,k}$ has no finite forbidden set characterization.

Exercise 4.11.3 Prove $(*)$ from Theorem 4.8.1; that is, prove the following.

> If a tree T has i internal vertices of degree at least 3,
>
> then T has at least $i + 2$ leaves.

Exercise 4.11.4 Show that the following problem is FPT by the method of reduction to a problem kernel:

SET BASIS (Garey and Johnson [337])
Instance: A collection C of subsets of a finite set S.
Parameter: A positive integer k.
Question: Is there a collection B of subsets of S with $|B| = k$ such that, for every set $A \in C$, there is a subcollection of B whose union is exactly A?

Exercise 4.11.5 (Bodleander [77]) Construct in polynomial time a kernel of size $O(k \log k)$ for the following problem.

WEIGHTED MARBLES
Input: A sequence of marbles, each with an integer weight and a color.
Parameter: A positive integer k.
Question   Can we remove marbles of a total cost $\leq k$, such that for each color, all marbles of that color are consecutive?

(Hint: Consider the following reduction rules: Rule 1: If we have two consecutive marbles of the same color, replace it them by one with the sum of the weights. Call a color *good* if there is only one marble with this color. Rule 2: Suppose two successive marbles both have a good color. Give the second the color of the first. Apply these exhaustively. No two successive marbles will be of the same color, and no two successive marbles will have a good color. The number of marbles is at most twice $(+1)$ the number with a bad color. This gives Rule 3: If there are at least $2k + 1$ bad colored marbles, say No. Finally for a kernelization we also apply exhaustively Rule 4: If a marble has weight $> k + 1$, give it weight $k + 1$.)

Exercise 4.11.6 (Bodlaender, Jansen, and Kratsch [92])   Prove that the following has a polynomial kernel ($O(\ell^4)$ is not difficult). Recall that, in Exercise 3.8.11, the reader showed that this problem is FPT.

LONG CYCLE PARAMETERIZED BY A MAX LEAF NUMBER

Instance: A graph G, and two integer k and ℓ.
Parameter: A positive integer $\ell = |X|$.
Promise: The maximum number of leaves in any spanning tree is ℓ.
Question: Is there a cycle in G of length at least k?

In this promise problem if the max leaf conditions are not satisfied, then the output can be arbitrary.

Exercise 4.11.7 (Bodlaender, Jansen, and Kratsch [92]) As in Exercise 4.11.6, prove that the parameterized problems (with the obvious definitions), below have polynomial kernels when parameterized by MAX LEAF NUMBER: LONG PATH, HAMILTON PATH, k-DISJOINT CYCLES, k-DISJOINT PATHS.

Exercise 4.11.8 Prove that the following problem is FPT.

GROUPING BY SWAPPING (Garey and Johnson [337])

Instance: A finite alphabet Σ, a string $x \in \Sigma^*$, and a positive integer k.
Parameter: A positive integer k.
Question: Is there a sequence of k or fewer adjacent symbol interchanges that transforms x into a string x' in which all occurrences of each symbol $a \in \Sigma$ are in a single block?

Exercise 4.11.9 (Niedermeier [546]) Give a problem kernel for the following problem.

MATRIX DOMINATION

Instance: An $n \times n$ matrix with entries in $\{0, 1\}$, X.
Parameter: A positive integer k.
Question: Find a set C of at most k 1's such that all the other nonzero entries are in the same row or column as at least one member of C.

Exercise 4.11.10 (Ganian [336]) A generalization of VERTEX COVER is the concept of TWIN COVER , which for a graph G, is $X \subseteq V(G)$, such that for each edge $e = ab$, either
 (i) $a \in X$ or $b \in X$, or
(ii) a and b are *twins* meaning that all other vertices are adjacent to *both* a and b or *neither* a nor b.
1. Give an example of a graph with low twin cover but which has not small vertex cover.
2. Prove that there is an FPT algorithm for k-TWIN COVER running in time $O(|E||V| + k|V| + \alpha^k)$ where α is the constant from the fastest running FPT algorithm for VERTEX COVER.
3. The problem GRAPH MOTIF below arises in the context of metabolic networks in computational biology. (See, for example, Lacroix, Fernandes, and Sagot [474].) Show that GRAPH MOTIF is FPT in time $O(|V||M| + 2^{2+2^k} \cdot (|E|\sqrt{|V|} + |V|))$ on graphs of twin cover bounded by k.

GRAPH MOTIF

Instance: A vertex colored (adjacent vertices can have the same coloring) graph
           $G = (V, E)$ and a multiset M of colors.
Question: Does there exist a connected subgraph H of G, such that the multiset of
           colors occurring in H is identical to M?

Exercise 4.11.11 The following problem is of some interest due to its potential
connections to combinatorial cryptosystems (Fellows and Koblitz [293, 295]). By
reducing to a problem kernel, show that it is FPT.

UNIQUE HITTING SET

Instance: A set X and k subsets $X_1, \ldots, X_k$ of X.
Parameter: A positive integer k.
Question: Is there a set $S \subseteq X$ such that for all i, $1 \le i \le k$, $|S \cap X_i| = 1$?

Exercise 4.11.12 Show that the following two (similar) problems are FPT by the
problem kernel method.

CONSECUTIVE BLOCK MINIMIZATION (Garey and Johnson [337])

Instance: An $m \times n$ matrix M of 0's and 1's and a positive integer k.
Parameter: A positive integer k.
Question: Is it possible to permute the columns of M to obtain a matrix M' that
           has at most k blocks of consecutive 1's? (A *block* of consecutive 1's is
           an interval of a row such that all entries in that interval are 1's.)

RECTILINEAR PICTURE COMPRESSION (Garey and Johnson [337])

Instance: An $n \times n$ matrix of 0's and 1's and a positive integer k.
Parameter: A positive integer k.
Question: Are there k rectangles that cover all the 1's?

Exercise 4.11.13 (Parametric duality) Classically, a problem that takes as input a
graph G and a positive integer $k \le n$ where n is the size of G is "the same" as the
problem where the integer part of the input is presented as $n - k$. Parametrically,
however, such "dual" problems are fundamentally different and may have different
complexities. Show that the following parametric dual of the usual graph coloring problem is FPT by the method of reduction to a problem kernel. (Due to Jan
Telle.)

DUAL OF COLORING

Instance: A graph G.
Parameter: A positive integer k.
Question: Can G be properly colored by $n - k$ colors, where n is the number of
           vertices of G?

The VERTEX COVER problem is parametrically dual to the INDEPENDENT SET
problem. An *irredundant set* of vertices in a graph $G = (V, E)$ is a set of vertices $V' \subseteq V$ such that for every vertex $u \in V'$, there is a vertex v in the neighborhood of u, but not in the neighborhood of any other vertex of V'. Prove that

the following problem is FPT by the method of reduction to a problem kernel.

DUAL OF IRREDUNDANT SET

Instance: A graph G.
Parameter: A positive integer k.
Question: Does G have an irredundant set of size $n - k$?

Exercise 4.11.14 (V. Raman) If E is a 3SAT expression having n clauses, then there is always a truth assignment that satisfies at least $\lceil n/2 \rceil$ clauses. By the method of reduction to a problem kernel, show that the following problem is FPT.[8]

MORE THAN HALF MAX-3SAT

Instance: A 3SAT expression E.
Parameter: A positive integer k.
Question: Is there a truth assignment that satisfies at least $\lceil n/2 \rceil + k$ clauses?

Exercise 4.11.15 (Mahajan and Raman [504]) Extend the result of Exercise 4.11.14 to the same problem except 3SAT is replaced by SAT.

Exercise 4.11.16 Show that the following problems are FPT by a combination of search tree and problem kernel techniques
(Hint: First, reduce part of the problem to VERTEX COVER.)

MATRIX DOMINATION (Garey and Johnson [337])

Instance: An $n \times n$ matrix M with entries from $\{0, 1\}$.
Parameter: A positive integer k.
Question: Is there a set C of k or fewer nonzero entries in M that dominate all others (in the sense that every nonzero entry in M is in the same row or in the same column as some element of C)?

NEARLY A PARTITION

Instance: A finite set X, a family F of subsets of X,.
Parameter: A positive integer k.
Question: Is there a subfamily $F' \subseteq F$ with $|F'| \leq k$ such that $F - F'$ is a partition of X?

Exercise 4.11.17 (VERTEX COVER ABOVE LP—Lokshtanov, Narayanaswamy, Raman, Ramanujan, and Saurabh [497]) The problem concerns a new programme, which looks like it should have other applications since Inter Linear Programming is a well-developed technique. We call this *parameterizing above the LP solution.*

---

[8]This, and its generalization in the next problem, is an example of what is called *parameterizing above guaranteed values* (an area of great interest we lack space to treat properly in this book). We will briefly discuss probabilistic methods for this kind of problem in Sect. 8.7.

By an *optimum solution to* LPVC(G), we mean a feasible solution with $x(v) \geq 0$ for all $v \in V$ minimizing the objective function $w(x) = \sum_{v \in V} x(u)$. By Theorem 4.3.2, for VERTEX COVER, we can have a feasible solution $x(v) \in \{0, \frac{1}{2}, 1\}$ for all $v \in V(G)$. Let vc*(G) denote the value of an optimum solution to LPVC(G).

VERTEX COVER ABOVE LP

Instance: An undirected graph G and $\lceil vc^*(G) \rceil$.
Parameter: $k - \lceil vc^*(G) \rceil$.
Question: Does G have a vertex cover of size k?

We drop the $\lceil, \rceil$ for ease of notation. Modify the proof of Theorem 4.3.2 to show this problem is $O^*(2.6181^{k-vc^*(G)})$ FPT.
(Hint: Do this modified proof as follows. Let G be given and we consider the LP solution given in Theorem 4.3.2. Define $V_i^x = \{u \in V \mid x(u) = i\}$ for each $i \in \{0, \frac{1}{2}, 1\}$ and define $x \cong i$ for $i \in \{0, \frac{1}{2}, 1\}$ if $x(u) = i$ for all $u \in V$. Let I be an independent set in G. Then define the *surplus* of I as $S(I) = |N(I)| - |I|$, for a set F of independent set in G let $S(F) = \min_{I \in F} S(I)$. Finally the surplus of G this quantity calculated for the family of all independent sets in G.

1. Prove the following rule is safe.
   Rule 1. Compute an optimal solution x to LPVC(G) such that $\frac{1}{2}$ is the unique optimal solution to LPVC($G[V_{\frac{1}{2}}^x]$). Delete the vertices in $V_0^x \cup V_1^x$ from the graph after including the vertex cover developed, and reduce k by $|V_1^x|$.

2. Prove that if Z be independent with $S(Z) = 1$ and such that for every $Y \subseteq Z$, $S(Y) \geq S(Z)$. Then,
   a. If $G[N(Z)]$ is not an independent set, then there exists a minimum vertex cover in G that includes all of $N(Z)$ and excludes all of Z.
   b. If $G[N(Z)]$ is an independent set, let G' be the graph obtained from G by removing $Z \cup N(Z)$ and adding a vertex z, followed by making z adjacent to every vertex $v \in G \setminus (Z \cup N(Z))$ which was adjacent to a vertex in $N(Z)$. Then G has a vertex cover of size at most k if and only if G' has a vertex cover of size at most $k - |Z|$.

3. Prove that the following rules are safe.
   Rule 2. If there is a set Z such that $S(Z) = 1$ and $G[N(Z)]$ are not an independent set, then reduce the instance as follows. Include $N(Z)$ in the vertex cover, delete $Z \cup N(Z)$ from the graph, and decrease k by $|N(Z)|$.
   Rule 3. If there is a set Z such that $S(Z) = 1$ and the graph induced by $G[N(Z)]$ is an independent set, then reduce the instance as follows: Remove Z from the graph, identify the vertices of $N(Z)$, and decrease k by $|Z|$.

4. Prove that if we have an instance (G, k) of VERTEX COVER ABOVE LP to which Rule 1 does not apply, we can test if Rule 2 applies to this instance in polynomial time.

5. Prove that if (G, k) is an instance of VERTEX COVER ABOVE LP on which Rules 1 and 2 do not apply, we can test if Rule 3 applies on this instance in polynomial time.

6. For whatever graph $R(G)$ we get after exhaustively applying the rules above, finish the algorithm using simple branching akin to that of Theorem 3.2.1. Now analyse its running time. To do this, we define a measure $\mu = \mu(G, k) = k - vc^*(G)$.

7. Prove that the pre-processing rules do not increase this measure.

8. Prove that the branching rules do not increase the measure.

9. Observe that branching results in a $(\frac{1}{2}, 1)$ vector and hence, resulting recurrence is $T(\mu) \leq T(\mu - \frac{1}{2}) + T(\mu - 1)$. Prove that this solves to $(2.6181)^\mu = (2.6181)^{k-vc^*(G)}$.)

We remark that with more complex branching rules (as treated in, say, Theorem 3.2.2) it is possible to improve this constant to give the following.

Theorem 4.11.1 (Lokshtanov, Narayanaswamy, Raman, Ramanujan, and Saurabh [497]) VERTEX COVER ABOVE LP *can be solved in time* $O^*([2.3146]^{k-vc^*(G)})$.

Lokshtanov et al. also used Theorem 4.11.1 together with some parametric reductions to give the first algorithm for GRAPH BIPARTITIZATION[9] treated by us in Chap. 6 which does *not* use iterative compression.

Exercise 4.11.18 (Bryant) Show that HITTING SET FOR SIZE THREE SETS is fixed-parameter tractable by the method of reduction to a problem kernel. (The problem arises in computing evolutionary phylogenies. The relationship of triples of species can be computed first, and then this information combined. In Exercise 3.8.1, we have seen that the problem is FPT by the method of bounded search trees.)

HITTING SET FOR SIZE THREE SETS (Garey and Johnson [337])

Instance: A collection C of 3-element subsets of a set S.
Parameter: A positive integer k.
Question: Does S contain a *hitting set* for C of size at most k (that is, a subset $S' \subseteq S$ with $|S'| \le k$, and such that S' contains at least one element from each set in C)?

Exercise 4.11.19 (H.T. Wareham) The following "Steiner problem" in hypercubes arises in computing maximum parsimony phylogenies for k species , based on binary character information. (The q-dimensional hypercube is the graph with vertex set consisting of all length q binary vectors, with two vertices adjacent if and only if the corresponding vectors differ in a single component.) Prove that the problem is FPT by reducing to a problem kernel.

STEINER PROBLEM IN HYPERCUBES

Instance: k length q binary sequences $X_1, \ldots, X_k$ and a positive integer M (given in binary).
Parameter: A positive integer k.
Question: Is there a subgraph S of the q-dimensional binary hypercube that includes the vertices $X_1, \ldots, X_k$, such that S has at most M edges?

Exercise 4.11.20 Suppose that Π is a parameterized problem having input (x, k), where k is the parameter. Let n denote the size of x. Say that Π is *kernelizable* if there is a parameterized reduction from Π to Π that runs in time polynomial in both n and k (and therefore is also a polynomial-time reduction, but with conditions on how the parameter is handled) that transforms (x, k) into (x', k'), where $k' = f(k)$ and $|x'| \le g(k)$ for some functions f and g. Prove that every recursive (computable) kernelizable problem is FPT. Describe a candidate problem for showing that the kernelizable problems are a proper subset of FPT.

---

[9]Can we remove at most k edges and get a bipartite graph?

4.12   Historical Notes

Pre-processing has been part of computer science since the dawn of the subject. The recognition of the method as more than a heuristic, but having a guaranteed performance in the parametric situation is due to Downey and Fellows [240, 243, 245], though the Buss kernel for VERTEX COVER can be found in Buss–Goldsmith [115]. Historically and for many applications, the most important kernel in parameterized complexity, the (1975!) Nemhauser–Trotter kernel can be found in Nemhauser and Trotter [544], the proof using linear programming is due to Bar-Yehuda and Even [54]. As mentioned in the text, VERTEX COVER has seen many algorithmic improvements, often by adding numerous cases to the search tree, beginning with Balasubramanian, Fellows, and Raman (Theorem 3.2.2) and moving to Chen, Kanj, and Jia [146] running in time $O(1.286^k|G|)$. The recent algorithm by Chen, Kanj and Xia [148], runs in time $O(1.2738^k + kn)$, the current best running time. Langston, Perkins, Saxton, Scharff, and Voy [480] gave impressive applications of some of these algorithms to significant problems in computational biology. There was a quite a bit of earlier work along these lines, such as by: Hallett, Gonnet, and Stege [376]; and Cheetham, Dehne, Rau-Chaplin, Stege, and Taillon [138]. A nice account of the impact of parameterized complexity on computational biology and, more generally interdisciplinary problem solving, can be found in Stege [628].

Using the VERTEX COVER kernelizations within algorithms has been rather successful. An excellent recent example of this is the following problem.

FAULT COVERAGE IN MEMORY ARRAY
Instance: A large reconfigurable rectangular memory array with defective elements with k_1 spare rows and k_2 spare columns.
Parameter: Positive integers k_1, k_2.
Question: Can the array structure be repaired by replacing some of the columns and rows by the spares?

Because of its obvious importance, this problem has attracted much interest in the literature, with most solutions being heuristic. A faster, simpler, and provably correct solution was given by Chen and Kanj [145] using the method of kernelization, filtering through the VERTEX COVER Nemhauser–Trotter kernelization. The Chen–Kanj algorithm runs in time $O(1.26^{k_1+k_2}n)$, the previous best algorithm (having had no explicit analysis) from 1988 being by Hasan and Liu [383] running in time $\theta(2^{k_1+k_2} + m\sqrt{n})$ (for an $m \times n$ array).

The influential example of "practice vs. theory" of TRAIN COVERING BY STATIONS is found in Weihe [661]. The first articulation of industrial strength FPT as the main goal of parameterized complexity can be found in Downey, Fellows and Stege [253]. For more on this evolution of our outlook, we refer the reader to Downey [237]. The formal definition of a kernel is due to the authors in [240, 245]. The fact that FPT = kernelizability came from a conversation of the authors with Niel Kolblitz and is essentially in Cai, Chen, Downey, and Fellows [122], where it is called the "advice view", as reported in Downey [237]. The Downey, Doyle, and Fellows algorithm for k-LEAF SPANNING TREE in the present section is taken from [247]. It will be vastly improved in subsequent sections. The Leizhen Cai material on hereditary properties is taken from Leizhen Cai [117]. The material on phylogenic trees and the MAF problem is due to Hallett and McCartin [377], building on earlier work of Allen and Steel [28]. The UNIQUE GAMES CONJECTURE is due to Khot and Regev [440]. The exercises are due to several authors, including Gramm, Niedermeier, and Rossmanith (from [357]) who were real pioneers of applications to biological "stringology", as well as Telle, Raman, Bryant, and several by Wareham, whose thesis and Hallett's were two of the first in parameterized computational biology.

4.13 Summary

The method of kernelization is introduced. Several methods of speeding up the algorithms are introduced and applied in settings including VERTEX COVER, k-LEAF SPANNING TREE, and MAXIMUM AGREEMENT FOREST. Heuristics are discussed including the work of Weihe on TRAIN COVERING BY STATIONS, and using parameterized algorithms as heuristics in practice.

More on Kernelization

5

5.1 Turing Kernelization

Recent work has shown that deriving kernels often needs a more general notion of kernelization than the one that was the focus of the preceding chapter. Indeed, if anything, Turing kernelization should be considered as based on potentially stronger practical considerations. We motivate this new notion by an example.

5.1.1 Rooted k-LEAF OUTBRANCHING

We first will study a problem of interest in its own right. In directed graphs, we generalize the notion of a spanning tree to that of an *out tree* which is an oriented tree with one root (the only vertex in the tree of indegree 0) spanning the digraph.

ROOTED k-LEAF OUTBRANCHING
Instance: A directed graph $D = (V, E)$, and a distinguished vertex r of G.
Parameter: An integer k.
Question: Is there an out tree in D with at least k leaves and having r as the root? (Such a tree is termed a *rooted outbranching* of D.)

Considered classically, this problem was shown to be NP-complete by Alon, Fomin, Gutin, Krivelevich, and Saurabh [32]. A closely related problem is k-LEAF OUTBRANCHING which is the same, except that there is no specification of a root r.

Alon et al. described the first FPT algorithms for both problems, based on the method of bounded search trees. The first poly(k) *kernelization* (in the sense of the last chapter) for the rooted problem was given by Fernau, Fomin, Lokshtanov, Raible, Saurabh, and Villanger [306]. The Fernau et al. kernel was an $O(k^3)$ vertex kernelization, obtained by means of reduction rules and detailed combinatorial analysis. Daligault and Thomassé [193] described an improved kernelization to a digraph on a quadratic number of vertices, based on a method called *s–t numberings*. These methods are interesting, as they may prove useful for other parameterized problems concerning digraphs.

R.G. Downey, M.R. Fellows, *Fundamentals of Parameterized Complexity*,
Texts in Computer Science, DOI 10.1007/978-1-4471-5559-1_5,
© Springer-Verlag London 2013

5.1.2 *s–t* and *r–r* **Numberings**

In the setting of *unrooted digraphs*, notions of connectivity are always interpreted in the directed sense (unless otherwise specified). For example, a digraph D is *2-connected* if there is no cutset of size 1, that is, any two distinct vertices of D are connected by at least two vertex-disjoint *directed* paths.[1]

In the setting of *rooted digraphs*, p-connectivity is defined slightly differently. If $uv \in E(D)$ we say u is an *inneighbor* of v, and v an *outneighbor* of u. The definitions of the *indegree* and *outdegree* of a vertex then follow straightforwardly.

We are concerned in this section with *rooted digraphs*. These are loopless digraphs with a single vertex r of indegree 0, the *root*, such that there is no arc xy with $x \neq r$ if $ry \in E(D)$. For a rooted digraph, we say it is *connected* if each vertex is reachable by a (directed) path beginning at r.

A *cut* is a set $C \subseteq V(D) - \{r\}$ for which there is a vertex $v \notin C$ which is not the end of a directed path from r to v in $V - C$. For $p \in \mathbb{N}$, p-*connectedness* is defined similarly to the unrooted case.

A cut of size 1 is called a *cutvertex*. As they rely only on facts about flows in digraphs for their proofs, Menger's Theorems still hold in the rooted case. A rooted digraph is p-*connected* if and only if there are at least p vertex-disjoint paths from r to any vertex that is not r and not one of its outneighbors.

The following definition is an adaptation of a definition of *s–t* numbering for undirected graphs by Lempel, Even, and Cederbaum [486].

Definition 5.1.1 (Cheriyan and Rief [157]) Let $D = (V, E)$ be a 2-connected digraph having at least three vertices. Let s, t be two distinct vertices. An *s–t num-bering* is a linear ordering σ of V with $\sigma(s) = 1$, $\sigma(t) = |V| = n$ such that for each $x \neq s, t$, there exist two inneighbors u and v of x such that $\sigma(u) < \sigma(x) < \sigma(v)$.

Similarly for rooted digraphs we have the following.

Definition 5.1.2 (Daligault and Thomassé [193]) Let D be a 2-connected rooted digraph. An *r–r numbering* is a linear ordering σ of $V(D) - \{r\}$ such that for each $x \neq r$, either x is an outneighbor of r, or there exist two inneighbors u and v of x such that $\sigma(u) < \sigma(x) < \sigma(v)$.

Cheriyan and Reif proved the following useful result.

Theorem 5.1.1 (Cheriyan and Rief [157]) *D is 2-vertex connected iff for each pair of vertices s, t, D has an $s–t$ numbering.*

For our purposes, we only need the result for *r–r* labelings, and we will leave the proof of Theorem 5.1.1 to the reader in the exercises (Exercise 5.3.2). Its proof is very similar to the one below (Daligault and Thomassé's proof is based on the proof of Theorem 5.1.1).

---

[1]If you are not familiar with this, read the material on Menger's Theorems in Chap. 35.

Fig. 5.1 Example of a
2-connected rooted digraph

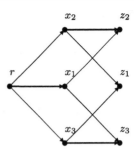

Theorem 5.1.2 (Daligault and Thomassé [193])   *Let D be a 2-connected rooted digraph. Then D has an r–r numbering.*

Proof We use induction on the number of edges in D. First suppose that D has a vertex x of indegree ≥ 3. We know that there are two vertex-disjoint paths from r to x. These can be refined to two such paths which only have one inneighbor of x each. Now make a new digraph D' by removing the edge yx for the inneighbor y of D not involved in these paths.

We argue that D' is 2-connected. To see this, take $v \neq r$ not a neighbor of r in D'. As D is 2-connected we would have two or more vertex-disjoint paths from r to v in D. Only one of those paths could involve xy. Hence there are at least two vertex-disjoint paths from r to x in D'. Therefore, any path involving xy in D from r to v can have the segment from r to x in the path replaced by one of these paths. Hence D' is 2-connected.

By induction, D' has an r–r numbering and this is also an r–r numbering of D. An illustration of the implicit algorithm of this proof is given in Fig. 5.1.

Thus we may assume that D has all vertices except r of indegree 2. By the Handshaking Lemma, there must exist a vertex w in D of outdegree at most 1, since the sum of the indegrees is twice the number of vertices minus 1 (as r has indegree 0). The sum of the outdegrees must equal this and it has 2 for r, and the remainder must be $2|V| - 2$, hence some term must be less than 2.

The first case is that the outdegree of w is 0. Now take an r–r numbering σ' of $D - \{w\}$. Let u_1, u_2 be the two inneighbors of w and suppose that $\sigma'(u_1) < \sigma'(u_2)$. Evidently, we can insert w between $\sigma'(u_1)$ and $\sigma'(u_2)$ and this is an r–r numbering σ of D.

The last case is that the outdegree of w is 1. Let x be the single outneighbor of w and let y be x's other inneighbor. This time let D' be the graph obtained by contracting the edge wx (i.e. making $w = x$). D' is still 2-connected by a similar argument to the one used above. Hence we have an r–r numbering σ' of D'. Now we make σ by replacing the vertex representing the contracted edge by x, and inserting w between y and x. We suppose that y is after the vertex representing the contraction in σ'. Consider the inneighbor s of w such that $\sigma'(s)$ is as small as possible. As σ' is an r–r numbering of D', s precedes the contracted vertex in σ'. We can insert w just after s, and this is an r–r numbering of D.   □

Fig. 5.2 The base graph

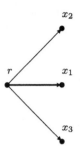

After deconstructing the digraph according to the algorithm implicit in the proof above (Exercise 5.3.1), one obtains the digraph in Fig. 5.2.

So we could start with x_1, x_2, x_3 and put in z_1. This needs to be between x_1 and x_2 and according to the method in the proof, should be placed as close to x_1 and possible, giving x_1, z_1, x_2, x_3. Now we place z_3, and it must slot between x_2 and x_3 and will between x_2 and x_3 and as close to x_2 as possible, so x_1, z_1, x_2, z_3, x_3 and then z_2 giving $x_1, z_1, x_2, z_2, z_3, x_3$.

As noted in Exercise 5.3.3, the r–r numbering σ decomposes the graph into two acyclic subgraphs covering D, one consisting of the forward edges of D generated by σ and the other the backward ones. We need the corollary and easy lemmas below.

Corollary 5.1.1 *If D is rooted and 2-connected then it has an acyclic connected covering subdigraph with at least half the edges of $D - \{r\}$.*

Lemma 5.1.1 (Folklore)
(1) *Any undirected graph with n vertices and m edges has a vertex cover of size $\leq \frac{n+m}{3}$.*
(2) *Suppose that $G = (A \sqcup B, E)$ is bipartite with every vertex in A of degree 2. Then B contains a dominating set of size $\leq \frac{|A|+|B|}{3}$.*

The proof of Lemma 5.1.1 is left to the reader in Exercise 5.3.4. It is not difficult. The following is the decisive lemma.

Lemma 5.1.2 (Daligault and Thomassé [193]) *Let D be a rooted acyclic digraph with ℓ vertices of indegree at least 2 and with a root of outdegree $d(r) \geq 2$. Then D has an outbranching with $\geq \frac{\ell+d(r)-1}{3} + 1$ many leaves.*

Proof Let $n = |V|$. For each vertex v of indegree ≥ 3, delete incoming edges until the indegree is 2. Since the root has indegree 0, this leaves a set Z of $n - \ell - 1$ vertices of indegree 1, with a set Y of inneighbors of size $\leq n - \ell - 1$. Let A_1 be the set of vertices of indegree 2 dominated by Y and A_2 the set of vertices of indegree 2 not dominated by Y. Z has all the outneighbors of r and hence Y contains r which has outdegree ≥ 2. Therefore, $|Y| + d(r) - 1 \leq |Z \cup A_1|$.

Now, as D is acyclic, it has a vertex s of outdegree 0. Let $B = V - Y - \{s\}$. Then as $A_2 = V - A_1 - Z - \{r\} = V - (A_1 \cup Z) - \{r\}$, it follows that $|A_2| \leq |V| - |A_1 \cup Z| - 1$. Hence $|B| = |V - Y - \{s\}| \geq |V| - |Z \cup A_1| - d(r) - 1 = |A_2| + d(r) - 1$.

Now $|Y| \leq n - \ell - 1$. Therefore $|B| \geq |V - Y - \{s\}| \geq n - (n - \ell - 1) - 1 = \ell$. We wish to apply the lemmas above. The idea of Daligault and Thomassé is to define a bipartite graph and then use the bounds of the lemmas. To this end, let A_2' be a copy of A_2 and B' a copy of B. Let G be the bipartite graph $(A_2' \sqcup B', E')$ with the edges ba for $a \in A_2'$ and $b \in B'$ iff $ba \in E(D)$. By construction, note that $d(a) = 2$ for all $a \in A_2'$. Thus we can apply Lemma 5.1.1(2) to get a G-dominating set $X' \subseteq B'$ of size $\frac{|A_2'| + |B'|}{3}$ and this in turn gives us a dominating set $X \subseteq B$ of size $\frac{|A_2| + |B|}{3} \leq \frac{2|B| - (d(r) - 1)}{3}$ dominating A_2 in D. Then the set $C = X \cup Y$ *strongly dominates* $V - \{r\}$ in the sense that every vertex of $V - \{r\}$ has a neighbor in C. Note that $|C| \leq |X| + |Y| \leq \frac{2|B| - (d(r) - 1)}{3} + |Y| = |B| + |Y| - \frac{|B| - (d(r) - 1)}{3}$. As $|Y| + |B| = n - 1$, and $|B| \geq \ell$, $|C| \leq n - 1 - \frac{\ell - (d(r) - 1)}{3}$. As D is acyclic, any strong dominating set is a connected dominating set (Exercise 5.3.5), hence there is an outbranching of D with a subset of C as internal vertices, and hence with at least $\frac{\ell - (d(r) - 1)}{3} + 1$ many leaves.   □

Daligault and Thomassé point out that the bound of Lemma 5.1.2 is tight up to one leaf (Exercise 5.3.7).

Corollary 5.1.2 *If D is a 2-connected rooted digraph with ℓ vertices of indegree at least 3, then D has an outbranching with at least $\frac{\ell}{3}$ leaves.*

Proof Apply Lemma 5.1.2 to the digraph with the largest number of vertices of indegree 2 from the subdigraph consisting of the forward edges generated by the r–r ordering σ, and the one generated by the forward edges of the reversal of σ. Since there are at least ℓ vertices of indegree ≥ 3, one of these two subdigraphs must satisfy the hypotheses of Lemma 5.1.2.   □

5.1.3 A Second Combinatorial Bound

To get our quadratic kernelization, following the analysis of Daligault and Thomassé we need one more combinatorial lower bound for the number of leaves. Let $D = (V, E)$ be a digraph. Following Daligault and Thomassé, we say that uv is a 2-circuit if $vu, uv \in E$. We say that an edge is *simple* if it is not part of a 2-circuit. Finally v is *nice* if it incident to a simple inedge.

Theorem 5.1.3 (Daligault and Thomassé [193]) *Let D be 2-connected and rooted. Assume that D has ℓ nice vertices. Then D has an outbranching with at least $\frac{\ell}{24}$ many leaves.*

Proof Let σ be an r–r numbering of D. Now for each nice vertex v incident with a simple inedge e and with indegree at least 3, delete from the graph all inedges different from e. This ensures that v has only one incoming forward edge and one incoming backward edge. For every other vertex of indegree at least 3, delete all incoming arcs until there is only one incoming forward edge and one incoming backward edge. Notice that the result is that D' remains a rooted 2-connected digraph and that σ is an r–r numbering. Moreover, the number of nice vertices has not increased.

Let T_f, T_b denote the forward (respectively, backward) edges of D, and as we have earlier observed, each is a spanning tree of D, partitioning the edges of $D - \{r\}$. For either tree T say that uv is *transverse* if $u \neq r$ and u and v are incomparable in the other tree in the sense that v is not an ancestor of u, and hence u is not an ancestor of v in T.

Now we choose $T = T_f$, say, with the most transverse edges. Take a plane representation of T_b. We partition the transverse edges uv of T_f into two sets, L and R, those vertices left and those right in the representation, since the plane representation will orient incomparable vertices. Suppose $|L| \geq |R|$ without loss of generality.

Then $T_b \cup L$ is an acyclic digraph by definition of L. Furthermore it has $|L|$ vertices of indegree 2, since the heads of the edges of L are pairwise distinct. Hence by Lemma 5.1.2, $T_b \cup L$ has an outbranching with at least $\frac{|L|+d(r)-1}{3} + 1$ many leaves.

It remains to estimate $|L|$. Take a nice vertex v that is not an outneighbor of r. Suppose that it has the simple inedge uv, and, without loss of generality, uv belongs to T_f. Let w be the outneighbor of v on the path from v to u in T_b. As uv is simple, $w \neq u$. No path in T_f goes from w to v, and hence uv is transverse. Hence every nice vertex is incident to a transverse edge. It follows that there are at least $\frac{\ell-d(r)}{2}$ transverse edges in D and hence at least $\frac{\ell-d(r)}{4}$ in T_f. Therefore $|L| \geq \frac{\ell-d(r)}{8}$.

Thus D has an outbranching with at least $\frac{|L|+d(r)-1}{3} \geq \frac{\frac{\ell-d(r)}{8}+1+d(r)-1}{3} \geq \frac{\ell}{24}$ many leaves. $\qquad\square$

5.1.4   A Quadratic Kernel

Daligault and Thomassé use the following reduction rules. In a digraph D, a *bipath* is a path $P = \{x_1, \ldots, x_t\}$ with $t \geq 3$ for which the set of edges incident with $\{x_2, \ldots, x_{t-1}\}$ is exactly $\{x_i x_{i+1}, x_{i+1} x_i \mid i \in \{1, \ldots, t-1\}\}$.

Reduction Rules
1. If there is a vertex not reachable from r in D, say *no*.
2. Let x be a cutvertex of D. Delete x and add edges vz for all $v \in N^-(x)$ (i.e. with $vx \in E$) and xz for all $z \in N^+(x)$.
3. For any bipath of length 4 contract two consecutive internal vertices.
4. Let $x \in V$. If there is a $y \in N^-(x)$ with $N^-(x) - \{y\}$ separating y, from r delete yx.

Standard techniques show that reachability in digraphs can be constructively determined in linear time. We say that a rule is *safe* if it preserves the fact that the reduced graph D' obtained from D has the same maximum rooted leaf outbranching number as D.

Lemma 5.1.3 *Rule* 2 *is safe, and can be implemented in polynomial time.*

Proof Let x be a cutvertex, and apply Rule 1. Assume that T is an outbranching of D rooted at r with k leaves. Then x is not a leaf, and let f be the parent of x in T. Construct T' by contracting fx, and we can see that T' is an outbranching of D' with k leaves.

Conversely, suppose that T' is an outbranching of D' with k leaves, rooted at r. We know that $N^-(x)$ is a cut in D' as x was a cutvertex in D. We observe that there is a nonempty collection of vertices $\{y_1, \ldots, y_t\} \subseteq N^-(x)$ which are not leaves in T'. Choose y_j with not an ancestor of any $y_i \in \{y_1, \ldots, y_t\} - \{y_j\}$. Now consider T as the graph obtained by adding x to T' as an isolated vertex. Then add the edge $y_j x$ and the edges $y_i x$ for $i \neq j$ with $y_i \in \{y_1, \ldots, y_t\}$ such that $y_i z \in N^+(x)$. Finally add the edge xz. As y_j is not an ancestor of y_i for all $\{y_1, \ldots, y_t\} - \{y_j\}$, there is no cycle in T, so it is an outbranching rooted at r with at least k leaves. Note that for this rule we only need to have the ability to find cutvertices, well known to be in polynomial time.   □

Lemma 5.1.4 *Rule* 3 *is safe and can be implemented in polynomial time.*

Proof Let u, v, w, x, z be the vertices of the bipath P in order, with vw contracted to obtain D'. Then let T be the k-leaf outbranching of D. First note that the rooted digraph obtained from T by contracting w to its parent yields the appropriate T'.

Conversely suppose that T' is an outbranching of D'. Suppose that x is the parent of the contracted vertex c in T'. Then replacing xc by xw and wv is the desired outbranching of D. The case where the parent c is u is analogous.   □

Lemma 5.1.5 *Rule* 4 *is safe and can be implemented in polynomial time.*

Proof Let x and y be as in Rule 4. Note that every outbranching of D' is one of D. Now let T be an outbranching of D containing xy. There is an ancestor $z \in N^-(x) - \{y\}$ of x. Then replacing yx in T by zx results in T', an outbranching of D' with at least k leaves.   □

The reduced graph is obtained by applying the rules above exhaustively.

Lemma 5.1.6 *Suppose that D is a reduced rooted graph with a vertex of indegree at least k. Then D is a yes instance of* Rooted k-Leaf Outbranching.

Proof Let x have $\ell \geq k$ inneighbors $\{u_1, \ldots, u_\ell\}$. By Rules 2 and 4, for each $i \in \{1, \ldots, \ell\}$, $N^-(x) - \{u_i\}$ does not cut u_i from r. Hence there must be a path P_i from r to u_i outside of $N^-(x) - \{u_i\}$. The rooted digraph $\bigcup_i P_i$ is connected and each u_i has outdegree 0 in D'. Therefore D' is outbranching and has at least k leaves which can be extended into an outbranching of D.   □

Here is our kernel result.

Theorem 5.1.4 (Daligault and Thomassé [193]) *A reduced digraph D with $|V| \geq (3k-2)(30k-2)$ has an outbranching with at least k leaves.*

Proof Call v *good* if it has indegree at least 3, or if one of its inedges is simple. Notice that a *bad* (i.e. not good) vertex u has exactly two inneighbors and these are also outneighbors of u.

Suppose that D has at least $6k + 24k - 1$ many good vertices. This is because it must either have at least $6k$ vertices of indegree ≥ 3 so Corollary 5.1.2 will apply, or it must have at least $24k$ nice vertices, so Theorem 5.1.3 will apply.

Thus we suppose there are at most $30k - 2$ good vertices in D. D is reduced so has no bipath of length 4. Define a weak bipath to be a maximal connected set of bad vertices. If $P = \{x_1, \ldots, x_t\}$ is a weak bipath, then the inneighbors of each internal x_i are exactly x_{i-1} and x_{i+1}. x_1 and x_t will be outneighbors of a good vertex. Thus we can associate with each weak bipath P of length t a set A_P of $\lceil \frac{t}{3} \rceil$ outedges towards good vertices. For each three consecutive vertices $x_i x_{i+1} x_{i+2}$ of P, $2 \leq i \leq t$, we see that $\{x_{i-1}, x_i, x_{i+1}, x_{i+2}, x_{i+3}\}$ is not a bipath by Rule 2, and hence there is an edge $x_j z$ for $j \in \{i, i+1, j+2\}$ and $z \notin P$. Also z must be good, since edges between bad vertices lie in their own weak bipaths. This is the set A_P of the desired size.

If this is a *No* instance, then by Lemma 5.1.6, any vertex of indegree less than k is reduced under Rule 4, and thus it follows that there are at most $3(k-1)(30k-2)$ many bad vertices in D.   □

Corollary 5.1.3 (Daligault and Thomassé [193])   ROOTED k-LEAF OUTBRANCHING *has a quadratic kernel.*

It is an open question whether it is possible to have a linear kernel for ROOTED k-LEAF OUTBRANCHING. Currently known techniques do not seem adequate to address this question.

Theorem 5.1.3 also allows us to demonstrate that some related problems have small kernels. Here is one illustration.

ROOTED k-LEAF OUT TREE

Instance: A directed graph $D = (V, E)$ and a root r.
Parameter: An integer k.
Question: Is there a tree in D with at least k leaves with root r?

The unrooted version is, of course, called k-LEAF OUT TREE. The point here is that there is no need for the tree to be an out branching.

Corollary 5.1.4   ROOTED k-LEAF OUT TREE *has a quadratic kernel.*

Proof If the size is more than $3(k-1)(30k-2)$ say "yes"; an out branching is an out tree. Else use complete search. □

5.1.5 k-LEAF OUTBRANCHING and Turing Kernels

The reader might wonder: "What about the apparently more natural problem of k-LEAF OUTBRANCHING where no root is specified?" As we will see in Chap. 30, we believe that *it has no polynomial size kernel*.[2]

The breakthrough was made by Fernau, Fomin, Lokshtanov, Raible, Saurabh, and Villanger [306].

Theorem 5.1.5 (Fernau et al. [306]) k-LEAF OUTBRANCHING *admits a polynomial-sized* Turing Kernel, *and hence is FPT. Specifically, for each vertex v of D, we generate an instance of* ROOTED k-LEAF OUTBRANCHING *with v as the root, and try them one at a time. If all say no then answer "no". If one says yes say "yes".*

The kernelization above is, confusingly, in the cited article, termed a "many–one" kernel.[3] As we mention in the footnote, it should be called a *disjunctive truth table* kernel. We analyze the above to extract a definition. An *oracle Turing machine computation* (such as in a Cook reduction) allows for many queries. The idea for a *Turing kernel* is the idea that each query to the oracle should be about an instance of size bounded by a function of the parameter.

In the following, think of $\langle \tau, k \rangle$ as the *query* and σ as the problem. The definition could be generalized to allow for the parameter to increase in length and for polynomially many oracle query (parameterized) languages.

Definition 5.1.3 (Turing kernelization) A *Turing kernel* consists of
(i) Two parameterized languages L_1 and L_2 (typically $L_1 = L_2$) with $L_i \subset \Sigma^* \times \mathbb{N}$ (for $i \in \{1, 2\}$), and a polynomial g.
(ii) Plus an oracle Turing procedure Φ, running in polynomial time on L_1, with oracle L_2, such that $\Phi_2^L = L_1$ on input $\langle \sigma, k \rangle$, only queries to L_2 of the form "Is $\langle \tau, k' \rangle$ in L_2?" for $|\tau|, k' \leq g(k)$.

---

[2]Specifically, what is proven there is that if this problem has a polynomial size kernel then the polynomial-time hierarchy collapses to two or fewer levels.

[3]As we saw in Chap. 1, in classical computability theory, and standard structural complexity theory, a many–one reduction is a (computable) (or polynomial) function $f : L \to \widehat{L}$ such that $x \in L$ iff $f(x) \in \widehat{L}$. The "many" is that it is possible that for various x, y $f(x) = f(y)$ as opposed to a 1–1 reduction where we insist that f is injective. If anything, the Fernau et al. reduction is really "one–many" since one question is being reduced to many. It is what is called a *disjunctive truth table* since it is the form $x \in L$ iff $\widehat{L}$ satisfies at least one of a polynomially computed (boolean) disjunction.

The idea is that on input $\langle \sigma, k \rangle$ Φ works like a normal polynomial-time oracle machine except on oracle queries, it converts every query to a query of a kernel *determined by the query*. Notice that since Φ runs in polynomial time, the whole process runs in time $\leq p(|\langle \sigma, k \rangle|)$, for some polynomial p.

In the case of k-LEAF OUTBRANCHING:

(i) L_1 is the collection of pairs $\langle G, k \rangle$ consisting of digraphs with k or more leaf outbranchings.

(ii) L_2 is the collection of pairs $\langle \widehat{G}, k \rangle$ consisting of kernels of *rooted* digraphs (the input $\widehat{G}$ would specify a root r) with k or more leaf outbranchings.

The procedure Φ would take an instance $\langle G, k \rangle$ and asks queries of the form "Does (G, r) have a k-leaf outbranching rooted at r?". It converts this into a kernelized query using Daligault and Thomassé technique, of the rooting of G at r, and queries L_2.

As we will see in Chap. 30, we now have methods to demonstrate that a problem likely does not have a polynomial-sized kernel (that is, of size bounded by a polynomial function of the parameter), assuming the mathematical universe is as we think it is. For Turing poly(k) kernelization, we have no corresponding lower-bound methods tied to familiar conjectures in classical complexity.

However, in Chap. 30, we report new completeness programme that might yield credible lower-bound technology for Turing kernelization. In fairness, the conjectures that P is not equal to NP, or that FPT is not equal to $M[1]$ or $W[1]$, are underwritten by completeness programs, in the context where provable inequalities seem untouchable. So why not add further plausible, but probably also untouchable, conjectures, in order to establish plausible lower-bound technology for poly(k) Turing kernelization—which is so well-motivated by practical considerations.

In the next section, we return to a discussion of many:1 kernelization, and a particularly important advanced topic, for what is arguably the single most important (in terms of applications) optimization problem. Results in this area have had powerful consequences in the analysis of real datasets in a surprisingly wide range of applications.

5.2   Crown Reductions

In this section we will look at an idea introduced by Chor, Fellows, and Juedes [159] which can also be used to give a size $3k$ kernel for VERTEX COVER. The reader might wonder why we are doing this given that we have two other ways *already* to get a size $2k$ one. The reasons are that (i) the new method of crown reductions is of independent interest, and (ii) because it is a simple technique based on *local* rules it often performs much better in practice than the Nemhauser–Trotter method. Such computational experiments are reported in Abu-Khzam, Collins, Fellows, Langston, Suters, and Symons [9]. The idea of a crown also has reflections in recent work on kernelization using what are called *protrusions*.

The starting point is the first rule in the Buss kernel for VERTEX COVER. In the Buss kernel, we delete any isolated vertex and always take its neighbor for the vertex cover since it is just as good. The idea is that if we have any independent set

Fig. 5.3 A crown

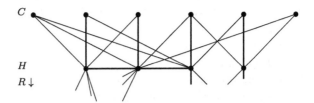

of vertices "resembling" a degree one vertex, in that they look like the "points of a crown", then we might as well take the neighbors and cover just as well. Figure 5.3 illustrates a crown.

Definition 5.2.1 (Crown decomposition) A crown decomposition of a graph $G = (V, E)$ is a partition $V = C \sqcup H \sqcup R$, where $C, H \neq \emptyset$ and
1. C is an independent set of vertices.
2. There is no edge between C and R.
3. There exists an injective (matching) map $f : H \to C$. That is, if $f(h) = c$, then $hc \in E$.

Notice that if G is connected then for all $c \in C$, there is an $h \in H$ with $hc \in E$. Therefore, the following is more or less immediate.

Proposition 5.2.1 *If G is connected and $V = C \sqcup H \sqcup R$ is a crown decomposition, then G has a size k vertex cover iff $G - (H \sqcup C)$ has a size $k - |H|$ vertex cover.*

There are several uses of similar crown-like procedures for data reduction, and they are problem dependent. We will use crowns to obtain two kernelizations for VERTEX COVER, as they are both of independent interest. First we will give a $4k$ kernelization, and then with a more intricate argument get a $3k$ one. Both use matching in general graphs. The reader should recall that in a general graph, a matching is a disjoint collection of edges in the graph. The greedy algorithm will find a maximal matching in a general graph, but we are interested in maximum (sized) matchings. For this, there are efficient algorithms for this going back to Edmonds [267]. If the reader is unfamiliar with this area, he or she should read Chap. 34. One notable fast algorithm for finding matchings in general graphs is due to Micali and Vazirani [531] and runs in time $O(\sqrt{|V|}|E|)$.

The following simple algorithm finds a crown decomposition or gives a certificate that none exists. The argument works as follows. If we take a maximal matching M in a graph G, then $C = V(G) - V(M)$ must be an independent set. The following lemma is implicit in the proof of Theorem 5.2.1, and hence we will leave it as an exercise for the reader in Exercise 5.3.8.

Lemma 5.2.1 (Chor, Fellows, and Juedes [159]) *Suppose that G has an independent set I with $|N(I)| < |I|$. Then a crown decomposition (C, H, R) for G with $C \subseteq I$ can be constructed in time $O(|V| + |E|)$.*

Proposition 5.2.2 (Sloper and Telle [624]) *Consider the algorithm below on input* $G = (V, E)$.

Algorithm 5.2.1
1. Find a maximal matching M for G.
2. If $|V(M)| > 2k$ say "no" and stop.
3. If $|V(G) - V(M)| \leq 2k$ then output (G, k) and stop.
4. Else use Lemma 5.2.1 to construct a crown decomposition $V = C \sqcup H \sqcup R$, and let $G = G[V - (C \sqcup H)]$, and $k := k - |H|$.
5. Repeat.

Then the algorithm terminates (correctly) with the answer no or produces a kernel of size $\leq 4k$ for k-VERTEX COVER.

Proof Termination is clear. Suppose that we terminate with a "no". We only get a "no" if there is a maximal matching M in G with $|V(M)| > 2k$. Since we would need at least one endpoint from each edge for any vertex cover, there cannot be one with only k vertices. Proposition 5.2.1 says that a "yes" instance is preserved under the crown reduction. The algorithm will halt when $|V(G)| = |V(M)| + |V(G) - V(M)| \leq 2k + 2k = 4k$.   □

Now we turn to the $3k$ kernelization.

Theorem 5.2.1 (Chor, Fellows, and Juedes [159]) *Crowns can be used to compute a $3k$ kernelization for* VERTEX COVER *in polynomial time.*

Proof Let $G = (V, E)$ be given and k be the parameter. First compute a maximal matching M in G. If $|M| > k$ simply say "no" as any vertex cover needs more than k vertices. Else, M determines a set of vertices $I = \{v \mid \forall w \in V(G)[vw \notin M]\}$, which will be an independent set. If $|I| > k$ we stop, since we would need more than k vertices for a vertex covering. Assuming that I is nonempty, we consider the bipartite graph induced in G by M. This is $G[I, V - I] = \{vw \mid v \in I \wedge vw \in N(I)\}$. The idea here is that this bipartite graph should be a crown, if it is not one already. The problem is that to obtain a crown, we need to make sure that all the vertices in the potential H (i.e. $N(I)$) are matched in I, and some may not be. Therefore, we compute a maximum matching M' in the bipartite graph $G[I, V - I]$.

Clearly, if every vertex of $N(I)$ is matched (so that M' is a $N(I)$-perfect matching) then we have our crown, $C = I$ and $H = N(I)$. If not, then the idea is to select a subset I' of I which has an $N(I')$-perfect matching. Let I_0 be the collection of vertices unmatched by M'. If $I_0 = \emptyset$ we stop, and output the current graph as the kernel. Else, compute $H_0 = N(I_0)$. If all the vertices of H_0 are now matchable we can stop as we have a crown, $C = I_0$, $H = H_0$. (Of course this would really mean that $C = \{x\}$ and $H = \{y\}$ for some isthmus $xy \in G$.) Otherwise, something in $N(I_0)$ is unmatchable by I_0. Here we have the key observation:

- *For such $y \in N(I_0)$, there must be a corresponding $x \in I$ with xy in the matching M'.*

Otherwise, we could add an edge xy to M' and make a bigger matching, and this would contradict the fact that M' is maximal. We let $I_1 = I_0 \cup N_{M'}(H_0)$ where $N_{M'}$ denotes the I-neighbors determined by the M' matching. More generally, we repeat the following steps until we get $I_{n+1} = I_n$;

$$H_n = N(I_n) \quad \text{and} \quad I_{n+1} = I_n \cup N_{M'}(H_n).$$

The claim is that
 (i) There is some n where we stop changing.
 (ii) $C = I_n$ and $H = H_n$ determine a crown.
Why should (i) be true? By induction, at each stage, if we don't have every element of H_n matched, as above, there must be some element y of H_n not matched by $N_{M'}(I_n)$ *and since the matching M' is maximal, it must be matched by something in I under M'*. Otherwise, from an element x of I_0 we would have an (odd) alternating path from x to y which we could use to construct a new bigger matching.[4] Since I is finite this process must stop, and hence (i) holds. But, by construction, we have made sure that everything in H is then matched so that (ii) holds as well. $C = I_n, H = H_n$ induce a crown, $C \sqcup H \sqcup (V - (C \cup H))$.

To kernelize, we will now use the following algorithm.

Algorithm 5.2.2
1. Find a maximal matching M of G. Let I be the set of unmatched vertices.
2. If $|M| > k$ say "no".
3. If $|I| > k$ say "no".
4. If all vertices are matched then stop, else.
5. Find a maximum sized matching M' between I and $N(I)$ in $G[I, N(I)]$. If this forms a crown, remove the $V(G[I, N(I)])$ and set $k := k - |G[I, N(I)]|$. Go to (1).
6. Else, let I_0 be the set of M'-unmatched vertices of I. If $I_0 \neq \emptyset$, compute until we get $I_{n+1} = I_n$;

$$H_n = N(I_n) \quad \text{and} \quad I_{n+1} = I_n \cup N_{M'}(H_n).$$

7. For the fixed point $C = I_n, H = H_n$, set $k = k - |C \sqcup H|$ and remove $C \sqcup H$ from G. Go to (1).

The termination and validity of the procedure are clear. It remains to argue that the procedure yields a kernel with at most $3k$ vertices. Since the size of the matching M is at most k, and the size of I is at most k, we will either reduce, or have a situation with $I_0 = \emptyset$. The case that $I_0 = \emptyset$ means that every element of I is matched in $N(I)$. Then $|V(G)| \leq 2k + k$, with $2k$ for the matching and k for I. □

---

[4]This kind of argument is carefully discussed in Chap. 34 when we look at e.g. Berge's Criterion. If the reader is new to this kind of argument he or she should pause and read that part of Chap. 34 at this point.

5.3   Exercises

Exercise 5.3.1   Write an algorithm to construct an $r–r$ labeling based around the proof of Theorem 5.1.2.

Exercise 5.3.2   Prove Theorem 5.1.1 based on the method of Theorem 5.1.2.
(Hint: First prove that if D is 2-connected then it has a subdigraph D' such that the indegrees of s and t are both 0, and for each $v \in D' - \{s, t\}$, there are two internally disjoint paths; one from s to v, and one from t to v. Then prove that D' has a vertex of outdegree at most one.)

Exercise 5.3.3 (Whitty [664])   A *branching* is a spanning acyclic subdigraph such that one vertex (called the root) has indegree 0, and all others have indegree 1. Let D be a digraph and specify a vertex z. Prove that if G is 2-vertex connected then there are two branchings $B_1 = (V, E_1)$, $B_2 = (V, E_2)$ rooted at z such that $E_1 \cap E_2 = \emptyset$, and for any v the two paths from z to v in B_1, B_2, respectively, are (internally) disjoint.
(Hint: Use Exercise 5.3.2.)

Exercise 5.3.4   Prove Lemma 5.1.1.
(Hint: For (1) use induction on $n + m$. For (2) use (1).)

Exercise 5.3.5   Prove the claim in Lemma 5.1.2 that if D is acyclic, any strong dominating set is a connected dominating set.

Exercise 5.3.6 (Leizhen Cai, Verbin, and Yang [128])   The *firefighting* problem takes a graph G with and a vertex $s \in V(G)$, where a fire starts burning at time $t = 0$. At each step $t \geq 1$, a firefighter can (permanently) protect a vertex, and at east step the fire spreads to all neighboring vertices of burning vertices.
    Consider the problem

SAVING k VERTICES
Instance: A graph $G = (V, E)$.
Parameter: An integer k.
Question: Is there a strategy to save at least k vertices when a fire breaks out?

Prove that if G is a tree then SAVING k VERTICES is FPT with a problem kernel of size $O(k^2)$.

Exercise 5.3.7   Prove that the rooted digraph D_k with $d(r) = 3$ (with D_6 in Fig. 5.4), is 2-connected, has $3k - 2$ vertices of indegree at least 2, and D_k has the maximum outbranching of D_k has $k + 2$ many leaves.

Exercise 5.3.8 (Chor, Fellows, and Juedes [159])   Prove Lemma 5.2.1.
(Hint: Use the method in the proof of Theorem 5.2.1.)

Fig. 5.4 The digraph D_6

Exercise 5.3.9 (Chor, Fellows, and Juedes [159]) The original use of crown reductions in [159] was to attack the problem below, which is related to computational biology. It is an example of parametric duality.

$n - k$ CLIQUE COVER

Instance: A graph $G = (V, E)$.
Parameter: An integer k.
Question: Are there $|G| - k$ many cliques in G that cover all the vertices?

Use methods similar to those described for the VERTEX COVER problem to show that $n - k$ CLIQUE COVER is kernelizable to a graph with at most $2k$ vertices.
(Hint: Assume that G is connected and consider how matching will interact with the independent set created by the matching.)

Exercise 5.3.10 (Chor, Fellows, and Juedes [159]) Show that the problem $n - k$ GRAPH COLORING below is FPT using the previous exercise.

$n - k$ GRAPH COLORING

Instance: A graph $G = (V, E)$.
Parameter: An integer k.
Question: Can G be (vertex) colored with at most $|G| - k$ many colors?

Exercise 5.3.11 (Kneis, Langer, and Rossmanith [452]) Prove that the following has a $O^*(1.5^t)$ algorithm for graphs of maximum degree 3.

PARTIAL VERTEX COVER

Instance: A graph $G = (V, E)$.
Parameter: Integers t and k.

Question: Is there $C \subseteq V$ of size k covering at least t edges?

For a bonus prove that this can be improved to $O^*(1.26^t)$.

Exercise 5.3.12 (Bodlaender, Jansen, and Kratsch [92]) Prove that the following problem admits a polynomial-time kernelization algorithm, where the number of vertices of a kernelized instance is $O(\ell^2)$.

LONG CYCLE PARAMETERIZED BY A VERTEX COVER
Instance: A graph G, and a vertex cover $X \subseteq V(G)$ and an integer k.
Parameter: $\ell = |X|$.
Question: Is there a cycle in G of length at least k?

Exercise 5.3.13 (Gutin) Prove that k-LEAF OUTBRANCHING and the rooted versions both admit a kernel with $O(k)$ vertices for acyclic digraphs.

5.4   Historical Notes

The parameterized complexity of ROOTED k-LEAF OUTBRANCHING was first discussed by Alon, Fomin, Gutin, Krivelevich, and Saurabh [32], where it was proven NP-complete and the parameterized version to be FPT. The first kernelization for this problem was given by Fernau, Fomin, Lokshtanov, Raible, Saurabh, and Villanger [306]. The quadratic kernel we described is due to Daligault and Thomassé [193]. The key insight of Fernau et al. [306] for the unrooted case was that it gave a poly(k) Turing kernel. The idea of a Turing kernel had already been discussed in Downey and Fellows [243] (in relation to the Robertson–Seymour Theorem, as we later see). However, the realization of Fernau et al. [306] is that the lower-bound methods of Chap. 30 applied to demonstrate that the unrooted case has *no* polynomial kernel unless co-NP $\subseteq$ NP/Poly. (More on this in Chap. 30.) The definition of a Turing kernel given here is more or less that same as those in the literature. The current fastest algorithm for k-LEAF OUTBRANCHING is due to Daligault, Gutin, Kim, and Yeo [192], and runs in time $O^*(3.72^k)$.

Crown reductions were introduced in Chor, Fellows, and Juedes [159]. They often work much faster than the Nemhauser–Trotter kernel in practice, as reported by Langston and his group.

5.5   Summary

We introduce further techniques for kernelization including Turing kernels. Particular problems addressed include ROOTED k-LEAF OUTBRANCHING and k-LEAF OUTBRANCHING. Crown reductions for kernelization are introduced.

Iterative Compression, and Measure and Conquer, for Minimization Problems

6

6.1   Iterative Compression: The Basic Technique

Iterative compression is a technique used for minimization problems. It came to prominence with Reed, Smith, and Vetta's solution to a problem that had been a longstanding one in parameterized complexity: EDGE BIPARTIZATION. This problem is specified as follows.

EDGE BIPARTIZATION
Instance: A graph $G = (V, E)$.
Parameter: An integer k.
Question: Can I delete k or fewer edges and turn G into a bipartite graph?

We will soon discuss the significance of this problem, and how it was solved using the new technique of iterative compression, which for many years was the *only* technique applicable to EDGE BIPARTIZATION. As we mentioned in Exercise 4.11.17 Narayanaswamy, Raman, Ramanujan, and Saurabh [497] use VERTEX COVER ABOVE LP to give an alternate route to parametric tractability.

However, we will begin our discussion by attacking VERTEX COVER with this technique. Initially, it seemed that applying iterative compression for problems like VERTEX COVER was not competitive with other known techniques for minimization problems. Yet, recent work has shown that *combining* iterative compression with other combinatorial techniques can achieve much sharper results, particularly for kernelization. We will take note of this when we re-examine FEEDBACK VERTEX SET in Sect. 6.5, where we will look at the current approach for this problem. Iterative compression is also applied by the state-of-the-art algorithm for VERTEX COVER, which is so intricate that detailing it would take us beyond the scope and planned length of the present book. The message is clear: iterative compression is a major tool that you, as an algorithm designer, need to know to address minimization problems.

Designers of an algorithm for a minimization problem like VERTEX COVER, recognize that it is *enough* to do the following iterative compression routine.

R.G. Downey, M.R. Fellows, *Fundamentals of Parameterized Complexity*,
Texts in Computer Science, DOI 10.1007/978-1-4471-5559-1_6,
© Springer-Verlag London 2013

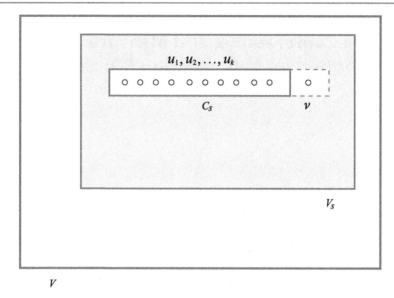

V

Fig. 6.1 Note that $C_s \cup \{v\}$ is a vertex cover of size $(k+1)$ for $V_s \cup \{v\}$

Algorithm 6.1.1 (Iterative Compression Routine) Given a size $k+1$ solution to the minimization problem, the algorithm either gives a certificate that there is no size k solution or produces one of size k.

The routine is to add the vertices of the graph, one at a time; if the relevant problem is of the compressible type, we will have a running solution for the induced subgraph of size k or at some stage get a certificate that there is no such solution for G. The algorithm modifies the current solution at each compression stage to make another one for the next stage.

Here is the iterative compression routine applied to VERTEX COVER. Importantly, in the algorithm below, if there is *no* size k solution at some stage, this certified "no" is a hereditary property. Therefore, as we only *add* vertices and edges, the instance cannot become a "yes"; so we are safe to abort.

Algorithm 6.1.2 (Iterative Compression for Vertex Cover)
1. Initially, set $C_0 = \emptyset$, and $V_0 = \emptyset$.
2. Until $V = V_s$ or the algorithm has returned "no", set $V_{s+1} = V_s \cup \{v\}$, where v is some vertex in $V - V_s$. The inductive hypothesis is that C_s is a vertex cover of $G[V_s]$ of size $\leq k$. Consider $G[V_{s+1}]$.
3. Clearly $C_s \cup \{v\}$ is a size $\leq k+1$ vertex cover of $G[V_{s+1}]$. See Fig. 6.1. If it has size $\leq k$, go to the next step. If not, we will seek to compress it. Let $\hat{C} = C \cup \{v\}$. Consider all possible partitions of $\hat{C}$ into two sets, $D \sqcup Q$. The idea is that the vertices of D will be *discarded*, and the vertices of Q *kept* in the search for the

Fig. 6.2 Note that $G[D]$ must be independent and $|H| < |D|$ for a successful compression

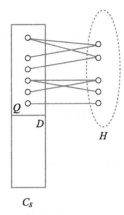

"next" vertex cover (of size k). We attempt to construct a suitable replacement H for the discarded part D of the vertex cover $C_s \cup \{v\}$.

For each such partition, and for each edge $e = uv$ not covered (i.e. $u, v \notin Q \cup H$), if $u, v \in D$, then there is no possible vertex cover of $G[V_{s+1}]$ that contains no vertices of D, the vertices to be discarded from $C_s \cup \{v\}$.

Otherwise, H must cover all the edges that have one endpoint in D and the other outside of Q. Thus, if uv is not covered by $H \cup Q$, find a vertex w, one of u or v not in D, and set $H := H \cup \{w\}$.

4. At the end of this if $H \cup Q$ has size $\leq k$, set $C_{s+1} = H \cup Q$ and go to step $s + 2$. Whenever $|H \cup Q|$ exceeds k then stop, and move to a new partition.
5. If we try all partitions, and fail to move to step $s + 2$, then stop and return "no".

Figure 6.2 can be useful here.

There are at most 2^{k+1} different partitions to consider in each step, so the algorithm requires time at most $O(2^k|G|)$.

6.2 EDGE BIPARTIZATION

Historically, Reed, Smith, and Vetta's proof that the EDGE BIPARTIZATION problem is FPT provided the first application of the method of iterative compression. Until the recent work of Lokshtanov, Narayanaswamy, Raman, Ramanujan, and Saurabh in [497] (as reported in Exercise 4.11.17), iterative compression is the only method known to demonstrate that EDGE BIPARTIZATION is FPT. The idea in the algorithm below is similar to that used for VERTEX COVER, except that we add edges rather than vertices. The algorithm we present is due to Guo, Gramm, Hüffner, Niedermeier, and Wernicke [365], and runs in time $O^*(2^k)$, improving on the original $O^*(3^k)$ algorithm of Reed, Smith, and Vetta. It is the state-of-the-art approach at present, especially when combined with the "algorithm engineering" heuristics of Hüffner [404].

Lemma 6.2.1 *Let G be a graph with an edge bipartization set X, and let C be a cycle in G. Then $|E(C) \cap X|$ and $|E(C)|$ have the same parity.*

Proof For each edge $uv \in X$, note that both u and v are in the same side the induced bipartite graph, otherwise they would not be needed for a minimal bipartization. Consider C as a cycle in G. The edges in $E(C) - X$ are all between two sides of $G - X$, and, as we just argued, those in $E(C) \cap X$ are all in one side. But for C to be a cycle, $E(C) - X$ must have an even number of edges, and since $|E(C)| = |E(C) - X| + |E(C) \cap X|$, it must be the case that $|E(C)|$ and $|E(C) \cap X|$ have the same parity.   □

Definition 6.2.1 Let $X \subseteq E$ for a graph $G = (V, E)$. Let $V(X) = \{v \mid \exists w(vw \in X)\}$. Then a black/white two-coloring $\Phi : V(X) \to \{B, W\}$ is called a *valid coloring* of $V(X)$ if for each $vw \in X$, $\Phi(v) \neq \Phi(w)$.

Now we have the decisive lemma allowing for iterative compression.

Lemma 6.2.2 *Suppose that X is a minimum edge bipartization of $G = (V, E)$. Suppose that Y is any set of edges disjoint from X. Then the following are equivalent.*

1. *Y is an edge bipartization set for G.*
2. *There is a valid two-coloring Φ of V such that Y is an edge cut in $G - X$ between $\Phi^{-1}(B)$ and $\Phi^{-1}(W)$. That is, all paths between the black and the white vertices include an edge in Y.*

Proof (2) implies (1). It suffices to prove that $E(C) \cap Y \neq \emptyset$ for any odd cycle C. Let $E(C) \cap X = \{u_0 v_0, u_1 v_1, \ldots, u_q v_q\}$, which is odd by Lemma 6.2.1, and numbered with vertices $u_i v_{(i+1) \bmod q}$ being connected by a path in $G - X$. Since Φ is a valid coloring, we see that $\Phi(v_i) \neq \Phi(u_i)$ for all $0 \leq i \leq q$. As q is odd, there must be some pair $v_i u_{(i+1) \bmod q}$ with $\Phi(v_i) \neq \Phi(u_{(i+1) \bmod q})$. But removal of Y destroys all paths between black and white vertices, and hence $E(C) \cap Y \neq \emptyset$.

(1) implies (2). Let $C_X : V \to \{B, W\}$ be a two-coloring of the bipartite graph $G - X$, and similarly G_Y a two-coloring of $G - Y$. Now define $\Phi : V \to \{B, W\}$ with $\Phi(v) = B$ by having if $C_X(v) = C_Y(v)$ and $\Phi(v) = W$, otherwise. The claim is that given the restriction of Φ to $V(X)$, $\hat{\Phi}$ is a valid coloring with Y an edge cut in $G - X$ between the white and the black vertices of $\hat{\Phi}$. Figure 6.3 might be useful here.

To see that $\hat{\Phi}$ is a valid coloring, let uv be any edge of X. If all the paths from u to v in $G - X$ were odd, then uv would be redundant, and X not minimal. Hence there must be at least one even path from u to v in $G - X$. Thus $C_X(u) = C_Y(v)$. However, in $G - Y$, since X and Y are disjoint, uv is an edge, and hence $C_Y(u) \neq C_X(v)$. Therefore $\hat{\Phi}(v) \neq \hat{\Phi}(v)$.

Now, C_Y and C_X must change values as they move from one vertex to its neighbor in $G - (Y \cup X)$; Φ is constant along any path in $G - (Y \cup X)$. We conclude that there can be no path from white $\Phi(v)$ to a black $\Phi(u)$ in $G - (Y \cup X)$, and hence Y must be an edge cut between the $\hat{\Phi}$ black and white vertices in $G - X$, as required.   □

Fig. 6.3 $1 \Rightarrow 2$. The set Y intercepts the paths shown

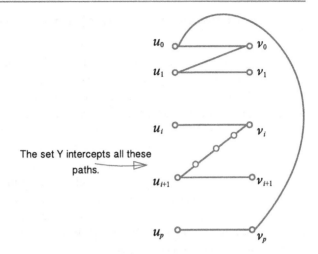

The set Y intercepts all these paths.

Theorem 6.2.1 (Guo et al. [365]) EDGE BIPARTIZATION *is solvable in time* $O(2^k k m^2)$ *for a graph with m edges.*

Proof The method is quite similar to that used for VERTEX COVER. Let $E(G) = \{e_1, \ldots, e_m\}$. At step i we consider $G[\{e_1, \ldots, e_i\}] := G_i$, and either construct from X_{i-1} a minimal bipartization set X_i or return a "no" answer. $X_1 = \emptyset$. Consider step i, and suppose that $|X_{i-1}| \leq k$. If X_{i-1} is a bipartization set for G_i, we are done, as we can keep $X_{i-1} = X_i$, and move on to step $i + 1$. Else, we consider $X_{i-1} \cup \{e_i\}$, which will clearly be a minimal edge bipartization set for G_i. If $|X_{i-1} \cup \{e_i\}| \leq k$ then set $X_i = X_{i-1} \cup \{e_i\}$ and move on to step $i + 1$. If $|X_{i-1} \cup \{e_i\}| = k + 1$, we seek an X_i that will have $\leq k$ edges, or we report "no". The plan is to use Lemma 6.2.2 with "$X = X_{i-1}$" and "$Y = X_i$" to seek new bipartizations. This lemma needs Y to be disjoint from X. This is achieved by a simple reduction at this step. For each $uv \in X_{i-1}$, we delete this edge, and replace it in G with 3 edges, uw_1, w_1w_2, w_2v (w_1, w_2 are distinct for each pair u, v). Then in X_{i-1} we include one of these edges for each uv, for instance w_1w_2. Notice that if uv is necessary for a minimal edge bipartization before this step, then we can use either of the other two in its stead for G_i. Now we proceed analogously to VERTEX COVER. Namely, we enumerate all valid colorings Φ of $V(X_{i-1})$, and determine a minimum edge cut between the white and black vertices of size $\leq k$. Each of the minimum cut problems can be solved using bipartite matching and hence in time $O((k + 1)i)$, as this has $k + 1$ rounds. If no such cut is found in any of the partitions we say "no". If we find a Y, set $G_i = Y$, and move to step $i + 1$. The total running time is thus $O(\sum_{i=1}^{m} 2^{k+1}ki) = O(2^k k m^2)$. $\qquad\square$

6.3   Heuristics and Algorithmic Improvements

It is possible to improve the above algorithms for vertex minimization problems using both heuristics and guaranteed improvements. For example, in the case of EDGE BIPARTIZATION, Hüffner [404] observed that a primary source of running time is the fact that when we do the bipartite matching to seek a smaller Y in the valid colorings, we seem to need to run something like Ford–Fulkerson or Edmonds–Karp at each step. However, if we use $(3, k)$-ary Gray codes (Guan [363]), such colorings can be enumerated with adjacent ones differing by only one element. Then the previous bipartite matching can be recycled and modified to see if a target Y is found. For example, consider the previous flow. If that flow involved the previously included vertex, which is now deleted, then zero the flow on the path, i.e. "drain the flow" and otherwise, keep things the same. It is not difficult to see that this works, and hence we get the following result.

Corollary 6.3.1 (Guo et al. [365]) EDGE BIPARTIZATION *is solvable by iterative compression in time* $O(2^k m^2)$.

Another method to speed up algorithms for edge minimization problems has been achieved by what Hüffner calls *filtration of valid colorings*. The idea here is to drastically omit searches. For example, suppose that uv is an edge of X_{i-1}. After the initial transformation, they are connected by an edge $w_1 w_2$. Since they are directly connected, it is not possible to find a cut in $G - X$ that excludes them. We can only look at valid colorings that assign to these vertices the same coloring.

The worst case running time does not improve. Yet with the modified algorithm derived from these methodologies, in situations like dense graphs, experimental improvements have been observed. Hüffner [404] describes engineering applications for a slightly slower algorithm (running in time $O(3^k m^2)$) more akin to the one of Reed, Smith, and Vetta, but his analysis is still relevant here. In particular, Hüffner ran experiments on synthetic data, as well as upon data obtained from data from the human genome, namely Wernicke [663]. Even with the slower algorithm, the heuristic improvements and the use of Gray codes resulted in quite significant improvements to the running times. In some cases *only the heuristically improved algorithms* actually *halted* in allowable time. There have been no more recent experiments.

In Sect. 6.5, we will again look at FEEDBACK VERTEX SET in a classic application of iterative compression *as a kernelization technique*. The algorithm also benefits from another reduction technique, "measure and conquer", which should be part of the toolkit of the designer of exact algorithms. We will examine this supportive technique in the next section, and encourage its application in Sect. 6.6 ("Exercises"), as well as give additional context for its development in Sect. 6.7 ("Historical Remarks").

6.4   Measure and Conquer

The "measure and conquer" reduction technique was developed by Fomin, Grandoni, and Kratsch [316]. This is a very general technique for speeding up branching algorithms (extending the interleaving method of Niedermeier and Rossmanith [547] in Sect. 4.4), and, as we mentioned, it should be in the toolkit of all workers in exact algorithms. The idea behind the "measure and conquer" technique is to use a "nonstandard measure" for the complexity of the problem, which gives weights to the branchings generated by the subcase analysis, and thus to derive recurrence relations for the cases in terms of their respective weights.

We will follow the analysis of van Rooij and Bodlaender [653]. We will examine an exact algorithm for DOMINATING SET. Although this is not FPT, as we later see, the method of measure and conquer is well-portrayed with this problem. The reader should recall that (for the unparameterized version of the problem) we are seeking the smallest collection of vertices that dominates all the other ones; meaning that either they are in the dominating set or they are adjacent to an element of the dominating set. The method of measure and conquer, soon to be described, was used to improve the analysis of exact algorithms for DOMINATING SET [316], INDEPENDENT SET (Fomin, Grandoni, and Kratsch [317]), DOMINATING CLIQUE (Kratsch and Liedloff [467]), and many other places. The van Rooij–Bodlaender paper also used the method for the *design* of the algorithm, which is why we use that as an archetype for these investigations.

We begin with an initial reduction to the following problem.

SET COVER
Instance: A collection of nonempty sets $\{S_i \mid S_i \in C\}$ and an integer k.
Question: Is there a subcollection of k of the sets $\{S_{i_1}, \dots, S_{i_k}\}$ such that $\bigcup_{j=1}^{k} S_{i_j} \supseteq \bigcup \{S_i \mid S_i \in C\}$?

Of course, the parameterized version would have k as a parameter, and the optimization version is to find the least such k (and the witness sets). Of importance to us is the complexity of the (parameterized) reduction of the lemma below.

Lemma 6.4.1 (Karp [431])   DOMINATING SET $\leq_m^P$ SET COVER.

Proof Take an instance $G = (V, E)$ of DOMINATING SET. For each vertex $v \in V$, make a set $S_v = \{v\} \cup N[v]$. Then set covers are in 1–1 correspondence to dominating sets. $\square$

This reduction is definitely a parameterized one in that all parameters are preserved. We will discuss this in more detail in Part V, in Chap. 20. The algorithm for SET COVER below uses what is called the *branch and reduce* method going back to Davis and Putnam [198]. The method coordinates a collection of branching rules and reduction rules. The reduction rules generate simpler equivalent instances, and when we cannot find any further applications of the reduction rules, we branch. This is rather similar to the methods we met in the kernelization sections. Again note that we wish to minimize the *branching*. The original use of measure and conquer was to re-analyse the branch and reduce technique to obtain better upper bounds for the running times. Now we give the van Rooij–Bodlaender algorithm method.

Algorithm 6.4.1 Apply the following reduction and branching minimization rules.
1. *The Base Case:* If all the sets have size 2, then the problem is essentially the same as a maximum matching, thinking of the sets now as edges. This can be done in polynomial time (Chap. 34).
2. *Splitting Connected Components:* Compute the set cover of each connected component separately for each component.
3. *Subset Rule:* If the instance has $S_i \subset S_j$, remove S_i and in each cover that contains S_i replace it by one with S_j.
4. *Subsumption Rule:* If the collection of sets in which an element e' occurs is a subset of the collection of sets in which some e occurs, then remove the element e, because in any set cover, covering e' covers e.
5. *Singleton Set Rule:* If a set of size 1 remains after the application of the preceding rules, then add this set to the set cover. This is because the element of the set must occur uniquely in this set; if not, then it would have been a subset of another, and removed by the Subset Rule.
6. *Avoiding Unnecessary Branchings Rule (for frequency-two elements):* For any set S in the problem instance, let r_2 be the number of frequency-two elements it contains, where the *frequency* of an element is the number of sets in which the element lies. Let m be the number of elements in the union of sets containing the other occurrences of these frequency-two elements, excluding any element already in S. If for any set S, $m < r_2$, include S in the set cover.

   This Rule 6 is seen to be safe because of the following. If we branch on S and include it in the set cover, we would cover all $|S|$ elements with one set. If the set cover does not use S, it then must include all the other occurrences of the frequency-two elements in it. They will now be singletons, and the preceding rule will pertain. Therefore we would cover $|S| + m$ many elements with r_2 many sets. But if $m < r_2$, clearly the first case is more parsimonious, as we will have $r_2 - 1 \leq m$ elements left to cover the rest.

Per Rule 6 to avoid branchings, select a subset of maximum cardinality, and create two subproblems either by including it in the minimum set cover and removing all newly covered elements, or removing it.

In the analysis of the Van Rooij and Bodlaender algorithm, we are giving weights to the set sizes and element frequencies, and are summing these over all items and sets. The idea is to enumerate the many subcases where the algorithm can branch and derive recurrence relations for each of these case in terms of their weights. In the analysis below we will see a variable measure of complexity.

Van Rooij and Bodlaender honed the measure and conquer algorithm above using *computer-aided algorithm design*. Our analysis of their algorithm starts by applying the variable measure of complexity and a set of polynomial time branching rules, and some polynomial time reduction rules relative to this measure. Initially, we will only have the trivial reduction and branch rules, namely the Base Case Rule 1 and Avoid Branching Rule 6. Next we will apply improvement steps to generate a potentially faster algorithm. The human designer adds the applicable new Rules 2–5 above, while the computer does extensive measure and conquer analysis to highlight possible points of further improvement. We will demonstrate later how this interactive methodology works to produce a faster algorithm.

This analysis is more or less the same as that by Fomin, Grandoni, and Kratsch [316]. Let v_i be the *weight* of an element of frequency i, and w_i similarly the weight of set of size i. Then we will assign a *variable measure of complexity* for a family $\mathcal{S}$, as

$$k_S = \sum_{S \in \mathcal{S}} w_{|S|} + \left| \bigcup \{S \mid S \in \mathcal{S}\} \right|.$$

The reader should note that $k_S \le d_S$ where $d_S = |\mathcal{S}| + |\bigcup\{S : S \in \mathcal{S}\}$. Originally, we would be solving an instance of set cover where elements of different frequencies and sets of different sizes contribute equally to the "dimension" $d = 2n$ of the family generated from the instance of DOMINATING SET (of size n), but by varying the variables v_i and w_i we can make elements of higher frequency or sets with more elements contribute more. Moreover, we can set $v_0 = w_0 = v_1 = w_1 = 0$ since all frequency-one elements are removed by the reduction rules. The following quantities turn out to be important.

$$\Delta w_i := w_i - w_{i-1} \quad \text{and} \quad \Delta v_i := v_i - v_{i-1}.$$

Notice that these quantities are nonnegative.

The next step is to derive the recurrence relations representing the problem instances whereupon the algorithm branches. Let $N(k)$ denote the number of subproblems generated to solve a problem of *measured* complexity k. Let Δk_{in} and Δk_{out} denote, respectively, the difference in measured complexity of the subproblems of including S and deleting S compared to the problem branched upon. Set $|S| = \sum_{i \ge 2} r_i$ where r_i is the number of elements of S of frequency i. Now we look at the sizes of Δk_{in} and Δk_{out}.

Case Δk_{in}. If we add S to the set cover, then S and its elements are removed. The reduction is $w_{|S|} + \sum_{i \ge 2} r_i v_i$. Also other sets will be smaller, and hence there is an additional reduction of at least $\min\{j \le |S| \mid \Delta w_j\} \sum_{i \ge 2} (i - 1) r_i$. For the analysis, the formulas are kept linear and so the quantity $\min\{j \le |S| \mid \Delta w_j\}$ is approximated as $\Delta w_{|S|}$.

Claim *The formula is kept correctly modeling the algorithm by placing constraints on the weights: $\Delta w_i \le \Delta w_{i-1}$, for all $i \ge 2$. The solution to the optimization problem is unchanged.*

The proof is left to the reader (Exercise 6.6.3).

Case Δk_{out}. In this case, as S is removed, all its elements have their frequency dropped by 1, giving a reduction of $w_{|S|} + \sum_{i \geq 2} r_i$. Also, any set which had contained an element of frequency 2 which was in S will also be included in the set cover, and these sets are distinct by the Subsumption Rule 4. Applying the Avoiding Branchings Rule 6, there are at least r_2 many such elements in each such set, and these must occur in another set in the instance, giving a reduction of additionally at least $r_2(v_2 + w_2 + \Delta w_{|S|})$. Notice that we also use Rule 2 (Splitting Connected Components) to make sure that not all frequency-two elements occur in the same set. In that case, all considered sets form a connected component of ≤ 5 sets, solvable in constant time.

We therefore get to the following recurrence relations. $3 \leq |S| = \sum_{i \geq 2} r_i$,

$$N(k) \leq N(k - \Delta k_{\text{out}}) + N(k - \Delta k_{\text{in}}), \quad \text{where}$$
$$\Delta k_{\text{out}} = w_{|S|} + \sum_{2 \leq i} r_i \Delta v_i + r_2(v_2 + w_2 + \Delta w_{|S|}) \quad \text{and}$$
$$\Delta k_{\text{in}} = w_{|S|} + \Delta w_{|S|}\left(\sum_{i \geq 2}(i - 1)r_i\right).$$

The problem is made finite by setting $w_i = v_i$ for $i \geq p$ with p sufficiently large. Then we consider the subcases $|S| = \sum_{i=2}^{p} + r_{>p}$ with $r_{>p} = \sum_{i>p} r_i$. This gives a finite set of recurrence relations since the cases where $|S| > p + 1$ are dominated by those with $|S| = p + 1$. The method is to use classical optimization for the value of p, and $p = 7$ was best.

Then the solution to the recurrence is of the form $N(k) = \alpha^k$, with α the smallest solution to the set of inequalities $\alpha^k \leq \alpha^{k - \Delta k_{\text{out}}} + \alpha^{k - \Delta k_{\text{in}}}$. Since $k \leq d$ where d is the number of sets in the family, the algorithm will have running time $O((\alpha + e)^d)$ where

$$O\big(\text{poly}(d)N(k)\big) = O\big(\text{poly}(d)\alpha^k\big) \leq O\big((\alpha + e)^d\big).$$

Then we let e be the error in the upward decimal rounding of α.

The overall conclusion, for any values $(0, v_2, v_3, \ldots)$ and $(0, w_2, w_3, \ldots)$, is that we can compute the running times with these weights. The goal is to choose the best weights so that the running time is minimized. This can be solved numerically, and the upper bound is $O(1.2302^d)$, for SET COVER and therefore $O(1.5134^k)$ for DOMINATING SET.

How does this work? The idea is to add the reduction rules and then seek the optimal solution. You begin with the trivial algorithm, with the formula for k_{out} as $w_{|S|} + \sum_{i \geq 1} r_i \Delta v_i$, $1 \leq |S| \leq \sum_{i=1}^{p} r_i + r_{>p} \leq p + 1 = 3$, the instance of the worst case being $\{1\} - \emptyset$. ($k_\in$ is not included, as it will not change, except that $r_1 \neq 0$ in the early stages.) This gives vectors $\overline{v} = (0.8808, 0.9901, \ldots), \overline{w} = (0.9782, \ldots)$ and a running time for SET COVER of $O(1.4519^k)$. If we then add the Singleton Set Rule 5 *Stop when all sets have size 1*, the formula for k_{out} becomes $w_{|S|} + \sum_{i \geq 1} r_i \Delta v_i$, $1 \leq |S| \leq \sum_{i=2}^{p} r_i + r_{>p} \leq p + 1 = 4$, the instance of the worst case being $\{1, 2\} - \emptyset$. $\overline{v} = (0.7829, 0.9638, \ldots), \overline{w} = (0.4615, \ldots)$ and the run time already improves to $O(1.2978^d)$. Next, per Rule 5, we can *include all frequency-one elements*, and get corresponding changes, eventually arriving at the given solution with k_{out} having the formula $w_{|S|} + \sum_{2 \le i} r_i \Delta v_i + r_2(v_2 + w_2 + \Delta w_{|S|})$, $3 \le |S| = \sum_{i=2}^{p} r_i + r_{>p} \le p + 1 = 8$ with instance part of the worst case $\{1, 2, 3\} - \{1, 4\}, \{2, 5\}, \{3, 6\}$, $\overline{v} = (0, 0, 219478, 0.671386, \ldots), \overline{w} = (0, 0.375418, 0.750835, \ldots)$ and $O(1.2302^d)$ for SET COVER. The stepwise addition of the new rules for *Subsumption, Avoiding Unnecessary Branchings, and Splitting Connected Components* were introduced in the van Rooij–Bodlaender paper, and the others were in the Fomin, Grandoni, and Kratsch paper on INDEPENDENT SET.

The construction thus works in steps. At each new step, we remove some subcases by using a larger smallest set, or removing small sets of elements, or the size of the formula for reduction k_{out} is increased. Then figure out a new worst case and re-factor the reduction rules.

However, you cannot reformulate a reduction rule that removes the current simplest worst case, as van Rooij and Bodlaender [653] point out, since they observe that the following problem is NP-complete (Exercise 6.6.5).

REDUCTION SET COVER

Instance: A SET COVER instance $\mathcal{S}$ and $S \in \mathcal{S}$ with the following properties: None of the reduction rules of the van Rooij–Bodlaender Algorithm apply. For all $\hat{S} \in \mathcal{S}$, $|\hat{S}| = 3$. Every element of S has frequency 2.

Question: Does there exist a minimum set cover of $\mathcal{S}$ containing S?

The reader might keep this question in mind as we examine a state-of-the-art reduction algorithm in the next section.

6.5 FEEDBACK VERTEX SET

This reduction algorithm, for FEEDBACK VERTEX SET (FVS), will use iterative compression as part of a subroutine. The reader should not be daunted by the length of the proof as it has a lot of interesting ideas. Moreover, it will strengthen the reader's intrepid spirit to see the state of the art in an area.

6.5.1 The Simplified Algorithm

The reduction algorithm for FVS begins by giving an algorithm for a simpler problem (DFVS), which sets it up for an application of iterative compression. Moreover, at the start, this algorithm gives the smallest kernelization for the DFVS problem. It was Dehne, Fellows, Langston, Rosamond, and Shaw [204] who pointed out the kernelization possibilities of iterative compression.

DISJOINT FEEDBACK VERTEX SET

Instance: A graph $G = (V, E)$ and a feedback vertex set F in G.

Parameter: An integer k.

Action: Construct a FVS F' of size k in G disjoint from F, or declare that no such FVS exists.

In Chap. 3, we saw a simple algorithm for this FEEDBACK VERTEX SET running in time $O((2k+1)^k \cdot n^2)$. This problem was recently proven to have an $O^*(3.83^k)$ algorithm by Cao, Chen, and Liu [130]. As well as iterative compression, the faster algorithm also uses a number of other techniques, as we will now see.

Returning to DISJOINT FEEDBACK VERTEX SET, if F is a FVS and we set $V_1 = G[V - F]$ then V_1 is a forest, and for all F if $G[F]$ is not a forest, then we cannot have a FVS $F' \subseteq G[V - F]$. Instances of DISJOINT FEEDBACK VERTEX SET are therefore of the form $(G, V_1, V_2; k)$ with V_1 and V_2 inducing forests. Following Cao, Chen, and Liu, we will call a FVS entirely contained in V_i a V_i-FVS. The algorithm to follow is in two parts. The first part is the simpler version and then we will improve the algorithm to get the $O^*(3.83^k)$ one.

Reduction Rules for DISJOINT FEEDBACK VERTEX SET:

- R1: Remove all degree-0 and degree-1 vertices.
- R2: For any degree-2 vertex v in V_1,
    - if the two neighbors of v are in the same connected component of $G[V_2]$, include v into the target V_1-FVS and set $G := G - \{v\}$ and $k := k - 1$;
    - else, move v from V_1 to V_2, $V_1 := V_1 - \{v\}$ and $V_2 := V_2 \cup \{v\}$.

Lemma 6.5.1

(1) *R1 is safe.*

(2) *R2 is safe.*

Proof (1) If both neighbors of v are in the same connected component of $G[V_2]$ then v and some vertices of V_2 form a cycle that must be broken. Hence the first case is safe.

(2) In the second case, it will suffice to show that if G has a V_1-FVS of size k, then G has one of size at most k not including v. So assume that F is a V_1-FVS containing v of size $\leq k$. Then note that if u is one of the neighbors of v and is in V_1 then $(F - \{v\}) \cup \{u\}$ works. Therefore we suppose that both neighbors u_1 and u_2 of v are in distinct connected components of $G[V_2]$. Since $G - F$ is acyclic there is at most one path in $G - F$ from u_1 to u_2. If there is no path, then adding v to $G - F$ cannot make a cycle, and hence $F - \{v\}$ is a V_1-FVS of size $\leq k - 1$ not containing v. If there is one path P, then P must contain one vertex w in V_1 as u_1 and u_2 are in distinct connected components of $G[V_2]$. Note that each cycle C of $(G - F) \cup \{v\}$ contains v (as F is a FVS) and also contain u_1 and u_2. We conclude that $C - \{v\}$ must be the unique path P from u_1 to u_2 in $G - F$. Therefore it contains w. Consequently, w must be contained in all cycles of $(G - F) \cup \{v\}$. This time we see that $(F - \{v\}) \cup \{w\}$ is a V_1-FVS of size bounded by k not containing v.   □

Lemma 6.6.1 can now be turned into an algorithm that gives a small kernel. The algorithm below works as follows. Given an instance $(G, V_1, V_2; k)$ of DISJOINT FEEDBACK VERTEX SET, keep all vertices of V_1 of degree 3 or greater, applying R1 and R2 if the degree drops to below 3, and repeatedly branch on the leaf of $G[V_1]$. The argument is that if $(G, V_1, V_2; k)$ has a V_1-FVS of size $\leq k$, then at least one of the computation paths of the branching algorithm will yield a V_1-FVS of size $\leq k$. Notice in the below that the vertex q will have at most one neighbor in $G[V_1]$. Since we have ensured each vertex in V_1 has degree ≥ 3, removing q from G will leave a neighbor of q in $G[V_1]$ of degree ≥ 2.

Algorithm 6.5.1 FindFVS$(G, V_1, V_2; k)$
Input: An instance $(G, V_1, V_2; k)$ of DISJOINT FEEDBACK VERTEX SET
Output: A V_1-FVS F if size $\leq k$ or a certificate for "no".
1. Initialize $F = \emptyset$.
2. While $|V_1| > 0$, do the following.
3. Pick a leaf q in $G[V_1]$.
4. (Case 1) If q is in the objective V_1-FVS F, add q to F and $k := k - 1$. If the neighbor v of q in $G[V_1]$ becomes degree 2, apply rule 2 to v. $G := G - \{q\}$.
5. (Case 2) If q is not in the objective V_1-FVS F, move q from V_1 to V_2.

Theorem 6.5.1 (Cao, Chen, and Liu [130]) *Let (G, V_1, G_2) be an instance of* DISJOINT FEEDBACK VERTEX SET *upon which neither reduction rules R1 nor R2 are applicable. If $|V_1| > 2k + c_2 - c_1$ there is no V_1-FVS in G with size $\leq k$, where c_i is the number of connected components in $G[V_i]$.*

Proof The main idea of this proof is to verify that if G has V_1-FVS of size $\leq k$, the computation path in the algorithm FindFVS that finds this V_1-FVS removes at most $2k + c_2 - c_1$ vertices from V_1 before V_1 becomes empty. This implies that V_1 has at most $2k + c_2 - c_1$ many vertices.

At each iteration of the algorithm, one of the following cases pertain: $q \in F$ or $q \notin F$.

Case 1. $q \in F$. Then q is removed. If the degree of the neighbor v of q is above 3 before removal, then q is the only vertex deleted and the degrees all remain ≥ 3. Otherwise, the degree of the neighbor v of q is 3, and hence q is moved from V_1 to V_2 by rule R2. The degree of any other vertex of V_1 is the same. So there are four subcases based on the neighbors u_1 and u_2 of v.
1. u_1 and u_2 are in the same tree in $G[V_2]$. v is added to F and $k := k - 1$.
2. u_1 and u_2 are in different trees in $G[V_2]$. Then v is moved to V_2 and c_2 decreases by 1.
3. u_1 and u_2 are in V_1. Then v is moved to V_2 and becomes a single vertex tree in $G[V_2]$, increasing c_2 by 1. But moving v to V_2 from V_1 splits u_1 and u_2 into different trees in $G[V_1]$, increasing c_1 by 1.
4. $u_1 \in V_1$ and $u_2 \in V_2$. Then v is moved to V_2 and nothing else changes.

Case 2. $q \notin F$. Then q is moved to V_2 increasing c_1 by at least 1, and no other vertex on V_1 is affected.

We examine the effects of the cases. In Case 1, one or two vertices are removed from V_1 and k is decreased by 1 or 2. In Case 2, exactly one vertex is removed, and c_1 is decreased by 1. Therefore Case 1 can only occur k times. The calculation of the number of invocations of Case 2 is complicated by the fact that in Subcase 3 of Case 1, c_2 is increased. If Case 1, Subcase 3 occurs x times, then Case 2 can occur at most $c_2 + x$ many times. The total number of vertices removed by this is $\leq 2k + c_2 + x$. Now, Subcase 3 does increase c_1, counteracting the effect of increasing c_2. In each step at most two vertices are removed from $G[V_1]$, and then to remove a whole tree from $G[V_1]$, the last step removes one or two vertices:

1. A trivial tree in $G[V_1]$ (a single vertex v with degree at least 3). If $v \in F$ so Case 1 is invoked, only one vertex v is removed from V_1 and $k := k - 1$. Else Case 2 is invoked and v is moved into $G[V_2]$, removing one vertex in V_1 and increasing the number of c_2 so that the total number loses at least 1 in both cases.
2. Both of the vertices u_1 and u_1 are leaves, so we can pick one of them arbitrarily, say u_1. Since this is not the final step, $u_1 \in F$, and hence u_2's fate is determined by Subcases 1 and 2. If Subcase 1, two vertices are removed from V_1 with $k := k - 2$, and the total loses 2. If Subcase 2, one vertex is removed from V_1, c_2 decreases by at least 2, and the total loses at least 1.

Therefore, whatever case pertains, at the final step of each tree, the total number loses at least 1. We know there are $c_1 + x$ many steps, where c_1 is the number of trees in $G[V_1]$, and x is the number of trees created in Subcase 3 of Case 1. The bound is therefore $2k + c_2 + x - (c_1 + x) = 2k + c_2 - c_1$, as required.

Finally, if $|V_1| > 2k + c_2 - c_1$, after $2k + c_2 - c_1$ many vertices have been disposed of, let q be a leaf left in $G[V_1]$. Then $G[V_2]$ has already become a single connected component, and adding any vertex with more than two neighbors will create a cycle, so q cannot be put into V_2. Since F, the V_1-FVS already has k vertices, and we cannot shift q to V_2, this is a "no" instance. $\qquad\square$

When we do the full algorithm, the reader should keep in mind that we always have $|V_2| = k + 1$ and therefore we will be able to invoke iterative compression. Moreover, also note that $c_2 \leq |V_2|$ and $c_1 > 0$, so that $2k + c_2 - c_1 \leq 3k$, thus we have:

Corollary 6.5.1 (Cao, Chen, and Liu [130]) DISJOINT FEEDBACK VERTEX SET *has a size $3k$ kernel.*[1]

---

[1] Strictly speaking the kernel is smaller than this, namely $3k - c_1 - \rho(V_1)$ where $\rho(V_1)$ is the size of a maximum matching of the subgraph induced by the set V_1' of vertices of V_1 of degree larger than 3. However, the present bound is sufficient for our purposes and we refer the reader to the journal version of [130] for analysis.

6.5.2 The 3-Regular Case

This section deals with an important subroutine of the Cao, Chen, and Liu algorithm.

Lemma 6.5.2 (Cao, Chen, and Liu [130]) *Let S be a set of vertices such that $G[S]$ is a forest. Then there is a spanning tree in G containing $G[S]$.*

The proof of Lemma 6.5.2 is essentially the same as that for Kruskal's Theorem for spanning trees (see Sect. 34.5), beginning with S. We leave it to the reader (Exercise 6.6.2). It can be done in time $O(m\alpha(n))$, where $\alpha(n)$ is a very slow growing function like the inverse to Ackermann's function.

Thus, let $(G : V_1, V_2; k)$ be an instance of DISJOINT FEEDBACK VERTEX SET so that $V_1 \sqcup V_2$ is a partition of G with the induced subgraphs $G[V_i]$ forests. By Lemma 6.5.2, there is a spanning tree T in G that contains $G[V_2]$. We call a spanning tree containing $G[V_2]$ a $T_{G[V_2]}$-tree. Let T be such a $T_{G[V_2]}$-tree. Every edge of $G - T$ has at least one vertex in V_1. Two edges of $G - T$ are said to be V_1-*adjacent* if they have a common vertex in V_1. We define a V_1-*adjacency matching* in $G - T$ to be a partition of the edges into two groups called *1-groups* and *2-groups*, respectively, where edges are in the same 2-group if they share a V_1 vertex in common. A maximum V_1-adjacency matching is one that *maximizes the number of 2-groups*.

Definition 6.5.1 Let T be a $T_{G[V_2]}$ tree in G. V_1-*adjacency number* with respect to T, $\mu(G, T)$ is the number of 2-groups. The V_1-*adjacency number* $\mu(G)$ is the largest possible $\mu(G, T)$ over all possible $T_{G[V_2]}$ trees.

The important subcase of the algorithm is a reduction to the case where the situation is 3-regular$_{V_1}$, meaning that every vertex in V_1 has degree exactly 3. Recall from classical graph theory that the *Betti* number $\beta(G)$ is the total number of edges in $G - T$ for some (any) spanning tree T. We need the following technical lemma.

Lemma 6.5.3 *Suppose that G is 3-regular$_{V_1}$. Let $f_{V_1}(G)$ denote the size of a minimum V_1-FVS. Then*

$$F_{V_1}(G) = \beta(G) - \mu(G).$$

Additionally, a minimum V_1-FVS F can be constructed in linear time from a $T_{G[V_2]}$-tree whose V_1-adjacency matching number is $\mu(G)$.

Proof The construction of F is straightforward. Take the relevant $T = T_{G[V_2]}$ tree with the V_1-adjacency matching M in $G - T$ of size $\mu(G)$. Let U be the set of edges in the 1-groups of M. For each $e \in U$, pick an end in V_1 and include this in F. For each 2-group of two adjacent edges, pick the common end and include this in F. Now every cycle must contain at least one edge of $G - T$, and every edge of $G - T$ has one end in F. Thus F is a V_1-FVS, and the number of vertices is $|F| = \beta(G) - \mu(G)$. Thus we also have $f_{V_1}(G) \le \beta(G) - \mu(G)$.

Now to prove the inequality the other way, take any V_1-FVS H, with $|H| = f_{V_1}(G)$. By Lemma 6.5.2, there is a spanning tree S containing the entire induced subgraph $G - H$. We will construct a V_1-adjacency matching with at least $\beta(G) - |H|$ many 2-groups. Since S contains $G - H$, each of the edges in $G - S$ has at least one end in H. Let E_2 be the set of edges of $G - S$ that have both ends in H and E_1 those with exactly one end in H.

We claim that each end of an edge in E_2 is shared by exactly one edge in E_1. Hence no edges in E_2 share a common end. To see this, let $uv \in E_2$ and both u and v in H. Let e_1, e_2 be two other edges incident with u. If u is not shared by any edge of E_1, then either both e_1 and e_2 are in S or one of the e_i is in E_2. If both of the e_i are in S, then since each edge in $G - S$ (including uv) has at least one end in $H - \{u\}$, the set $H - \{u\}$ would be a smaller V_1-FVS, a contradiction. If the edge $e_i = uw$ is in E_2, with $w \in H$, we again see that $H - \{u\}$ would be a smaller V_1-FVS, a contradiction. The claim is established.

Suppose that there are m_2 vertices in H incident to two edges in $G - S$. Then there are $|H| - m_2$ remaining edges incident to at most one edge in $G - S$. Counting the number of incidences between edges in H and $G - S$, we see

$$2|E_2| + \big(\beta(G) - |E_2|\big) \le 2m_2 + \big(|H| - m_2\big).$$

Thus $m_2 - |E_2| \ge \beta(G) - |H|$. We construct the desired V_1-adjacency matching in $G - S$. For each edge e in E_2, by the claim, we can make a 2-group consisting of e and an edge of E_1 sharing a vertex with e. As well as the edges of E_2 there are $m_2 - 2|E_2|$ vertices of H that are incident with two edges in E_1. For each such vertex v, make a 2-group consisting of the two edges of E_1 incident with v. This construction gives $|E_2| + (m_2 - 2|E_2|) = m_2 - |E_2|$ many disjoint 2-groups. The rest of the edges in $G - S$ can be made into 1-groups. This V_1-adjacency matching in $G - S$ has $m_2 - |E_2|$ many 2-groups. Since we have proven that $m_2 - |E_2| \ge \beta(G) - |H|$, we see

$$\mu(G) \ge \mu(G, S) \ge m_2 - |E_2| \ge \beta(G) - |H| = \beta(G) - f_{V_1}(G).$$

The result now follows and moreover the construction of F is clearly linear time. $\square$

We wish to prove the following theorem.

Theorem 6.5.2 (Cao, Chen, and Liu [130]) *There is an $O(n^2 \log^6 n)$ algorithm that takes an instance $(G : V_1, V_2; k)$ of* DISJOINT FEEDBACK VERTEX SET *with 3-regular$_{V_1}$ and either constructs a V_1-FVS of size $\le k$ or says "no" correctly.*

Given a 3-regular$_{V_1}$ instance $(G : V_1, V_2; k)$ of DISJOINT FEEDBACK VERTEX SET, we need only construct a $T_{G[V_2]}$-tree in the graph whose V_1-adjacency matching number is $\mu(G)$. To do this, we will appeal to some material from matching theory in matroids, as per Sect. 34.5. Recall that a (finite) matroid is a pair $(E, \mathcal{I})$ with $\mathcal{I}$ a collection of *independent sets*[2] such that

1. $I \in \mathcal{I}$ and $J \subseteq I$ implies $J \in \mathcal{I}$.
2. $I, J \in \mathcal{I}$ and $|I| > |J|$ implies that there is some $a \in I - J$ with $J \cup \{a\} \in \mathcal{I}$.

As we discuss in Appendix 1, Sect. 34.5, where the parity problem is treated in somewhat more detail, we will be concerned with a class of matroids formed under duality, called *cographic matroids*. These have sets S of edges of the graph and are regarded as independent if $G - S$ is connected. The reader is invited to verify that this is indeed a matroid.

The relevant problem for the algorithm is the following, which generalizes the problem of spanning tree parity.

MATROID PARITY PROBLEM

Instance: A matroid $(E, \mathcal{I})$ and a perfect pairing (i.e. a partition of E into pairs) $\{a_1 \overline{a_1}, \ldots, a_n \overline{a_n}\}$ of elements of E.

Task: Find the largest subset $P \in \mathcal{I}$, such that, for all i, either both $a_i \in P$ and $\overline{a_i} \in P$ or neither $a_i \in P$ nor $\overline{a_i} \in P$.

Using methods akin to Edmond's classic matching algorithm, and reduction to the problem for linear matroids, Gabow and Stallman [334] showed that *the matroid parity problem for cographic matroids is solvable in polynomial time. More precisely, it is solvable in time $O(mn \log^6 n)$ where m is the size of $\mathcal{E}$.*

Reduction to the Cographic Matroid Parity Problem Take a 3-regular$_{V_1}$ instance $(G : V_1, V_2; k)$ of DISJOINT FEEDBACK VERTEX SET. Without loss of generality assume G is connected (else work on components), for each $v \in V_1$, there is at most one edge from v to a connected component of $G[V_2]$, for otherwise, we can include v directly in the V_1-FVS.

Let $d_{V_1}(e)$ denote the number of edges V_1-adjacent to e. (Recall: sharing a vertex in V_1.) Then we can construct a *labeled subdivision* G_2 of G as follows.

1. Shrink each component of $G[V_2]$ to a single vertex and let the resulting graph be G_1.
2. Assign each edge in G_1 a distinguishing label.
3. For each edge labeled e_0 in G_1, let the V_1-adjacent edges to e_0 be labeled $e_1, \ldots, e_d$, with $d = d_{V_1}(e_0)$. Subdivide e_0 into d segment edges labeled $(e_0 e_1), (e_0 e_2), \ldots, (e_0 e_d)$. (We say that these edges are *from* e_0.) Call the resulting graph G_2.

Note that every edge of G_1 corresponds uniquely to an edge of G with at least one of its vertices in V_1, so we can regard the edge as in G or G_1. Also G_1 is simple and connected, implying the same of G_2. Since each edge of V_1 has degree 3, every edge of G_1 is subdivided into at least two segment edges in G_2.

In G_2, pair the segment edges labeled $(e_0 e_i)$ with the segment edge $(e_i e_0)$ for all segment edges since both will be there, and this is a perfect pairing of the edges in G_2. This edge pairing is a perfect pairing of the cographic matroid induced by G_2.

---

[2]Not, of course, to be confused with the notion of independent set in a graph.

Then the Gabow and Stallman's algorithm produces maximum edge subset P in the independent sets of the matroid, such that for each segment edge $(e_0 e_i)$, it either contains both $(e_0 e_i)$ and $(e_i e_0)$ or neither.

Lemma 6.5.4 *From the maximum edge subset P for the cographic matroid constructed above, a $T_{G[V_2]}$-tree for G whose V_1-adjacency number is $\mu(G)$ can be constructed in close-to-linear time[3] (more precisely, time $O(|E|\alpha(|V|))$).*

Proof Let P consist of the edges $\{(e_1 e_1'), (e_1' e_1), \ldots, (e_h' e_h)\}$. Note that $G - P$ is connected by definition of the cographic matroid. Thus P corresponds to an edge subset P' of size $2h$ in G_1, $P' = \{e_1, e_1', e_2, e_2', \ldots, e_h, e_h'\}$. The edges e_i and e_i' are V_1-adjacent. As $G_2 - P$ is connected so too are $G_1 - P$ and $G - P$. Furthermore, $G - P'$ contains $G[V_2]$. By Lemma 6.5.2, in linear time we can construct the $T_{V[G_2]}$-tree T_1 for the graph G fro $G - P'$. Now we make each pair $e_i e_i'$ a 2-group and the rest a 1-group; we get a V_1-adjacency matching with h many 2-groups in $G - T_1$.

It suffices to show that $\mu(G) = h$. To do this we only need to prove that no V_1-adjacency matching can have more than h 2-groups. Let T_2 be a $T_{G[V_2]}$-tree with q 2-groups $\{e_{i_1} e_{i_1}', \ldots, e_{i_q} e_{i_q}'\}$. As T_2 is contained in $\hat{G} = G - \{e_{i_1}, e_{i_1}', \ldots, e_{i_q}, e_{i_q}'\}$, $\hat{G}$ is connected, as is $G_1 - \{e_{i_1}, e_{i_1}', \ldots, e_{i_q}, e_{i_q}'\}$. Thence $G_2 - \{e_{i_1}, e_{i_1}', \ldots, e_{i_q}, e_{i_q}'\}$ is also connected. However, P is a solution to the matroid parity problem for the cographic matroid, and hence $h \geq q$, as required. $\square$

We now finish the proof of Theorem 6.5.2. For the 3-regular$_{V_1}$ instance $(G; V_1, V_2; k)$ of DISJOINT FEEDBACK VERTEX SET, first construct, as above, G_1 by shrinking each connected component of $G[V_2]$ to a single component. By the 3-regularity, the total number of edges of G_1 is $\leq 3|V_1|$. As above we construct a labeled subdivision graph G_2. By 3-regularity, each edge of G_1 is subdivided into at most 4 segments in G_2. Therefore the numbers n_2 and m_2 of vertices and edges, respectively, of G_2 are bounded by $O(|V_1|)$. Next we construct a cographic matroid from G_2, and apply the Gabow–Stallmann algorithm to get a maximum matroid parity solution. This gives the relevant set P and is done in time $O(m_2 n_2 \log^6 n_2) = O(n^2 \log^6 n)$ time. Using Lemma 6.5.4, construct the $T_{G[V_2]}$ tree T for the graph whose V_1-adjacency matching number is $\mu(G)$. Finally, using Lemma 6.5.3, we construct a minimum V_1-FVS F in linear time from T, and this solves the problem comparing k to the size of F. This concludes the proof.

Corollary 6.5.2 *There is an $O(n^2 \log^6 n)$ time algorithm which takes an instance $(G : V_1, V_2; k)$ of DISJOINT FEEDBACK VERTEX SET where all the vertices of V_1 have degree bounded by 3, and either constructs a V_1-FVS of size $\leq k$ or reports correctly that no such set exists.*

---

[3]It is not linear since $\alpha(n)$ is unbounded.

6.5.3 An Improved Running Time Using Measure and Conquer

Consider that $(G; V_1, V_2, k)$, and we are looking for a V_1-FVS of size k. We can look for certain structures within the graph to make the branching smaller. For example, G cannot contain edges where both ends have the same vertex (self-loops) as both $G[V_1]$ and $G[V_2]$ are forests. Also, if two vertices are connected by multiple edges then one is in V_1 and the other is in V_2, and hence we can by fiat include one in V_1 and the other in V_2. Thus we can assume that G has no self-loops nor multiple edges.

Definition 6.5.2 $v \in V_1$ is called *nice* if $\deg(v) = 3$ and all of v's neighbors are in V_2.

Now let p be the number of nice V_1 vertices, c_2 again the number of connected components in $G[V_2]$. The *measure* will be

$$m = k + \frac{c_2}{2} - p.$$

Lemma 6.5.5
(1) *If the measure ≤ 0, then there is no V_1-FVS of size $\leq k$ in G.*
(2) *If all vertices in V_1 are nice, then we can construct a minimal V_1-FVS in G in polynomial time.*

Proof Suppose that $m = k + \frac{c_2}{2} - p \leq 0$ and G has a V_1-FVS F of size $k' \leq k$. Let S be the set of $p - k'$ nice vertices not in F. Then $G[V_2 \cup S]$ must be a forest as F is a FVS disjoint from $V_2 \cup S$. But $G[V_2 \cup S]$ can be constructed from $G[V_2]$ and the $p - k'$ elements of S, via adding the $3(p - k')$ vertices adjacent to those in S (which are all nice). As $k' \leq k$, $p - k' \geq p - k \geq \frac{c_2}{2}$. Therefore, $3(p - k') = 2(p - k') + (p - k') \geq c_2 + (p - k')$. This is a contradiction, as $G[V_2 \cup S]$ is a forest. In order to keep it a forest, we can add at most $c_2 + (p - k') - 1$ edges to the structure that consists of the $G[V_2]$ of c_2 connected components and the $p - k'$ vertices of S.

To prove Lemma 6.5.5(2), note that if all the vertices of V_1 are nice, then $(G; V_1, V_2, k)$ is a 3-regular$_{V_1}$ and Theorem 6.5.2 gives a polynomial time algorithm for this. □

Here is the algorithm for DISJOINT FEEDBACK VERTEX SET.

Algorithm 6.5.2 (Feedback(G, V_1, V_2, k))
Input: An instance $(G; V_1, V_2; k)$ of DISJOINT FVS.
Output: A V_1-FVS of size $\leq k$ in G if one exists, or a certificate of "no" if none exists.
1. If $k < 0$ or ($k = 0$ and G is not a forest), then return a certificate of "no".
2. If $k \geq 0$ and G is a forest, then return $\emptyset$, let c_2 be the number of connected components in $G[V_2]$, and let p be the number of nice vertices.

3. a. If $p > k + \frac{c_2}{2}$, return "no".

   b. If $p = |V_1|$, solve the problem in polynomial time.

4. If $w \in V_1$ has degree ≤ 1, then return Feedback$(G, V_1 - \{w\}, V_2, k)$.

5. If $w \in V_1$ has two neighbors in the same tree in $G[V_2]$, $F_1 = $ Feedback$(G - \{w\}, V_1 - \{w\}, V_2, k - 1)$. If $F_1 = $"no" then return "no"; or return $F_1 \cup \{w\}$.

6. If a vertex $w \in V_1$ has degree 2, return Feedback$(G - \{w\}, V_1 - \{w\}, V_2 \cup \{w\}, k)$.

7. If a leaf $w \in G[V_1]$ is not a nice V_1 vertex and has ≥ 3 neighbors in V_2, set $F_1 = $ Feedback$(G - \{w\}, V_1 - \{w\}, V_2, k - 1)$.

   a. If $F_1 \neq $"no" return $F_1 \cup \{w\}$.

   b. If $F_1 = $"no", return Feedback$(G, V_1 - \{w\}, V_2 \cup \{w\}, k)$.

8. If the neighbor y of a leaf w in $G[V_1]$ has at least one neighbor in V_2, set $F_1 = $ Feedback$(G - \{y\}, V_1 - \{w, y\}, V_2 \cup \{w\}, k - 1)$.

   a. If $F_1 \neq $"no" return $F_1 \cup \{y\}$.

   b. If $F_1 = $"no", return Feedback$(G, V_1 - \{y\}, V_2 \cup \{y\}, k)$.

9. On entry to this step, we will claim the following: Now any tree of a single vertex in V_1 is nice, and we know that there is no tree of two vertices in $G[V_1]$. Hence for a tree T of at least 3 vertices in $G[V_1]$, we can fix any internal vertex as the root. Then we can find the lowest leaf w_1 in polynomial time. Since t has at least three vertices, it has a parent $w \in T$ in $G[V_1]$. Thus, pick a lowest leaf w_1 in any tree T in $G[V_1]$ with parent w. Let $w_1, \ldots, w_t$ be the children of w in T. Set $F_1 = $ Feedback$(G - \{w\}, V_1 - \{w, w_1\}, V_2 \cup \{w_1\}, k - 1)$.

   a. If $F_1 \neq $"no" return $F_1 \cup \{w\}$.

   b. If $F_1 = $"no", return Feedback$(G, V_1 - \{w\}, V_2 \cup \{w\}, k)$.

Lemma 6.5.6 *Algorithm Feedback(G, V_1, V_2, k) solves* DISJOINT FEEDBACK VERTEX SET *in time* $O(2^{k + \frac{c_2}{2}} n^2)$.

Proof We begin with correctness. Steps 1 and 2 of Feedback(G, V_1, V_2, k) are clearly correct. By Lemma 6.5.5, Step 3 is correct. Step 4 is correct by Theorem 6.5.1. After Step 4, all the vertices have degree at least 2.

If the vertex w has two neighbors in V_2 that belong to the same tree T in $G[V_2]$, then $T \cup \{w\}$ has at least one cycle. The only way to break cycles in $T \cup \{w\}$, is to include the vertex w in the objective V_1-FVS (since we are searching for a V_1-FVS). The objective V_1-FVS of size $\leq k$ exists in G iff the remaining graph $G - \{w\}$ has a V_1-FVS of size $\leq k - 1$ in the subset $V_1 - \{w\}$. Therefore Step 5 is correct and after that all the vertices have at most one neighbor in a tree in $G[V_2]$. Also because of this step, a degree-2 vertex in Step 6 cannot have *both* of its neighbors in the same tree in $G[V_2]$. By Theorem 6.5.1, Step 6 is also safe. After Step 6 we will assume that all vertices have degree at least 3.

Now we consider Steps 7–9. Consider a vertex $w \in V_1$. If w is in the objective V_1-FVS, there will be a V_1-FVS $F_1 \subseteq V_1 - \{w\}$ in $G - \{w\}$ with $|F_1| \leq k - 1$. If w is not in the V_1-FVS, then the objective V_1-FVS must be contained in $V_1 - \{w\}$, and moreover, $G[V_2 \cup \{w\}]$ will still be a forest, as no two neighbors of w in V_2 belong to the same tree in $G[V_2]$. Therefore Step 7 is safe.

Now on entering Step 8, every leaf $w \in G[V_1]$ has exactly two neighbors in V_2. Consider the vertex y of Step 8. If y is in the objective V_1-FVS, then there should be a V_1-FVS $F_1 \subseteq V_1 - \{y\}$ in $G - \{y\}$ with $|F_1| \leq k - 1$. If we remove y from G, w becomes degree 2 and both of its neighbors are in V_2 (by the case analysis, as Step 7 is not applicable). If y is not in the objective V_1-FVS, then analogously to Step 7, there must be an objective V_1-FVS contained in $V_1 - \{y\}$. Again note that $g[V_2 \cup \{y\}]$ is a forest since no two neighbors are in the same tree in $G[V_2]$. Thus Step 8 is safe.

On entering Step 9, we know the following:

1. $k > 0$ and G is not a forest. (Steps 1 and 2)
2. $p \leq k + \frac{c_2}{2}$ and not all vertices of V_1 are nice. (Step 3)
3. Any vertex in V_1 has degree at least 3. (Steps 4–6)
4. Any leaf in $G[V_1]$ is nice or has exactly two neighbors in V_2. (Step 7)
5. For any leaf w in $G[V_1]$, the neighbor $y \in V_1$ of w has no neighbors in V_2. (Step 8)

The hypotheses in this algorithm relevant to the case analysis are satisfied, and hence the we have the collection $w, w_1, \ldots, w_t$. The vertex w either is, or is not, in the objective V_1-FVS. If w is in the objective V_1-FVS, then there should $F_1 \subseteq V_1 - \{w\}$ in $G - \{w\}$ of size $\leq k - 1$. After removing w, w_1 becomes degree 2, and by Lemma 6.5.1 we can move w_1 from V_1 to V_2. If w in not in the objective V_1-FVS, then the objective V_1-FVS must be in $V_1 - \{w\}$. Thus Step 9 is safe. Therefore this algorithm is correct.

Complexity Analysis The recursive execution of this algorithm uses a search tree S. Only Steps 7–9 of this algorithm are branching. Let $S(m)$ denote the number of branches in the search tree when $m = k + \frac{c_2}{2} - p$, with c_2 the number of components of the forest $G[V_2]$, and again p is the number of nice V_1 vertices.

Step 7. The branch step 1 has $k := k - 1$, $c_2 := c_2$, $p :\geq p$. Thus $m :\leq k - 1 + \frac{c_2}{2} - p \leq m - 1$. The branch step 2 has no change except $c_2 :\leq c_2 - 2$. Thus $n :\leq m - 1$, again. Hence this gives the recurrence $S(m) \leq 2S(m - 1)$.

Step 8. The first case has $k := k - 1$, $c_2 := c_2 - 1$, $p :\geq p$, giving $m :\leq m - \frac{3}{2}$. The second case has the only change that $p := p + 1$. Hence for Step 8 the recurrence is $S(m) \leq S(m - \frac{3}{2}) + S(m - 1)$.

Step 9. The first case has $k := k - 1$, $c_2 := c_2 - 1$, $p :\geq p$, and hence $m :\leq m - \frac{3}{2}$, again. The second case has $k := k$, $c_2 := c_2 + 1$ (because of w), and $p :\geq p + 2$ (as w has at least two children which are leaves). This gives $m :\leq m - \frac{3}{2}$. The recurrence is $S(m) \leq 2S(m - \frac{3}{2})$.

Analyzing this recurrence, we observe that the worst case is Step 7, and we get a bound of $S(m) \leq 2^m$.

Finally, Steps 1–3 return an answer, Step 4 does not increase the measure m as w is not nice, Step 5 does not change m as $k = k - 1$ and $p :\leq p + 1$. However, Step 6 can increase the measure by $\frac{1}{2}$ as c_2 can increase by 1, but we can bypass the vertex w by putting it into V_2. This bypassing means that the measure does not change. This seems a cheat. However, in the proof of Lemma 6.5.1, we did not bypass w because it made the analysis for constructing the kernel easier. Since

$m = k + \frac{c_2}{2} - p \leq k + \frac{c_2}{2}$, the computation time on each path in S is $O(n^2)$. Thus this algorithm runs in time $O(2^{k+\frac{c_2}{2}} n^2)$ as required. □

Theorem 6.5.3 (Cao, Chen, and Liu [130]) FEEDBACK VERTEX SET *can be solved in time* $O(3.83^k n^2)$.

Proof The proof is to use iterative compression on the problem below which we claim can be solved in time $O(3.83^k n^2)$.

FVS REDUCTION
Instance: A graph G and a FVS F of size $k + 1$.
Parameter: An integer k.
Action: Either construct a FVS of size $\leq k$ or report that G does not have one.

To see that FVS REDUCTION is solvable in time $O(3.83^k n^2)$, let G be a graph and F_{k+1} the given FVS. As with VERTEX COVER, or GRAPH BIPARTITIZATION, for each subset Q of F_{k+1}, we work on the belief that the objective FVS F has intersection Q with F_{k+1}. The construction of Q is equivalent to constructing a FVS of size $\leq k - |Q|$ in $G - Q$. Let $|Q| = k - j$. Using Lemma 6.5.6, this can be done in time $O(2^{j+\frac{j+1}{2}} n^2) = O(2.83^k n^2)$. Applying this to each j for $0 \leq j \leq k$, and all subsets Q of this size $k - j$, we get an algorithm for FVS REDUCTION in time $\sum_{j=0}^{k} \binom{k+1}{k-j} O(2.82^k n^2) = O(3.83^k n^2)$.

Calling this algorithm on a graph of $\hat{n}$ many vertices, gives a running time of $\sum_{n \leq \hat{n}} O(3.83^k n^2) = O(3.83^k \hat{n}^3)$, as required. This concludes the proof of the algorithm for FEEDBACK VERTEX SET. □

6.6   Exercises

Exercise 6.6.1 (Folklore) Let $G = (V, E)$ be a graph and $w : V \to \mathbb{Q}^+$ be a function assigning weights to the vertices. We say the weighting function is *degree weighted* if there is a constant c such that $w(v) = c \cdot \deg(v)$. The *optimal vertex cover* C is the one of least overall weight $\sum_{v \in C} w(v)$. It has weight *OPT*.
1. Prove that $w(V) = \sum_{v \in V} w(v) \leq 2 \cdot OPT$.
2. Construct a 2-approximation algorithm for minimal-weight vertex cover, assuming arbitrary vertex weights.
   (Hint: First include the set V_0 of weight 0 vertices and let $G_1 = G[V - V_0]$. Consider $c = \min \frac{w(v)}{\deg(v)}$. Now define $w'(v) = w(v) - c \cdot \deg(v)$, and repeat with G_1, etc.)
3. Hence compute a 2-approximation solution to SET COVER.

Exercise 6.6.2 Prove Lemma 6.5.2.

Exercise 6.6.3 Prove Claim from Sect. 6.4.

Exercise 6.6.4 (Cao, Chen, and Liu [130]) Prove that DISJOINT FEEDBACK VERTEX SET is NP-complete when we relax the regularity to 4.

Exercise 6.6.5 Prove that REDUCTION SET COVER is NP complete.

Exercise 6.6.6 Prove that d-HITTING SET ($d \geq 2$) is FPT using iterative compression.

Exercise 6.6.7 Use iterative compression to show that FEEDBACK VERTEX SET IN TOURNAMENTS is FPT.

Exercise 6.6.8 (Guo, Moser, and Niedermeier [366]) Use iterative compression to show that the following problem is FPT.

CLUSTER DELETION
Instance: A graph $G = (V, E)$.
Parameter: An integer k.
Question: Can I delete k or fewer and turn G into a cluster graph; that is a graph whose components consist of cliques?

(Hint: Apply the following reduction rules to simplify the problem first, and then use compression. Given a graph G and a vertex set X such that $G[X]$ and $[V \setminus X]$ are cluster graphs, delete all $R := V \setminus X$ that are adjacent to more than one clique. Delete all R that are adjacent to some but not all vertices of a clique, and remove all components that are cliques.)

Exercise 6.6.9 (Gutin, Kim, Lampis, and Mitsou [370]) Prove that the following has an FPT algorithm running in time $O^*(2^{kB})$ using the EDGE BIPARTITIZATION algorithm of Theorem 6.2.1.

VERTEX COVER ABOVE TIGHT LOWER BOUND
Instance: A positive integer B, a graph G with m edges and of maximum degree B, and a nonnegative integer k.
Parameter: $k + B$.
Question: Is there a vertex cover of G with at most $\frac{m}{B} + k$ vertices?

6.7   Historical Remarks

As a technique, iterative compression dates back to the parameterized algorithm given by Reed, Smith, and Vetta in [576] for the problem ODD CYCLE TRANSVERSAL that, given a graph G and an integer k, asks whether there is a set S of at most k vertices meeting all odd cycles of G (see also [404]). Before [576], this was a popular open problem and iterative compression appeared to be just the correct approach for its solution. The FEEDBACK VERTEX SET algorithm here is taken from Cao, Chen, and Liu [130]. Earlier algorithms such as [204] and Guo, Gramm, Hüffner, Niedermeier, and Wernicke [365], methods used various annotated version of the FEEDBACK VERTEX SET. Other illustrative problems making direct use of this technique include Chen, Liu, Lu, O'Sullivan, and Razgon [150] algorithm for DIRECTED FEEDBACK VERTEX SET. The DIRECTED FEEDBACK VERTEX SET problem had famously resisted solution for nearly 20 years. Other problems making direct use of the technique include Razgon and O'Sullivan [574] ALMOST 2-SAT, and Hüffner, Komusiewicz, Moser, and Niedermeier [405] CLUSTER VERTEX DELETION. We remark that as well as giving an alternative route to proving parametric tractability of GRAPH BIPARTITIZATION, the methods of Exercise 4.11.17 allowed Narayanaswamy, Raman, Ramanujan, and Saurabh [497] to give a new method of proof for the parametric tractability of ALMOST 2-SAT. We also refer to the survey of Guo, Moser, and Niedermeier [366] on results using the iterative compression technique.

The idea of measure and conquer, which is a systematic upgrading of the notion of improving search trees and seems to reflect ideas from machine learning originated in Fomin, Grandoni, and Kratsch [316]. It has been extensively applied in, for example, Fomin, Grandoni, and Kratsch [317, 318], Fomin, Grandoni, Pyatkin, and Stepanov [319]. The algorithm and analysis for SET COVER and DOMINATING SET based on measure and conquer follows the paper of van Rooij and Bodlaender [653]. The current fastest algorithm for DOMINATING SET is $O^*(1.4864^n)$, and is due to Iwata [414]. For this running time, Iwata begins with the van Rooij–Bodlaender algorithm. Iwata noticed that certain worst case branching cannot occur consecutively in the branching algorithm, and devises a method for dealing with consecutive branching in the algorithm simultaneously. He does this by adding a state diagram, and then not only measures the cost, but considers the state of the transition diagram. Iwata calls this method "the potential method". The algorithm uses around 1,000 potential new variables, and suggests that running times could be improved by adding more. Presumably, this method could also be applied in the context of the measure and conquer part of the FEEDBACK VERTEX SET algorithm given here with a consequential improvement in running times. Space limitations preclude us from including this method. The first algorithms for FEEDBACK VERTEX SET were found in Downey and Fellows [247] and one is given in Chap. 3. The first small kernel for FEEDBACK VERTEX SET was given by Bodlaender [76]. The running times were improved about once every two years from 1999–2010 culminating in the algorithm given here. The exact history can be found in [130]. The realization that iterative compression could be used for kernelization was in Dehne, Fellows, Langston, Rosamond, and Shaw [204]. The [204] paper gave an $O(10.6^k n^3)$ algorithm. The algorithm and proof for FEEDBACK VERTEX COVER follow the methods of Cao, Chen, and Liu. The use of measure and conquer is one of the primary reasons that the running time improved the earlier algorithm of the Chen, Fomin, Liu, Lu, and Villanger [141] which ran in time $O(5^k k n^2)$ and also used iterative compression.

6.8 Summary

We introduce iterative compression. This method adds vertices stage by stage and either updates a solution or gives a certificate that no solution is possible. The method of measure and conquer is introduced. We show how the methods combine to give state-of-the-art algorithms.

Further Elementary Techniques

7

7.1 Methods Based on Induction

7.1.1 The Extremal Method

The *extremal method* inductively explores the boundary between "yes" and "no" instances of a parameterized problem. So far, the approach has mostly been used for *parameter monotone* maximization problems. By *monotone* is meant that if (G, k) is a no-instance of the parameterized problem, then necessarily $(G, k + 1)$ is also a no-instance. The *boundary* investigation referred to consists in exploring the question:

> If (x, k) is a thoroughly kernelized *yes-instance, where* $(x, k + 1)$ *is a no-instance, then what is the detailed structure of x?*

There is no logical reason why the approach cannot be generalized to minimization problems, although currently there are no striking examples. The method was first introduced by Estivill-Castro, Fellows, Langston, and Rosamond [278]. The method supports the thesis that the *proper way to find an efficient kernelization algorithm is* "PTIME EXTREMAL STRUCTURE THEORY". The extremal method is essentially about deriving efficient FPT kernelization algorithms in a mathematically disciplined way. Many examples are described in Prieto's dissertation [564], and in Mathieson, Prieto, and Shaw [518] and Prieto and Sloper [565].

By intensely studying the boundary inductively (e.g., minimal counterexample arguments with many inductive priorities) *the goal is to discover a powerful set of reduction rules $\mathcal{R}$, such that the following holds in the tightest possible way with respect to some kernelization bound $g(k)$:*

> **Boundary Lemma.** *If (x, k) is reduced with respect to $\mathcal{R}$, (x, k) is a yes-instance, and $(x, k + 1)$ is a no-instance, then $|x| \leq g(k)$.*

A nice illustration of the approach due to Prieto and Sloper [565] concerns the following problem.

R.G. Downey, M.R. Fellows, *Fundamentals of Parameterized Complexity*,
Texts in Computer Science, DOI 10.1007/978-1-4471-5559-1_7,
© Springer-Verlag London 2013

k-PACKING OF $K_{1,3}$

Instance: A graph $G = (V, E)$.
Parameter: A positive integer k.
Question: Can I find k vertex disjoint copies of $K_{1,3}$ in G?

Theorem 7.1.1 (Prieto and Sloper [565]) k-PACKING OF $K_{1,3}$ *is linear time FPT and has a kernel of size* $48k^2 - 8k$.

Proof sketch By an inductive study of the parameterized problem boundary, Prieto, and Sloper identified the following reduction rules that can be applied in polynomial time.

1. If an edge uv has both u and v of degree 2, then $(G, q) \in \mathcal{Y}(q)$ iff $(G - \{u, v\}, q) \in \mathcal{Y}(q)$.
2. If G has a high degree vertex, namely v with degree $\geq 4q - 1$, then $(G, q) \in \mathcal{Y}(q)$ iff $(G - \{v\}, q - 1) \in \mathcal{Y}(q - 1)$.

The safety of the first rule is immediate, since such edges are useless for a 3-star packing. The safety of the second rule comes from the fact that a 3-star packing needs at least $4q$ many vertices. Now suppose we are in general position in the induction. We have dealt with (G, q) and have moved on to $(G, q + 1)$. Take the packing S of the 3-stars from q and consider $G[V - S]$. If there is any vertex in this of degree 3 say "yes", and move on to $q + 2$, unless $q + 1 = k$ and then stop. If all the vertices of $G[V - S]$ have degree ≤ 2, then by Rule 2 above, we see that $N[S] \leq 4(q + 1)(4(q + 1) - 1)$. Per Rule 1, we observe that $I = V - (S \cup N(S))$ is an independent set, and hence each vertex in I has one adjacent to it in $N(S)$. Therefore $|I| \leq 2|N(S)|$. Thus, $|V| \leq |S| + |N(S)| + |I| \leq 4(q + 1) + 3 \cdot 4(q + 1)(4(q + 1) - 1) = 48(q + 1)^2 - 8(q + 1)$. That is, we have a bound $f(q + 1)$ for the kernel, as required. □

7.1.2 *k*-LEAF SPANNING TREE, Revisited

Extremal combinatorial results can be used in devising kernelization algorithms in a variety of ways. Although the method of *P-time Extremal Structure Theory* expositied in the previous section is aimed at formulating an *algorithmic* approach to extremal structure theory, specifically in tune with FPT objectives, one can still use "off-the-shelf" existential (i.e., non-algorithmic) extremal combinatorial results to good effect. Here is an example.

Let us reconsider the k-LEAF SPANNING TREE problem we met in Chap. 4, Theorem 4.8.1. We recall this problem.

k-LEAF SPANNING TREE
Instance: A graph $G = (V, E)$.
Parameter: A positive integer k.
Question: Is there a spanning tree of G with at least k leaves?

In Theorem 4.8.1 we found a kernel of size $O(4k + 2)(4k + 1)$. A kernel of at most $8k$ vertices for this problem is the procedure that repetitively applies the following reduction rules until none of them applies:

R1: If G contains a vertex v of degree 1 with a neighbor of degree 2 then set $(G, k) \leftarrow (G \setminus \{v\}, k)$.

R2: If G contains two vertices v and v' of degree 1 with the same neighbor, then set $(G, k) \leftarrow (G \setminus \{v\}, k - 1)$.

R3: If G contains a chain of length at least 4 then set $(G, k) \leftarrow (G', k)$ where G' is the graph obtained if we contract some of the edges of this chain. (A *chain* is a path of length at least 2, where all internal vertices have degree 2, and with endpoints of degree greater than 2.)

It is not hard to see that all of the reduction rules above sound—they produce smaller equivalent instances. A *reduced instance* (G', k'), is an instance of the problem where none of the above rules can be applied any more. What remains to prove is that if in a reduced instance (G', k'), G' has least $8k'$ vertices, then (G', k') is a "no"-instance.[1]

We denote by V_1, V_2, and $V_{\geq 3}$ the set of vertices of G' with degrees 1, 2, or ≥ 3, respectively. Because of the first rule, vertices in V_1 are adjacent only with vertices in V_3, and, from the second rule, we have $|V_1| \leq |V_3|$. Moreover, from the third rule, all the chains of G' have length at most 3.

We aim to prove that if $|V(G')| \geq 8k'$, then G' contains a spanning tree of k' leaves. For this, we construct, using G', an auxiliary graph H by removing all vertices of degree 1 and by replacing all chains of G' by chains of length 2. Notice that H does not have vertices of degree 1 and all its degree 2 vertices—we call them *subdivision vertices*—are internal vertices of some chain of length 2. We denote by Q_2 and $Q_{\geq 3}$ the vertices of H that have degree 2 or ≥ 3, respectively. Notice that $|V_{\geq 3}| = |Q_{\geq 3}|$ which, combined with the fact that $|V_1| \leq |V_3|$, implies that $|V_1| + |V_{\geq 3}| \leq 2 \cdot |Q_{\geq 3}|$. Moreover, each vertex in Q_2 corresponds to at most two vertices of V_2. Therefore $|V_2| \leq 2 \cdot |Q_2|$. We conclude that $|V(G')| = |V_1| + |V_2| + |V_{\geq 3}| \leq 2 \cdot |Q_2| + 2 \cdot |Q_{\geq 3}| = 2 \cdot |V(H)|$. As $|V(G')| \geq 8k'$, we conclude that $|V(H)| \geq 4k'$. We call two subdivided edges of H *siblings* if they have the same neighborhood. If a vertex of degree 2 has no sibling then we will subdivide it. Next we add edges between each set of siblings with same neighborhood, so that H is transformed to a graph H' with minimum degree 3. It is easy to see that if H' contains a spanning tree with at least k' leaves then also H contains a spanning tree with at least k' leaves. Specifically, if X_1 is a set of siblings we make X_1 a clique. We do the same for any other set of siblings of X_i. Suppose now X is a set of siblings adjacent with two vertices a and b. Then in H' this makes a clique on X whose all vertices are adjacent to a and to b. If a spanning tree of H uses has a leaf q adjacent to an edge $\{q, p\}$ of this clique (i.e. an edge that does not exist in H), then q is also adjacent to a and therefore we can substitute edge $\{q, p\}$ by edge aq

---

[1] We remark that the P-time extremal structure approach uncovers 14 more reduction rules, yielding a much tighter kernelization bound, but the argument is laborious and beyond the scope of this book. See [278].

that is an edge of H. This way, the spanning tree of H' is transformed to a spanning tree of H with the same number of leaves. As $|V(H')| \geq 4k'$, then, from the main result of Kleitman and West in [449], H' (and therefore H) contains a spanning tree of with least k' leaves. By the construction of H, it easily follows that this spanning tree of H can be extended to a spanning tree in G' with at least the same number of leaves and we are done.

The correctness of the above kernel is based on the purely existential extremal combinatorial result of [449]. Such results simply assert the existence of combinatorial structures and bounds. It is clear that FPT as a complexity phenomenon is deeply entwined with rich extremal combinatorial structure theory (e.g., *what is Graph Minors about?*). It is an interesting challenge to further develop polynomial time extremal structure theory—much remains to be explored!

7.2   Greedy Localization

The method was named in Dehne, Fellows, Langston, Rosamond, and Shaw [204]. The method itself was introduced in Chen, Friesen, and Kanj [142] and Jia, Zhang, and Chen [416].

The method of *Greedy Localization* is the branching approach combined with a step further: a non-deterministic guess of some potential part of the solution. We will briefly present the idea using the following problem (which can also be shown to be FPT by the extremal method, as the reader should show in Exercise 7.4.3).

k-PACKING OF TRIANGLES
Instance: A graph $G = (V, E)$.
Parameter: A positive integer k.
Question: Can I find k vertex disjoint copies of K_3 in G?

Here is our non-deterministic version of the FPT-algorithm for this problem. We will use the term *partial triangle* for any vertex or edge of G and we will treat it as a potential part of a triangle in the requested triangle packing. The first step is to find in G a maximal set of disjoint triangles $\mathcal{T} = \{T_1, \ldots, T_r\}$. This greedy step justifies the name of the technique and can be done in polynomial time. If $r \geq k$ then return YES and stop. If not, let $G_\mathcal{T}$ be the graph formed by the disjoint union of the triangles in $\mathcal{T}$ and observe that $|V(G_\mathcal{T})| \leq 3(k-1)$. The key observation is that if there exists a solution $\mathcal{T}' = \{T_1', \ldots, T_r'\}$ to the problem ($r \geq k$), then each T_i' should intersect some triangle in $\mathcal{T}$, (because of the maximality of the choice of $\mathcal{T}$). These intersections define a partial solution $\mathcal{P} = \{A_1, \ldots, A_p\}$ of the problem. The next step of the algorithm is to *guess* them: let S be a subset of $V(G_\mathcal{T})$ such that $G_\mathcal{T}[S]$ has at least k connected components (if such a set does not exist, the algorithm returns NO and stops). Let $\mathcal{P} = \{P_1, \ldots, P_p\}$ be these connected components and observe that each P_i is a partial triangle of G. We also denote by $V(\mathcal{P})$ the set of vertices in the elements of $\mathcal{P}$. Then, the algorithm intends to extend this partial solution to a full one using the following non-deterministic procedure BRANCH($\mathcal{P}$):

Algorithm 7.2.1

Procedure BRANCH$(P_1, \ldots, P_p)$

1. $A \leftarrow B \leftarrow \emptyset, i \leftarrow 1$
2. while $i \leq p$

   if there exists a partial triangle B of $G \setminus (V(\mathcal{P}) \cup A)$

   such that $G[P_i \cup B]$ is a triangle,

   then $A \leftarrow A \cup B, i \leftarrow i + 1$

   otherwise goto step 4
3. **return** YES and stop.
4. **let** A' be the set containing each vertex $v \in A$

   such that $P_i \cup \{v\}$ is a (partial) triangle.
5. **if** $A' = \emptyset$, **then return** NO and stop,

   otherwise *guess* $v \in A'$ and

   return BRANCH$(P_1, \ldots, P_i \cup \{v\} \ldots, P_p)$.

In Step 2, the procedure checks greedily for a possible extension of $\mathcal{P}$ to a full triangle packing. If it succeeds, it enters Step 3 and returns YES. If it fails at the ith iteration of Step 2, this means that one of the vertices used for the current extension (these are the vertices in A' constructed in Step 4) should belong to the extension of P_i (if such a solution exists). So we further *guess* which vertex of A should be added to P_i and we recurse with this new collection of partial triangles (if such a vertex does not exist, return NO). This algorithm has two non-deterministic steps: one is the initial choice of the set S and the other is the one in Step 5 of BRANCH$(P_1, \ldots, P_p)$. The first choice implies that BRANCH$(P_1, \ldots, P_p)$ is called $k^{O(k)}$ times (recall that $|V(G_\mathcal{T})| = O(k)$). Each call of BRANCH$(P_1, \ldots, P_p)$ has $|A'| \leq 2k$ recursive calls and the depth of the recursion is upper bounded by $2k$, therefore the size of the search tree is bounded by $k^{O(k)}$. Summing up, we have shown that k-PACKING OF TRIANGLES is in FPT by an algorithm with running time $O(2^{O(k \log k)} \cdot n)$.

The greedy localization technique has been used to solve a wide variety of problems. It was applied for the standard parameterizations of problems such as r-DIMENSIONAL MATCHING in [142], r-SET PACKING in [416], H-GRAPH PACKING in [288]. Similar ideas, combined with other techniques, have been used for various problems such as the (long-standing open) DIRECTED FEEDBACK VERTEX SET [150] and the BOUNDED LENGTH EDGE DISJOINT PATHS in [350]. The main drawback of this method is that the parameter dependence of the derived FPT-algorithm is typically $O(2^{O(k \log k)})$. Alternative methods for 3-DIMENSIONAL MATCHING are found in the next section.

7.2.1 3-DIMENSIONAL MATCHING

We finish this section with a problem of a different character

3-DIMENSIONAL MATCHING

Instance: Three disjoint sets A, B, C, a collection of triples $\mathcal{T} \subseteq A \times B \times C$ and an integer $k \geq 1$.

Parameter: A positive integer k.

Question: Is there a subset $\mathcal{M} \subseteq \mathcal{T}$ where $|\mathcal{M}| \geq k$ and such that no two triples in $\mathcal{M}$ share a common element (i.e. $\mathcal{M}$ is a matching of $\mathcal{T}$)?

Let $\mathcal{M}'$ be a maximal matching of $\mathcal{T}$. Clearly, such a matching can be found greedily in polynomial time. If $|\mathcal{M}'| \geq k$ then we are done (in this case, return a trivial YES-instance). Therefore, we assume that $|\mathcal{M}'| \leq k - 1$ and let S be the set of all elements in the triples of $\mathcal{M}'$. Certainly, $|S| \leq 3k - 3$ and, by the maximality of $\mathcal{M}'$, S will intersect every triple in $\mathcal{T}$. Suppose now that $\mathcal{T}$ contains a sub-collection $\mathcal{A}$ with at least $k + 1$ triples that agree in two coordinates. Without loss of generality (w.l.o.g.), we assume $\mathcal{A} = \{(x, y, v_i) \mid i = 1, \ldots, \rho\}$ for some $r \geq k + 1$. Our first reduction rule is to replace $(\mathcal{T}, k)$ by $(\mathcal{T}', k)$ where $\mathcal{T}' = \mathcal{T} \setminus \{(x, y, v_i) \mid i = k + 1, \ldots, \rho\}$. We claim that this reduction rule is *safe*, i.e. the new instance $(\mathcal{T}', k)$ is equivalent to the old one. Obviously, every matching of $\mathcal{T}'$ is also a matching of $\mathcal{T}$. Suppose now that $\mathcal{M}$ is a matching of $\mathcal{T}$ that is not any more a matching of $\mathcal{T}'$. Then we show that $\mathcal{M}$ can be safely replaced by another one that is in $\mathcal{T}'$. Indeed, if $\mathcal{M}$ is not a matching of $\mathcal{T}'$, then one of the triples in $\mathcal{M}$ is missing from $\mathcal{T}'$ and therefore is of the form $T = (x, y, v_j)$ for some $j \in \{k + 1, \ldots, \rho\}$. Notice that if one of the triples in $\mathcal{M} \setminus \{T\}$ intersects a triple in $\{(x, y, v_i) \mid i = 1, \ldots, k\}$, then this intersection will be a vertex in $\{v_i \mid i = 1, \ldots, k\}$. As $|\mathcal{M} \setminus \{T\}| < k$, one, say T' of the k triples in $\{(x, y, v_i) \mid i = 1, \ldots, k\}$ will not be intersected by the triples in $\mathcal{M} \setminus \{T\}$. Therefore, $\mathcal{M}^* = \mathcal{M} \setminus \{T\} \cup \{T'\}$ is a matching of $\mathcal{T}'$ and the claim follows as $|\mathcal{M}^*| \geq k$. Notice that the first rule simply "truncates" all but k triples agreeing in the same two coordinates.

We now assume that $\mathcal{T}$ is a collection of triples where no more than k of them agree in the same two coordinates. Suppose now that $\mathcal{T}$ contains a sub-collection $\mathcal{B}$ with more than $2(k - 1)k + 1$ triples, all agreeing to one coordinate. W.l.o.g. we assume that the agreeing coordinate is the first one. The second reduction rule removes from $\mathcal{T}$ all but $2(k - 1)k + 1$ of the elements of $\mathcal{B}$. Again using a pigeonhole argument, it follows that in the $2(k - 1)k + 1$ surviving triples of $\mathcal{B}$ there is a subset $\mathcal{C}$ of $2k - 1$ triples where each two of them disagree in both second and third coordinate. Again, if a discarded triple is used in a solution $\mathcal{M}$, then the $k - 1$ other triples cannot intersect more than $2k - 2$ triples of $\mathcal{C}$ and therefore a "surviving" one can substitute the discarded one in $\mathcal{M}$. Therefore, the second truncation is also safe and leaves an equivalent instance $\mathcal{T}'$ where no more than $2(k - 1)k + 1$ of them agree in the same coordinate. Recall now that the elements of the set S intersect all triples in $\mathcal{T}$. As $|S| \leq 3(k - 1)$, we obtain the result that, in the equivalent instance $(\mathcal{T}', k)$, $\mathcal{T}'$ has at most $3(k - 1) \cdot (2(k - 1)k + 1) = O(k^3)$ triples. We conclude that p-3-DIMENSIONAL-MATCHING has a polynomial size kernel.

The above kernelization was proposed by Fellows, Knauer, Nishimura, Ragde, Rosamond, Stege, Thilikos, and Whitesides [292]. When combined with the color coding technique in Chap. 8, the greedy localization technique gave algorithms of total complexity $2^{O(k)} + O(n)$ for a wide variety of packing and matching problems.

7.3   Bounded Variable Integer Programming

One method which initially seemed to promise a lot is based on integer linear programming. We have already seen a related application of the linear programming methodology in the Nemhauser–Trotter kernelization of VERTEX COVER. However, here we will look at applications of the following result.

Theorem 7.3.1 (Lenstra [487], then Kannan [429])   *The following problem is* $O(p^{2p}L)$ *(counting arithmetical operations)*:

INTEGER PROGRAMMING FEASIBILITY
Instance: *A collection of linear integer inequalities with p variables and L bits of input.*
Parameter: *A positive integer p.*
Question: *Does there exist an integer solution satisfying the constraints?*

The proof of this result is beyond the scope of this book and we will simply invoke it as a black box. Recently, Fellows, Gaspers, and Rosamond [287] observed that this FPT result could be widely applied in the situation where we parameterized *by the number of numbers*. Here is an example.

VAR-SUBSET SUM
Instance: *A multiset A of integers and integer s, k. where k is the number of distinct integers in A.*
Parameter: *A positive integer k.*
Question: *Is there a multiset $X \subseteq A$ such that $\sum_{a \in X} a = s$?*

Theorem 7.3.2 (Fellows, Gaspers, and Rosamond [287])   VAR-SUBSET SUM *is FTP.*

Proof Let $a_1, \ldots, a_k$ denote the distinct elements of A. Let m_i denote the multiplicity of a_i. We create an Integer Programming Instance with variables $x_1, \ldots, x_k$, and with the following constraints:

$$x_i \leq m_i \wedge x_i \geq 0 \quad \text{for } i \in [k],$$

$$\sum_{i=1}^{k} x_i \cdot a_i = s.$$

Clearly this describes the problem, and hence it is FPT by Theorem 7.3.1.   □

To our knowledge, the first applications of Theorem 7.3.1 was due to Gramm, Niedermeier, and Rossmanith [357], and concerns our old friend CLOSEST STRING from Chap. 3. We recall this problem was the following.

CLOSEST STRING

Input: A set of k length L strings $s_1, \ldots, s_k$ over an alphabet Σ and an integer
              $d \geq 0$.

Parameter: Positive integers d and k.

Question: Find the center string, if any. That is find $s = s_j$ such that $d_H(s, s_i) \leq d$
              for all $i = 1, \ldots, k$.

The idea used by Gramm, Niedermeier and Rossmanith was the following. Given a set of k strings of length L, we will think of these as a $k \times L$ matrix of characters. Thinking of this, we can refer to the columns of the CLOSEST STRING problem as being the columns of this matrix. For any permutation $\pi : L \to L$, s is a solution to the original problem if and only if $\pi(s)$ is a solution to the problem involving the permuted strings $\pi(s_j)$ for $j = 1, \ldots, k$.

The reader should note that columns can be identified via *isomorphism*. The distance of the closest string is measured columnwise; it is always possible to *normalize* the solution, so that a is the letter occurring most often in a column, then b the next most, etc. Finally we regard two instances as being isomorphic if there is a bijection between the columns of the instances so that matched pairs of columns are isomorphic.

Lemma 7.3.1 (Gramm, Niedermeier and Rossmanith [357]) *An instance of* CLOSEST STRING *over* Σ, k *with* $|\Sigma| > k$ *is isomorphic to an instance with* $|\Sigma| = k$.

Proof If $|\Sigma| > k$, then we note that in each column there are at most k symbols of Σ. Since the columns are independent of each other, it suffices to represent the structure of a column by an isomorphic instance, and for this we only need k symbols. $\qquad\qquad\square$

The number of columns for k strings depends only on k and hence is bounded by $k!$. Using the column types, Gramm, Niedermeier and Rossmanith [357] observed that CLOSEST STRING can be formulated by an integer linear program needing at most $k! \cdot k$ variables $x_{t,\varphi}$ where t denotes a column type represented by a number between 1 and $k!$, and $\varphi \in \Sigma$, with $|\Sigma| = k$. This programme wishes to minimize

$$\max_{1 \leq i \leq k} \sum_{1 \leq t \leq k!} \sum_{\varphi \in (\Sigma \setminus \{\varphi_{t,i}\})} x_{t,\varphi}.$$

Here $\varphi_{t,i}$ denotes the ith entry in column type t. We then need the following constraints to formulate the problem.

1. $x_{t,\varphi} \in \mathbb{Z}^+$.
2. If n_t denotes the number of columns of type t in the instance, then $\sum_{\varphi \in \Sigma} x_{t,\varphi} = n_t$ for each column type t.

Gramm, Niedermeier and Rossmanith [357] point out that Theorem 7.3.1 does not give the *optimal* solution, only whether there is a feasible solution. The solution is then reformulated as a decision problem, by specifying

$$\max_{1 \le i \le k} \sum_{1 \le t \le k!} \sum_{\varphi \in (\Sigma \setminus \{\varphi_{t,i}\})} x_{t,\varphi} \le d.$$

As is noted in [357], the number of variables becomes huge and additionally the running time, whilst *linear*, has infeasible constants. Thus the methodology above has remained a classification tool, and it is not used in practice, to the knowledge of the authors.

7.4 Exercises

Exercise 7.4.1 (Prieto and Sloper [565]) Show that the following is FPT using the extremal method.

k-PACKING OF $K_{1,s}$

Instance: A graph $G = (V, E)$.
Parameter: Positive integers s and k.
Question: Can I find k vertex disjoint copies of $K_{1,s}$ in G?

Exercise 7.4.2 (Mathieson, Prieto, and Shaw [518])   Show that the following problem is FPT using the extremal method.

EDGE DISJOINT PACKING OF TRIANGLES
Instance: A graph $G = (V, E)$.
Parameter: A positive integer k.
Question: Can I find k edge disjoint copies of K_3 in G?

Exercise 7.4.3 (Fellows, Heggernes, Rosamond, Sloper, and Telle [289])   Show that the following is FPT using the extremal method.

k-PACKING OF TRIANGLES
Instance: A graph $G = (V, E)$.
Parameter: A positive integers k.
Question: Can I find k vertex disjoint copies of K_3 in G?

Exercise 7.4.4 Show that the following problem is FPT. You can use the extremal method (Prieto [564]). It has a kernel of size $5/3 \cdot k$ by Dehne, Fellows, Fernau, Prieto, and Rosamond [203].

NON-BLOCKER
Instance: A graph $G = (V, E)$.
Parameter: A positive integer k.

Question: Is there a set S of k vertices such that every member has a neighbor not
              in S?

Exercise 7.4.5 Consider the following problem.

$\leq k$-LEAF SPANNING TREE
Instance: A graph $G = (V, E)$.
Parameter: A positive integer k.
Question: Is there a spanning tree of G with at most k leaves?

Prove that the problem is NP-complete for $k = 2$.

Exercise 7.4.6 (Sloper and Telle [624]) Using the method of greedy localization,
prove that the following is FPT:

k-PACKING OF k-CYCLES
Instance: A graph $G = (V, E)$.
Parameter: A positive integer k.
Question: Does G contain k vertex disjoint cycles on k vertices?

Exercise 7.4.7 (Fellows, Gaspers, and Rosamond [287])   Use the method of Theorem 7.3.2 to show that the following are FPT.

VAR-PART
Instance: A multiset A of integers and a positive integer k, where k is the number
              of distinct integers in A.
Parameter: k.
Question: Is there a multiset $X \subseteq A$ such that $\sum_{a \in X} a = \sum_{b \in A-X} b$?

VAR-NUM 3DM
Instance: Multisets A, B, C of n integers each, and positive integers s and k,
              where k is the number of distinct integers in $A \cup B \cup C$.
Parameter: k.
Question: Are there n triples $S_1, \ldots, S_n$ each with one element of A, B, C such
              that for all $i \in [n]$, $\sum_{a \in S_i} a = s$?

Exercise 7.4.8 (Niedermeier [546]) Prove that the following is FPT using bounded
integer programming. We have a cycle of n vertices, upon which are stacked multiple rings, each with capacity c. The goal is to realize communication channels
subject to the following conditions.
1. The communication channel between two vertices must use the same ring.
2. Through each edge of a ring there may be at most c communication channels.
3. If a vertex of the cycle is the endpoint of a communication channel for a ring,
   then it must be the "ATM" for this vertex pair/ring.

RING GROOMING
Instance: A setup as above, pairs $(j_1, j_2), \ldots, (j_s, k_s)$ of demands from $[n]$, representing communication demands, and the endpoints must be different.
Parameter: A positive integer k.
Question: Can the demands be met using at most k ATM's?

7.5 Historical Notes

The method has been extensively used as a FPT design paradigm, especially in systematic kernelization. For example, it was used in Prieto [564], Mathieson, Prieto, and Shaw [518] and Prieto and Sloper [565]. Greedy localization was introduced in Chen, Friesen, and Kanj [142] and Jia, Zhang, and Chen [416]. It is identified as a singularly appropriate method for FPT design in Dehne, Fellows, Langston, Rosamond, and Shaw [204]. The kernelization for 3-DIMENSIONAL MATCHING was proposed by Fellows, Knauer, Nishimura, Ragde, Rosamond, Stege, Thilikos, and Whitesides [292]. This technique combined with the color coding technique of the next chapter, produces algorithms of total complexity $2^{O(k)} + O(n)$ for a wide variety of packing and matching problems. Related results include Chen, Fernau, Kanj, and Xia [140] Jia, Zhang, and Chen [416], and Abu-Khzam [8]. Further examples have been given in the exercises. INTEGER PROGRAMMING FEASIBILITY was shown to be FPT parameterized by the number of variables in Lenstra [487], then Kannan [429] improved the running time to the stated one. The method is rather difficult, and beyond the scope of the present book. The first applications of this method to combinatorial problems, to demonstrate parametric tractability, was Gramm, Niedermeier, and Rossmanith [357], except possibly Alon, Azar, Woeginger, and Yadid [30], which can be interpreted as an FPT result. This was how matters stood until Fellows, Gaspers, and Rosamond [287] observed that the method could be widely applied in situations where we parameterize by the number of numbers.

7.6 Summary

We look at a number of elementary techniques including the extremal method, greedy localization, and bounded variable integer linear programming.

Color Coding, Multilinear Detection, and Randomized Divide and Conquer

8

8.1 Introduction

The first technique introduced in this chapter is *color coding*, one of the most beautiful ideas in parameterized algorithm design. In Sect. 8.4 we will show how this method can be, in some sense, simulated algebraically to improve running times. Color coding has been extensively applied, not only to FPT algorithm design, but also in the construction of FPT-*reductions* for tight lower bounds. We see one example of this reduction methodology with the work of Marx [516], which we consider in Chap. 31, Theorem 31.5.2. Color coding was introduced by Alon, Yuster, and Zwick in [36]. This technique was first applied to special problems of the following type.

k-SUBGRAPH ISOMORPHISM

Instance: A graph $G = (V, E)$ and a graph $H = (V^H, E^H)$ with $|V^H| = k$.
Parameter: An integer k.
Question: Is H isomorphic to a subgraph in G?

If H is, for example, a clique, then the problem seems not FPT, as it is hard for the class $W[1]$. But in certain special cases of H, we can show tractability using the technique of color coding, as we now see.

The idea behind color coding is that, in order to find the desired set of vertices V' in G, such that $G[V']$ is isomorphic to H, we randomly color all the vertices of G with k colors and expect that, with some high degree of probability, all vertices in V' will obtain different colors. In some special cases of the SUBGRAPH ISOMORPHISM problem, dependent on the nature of H, this will simplify the task of finding V'.

If we color G uniformly at random with k colors, a set of k distinct vertices will obtain different colors with probability $(k!)/k^k$. This probability is lower-bounded by e^{-k}, so we need to repeat the process e^k times to have sufficiently high probability of obtaining the required coloring.

We can de-randomize this kind of algorithm using *hashing*, but at the cost of extending the running time. We need a list of colorings of the vertices in G such that, for *each* subset $V' \subseteq V$ with $|V'| = k$ there is at least one coloring in the list

R.G. Downey, M.R. Fellows, *Fundamentals of Parameterized Complexity*,
Texts in Computer Science, DOI 10.1007/978-1-4471-5559-1_8,
© Springer-Verlag London 2013

by which all vertices in V' obtain different colors. Formally, we require a k-perfect family of hash functions from the set of vertices in G $\{1, 2, \ldots, |V|\}$, onto the set of colors $\{1, 2, \ldots, k\}$.

8.1.1 k-PATH

To make this more specific, we consider the problem of finding a path in a graph of length k, i.e. $H = P_k$.

k-PATH

Instance: A graph $G = (V, E)$.
Parameter: An integer k.
Question: Does G contain a path of at least k vertices?

The above problem can be solved in $2^{O(k \cdot \log k)} \cdot n$ steps using dynamic programming techniques. This dynamic programming technique confirmed the Papadimitriou and Yannakakis conjecture, in [553], that it is possible to check—in polynomial time—whether an n-vertex graph contains a path of length $\log n$. As we see in Sect. 8.4, we will also be able to improve this running time to $O^*(2^k)$ using algebraic methods to replace the dynamic programming.

8.1.2   Dynamic Programming

Dynamic programming simplifies a global problem by breaking it down into a sequence of smaller decision subproblems over time. These often have certain overlapping values and this is adjusted, but crucially the local values can be combined in a controllable way. Typically, value functions are $V_1, V_2, \ldots, V_n$, with V_i the value of the system at step i. There is usually some kind of recursion in the process. As we see below, in the case of color coding, the dynamic programming will be determined by the color classes and their connections. This method will be used extensively when we look at treewidth and branchwidth (see Chap. 10 and Sect. 16.3).

8.2   The Randomized Technique

To solve k-PATH, begin by considering a function $\chi : V(G) \to \{1, \ldots, k\}$ coloring the vertices of G with k distinct colors. Given a path P of G, we denote by **col**(P) the set of colors of the vertices in P. We call a path P of G χ-*colorful*, or simply *colorful*, if all its colors are pairwise distinct. We also use the term i-*path* for a path of i vertices. We now define the following variant of the original problem:

k-COLORFUL PATH

Instance: A graph $G = (V, E)$ and a coloring $\chi : V(G) \to \{1, \ldots, k\}$.
Parameter: An integer k.
Question: Does G contain a colorful path of at least k vertices?

k-COLORFUL PATH can be solved by the following dynamic programming procedure. First we fix a vertex $s \in V(G)$ and then for any $v \in V(G)$, and $i \in \{1, \ldots, k\}$, we define

$$\mathcal{C}_s(i, v) = \{R \subseteq \{1, \ldots, k\} \mid G \text{ has a colorful } i\text{-path } P \text{ from}$$
$$s \text{ to } v \text{ such that } \mathbf{col}(P) = R\}.$$

Notice that $\mathcal{C}_s(i, v)$ stores sets of colors in paths of length $i - 1$ between s and v, instead of the paths themselves. Clearly, G has a colorful k-path starting from s iff $\exists v \in V(G) : \mathcal{C}_s(k, v) \neq \emptyset$. The dynamic programming is based on the following relation:

$$\mathcal{C}_s(i, v) = \bigcup_{v' \in N_G(v)} \{R \mid R \setminus \{\chi(v)\} \in \mathcal{C}_s(i - 1, v')\}.$$

Notice also that $|\mathcal{C}_s(i, v)| \leq \binom{k}{i}$ and, for all $v \in V(G)$, $\mathcal{C}_s(i, v)$ can be computed in $O(m \cdot \binom{k}{i} \cdot i)$ steps (here, m is the number of edges in G). For all $v \in V(G)$, one can compute $\mathcal{C}_s(k, v)$ in $O(\sum_{i=1,\ldots,k} m \cdot \binom{k}{i} \cdot i) = O(2^k \cdot k \cdot m)$ steps. We conclude that one can check whether G colored by χ has a colorful path of length k in $O(2^k \cdot k \cdot m \cdot n)$ steps (just apply the above dynamic programming for each possible starting vertex $s \in V(G)$). Applying this method to randomly chosen colorings gives the desired randomized algorithm.

8.3 De-randomizing

As we mentioned above, we will use what are called hash functions to de-randomize the algorithms obtained by color coding.

Definition 8.3.1 A *family of k-perfect hash functions* is a family $\mathcal{F}$ of functions from $\{1, \ldots, n\}$ onto $\{1, \ldots, k\}$ such that for each $S \subseteq \{1, \ldots, n\}$ with $|S| = k$, there exists an $\chi \in \mathcal{F}$ that is bijective when restricted to S.

Clearly G contains a k-path if and only if there is a $\chi \in \mathcal{F}$ such that G, colored by χ, contains a χ-colorful k-path. This equivalence reduces the k-PATH problem to the k-COLORFUL PATH problem: just run the above dynamic programming procedure for k-COLORFUL PATH for all colorings in $\mathcal{F}$. We conclude that k-PATH is fixed-parameter tractable, by an FPT algorithm with running time $O^*((2|\mathcal{F}|)^k)$.

Research on the construction of small-size families of k-perfect hash functions dates back to the work of Fredman, Komlós, and Szemerédi [331] (see also Slot and van Emde Boas [625], Schmidt and Siegel [603] and Naor, Schulman, and Srinivasan [541]). The most "efficient" families have been described by Alon, Yuster, and Zwick in [36] by means of sophisticated probabilistic methods that are beyond the scope of this book, although we will describe their results below. Their results are a bit impractical due to the large constants involved, but there are efficiently described families which work in practice due to Chen, Lu, Sze, and Zhang [151], which we also describe below.

We first describe a simpler but less efficient construction due to Dennenberg, Gurevich, and Shelah [219] and improved slightly by Koblitz [455], whose proof we give here. This is sufficient to provide a basis for the method. Improved FPT algorithms are possible by using more efficient families of hash functions. It will need an estimation theorem from the number theory literature as part of its proof.

Let $l_1(x) = \log_2(x)$, and $l_r(x) = \log_2(l_{r-1}(x))$ for $r > 1$, be the rth repeated log function. Let $e_1(x) = 2^x$ and $e_r(x) = 2^{e_{r-1}(x)}$ be the rth repeated exponential functions, i.e., the inverse functions of $l_1(x)$ and $l_r(x)$, respectively. For integers $u_1 > u_2 > \cdots > u_r > 0$, let $\mathrm{mod}(u_1, \ldots, u_r)$ be the function defined by reduction mod u_1 followed by reduction mod u_2 followed by $\ldots$ followed by reduction mod u_r.

Lemma 8.3.1 (Dennenberg, Gurevich, and Shelah [219], Koblitz [455]) *Let r be an arbitrary positive integer and let $n_0 = e_r(8)$. Let $X \subset \{1, \ldots, n\}$ where $n \geq n_0$ and $|X| = k$. Then, there exist integers $n > u_1 > u_2 > \cdots > u_r > 0$ such that $u_r < \frac{1}{2}k^2 l_r(n) + k^{2+\varepsilon}$ and the function $\mathrm{mod}(u_1, \ldots, u_r)$ is an injective map on X.*

Proof Let $B(u)$ denote the least common multiple of all positive integers less than u: $B(u) = \mathrm{l.c.m.}(1, 2, \ldots, u - 1)$. We use the fact that if u is an integer ≥ 8, then $B(u) > 2^u$. To see this, note that the Chebyshev ψ-function is related to $B(u)$ as follows: $\exp(\psi(x)) = B([x] + 1)$, and so it suffices to know that $\psi(x) > (\ln 2)x$ for $x \geq 7$. By the main result of Rosser and Schoenfield [595] we have $\psi(x) > (1 - \frac{8.6853}{\ln^2 x})x$, and this is greater than $(\ln 2)x$ for $x \geq 175$. For $175 \geq u \geq 8$, the inequality $B(u) > 2^u$ is quickly checked by hand.

Define u_1 as the smallest positive integer such that $\mathrm{mod}(u_1)$ is an injective map on X. Let A be the least common multiple of all differences $|x - y|$ as x and y range over pairs of distinct elements of X. Then,

$$A \leq \prod_{x, y \in X, x > y} (x - y) < n^{k(k-1)/2}.$$

But $B(u_1)$ must divide A, since otherwise there would be a $j < u_1$ that does not divide A, and hence does not divide any of the differences $x - y$, contrary to the definition of u_1. Thus, $B(u_1) \leq A$, and so $2^{u_1} < n^{k(k-1)/2}$. Taking $\log_2$ of both sides gives

$$u_1 < k(k - 1)/2 \log_2(n) < \frac{1}{2}k^2 \log_2(n).$$

Now repeat the argument in the last paragraph with the set X replaced by its image X_1 under reduction modulo u_1, with n replaced by u_1, etc. One finds u_2 such that $\mathrm{mod}(u_1, u_2)$ is an injective map on X and such that

$$u_2 < \frac{1}{2}k^2 \log_2(u_1) < \frac{1}{2}k^2 \log_2(k^2 \log_2(n)) = \frac{1}{2}k^2 \log_2(\log_2(n)) + k^2 \log_2(k).$$

The lemma then follows by induction on r. ☐

Corollary 8.3.1 *Choose r to be the maximum possible value such that $n \geq e_r(8)$. Then, $\mathrm{mod}(u_1, \ldots, u_r)$ is a function from $\{1, \ldots, n\}$ to the set $J(k)$ consisting of the first $Ck^{2+\varepsilon}$ non-negative integers, where C is independent of k and n.*

Let $\mathcal{H}(n, k)$ denote the family of hash functions $h : \{1, \ldots, n\} \to J(k)$ based on all possible sequences $(u_1, \ldots, u_r)$. The size of this collection can be bounded:

$$|\mathcal{H}(n, k)| \leq k^{2\log^* n}(\log n)^2 \leq Ck^{\log k}(\log n)^3$$

for some constant C independent of k and n.

Note that for $h \in \mathcal{H}(n, k)$ and $x \in \{1, \ldots, n\}$, the value of $h(x)$ can be computed in time $O(\log n)^3$.

8.3.1 MULTIDIMENSIONAL MATCHING

We will apply the results of the previous section, and the method of color coding to another problem.

MULTIDIMENSIONAL MATCHING

Input: A set $M \subseteq X_1 \times \cdots \times X_r$, where the X_i are disjoint sets.
Parameter: The positive integer r and a positive integer k.
Question: Is there a subset $M' \subseteq M$ with $|M'| = k$, such that no two elements of M' agree in any coordinate?

Theorem 8.3.1 (Downey, Fellows, and Koblitz [247, 249]) MULTIDIMENSIONAL MATCHING *is fixed-parameter tractable.*

Proof Let N denote the number of distinct coordinates that appear in the r-tuples of M, and assume that these coordinates are $\{1, \ldots, N\}$. Let $K = kr$. Let n denote the total size of the description of M.

Define a *solution schema* S for the problem to be a set of k r-tuples with all coordinates distinct, and with the coordinates chosen from $J(K)$. If α is an r-tuple in M, $\alpha = (x_1, \ldots, x_r)$, let $h(\alpha)$ denote the image r-tuple, $h(\alpha) = (h(x_1), \ldots, h(x_r))$.

Algorithm 8.3.1
1. For each $h \in \mathcal{H}(N, K)$ and each solution schema S:
2. Compute $h(\alpha)$ for each $\alpha \in M$.
3. Determine whether the solution schema S is *realized*, meaning that every r-tuple in S occurs as the image $h(\alpha)$ of some $\alpha \in M$. If so, answer "yes".
4. If no hash function h realizes a solution schema, then answer "no".

The correctness of the algorithm is nearly immediate. If M has a k-matching, then this involves $K = kr$ distinct coordinates, and by the lemma, there must be some $h \in \mathcal{H}(N, K)$ that maps these invectively to $J(K)$, and the image under h is the solution schema that causes the algorithm to output "yes". Conversely, if there is an h that realizes a solution schema S, then choosing one r-tuple in each of the k preimages yields a matching.

The running time of the algorithm is bounded by $O(K!K^{3K+1}n(\log n)^6)$.   □

The running time for the algorithm above can be improved simply by using a better family of hash functions.

By means of a variety of sophisticated methods, Alon, Yuster, and Zwick [36] have proved the following two theorems.

Theorem 8.3.2 (Alon, Yuster, and Zwick [36]) *There is a k-perfect family of hash functions from $\{1, \ldots, n\}$ to $\{1, \ldots, k\}$ of size $2^{O(k)} \log n$.*

Theorem 8.3.3 (Alon, Yuster, and Zwick [36]) *There is a k-perfect family of hash functions from $\{1, \ldots, n\}$ to $\{1, \ldots, k^2\}$ of size $k^{O(1)} \log n$.*

The problem with these results is that the hidden constant in the O-notation of this bound is quite big. Instead, we may use the following recent results.

Theorem 8.3.4 (Chen, Lu, Sze, and Zhang [151])
1. *There is a family of k-perfect hash functions $\mathcal{F}$ where $|\mathcal{F}| = O(6.4^k \cdot n)$.*
2. *Moreover, this collection can be constructed in $O(6.4^k \cdot n)$ steps.*

These families of k-perfect hash functions are actually constructible in practice and the hidden constants are not unreasonable. Unfortunately the proofs are too intricate to be included here, although they are quite elementary. We refer the reader to [151] for details.

Using these families we can obtain

Corollary 8.3.2 k-PATH *is solvable in time $O^*(12.8^k)$.*

Chen et al. [151] also gives lower bounds on the size of a family of k-perfect hash functions, indicating the limits of the color-coding technique. In the next sections we will investigate further developments, firstly using group algebras to improve running times, and then looking at a related method of Chen et al. [151], which combines the above with a divide and conquer technique.

8.4   Speeding Things up with Multilinear Detection

The above was how matters stood until the dramatic insight of Koutis [460], which was: that by using algebraic methods, the run times could be improved greatly in the randomized case. These methods use only a polynomial number of random bits. Currently there is no known de-randomization method for Koutis' approach.

Koutis' insight was the following. He noted that color coding worked in two parts. First there was the randomized coloring of the objects, and then the solution by dynamic programming. As we will see below, dynamic programming can be considered the expansion of a polynomial into a sum of product form. *If this is done in the right algebra*, then you can throw away certain squared terms, as they have no effect on the outcome, and looking for "colorful" solutions corresponds to detecting multilinear forms. As noted by Koutis, addition corresponds roughly to set union, and multiplication corresponds to joining partially colored structures. *If a join uses a color more than once, it can be discarded; and this is why squares are irrelevant to solutions.* Koutis and then Williams used this method, analyzing the particular dynamic programming, to find the correct polynomial and then solve pretty well any color-coding problem using algebraic methods.

Koutis' methods were improved somewhat by Williams [666], and Koutis–Williams [461] and these analyses we will follow. We now turn to some formal details.

8.4.1 Group Algebras over $\mathbb{Z}_2^k$

We will consider the group $\mathbb{Z}_2^k$ of length k binary vectors with group multiplication being entry-wise addition modulo 2. For a ring R, the group algebra $R[\mathbb{Z}_2^k]$ is the set of all linear combinations $\sum_{v \in \mathbb{Z}_2^k} r_v v$ with $r_v \in R$. We define addition as inherited from R: $\sum_{v \in \mathbb{Z}_2^k} r_v v + \sum_{w \in \mathbb{Z}_2^k} r_v' v = \sum_{v \in \mathbb{Z}_2^k} (r_v + r_v')v$; scalar multiplication similarly: $\alpha \sum_{v \in \mathbb{Z}_2^k} r_v v = \sum_{v \in \mathbb{Z}_2^k} (\alpha r_v)v$. Finally, multiplication is defined via the usual convolution familiar from polynomial multiplication:

$$\left(\sum_{v \in \mathbb{Z}_2^k} r_v v \right) \cdot \left(\sum_{v \in \mathbb{Z}_2^k} r_v' v \right) = \sum_{v,w \in \mathbb{Z}_2^k} (r_v r_w')(vw).$$

It is easy to verify the axiom that if R is a commutative ring then so is $R[\mathbb{Z}_2^k]$ (Exercise 8.8.10).

The algorithm below works with the setting $R = GF(2^\ell)$ where this denotes the Galois field with 2^ℓ many elements. Let $\mathbf{0}$ denote the all-zero vector in $\mathbb{Z}_2^k$. Algebraic facts we need are that if $v \in \mathbb{Z}_2^k$ then $v^2 = \mathbf{0}$.

Theorem 8.4.1 (Williams [666], improving Koutis [460]) *Let $P = P(x_1, \ldots, a_n)$ be a polynomial of degree at most k represented by an arithmetical circuit of size $s(n)$ with $+$ gates of unbounded fan-in and $\times$ gates of fan-in at two. Choose a suitable field $F = GF[2^\ell]$ of characteristic 2, as described below. There is a randomized algorithm that runs on every such P in time $O^*(2^k s(n))$ and outputs yes with high probability if there is a multilinear term in the sum-product expansion of P, and always answers no if there is no such term.*

Before we prove Theorem 8.4.1, we will look at how to apply the result for the k-PATH problem. Let G be a graph with vertex set $1, \ldots, n$ and with m edges. For the vertices, define corresponding variables $x_1, \ldots, x_n$, and A the adjacency of matrix of G. Now define a matrix $B[i, j] = A[i, j]x_i$. Let $\mathbf{1}$ be the row vector of n-ones, and $\overline{x}$ the column vector defined as $\overline{x}[i] = x_i$. Define the k-*walk polynomial* as $P_k(x_1, \ldots, x_n) = \mathbf{1} \cdot B^{k-1}\overline{x}$. Then the reader can verify (Exercise 8.8.11) that

$$P_k(x_1, \ldots, x_n) = \sum_{i_1, \ldots, i_k \text{ a walk in } G} x_{i_1} \cdots x_{i_k}.$$

Then G has a k path iff $P_k(x_1, \ldots, x_n)$ has a multilinear term when expanded in sum-product form. The reason that Theorem 8.4.1 gives an efficient randomized algorithm is that P_k can be implemented using a circuit of size $s(n) = O(k(m+n))$.

Proof The proof of Theorem 8.4.1 relies on the following algorithm.

Algorithm 8.4.1 Pick n vectors $v_1, \ldots, v_n$ randomly from $\mathbb{Z}_2^k$. For each multiplication gate g_i of the circuit for P, uniformly pick at random $w_i \in F \setminus \{0\}$. Insert a new gate that multiplies the output of g_i by w_i and feeds that outcome into those gates which read g_i in the circuit for P. Let P' be the polynomial for the new circuit. Output yes if $P'(\mathbf{0} + v_1, \ldots, \mathbf{0} + v_n) \neq 0$, where $\mathbf{0}$ denotes the all-zero vector.

Williams in [666] notes that in the case of the k-PATH polynomial, the algorithm above corresponds to picking random $w_{i,j,c} \in F$ for all $c \in \{1, \ldots, k-1\}$ and $i, j \in \{1, \ldots, n\}$ and defining $B_c[i, j] = w_{i,j,c}B[i, j]$. Then evaluating $P_k'(x_1, \ldots, x_n) = \mathbf{1} \cdot B_{k-1} \ldots B_1(x_1, \ldots, x_n)$ for randomly chosen vectors.

The proof of correctness of Algorithm 8.4.1 works by the key observation of Koutis [460]. *If* $v_i \in \mathbb{Z}_2^k$, *then*

$$(\mathbf{0} + v_i)^2 = \mathbf{0}^2 + 2v_i + v_i^2 \cong 0 \pmod{2}.$$

Thus all squares in P vanish in $P'(\mathbf{0} + v_1, \ldots, \mathbf{0} + v_n)$ as F has characteristic 2. Hence if P has no multilinear term then it must evaluate to 0.

It remains to show that if P has a multilinear term in its sum-product expansion, then P' evaluates to a nonzero value with probability at least $\frac{1}{5}$. We may assume that every multilinear term has degree at least k, and if it has one, it has one of degree exactly k. If not, for $j = 1, \ldots, k$ in turn, multiple the circuit for P by j new variables $x_{n+1}, \ldots, x_{n+j}$, and get a new polynomial P^j, which is the input to the randomized algorithm. If P has a term of degree $k' < k$, then $P^{k-k'}$ will have a term of degree exactly k. We can also assume that each multilinear form in the sum-product expansion of P has the form $cx_{i_1} \ldots x_{i_d}$, where $d \geq k$, and $c \in \mathbb{Z}$. For such a term, there corresponds a collection of multilinear terms in P', of the form $w_1 \ldots w_{d-1} \prod_{i=1}^{d}(\mathbf{0} + v_{i_j})$. Since each sequence of gates $g_1, \ldots, g_{d-1}$ is distinct, each $w_1, \ldots, w_{d-1}$ is distinct for each term in the collection.

Lemma 8.4.1 (Koutis [460]) *If $v_1, \ldots, v_q$ are linearly independent over GF[2], then $\prod_{i=1}^{q}(\mathbf{0} + v_i) = 0$ in $F[\mathbb{Z}_2^k]$.*

Proof If $v_1, \ldots, v_q$ are dependent, take the collection C, which sum to the zero vector mod 2 (and note that in $F[\mathbb{Z}_2^k]$), this means $\prod_{i \in C} v_i = \mathbf{0}$. If $C' \subseteq C$, and we multiply both sides by $\prod_{j \in (C \setminus C')} v_j$, we see $\prod_{j \in C'} v_j = \prod_{j \in (C \setminus C')} v_j$. Thence, $\prod_{i \in C}(\mathbf{0} + v_i) = \sum_{C' \subseteq C}(\prod_{j \in C'} v_j) = 0 \mod 2$, and hence $\prod_{i=1}^{q}(\mathbf{0} + v_i) = 0$ over $F[\mathbb{Z}_2^k]$, as F has characteristic 2. $\square$

Lemma 8.4.2 Koutis–Williams [666] *If $v_1, \ldots, v_q$ are linearly independent over GF[2], then $\prod_{i=1}^{q}(\mathbf{0} + v_i) = \sum_{v \in \mathbb{Z}_2^k} v$.*

Proof If $v_1, \ldots, v_q$ are independent, the $\prod_{i=1}^{q}(\mathbf{0} + v_i)$ is a sum of all vectors in the span of $v_1, \ldots, v_q$; as all vectors in this span are of the form $\prod_{j \in C'} v_j$ for some $C' \subseteq \{1, \ldots, q\}$, and by independence there is a unique way to generate each such vector. $\square$

Since and $q > k$ vectors are dependent, from these two lemmas we conclude that $P'(\mathbf{0} + v_1, \ldots, \mathbf{0} + v_n)$ will either evaluate to 0, or some $c \sum_{v \in \mathbb{Z}_2^k} v$ for some $c \in F$. To finish the argument, we need to show that with probability $\geq \frac{1}{5}$, if P has a multilinear term, this will have $c \neq 0$.

The probability that $v_{\ell_1}, \ldots, v_{\ell_k}$ chosen for the variables in a multilinear term of P chosen at random are independent is at least $\frac{1}{4}$ (since the probability that a random $k \times k$ matrix over $GF[2]$ has full rank is at least $0.28 > \frac{1}{4}$) (Exercise 8.8.12). Hence in $P'(\mathbf{0} + v_1, \ldots, \mathbf{0} + v_n)$ there is at least one multilinear term corresponding to an independent set that is at least $\frac{1}{4}$.

To finish the argument, suppose that C is the collection of multilinear terms of P which correspond to k independent vectors in $P'(\mathbf{0} + v_1, \ldots, \mathbf{0} + v_n)$. Note that $c = \sum_i c_i$ for those $c_i \in F$ corresponding to the ith multilinear term of C. We claim that $\Pr(c = 0 | C \neq \emptyset) \leq \frac{1}{8}$. Note that each c_i comes from a product of $k - 1$ elements $w_{i,1}, \ldots, w_{i,k-1}$, corresponding to some gates $g_{i,1}, \ldots, g_{i,k-1}$ of the circuit.[1] Thinking of the w_i as variables, then the sum $Q(w_1, \ldots, w_{s(n)}) = \sum_i c_i$ is a degree k polynomial over F. Since $C \neq \emptyset$, the polynomial Q is not identically 0. At this point, we will invoke a classical result in computational complexity.[2] It gives a quantitative version of the intuition that in a big field F, if we randomly choose elements and evaluate a polynomial Q at those elements, we would be unlikely to evaluate to zero.

---

[1] Williams [666] remarks that in the case of k-PATH, each c_i is a sum of products of the form $y_{i_1,i_2,1} y_{i_2,i_3,2} \cdots y_{i_{k-1},i_k,k-1}$.

[2] This was a key observation of Williams [666]. Namely, we could use the Schwartz–Zippel Lemma by running the construction in a large characteristic 2 field.

Lemma 8.4.3 (Schwartz–Zippel [606]) *Let F be any field and choose a nonzero degree d polynomial Q in $F[x_1, x_2, \ldots, x_n]$. Let F' be a finite subset of F and let $r_1, \ldots, r_n$ be chosen randomly from F'. Then $\Pr[Q(r_1, r_2, \ldots, r_n) = 0] \leq \frac{d}{|F'|}$.*

Proof The proof is quite short. We use induction on n. For $n = 1$, Q can have at most d roots and we are done. For an induction, assume the result for $n - 1$, and re-write $Q(x_1, \ldots, x_n)$ as

$$Q(x_1, \ldots, x_n) = \sum_{i=0}^{d} x_1^i \, Q_i(x_2, \ldots, x_n).$$

By the induction hypothesis, there is some largest i such that Q_i is not identically 0. Note $\deg Q_i \leq d - i$.

Randomly pick $r_2, \ldots, r_n$ from F'. By induction, $\Pr[Q_i(r_2, \ldots, r_n) = 0] \leq \frac{d-i}{|F'|}$. If $Q_i(r_2, \ldots, r_n) \neq 0$, then $Q(x_1, r_2, \ldots, r_n)$ has degree i so $\Pr[Q(r_1, r_2, \ldots, r_n) = 0 | Q_i(r_2, \ldots, r_n) \neq 0] \leq \frac{i}{|F'|}$. The proof is finished with a simple calculation. $\Pr[Q(r_1, r_2, \ldots, r_n) = 0] = \Pr[(Q(r_1, r_2, \ldots, r_n) = 0) \wedge (Q_i(r_2, \ldots, r_n) = 0)] + \Pr[(Q(r_1, r_2, \ldots, r_n) = 0) \wedge (Q_i(r_2, \ldots, r_n) \neq 0)]$. Now, $\Pr[(Q(r_1, r_2, \ldots, r_n) = 0) \wedge (Q_i(r_2, \ldots, r_n) = 0)] \leq \Pr[Q_i(r_2, \ldots, r_n) = 0] \leq \frac{d-i}{|F'|}$. Also $\Pr[(Q(r_1, r_2, \ldots, r_n) = 0) \wedge (Q_i(r_2, \ldots, r_n) \neq 0)] = \Pr[(Q_i(r_2, \ldots, r_n) \neq 0)] \Pr[Q(r_1, r_2, \ldots, r_n) = 0 | (Q_i(r_2, \ldots, r_n) \neq 0)] \leq \frac{i}{|F'|}$. Thus,

$$\Pr[Q(r_1, r_2, \ldots, r_n) = 0] \leq \frac{d - i}{|F'|} + \frac{i}{|F'|} \leq \frac{d}{|F'|}. \qquad \square$$

Now in the proof at hand, we apply the Lemma 8.4.3 in the case that $F = F'$, the random assignment results in 0 with probability at most $\frac{k}{|F|} \leq \frac{1}{8}$, if we choose $\ell \geq 3 + \log k$. Since $C \neq \emptyset$ with probability at least $\frac{1}{4}$, the overall probability of a "yes" is at least $\frac{1}{4}(1 - \frac{1}{8}) > \frac{1}{5}$.

To conclude the proof, we calculate the running time. The evaluation of the polynomial P' takes $O(s(n))$ many arithmetical operations. To do this in $F[\mathbb{Z}_2^k]$, we will interpret elements as vectors in F^{2^k}. Using component-wise sum, addition in F^{2^k} takes at most time $O(2^k \log |F|)$. For multiplication here we choose, as above, $\ell = 3 + \log k$. Elements of $F = GF[2^\ell]$ are represented as univariant (over x) degree ℓ polynomials using the standard construction of $GF[2^\ell]$ as a field extension. It is possible using fast Fourier transforms[3] to show that the running time is bounded by

---

[3]In particular, in $\mathbb{C}[x]$ to multiply polynomials u, v, say, by H_k, the discrete Fourier transform over $\mathbb{Z}_2^k$ runs in time $O(k2^k M(\ell))$ where $M(\ell)$ is the runtime for multiplying two polynomials in $GF[2][x]$. Note that $M(\ell) \leq O(\ell^2)$. Taking the pointwise product of the two resulting vectors to get the vector w, which is multiplied by $H_k = H_k^1$, giving a vector z containing 2^k polynomials each of degree at most 2ℓ. These are reduced modulo an irreducible degree ℓ polynomial over $GF[2]$ in time $O(2^\ell M(\ell))$, mapping the computations in $\mathbb{C}[x]$ down to $GF[2^\ell]$. This give the desired time of $O^*(2^k s(n))$.

$O^*(2^k s(n))$. In some sense, the methods are not important, in that they are applied as a black box: it is well-known that multiplication in the group algebra $GF[2^\ell][\mathbb{Z}_2^k]$ can be achieved efficiently. Koutis [460] gives another efficient method, based upon representation theory, of running the algebra, and only uses space $O(\text{poly}(n, k))$. $\square$

Koutis and Williams [461] prove the following generalized version of Theorem 8.4.1 which promises many applications. In the following, X will be a set of variables.

Theorem 8.4.2 (Koutis and Williams [461]) *Let $P(X, z)$ be a polynomial representing a commutative arithmetical circuit C. Then there is a randomized Monte-Carlo algorithm with success probability $\frac{1}{5}$, which will decide in time $O^*(2^k t^2 |C|)$ and space $O(t|C|)$ whether P contains a term of the form $z^t Q(X)$ where $Q(X)$ is multilinear, monomial, and has degree at most k.*

The proof of Theorem 8.4.2 is similar to that of Theorem 8.4.1, but is omitted for space considerations.

We finish this section with one application of Theorem 8.4.2, giving an attractive improvement of Theorem 8.3.4.

Theorem 8.4.3 (Koutis and Williams [461]) r-MULTIDIMENSIONAL MATCHING *is solvable in randomized time $O^*(2^{(r-1)k})$, with polynomially many random bits.*

Proof Encode each element u of $X_2 \cup X_r$ by a variable x_u. Encode each $r - 1$-tuple $t = (u_2, \ldots, u_r) \in X_2 \times \cdots \times X_r$ by the monomial $M_t = \prod_{i=2}^r x_{u_i}$. Let $X_1 = \{u_{1,1}, \ldots, u_{1,n}\}$, and let T_j denote the subset of r-tuples whose first coordinate is $u_{1,j}$. The relevant polynomial is

$$P(X, z) = \prod_{j=1}^n \left(1 + \sum_{t \in T_j} (z \cdot M_t) \right),$$

where z is a new indeterminant. Then the coefficient of z^k is a polynomial $Q(X)$ which contains a multilinear $((m - 1)k)$-term iff there is a k matching. Now we apply Theorem 8.4.2. $\square$

8.5   Randomized Divide and Conquer

A related method called *randomized divide and conquer* is due to Chen, Kneis, Lu, Molle, Richter, Rossmanith, Sze, and Zhang [149]. It is described as follows.[4]

A fairly large class of NP-hard problems can be described as finding a subset S of a "base set" U, such that the subset S satisfies certain pre-specified conditions. For example, the k-PATH problem is seeking a subset S of k vertices in the vertex

---

[4]Thanks to Jianer Chen for help with this account.

set U of a given graph that makes a simple path, and the 3-D MATCHING problem is seeking a subset S of $3k$ different elements in a set U that makes k disjoint triples in the given triple set. Some of these problems can be solved based on the classical divide and conquer method, as follows. First *properly* partition the base set U to divide the given instance into two disjoint sub-instances, then recursively construct solutions to the sub-instances, finally construct a solution to the original instance from the solutions to the sub-instances.

In solving the k-PATH problem, i.e., finding a simple path P of k vertices in a graph $G = (U, E)$, we can first partition the vertex set U into two subsets U_1 and U_2, such that U_1 *contains all vertices in the first half of the path P, and U_2 contains all vertices in the second half of the path P*. Then, we recursively construct simple paths of $k/2$ vertices in the induced subgraphs $G[U_1]$ and $G[U_2]$. Finally, we concatenate two simple paths of $k/2$ vertices, one in $G[U_1]$ and one in $G[U_2]$, to obtain a simple path of k vertices in the original graph G.

Similarly, to find a matching of k triples in a triple set T, we can first partition the base set U into two subsets U_1 and U_2 *such that $3k/2$ elements in U_1 make a matching M_1 of $k/2$ triples, and $3k/2$ elements in U_2 make a matching M_2 of $k/2$ triples*. Then, obviously the union of M_1 and M_2 is a matching of k triples in T.

Note that the essential difference between the above processes and the traditional polynomial-time divide and conquer algorithms (e.g., for sorting) is that in these above processes, the partitioning of the base set U is not arbitrary. Currently there is no known polynomial-time algorithm that guarantees the desired partition of the base set U.

On the other hand, if we know in advance that the size k of the objective subset S is small, then a simple probabilistic process can be used that significantly speeds up the process of partitioning the base set U into the desired subsets. For example, a random partition of the vertex set U of a graph G into two subsets U_1 and U_2 has a probability at least $1/2^k$ such that the first half of the objective path P of k vertices is in $G[U_1]$ and the second half of P is in $G[U_2]$. Similarly, if a triple set T contains a matching of k triples, then a random partition of the base set U into two subsets U_1 and U_2 has a probability at least $1/2^{3k}$ such that $3k/2$ elements in U_1 make a matching of $k/2$ triples and $3k/2$ elements in U_2 make a matching of $k/2$ triples. Therefore, if we try sufficiently many times of the random partitions ($O(2^k)$ times for the k-*path* problem, and $O(2^{3k})$ times for the 3-D MATCHING problem), then we will have a high probability to obtain the desired partition of the base set U.

Based on this observation, we have the following basic framework for developing parameterized algorithms for this kind of NP-hard parameterized problems.

Algorithm 8.5.1
1. Loop $O(c^k)$ times.
2. Pick a random partition of the base set U that divides the given instance I into two disjoint sub-instances I_1 and I_2.
3. Recursively solve the sub-instances I_1 and I_2.
4. Construct a solution to the original instance I from the solutions to I_1 and I_2.

A careful analysis [149] shows that the above process runs in time $O(c^{2k}n^{O(1)})$. In particular, based on this framework, the k-PATH problem can be solved in time $O(4^k n^{O(1)})$ and the 3-DIMENSIONAL MATCHING problem can be solved in time $O(4^{3k}n^{O(1)})$.

The above randomized algorithms can be de-randomized based on the concept of *universal sets*. Formally, an (n, k)-*universal set* $\mathcal{U}$ is a collection of partitions of the base set U of n elements such that for *any* partition (S_1, S_2) of *any* subset S of k elements in U, there is a partition in $\mathcal{U}$ that divides the set U into two subsets U_1 and U_2 such that $S_1 \subseteq U_1$ and $S_2 \subseteq U_2$. An (n, k)-universal set can be constructed in time $O(c^{k+o(k)}n^{O(1)})$; see [149] for a detailed construction. Therefore, by replacing (in the above randomized framework) the iterating random partitions with an enumeration of the partitions in an (n, k)-universal set, we can obtain a deterministic algorithm of running time $O(c^{2k+o(k)}n^{O(1)})$.

8.6   Divide and Color

A new randomization method was introduced by Alon, Lokshtanov and Saurabh [35], which gave a randomized sub-exponential time, polynomial space, parameterized algorithm for the k-WEIGHTED FEEDBACK ARC SET IN TOURNAMENTS (k-FAST) problem. Recall that a *tournament* is a directed graph where every pair of vertices is connected by exactly one arc, and a *feedback arc set* is a set of arcs whose removal makes the graph acyclic. The problem is then specified as follows.

k-WEIGHTED FEEDBACK ARC SET IN TOURNAMENTS (k-FAST)

Instance: A tournament $T = (V, A)$, a weight function $w : A \to \{x \in \mathbb{R} : x \geq 1\}$.
Parameter: An integer k.
Question: Is there an arc set $S \subseteq A$ such that $\sum_{e \in S} w(e) \leq k$ and $T \setminus S$ is acyclic?

Alon, Lokshtanov, and Suarabh demonstrated that the algorithm can be de-randomized by slightly increasing the running time. To perform this de-randomization, Alon, Lokshtanov, and Saurabh constructed a new kind of universal hash functions, now called *universal coloring families*.

Definition 8.6.1 (Alon, Lokshtanov, and Saurabh [35])   For integers m, k and r, a family $\mathcal{F}$ of functions from $[m]$ to $[r]$ is called a universal (m, k, r)-coloring family if for any graph G on the set of vertices $[m]$ with at most k edges, there exists an $f \in \mathcal{F}$ which is a proper vertex coloring of G.

We will present these results here, following their presentation. These methods would seem to have wider applications.

By way of motivation, as pointed out by Alon, Lokshtanov, and Saurabh, in a competition where everyone plays against everyone it is uncommon that the results are acyclic and hence one cannot rank the players by simply using a topological ordering. A natural ranking is one that minimizes the number of upsets, where an upset is a pair of players such that the lower ranked player beats the higher ranked one. The problem of finding such a ranking given the match outcomes is the FEEDBACK ARC SET problem restricted to tournaments.

The technique is also a version of randomized divide and conquer, and will combine color coding with a divide and conquer algorithm with the simple k^2 kernelization below. In Exercise 8.8.15, we will mention a more recent $k\sqrt{k}$ kernelization for this problem.

For an arc weighted tournament, we define the weight function $w^* : V \times V \to \mathbb{R}$, such that $w^*(u, v) = w(uv)$ if $uv \in A$ and 0 otherwise. Given a directed graph $D = (V, A)$ and a set F of arcs in A, define $D\{F\}$ to be the directed graph obtained from D by reversing all arcs of F. In our arguments, we will need the following characterization of minimal feedback arc sets in directed graphs.

Proposition 8.6.1 *Let $D = (V, A)$ be a directed graph and F be a subset of A. Then F is a minimal feedback arc set of D if and only if F is a minimal set of arcs such that $D\{F\}$ is a directed acyclic graph.*

Given a minimal feedback arc set F of a tournament T, the ordering σ corresponding to F is the unique topological ordering of $T\{F\}$. Conversely, given an ordering σ of the vertices of T, the feedback arc set F corresponding to σ is the set of arcs whose endpoint appears before their startpoint in σ. The cost of an arc set F is $\sum_{e \in F} w(e)$, and the cost of a vertex ordering σ is the cost of the feedback arc set corresponding to σ.

For a pair of integer row vectors $\hat{p} = [p_1, \dots, p_t]$, $\hat{q} = [q_1, \dots, q_t]$, we say that $\hat{p} \leq \hat{q}$ if $p_i \leq q_i$ for all i. The transpose of a row vector $\hat{p}$ is denoted by $\hat{p}^\dagger$. The t-sized vector $\hat{e}$ is $[1, 1, \dots, 1]$; $\hat{0}$ is $[0, 0, \dots, 0]$; and $\hat{e}_i$ is the t-sized vector with all entries 0 except for the ith, which is 1. Let $\tilde{O}(\sqrt{k})$ denote, as usual, any function which is $O(\sqrt{k}(\log k)^{O(1)})$. For any positive integer m put $[m] = \{1, 2, \dots, m\}$.

8.6.1 The Algorithm and the Kernel

Here is the algorithm.

Algorithm 8.6.1
1. Kernelize to a graph with at most $O(k^2)$ vertices.
2. Randomly color the vertices of our graph with $t = \sqrt{8k}$ colors, and define the arc set A_c to be the set of arcs whose endpoints have different colors.
3. Check whether there is a weight k feedback arc set $S \subseteq A_c$.

Lemma 8.6.1 (Dom, Guo, Hüffner, Niedermeier, and Truß [225]) *k-FAST has a kernel with $O(k^2)$ vertices.*

Proof We give two simple reduction rules.

1. (Sunflower Rule) If an arc e is contained in at least $k + 1$ triangles,[5] reverse the arc and reduce k by $w(e)$.
2. If a vertex v is not contained in any triangle, delete v from T.

The first rule is safe because any feedback arc set that does not contain the arc e must contain at least one arc from each of the $k + 1$ triangles containing e, and thus must have weight at least $k + 1$. The second rule is safe because the fact that v is not contained in any triangle implies that all arcs between $N^-(v)$ and $N^+(v)$ are oriented from $N^-(v)$ to $N^+(v)$. Hence for any feedback arc set S_1 of $T[N^-(v)]$ and feedback arc set S_2 of $T[N^+(v)]$, $S_1 \cup S_2$ is a feedback arc set of T.

Finally, we show that any reduced yes instance T has at most $k(k + 2)$ vertices. Let S be a feedback arc set of T with weight at most k. The set S contains at most k arcs, and for every arc $e \in S$, aside from the two endpoints of e, there are at most k vertices that are contained in a triangle containing e, because otherwise the first rule would have applied. Since every triangle in T contains an arc of S and every vertex of T is in a triangle, T has at most $k(k + 2)$ vertices. $\qquad\square$

8.6.2 Probability of a Good Coloring

We now proceed to analyze the second step of the algorithm. What we aim for, is to show that if T does have a feedback arc set S of weight at most k, then the probability that S is a subset of A_c is at least $2^{-c\sqrt{k}}$ for some fixed constant c. We show this by showing that if we randomly color the vertices of a k edge graph G with $t = \sqrt{8k}$ colors, then the probability that G has been properly colored is at least $2^{-c\sqrt{k}}$.

Lemma 8.6.2 (Alon, Lokshtanov and Saurabh [35]) *If a graph on q edges is colored randomly with $\sqrt{8q}$ colors, then the probability that G is properly colored is at least $(2e)^{-\sqrt{q/8}}$.*

Proof Arrange the vertices of the graph by repeatedly removing a vertex of lowest degree. Let $d_1, d_2, \ldots, d_s$ be the degrees of the vertices when they have been removed. Then for each i, $d_i(s - i + 1) \le 2q$, since when vertex i is removed, each vertex had degree at least d_i. Furthermore, $d_i \le s - i$ for all i, since the degree of the vertex removed cannot exceed the number of remaining vertices at that point. Thus $d_i \le \sqrt{2q}$ for all i. In the coloring, consider the colors of each vertex one by one, starting from the last one, that is vertex number s. When vertex number i is colored, the probability that it will be colored by a color that differs from all those of its d_i neighbors following it, is at least $(1 - \frac{d_i}{\sqrt{8q}}) \ge (2e)^{-d_i/\sqrt{8q}}$ because $\sqrt{8q} \ge 2d_i$.

---

[5] A triangle in what follows means a directed cyclic triangle.

Hence the probability that G is properly colored is at least

$$\prod_{i=1}^{s}\left(1-\frac{d_i}{\sqrt{8q}}\right) \geq \prod_{i=1}^{s}(2e)^{-d_i/\sqrt{8q}} = (2e)^{-\sqrt{q/8}}.$$

$\square$

8.6.3   Solving a Colored Instance

Given a t-colored tournament T, as with standard color coding we will say that an arc set F is *colorful* if no arc in F is monochromatic. An ordering σ of T is colorful if the feedback arc set corresponding to σ is colorful. An optimal colorful ordering of T is a colorful ordering of T with minimum cost among all colorful orderings. We now give an algorithm that takes a t-colored arc weighted tournament T as input and finds a colorful feedback arc set of minimum weight, or concludes that no such feedback arc set exists.

Proposition 8.6.2 *Let $T = (V_1 \cup V_2 \cup \cdots \cup V_t, A)$ be a t-colored tournament. There exists a colorful feedback arc set of T if and only if $T[V_i]$ induces an acyclic tournament for every i.*

We say that a colored tournament T is *feasible* if $T[V_i]$ induces an acyclic tournament for every i. Let $n_i = |V_i|$ for every i and let $\hat{n}$ be the vector $[n_1, n_2, \ldots, n_t]$. Let $\sigma = v_1 v_2 \ldots v_n$ be the ordering of V corresponding to a colorful feedback arc set F of T. For every color class V_i of T, let $v_i^1 v_i^2 \ldots v_i^{n_i}$ be the order in which the vertices of V_i appear according to σ. Observe that since F is colorful, $v_i^1 v_i^2 \ldots v_i^{n_i}$ must be the unique topological ordering of $T[V_i]$. We exploit this to give a dynamic programming algorithm for the problem.

Lemma 8.6.3 (Alon, Lokshtanov and Saurabh [35]) *Given a feasible t-colored tournament T, we can find a minimum weight colorful feedback arc set in $O(t \cdot n^{t+1})$ time and $O(n^t)$ space.*

Proof For an integer $x \geq 1$, define $S_x = \{v_1, \ldots, v_x\}$ and $S_x^i = \{v_1^i, \ldots, v_x^i\}$. Let $S_0 = S_0^i = \emptyset$. Notice that for any x there must be some x' such that $S_x \cap V_i = S_{x'}$. Given an integer vector $\hat{p}$ of length t in which the ith entry is between 0 and n_i, let $T(\hat{p})$ be $T[S_{p_1}^1 \cup S_{p_2}^2 \cup \cdots \cup S_{p_t}^t]$.

For a feasible t-colored tournament T, let $\mathrm{FAS}(T)$ be the weight of the minimum weight colorful feedback arc set of T. Observe that if a t-colored tournament T is feasible, then so are all induced subtournaments of T, and hence the function FAS is well defined on all induced subtournaments of T. We proceed to prove that the following recurrence holds for $\mathrm{FAS}(T(\hat{p}))$.

$$\mathrm{FAS}\big(T(\hat{p})\big) = \min_{i\,:\,\hat{p}_i>0}\left(\mathrm{FAS}\big(T(\hat{p}-\hat{e}_i)\big) + \sum_{u \in V(T(\hat{p}))} w^*\big(v_{\hat{p}_i}^i, u\big)\right) \qquad (8.1)$$

First we prove that the left hand side is at most the right hand side. Let i be the integer that minimizes the right hand side. Taking the optimal ordering of $T(\hat{p} - \hat{e}_i)$

and appending it with $v_{\hat{p}_i}^i$ gives an ordering of $T(\hat{p})$ with cost at most $\text{FAS}(T(\hat{p} - \hat{e}_i)) + \sum_{u \in V(T(\hat{p}))} w^*(v_{\hat{p}_i}^i, u)$.

To prove that the right hand side is at most the left hand side, take an optimal colorful ordering σ of $T(\hat{p})$ and let v be the last vertex of this ordering. There is an i such that $v = v_{\hat{p}_i}^i$. Thus σ restricted to $V(T(\hat{p} - \hat{e}_i))$ is a colorful ordering of $T(\hat{p} - \hat{e}_i)$. The total weight of the edges with startpoint in v and endpoint in $V(T(\hat{p} - \hat{e}_i))$ is exactly $\sum_{u \in V(T(\hat{p}))} w^*(v_{\hat{p}_i}^i, u)$. So the cost of σ is at least the value of the right hand side of the inequality, completing the proof.

Recurrence (8.1) naturally leads to a dynamic programming algorithm for the problem. We build a table containing $\text{FAS}(T(\hat{p}))$ for every $\hat{p}$. There are $O(n^t)$ table entries; and for each entry it takes $O(nt)$ time to compute it, giving the $O(t \cdot n^{t+1})$ time bound. □

In fact, the algorithm provided in Lemma 8.6.3 can be made to run slightly faster by pre-computing the value of $\sum_{u \in V(T(\hat{p}))} w^*(v_{\hat{p}_i}^i, u)$ for every $\hat{p}$ and i using dynamic programming, and storing these pre-computed values in a table. This pre-computing process would let us reduce the time to compute a table entry using Recurrence (8.1) from $O(nt)$ to $O(t)$, yielding an algorithm that runs in time and space $O(t \cdot n^t)$.

Lemma 8.6.4 (Alon, Lokshtanov and Saurabh [35]) *For a tournament of size* $O(k^2)$, *k-FAST can be solved in expected time* $2^{O(\sqrt{k} \log k)}$ *and* $2^{O(\sqrt{k} \log k)}$ *space.*

Proof Our algorithm proceeds as described in Algorithm 8.3.1. The correctness of the algorithm follows from Lemma 8.6.3. Combining Lemmata 8.6.1, 8.6.2, 8.6.3 yields an expected running time of $O((2e)^{\sqrt{k/8}}) \cdot O(\sqrt{8k} \cdot (k^2 + 2k)^{1+\sqrt{8k}}) \leq 2^{O(\sqrt{k} \log k)}$ for finding a feedback arc set of weight at most k if one exists. The space required by the algorithm is $O((k^2 + 2k)^{1+\sqrt{8k}}) \leq 2^{O(\sqrt{k} \log k)}$. □

The dynamic programming algorithm from Lemma 8.6.3 can be turned into a divide and conquer algorithm that runs in polynomial space, at a small cost in the running time.

Lemma 8.6.5 (Alon, Lokshtanov and Saurabh [35]) *Given a feasible t-colored tournament* T, *we can find a minimum weight colorful feedback arc set in time* $O(n^{1+(t+2) \cdot \log n})$ *in polynomial space.*

Proof By expanding recurrence (8.1) $\lfloor n/2 \rfloor$ times and simplifying the right hand side, we obtain the following recurrence.

$$\text{FAS}(T(\hat{p})) = \min_{\substack{\hat{q} \geq \hat{0} \\ \hat{q}^\dagger \cdot \hat{e} = \lceil n/2 \rceil}} \left\{ \text{FAS}(T(\hat{q})) + \text{FAS}(T \setminus V(T(\hat{q}))) + \sum_{\substack{u \in V(T(\hat{q})) \\ v \notin V(T(\hat{q}))}} w^*(v, u) \right\}.$$

$$(8.2)$$

Recurrence (8.2) immediately yields a divide and conquer algorithm for the problem. Let $\mathcal{T}(n)$ be the running time of the algorithm restricted to a subtournament of T with n vertices. For a particular vector $\hat{q}$ it takes at most n^2 time to find the value of $\sum_{u \in V(T(\hat{q})), v \notin V(T(\hat{q}))} w^*(v, u)$. It follows that $\mathcal{T}(n) \leq n^{t+2} \cdot 2 \cdot \mathcal{T}(n/2) \leq 2^{\log n} \cdot n^{(t+2) \cdot \log n} = n^{1+(t+2) \cdot \log n}$. $\qquad\square$

Theorem 8.6.1 (Alon, Lokshtanov and Saurabh [35]) *k-FAST (for a tournament of size $O(k^2)$) can be solved in expected time $2^{O(\sqrt{k} \log^2 k)}$ and polynomial space. Therefore, k-FAST for a tournament of size n can be solved in expected time $2^{O(\sqrt{k} \log^2 k)} + n^{O(1)}$ and polynomial space.*

8.6.4   Derandomization with Universal Coloring Families

Recall the definition of a universal coloring family in Definition 8.6.1.

Theorem 8.6.2 (Alon, Lokshtanov and Saurabh [35]) *There exists an explicit universal $(10k^2, k, O(\sqrt{k}))$-coloring family $\mathcal{F}$ of size $|\mathcal{F}| \leq 2^{O(\sqrt{k})}$.*

For simplicity we omit all floor and ceiling signs whenever these are not crucial. No attempt is made to optimize the absolute constants in the $O(\sqrt{k})$ or in the $O(\sqrt{k})$ notation, so likely this can be improved. Whenever this is needed, we assume that k is sufficiently large.

Proof Let $\mathcal{G}$ be an explicit family of functions g from $[10k^2]$ to $[\sqrt{k}]$ so that every coordinate of g is uniformly distributed in $[\sqrt{k}]$, and every two coordinates are pairwise independent. There are known constructions of such a family $\mathcal{G}$ with $|\mathcal{G}| \leq k^{O(1)}$. Indeed, each function g represents the values of $10k^2$ pairwise independent random variables distributed uniformly in $[\sqrt{k}]$ in a point of a small sample space supporting such variables; a construction is given, for example, in Alon and Babai and Itai [31]. The family $\mathcal{G}$ is obtained from the family of all linear polynomials over a finite field with some $k^{O(1)}$ elements, as described in [31].[6]

We can now describe the required family $\mathcal{F}$. Each $f \in \mathcal{F}$ is described by a subset $T \subset [10k^2]$ of size $|T| = \sqrt{k}$ and by a function $g \in \mathcal{G}$. For each $i \in [10k^2]$, the value of $f(i)$ is determined as follows. Suppose $T = \{i_1, i_2, \ldots, i_{\sqrt{k}}\}$, with $i_1 < i_2 < \cdots < i_{\sqrt{k}}$. If $i = i_j \in T$, define $f(i) = \sqrt{k} + j$. Otherwise, $f(i) = g(i)$. Note that the range of f is of size $\sqrt{k} + \sqrt{k} = 2\sqrt{k}$, and the size of $\mathcal{F}$ is at most

$$\binom{10k^2}{\sqrt{k}} |\mathcal{G}| \leq \binom{10k^2}{\sqrt{k}} k^{O(1)} \leq 2^{O(\sqrt{k} \log k)} \leq 2^{O(\sqrt{k})}.$$

---

[6]We will simply use this as a black box, but suffice it to say that the construction techniques are similar to the other families of hash functions.

To complete the proof we have to show that for every graph G on the set of vertices $[10k^2]$ with at most k edges, there is an $f \in \mathcal{F}$ which is a proper vertex coloring of G. Fix such a graph G.

The idea is to choose T and g in the definition of the function f that will provide the required coloring for G as follows. The function g is chosen at random in $\mathcal{G}$, and is used to properly color all but at most $\sqrt{k}$ edges. The set T is chosen to contain at least one endpoint of each of these edges, and the vertices in the set T will be re-colored by a unique color that is used only once by f. Using the properties of $\mathcal{G}$ we now observe that, with positive probability, the number of edges of G which are monochromatic is bounded by $\sqrt{k}$.

Claim *If the vertices of G are colored by a function g chosen at random from $\mathcal{G}$, then the expected number of monochromatic edges is $\sqrt{k}$.*

Proof Fix an edge e in the graph G and $j \in [\sqrt{k}]$. As g maps the vertices in a pairwise independent manner, the probability that both the end points of e get mapped to j is precisely $\frac{1}{(\sqrt{k})^2}$. There are $\sqrt{k}$ possibilities for j and hence the probability that e is monochromatic is given by $\frac{\sqrt{k}}{(\sqrt{k})^2} = \frac{1}{\sqrt{k}}$. Let X be the random variable denoting the number of monochromatic edges. By linearity of expectation, the expected value of X is $k \cdot \frac{1}{\sqrt{k}} = \sqrt{k}$. $\square$

Returning to the proof of the theorem, observe that by the above claim, with positive probability, the number of monochromatic edges is upper bounded by $\sqrt{k}$. Fix a $g \in \mathcal{G}$ for which this holds, and let $T = \{i_1, i_2, \ldots, i_{\sqrt{k}}\}$ be a set of $\sqrt{k}$ vertices containing at least one endpoint of each monochromatic edge. Consider the function f defined by this T and g. As mentioned above, f colors each of the vertices in T by a unique color, which is used only once by f, and hence we only need to consider the coloring of $G \setminus T$. However, all edges in $G \setminus T$ are properly colored by g, and f coincides with g on $G \setminus T$. Hence f is a proper coloring of G, completing the proof of the theorem. $\square$

The next result was also proven in [35], and is spiritually similar; and as those authors state, should have further applications.

Theorem 8.6.3 (Alon, Lokshtanov and Saurabh [35]) *For any $n > 10k^2$ there exists an explicit universal $(n, k, O(\sqrt{k}))$-coloring family $\mathcal{F}$ of size $|\mathcal{F}| \leq 2^{O(\sqrt{k})} \log n$.*

Proof Let $\mathcal{F}_1$ be an explicit $(n, 2k, 10k^2)$-family of hash functions from $[n]$ to $10k^2$ of size $|\mathcal{F}_1| \leq k^{O(1)} \log n$. This means that for every set $S \subset [n]$ of size at most $2k$ there is an $f \in \mathcal{F}_1$ mapping S in a one-to-one fashion. The existence of such a family is well known, and follows, for example, from constructions of small spaces supporting n nearly pairwise independent random variables taking values in $[10k^2]$. Let $\mathcal{F}_2$ be an explicit universal $(10k^2, k, O(\sqrt{k}))$-coloring family, as described in Theorem 8.6.2. The required family $\mathcal{F}$ is simply the family of all compositions of a function from $\mathcal{F}_2$, followed by one from $\mathcal{F}_1$. It is easy to check that $\mathcal{F}$ satisfies the assertion of Theorem 8.6.3.   □

Combining the algorithm from Theorem 8.6.1 with the universal coloring family given by Theorem 8.6.2 yields a deterministic sub-exponential time, polynomial space, algorithm for k-FAST.

Theorem 8.6.4 (Alon, Lokshtanov and Saurabh [35]) k-FAST *can be solved in time* $2^{O(\sqrt{k})} + n^{O(1)}$ *and polynomial space.*

8.7   Solving MAX-r-SAT Above a Tight Lower Bound

In this section[7] we will examine a further example of using randomization/derandomization in parametric algorithms. In Exercise 4.11.14, we met parameterization above guaranteed values. We further explore this concept here.[8]

The Maximum r-Satisfiability Problem (MAX-r-SAT) is a classic optimization problem with a wide range of real-world applications. The task is to find a truth assignment to a multiset of clauses, each with exactly r literals, that satisfies as many clauses as possible. Even MAX-2-SAT is NP-hard by Garey, Johnson, and Stockmeyer [338] and hard to approximate as proven by Hastad [384], in strong contrast with 2-SAT, which is solvable in linear time.

It is always possible to satisfy a $1 - 2^{-r}$ fraction of a given multiset of clauses with exactly r literals each; a truth assignment that meets this lower bound can be found in polynomial time by Johnson's algorithm [419] (Exercise 8.8.14). This lower bound is *tight* in the sense that it is optimal for an infinite sequence of instances. Mahajan, Raman, and Sikdar [505] asked what the parameterized complexity is of the following problem.

MAX-r-SAT Above Tight Lower Bound
Input: A CNF formula F with m clauses with r literals with r fixed.
Parameter: A positive integer k.
Problem: Decide whether at least $(1 - 2^{-r})m + k2^{-r}$ clauses of F can be satisfied by a truth assignment.

As we will now show, Alon, Gutin, Kim, Szeider, and Yeo [33] solved MAX-r-SAT Above Tight Lower Bound by proving that this problem is FPT and moreover admits a polynomial-size kernel. The method that we will use was first introduced in [33]. This method was also applied to a number of constraint satisfaction and graph problems in [33], and, for instance, Gutin, van Iersel, Mnich, and Yeo [372]. Some such examples can be found in the exercises.

---

[7]Thanks to Gregory Gutin for material for this section.

[8]This section does require a bit of knowledge of classical probability theory, and could be omitted on a first reading.

Let F be an r-CNF formula with clauses $C_1, \ldots, C_m$ in the variables $x_1, x_2, \ldots, x_n$, let sat(F, τ) denote the maximum number of clauses of F satisfied by a truth assignment τ, and let var(C_i) be the set of variables in C_i.

For F, consider

$$X = \sum_{C \in F} \left[1 - \prod_{x_i \in \text{var}(C)} (1 + \varepsilon_i x_i) \right],$$

where $\varepsilon_i \in \{-1, 1\}$ and $\varepsilon_i = -1$ if and only if x_i is in C (as a literal).

Lemma 8.7.1 *For a truth assignment τ, we have $X = 2^r (\text{sat}(\tau, F) - (1 - 2^{-r})m)$.*

Proof Observe that $\prod_{x_i \in \text{var}(C)} (1 + \varepsilon_i x_i)$ equals 2^r if C is falsified and 0, otherwise. Thus, $X = m - 2^r (m - \text{sat}(\tau, F))$ implying the claimed formula. $\qquad \square$

After algebraic simplification $X = X(x_1, x_2, \ldots, x_n)$ can be written as

$$X = \sum_{I \in \mathcal{S}} X_I, \tag{8.3}$$

where $X_I = c_I \prod_{i \in I} x_i$, each c_I is a nonzero integer and $\mathcal{S}$ is a family of nonempty subsets of $\{1, \ldots, n\}$ each with at most r elements.

The question we address is that of deciding whether or not there are values $x_i \in \{-1, 1\}$ so that $X = X(x_1, x_2, \ldots, x_n) \geq k$. The idea is to use a probabilistic argument and show that if the above polynomial has many nonzero coefficients, that is, if $|\mathcal{S}|$ is large, this is necessarily the case, whereas if it is small, the question can be solved by checking all possibilities of the relevant variables.

In what follows, we assume that each variable x_i takes its values randomly and independently in $\{-1, 1\}$ and thus X is a random variable. We will use the following lemma whose proof we omit because of space considerations. This lemma assumes some knowledge of classical probability theory.

Lemma 8.7.2 (Alon et al. [33]) *Let Y be a real random variable and suppose that its first, second, and fourth moments satisfy $\mathbb{E}(Y) = 0$, $\mathbb{E}(Y^2) = \sigma^2 > 0$ and $\mathbb{E}(Y^4) \leq b\sigma^4$. Then*

$$\mathbb{P}\left(Y \geq \frac{\sigma}{2\sqrt{b}} \right) > 0.$$

We also use the following lemma, which is a special case of what is called the *Hypercontractive Inequality* first proved in Bonami [104], and for which, again, space considerations preclude inclusion of a proof.

Lemma 8.7.3 *Let $f = f(y_1, \ldots, y_n)$ be a polynomial of degree r in n variables $y_1, \ldots, y_n$ each with domain $\{-1, 1\}$. Define a random variable Y by choosing a vector $(\varepsilon_1, \ldots, \varepsilon_n) \in \{-1, 1\}^n$ uniformly at random and setting $Y = f(\varepsilon_1, \ldots, \varepsilon_n)$. Then $\mathbb{E}(Y^4) \leq 9^r \mathbb{E}(Y^2)^2$.*

Returning to the random variable $X = X(x_1, x_2, \ldots, x_n)$ defined in above, we prove the following:

Lemma 8.7.4 *Let $X = \sum_{I \in \mathcal{S}} X_I$, where $X_I = c_I \prod_{i \in I} x_i$ is as above and assume it is not identically zero. Then $\mathbb{E}(X) = 0$, $\mathbb{E}(X^2) = \sum_{I \in \mathcal{S}} c_I^2 \geq |\mathcal{S}| > 0$ and $\mathbb{E}(X^4) \leq 9^r \mathbb{E}(X^2)^2$.*

Proof Since the x_i's are mutually independent, $\mathbb{E}(X) = 0$. Note that for $I, J \in \mathcal{S}$, $I \neq J$, we have $\mathbb{E}(X_I X_J) = c_I c_J \mathbb{E}(\prod_{i \in I \Delta J} x_i) = 0$, where $I \Delta J$ is the symmetric difference of I and J. Thus, $\mathbb{E}(X^2) = \sum_{I \in \mathcal{S}} c_I^2$. By Lemma 8.7.3, $\mathbb{E}(X^4) \leq 9^r \mathbb{E}(X^2)^2$. □

Theorem 8.7.1 (Alon et al. [33]) *The problem* MAX-r-SAT ABOVE TIGHT LOWER BOUND *can be solved in time* $O(m) + 2^{O(k^2)}$.

Proof By Lemma 8.7.1 the problem is equivalent to that of deciding whether or not there is a truth assignment to the variables $x_1, x_2, \ldots, x_n$, so that

$$X(x_1, \ldots, x_n) \geq k. \tag{8.4}$$

Note that in particular this implies that, if X is the zero polynomial, then any truth assignment satisfies exactly a $(1 - 2^{-r})$ fraction of the original clauses. By Corollary 8.7.2 and Lemma 8.7.4, $\mathbb{P}(X \geq \frac{\sqrt{\mathbb{E}(X^2)}}{2\sqrt{b}}) > 0$, where $b = 9^r$ and $\mathbb{E}(X^2) = \sum_{I \in \mathcal{S}} c_I^2 \geq |\mathcal{S}|$; the last inequality follows from the fact that each c_I^2 is a positive integer. Therefore $\mathbb{P}(X \geq \frac{\sqrt{|\mathcal{S}|}}{2 \cdot 3^r}) > 0$. Now, if $k \leq \frac{\sqrt{|\mathcal{S}|}}{2 \cdot 3^r}$ then (8.4) and there is an assignment for which the answer to the problem is YES. Otherwise, $|\mathcal{S}| = O(k^2)$.

For any fixed r, the representation of a problem instance of m clauses as a polynomial, and the simplification of this polynomial, can be performed in time $O(m)$. If the number of nonzero terms of this polynomial is larger than $4 \cdot 8^{2r} k^2$, then the answer to the problem is YES. Otherwise, the polynomial has at most $O(k^2)$ terms and depends on at most $O(k^2)$ variables, and its maximum can be found in time $2^{O(k^2)}$. □

Corollary 8.7.1 (Alon et al. [33]) *The problem* MAX-r-SAT Above Tight Lower Bound *admits a kernel of size polynomial in k.*

Proof Let (I, k) be an instance of MAX-r-SAT ABOVE TIGHT LOWER BOUND. As in the proof of Theorem 8.7.1 we represent (I, k) as a simplified polynomial X of form (8.3), and this representation can be found in polynomial time. If the number of nonzero terms of P is larger than $4 \cdot 3^{2r} k^2$, then we take an arbitrary fixed YES-instance of the problem as the kernel; in this case we have a kernel of constant size. It remains to consider the case where the number of nonzero terms of X is at most

$4 \cdot 3^{2r} k^2$. In this case X has at most $O(k^2)$ terms and depends on at most $O(k^2)$ variables.

We may consider the pair (P, k) as an instance of the problem MAX-r-POLY: given is a polynomial X of the form (8.3) and an integer k; decide if the polynomial has a solution for which $X \geq k$. Observe that (I, k) and (P, k) are decision-equivalent instances of MAX-r-SAT ABOVE TIGHT LOWER BOUND and MAX-r-POLY, respectively. Clearly the problem MAX-r-POLY is in NP. Consequently, since MAX-r-SAT ABOVE TIGHT LOWER BOUND is NP-hard, there exists a polynomial-time many-to-one reduction from MAX-r-POLY to MAX-r-SAT ABOVE TIGHT LOWER BOUND. Thus we can find in polynomial time an instance (I', k') of MAX-r-SAT ABOVE TIGHT LOWER BOUND, which is decision-equivalent with (P, k), and in turn with (I, k). The size of (P, k) (under reasonable encoding) is polynomial in k, hence also the size of (I', k') (because the reduction from MAX-r-POLY to MAX-r-SAT ABOVE TIGHT LOWER BOUND is polynomial). Thus (I', k') is a kernel of MAX-r-SAT ABOVE TIGHT LOWER BOUND of polynomial size. $\qquad\square$

8.8 Exercises

Exercise 8.8.1 (Downey and Fellows [247]) Use the hashing technique to show that DISJOINT r-SUBSETS is fixed-parameter tractable.

DISJOINT r-SUBSETS

Instance: A collection $\mathcal{F}$ of r subsets of a set X and a positive integer k, which must answer whether there are k subsets in $\mathcal{F}$ that are disjoint.

Parameter: The pair (r, k).

Question: Are there k subsets $S_1, \ldots, S_k \in \mathcal{F}$ such that $\forall i, j \in \{1, \ldots, k\}$ with $i \neq j$, both $S_i - S_j$ and $S_j - S_i$ are nonempty?

Exercise 8.8.2 Use color coding to show that ANTICHAIN OF r-SUBSETS is fixed-parameter tractable.

DISJOINT r-SUBSETS

Instance: A collection $\mathcal{F}$ of r subsets of a set X and a positive integer k.

Parameter: The pair (r, k).

Question: Are k subsets $S_1, \ldots, S_k \in \mathcal{F}$ such that $\forall i, j \in \{1, \ldots, k\}$ with $i \neq j$, both $S_i - S_j$ and $S_j - S_i$ are nonempty?

Exercise 8.8.3 (Downey and Fellows [247]) Use the method of color coding to show that the following is FPT.

POLYCHROME MATCHING

Instance: An r-edge-colored graph G.

Parameter: A positive integer k.

Question: Is there is a partial matching in G consisting of r edges, with one edge of each color?

Exercise 8.8.4 (Reidl, Rossmanith, Sánchez, Sikdar) Construct a randomized FPT algorithm for the following.

VERTEX DISJOINT 3-CYCLES
Input: A graph $G = (G, E)$.
Parameter: A positive integer k.
Question: Does G contain at least k vertex disjoint triangles?

Exercise 8.8.5 (Reidl, Rossmanith, Sánchez, Sikdar) The algorithm for k-PATH uses k colors. Consider the following idea which uses only two colors. Randomly color the vertices into two colors, red and black, say. Assume that the graph has a k-path. Show that the probability that the k-path is split into two roughly equal parts is about 2^{-k}. Use this idea and recursion to design a randomized $O^*(4^k)$ algorithm for k-PATH.

Exercise 8.8.6 (Downey and Fellows [247]) Define a *pseudo-homomorphism* from a graph G to a graph H to be a map $f : V(G) \to V(H)$ that is onto, and that satisfies: whenever xy is an edge of H, then there is a vertex $u \in f^{-1}(x)$ and there is a vertex $v \in f^{-1}(y)$ such that $uv \in E(G)$. Use the method of hashing to show that this problem, which takes as inputs G and H, and determines whether there is a pseudo-homomorphism from G to H, is fixed-parameter tractable for the parameter H.

Exercise 8.8.7 (Becker, Bar-Yehuda and Geiger [58]) This is another example of an application of randomization. Given a graph G, consider the following rules.
Reduction rule 1: remove vertices of degree at most 1.
Reduction rule 2: bypass vertices of degree 2 (this may lead to multiple edges).
1. Prove that if S is a feedback vertex set of G with min degree at least 3, then $|E_S| \geqslant |E|/2$ where E_S contains the edges uv with u or v in S.
   (Hint: Use a double degree counting argument.)
2. For a parameter k, prove that in time $O^*(4^k)$ we can compute a set of k vertices that is a feedback vertex set with probability at least $1 - 1/\varepsilon$.

Exercise 8.8.8 (Leizhen Cai, Verbin, and Yang [128]) Recall the *firefighting* problem from Exercise 5.3.6, which takes a graph G, and a vertex $s \in V(G)$ where a fire starts burning at time $t = 0$. At each step $t \geq 1$, a firefighter can (permanently) protect a vertex, and at east step the fire spreads to all neighboring vertices of burning vertices.
   Consider the problem

SAVING ALL BUT k VERTICES
Instance: A graph $G = (V, E)$.
Parameter: An integer k.
Question: Is there a strategy to save at least $n - k$ vertices when a fire breaks out?

Prove that SAVING ALL BUT k VERTICES is FPT.

Exercise 8.8.9 (Koutis [459]) Use color coding to show that k-RESOLUTE AS-SIGNMENTS is fixed-parameter tractable.

k-RESOLUTE ASSIGNMENTS
Instance: A Boolean formula φ.
Parameter: A positive integer k.
Question: Decide if either φ is false or there is a subset $k' \leq k$ of variables, such that any truth assignment of the remaining $n - k$ variables makes φ true.

Exercise 8.8.10 Verify that if R is a commutative ring then so is $R[\mathbb{Z}_2^k]$.

Exercise 8.8.11 Verify the claim made after the statement of Theorem 8.4.1; namely if $P_k(x_1, \ldots, x_n) = \sum_{i_1, \ldots, i_k}$ a walk in G $x_{i_1} \ldots x_{i_k}$. Then G has a k path iff $P_k(x_1, \ldots, x_n)$ has a multilinear term when expanded in sum-product form.

Exercise 8.8.12 Prove that the probability that a random $k \times k$ matrix over $GF[2]$ has full rank is $\geq \frac{1}{4}$.

Exercise 8.8.13 (Koutis and Williams [461]) Use the method of multivariable detection to show that the following are solvable in randomized time $O^*(2^k)$.

k-TREE
Instance: A graph $G = (V, E)$ and a k vertex tree T.
Parameter: T.
Question: Is T isomorphic to a subgraph in G?

t-DOMINATING SET
Instance: A graph $G = (V, E)$.
Parameter: An integer t.
Question: Find a minimal set S of vertices that dominate at least t vertices of G.

Exercise 8.8.14 (Johnson [419]) Prove that it is always possible to satisfy a $1 - 2^{-r}$ fraction of a given multiset of clauses with exactly r literals each and that a truth assignment that meets this lower bound can be found in polynomial time.

Exercise 8.8.15 (Bessy, Fomin, Gaspers, Paul, Perez, Saurabh, and Thomassé [62]) Prove that FAST has a size $O(k\sqrt{k})$ kernel using the following rule: A transitive (or acyclic) module in a tournament T is a subset of vertices M such that for every vertex $u \notin M$, either $M \subseteq N^+(u)$ or $M \subseteq N^-(u)$ and the subtournament $T[M]$ is transitive.
Rule 3 (Transitive/acyclic module rule) If M is a transitive module such that T contains at most $|M|$ arcs from $N^+(M)$ to $N^-(M)$, then reverse all these arcs and decrease k accordingly.
(Hint: The intuition for the safeness of the rule is that each of these (to be) reversed arcs can be matched to distinct vertices of M to obtain a packing of vertex disjoint triangles.)

The proof that the kernel is the correct size, consists in analyzing the structure of a solution with respect to the resulting permutation. At most $2k$ vertices are affected, i.e., incident to a reversed arc. Now observe that the vertices (between two affected vertices in the permutation) form a transitive module. To estimate the number of non-affected vertices, you do counting using the span of the reversed arcs and Cauchy–Schwartz.)

Exercise 8.8.16 (Gutin)   Extend Theorem 8.7.1 and Corollary 8.7.1 for the BOOLEAN MAX-r-CONSTRAINT SATISFACTION PROBLEM described in the Historical Notes below.

Exercise 8.8.17 (Gutin, Kim, Szeider, and Yeo [371]) Consider the following problem.

2-LINEAR ORDERING ABOVE AVERAGE (2-LINEAR ORDERING-AA)

Instance: A digraph $D = (V, A)$, where each arc ij has an integral positive weight w_{ij}.

Parameter: A positive integer k.

Question: Is there an acyclic subdigraph H of D of weight at least $W/2 + k$, where $W = \sum_{ij \in A} w_{ij}$?

Prove that 2-LINEAR ORDERING-AA admits a kernel with $O(k^2)$ arcs. (This problem is not easy, and requires some knowledge of probability theory similar to that of Sect. 8.7.)

(Hint: Apply the following reduction rule to eliminate all directed cycles of length 2: For each directed cycle (i, j, i), if $w_{ij} = w_{ji}$, delete the cycle; if $w_{ij} > w_{ji}$, delete the arc ji and replace w_{ij} by $w_{ij} - w_{ji}$; and if $w_{ji} > w_{ij}$, delete the arc ij and replace w_{ji} by $w_{ji} - w_{ij}$.

Consider an ordering $\alpha : V \to \{1, \ldots, |V|\}$ and $X = \frac{1}{2} \sum_{ij \in A} x_{ij}$, where $x_{ij} = w_{ij}$ if $\alpha(i) < \alpha(j)$ and $x_{ij} = -w_{ij}$, otherwise. Show that $X = \sum \{w_{ij} : ij \in A, \alpha(i) < \alpha(j)\} - W/2$ and thus the answer to 2-LINEAR ORDERING-AA is YES if and only if there is an ordering α such that $X \geq k$.

Let $\alpha : V \to \{1, \ldots, |V|\}$ be a random ordering. Then X will be a random variable. Show that X is symmetric (i.e., $\Pr(X = a) = \Pr(X = -a)$ for each a). Prove that for the second moment of X we have $\mathbb{E}(X^2) \geq W^{(2)}/12$, where $W^{(2)} = \sum_{ij \in A} w_{ij}^2$.

Let Y be a discrete symmetric random variable. It is easy to see that $\Pr(Y \geq \mathbb{E}(Y^2)) > 0$. Using this inequality and the lower bound for $\mathbb{E}(X^2)$, prove that either the answer to 2-LINEAR ORDERING-AA for (the reduced digraph) D is YES or D has $O(k^2)$ arcs.)

8.9   Historical Notes

Color coding was introduced by Alon, Yuster, and Zwick in [36]. As mentioned in the chapter, there has been a lot of work on constructing small families of hash functions efficiently. Color coding has been used extensively in parameterized algorithm design. Applications of the same technique can be found in: Fellows, Knauer, Nishimura, Ragde, Rosamond, Stege, Thilikos, and Whitesides [292]; Demaine, Hajiaghayi, and Marx [216]; Misra, Raman, Saurabh, and Sikdar [532]; Betzler, Fellows, Komusiewicz, and Niedermeier [63] (see also Hüffner, Wernicke, and Zichner

[406] who studied algorithm engineering issues for color coding); Shlomi, Segal, Ruppin, and Sharan [617], Scott, Ideker, Karp, and Sharan [610] (both of whom studied color coding applications in computational biology); and Alon, Yuster, and Zwick [37] (for an overview of the area of color coding). Novel developments of the color-coding idea also appeared recently in Fomin, Lokshtanov, Raman, and Saurabh [321] and Alon, Lokshtanov, and Saurabh [35]. *Randomized divide and conquer* is a technique due to Chen, Kneis, Lu, Molle, Richter, Rossmanith, Sze, and Zhang [149]. At this time it is not widely applied. Randomized methods using multilinear detection are due to Koutis [460]. The improvements in the running time obtained using the Schwartz–Zippel Lemma were obtained by Williams [666], and this method was further refined in Koutis–Williams [461]. In [461], it is shown that k-monomial detection cannot be solved faster using cleverer algebra, since they use *communication complexity* to demonstrate that, for any commutative algebra $\mathcal{A}$ used to evaluate a circuit C, the lengths of elements in $\mathcal{A}$ must be at least $\Omega(\frac{2^k}{k})$ in order to perform k-monomial detection. We are not aware of any implementations of the Koutis–Williams techniques. There are also no known de-randomization methods for these techniques, and this is an important area of possible research. As we have seen, Koutis and Williams give a randomized algorithm for k-PATH that runs in time $O^*(2^k)$, by reducing it to a Multilinear Monomial testing algorithm. The best known deterministic algorithm for Multilinear Monomial testing has running time $O^*(2e)^k$. Very recently Lokshtanov and Saurabh have announced an algorithm for Multilinear Monomial testing that runs in time $O^*(4.32^k)$. For the special case of k-PATH, Lokshtanov and Saurabh[9] used fast computation of what are called representative sets for a uniform matroid to give a deterministic algorithm that runs in time $O^*(2.82^k)$.

Cygan, Nederlof, Pilipczuk, Pilipczuk, Rooij, and Wojtaszczyk [190] applied the Khoutis-Williams methodology to giving single exponential randomized algorithms for treewidth problems. Other uses of probabilistic techniques can be found throughout the literature; the material on performance guarantees above tight values is quite representative. As we mentioned in Exercise 4.11.14, the first such result was due to Raman. The material here is due to Alon, Gutin, Kim, Szeider, and Yeo [33].

With a little extra work, Alon, Gutin, Kim, Szeider, and Yeo [33] showed that MAX-r-SAT ABOVE TIGHT LOWER BOUND admits a kernel with $O(k^2)$ clauses and variables. In fact, it is possible to prove Crowston, Fellows, Gutin, Jones, Rosamond, Thomassé and Yeo [184] that MAX-r-SAT ABOVE TIGHT LOWER BOUND admits a kernel with at most $r(2k-1)$ variables.

It is not hard to extend Theorem 8.7.1 and Corollary 8.7.1 to a more general constraint satisfaction problem. Let r be a fixed positive integer; let Φ be a set of Boolean functions, each involving at most r variables; and let $\mathcal{F} = \{f_1, f_2, \ldots, f_m\}$ be a collection of Boolean functions, each being a member of Φ, and each acting on some subset of the n Boolean variables $x_1, x_2, \ldots, x_n$. The BOOLEAN MAX-r-CONSTRAINT SATISFACTION PROBLEM (corresponding to Φ), when Φ is clear

---

[9]See http://arxiv.org/abs/1304.4626.

from the context, is the problem of finding a truth assignment to the variables so as to maximize the total number of functions satisfied. As the parameter we can take the number of functions satisfied minus the expected value of this number (Exercise 8.8.16).

Feedback arc sets in tournaments are well studied. These include the combinatorial point of view in papers such as: Erdös and Moon [276]; Fernandez de la Vega [304]; Jung [424]; Reid [578]; Seshu and Reid [614]; and Spencer [626, 627], for examples. They have also long been studied for algorithmic reasons. The problem has many applications such as in psychology, in relation to *ranking by paired comparisons*; here you wish to rank some items by an objective, but you do not have access to the objective function, only to pairwise comparisons of the objects in question. An example for this setting is measuring people's preferences for food. The weighted generalization of the problem, WEIGHTED FEEDBACK ARC SET IN TOURNAMENTS is also applied in *rank aggregation*; here we are given several rankings of a set of objects, and we wish to produce a single ranking that on average is as consistent as possible with the given ones, according to some chosen measure of consistency. This problem has been studied in the context of voting such as in Borda [107], and Condorcet [201]; in machine learning (e.g. Cohen, Schapire, and Singer [161]); and in search engine ranking (e.g. Dwork, Kumar, Naor, and Sivakumar [264, 265]).

The problem of finding a feedback arc set of minimum size in an unweighted tournament is NP-hard by Alon [29]. However, even the weighted version of the problem admits a polynomial-time approximation scheme, as proven by Kenyon-Mathieu and Schudy [438]. Venkatesh Raman demonstrated the problem to be fixed-parameter tractable in [571].

Before the use of divide and color, the fastest previously known parameterized algorithm for k-FAST by Raman and Saurabh [571] ran in time $O(2.415^k \cdot k^{4.752} + n^{O(1)})$, and it was an open problem of Guo, Moser, and Niedermeier [366] whether k-FAST can be solved in time $2^k \cdot n^{O(1)}$. We give a randomized and a deterministic algorithm both running in time $2^{O(\sqrt{k}\log^2 k)} + n^{O(1)}$. The Alon, Lokshtanok, and Saurabh algorithm runs in sub-exponential time, a trait uncommon to parameterized algorithms. Likely this technique will have further applications.

8.10   Summary

We introduce the important technique of color coding. We introduce the idea of how to de-randomize the color-coding technique using hash functions and universal sets. We also show how this relates to randomized divide and conquer.

Optimization Problems, Approximation Schemes, and Their Relation to FPT

9.1   Optimization

Optimization problems are at the heart of complexity theory. In practice, we usually wish to optimize some objective function (e.g., find a least-cost tour) rather than solve a related decision problem. It is often acceptable to find an approximate solution to some desired performance ratio, especially when a computational problem has no efficient algorithm to find an exactly optimal solution. In this section, we will see that parameterized complexity and efficient approximability are deeply related.

This will be a first foray into this relationship. Later in Chap. 29, we will see that parameterized complexity allows us to establish a *tight* relationship between optimization problems and properness of problem class hierarchies, via a program called XP-*optimality*.

In this chapter, we point out some very interesting *basic* connections between the concept of parameterized tractability and of having a polynomial-time approximation scheme. This connection will allow us to tie in with the literature on these well-studied concepts. We refer the reader to Garey and Johnson [337], especially Chap. 6, for more details on polynomial-time approximation algorithms. We will also relate fixed-parameter tractability to a syntactic class called MAX SNP and another called MIN $F^+ \Pi_1(h)$. The relationship between these classes and FPT enable us to give sufficient conditions to say that a problem is FPT *from its syntactic description alone*.

9.2   Optimization Problems

The following definitions are classical and can be found in Garey and Johnson [337], for example.

R.G. Downey, M.R. Fellows, *Fundamentals of Parameterized Complexity*,
Texts in Computer Science, DOI 10.1007/978-1-4471-5559-1_9,
© Springer-Verlag London 2013

Definition 9.2.1 An *NP optimization problem* Q is either a *minimization problem* or a *maximization problem* and is given as a 4-tuple (I_Q, S_Q, f_Q, opt_Q), where

1. I_Q is the set of *input instances*. I_Q is recognizable in polynomial time.
2. $S_Q(x)$ is the set of *feasible solutions* for the input $x \in I_Q$ such that there is a polynomial p and a polynomial-time computable relation Π (p and Π only depend on Q) such that for all $x \in I_Q$, $S_Q(x)$ can be expressed as $S_Q(x) = \{y : |y| \le p(|x|) \wedge \Pi(x, y)\}$.
3. $f_Q(x, y) \in \mathbb{N}$ is the *objective function* for each $x \in I_Q$ and $y \in S_Q(x)$. The function f_Q is computable in polynomial time.
4. $opt_Q \in \{\max, \min\}$.

An *optimal solution* for an input instance $x \in I_Q$ is a feasible solution $y \in S_Q(x)$ such that $f_Q(x, y) = opt_Q\{f_Q(x, z) : z \in S_Q(x)\}$. We write

$$opt_Q(x) = opt_Q\{f_Q(x, z) : z \in S_Q(x)\}.$$

An algorithm A is an *approximation algorithm* for Q if, given any input instance x in I_Q, A finds a feasible solution $y_A(x)$ in $S_Q(x)$. A maximization problem (resp. minimization problem) Q is *polynomial-time approximable to a ratio $r(n)$* if there is a polynomial-time approximation algorithm A for Q such that the *relative error*

$$\mathrm{Re}_A(x) = \frac{opt_Q(x) - f_Q(x, y_A(x))}{f_Q(x, y_A(x))}$$

$$\left(\text{resp. } \mathrm{Re}_A(x) = \frac{f_Q(x, y_A(x)) - opt_Q(x)}{opt_Q(x)} \right)$$

is bounded by $r(|x|) - 1$ for all input instances $x \in I_Q$. An optimization problem Q has a *polynomial-time approximation scheme* if there is an algorithm A that takes a pair $x \in I_Q$ and $\varepsilon > 0$ as input and outputs a feasible solution in $S_Q(x)$ whereby the relative error is bounded by ε for all input instances $x \in I_Q$ and the running time of A is bounded by a polynomial of $|x|$ for each fixed ε. If the running time of the algorithm A is bounded by a polynomial of $|x|$ and $1/\varepsilon$, then the optimization problem Q has a *fully polynomial-time approximation scheme*.

9.3　How FPT and Optimization Problems Relate: Part I

In this section, we will demonstrate an interesting relationship between the fixed-parameter tractability and approximability of NP optimization problems. We will show the FPT for some well-known NP optimization classes, and investigate the approximability of NP optimization problems based on their fixed-parameter tractability.

This relationship has more than passing interest. Aside from the original work in Garey and Johnson [337], there have been few advances enabling one to prove that an approximation problem has no good approximation algorithm, even assuming $P \neq NP$. The celebrated work of S. Arora, C. Lund, R. Motwani, M. Sudan, and M. Szegedy [47] did demonstrate that many NP optimization problems do not have good approximation algorithms unless $P \neq NP$. We shall see that the FPT status of an NP optimization problem often implies nonapproximability of the problem. This result applies to many problems *not* covered by the Arora et al. results. Thus, the study of fixed-parameter complexity provides a new and potentially powerful approach to proving nonapproximability of NP optimization problems. Our concern will be parameterized problems associated with *optimization problems*. To this end, for a relation R from $\{\geq, \leq, =\}$, for example, we can associate a natural decision problem.

DECISION PROBLEM ASSOCIATED WITH AN NP OPTIMIZATION PROBLEM $Q = (I_Q, S_Q, f_Q, opt_Q)$.

Input: $x \in I_Q$.
Parameter: A positive integer k.
Question: Does $R(opt_Q(x), k)$ hold?

We write Q_R for the decision problem associated with R. The complexity of these parameterized problems ($Q_=$; $Q_{\leq}$; and $Q_{\geq}$) have been studied in the literature. For example, Leggett and Moore [483] showed that for many NP optimization problems Q, the problems $Q_{\leq}$ and $Q_{\geq}$ are in $NP \cup co\text{-}NP$, whereas the problem $Q_=$ is not in $NP \cup co\text{-}NP$ unless $NP = co\text{-}NP$. With the following elementary argument, Liming Cai and Chen [121] demonstrated that the problems $Q_=$, $Q_{\leq}$, and $Q_{\geq}$ have the same complexity *from the viewpoint of fixed-parameter tractability*.

Lemma 9.3.1 (Cai and Chen [121]) *Let Q be an arbitrary NP optimization problem. Then the three parameterized problems $Q_=$, $Q_{\leq}$, and $Q_{\geq}$ are either all (strongly) (uniformly) fixed-parameter tractable, or all (strongly) (uniformly) fixed-parameter intractable.*

Proof We argue for uniformly FPT; the others are similar. Suppose that the parameterized problem $Q_{\leq}$ is fixed-parameter tractable. Thus, there is an algorithm $A_{\leq}$ that solves the problem $Q_{\leq}$ in time $O(f(k)n^c)$, where f is a function and c is a constant independent of the parameter k. To show that the parameterized problem $Q_=$ is fixed-parameter tractable, note that the condition $k = opt_Q(x)$ is equivalent to the condition: $(k \leq opt_Q(x))$ and $((k+1) > opt_Q(x))$. Therefore, given an instance $\langle x, k \rangle$ for the parameterized problem $Q_=$, we can run the algorithm $A_{\leq}$ on the input $\langle x, k \rangle$ and on the input $\langle x, k+1 \rangle$ to decide whether $k = opt_Q(x)$. The total running time is obviously bounded by $O((f(k) + f(k+1))n^c)$. Thus, the problem $Q_=$ is fixed-parameter tractable.

Suppose that the problem $Q_=$ is fixed-parameter tractable. Then there is an algorithm $A_=$ that solves the problem $Q_=$ in time $O(f'(k)n^{c'})$, where f' is a computable function and c' is a constant independent of the parameter k. Now, given an instance $\langle x, k \rangle$ of the parameterized problem $Q_{\leq}$, we run the algorithm $A_=$ on k inputs $\langle x, 0 \rangle, \langle x, 1 \rangle, \ldots, \langle x, k-1 \rangle$. It is easy to see that $\langle x, k \rangle$ is a no-instance of the parameterized problem $Q_{\leq}$ if and only if the algorithm $A_=$ concludes yes for at least one of those k inputs. Note that the running time of the above process is bounded by $O(kf'(k)n^{c'})$. Thus, the problem $Q_{\leq}$ is fixed-parameter tractable.

Therefore, the parameterized problem $Q_=$ is fixed-parameter tractable if and only if the parameterized problem $Q_{\leq}$ is fixed-parameter tractable. In a similar way, we can show that the parameterized problem $Q_{\geq}$ is fixed-parameter tractable if and only if the parameterized problem $Q_=$ is fixed-parameter tractable.   □

The key result tying FPT with approximability is the following theorem, which extends earlier work of Liming Cai [116] on pseudo-polynomial time and FPT.

Theorem 9.3.1 (Bazgan [57], Cai [116], Liming Cai and Chen [121]) *If an NP optimization problem has a fully polynomial-time approximation scheme, then it is fixed-parameter tractable.*

Proof Suppose that an NP optimization problem $Q = (I_Q, S_Q, f_Q, opt_Q)$ has a fully polynomial-time approximation scheme A. Then, algorithm A takes as input, both an instance $x \in I_Q$ of Q and an accuracy requirement $\varepsilon > 0$, and then outputs a feasible solution in time $O(p(1/\varepsilon, |x|))$ such that the relative error $\mathrm{Re}_A(x)$ is bounded by ε, where p is a two-variable polynomial.

First, we assume that Q is a maximization problem. By Lemma 9.3.1, we only need to show that the problem $Q_{\leq}$ is fixed-parameter tractable. Given an instance $\langle x, k \rangle$ for the parameterized problem $Q_{\leq}$, we run algorithm A on input x and $1/2k$. In time $O(p(2k, |x|))$, algorithm A produces a feasible solution $y \in S_Q(x)$ such that

$$\mathrm{Re}_A(x) = \frac{opt_Q(x) - f_Q(x, y)}{f_Q(x, y)} \leq \frac{1}{2k} < \frac{1}{k}.$$

If $k \leq f_Q(x, y)$, then certainly $k \leq opt_Q(x)$ since Q is a maximization problem. On the other hand, if $k > f_Q(x, y)$, then $k - 1 \geq f_Q(x, y)$. Combining this with the inequality $(opt_Q(x) - f_Q(x, y))/f_Q(x, y) < 1/k$, we get $k > opt_Q(x)$ immediately. Therefore, $k \leq opt_Q(x)$ if and only if $k \leq f_Q(x, y)$. Since the feasible solution y can be constructed by the algorithm A in time $O(p(2k, |x|))$, we conclude that the NP optimization problem Q is fixed-parameter tractable. The case when Q is a minimization problem can be proved similarly by showing that the problem $Q_{\geq}$ is fixed-parameter tractable.   □

Actually, the proof above demonstrates the FPT of any optimization problem with a polynomial-time approximation algorithm $A(x, \varepsilon)$ *and with the additional property that there exists a constant c independent of ε such that $A(x, \varepsilon)$ runs in time $O(|x|^c)$ for a fixed ε.*

Theorem 9.3.1 allows us to establish that many problems are FPT. For instance, BOUNDED KNAPSACK (below) is FTP. This follows by the algorithm of Ibarra and Kim [408].

BOUNDED KNAPSACK

Input: $U \subseteq \mathbb{N}$ with sizes $s(a) \in \mathbb{Z}$ for $a \in U$, and $B \leq k$.
Parameter: A positive integer k.
Question: Is there some $S \subseteq U$ with $\sum_{a \in S} s(a) = B$?

Similarly, the approximation algorithm of Lipton and Tarjan [493] shows that the problem PLANAR INDEPENDENT SET below is FPT. Of course we met this problem in Chap. 3, but the point here is that we can deduce its parameterized tractability simply by appealing to the approximation literature.

PLANAR INDEPENDENT SET

Input: A planar graph G.
Parameter: A positive integer k.
Question: Does G have an independent set of size k?

A final example: by the randomized fully polynomial-time approximation scheme of Brightwell and Winkler [111, 112] the following problem is randomized FPT (with the obvious meaning).

LINEAR EXTENSION COUNT

Input: A partially order set P.
Parameter: A positive integer k.
Question: Does P have at least k linear extensions? [A linear extension L of a poset $(\{a_1, \ldots, a_n\}, \leq)$ is a linear ordering $(\{a_1, \ldots, a_n\}, \leq^*)$ such that if $a_i \leq a_j$ in P, then $a_i \leq a_j$ in L.]

The problem of counting linear extension is shown (in Brightwell and Winkler [112]) to be #P complete.

Theorem 9.3.1 implies that many other natural problems are FPT. One powerful application comes from the contrapositive.

Corollary 9.3.1 (Bazgan [57], Cai and Chen [121]) *If an NP optimization problem is fixed-parameter intractable, then it has no fully polynomial-time approximation scheme. In particular, under the current working hypothesis that $W[1] \neq$ FPT—the focus of Chap. 21 on hardness theory for parameterized complexity—the NP optimization problems that are $W[1]$-hard under the uniform reduction have no fully polynomial-time approximation scheme.*

In Part III directly after this chapter, we will sharpen Corollary 9.3.1 by using another class called $M[1]$ and discover that we can get very tight limits on our ability to efficiently compute approximate solutions. We take this up in Sect. 29.3.

Now, following Liming Cai and Chen [121], we point out that Corollary 9.3.1 complements a classical result of Garey and Johnson (Theorem 6.8 in [337], p. 140) that a strongly NP-hard optimization problem Q with $opt_Q(x) < p(|x|, num(x))$ for some two-variable polynomial p has no fully polynomial-time approximation scheme unless $P = NP$, where $num(x)$ is the largest integer appearing in the input instance x. The consequence $W[1] = $ FPT in Corollary 9.3.1 is weaker than $P = NP$, but the assumption in Corollary 9.3.1 requires neither strong NP-hardness nor the condition $opt_Q(x) < p(|x|, num(x))$.

There are a number of applications of Theorem 9.3.1. One not covered by the classical literature is that of the so-called VC-DIMENSION from computational learning theory, which has been studied extensively. As we see in Chap. 22, the VC-DIMENSION is $W[1]$-hard, and so cannot have a fully polynomial-time approximation scheme unless $W[1] = $ FPT. However, Papadimitriou and Yannakakis [552] have shown strong evidence that this problem is in NP but neither in P nor NP-complete. Therefore, Theorem 6.8 of Garey and Johnson [337] is not applicable. The conclusion is that VC-DIMENSION is unlikely to have a polynomial-time approximation scheme even though it seems easier than NP-complete problems.

9.4   The Classes MAX SNP, MIN $F^+\Pi_1(h)$ and FPT

Cai and Chen [121] proved that large classes of important optimization problems have corresponding parameterized problems which are fixed-parameter tractable. These problems are defined from subclasses of the collection of NP-complete which are described syntactically.

Many important natural and well-studied NP optimization problems belong to the class MAX SNP, introduced by Papadimitriou and Yannakakis [552]. The class comes from consideration of the syntactic form of NP problems via Fagin's Theorem below, known in finite model theory as NP $= \Sigma_1^1$. More precisely, the exact statement of the theorem is the following. (For a more detailed discussion of second-order logic, we refer the reader to Chap. 13, where the specific case of graphs is discussed in detail.)

Theorem 9.4.1 (Fagin's Theorem [281])   NP *is the class of all problems that can be put in the form*:

$$\left\{I : \exists S \in 2^{[n]^k} \forall x \in [n]^p \exists y \in [n]^q \phi(I, S, x, y)\right\}.$$

$I \subseteq [n]^m$ *is the input relation, x and y are tuples ranging over $[n] = \{1, 2, \ldots, n\}$, and ϕ is quantifier free.*

For instance, SATISFIABILITY can be written as $\exists T \forall c \exists x[(P(c, x) \wedge x \in T) \vee (N(c, x) \wedge x \notin T)]$.

Proof sketch   We sketch the proof, leaving the details to the reader (Exercise 9.5.2). First, to get $\Sigma_1^1 \subseteq$ NP, let $\mathcal{C} = \{\mathcal{A} : \mathcal{A} \models \exists R\psi\}$, with ψ first order, as above. Then, given $\mathcal{A}$, we can guess relations R and verify that $(\mathcal{A}, R) \models \psi$, which is polynomial time. Conversely, NP $\subseteq \Sigma_1^1$, as follows. Given an NP language L, and hence a nondeterministic polynomial-time machine M, we have $\mathcal{A} \in L$ iff M accepts the encoding of $(\mathcal{A}, <)$. Here one needs only construct a first-order formula $\psi(<, X)$ which means "X encodes an accepting computation of M on the encoding of $(\mathcal{A}, <)$." Then $L = \{\mathcal{A} : \mathcal{A} \models \exists < \exists X \, (\text{"} < \text{ is a total order"} \wedge \psi)\}$. $\quad\square$

Some problems have a form that is *simpler* than the basic Fagin Theorem. We can write 3SAT with one less alternation of quantifiers by assuming that the input consists of four relations C_0, C_1, C_2, and C_3. C_i consists of all clauses with exactly j negative literals, and $(x_1, x_2, x_3) \in C_j$ means that there is a clause with $x_1, \ldots, x_j$ occurring negatively and $x_{j+1}, \ldots, x_3$ positively. 3SAT is then

$$\exists T \forall (x_1, x_2, x_3) \left[\bigwedge_{j=0}^{3} \left((x_1, x_2, x_3) \in C_j \rightarrow \left(\bigwedge_{i \leq j} x_i \notin T \wedge \bigwedge_{i > j} x_i \in T \right) \right) \right].$$

Problems such as 3SAT, which can be expressed as $\exists S \forall x \, \psi(x, G, S)$, constitute a class called *simple* NP or SNP. For each $\Pi \in$ NP, one can define the *maximization version* MAX Π as

$$\max_S \left| \{x : \exists y \psi(x, y, G, S)\} \right|.$$

The immediate relevance of all the above NP relations to our studies of FPT comes from the following consequential definition.

Definition 9.4.1 (Papadimitriou and Yannakakis [552]) A maximization problem $Q = (I_Q, S_Q, f_Q, opt_Q)$ is in the class MAX SNP if its optimum $opt_Q(X)$, $X \in I_Q$, can be expressed as

$$opt_Q(X) = \max_S \left| \{v : \psi(v, X, S)\} \right|,$$

where the input instance $X = (U, P^1, \ldots, P^b)$ is described by a finite structure over the finite universe U; and P^i is a predicate of arity r_i for some integer $r_i \geq 1$; S is also a finite structure over the universe U; v is a vector of fixed arity of elements in U; and ψ is a quantifier-free formula.

Despite the fact that Papadimitriou and Yannakakis [552] proved that any problem in MAX SNP can be approximated to within some constant ratio, Arora et al. [47] proved the beautiful result *that* MAX SNP-*complete*[1] *optimization problems have no fully polynomial-time approximation scheme unless* P = NP. Because of this, we cannot use Theorem 9.3.1 to demonstrate that MAX SNP problems are FPT. Nevertheless, Liming Cai and Chen were able to establish the fixed-parameter tractability of all problems in MAX SNP.

---

[1] The appropriate reduction here is called an *L*-reduction. See Exercise 9.5.4.

Theorem 9.4.2 (Cai and Chen [121]) *All maximization problems in the class* MAX SNP *are fixed-parameter tractable.*

Proof Let c be a fixed constant. Consider the following MAX c-SAT problem.

MAX c-SAT
Input: A set of Boolean formulas $\phi_1, \phi_2, \ldots, \phi_m$, each with at most c variables.
Question: Find a truth assignment that satisfies the maximum number of the formulas.

We first show that the corresponding parameterized problem (MAX c-SAT)$_=$ is fixed-parameter tractable for every constant c. By Papadimitriou and Yannakakis [552], for any input instance $x = \{\phi_1, \ldots, \phi_m\}$ of the problem MAX c-SAT, the optimal value of x is at least $m/2^c$. Moreover, since each formula in x has at most c variables, the total number of variables appearing in x is bounded by cm. Therefore, the following algorithm demonstrates the fixed-parameter tractability of the parameterized problem (MAX c-SAT)$_=$: Given a pair $\langle x, k \rangle$, where $x = \{\phi_1, \ldots, \phi_m\}$ is an instance of the problem MAX c-SAT and k is an integer, we first compare the values k and $m/2^c$. If $k < m/2^c$, then $\langle x, k \rangle$ is obviously a no-instance for the parameterized problem (MAX c-SAT)$_=$. If $k \geq m/2^c$, then we check all assignments to the variables in x to verify whether k is the maximum number of formulas in x that can be satisfied by a single assignment. Since the total number of variables in x is bounded by cm, there are at most 2^{cm} different assignments to the variables in x. Therefore, the above algorithm runs in time $O(2^{cm}m)$, which is bounded by $O(2^{O(k)}m)$ since $m = O(k)$. This shows that the parameterized problem (MAX c-SAT)$_=$ is fixed-parameter tractable.

Every MAX SNP problem Q can be reduced to the problem MAX c-SAT for some constant c in the following sense, according to Papadimitriou and Yannakakis [552]. Given an input instance $x \in I_Q$ of Q, there is a polynomial-time algorithm that constructs an input instance x' of MAX c-SAT, such that the optimal value for the instance x of the problem Q is equal to the optimal value for the instance x' of the problem MAX c-SAT. Therefore, the membership of the input instance $\langle x, k \rangle$ for the parameterized problem $Q_=$ is identical to the membership of the input instance $\langle x', k \rangle$ for the parameterized problem (MAX c-SAT)$_=$. Since x' can be constructed from x in polynomial time, and the parameterized problem (MAX c-SAT)$_=$ is fixed-parameter tractable, we conclude that the optimization problem Q in MAX SNP is also fixed-parameter tractable. Q is an arbitrary problem in the class MAX SNP, and the result agrees with the Cai and Chen result Theorem 9.4.2 on the FPT status of all optimization problems in MAX SNP. $\qquad\square$

Liming Cai and Chen also showed that an important class of minimization problems introduced by Kolaitis and Thakur [443] are fixed-parameter tractable.

Definition 9.4.2 (MIN F$^+\Pi_1(h)$) Define MIN F$^+\Pi_1(h)$, $h \geq 2$, to be the class of all minimization problems Q whose optimum can be expressed as

$$opt_Q(X) = \min_S \big\{ |S| : \forall v \psi(v, S, X) \big\},$$

where $S = (U, P_0)$; and P_0 is a predicate of arity 1 over the finite universe U. $|S|$ denotes the weight of the predicate P_0, i.e., the number of elements of the universe U on which the predicate P_0 has truth value. $\psi(v, S, X)$ is a quantifier-free CNF formula in which all occurrences of P_0 are positive. P_0 occurs at most h times in each clause. Let MIN $F^+ \Pi_1 = \bigcup_{h \geq 2}$ MIN $F^+ \Pi_1(h)$.

The class MIN $F^+ \Pi_1$ contains well-studied minimization problems, including VERTEX COVER and a large number of *vertex deletion* and *edge deletion* graph problems. It is known that all minimization problems in MIN $F^+ \Pi_1$ are polynomial-time approximable to a constant ratio [443]. Theorem 9.3.1 is not applicable to the class MIN $F^+ \Pi_1$ since the problem VERTEX COVER is MAX SNP-hard [552], so it does not have a fully polynomial-time approximation scheme unless $P = NP$.

Theorem 9.4.3 (Cai and Chen [121]) *All minimization problems in the class* MIN $F^+ \Pi_1$ *are fixed-parameter tractable.*

Proof Let Q be a minimization problem in the class MIN $F^+ \Pi_1(h)$, where $h \geq 2$ is a fixed integer. Then the optimum $opt_Q(X)$ for any input instance X of Q can be expressed as

$$opt_Q(X) = \min_S \big\{ |S| : \forall v \psi(v, S, X) \big\},$$

where $S = (U, P_0)$. Here P_0 is a predicate of arity 1 over the finite universe U; $v = (v_1, \ldots, v_d)$ is a vector over U of arity d for some fixed d; $\psi(v, S, X)$ is a quantifier-free CNF formula in which all occurrences of P_0 are positive; and P_0 occurs at most h times in each clause.

We show below that the parameterized problem $Q_\leq$ is fixed-parameter tractable. Fix an input instance X of Q. Consider the formula

$$F(X, S) = \bigwedge_{v \in U^d} \psi(v, S, X).$$

Note that the formula $F(X, S)$ is a quantifier-free CNF formula in which the only unknowns are $P_0(u)$, where $u \in U$. Moreover, according to the definition of ψ, each clause of $F(X, S)$ contains at most h unknowns $P_0(u)$, $u \in U$ and all of them occur positively. Since the number of elements of the universe U is bounded by the length $|X|$ and d is a fixed integer, the formula $F(X, S)$ can be constructed from the input instance X in polynomial time.

Now, given an instance $\langle X, k \rangle$ of the parameterized problem $Q_{\leq}$, we first construct the formula $F(X, S) = \{C_1, C_2, \ldots, C_s\}$, and then execute the algorithm below.

Algorithm 9.4.1 (WEIGHTEDCNF(F, k))

WEIGHTEDCNF(F, k)

Input: A CNF formula $F = \{C_1, \ldots, C_s\}$ and an integer k, where each C_i is
             a clause with at most h unknowns, all of the form $P_0(u)$, $u \in U$, and
             appearing positively.

Question: Is there a finite structure $S = (U, P_0)$, $k > |S|$, and S satisfies F?

1. IF $k \leq 0$, then REJECT.
2. IF $F = \phi$, then ACCEPT.
3. Suppose that $C_1 = \{P_0(u_1), P_0(u_2), \ldots, P_0(u_c)\}$, where $u_j \in U$ and $c \leq h$.
4. FOR $j = 1$ to c
       Make $P_0(u_j) = 1$;
       Let F' be the CNF formula obtained from the formula F by
          deleting all clauses that contain the occurrence $P_0(u_j)$;
       Recursively call WEIGHTEDCNF($F', k - 1$).

It is easy to verify the correctness of the algorithm WEIGHTEDCNF(F, k). If the algorithm WEIGHTEDCNF(F, k) (accepts), then there is a finite structure $S_0 = (U, P_0)$, $|S_0| < k$, and $F(X, S_0) = \forall v \psi(v, S_0, X) = 1$. Therefore,

$$opt_Q(X) = \min_S \{|S| : \forall v \psi(v, S, X)\} \leq |S_0| < k.$$

This $\langle X, k \rangle$ is a no-instance of the parameterized problem $Q_{\leq}$. Otherwise, we should have

$$opt_Q(X) = \min_S \{|S| : \forall v \psi(v, S, X)\} = \min_S \{|S| : F(X, S)\} \geq k.$$

This $\langle X, k \rangle$ is a yes-instance of the parameterized problem $Q_{\leq}$. Thus, the problem $Q_{\leq}$ can be solved through the algorithm WEIGHTEDCNF.

We claim that the running time of the algorithm WEIGHTEDCNF(F, k) is bounded by $O(h^k|F|)$. This is certainly true when $k \leq 1$. If $k \geq 2$, then the loop body of Step 4 is executed at most h times; each time involves a recursive call WEIGHTEDCNF($F', k - 1$), where $|F'| \leq |F|$. By the inductive hypothesis, the running time of each call of WEIGHTEDCNF($F', k - 1$) is bounded by $O(h^{k-1}|F|)$.

Note that the length of the formula F is bounded by a polynomial of the length of the input X which is independent of the parameter k, and h is a constant. Consequently, the parameterized problem $Q_{\leq}$ is fixed-parameter tractable.   □

9.5 Exercises

Exercise 9.5.1 Consider the problem BIN PACKING below.

BIN PACKING

Input: $U = \{u_1, \ldots, u_n\}$ and a size function $s(u_i) : U \mapsto \mathbb{Q}$ with $0 \leq s(u_i) \leq 1$.

Question: Find a partition of U into disjoint subsets $U_1, \ldots, U_k$ such that for each j, $\sum_{u_i \in U_j} \leq 1$, and k is least.

Prove that the "first fit" algorithm, which begins with all the U_t being empty, and then for $\leq n$ in order, will put u_i into the first acceptable U_t, and is a polynomial-time approximation algorithm with performance ratio $\leq \frac{1}{2}$. Actually, [420] proved that the performance of this algorithm is always at least as good as $\frac{17}{10} opt(I) + 2$.

Exercise 9.5.2 Prove Fagin's Theorem in detail.

Exercise 9.5.3 (Stockmeyer) Prove that "SO = PH"; that is, prove that $\bigcup_n \Sigma_n^P$ coincides with the polynomial-time hierarchy.

Exercise 9.5.4 (Papadimitriou and Yannakakis [552]) For two optimization problems Π and Π', we say that Π L-reduces to Π' iff there exist two polynomial algorithms f and g and positive constants α and β such that for each instance I of Π, $f(I)$ is an instance of Π' with $opt(f(I)) \leq \alpha opt(I)$; and given any solution of $I' = f(I)$ with cost c', g produces a solution to I with cost c such that $|c - opt(I)| \leq \beta |c' - opt(I')|$.
1. Prove that L-reductions compose.
2. Prove that if Π L-reduces to Π' and are both maximization problems, then if there is a polynomial-time approximation algorithm for Π' with worst-case error ε', then there is also a polynomial-time approximation algorithm for Π with worst-case error $\alpha\beta\varepsilon'$.

Exercise 9.5.5 (Papadimitriou and Yannakakis [552]) Prove that the following are complete for MAX SNP under L-reductions. (See Exercise 9.5.4.)

MAX 3SAT

Input: A Boolean formula X in 3CNF form.
Problem: Find an assignment satisfying the most number of clauses.

MAX 3SAT-B

Input: A Boolean formula X in 3CNF form where the number of occurrences of variables are bounded by a constant B.
Problem: Find an assignment satisfying the most number of clauses.

INDEPENDENT SET-B

Input: A graph with the maximum degree $\leq B$.
Question: Find the largest independent set.

DOMINATING SET-B

Input: A graph with the maximum degree $\leq B$.
Question: Find the smallest dominating set.

MAX CUT

Input: A graph G.
Question: Find a partition of the nodes of G into sets S and $\overline{S}$ such that the number of edges from S to $\overline{S}$ is greatest.

MAX k-COLORABLE SUBGRAPH

Input: A graph G.
Question: Find the largest k-colorable subgraph. (Here, k is a fixed parameter.)

Exercise 9.5.6 (Papadimitriou and Yannakakis [552]) Prove that every problem in MAX SNP can be approximated to within some constant ratio.

(Hint: It is enough to consider the case of MAX 3SAT by the above. Approximate to within $\frac{1}{2}$ by repeatedly choosing the truth value that satisfies the most as yet unsatisfied clauses.)

9.6   Historical Notes

The results of this section, are taken from material by Bazgan [57], Liming Cai [116], and Liming Cai and Chen [121]. Fagin's Theorem was proven in Fagin [281]. (See Fagin [282] for more of the history of the area.) There is a great deal of material now available on finite model theory and its relation to complexity classes. We refer the reader to Ebbinghaus and Flum [266]; Immermann [409]; and Libkin [489]. Downey, Fellows and Regan [255] have material relating descriptive complexity to the parametric complexity classes of this section, some of which is represented in Part V. The approach of using descriptive complexity as a basis for parameterized complexity has been championed by Flum and Grohe in papers by them and with co-authors beginning with [309] and continuing with others such as [154] and [310]. The material on L-reductions is drawn from Papadimitriou and Yannakakis [552] where the reductions in Exercises 9.5 can be found.

Flum and Grohe [312] base their whole approach to parameterized complexity around logic, particularly model checking and logical definability.

9.7   Summary

This chapter introduces the basic connections between parameterized complexity and optimization problems. It looks at Fagin-type definability and observes that a wide class of problems can be shown to be FPT via descriptive complexity, through classes of problems such as SNP and MIN $F^+(X)$.

Part III
Techniques Based on Graph Structure

Treewidth and Dynamic Programming

10

10.1 Basic Facts About Treewidth

10.1.1 Introduction

One of the major outgrowths from the beautiful work of Robertson and Seymour, for example [582, 588, 589], was the focus on a new parameter, which measures how "tree-like" a graph is in a precisely defined sense. We could then lift parametric results from trees to graphs that are "tree-like".

Concepts equivalent to tree decomposition were subsequently, and apparently independently, discovered in other contexts such as the artificial intelligence literature. As we see in the next section, this parametric method of "tree-like" data analysis also enables a uniform approach to linear-time algorithms on a wide class of graphs. Furthermore, this parameter is also the basis of modern algorithmic developments like "bidimensionality" and "protrusions" we will meet later in this book. Finally, the focus on *topological* graph theory had a profound impact on combinatorics.

Whilst determining whether a graph has treewidth k is FPT, the algorithms to do this are impractical. Because of this, it was thought that treewidth would have no impact on practical algorithmics, besides a classification tool. Yet there is a growing use of treewidth in practice, particularly via heuristics and where the certificate of the treewidth is either given or close to the surface. For example, Koster, van Hoesel and Kolen [458] applied this to constraint satisfaction problems such as frequency assignment problems from practical data. Inference problems and Bayesian belief networks come readily equipped with a tree structure, and admit analysis by treewidth-based algorithms. (See, for example, S.L. Lauritzen and D.J. Spiegelhalter [482] or Alber, Dorn, and Niedermeier [23].)

We remark that algorithms for (truly) small treewidth are of interest, and such graphs arise quite regularly to depict natural phenomena. For example, Thorup [645] showed that the flow control graph of go-to free programs like *Pascal*. Yamagucki, Aoki, and Mamitsuka [667] computed the treewidth of 9712 chemical compounds from the LIGAND database and found that all but one had treewidth at most 3 and the remainder had treewidth 4.

R.G. Downey, M.R. Fellows, *Fundamentals of Parameterized Complexity*,
Texts in Computer Science, DOI 10.1007/978-1-4471-5559-1_10,
© Springer-Verlag London 2013

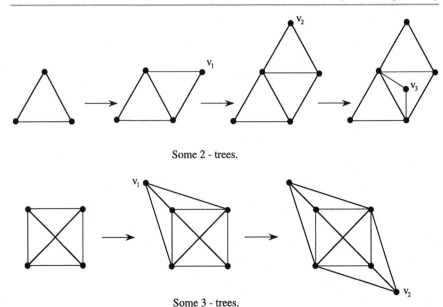

Some 2 - trees.

Some 3 - trees.

Fig. 10.1 Examples of 2- and 3-trees

10.1.2 The Basic Definitions

The simplest definition of treewidth for *graphs* comes from *partial k-trees*.

Definition 10.1.1 (*k*-tree)
(a) The class of *k*-trees is the smallest class obeying (i) and (ii) below.
  (i) K_{k+1}, the complete graph on $k + 1$ vertices, is a *k*-tree.
  (ii) If G is a *k*-tree and H is a subgraph of G isomorphic to K_k, then the
       graph G' constructed from G by first adding a new vertex v to G and
       then adding edges to make $H \cup \{v\}$ a copy of K_{k+1}, is a *k*-tree.
(b) If G_1 is a subgraph of a *k*-tree, then G_1 is called a *partial k-tree*.

In Fig. 10.1 we give examples of 2- and 3-trees.

Definition 10.1.2 (Treewidth) We say that a graph G has *treewidth k* if k is
least such that G is a partial k tree.

For algorithmic purposes, Definition 10.1.2 is not the most useful. An equivalent
characterization of treewidth is provided below.

Fig. 10.2 Example of tree decomposition of width 2

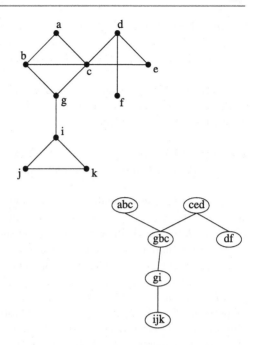

> **Definition 10.1.3** (Tree decomposition)
> (a) A *tree decomposition* of a graph $G = (V, E)$ is a tree $\mathcal{T}$ together with a collection of subsets T_x (called *bags*) of V labeled by the vertices x of $\mathcal{T}$ such that $\bigcup_{x \in \mathcal{T}} T_x = V$ and the following connectivity properties (1) and (2) hold:
>   (1) For every edge uv of G there is some x such that $\{u, v\} \subseteq T_x$.
>   (2) (Interpolation property) If y is a vertex on the unique path in $\mathcal{T}$ from x to z then $T_x \cap T_z \subseteq T_y$.
> (b) The *width* of a tree decomposition is the maximum value of $|T_x| - 1$ taken over all the vertices x of the tree $\mathcal{T}$ of the decomposition.

We remark that Definition 10.1.3 of treewidth is not only algorithmically more useful, but easily allows for generalization to multigraphs, hypergraphs, matroids, and the like.

If a tree decomposition of a graph G gives a path, then we say that the tree decomposition is a *path decomposition*, and use *pathwidth* in place of treewidth.

Figure 10.2 gives an example of a tree decomposition of width 2.

The axioms above imply that certain connectivity conditions will occur on the graphs.

Lemma 10.1.1 *Let* $\{T_x : x \in \mathcal{T}\}$ *be a tree decomposition of* G.
(i) *Let* $x \in V(G)$. *Then the collection of* $y \in \mathcal{T}$ *with* $x \in T_y$ *forms a subtree of* T.
(ii) *Suppose that* C *is a clique of* G. *Then there is some* $x \in \mathcal{T}$ *with* $C \subseteq T_x$.

We leave this proof as an exercise (Exercise 10.2.1 below). The following theorem and its proof demonstrates that treewidth and pathwidth, the two basic representations of width, are identical.

Theorem 10.1.1 (Arnborg, Gavril [341], Rose, Tarjan, and Lueker [594]) *A graph has treewidth* k *iff* G *has a tree decomposition of width* k.

Proof Let G be a graph with a tree decomposition $\{T_x : x \in \mathcal{T}\}$ of width k. Thus $\max_{x \in \mathcal{T}} |T_x| = k+1$. We prove that G is a partial k-tree, and in fact G is a subgraph of a k-tree $K(G)$ in which every $\leq k$ element subset of a T_x is part of a k-clique. We use induction on trees $\mathcal{T}$. Let x be a leaf of $\mathcal{T}$. Let $\mathcal{T}'$ be the result of removing x from $\mathcal{T}$, and let G' be a graph corresponding to $\mathcal{T}'$. By hypothesis, G' is a subgraph of a k-tree $K(G')$ with the desired properties. Let y be attached to x in $\mathcal{T}$. Let $T_x \cap T_y = T'_x$ and $T_x - T'_x = T''_x$. If $T''_x = \emptyset$, then $T_x \subseteq T_y$ and hence $G = G'$. So suppose that $T''_x \neq \emptyset$. By property 2. of Definition 10.1.3, for all $z \neq x$, $T''_x \cap T_z = \emptyset$. It follows that T'_x is a proper subset of T_x, and hence as the width is k, T'_x is a subset of at most k elements of T_y. By hypothesis, this means that T'_x is a subset of part of a k-clique C of $K(G')$. The idea is now to simply add T''_x one node at a time. We will use Algorithm 10.1.1.

Algorithm 10.1.1
1. Initially we let $K = K(G')$.
   While $T''_x \neq \emptyset$ we repeat the following, steps 2–5.
2. Pick a vertex v of T''_x.
3. Add v to K and also add all edges between vertices of C and v.
4. Let $T''_x \leftarrow T''_x - \{v\}$.
5. If there is any vertex $c \in C - T_x$, let $C \leftarrow [C - \{c\}] \cup \{v\}$.

By induction on this algorithm, one can see that after each iteration, C is a k-clique of K. Each step emulates the definition of a k-tree. Hence as $K(G')$ is a k-tree, so too is K_{fin} the result of Algorithm 10.1.1 applied to $K(G')$. By induction, we also can see that all of T''_x is added together with all possible edges that can exist between vertices in $T_x = T'_x \cup T''_x$. Thus K_{fin} contains G and has the desired properties.

Suppose, however, that G is a partial k-tree. Since it suffices to consider k-trees, we will first suppose that G is a k-tree. We prove by induction on $|V(G)|$ that G will have a tree decomposition of width k. We use the stronger hypothesis that if G' is a k-tree with $< |V(G)|$ vertices, then G' has a tree decomposition $\{T_x : x \in \mathcal{T}\}$ where $|T_x| \leq k+1$ for all $x \in \mathcal{T}$, and each k-clique of G' occurs in some T_x. The result is clear if $G = K_{k+1}$. So if G has more than $k+1$ nodes, let v be the last node added in the formation of G. (That is, v is a *simplicial* node.) Let $N(v)$ be

Fig. 10.3 Example of tree decomposition of width 3

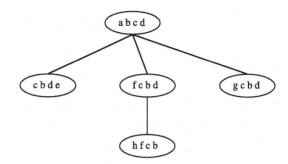

the neighbors of v in G. Let G' be the result of the removal of v and the k edges vu with $u \in N(v)$. Then we can apply the induction hypothesis to G' and obtain a tree decomposition $\{T'_x : x \in T'\}$ of G' with the desired properties. Now $N(v)$ is a k-clique and hence, by hypothesis, there is some $y \in T'$ with $N(v) \subseteq T'_y$. Create a new tree T from T' by attaching a new leaf $\ell = \ell(v)$ at y. Let $T_\ell = N(v) \cup \{v\}$. By hypothesis, T will be the relevant tree decomposition of G.   □

Definition 10.1.4 (Bounded treewidth)   We say that a class C of graphs has *bounded treewidth* if there is some k such that for all $G \in C$, G has treewidth $\leq k$.

10.2 Exercises

Exercise 10.2.1   Prove Lemma 10.1.1.

Exercise 10.2.2 (Per W. Merkle)   An ordering $I = \langle I_1, \ldots, I_n \rangle$ of a set of subsets of V is called an *acyclic hypergraph* iff for all $j \leq n$, there is a $q < j$ such that $I_j \cap (\bigcup_{i < j} I_i) \subseteq I_q$. (This is also called the *running intersection property* per Lauritzen and Spiegelhalter [482].) Given a finite graph, we say that an acyclic hypergraph I is an acyclic hypergraph decomposition of G iff $V = V(G)$ and for each edge e of G there is some i such that the vertices of e are in I_i.

Prove that G has a tree decomposition of width w iff G has an acyclic hypergraph decomposition I such that for all $i \leq n$, $|I_i| \leq w$.

(Hint: Prove that one can construct the tree decomposition from I to have the I_i as the actual bags, and conversely.)

Exercise 10.2.3   Apply the algorithm implicit in Theorem 10.3 to embed the graph described by the tree decomposition of Fig. 10.3 into a 3-tree.

Exercise 10.2.4   Prove that if a graph G has a size k vertex cover, then it must also have treewidth $\leq k$.

Exercise 10.2.5 A *Halin graph* is constructed as follows: take a tree without vertices of degree 2. Then, embed it in the plane, and add a cycle through its leaves, in order (i.e., the result remains planar). Prove that a Halin graph must have treewidth 3.

Exercise 10.2.6 A graph is called *outerplanar* iff there exists a plane embedding G of the graph such that G has a cycle C with the property that every edge of G is either a member of C or a chord of C. Prove that if G is outerplanar then it has treewidth 2.

10.3   Algorithms for Graphs of Bounded Treewidth

Historically, algorithm authors often tried to get around classical intractability by looking at their particular problems represented on special classes of graphs. They often discovered that intractable problems became tractable if the problems were restricted to, say, "outerplanar" graphs. Such restriction is not purely an academic exercise since, in many practical situations, the graphs that arise do not, in fact, demonstrate the full pathology of the class of all graphs. (Consider, for instance, a TRAVELING SALESMAN concerned with a given city. The graph will almost certainly be of small maximum clique size.) The table (from Van Leeuwen [652]) below, lists some families of graphs that have been studied and have treewidth.

Families of graphs	Bound on treewidth
Trees	1
Almost trees (k)	$k + 1$
Partial k-trees	k
Bandwidth k	k
Cutwidth k	k
Planar of radius k	$3k$
Series parallel	2
Outerplanar	2
Halin	3
k-Outerplanar	$3k - 1$
Chordal with maximum clique size k	$k - 1$
Undirected path with maximum clique size k	$k - 1$
Directed path with maximum clique size k	$k - 1$
Interval with maximum clique size k	$k - 1$
Proper interval with maximum clique size k	$k - 1$
Circular arc with maximum clique size k	$2k - 1$
Proper circular arc with maximum clique size k	$2k - 2$

Fig. 10.4 A tree
decomposition

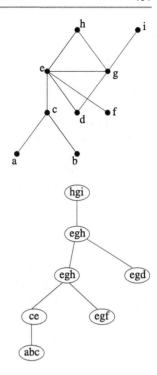

We present this table only for background information. It is not really important
if the reader is unfamiliar with some of the graph families above. The reader is
invited to verify a few of these treewidth bounds in the Exercises in Sect. 10. We
consider the cases of CUTWIDTH and BANDWIDTH in some detail in Sect. 12.7.

Important, however, is the fact that treewidth can be represented by these graph
families. Demonstration of tractability for graphs of bounded treewidth will demonstrate tractability for all these families. *Furthermore, and perhaps even more importantly, as we see in the next section, tree decomposition enables general automata-
theoretic methods for demonstrating tractability of a wide class of properties for
graphs of bounded treewidth.*

We now sketch how to run algorithms on graphs of bounded treewidth. This
can be viewed as "*dynamic programming*", a very valuable algorithmic technique.
A good introduction to this technique is given by Bodlaender [70]; helpful intro-
ductions to other algorithmic aspects of treewidth are also provided by Bodlaender
[73, 75].

We can apply the dynamic programming technique of graphing bounded
treewidth to refine algorithms for the INDEPENDENT SET. For general graphs G,
the problem is NP-complete and we also think it is fixed parameter intractable, as
we will see that it is $W[1]$ complete.

In the case of graphs of bounded treewidth, we can give a linear-time algorithm.
So suppose that we have a tree decomposition of G. Consider the one in Fig. 10.4.

We will use *tables* to grow the independent set up the tree, starting at the leaves of the decomposition. Notice that once a vertex leaves the bags, it will never come back, and hence we do not really need to keep track of its effect. Thus we can focus on working with tables that correspond to all the subsets of the current bag. We need only consider the size of independent sets I growing relative to the information flow across the boundary of the current bag.

Starting at the leaf corresponding to the triangle abc, we could generate the subset table below. Here, the box with the entry ab denotes the subset $\{a, b\}$, meaning that the independent set should contain both a and b, and would have cardinality 2 entered in the cell below. The box with bc corresponds to subset $\{b, c\}$, and since $\{b, c\}$ is *not* an independent set, no cardinality is denoted by a line in the cell below.

Ø	a	b	c	ab	ac	bc	abc
0	1	1	1	2	–	–	–

The subset table for the next box would have only four columns since there are only four subsets of $\{c, d\}$ and we consider maximal independent sets I containing the specified subsets.

Ø	c	e	ce
2	1	3	–

The first entry has value 2 since this means that neither of c or e is included in this independent set I, and we take the maximum entry from the previous table compatible with $\{c, e\} \cap I = \emptyset$, namely 2 for the $\{a, b\}$ column.

The second entry indicates that $c \in I$ and $e \notin I$, that is, that the intersection of the bag $\{c, e\}$ with I is the singleton set $\{c\}$. The maximum entry in the second row of the previous table, over compatible columns, yields an entry of 1.

The third entry corresponds to $I \cap \{c, e\} = \{e\}$. This is compatible with $\{a, b\} \subseteq I$. In particular, e is not adjacent to a or b, yielding an entry of $2 + 1 = 3$.

The fourth entry corresponds to $I \cap \{c, e\} = \{c, e\}$, but this is impossible since c and e are adjacent, hence the line entry.

The rest of the tables are filled in similarly. With pointers you can also keep track of the relevant independent set.

The next table is for a join node of the tree decomposition. As above, each column corresponds to the intersection of I with the bag $\{e, g, h\}$. Now we must check compatibility with respect to the *two* tables, corresponding to the bags $\{c, e\}$ and $\{e, g, f\}$. The entry is the maximum sum from the two tables, over compatible columns, and numerically adjusting for either repetitions, or for adding new vertices to I.

The final tree of tables corresponding to the tree decomposition is shown in Fig. 10.5.

The maximum value of 4 for the column corresponding to $I \cap \{e, g, h\} = \{h\}$ can be obtained by noting that this intersection information is jointly compatible with the column corresponding to Ø in the table for the bag $\{c, e\}$, with value 2, and with

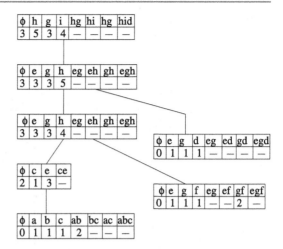

Fig. 10.5 Dynamic programming

the column corresponding to $\{f\}$ in the table for the bag $\{e, g, f\}$, with value 1. These are disjoint, so no repetitions are involved, and the new vertex h joins I, yielding $2 + 1 + 1 = 4$.

Now the data structures above are the most naive implementation possible, and using many techniques we can speed things up. Later we will discuss one method using matrices and distance products as data structures. We do this in Sect. 16.4, when we consider dynamic programming on graphs of bounded branchwidth, which is a similar parameter to treewidth. Other sophisticated techniques are mentioned in Chap. 16.

A sly feature of dynamic programming by this method of growing the independent set using coordinated tables, is that we need a tree decomposition for the graph G. It turns out that for a fixed t there is a *linear-time algorithm* that determines if G has width t, and then finds the tree decomposition of G—should the graph actually have one. However, the algorithm has really terrible constants and there is no actually feasible tree decomposition for this problem. For more on this we refer the reader to Bodlaender [73], and his web site. John Fouhy [329] in his M.Sc. Thesis looked at computational experiments for treewidth heuristics. (See www.mcs.vuw.ac.nz/~downey/students.html.) Chapter 11 reports on many other treewidth heuristics. There is much work to be done in this field.

10.4 Exercises

Exercise 10.4.1 Construct a tree decomposition for the graph of treewidth 2 in Fig. 10.6.

Exercise 10.4.2 Using dynamic programming on the graph of Exercise 10.4.1, compute

Fig. 10.6 A graph of
treewidth 2

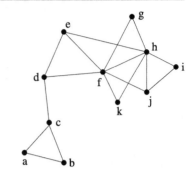

Fig. 10.7 A graph of
treewidth 3

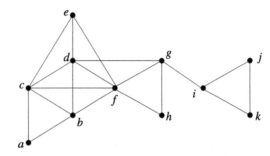

(i) A maximum independent set.
(ii) A minimum dominating set.
(iii) A minimum vertex cover.

Exercise 10.4.3 Construct a tree decomposition for the graph of treewidth 3 in Fig. 10.7.

Exercise 10.4.4 Using dynamic programming on the graph of Exercise 10.4.3, compute
(i) A maximum independent set.
(ii) A minimum dominating set.
(iii) A minimum vertex cover.

Exercise 10.4.5 A graph G is called *series parallel* if it can be generated by the following operations. Take a loop or an edge and interpolate a vertex v. Take an edge e and create a parallel edge (i.e., with the same vertices) e'. Prove that if G is series parallel, then G has treewidth 2.

Exercise 10.4.6 A graph G is said to be *almost tree* (k) iff it has at most k cycles. Prove that if G is almost tree (k), then G has treewidth $\leq k + 1$.

Treewidth methods can be used to replace the *ad hoc* combinatorial methods previously employed by many authors and which usually entailed heroic case analysis. The quintessential example of this phenomenon will be Courcelle's theorem (Theorem 13.1.1), which demonstrates for graphs of bounded treewidth, a linear-time algorithm for recognition of any property of graphs statable in a very powerful fragment of second-order arithmetic (e.g., HAMILTONICITY). All of these arguments are based on the fact that the following problem is fixed parameter tractable.

TREEWIDTH
Input: A graph G.
Parameter: A positive integer k.
Question: Does G have treewidth k?

Theorem 10.4.1 (Bodlaender's Theorem, Bodlaender [74]) TREEWIDTH *is strongly* FPT. *In fact, for each k, there is a linear-time algorithm for recognition of graphs of treewidth k. Furthermore, given a graph G of treewidth k, the algorithm will produce a tree decomposition of G of width k in linear time.*

Bodlaender's theorem is proven in Bodlaender [74], via Bodlaender and Kloks [94]. Modulo the latter result, we give a complete proof of Bodlaender's theorem in the next section.

10.5 Bodlaender's Theorem

Our proof that there is a linear-time FPT algorithm which
 (i) determines if a graph G has treewidth k and
(ii) produces a tree decomposition of width k if the answer to (i) is "yes", relies
     upon the following theorem.

Theorem 10.5.1 (Bodlaender and Kloks [94])
1. *There is an $O(|G|)$ FPT algorithm which has parameter k and p with $k \leq p$ which, given a treewidth p tree decomposition of G determines if G has treewidth k.*
2. *Furthermore, if the treewidth of G is k, there is a linear-time algorithm which will output a treewidth k tree decomposition for G.*

There are several different approaches to proving Theorem 10.5.1(1). We can deduce it by previewing later chapters of this book. By Robertson and Seymour [589], as we see in Chap. 17, there is a finite set of graphs $H_1, \ldots, H_q$ such that a graph G of treewidth p has treewidth k iff H_i is not a minor of G for $i \in \{1, \ldots, q\}$.[1] Actually, since we know a bound on the treewidth of the H_i, it is possible to calculate the set $H_1, \ldots, H_q$, as we will see in Chap. 19. We will soon see (after Chap. 12 on Automata and Bounded Treewidth), that the minor relation is definable in monadic second-order logic (see Courcelle's theorem (Theorem 13.1.1)), we can construct an

---

[1]We say a graph H is a minor of a graph G if H can be obtained from G by deletion and contractions of edges. We refer to Definition 17.1.7.

automaton which runs in linear time and accepts only parse trees of width k from the input of (parse trees in linear time derived from) treewidth p tree decompositions.[2] The algorithm of Bodlaender and Kloks, which produces the tree decomposition if there is one, is concrete and elementary, but too involved to easily fit in the book. (We will give some further notes on the development of Bodlaender's theorem in the Historical Notes at the end of this chapter.)

The main idea of the proof of Bodlaender's theorem is that given G, we first partition the vertices into vertices of *low* and of *high* degree (soon to be described in Definition 10.5.1). There are then two cases.

Case 1. There are many vertices of low degree adjacent to each other. In this case, it is shown that a maximal matching contains *many* edges. One then *contracts*[3] all edges in a maximal matching and generates a new derived graph G'. Now, recursively get a treewidth k decomposition for G', or get a certificate that the treewidth of G' and therefore G exceeds k. In the former case it is possible to generate a treewidth $2k + 1$ decomposition for G from that of G' using the Bodlaender–Kloks Algorithm.

Case 2. There are only few vertices of low degree adjacent to another vertex of low degree. In this case, we will *enrich* the graph by adding certain new edges to form a new graph G' of the same treewidth as G (if it is t). This new graph G' will have many vertices which are *simplicial* in the sense that their neighbors form cliques and can be removed to form G''. Given a tree decomposition of G'', one can easily get one for G.

It will be easy to see that the algorithm runs in linear time.

Let us first consider some details of the algorithm parameters.

Lemma 10.5.1 (Folklore) *Suppose that G has treewidth k. Let $\{X_i : i \in \mathcal{T}\}$ be a tree decomposition for G. Then the following properties hold*:

(i) $|E| < k|V| - \frac{1}{2}k(k + 1)$.

(ii) *If $W \subseteq V$ is a clique, for some $i \in \mathcal{T}$, $W \subseteq X_i$.*

(iii) *If W_1 and W_2 induce a complete bipartite subgraph of G, then for some $i \in \mathcal{T}$, $W_1 \subseteq X_i$.*

(iv) *Suppose that $v, w \in X_i$ and $v \neq w$. Then $\{X_i : i \in \mathcal{T}\}$ is a tree decomposition for $G + vw$.*

(v) *Suppose that x and y are vertices of G with at least $k + 1$ common neighbors. Then the treewidth of $G + xy$ is k and $\{X_i : i \in \mathcal{T}\}$ is also a tree decomposition for $G + xy$.*

---

[2]The reader might notice an apparent circularity in all of this. Is not Bodlaender's theorem actually *used* in the proof of Courcelle's theorem? The proof of Courcelle's theorem actually demonstrates that, for a fixed p, monadic second-order statements have linear-time recognition from a given treewidth p decomposition. This is independent of Bodlaender's theorem, which can now be used to add that "hence monadic second-order statements have linear-time recognition algorithms".

[3]Recall that a contraction of an edge uv in a graph G is the operation of identifying u with v. We refer to Fig. 17.2.

Proof See Exercise 10.6.1. □

A tree decomposition $\{X_i : i \in \mathcal{T}\}$ of G is called *smooth* iff for all $i \in \mathcal{T}$, X_i has $k + 1$ elements and for all nodes p, q of $\mathcal{T}$ with p the child of q, $|X_p \cap X_q| = k$. Part (i) of the following lemma is well known and has a proof similar to the proof of Theorem 12.7.1 (see Exercise 10.6.2).

Lemma 10.5.2
(i) *Any tree decomposition can be converted into a smooth one for the same graph by applying the following operations to nodes p, q in $\mathcal{T}$ with p the child of q or q the child of p.*
   (a) *If $X_p \subseteq X_q$, contract pq let the bag be X_q.*
   (b) *If $X_p \not\subseteq X_q$, but $|X_q| < k + 1$, take any vertex from $X_p - X_q$ and add it to X_q.*
   (c) *If $|X_p| = |X_q| = k + 1$ and $|X_p - X_q| > 1$, interpolate new bags between X_p and X_q.*
(ii) *(Bodlaender) If $\{X_i : i \in \mathcal{T}\}$ is a smooth tree decomposition of G of width k, then $\mathcal{T}$ has exactly $|V(G)| - k$ nodes.*

Proof See Exercise 10.6.2. □

Algorithm Parameters Choose rationals $0 < c_1, c_2 < 1$ such that

$$c_2 = \frac{1}{4k^2 + 12k + 16} - \frac{c_1 k^2 (k+1)}{2} > 0.$$

Also, define

$$d = \max\left\{k^2 + 4k + 4, \left\lceil \left(\frac{2k}{c_1}\right) \right\rceil \right\}.$$

Definition 10.5.1 (High, low, and friendly degrees of a vertex; per Bodlaender [74]) If the degree of a vertex v exceeds d, then we say that v has high degree; if it is less than or equal to d, it has low degree; and finally if v has both low degree and is also adjacent to one of low degree, we say that it is *friendly*.

Lemma 10.5.3
(i) *If G has treewidth k, then it contains fewer than $c_1|V|$ high-degree vertices.*
(ii) *If there are f friendly vertices in G, then any maximal matching contains at least $\frac{f}{2d}$ edges.*

Proof (i) is clear. (ii) Let M be a maximal matching. Note that any friendly vertex must either be an endpoint of an edge of M or adjacent to an endpoint of an edge of M. Now, associated with each edge e of M there can be at most $2d$ friendly vertices which are either endpoints of e or adjacent a low-degree friendly endpoint of e. Hence, $M > \frac{f}{2d}$. □

Definition 10.5.2 (*M*-derived graph)

1. Let M be a maximal matching of G. The *M*-*derived graph* is the one obtained by contracting all the edges of M.
2. The *M*-*derivation function* is defined via $g_M(v) = v$ if v is not a vertex of an edge of M, and $g_M(v) = g_M(w)$ is the vertex v' resulting from the contraction of vw of M.

We need one further lemma regarding algorithm parameters for the *M*-derived graph of G. Its proof is left to the reader (Exercise 10.6.3).

Lemma 10.5.4

1. *If* $\{X_i : i \in \mathcal{T}\}$ *is a treewidth* k *decomposition of* G', *where* G' *denotes the M*-*derived graph of* G, *then* $\{Y_i : i \in \mathcal{T}\}$ *defined via* $Y_i = \{v \in V : g_M(v) \in X_i\}$ *is a treewidth* $\leq 2k + 1$ *decomposition of* G.
2. *The treewidth of* G' *is less than or equal to that of* G.

For a graph G, we define the *improved graph* $\widehat{G}$ of G to be that obtained by adding an edge uv for all pairs u, v of V with at least $k + 1$ common neighbors in G.

Now we get to the key theorem. We need this preliminary definition.

Definition 10.5.3 ($\mathcal{T}$-simplicial vertex)

1. We say that a vertex v of G is *simplicial in* G if its set of neighbors form a clique in G.
2. A vertex w is called $\mathcal{T}$-*simplicial vertex* if it is simplicial in the improved graph $\widehat{G}$ of G, it is of low degree and it is not friendly in G.

We also need this decisive lemma, upon which Bodlaender's theorem and linear-time FPT algorithm are based.

Lemma 10.5.5 *Let* $\{X'_i : i \in \mathcal{T}'\}$ *be a tree decomposition of the graph* G' *obtained by removing all* $\mathcal{T}$-*simplicial vertices (and adjacent edges) from the improved graph* $\widehat{G}$ *of* G. *Then, for all* $\mathcal{T}$-*simplicial vertices* v, *there is some node* i_v *such that the set of neighbors of* v *in* G *all lie in* X'_{i_v}.

Proof See Exercise 10.6.4. $\square$

We are now prepared to follow the central argument of Bodlaender's theorem into the actions of his algorithm.

Theorem 10.5.2 (Bodlaender [74]) *For every graph* G, *one of the following holds*:

(i) G *contains at least* $\frac{|V|}{4k^2 + 12K + 16}$ *friendly vertices.*
(ii) *The improved graph* $\widehat{G}$ *of* G *contains at least* $c_2|V|$ $\mathcal{T}$-*simplicial vertices.*
(iii) *The treewidth of* G *exceeds* k.

Before we verify Theorem 10.5.2, we can now state Bodlaender's algorithm.

Algorithm 10.5.1 (Bodlaender's algorithm) For graphs larger than some fixed constant (which is computable), the following algorithm determines if a graph G has treewidth $\leq k$ and, if so, produces a treewidth k tree decomposition for G.
Step 1. Check if $|E| \leq k|V| - \frac{1}{2}k(k+1)$. If the answer is *no*, output that the treewidth of G exceeds k (Lemma 10.5.1(i)). If the answer is *yes*, go to step 2.
Step 2. There are two cases, depending on the number f of friendly vertices.
Case 1. $f \geq \frac{|V|}{4k^2+12K+16}$.
Action:
1. Find a maximal matching M in G.
2. Compute the derived graph (G', E') by contracting edges of M.
3. Recursively apply the algorithm to G'. If the treewidth of G' exceeds k, stop and output that the treewidth of G exceeds k (Lemma 10.5.4(ii)). Otherwise, the recursive call yields a treewidth k tree decomposition $\{X_i' : i \in \mathcal{T}'\}$ for G'. Now apply Lemma 10.5.4(i) to get a treewidth $2k+1$ tree decomposition $\{Y_i : i \in \mathcal{T}'\}$ of G.
4. Now apply the Bodlaender–Kloks algorithm to, for instance, the treewidth $2k+1$ tree decomposition of G, to either return that G has treewidth exceeding k or return a treewidth k tree decomposition of G.
Case 2. $f < \frac{|V|}{4k^2+12K+16}$.
Action:
1. Compute the improved graph $\widehat{G}$ of G. If there is a $\mathcal{T}$-simplicial vertex of degree $\geq k+1$, then stop and output that the treewidth of G exceeds k since $\widehat{G}$ contains a $k+2$-clique.
2. Otherwise, put all $\mathcal{T}$-simplicial vertices into some set S. Now, compute the graph G' obtained by deleting all $\mathcal{T}$-simplicial vertices and adjacent edges from G.
3. If $|S| < c_2|V|$, then stop and output that the treewidth of G exceeds k (Theorem 10.5.2(ii)).
4. Otherwise, recursively apply the algorithm to G'. If the treewidth of G' is greater than G, then the treewidth of G is greater than k, as G' is a subgraph of G. Otherwise, the algorithm will return a treewidth k tree decomposition $\{X_i' : i \in \mathcal{T}'\}$ for G'.
5. In that case, for all $v \in S$, find an $i_v \in \mathcal{T}$ such that the neighbors $N_G(v)$ of v in G all lie in bag X_{i_v}' (Lemma 10.5.5). We then add a new node j_v to $\mathcal{T}$ adjacent to i_v in $\mathcal{T}$ with $X_{j_v} = \{v\} \cup N_G(v)$.

The fact that the algorithm is correct is immediate by the lemmata preceding it. Note that we either recursively apply the algorithm on a graph with $(1 - \frac{1}{2d(4k^2+12k+16)})|V|$ vertices or on a graph of $(1 - c_2)|V|$ vertices. Since both $1 - c_2$ and $1 - \frac{1}{2d(4k^2+12k+16)}$ are greater than or equal to 0 and less than 1, it follows that the algorithm must run in linear time, given that the called nonrecursive parts have linear-time implementations, which they do.

Proving Bodlaender's Theorem 10.5.2 remains to be done. We do this in the following four lemmas. We say that a vertex v is *localized* with respect to a tree decomposition $\{X_i : i \in \mathcal{T}\}$ if it has low degree, it is not friendly, and for some bag X_i, all of the neighbors of v are in X_i.

Lemma 10.5.6 *Suppose that $\{X_i : i \in \mathcal{T}\}$ is a smooth tree decomposition of G of width k. Then, (i) and (ii) below hold.*

(i) *To each leaf i of $\mathcal{T}$ one can associate a vertex v_i which is either friendly or localized, and furthermore for all $j \neq i$, $v_i \notin X_j$.*

(ii) *To each path P of the form $i_0, \ldots, i_{k^2+3k+2}$ in $\mathcal{T}$ with i_j of degree 2 for $j \in \{0, \ldots, k^2 + 3k + 2\}$, there is a vertex $v_P \in \bigcup_{j=0}^{k^2+3k+2} X_{i_j}$ such that v_P is either friendly or localized, and for all q not on the path P, $v_P \notin X_q$.*

Proof (i) Let i be a leaf of $\mathcal{T}$, and j a neighbor. Let v be the unique vertex in $X_i - X_j$. Note that the only neighbors of v are in X_i so it has low degree. If all the neighbors of v have high degree, then it is localized. Otherwise it is friendly.

(ii) $|\bigcup_{j=0}^{k^2+3k+2} X_{i_j}| = k^2 + 4k + 4$ (by smoothness) and hence $\leq d$. Hence, all the vertices in $(\bigcup_{j=0}^{k^2+3k+2} X_{i_j}) - (X_{i_0} \cup X_{i_{k^2+3k+2}})$ are of low degree. Suppose that all of them are unfriendly, and hence only adjacent to high-degree vertices in $X_{i_0} \cup X_{i_{k^2+3k+2}}$. Let $w_1, \ldots, w_r$ list the high-degree vertices in X_{i_0}. For $s \in \{1, \ldots, r\}$, we know by smoothness that there is a number n_s such that $w_s \in X_{i_0}, \ldots, X_{i_{n_s}}$. We may suppose that $n_1 \leq n_2 \leq \cdots \leq n_r$. If some vertex v belongs to exactly one X_{i_j} on the path P, it must be of low degree and hence localized. If some low-degree vertex belongs to only bags from $X_{i_{n_j}+1}, \ldots, X_{i_{n_j+1}}$, then it is localized since all of its neighbors must belong to $X_{i_{n_j+1}}$. All vertices in $\bigcup_{j=0}^{k^2+3k+2} X_{i_j}$ not of these two types must belong to at least one of the distinguished bags

$$X_{i_0}, X_{i_{n_1}}, \ldots, X_{i_{n_r}}, X_{i_{k^2+3k+2}}.$$

This is only a total of $k^2 + 4k + 3$ vertices, and the result follows. $\square$

Now we demonstrate that we can nontrivially apply the lemma. We define a *leaf-path collection* to be a set of leaves of $\mathcal{T}$ plus a set of paths of $\mathcal{T}$ of length at least $k^2 + 3k + 4$. All the nodes of this leaf-path collection, except the endpoints, have degree 2 and do not belong to any other path in the collection.

Lemma 10.5.7

(i) *If $\mathcal{T}$ is any tree with n nodes then it must have a leaf-path collection of size $\geq \frac{n}{2k^2+6k+8}$ (i.e., it contains that many leaves and paths).*

(ii) *Consequently, if $\{X_i : i \in \mathcal{T}\}$ is a smooth tree decomposition of width k, then G has at least $\frac{|V|}{2k^2+6k+8} - 1$ vertices that are localized or friendly.*

Proof (i) This is a calculation. Let n_3 be the number of nodes in $\mathcal{T}$ of degree ≥ 3, n_1 the number of leaves, and n_2 the number of nodes of degree 2. We note that all nodes of degree 2 belong to $< n_1 + n_3$ connected components of the forest obtained from $\mathcal{T}$ by removing all nodes of degree ≥ 3. Each such component has at most $k^2 + 3k + 3$ nodes if it does not contribute to some leaf-path collection. Therefore, there are fewer than $(n_1 + n_3)(k^2 + 3k + 3)$ nodes of degree 2 not on a path in the leaf-path collection $\mathcal{C}$ of maximum size. The number of paths in $\mathcal{C}$ is at least:

$$\frac{n_2 - (n_1 + n_3)(k^2 + 3k + 3)}{k^2 + 3k + 4}.$$

Thus, the size of leaf-path collection $\mathcal{C}$ is at least:

$$\frac{n_2 - (n_1 + n_3)(k^2 + 3k + 3)}{k^2 + 3k + 4} + n_1 \geq \frac{\frac{1}{2}n}{k^2 + 3k + 4},$$

giving part (i). For (ii), by Lemma 10.5.2(ii), $\mathcal{T}$ has $|V| - k$ many nodes and, hence, an application of (i) and Lemma 10.5.6 yields the result.   □

To fully prove Bodlaender's theorem, we need one more definition and two more lemmas.

Definition 10.5.4 (Important high-degree vertices) We say that a set of high-degree vertices $Y \subseteq V$ is *important* with respect to the decomposition $\{X_i : i \in \mathcal{T}\}$, if $Y \subseteq X_i$ for some $i \in \mathcal{T}$ and for any other set Y' of high-degree vertices with $Y' \subseteq X_j$ for some j, Y is not a proper subset of Y'.

Lemma 10.5.8 *Let $\{X_i : i \in \mathcal{T}\}$ be a decomposition of G of width k. The number of important sets is at most the number of high-degree vertices of G.*

Proof Let H be the set of high-degree vertices of G. Then, $\{X_i \cap H : i \in \mathcal{T}\}$ is a tree decomposition of the subgraph induced by H. Each important set Y is a set of the form $X_i \cap H$ not contained in any other $X_j \cap H$. Repeatedly contract the edges ii' with $X_i \cap H \subseteq X_{i'} \cap H$, forming a new node containing all the vertices of $X_{i'}$. Clearly, the resulting tree decomposition will have size at most H and will contain all the important sets of G.   □

Fix a tree decomposition $\{X_i : i \in \mathcal{T}\}$ of G. Since localized vertices are not friendly, there is a function f, called a *localized to important function*, which associates each localized vertex v with an important set Y such that $N_G(v) \subseteq Y$.

Lemma 10.5.9 *Let f be such a localized to important function for a smooth decomposition $\{X_i : i \in \mathcal{T}\}$ of width k for G. Then, for any important set Y, there are at most $\frac{1}{2}k^2(k + 1)$ localized vertices in $f^{-1}(Y)$ which are not $\mathcal{T}$-simplicial.*

Proof To each non-$\mathcal{T}$-simplicial localized vertex v, associate a pair of its neighbors that are nonadjacent in the improved graph. Notice that we cannot so associate more than k vertices with the same pair—lest the graph have $\geq k+1$ common low-degree vertices, which would necessarily have an edge between them in the improved graph. By simply counting, we see that the number of $\mathcal{T}$-simplicial vertices with $f(v) = Y$ is at most $\frac{1}{2}|Y|(|Y|-1)$, which is $\leq \frac{1}{2}k^2(k+1)$. □

Proof of Bodlaender's theorem, completed To complete the proof of Theorem 10.5.2, suppose that G contains fewer than $\frac{|V|}{4k^2+12k+16}$ friendly vertices and that the treewidth of G is at most k. By Lemma 10.5.3(i), there are at most $c_1|V|$ high-degree vertices in G. Therefore, by Lemma 10.5.8, the number of important sets for any smooth tree decomposition $\{X_i : i \in \mathcal{T}\}$ of G of width k is $\leq c_1|V|$. Therefore, by Lemma 10.5.9, at most $\frac{1}{2}k^2(k+1)(c_1|V|-1)$ localized vertices are not $\mathcal{T}$-simplicial. Using Lemma 10.5.7(ii), we see that

$$\frac{|V|}{2k^2+6k+8} - 1 - \frac{|V|}{4k^2+12k+16} - \frac{1}{2}k^2(k+1)(c_1|V|-1)c_2|V|$$

vertices are $\mathcal{T}$-simplicial, which establishes Theorem 10.5.2. □

10.6　Exercises

Exercise 10.6.1 Prove Lemma 10.5.1.
(Hint: For (i) consider a k-tree, for (ii) and (iii) use the interpolation property in the definition of tree decomposition, (iv) is easy, and for (v), argue that x and y must occur in some bag together since otherwise we could apply (iii) and then (ii) to get a $k+2$-clique in a treewidth k graph.)

Exercise 10.6.2 Prove Lemma 10.5.2.
(Hint: For (ii), use induction on $|V(\mathcal{T})|$.)

Exercise 10.6.3 Prove Lemma 10.5.4.

Exercise 10.6.4 Prove Lemma 10.5.5.
(Hint: Use Lemma 10.5.1.)

10.7　Historical Notes

The material of this section is drawn from the 1966 paper wherein Bodlaender [74] presented his theorem. The problem of deciding if a graph has treewidth k is NP-complete if k is allowed to vary, was discussed a decade later in Arnborg, Corneil, and Proskurowski [44]. In 1987 [44], those authors also showed that there was an $O(n^{k+2})$ for treewidth k graphs. Robertson and Seymour [588] gave the first FPT (it was $O(n^2)$) algorithm in 1995. Their algorithm was based upon the minor well-quasi-ordering theorem, and thus was highly nonconstructive and nonelementary and had huge constants. Algorithms for recognition of bounded treewidth graphs have been improved, approximated, and parallelized since 1987 in the work of, for instance, Lagergren [475], Reed [575], Perkovíc and B. Reed [558], Fellows and Langston [297], and Matousek and Thomas [519].

Linear-time algorithms had been demonstrated in 1991 for recognition of tree-width k graphs $k \leq 3$ (e.g., Matousek and Thomas [519]). For a general k, linear-time recognition of treewidth k graphs from among treewidth $m \leq k$ graphs given by treewidth m decompositions is due independently to Lagergren and Arnborg, Abrahamson and Fellows, Bodlaender and Kloks [94] (where it first appears in print), although it follows also by Courcelle's theorem as we noted earlier. Bodlaender and Kloks gave the explicit algorithm which allows us to actually derive the relevant tree decomposition needed for Bodlaender's theorem. Their algorithm and Bodlaender's theorem are both double exponential in k, which is the best known constant. It would be very interesting if this could be reduced to a single exponential.

In the next chapter, we discuss heuristics for treewidth to optimize algorithm running time while finding tree decompositions.

10.8 Summary

This chapter introduces the reader to the key concepts of treewidth and tree decomposition. It is shown how to run dynamic programming on tree decompositions. Bodlaender's algorithm is proven.

Heuristics for Treewidth

11

11.1 Introduction

Treewidth has become a central concept in graph algorithm design, with a wide range of applications. It is important to be able to find good tree decompositions quickly. Much the same could be said for pathwidth. Algorithms and heuristics for finding good path decompositions are considered in Chap. 31 and Remark 31.4.1. In Chap. 16, we will look at algorithms and heuristics for other width metrics. As a first step, we closely examine in this chapter heuristics for treewidth.

In the previous chapter, we met Bodlaender's Theorem for tree decompositions. Several tries at implementing the algorithm have met with failure. While some algorithms that have a bad worst-case running time analysis *do* work well in practice, Bodlaender's algorithms fpr finding optimal tree and path decompositions do not seem to be among these!

As we have seen, Bodlaender's algorithms recursively invoke a linear-time FPT algorithm. The main results are due to Bodlaender and Kloks [94]. The main part of the algorithms work by "squeezing" a given width p tree decomposition of a graph G down to a width k tree (path) decomposition of G, if this is possible. Otherwise, there is a certification that G has treewidth (pathwidth) greater than k. A small improvement to the Bodlaender/Kloks algorithm would substantially improve the performance of Bodlaender's algorithms.

Perkovíc and Reed [558] have recently improved upon Bodlaender's work, giving a streamlined algorithm for k-TREEWIDTH that recursively invokes the Bodlaender/Kloks algorithm no more than $O(k^2)$ times, while Bodlaender's algorithms may require $O(k^8)$ recursive iterations. However, even these improved results are still far from practical.

Efficient polynomial time algorithms can find optimal treewidth and pathwidth, or good approximations of these, for some graph classes. Examples are chordal bipartite graphs, interval graphs, permutation graphs, circle graphs, and co-graphs.

In this chapter we will look at some classes of heuristics that give approximate tree decompositions and work pretty well in practice.

R.G. Downey, M.R. Fellows, *Fundamentals of Parameterized Complexity*,
Texts in Computer Science, DOI 10.1007/978-1-4471-5559-1_11,
© Springer-Verlag London 2013

11.2   Separators

A natural guess for an efficient algorithm, and one that is parallelizable, would be to take a (vertex) separator S of G, split G into two subgraphs, and solve the problem recursively. *Sometimes* this works. As we see in Chap. 31, Sect. 31.3, separators can also be used for *parameterized approximation* algorithms for treewidth, and yielding a $3k + 3$ approximation algorithm for treewidth.

Definition 11.2.1 We say a separator S of G is *safe* if

$$\text{tw}(G) = \max\{\text{tw}(G[S \cup C] + \text{clique}(S)) \mid C \text{ is a component of } G \setminus S\}.$$

The following lemma indicates why this is an interesting concept.

Lemma 11.2.1 (Bodlaender and Koster [95]) *Suppose that S is any separator of G. Then* $\text{tw}(G) \leq \max\{\text{tw}(G[S \cup C] + \text{clique}(S)) \mid C \text{ is a component of } G \setminus S\}$.

Proof The proof actually shows how to construct the decomposition. Let $t = \max\{\text{tw}(G[S \cup C] + \text{clique}(S)) \mid C \text{ is a component of } G \setminus S\}$. Let $C_1, \ldots, C_r$ denote the components of $G \setminus S$. For each component C_j let $\{X_i^j \mid i \in T_j\}$ denote a tree decomposition of $G[S \cup C_j] + \text{clique}(S)$ of width $\leq t$. Since $\text{clique}(S)$ is a clique, by Lemma 10.1.1, there is some bag $X_{i_{i_j}}^i \supseteq S$. Now take the disjoint union of the T_j, and add one additional bag $X = S$, and make the $X_{i_j}^i$ as children of this bag for $i = 1, \ldots, r$. This is a tree decomposition of G of width $\leq t$. $\square$

Essentially the same arguments enable us to establish the following. The proof is left to the reader in Exercise 11.5.1.

Lemma 11.2.2 (Bodlaender and Koster [95]) *Let S be any separator in G. Suppose that $I_1, \ldots, I_r$ induce components of $G \setminus S$. Let $H_j = (W_j \cup S, E_j)$ be a triangulation of $G[S \cup W_j] + \text{clique}(S)$ with maximum clique size k. Then if we let $H = (V, F)$ where $F = \bigcup_{i \leq j \leq r} E_j$, H is a triangulation of G with maximum clique size at most k.*

The following is immediate by the lemmas above.

Lemma 11.2.3 *Suppose that S is a clique separator; namely, a separator which induces a clique in G. Then S is safe.*

Hence every separator consisting of a single vertex is safe. Similar techniques show that any separator of size 2 which is also minimal is safe. Considerations about taking minors allow us to establish the following.

Lemma 11.2.4 (Bodlaender and Koster [95]) *Suppose that S is a minimal separator of size k such that G \ S has at least k connected components. Then S is safe.*

These simple results allow us to be able to give sufficient conditions for safe separators for small treewidth.

Theorem 11.2.1 (Bodlaender and Koster [95])
1. *Every separator of size 1 is safe.*
2. *Every minimal separator of size 2 is safe.*
3. *If S is a separator of size 3, and S has two connected components, then S is safe.*
4. *Suppose that $S = \{v, w, x, y\}$ is a separator of size 4, and v is adjacent to w, x, and y, and G has no separator of size 2 and none of size 3 containing v. Then if every connected component of G \ S has at least two vertices, S is safe.*

The general algorithmic strategy that the above structural lemmas support is to find safe separators S in G and replace G by the collection of graphs $G[S \cup C] + \text{clique}(S)$ for all components C of $G \setminus S$. That is, safe separators allow a divide-and-conquer approach.

The proof of Lemma 11.2.1 tells us how to assemble a tree decomposition for G from tree decompositions for the graphs $G[S \cup C] + \text{clique}(S)$. This process can be repeated until there are no safe separators left.

The central algorithmic problem for this strategy is that of finding safe separators. Clearly, ones of size 1 are easy to find via depth-first search. Minimal ones of size 2 can be found using network flow in linear time. Clique separators (of no guaranteed size) can be found in time $O(mn)$ by an algorithm due to Tarjan [637].

Experimental analyses of implementations can be found in, for example, Bodlaender and Koster [95], and Fernau [305], as well as Fouhy's Thesis [329] and Koster, Bodlaender and van Hoesel [457]. The most successful implementations combine several different algorithmic strategies; separators (as discussed above), including branch and bound, and (as discussed below) heuristics based on perfect elimination orderings and reduction rules. Perhaps it is not surprising that the best-performing heuristics combine several approaches—this seems to be true in general for computationally hard problems.

11.3 Perfect Elimination

Another class of heuristics are based upon the k-tree version of the definition of tree decomposition given in Theorem 10.1.1. The following is a re-interpretation of this, and is similar, since family of graphs of pathwidth at most k consists of precisely the subgraphs interval graphs of maximum clique size at most k.

Definition 11.3.1
1. An *elimination ordering* of G is an ordering of the vertices $f : V \rightarrow \{1, \ldots, n\}$. It is called *perfect* if for all $v \in V$,

$$\{w \mid vw \in E \wedge f(w) > f(v)\} \quad \text{is a clique.}$$

2. G is called the *intersection graph of subtrees of a tree* if and only if there is a tree T, and for each $v \in V$ there is a subtree T_v of T such that for all $v \neq w \in V$, $vw \in E$ if and only if T_v and T_w have at least one vertex in common.

We will be using triangulations of graphs G. Given an elimination ordering f of G, we can construct a triangulation H of G by beginning with $H := G$ and adding edges by seeing w, v with $wv \notin E(H)$, $wx, vx \in E(H)$ and $f(w), f(v) > f(x)$, and *adding wv to H* inductively. This is called the *fill-in* generated by G, f. The following is a re-interpretation of Theorem 10.1.1, and its proof is left to the reader in Exercise 11.5.2.

Theorem 11.3.1 (Arnborg, Gavril [341], Rose, Tarjan, and Lueker [594]) *"A triangulation graph H of G has treewidth no more than k after an elimination ordering f of G generates a fill-in tree decomposition of width no more than k." More precisely,*
1. *The following are equivalent for a graph G.*
   a. *G is triangulated.*
   b. *G has a perfect elimination ordering.*
   c. *G is the intersection graph of subtrees of a tree.*
   d. *There is a tree decomposition $\{T_x : x \in T\}$ such that each bag T_x is a clique.*
2. *Let G be a graph and $k \leq n$. The following are equivalent.*
   a. *$\mathrm{tw}(G) \leq k$.*
   b. *G has a triangulation H with maximum clique size $\leq k + 1$.*
   c. *There is an elimination ordering whose fill-in has maximum clique size $\leq k + 1$.*
   d. *There is an elimination ordering f such that no vertex $v \in V$ has more than k numbers in the fill-in H of G, f.*

The above gives us a general algorithmic strategy for treewidth heuristics based on *graph triangulations*.

Algorithm 11.3.1 (PERMUTATIONTOTREEDECOMPOSITION) (Bodlaender and Koster [95])
Input: G and an elimination ordering $\{v_2, \ldots, v_n\}$.
If $n = 1$ return a tree decomposition with one bag $T_{v_1} = \{v_1\}$.
Else, follow these steps.

1. Compute the graph G' obtained by eliminating v_1 from G.
   Call PERMUTATIONTOTREEDECOMPOSITION on G', $\{v_1, \ldots, v_n\}$, recursively, and let $\{T_w : w \in T'\}$ (the underlying set for T' will be indexed by vertices $v \in \{v_2, \ldots, v_n\}$) be the returned tree decomposition.
2. Let $j = \min\{i \mid v_1 v_i \in E(G)\}$. Construct a bag $T_{v_1} = N_G[v_1]$. Adjoin this bag to T' by connecting it to T_{v_j}.

11.4 Reduction Rules

Reduction rules are another important general algorithmic strategy that seem to play an important role in the most successful heuristics. In the theorem below, we maintain a variable *low* that captures information on a lower bound for the treewidth of G.

Theorem 11.4.1 (Bodlaender, Koster, and van der Eijkhof [99]) *The following reduction rules are safe for treewidth.*
1. Simplicial rule. *Remove a vertex v whose neighbors form a clique and set low $:= \max\{\deg(v), low\}$. (Special cases are degree 1 (twig) and degree 0 (islet).)*
2. Almost simplicial rule. *We say that $S \subseteq V$ forms an* almost clique *if there is a single vertex w such that $S \setminus \{w\}$ is a clique. If the neighbors of v form an almost clique, remove v, and turn the neighbors of v into a clique. (Special cases are where v has degree 2 (series rule) or degree 2 (triangle rule).)*
3. Buddy rule. *Suppose that v and w have the same neighbors and both have degree 3, and also that low ≥ 3. Turn the neighbors of v and w into a clique and remove both v and w.*
4. Generalized buddy rule (van der Eijkhof, Bodlaender, and Koster [651]) *If vertices $v_1, \ldots, v_k$ have the same neighbors and each of the neighbors has degree $k + 1$, and low $\geq k$, make the neighbors into a clique, and remove $v_1, \ldots, v_k$.*
5. Cube rule. *Suppose that a, b, c, d, v, w, x with a, b, c having degree 3 in G and with edges $ad, av, aw, bd, bv, bx, cw, cx \in E$, and no other edges to the "cube" except edges from the rest of G to v, w, x and d. Let G' be constructed by making v, w, x, d mutually adjacent and removing a, b and c.*
   Then we claim the following:
   a. $\text{tw}(G) = \max\{3, \text{tw}(G')\}$.
   b. *There is an elimination ordering a, b, c, f of G where f is an elimination ordering of G' to treewidth at most $\max\{3, \text{tw}(G')\}$.*
6. Extended cube rule. *Suppose that low ≥ 3 and G contains vertices a, b, c, d, v, w, x with only edges $va, vb, ad, bd, aw, bx, wc, dc, xc$, and all other edges in G incident only with v, x or w. Then remove d, c, b, a and replace this "cube" with a triangle with edges vw, wx, vx.*

Proof We prove the safety of these reduction rules as follows.

1. *Simplicial rule.* By the way that k-trees are formed, simplicial vertices and their neighbors must be in a bag together, and hence since we have set $low :=$ $\max\{\deg(v), low\}$, we are safe.

2. *Almost simplicial rule.* Consider the graph G' obtained by the almost simplicial rule. We claim the following.

a. $\text{tw}(G') \le \text{tw}(G) \le \max\{d, \text{tw}(G')\}$ where $d = \deg(v)$.

b. If f is an elimination ordering of G' of treewidth at most $\max\{d, \text{tw}(G')\}$, then the ordering given by putting v first and following with f is an elimination ordering of G of treewidth at most $\max\{d, \text{tw}(G')\}$.

To establish these claims: let w be a neighbor of v such that the remainder form a clique. Since we obtain G' from G essentially by contracting the edge vw, we can easily prove by induction that $\text{tw}(G') \le \text{tw}(G)$. For the second part, suppose that f is an elimination ordering of G' of treewidth $t = \max\{d, \text{tw}(G')\}$. Let H be the fill-in of G' given f. Add v and its formerly adjacent edges to H and note that v is simplicial in the resulting graph $\widehat{H}$. Clearly this triangulation of G induces a tree decomposition of G with treewidth $\le \max\{t, d\}$. To see that the rule is safe, note that since $d \le low$, keeping low unchanged is safe, as $\max\{\text{tw}(G), low\} = \max\{\text{tw}(G'), low'\}$.

3. *Buddy rule.* Let G' be the result of the buddy rule. We claim that:

a. $\text{tw}(G') \le \text{tw}(G)$ and $\text{tw}(G) \le \max\{3, \text{tw}(G)\}$.

b. The ordering v, w, f with f an elimination ordering of G' or treewidth at most $\max\{3, \text{tw}(G')\}$ is an elimination ordering of G of treewdith at most $\max\{3, \text{tw}(G')\}$.

To establish these claims, suppose that x, y, z are the neighbors of v and w. Contract the edges vx and wy to obtain G'. An elementary induction shows that the treewidth is not increased. Adding v and w to the fill-in H of G' induced by f gives the result (as both of these will be simplicial). Safety follows immediately.

4. *Generalized buddy rule.* This is Exercise 11.5.4.

5. *Cube rule.* As $\text{tw}(G[\{a, b, c, d, v, w\}]) \ge 3$, $\text{tw}(G) \ge 3$. As edges are contracted, $\text{tw}(G') \le \text{tw}(G)$. As before, take the fill-in H and an elimination ordering f of treewidth $\text{tw}(G')$. Now if we consider a, b, c, f as an elimination ordering of G, we note the fill-in of this eliminations ordering is already present in H; hence the result follows.

6. *Extended cube rule.* This is safe, as the extended cube rule can be obtained by first applying the cube rule and then the simplicial vertex rule, noting that d becomes simplicial after the cube rule is applied.   □

Heuristics based on various combinations of the general strategies discussed in this chapter have been applied quite successfully. With algorithms engineering care, the running time of the reduction rules is $O(n^2)$.

Other approaches have been tried, including *simulated annealing, greedy triangulation*, and *greedy sparse subgraphs*—those we do not cover here. We refer the reader to Bodlaender and Koster [95, 97, 98], Bodlaender, Heggernes, and Villanger [90] and Bodlaender, Jansen, and Kratsch [93] for the latest word on practical computational approaches to this central problem.

11.5 Exercises

Exercise 11.5.1 Prove Lemma 11.2.2.

Exercise 11.5.2 Prove Theorem 11.3.1 using the methods of Theorem 10.1.1.

Exercise 11.5.3 Let $\{X_i \mid i \in T\}$ be a tree decomposition of G. Show that for each edge ij of T, one of the following holds.
1. $X_i \cap X_j$ is a separator of G.
2. All of the vertices of the component of $T \setminus \{e\}$ containing X_i are subsets of X_j.
3. All of the vertices of the component of $T \setminus \{e\}$ containing X_j are subsets of X_i.

Exercise 11.5.4 Prove that the *generalized buddy rule* is safe.

11.6 Historical Notes

Heuristics for efficient computation of "good" tree decompositions have been with the subject since its beginnings. Recent perspectives on this core algorithmic problem can be found by referring to Bodlaender and Koster [97], [98] and recent surveys, such as Bodlaender and Koster [96] and Bodlaender [78].

11.7 Summary

We described some of the main algorithmic strategies for the practical computation of optimal or near-optimal tree decompositions, based on separators, elimination orderings, and reduction rules.

Methods via Automata and Bounded Treewidth

12

12.1 Introduction

Finite-state automata that process either:

- strings of symbols (linear automata), or
- rooted trees of symbols (tree automata)

support an important, general, FPT methodology (which continues to develop with significant vigor and exciting research horizons) based on the following two-step algorithmic strategy.

(1) Compute a representation for the graph (or hypergraph, or matroid, etc.) of bounded width, as a labeled binary tree, termed a *parse tree*, for the object. This can be accomplished in FPT time for a variety of objects, and for a number of important structural width metrics, with the parameter being the width bound.

(2) Use a finite-state tree automaton to recognize precisely the parse trees that represent the objects (e.g., graphs of bounded treewidth) that have the property of interest (e.g., Hamiltonicity).

This turns out to be a powerful paradigm for FPT algorithm design for natural parameterized problems concerning a wide range of mathematical objects, often deployed in conjunction with strategies to "kernelize" to objects of bounded width (meaning that in FPT time, we are able to restrict attention to problem instances of bounded width).

12.2 Deterministic Finite Automata

Since we will be using analogs of methods used for languages accepted by finite automata, in the next few sections we will give a (brief) self-contained account of the aspects of classical formal language theory which we need. We begin with finite automata. The reader who has had a solid course in automata theory might still find the material on tree automata beginning with Sect. 12.6 new. The plan

R.G. Downey, M.R. Fellows, *Fundamentals of Parameterized Complexity*,
Texts in Computer Science, DOI 10.1007/978-1-4471-5559-1_12,
© Springer-Verlag London 2013

is to successively extend the notion of automata to culminate with the use of tree automata to decide properties of mathematical objects that can be structurally parsed to a rooted tree-of-symbols representation.

Definition 12.2.1 (Deterministic finite automaton)   A *deterministic finite automaton* is a quintuple $M = \langle K, \Sigma, \delta, S, F \rangle$, where

K	is a finite set called the set of states,
$S \subseteq K$	is called the set of *initial* states,
$F \subseteq K$	is called the set of *accepting* states,
Σ	is a finite set of *symbols*, termed the *alphabet*, and
$\delta : K \times \Sigma \mapsto K$	is a function called the *transition* function.

We will usually denote states by $\{q_i : i \in G\}$ for some set G and have $S = \{q_0\}$. As the reader will recall, we think of automata as physical machines that act to accept or reject strings from Σ^*. We consider the machine M as starting on the leftmost symbol of a string $\sigma \in \Sigma^*$, in state q_0. The transition function δ induces a rule "If M is in state q_i reading a ($\in \Sigma$), move one symbol right and change the internal state of M to $\delta(q_i, a)$." For each $\gamma \in \Sigma^*$, we may write this action as

$$\langle q_i, a\gamma \rangle \vdash \langle \delta(q_i, a), \gamma \rangle.$$

Here, the notation $\vdash$ is read as "yields in one step". That is, the first symbol of the (remainder) of the string of symbols being analyzed, is processed, leading to a new "current state".

We say that M *accepts* a string of symbols σ if, after starting M on input σ in state q_0 and then applying $\vdash$ $|\sigma|$ times, M is in one of the "accept" states in F. (More formally, one can define M to accept σ if, for some $f \in F$,

$$\langle q_0, \sigma \rangle \vdash^* \langle f, \lambda \rangle,$$

where λ denotes the empty string, and $\vdash^*$ denotes the transitive closure of $\vdash$.) We let

$$L(M) = \left\{ \sigma \in \Sigma^* : M \text{ accepts } \sigma \right\}.$$

Example 12.2.1 Consider M to be specified by the information: $K = \{q_0, q_1\}$, $\Sigma = \{0, 1\}$, $S = \{q_0\}$, $F = \{q_0\}$ with the transition function given by the table:

State (q)	Symbol (a)	$\delta(q, a)$
q_0	0	q_0
q_0	1	q_1
q_1	1	q_0
q_1	0	q_1

Fig. 12.1 Transition diagram
for Example 12.2.1

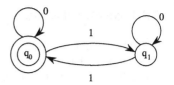

The string 00110 is then processed:

$$\langle q_0, 00110 \rangle \vdash \langle q_0, 0110 \rangle$$
$$\vdash \langle q_0, 110 \rangle$$
$$\vdash \langle q_1, 10 \rangle$$
$$\vdash \langle q_0, 0 \rangle$$
$$\vdash \langle q_0, \lambda \rangle.$$

As $q_0 \in F$, M accepts 00110 and, hence, $00110 \in L(M)$. On the other hand, one can easily see that $\langle q_0, 0010 \rangle \vdash^* \langle q_1, \lambda \rangle$ and, since $q_1 \notin F$, $0010 \notin L(M)$.

The traditional way of representing automata is via *transition (or state) diagrams*. Here, there is one vertex for each of the states of M and arrows from one state to another labeled by the possible transitions given by δ. Accept states are usually represented by a "double circle" at the node. For instance, the machine M of the above example would be represented by the diagram in Fig. 12.1. In this simple machine, q_0 has a double circle around it since it is an accept state, whereas q_1 has only a single circle around it since it is not an accept state.

The crucial fact we will need in our investigations is that, for a fixed M, one can quickly decide if $\sigma \in M$; that is, we note the following trivial observation.

DETERMINISTIC FINITE AUTOMATA ACCEPTANCE
Input: A string $\sigma \in \Sigma^*$.
Parameter: M, a deterministic finite automata over Σ^*.
Question: Is $\sigma \in L(M)$?

Proposition 12.2.1 DETERMINISTIC FINITE AUTOMATA ACCEPTANCE *is solvable in time* $O(|\sigma|)$, *and hence is (strongly uniformly)* FPT.

The following exercises should provide the reader unfamiliar with automata with a little warm-up for the next sections, where we will that linear automata can be used to prove FPT results for graphs of bounded pathwidth, and the generalization to tree automata can be used to prove FPT results for bounded treewidth.

12.2.1 Historical Notes

The study of what we now call finite automata essentially began with McCulloch and Pitts [524]. Kleene [448] introduced the model we now use. Many others had similar ideas and introduced various models. For instance, the reader is referred to Huffman [403], Moore [537], and Mealy [526]. Hopcroft and Ullman [401] discuss the development of finite automata in more detail.

12.2.2 Exercises

Exercise 12.2.1 Prove that for the machine M of Example 12.2.1,

$$L(M) = \{\sigma : \sigma \text{ contains an even number of ones}\}.$$

Exercise 12.2.2 Construct automata to accept the following languages $L_i \subseteq \{0, 1\}^*$:
1. $L_1 = \{\sigma : \sigma \text{ contains at most 5 ones}\}$.
2. $L_2 = \{\sigma : \sigma \text{ contains no 1 followed by a 0}\}$.
3. $L_3 = \{\sigma : \text{the number of ones in } \sigma \text{ is a multiple of 3}\}$.

Exercise 12.2.3 Prove that if L_1 and L_2 are languages accepted by deterministic finite automata, so is $L_1 \cup L_2$.

12.3   Nondeterministic Finite Automata

The basic idea behind nondeterminism is that from any position, there are a number of possible computation paths. For automata nondeterminism is manifested as the generalization the transition relation δ from a function to a multifunction. Thus, δ now becomes a relation so that from a given $\langle$state, symbol$\rangle$ there may be several possibilities for δ(state, symbol). Formally, we have the following definition.

Definition 12.3.1 (Nondeterministic finite automaton) A *nondeterministic finite automaton* is a quintuple $M = \langle K, \Sigma, \Delta, S, F \rangle$, where K, S, F, and Σ are the same as for a deterministic finite automaton, but $\Delta \subseteq K \times \Sigma \times K$ is a relation called the *transition* relation.

We can interpret the action of a machine M being in state q_i reading a symbol a, as "being able to move to any one of the states q_k with $\langle q_i, a, q_k \rangle \in \Delta$". Abusing notation slightly, we write

$$\langle q_i, a\gamma \rangle \vdash \langle q_k, \gamma \rangle,$$

but now we will interpret $\vdash$ to mean "can *possibly* yield". Again, we will let $\vdash^*$ denote the reflexive transitive closure of $\vdash$ and declare that

$$\sigma \in L(M) \quad \text{iff} \quad \langle q_0, \sigma \rangle \vdash^* \langle f, \lambda \rangle,$$

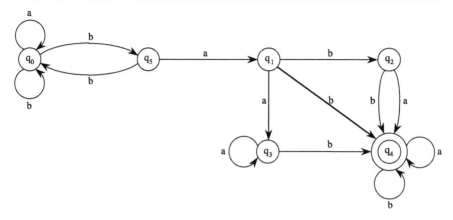

Fig. 12.2 Transition diagram for Example 12.3.1

for some $f \in F$. The interpretation of the definition of $\vdash^*$ is that we will accept σ provided that there is some way to finish in an accepting state; that is, *we will accept σ provided that there is some computation path of M accepting σ*. Of course, there may be many possible computation paths for the input σ. Perhaps only one leads to acceptance. Nevertheless, we would accept σ.

Example 12.3.1 Specify M by the following sets: $K = \{q_0, q_1, q_2, q_3, q_4, q_5\}$, $\Sigma = \{a, b\}$, $S = \{q_0\}$, $F = \{q_4\}$, and $\Delta = \{\langle q_0, a, q_0 \rangle, \langle q_0, b, q_0 \rangle, \langle q_0, b, q_5 \rangle, \langle q_5, b, q_0 \rangle, \langle q_5, a, q_1 \rangle, \langle q_1, b, q_2 \rangle, \langle q_1, a, q_3 \rangle, \langle q_1, b, q_4 \rangle, \langle q_2, a, q_4 \rangle, \langle q_2, b, q_4 \rangle, \langle q_3, b, q_4 \rangle, \langle q_3, a, q_3 \rangle, \langle q_4, a, q_4 \rangle, \langle q_4, b, q_4 \rangle\}$. See Fig. 12.2.

Note that upon reading, say, *bab*, M could finish in states q_0, q_2, q_4, or q_5. The reader should verify that

$$L(M) = \{\sigma \in \{a, b\}^* : bab \text{ or } ba(a \ldots a)ab \text{ is a substring of } \sigma\}.$$

At first glance, it would appear that nondeterministic automata are much more powerful than deterministic ones. This is not the case, as we see by the classic theorem of Rabin and Scott below. We say that two automata M_1 and M_2 are equivalent if $L(M_1) = L(M_2)$.

Theorem 12.3.1 (Rabin and Scott's theorem [569]) *Every nondeterministic finite automata is equivalent to a deterministic one. Furthermore, if M is a (nondeterministic) machine with n states, then we may compute in $O(2^n)$ steps an equivalent deterministic machine M' with at most 2^n states. Here, the O only depends on $|\Sigma|$.*

Proof The construction to follow is often referred to as the *subset construction*. Let $M = \langle K, \Sigma, S, F, \Delta \rangle$ be the nondeterministic automaton. The idea is quite simple. We will consider the states of the new machine *we* create from M as corresponding to *sets* of states from M. We will assume that $S = \{q_0\}$. Now, on a given symbol

$a \in \Sigma^*$, M can go to any of the states q_k with $\langle q_0, a, q_k \rangle \in \Delta$. Let $\{q_{k_1}, \ldots, q_{k_p}\}$ list the states with $\langle q_0, a, q_{k_j} \rangle \in \Delta$. The idea is that we will create a state $Q_1 = \{q_{k_1}, \ldots, q_{k_p}\}$, consisting of the *set of states of M reachable from q_0 on input a*. Consider Q_1. For each symbol $b \in \Sigma^*$, and for each $q_{k_j} \in Q_1$, compute the set D_{k_j} of states reachable from q_{k_j} via b, and we let Q_2 denote $\bigcup_{q_{k_j} \in Q_1} D_{k_j}$. Formally,

$$Q_0 = S = \{q_0\},$$
$$Q_1 = \{q \in K : \langle q_0, a \rangle \vdash \langle q, \lambda \rangle\},$$

and

$$Q_2 = \bigcup_{q \in Q_1} \{r \in K : \langle q, b \rangle \vdash \langle r, \lambda \rangle\}.$$

The idea continues inductively as we run through all the members of Σ and longer and longer input strings. We continue until no new states are generated. Notice that path lengths are bounded by $n = |K|$, and that the process will halt after $O(2^n)$ steps. The accept states of the machine so generated consists of those subsets of K that we construct and contain at least one accept state of M. It is routine to prove that this construction works (see Exercise 12.3.3). □

The algorithm implicit in the proof above is often called "Thompson's construction" [644]. Notice that we may not necessarily construct all the 2^n subsets of K as states. The point here is that many classical proofs of Rabin and Scott's theorem simply say to generate all the 2^n subsets of K. Then, let the initial state of M' be the set of start states of M, the accept states of M' are the subsets of K containing at least one accept state of M, and, if $F \subseteq K$ and $a \in \Sigma$, define

$$\delta(F, a) = \bigcup_{q \in F} \{r \in K : \langle q, a \rangle \vdash \langle r, \lambda \rangle\}.$$

The difference is that the latter proof often constructs many *useless* states since many of the $F \subseteq K$ simply *cannot* be accessed upon *any* input string. But no matter what, if we use the latter proof, we will indeed take $O(2^n)$ steps.

Example 12.3.2 We apply Thompson's [644] construction to the automaton from Example 12.3.1.

The initial state of Example 12.3.1 is $Q_0 = \{q_0\}$. We obtain the transition diagram of Fig. 12.3.

Notice that the deterministic automaton generated by Thompson's construction has only eight states, whereas the worst possible case would be $2^5 = 32$ states. We remark that it is possible for Thompson's construction and, indeed *any* construction to generate *necessarily* $\Omega(2^n)$ states. This is because there exist examples of n state nondeterministic automata where any deterministic automata accepting the same strings necessarily need $\Omega(2^n)$ states. We give one example of this phenomenon in the next section (Example 12.4.3).

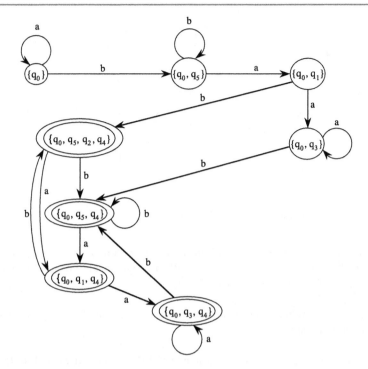

Fig. 12.3 The deterministic automaton for Example 12.3.1

As with deterministic automata, we can extract the following algorithmic problem.

NONDETERMINISTIC FINITE AUTOMATA ACCEPTANCE
Input: A string $\sigma \in \Sigma^*$.
Parameter: A nondeterministic finite automaton M.
Question: Does M accept σ?

Theorem 12.3.1 has the following important algorithmic consequence.

Corollary 12.3.1 NONDETERMINISTIC FINITE AUTOMATA ACCEPTANCE *can be solved in time $O(n)$ and (strongly uniformly) FPT.*

It is convenient to extend the model of a nondeterministic automaton to *nondeterministic finite automata with λ moves*. Here, one allows "spontaneous" movement from one state to another without consuming a symbol.

Example 12.3.3 Consider the automaton M given by Fig. 12.4.

For this automaton, on input 110 it is possible for the machine to follow a computation path to *any* of the four states except q_1. For instance, to get to state q_0, M could read 1 in state q_0, move to state q_1, spontaneously move into state q_2 via the

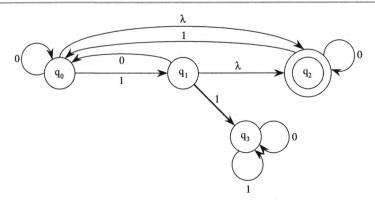

Fig. 12.4 A nondeterministic automaton with λ moves

empty transition from q_1 to q_2, travel the 1 transition from q_2 back to q_0, and then finally travel the 0 transition from q_0 to q_0. We invite the reader to check that all the other states are accessible on input 110.

It is easy to show that there is no real increase in computational power for the new class of automata.

Theorem 12.3.2 (McNaughton and Yamada [525], Rabin and Scott [569]) *Every nondeterministic finite automaton with λ moves is equivalent to one without λ moves. Moreover, there is a construction, which when applied to M, a nondeterministic finite automaton with λ moves and n states, yields an equivalent nondeterministic automaton M' with $O(n)$ states in $O(n)$ steps.*

Proof Let M be a given nondeterministic finite automaton with λ moves. We construct an equivalent M' without λ moves. For a state q of M, define the equivalence class of q via

$$E(q) = \left\{ p \in K(M) : \forall w \in \Sigma^* \big(\langle q, w \rangle \vdash^* \langle p, w \rangle \big) \right\};$$

that is, $E(q)$ denotes the collection of states accessible from q with no head movement. In Example 12.3.3, $E(q_0) = \{q_0, q_2\}$. As with the proof of Theorem 12.3.1, for M', we replace states by sets of states of M. This time the set of states are the equivalence classes $E(q)$ for $q \in M$. Of course, the accept states are just those $E(q)$ containing accept states of M, and the new transition on input, a, takes $E(q)$ to $E(r)$, provided that there is some q_i in $E(q)$ and some q_j in $E(r)$ with Δ taking q_i to q_j on input a. It is easy to see that this construction works (Exercise 12.3.4). □

When applied to the automaton of Example 12.3.3, the method used in the proof of Theorem 12.3.2 yields the automaton in Fig. 12.5.

Definition 12.3.2 If L is the set of strings accepted by a finite automaton, we say that L is *finite state*.

Fig. 12.5 The automaton corresponding to Example 12.3.3

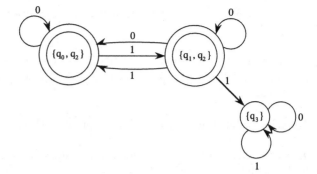

We can use these results to establish a number of closure properties for finite-state languages.

Theorem 12.3.3 *Let $\mathcal{L}$ denote the collection of languages accepted by finite automata. Then $\mathcal{L}$ is closed under the following:*

(i) *union*
(ii) *intersection*
(iii) *star (i.e., if $L \in \mathcal{L}$, then so is L^*, the collection of strings that can be obtained by concatenating strings in L)*
(iv) *complementation*
(v) *concatenation (i.e., if $L_1, L_2 \in \mathcal{L}$, then so is $L_1 L_2 = \{xy : x \in L_1 \wedge y \in L_2\}$).*

Proof Let $M_i = \langle K_i, \Sigma_i, S_i, \Delta_i, F_i \rangle$ for $i = 1, 2$, be automata over Σ. By renaming, we can suppose that $K_1 \cap K_2 = \emptyset$.

For (i), we simply add a new start state S' with a λ move to each of the start states of M_1 and M_2. The accept states for $M_1 \cup M_2$ would be $F_1 \cup F_2$.

For (iv), we need to prove that $\overline{L(M_1)}$ is finite state. This is easy. Let $M' = \langle K_1, \Sigma, S_1, \delta_1, K_1 - F_1 \rangle$. Clearly, $\overline{L(M_1)} = L(M')$.

For (ii), we can use the fact that $L(M_1) \cap L(M_2) = \overline{\overline{L(M_1)} \cup \overline{L(M_2)}}$.

We leave (iii) and (v) for the reader (Exercise 12.3.6).   □

12.3.1 Historical Notes

Nondeterministic Finite Automata were introduced by Rabin and Scott [569]. Rabin and Scott established the equivalence of nondeterministic finite automata with deterministic finite automata. McNaughton and Yamada consider automata with λ moves. Thompson's construction is taken from Thompson [644].

12.3.2 Exercises

Exercise 12.3.1 Compute a deterministic finite automaton equivalent to the automaton of Example 12.3.3.

Exercise 12.3.2 Describe the language L accepted by the automaton of Example 12.3.3.

Exercise 12.3.3 Carefully complete the details of the proof of Theorem 12.3.1 by proving that if $\langle q_0, \sigma \rangle \vdash^*_M \langle r, \lambda \rangle$, then $\langle Q_0, \sigma \rangle \vdash^*_{M'} \langle R, \lambda \rangle$ for some state R of M' containing r.

Exercise 12.3.4 Carefully formalize the proof of Theorem 12.3.2.

Exercise 12.3.5 Prove that for any deterministic finite automaton $M = \langle K, \Sigma, S, F, \delta \rangle$ for all $\sigma, \tau \in \Sigma^*$ and $q \in K$,

$$\delta(q, \tau\sigma) = \delta\big(\delta(q, \tau), \sigma\big).$$

Exercise 12.3.6 Prove (iii) and (v) of Theorem 12.3.3; that is, prove that the collection of finite-state languages over Σ is closed under star and concatenation.

Exercise 12.3.7 It is possible to generalize the notion of a nondeterministic automaton as follows. Now M is a quintuple $\langle K, \Sigma, \delta, S, F \rangle$ with all as before save that δ is a multifunction from a subset of $K \times \Sigma^* \mapsto K$ giving a rule:

$$\langle q_i, \gamma\sigma \rangle \vdash \big(\delta(q_i, \gamma), \sigma\big).$$

The difference is that in the standard definition of an automaton, we can only have $a \in \Sigma$, whereas now we can have $\gamma \in \Sigma^*$. (λ moves can be thought of as a special case of this phenomenon with $\gamma = \lambda$.) We will call such automata nondeterministic automata with *string* moves. Prove that any nondeterministic finite automaton with string moves, M, is equivalent to a standard nondeterministic finite automaton, M'. What is the relationship between the number of states of M and M'?

Exercise 12.3.8
1. Suppose that L, L_1, and L_2 are finite state. Prove that the following are decidable if we are given finite automata M, M_1, and M_2 with $L = L(M), L_1 = L(M_1)$, and $L_2 = L(M_2)$.
   (a) Is $L = \emptyset$?
   (b) Is $L = \Sigma^*$?
   (c) Is $L_1 \subseteq L_2$?
   (d) Is $L_1 = L_2$?
2. Prove that (a) and (b) are solvable in $O(2^n)$ for a fixed Σ.

12.4   Regular Languages

Finite automata provide a *semantical* approach to formal language theory, with implicit algorithmic consequences. There is another well-known *syntactical* approach to finite-state languages. It has been extensively studied due to its importance in practical applications such as the GREP family of commands in UNIX (and surely because of the beauty of the resulting theory). We refer, of course, to the class of regular languages. The reader should recall that a regular language is one that has a representation via a finite string of symbols taken from the members of Σ, together with $\cup$ (union),[1] concatenation [which we represent by juxtaposition for regular expressions (see below) and by $\frown$ for strings], $*$ (Kleene star), and $\emptyset$. Formally, we have the following definitions.

Definition 12.4.1 (Regular expression) A *regular expression* over Σ is an expression in the alphabet $\Sigma, \cup, (,)$, and $\emptyset$ defined by induction on logical complexity as follows:
1. $\emptyset$ and each member of Σ is a regular expression.
2. If α and β are regular expressions, so are $(\alpha\beta)$ and $(\alpha \cup \beta)$.
3. If α is a regular expression, so is $(\alpha)^*$.
4. Nothing else.

Naturally, we drop the brackets where the meaning is clear.

Definition 12.4.2 (Regular language) We associate a language $L(\alpha)$ with a regular expression α as follows.
1. If $\alpha = \emptyset$, then $L(\alpha) = \emptyset$.
2. If $\alpha = a$ with $a \in \Sigma$, then $L(\alpha) = \{a\}$.
3. If $\alpha = \gamma\beta$, then $L(\alpha) = L(\gamma)L(\beta)$ $[=_{\text{def}} \{xy : x \in L(\gamma) \text{ and } y \in L(\beta)\}]$.
4. If $\alpha = \gamma \cup \beta$, then $L(\alpha) = L(\gamma) \cup L(\beta)$.
5. If $\alpha = \beta^*$, then $L(\alpha) = (L(\beta))^*$ $[=_{\text{def}}$ the set of all strings obtainable from $L(\beta).]$
   We call a language L *regular* if $L = L(\alpha)$ for some regular expression α.

Example 12.4.1 Let $L = \{0, 1\}$.

$$L\big(0(0 \cup 1)^*\big) = L(0)L\big((0 \cup 1)^*\big) \quad \text{by (iii)}$$
$$= \{0\}\big(L(0 \cup 1)\big)^* \quad \text{by (ii) and (v)}$$
$$= \{0\}\big(L(0) \cup L(1)\big)^* \quad \text{by (iv)}$$
$$= \{0\}\big(\{0\} \cup \{1\}\big)^* \quad \text{by (ii)}$$
$$= \{0\}\{0, 1\}^*$$
$$= \big\{x \in \{0, 1\}^* : x \text{ starts with a } 0\big\}.$$

The relationship with the work of the previous sections is described by the following classic theorem.

---

[1]Often union is represented by |.

Theorem 12.4.1 (Kleene [448]) *For a language L, L is finite state iff L is regular. Furthermore, from a regular expression of length n, one can construct a deterministic finite automaton with $O(2^n)$ states accepting L.*

Proof We show how to construct automata to accept regular languages. Clearly, $\emptyset$ and $a \in \Sigma$ are fine. Otherwise, a regular language is built from one of lower logical complexity via one of parts (iii)–(v) of Definition 12.4.2. But now the result follows by the closure properties of Theorem 12.3.3. The complexity $O(2^n)$ comes from looking at the complexity of the inductive definitions of the machines for union, star, and concatenation.

Conversely, suppose that L is finite state. Let $L = L(M)$ with $M = \langle K, \Sigma, \delta, S, F \rangle$ a deterministic finite automaton. We need a regular language R with $R = L(M)$. Let $K = \{q_1, \ldots, q_n\}$, and $S = \{q_1\}$. For $i, j \in \{1, \ldots, n\}$, and $1 \le k \le n+1$, let $R(i, j, k)$ denote the set of all strings in Σ^* derivable in M from state q_i to state q_j without using any state q_m for $m \ge k$.

That is, we let

$$R(i, j, k) = \{\sigma : \langle q_i, \sigma \rangle \vdash^* \langle q_j, \lambda \rangle \wedge \forall \gamma \big[\langle q_i, \sigma \rangle \vdash^* \langle q_m, \gamma \rangle \to$$
$$\big(m < k \vee (\gamma = \lambda \wedge m = j) \vee (\gamma = \sigma \wedge m = i) \big) \big] \}.$$

Notice that $R(i, j, n + 1) = \{\sigma : \langle q_i, \sigma \rangle \vdash^* \langle q_j, \lambda \rangle\}$. Thus,

$$L(M) = \bigcup \{ R(i, j, n + 1) : q_i \in S \wedge q_j \in F \}.$$

Since the union of a finite number of regular languages is again regular, Theorem 12.4.1 will follow once we verify the lemma below.

Lemma 12.4.1 *For all i, j, k, $R(i, j, k)$ is regular.*

Proof of Lemma 12.4.1 We prove Lemma 12.4.1 by induction upon k, the hypothesis being that $R(i, j, p)$ is regular for all $p \le k$.

For $k = 1$,

$$R(i, j, 1) = \{\sigma : \delta(q_i, \sigma) = q_j\}.$$

Then $R(i, j, 1)$ is finite and hence regular. Now, we claim that

$$R(i, j, k + 1) = R(i, j, k) \cup R(i, k, k) R(k, k, k)^* R(k, j, k).$$

To see that the claim holds, note that $R(i, j, k + 1)$ is the collection of strings σ with $\langle q_i, \sigma \rangle$ moving to $\langle q_j, \lambda \rangle$ without using states with indices $\ge k + 1$. Now to go from state q_i to state q_j without involving q_m for $m \ge k + 1$, we must choose one of the following options.

1. Go from q_i to q_j and only involve states q_p for $p < k$. This is the set $R(i, j, k)$.
2. Otherwise go from q_i to q_j and necessarily involve q_k. Therefore, we must choose one of the three sub-options below.

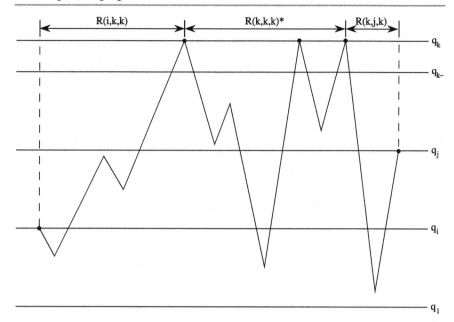

Fig. 12.6 Case (b) of Lemma 12.4.1

(a) First go from q_i to q_k using only q_p for $p < k$. [This is $R(i, k, k)$.]
(b) Go from q_k to q_k some number of times, using only states q_p for $p < k$. [This is $R(k, k, k)$ each time and hence gives $R(k, k, k)^*$.]
(c) Finally, go from q_k to q_j only using states q_p for $p < k$. [This is $R(k, j, k)$.]
Figure 12.6 may be helpful in visualizing (b) above. Thus the contribution of (b) is $R(i, k, k)R(k, k, k)^* R(k, j, k)$. Now we can apply the induction hypothesis to each of the expressions in the decompositions of $R(i, j, k + 1)$ of the claim. This observation establishes the claim, and hence the Lemma and finishes the proof of Theorem 12.4.1.   □

Example 12.4.2 Figure 12.7 demonstrates the construction of a nondeterministic automaton accepting a given regular expression. Let $L = L((ab \cup abb)^*)$.

Example 12.4.3   Armed with Theorem 12.4.1, it is easy to write a language that takes exponential time for Thompson's construction. A deterministic automaton accepting the language

$$L = (0 \cup 1)^* 0 \underbrace{(0 \cup 1) \cdots (0 \cup 1)}_{n \text{ times}}$$

can be shown to require $O(2^n)$ states but is accepted by a nondeterministic automaton with $O(n)$ states (Exercise 12.4.1).

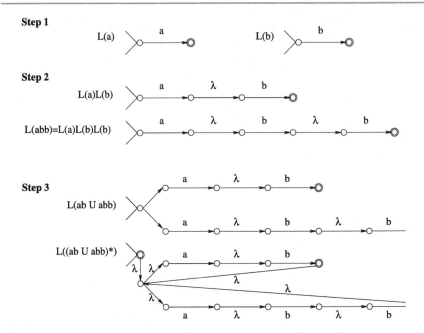

Fig. 12.7 Construction of an automaton accepting L

We remark that it is possible to construct a nondeterministic finite automaton with $k+1$ states for which *any* deterministic automaton requires 2^k+1 states (Leiss [484]) (see Exercise 12.4.2).

12.4.1 Historical Notes

Kleene's theorem is taken from Kleene [448]. The now-standard proof that finite-state languages are regular was formulated by McNaughton and Yamada [525].

12.4.2 Exercises

Exercise 12.4.1
1. Following the proof of Theorem 12.4.1, construct a nondeterministic finite automaton with $O(n)$ states accepting the language L of Example 12.4.3.
2. Prove that any deterministic finite automaton accepting such L requires $\Omega(2^n)$ states.

Exercise 12.4.2 (Leiss [484])   Let L be the language denoted by the regular expression

$$2(0 \cup 1)^* 12^* \big((0 \cup 1)2^*\big)^n.$$

Prove that L can be represented by a nondeterministic finite automaton with $n + 2$ states, but any deterministic finite automaton accepting L needs at least $2^{n+2} + 1$ states.

12.5   The Myhill–Nerode Theorem and the Method of Test Sets

The Myhill–Nerode theorem is due to Nerode [545], although Myhill has a similar result in [539]. (See the Historical Notes at the end of the section.) The Myhill–Nerode theorem is a beautiful, purely algebraic characterization of the computational notion of having a finite automaton accepting a language. Regular languages have another algebraic characterization in terms of well-quasi-ordering.

The theorem is of great importance in terms of practical algorithms for minimizing the number of states in a finite automaton. First we will need the theorem!

Definition 12.5.1 (Right congruence)
1. We call a relation R on a set Σ^* a *right congruence* (with respect to concatenation) iff
   (a) R is an equivalence relation on Σ^*;
   (b) for all $x, y \in \Sigma^*$, xRy iff for all $z \in \Sigma^*$, $xzRyz$.
2. For a language L, the *canonical right congruence* (induced by L) is the relation $\sim_L$ defined by

$$x \sim_L y \quad \text{iff} \quad \text{for all } z \in \Sigma^*, xz \in L \text{ iff } yz \in L.$$

Intuitively, right congruences must agree on all extensions. Similarly, we can define a *left* congruence by replacing "$xzRyx$" in Definition 12.5.1 by "$zxRzy$". Finally, we say that R is a congruence if it is both a left and a right congruence. An equivalence relation on Σ^* is said to have *finite index* if it has only a finite number of equivalence classes over Σ^*.

Theorem 12.5.1 (The Myhill–Nerode theorem)
1. *The following are equivalent over Σ^*:*
   (a) *L is finite state;*
   (b) *L is the union of a collection of equivalence classes of a right congruence of finite index over Σ^*;*
   (c) *$\sim_L$ has finite index.*
2. *Furthermore, any right congruence satisfying (ii) is a refinement[2] of $\sim_L$.*

---

[2]Recall that an equivalence relation E is said to be a refinement of an equivalence relation F iff the equivalence classes of F are unions of equivalence classes of E.

Proof (a) $\rightarrow$ (b). Suppose that L is regular and is accepted by the deterministic finite automaton $M = \langle K, \Sigma, S, F, \delta \rangle$. Without loss of generality, let $S = \{q_0\}$. We define $R = R_M$ via

$$x R y \quad \text{iff} \quad \text{there is a } q \in K \text{ such that } \langle q_0, x \rangle \vdash^* \langle q, \lambda \rangle \text{ and } \langle q_0, y \rangle \vdash^* \langle q, \lambda \rangle$$

(that is, on input x or y, we finish up in the same state q). It is evident that R_M is a right congruence since, for all $z \in \Sigma^*$,

$$\langle q_0, xz \rangle \vdash^* \langle q, z \rangle,$$

$$\langle q_0, yz \rangle \vdash^* \langle q, z \rangle.$$

Also, the index of R_M is finite since there are only finitely many states in K. The implication follows since $L = \bigcup \{\sigma \in \Sigma^* : \langle q_0, \sigma \rangle \vdash^* \langle q, \lambda \rangle\}$ with $q \in F$.

(b) $\rightarrow$ (c). Let R be a right congruence of finite index so that L is a union of some of equivalence classes of R. Since R has finite index, it is sufficient to argue that if $x R y$ then $x \sim_L y$, since this implies that each equivalence of $\sim_L$ is the union of some of the equivalence classes of R. Suppose that the implication does not hold. The there are words x, y, z with $x R y$, $xz \in L$ and $yz \notin L$. Since R is a right congruence, $xz R yz$. Let C denote the equivalence class of R to which both xz and yz belong. We now reach a contradiction, since for every equivalence class C of R, either $C \subseteq L$ or $C \cap L = \emptyset$.

(c) $\rightarrow$ (a). Assume that $\sim_L$ has finite index. We consider the deterministic finite automaton M described as follows.

1. The states of M are the equivalence classes of $\sim_L$.
2. The initial state is $[\lambda]_{\sim_L}$, the equivalence class of λ.
3. The transition function δ of M is defined by $\delta([x], a) = [xa]$ for $x \in \Sigma^*$ and $a \in \Sigma$.
4. The accepting states of M are those x with $[x] \subseteq L$.

We must first argue that M is well defined. The first two specifications are unproblematic.

The third specification is well defined since $\sim_L$ is a right congruence. We argue this by contradiction. If $\sim_L$ were not a right congruence, then there would be x, y, z with $x \sim_L y$ but $xz \nsim_L yz$, implying that for some word w we have (without loss of generality, as the cases are symmetric), $xzw \in L$ and $yzw \notin L$. But then, since concatenation is associative, zw witnesses that $x \nsim_L y$, a contradiction. If $[x] = [y]$, that is, $x \sim_L y$, then the right congruence property $\sim_L$ implies $xa \sim_L ya$, or equivalently, $[xa] = [ya]$.

The fourth specification is well defined because each equivalence class of $\sim_L$ is either contained in L or has empty intersection with L. If this were not the case, then there would be x, y with $x \sim_L y$, $x \in L$ and $y \notin L$. But then the empty string λ witnesses that $x = x\lambda \nsim_L y\lambda = y$, a contradiction.

Now we must M as specified above correctly recognizes the words of L. We argue by induction on word length that if

$$\langle [\lambda], x \rangle \vdash^* \langle q, \lambda \rangle$$

then $q = [x]_{\sim_L}$. For the word λ of length zero (the base case) there is nothing to prove. The induction step follows directly from the definition of the transition function.

We can now argue correctness. For any word $x \in \Sigma^*$ we have

$$\langle [\lambda], x \rangle \vdash^* \langle [x], \lambda \rangle$$

and by the fourth specification of M, either $x \in L$ and therefore $[x]$ is an accept state, or $x \notin L$ and $[x]$ is not an accept state. □

One of the beauties of the Myhill–Nerode theorem is in proving that concrete languages *are* finite-state. An unsatisfactory alternative—which one often sees—is to sketch a nondeterministic machine, usually as a "spaghetti-style transition diagram" and then exhort the reader to "see" that it works. As with some inferior styles of software engineering (which seems to us an apt comparison), this is error-prone, and certainly unsatisfying as a method of *proof*. If a concrete language is to be proved finite-state, and the proof begins by defining a finite-state machine M, then there is nothing to be done next, in a *real proof*, except to argue correctness by induction on word length, which is cumbersome, given the nature of the beast. The beauty of the Myhill–Nerode theorem is that the *one* induction on word length in the *proof* of the theorem is sufficient to transfer the issue of regularity to a powerful *abstract* perspective. The strategic deployment of *abstraction* is a key issue in effective software engineering.

We now turn to some examples of using the Myhill–Nerode theorem to establish that certain languages are finite state. We first investigate a more fine-grained variation of the canonical right congruence.

Definition 12.5.2 (Myhill's congruence[3])   Define $x \approx_L y$ to mean that for all $u, v \in \Sigma^*$,

$$uxv \in L \quad \text{iff} \quad uyv \in L.$$

It is easy to prove that $\approx_L$ has finite index if and only if $\sim_L$ has finite index (Exercise 12.5.1).

Example 12.5.1   If L is finite state, so is the reverse L^{rev} of L.

To see this, define $x R y$ to hold iff $x^{\text{rev}} \approx_L y^{\text{rev}}$. As $\approx_L$ is an equivalence relation, so is R. Furthermore, by the Myhill–Nerode theorem, $\sim_L$ has finite index and hence $\approx_L$ has finite index. This implies that R has finite index. Finally, to see that R is a right congruence, suppose that $x R y$ and $z \in \Sigma^*$. Now $x^{\text{rev}} \approx_L y^{\text{rev}}$ and since $\approx_L$ is a (left) congruence, we have $z^{\text{rev}} x^{\text{rev}} \approx_L z^{\text{rev}} y^{\text{rev}}$ and hence $(xz)^{\text{rev}} \approx_L (yz)^{\text{rev}}$. This implies that $xz R yz$ and hence R is a right congruence.

---

[3] Sometimes in the literature this is called the syntactic congruence.

Our result will then follow from the Myhill–Nerode theorem once we can establish that L^{rev} is a union of equivalence classes of R. We need to show that if $x R y$, then $x \in L^{\text{rev}}$ implies $y \in L^{\text{rev}}$. If $x \in L^{\text{rev}}$, then $x^{\text{rev}} \in L$. If $x R y$, then $x^{\text{rev}} \approx_L y^{\text{rev}}$ and hence $y^{\text{rev}} \in L$, which implies that $y \in L^{\text{rev}}$.

In Exercises 12.5.2–12.5.4, we invite the reader to apply the technique of Example 12.5.1 to other languages.

Another beautiful use of the Myhill–Nerode characterization is in efficiently demonstrating the *non-regularity* of languages by exhibiting infinitely many equivalence classes for $\sim_L$. This technique is often significantly more elegant than classical approaches using the "Pumping lemma" (Corollary 12.5.1 and Exercise 12.5.8). Here is an example.

Example 12.5.2 We will demonstrate that the language $L = \{a^n b^n : n > 1\}$ is not regular using the Myhill–Nerode theorem. For all $n < m$, $a^n \not\approx_L a^m$ since $a^n b^n \in L$ yet $a^m b^n \notin L$. Hence, $\sim_L$ has infinite index, and therefore L is not regular.

In fact, the classical Pumping lemma is an easy consequence of the Myhill–Nerode theorem as we now see.

Corollary 12.5.1 (Pumping lemma, Bar-Hillel, Perles, and Shamir [53]) *Suppose that L is regular and infinite. Then there exist strings x, y, and z, with $y \neq \lambda$, such that for all $n \geq 1$,*

$$x y^n z \in L.$$

Proof Define a collection C of strings inductively. Let $y_0 = \lambda$ and y_{n+1} denote the lexicographically least string extending y_n such that (i) $|y_{n+1}| = n + 1$ and (ii) y_{n+1} has an extensions in L. By the Myhill–Nerode theorem, there exist $n < m$ such that $y_n \sim_L y_m$. Let $x = y_n$ and $y_m = xy$. Since $\sim_L$ is a right congruence, by elementary induction we have the infinite set of equivalent words

$$x \sim_L xy \sim_L xy^2 \sim_L xy^3 \sim_L \cdots$$

Since x has an extension in L, there is a z such that $xz \in L$. The theorem follows. □

The Method of Test Sets One particularly useful strategy for showing that a formal language $L \subseteq \Sigma^*$ is finite-state, is the *method of test sets*.

In the context of graph algorithms, this is particularly powerful in delivering FPT results, for graphs of bounded width, e.g., pathwidth or treewidth—but also many other "width" measures for many kinds of mathematical objects—anything that supports FPT parsing into strings or trees of symbols. Crucial to the strategy in those settings is the translation back-and-forth between the formal language *tests* as expressed in the parsed setting, and the corresponding *tests* as viewed in the setting of the (unparsed) combinatorial objects of interest. (This will become clearer in the chapters ahead, where it will be seen that the translations of the tests between the parsed and unparsed settings provide much of the force of the strategy.)

The strategy is summarized as follows and consists of the steps:

(1) Define a finite number of "tests" (which can be either *abstract* or *concrete*) that capture the essential information about the property of interest. What we mean by this, specializing to the concrete situation of formal languages $L \subseteq \Sigma^*$, is that we have a finite set of *yes/no questions* with which we *interrogate* each word of Σ^*. We define words x and y to be T-equivalent, xTy, if they give the same answers to the tests. Since there are a finite number of questions, the number of equivalence classes is necessarily finite.

(2) Argue that R is a right congruence.

(3) Argue that if xRy, then x and y agree about membership in L.

By a set of *concrete tests*, in the setting of formal languages, we mean simply a list of words $\tau_1, \tau_2, \ldots, \tau_m$, where the interrogation (of $x \in \Sigma^*$) consists of the m questions:

Is $x\tau_i \in L$?

The following proposition shows that we can *always* view L-regularity in terms of (concrete) test sets. The interesting point in the chapters ahead is that viewing regularity in terms of *abstract test sets* seems more powerful, and useful, and enables us to prove concrete FPT results apparently not available in any other way.

Proposition 12.5.1 *Suppose that L is regular. Then there exists a finite set $\{\tau_1, \ldots, \tau_n\}$ of strings such that for all strings x and y, $x \sim_L y$ if and only if, for all τ_i, $x\tau_i \in L$, if and only if, $y\tau_i \in L$. The τ_i can be taken as representatives of the $\sim_L$ equivalence classes.*

Proof Let $\{\tau_1, \ldots, \tau_n\}$ represent the $\sim_L$ equivalence classes. Suppose that we have x, y such that for all τ_i, $x\tau_i \in L$ if and only if $y\tau_i \in L$. Suppose that $x \not\sim_L y$. Then it must be (w.l.o.g.) that for some string z, $xz \in L$ and $yz \notin L$. Now since $\{\tau_1, \ldots, \tau_n\}$ represent the $\sim_L$ equivalence classes, we know that $z \sim_L \tau_i$, for some i. But this is a contradiction, as $xz \in L$ if and only if $x\tau_i \in L$ and $yz \in L$ if and only if $y\tau_i \in L$, by the definition of $\sim_L$. $\square$

Usually the tests prescribed by the *method of test sets* (whether concrete or abstract) are computable by a known algorithm, and usually membership in L is also somehow known to be computable. In these circumstances, we can compute a finite-state machine recognizing L, by methods that delve a little more deeply into the algorithmic aspects of the proof of Theorem 12.5.1.

What information do we constructively need about a regular language L in order to be able to compute a finite-state machine recognizing L? Suppose we only know a description of a Turing machine whose set of accepted words is L. Is this enough? The answer is *No*.

Table 12.1 Table of test set answers

×	λ	0	1	00	01	10	11	000	001	010	011	100	101	110	111
													Test strings z		
λ	No	No	No	No	Yes	No	No	No	No	No	Yes	No	Yes	No	No
0	No	No	Yes	No	No	No	Yes	No	Yes	No	No	No	Yes	No	Yes
1	No	No	No	No	Yes	No	No	No	No	No	Yes	No	Yes	No	No
00	No	No	No	No	Yes	No	No	No	No	No	Yes	No	Yes	No	No
01	Yes	No	Yes	No	Yes	No	Yes	No	No	No	Yes	No	Yes	No	Yes
10	No	No	Yes	No	No	No	Yes	No	Yes	No	No	No	Yes	No	Yes
11	No	No	No	No	Yes	No	No	No	No	No	Yes	No	Yes	No	No
000	No	No	Yes	No	No	No	Yes	No	Yes	No	No	No	Yes	No	Yes
001	No	No	No	No	Yes	No	No	No	No	No	Yes	No	Yes	No	No
010	No	No	Yes	No	No	No	Yes	No	Yes	No	No	No	Yes	No	Yes
011	Yes	No	Yes	No	Yes	No	Yes	No	No	No	Yes	No	Yes	No	Yes
100	No	No	No	No	Yes	No	No	No	No	No	Yes	No	Yes	No	No
101	Yes	No	Yes	No	Yes	No	Yes	No	No	No	Yes	No	Yes	No	Yes
110	No	No	Yes	No	No	No	Yes	No	Yes	No	No	No	Yes	No	Yes
111	No	No	No	No	Yes	No	No	No	No	No	Yes	No	Yes	No	No

Proposition 12.5.2 *There is no algorithm to deduce a finite-state machine for a regular language L from a description of a Turing machine M, which comes with a promise that the language $L = L(M)$ recognized by M is regular.*

Proof If we had such an algorithm, then we could decide, on input M, whether M halts on the empty input (the famously undecidable HALTING PROBLEM), by trivially modifying M to M', so that either:

– if M does not halt and accept on the empty input λ, then M' does not halt on any input, and so $L(M') = \emptyset$, and
– if M halts on and accepts the empty input λ, then M' does likewise. □

However, if we are given a little more information:
– a Turing machine M deciding L
– a promise that $L(M) = L$ is regular
– a decision algorithm for the relation $\sim_L$
then we can compute (*compile*) a finite-state machine M, such that $L(M) = L$. This is left to the reader as an important(!) exercise.

Example 12.5.3 Here is an example of the method of test sets. We will pretend that we have been implicitly given a regular language over $\{0, 1\}^*$, and know that there is some automaton M with ≤ 4 states accepting L. Suppose that we obtained the table of answers of Table 12.1 from the decision procedure for L.

Fig. 12.8 The resulting
automaton

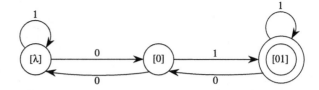

We note that

$$[\lambda] = [1] = [11] = [111] = [100] = [00] = [001],$$

$$[0] = [10] = [110] = [000] = [010],$$

$$[01] = [101] = [011].$$

Hence the three states of M' are $[\lambda]$, $[0]$, and $[01]$. Note that $[01]$ will be the accept state since $01 \in L$. The transition diagram is given in Fig. 12.8.

An important consequence of the proof of the Myhill–Nerode theorem is the fact that the automaton in described on the basis of $\sim_L$ has (uniquely) the least number of states of any deterministic automaton accepting L. Implicitly, the proof provides us with a *state minimization procedure;* that is, we can take an automaton M accepting L, and identify states equivalent under $\sim_L$. Again, since the number of states of L determines the length of the longest path in L, we only need to look at strings of length $\leq |K|$. One minimization algorithm derived from these observations is given below.

Algorithm 12.5.1 (State minimization algorithm)
Step 1. Make a table of entries $(p, q) \in K \times K$ for $p \neq q$.
Step 2. (*Round 1*) For each (p, q) with $p \in F \wedge q \notin F$, mark (p, q).
Step 3. Repeat the following in rounds until there is no change from one round to the next.
(*Round $i + 1$ for $1 \leq i$*) For all (p, q) not yet marked by the end of round i, if for some $a \in \Sigma$, $(\delta(p, a), \delta(q, a))$ is marked, then mark (p, q).
Step 4. Any unmarked squares representing (p, q) represent states that are equivalent and hence can be identified.

In Fig. 12.9, we give an application of the minimization algorithm.
We remark that in Round 2, the only unmarked (x, y) are (a, e), (b, c), (b, d), and (d, c). We mark (c, d) since $\delta(c, 0) = d$, $\delta(d, 0) = e$ and (d, e) was marked in Round 1. We mark (b, d) since $\delta(b, 0) = d$ and $\delta(d, 0) = e$, and, finally, we mark (e, a) since $\delta(e, 1) = e$ and $\delta(a, 1) = b$. In Round 3, there is no change from Round 2 and the algorithm terminates. Notice that the cost of the algorithm is at most $O(n^2)$ if M has n states (for a fixed Σ).

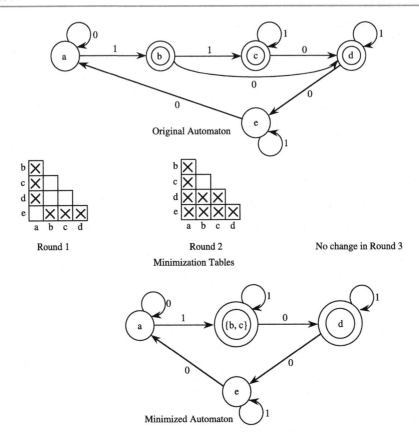

Fig. 12.9 The minimization algorithm

12.5.1 Historical Notes

The Myhill–Nerode theorem is due to Nerode [545]. Myhill [539] proved Theorem 12.5.1 but with the congruence $\approx_L$ of Definition 12.5.2 in place of $\sim_L$ in (iii), and proved that $\approx_L$ is the coarsest congruence (rather than *right* congruence) such that L is a union of equivalence classes of $\approx_L$. The algorithm to minimize the number of states of a finite automaton can be found in Huffmann [403] and Moore [537]. The Pumping lemma is due to Bar-Hillel, Perles and Shamir [53]. The Method of Test Sets is implicit in the Myhill–Nerode theorem but, as best we can tell, was only made explicit in Abrahamson and Fellows [5] where it applied in a general setting of languages consisting of parse trees of graphs of bounded treewidth. (We will discuss this point more fully in Sect. 12.7.)

12.5.2 Exercises

Exercise 12.5.1   Prove that the equivalence relation $\approx_L$ of Definition 12.5.2 has finite index iff $\sim_L$ does.

Exercise 12.5.2   Let L_1 and L_2 be finite state. Define $shuffle(L_1, L_2) = \{x : x \in shuffle(u, v) \text{ where } u \in L_1 \text{ and } v \in L_2\}$. Here, $shuffle(u, v) = \{y : y$ is obtained by some interleaving of the sequences of letters of u and $v\}$.
   Prove that $shuffle(L_1, L_2)$ is finite state using the technique of Example 12.5.1.
(Hint: Define R via xRy iff $(x \sim_{L_1} y \wedge x \sim_{L_2} y \wedge S_x = S_y)$, where $S_z = \{[u]_{\sim_{L_1}}, [v]_{\sim_{L_2}} : z \in shuffle(u, v)\}$.)

Exercise 12.5.3   Let $x, y \in \Sigma^*$. Define the *Hamming distance* between x and y to be

$$d(x, y) = \begin{cases} \text{the number of letters that } x \text{ and } y \text{ differ} & \text{if } |x| = |y|, \\ \infty & \text{otherwise.} \end{cases}$$

Use the method of Example 12.5.1 to prove that for any $r \in \mathbb{Z}$, if L is finite state, then so is

$$H_r(L) = \left\{ x \in \Sigma^* : \exists y \in L \big[d(x, y) \le r \big] \right\}.$$

(Hint: For $x \in \Sigma^*$, let $S_{x,i} = \{[u]_{\sim_L} : d(x, u) = i\}$. For $i \le r$, let $x R_i y$ iff $S_{x,i} = S_{y,i}$. Argue via $R = \bigcap_{0 \le i \le r} R_i$.)

Exercise 12.5.4   Use the method of Example 12.5.1 to prove that if L is finite state, then so is $(1/3)L = \{x \in \Sigma^* : \exists y \in \Sigma^* (xy \in L \wedge |xy| = 3|x|)\}$.
(Hint: Define $x R y$ to hold iff $x \approx_L y \wedge S_x = S_y$, where $S_z = \{[u]_{\sim_L} : |zu| = 3|z|\}$.)

Exercise 12.5.5   Apply the method of test sets to the data in Table 12.2.

Exercise 12.5.6   Apply the state minimization algorithm (Algorithm 12.5.1) to the automaton of Fig. 12.10.

Exercise 12.5.7   Use the technique of Example 12.5.2 to demonstrate the non-regularity of the following languages
   (i) $\{a^p : p$ is prime$\}$;
   (ii) $\{a^n b^m : n \ne m\}$.

Exercise 12.5.8 (The Pumping lemma)   In this exercise, we will use the classical approach to proving Pumping lemma: *If L is an infinite regular language, then there exist $x, y, z \in \Sigma^*$ with $y \ne \lambda$ such that*

$$\forall n > 0 \quad (xy^n z \in L).$$

Prove the Pumping lemma (Corollary 12.5.1) by using induction on the length of the regular expression α with $L = L(\alpha)$.

Table 12.2 Table for Exercise 12.5.5

	Test strings z														
×	λ	0	1	00	01	10	11	000	001	010	011	100	101	110	111
λ	No	No	Yes	No	Yes	No	Yes	No	Yes	No	Yes	No	Yes	No	Yes
0	No	No	Yes	No	Yes	No	Yes	No	Yes	No	Yes	No	Yes	No	Yes
1	Yes	No	Yes	No	Yes	No	Yes	No	Yes	No	Yes	No	Yes	No	Yes
00	No	No	Yes	No	Yes	No	Yes	No	Yes	No	Yes	No	Yes	No	Yes
01	Yes	No	Yes	No	Yes	No	Yes	No	Yes	No	Yes	No	Yes	No	Yes
10	No	No	Yes	No	Yes	No	Yes	No	Yes	No	Yes	No	Yes	No	Yes
11	Yes	No	Yes	No	Yes	No	Yes	No	Yes	No	Yes	No	Yes	No	Yes
000	No	No	Yes	No	Yes	No	Yes	No	Yes	No	Yes	No	Yes	No	Yes
001	Yes	No	Yes	No	Yes	No	Yes	No	Yes	No	Yes	No	Yes	No	Yes
010	No	No	Yes	No	Yes	No	Yes	No	Yes	No	Yes	No	Yes	No	Yes
011	Yes	No	Yes	No	Yes	No	Yes	No	Yes	No	Yes	No	Yes	No	Yes
100	No	No	Yes	No	Yes	No	Yes	No	Yes	No	Yes	No	Yes	No	Yes
101	Yes	No	Yes	No	Yes	No	Yes	No	Yes	No	Yes	No	Yes	No	Yes
110	No	No	Yes	No	Yes	No	Yes	No	No	No	Yes	No	Yes	No	Yes
111	Yes	No	Yes	No	No	No	Yes	No	Yes	No	No	No	Yes	No	Yes

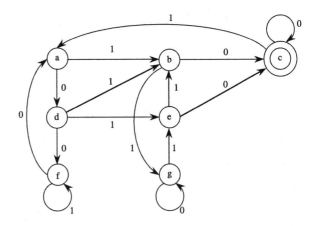

Fig. 12.10 Automaton for Exercise 12.5.6

Exercise 12.5.9 (Kozen [464]) For a language A and a function f, define

$$A_f = \big\{ y : \exists x \big(|y| = f(|x|) \wedge xy \in A \big) \big\}.$$

We call a function *regularity preserving* iff A is regular iff A_f is regular. A function $f : \mathbb{N} \mapsto \mathbb{N}$ is called *eventually periodic* iff there exists a $p > 0$ such that for almost all x, $f(x) = f(x + p)$. Finally, a language $U \subseteq \mathbb{N}$ is called eventually periodic iff there exists a q such that for almost all n,

$$n \in U \quad \text{iff} \quad n + p \in U.$$

Prove that the following are equivalent for $f : \mathbb{N} \mapsto \mathbb{N}$:
1. f is regularity preserving;
2. A is regular iff $A'_f =_{\text{def}} \{x : \exists y (y \in A \wedge f(|y|) = |x|)\}$;
3. f is eventually periodic modulo m for all $m \geq 1$ and $f^{-1}(\{n\})$ is eventually periodic for all $n \in \mathbb{N}$;
4. f preserves eventual periodicity.

(Hint: Use Myhill–Nerode theorem.)

12.6   Tree Automata

The purpose of this section is to generalize the class of objects accepted by automata to labeled rooted trees. To allow such a generalization, we will need to modify our notion of automata. Once we have done this modification, we find that the resulting theory follows the standard theory fairly closely. Our treatment will be a little nonstandard and will be basically tailored to the graph-theoretic applications in the sections to follow. In particular, as best we can, we will avoid the abstraction to universal algebra that many authors use. It suffices to say that it is possible to put all these results in a more general context such as syntactic monoids in texts such as those by Giécseg and Steinby [345], Mezei and Wright [530], or Kozen [463].

By *tree*, we will mean a (directed) rooted tree[4] with labels chosen from Σ at each of the nodes, and each node has fan-in bounded by some constant f. We will denote the collection of such trees over Σ as Σ^{**}. If $T \in \Sigma^{**}$, we will denote the root of T by r_T.

In Fig. 12.11, we give an example of a tree over $\{0, 1\}$. Although the example of Fig. 12.11 is a binary tree, any fixed fan-in f will do. The reader would do well to keep the binary example in mind, for most applications seem to need only binary trees. Notice that we can consider a string as a *unary* tree with the single leaf of a string such as $a_1 \ldots a_n$ being the leftmost symbol a_1 and a_n being the root. As we have seen, automata act on such unary trees by reading the symbols from leaf to root, changing states as they move down the tree. The extension to more general trees is natural. In a *leaf-to-root* tree automaton M, we will move from the leaves to the root, changing states as we are directed by the labels on the nodes. We know what such movement means for *unary* nodes, but we have not specified what such movement means for k-ary nodes for $k \leq f$. The idea is clear. Suppose that v is

---

[4]We will not indicate the directions of the edges on the trees since it will be totally clear from context which vertex is the root. Similarly there will always be an implicit ordering of the children of each node. Some authors such as Courcelle and Engelfriet [178] argue that this should all be set in the more general language of *terms* in *universal algebra*, but we have chosen to stay with the more concrete approach of trees which seem more intuitive, at least to us.

Fig. 12.11 An example of a
tree

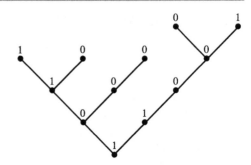

a node of a tree T with k inputs $i_1, \ldots, i_k$, and suppose that v is labeled $a \in \Sigma$. Inductively, we may suppose that corresponding to the k input lines $i_1, \ldots, i_k$, the machine would be in states $q_{j_1}, \ldots, q_{j_k}$ (i.e., "simultaneously"). Then, at node v the machine would move to the state specified by the $k + 1$-tuple $\langle q_{j_1}, \ldots, q_{j_k}, a \rangle$. (Here, of course, we are regarding the nodes v of the tree as being lexicographically ordered in some standard fashion by $\leq_L$. Thus, $i_1 <_L \cdots <_L i_k$, for example.)

Definition 12.6.1 (Deterministic finite tree automaton) A *Tree Automaton* over Σ is thus a quintuple $\langle K, \Sigma, \delta, S, F \rangle$ and a bound f (the *fan-in* bound) where K, S, F are as for a standard automaton (as in Definition 12.2.1) and δ is a function from (a subset of) $\bigcup_{1 \leq i \leq f}(\underbrace{K \times \cdots \times K}_{i \text{ times}} \times \Sigma)$ to K.

We remark that it is also possible to study *root-to-leaf* automata (Exercise 12.6.1).

We will write $\delta(x_1, \ldots, x_i, a)$ in place of $\delta(\langle x_1, \ldots, x_i \rangle, a)$. Notice that if the trees were all binary and all nonleaf nodes had fan-in 2 and $S = \{q_0\}$, then we could equally well specify δ simply by specifying $\delta(q_0, a)$ for all $a \in \Sigma$ and $\delta(q_i, q_j, a)$ for $q_i, q_j \in K$ and $a \in \Sigma$.

We will have an analogy of the "yield" relation $\vdash^*$ for standard deterministic finite automata. We call this an evaluation rule *eval*. Let $T_1, \ldots, T_n$ be labeled rooted trees. We denote by $T_1 \odot_a \cdots \odot_a T_n$ the tree obtained by adding a further node r and gluing the roots $r_{T_1}, \ldots, r_{T_n}$ to r and labeling r by a. The example in Fig. 12.12 makes this basic action clear.

It is obvious that all trees in Σ^{**} are obtainable from Σ and $\odot_a$ for $a \in \Sigma$. We regard $\lambda \odot_a = a$ for $a \in \Sigma$, so, in fact, all of Σ^{**} can be generated by λ and $\odot_a$, for $a \in \Sigma$.

Definition 12.6.2 (Tree automaton acceptance) If $M = \langle K, \Sigma, \delta, S, F \rangle$ is a finite deterministic tree automaton, the function $eval_M$ is defined via

1. $eval_M(q_j, a) = \delta(q_j, a)$ for $a \in \Sigma$ and $q_j \in S$ [and, without loss of generality, we will assume that $S = \{q_0\}$ and write $eval(a)$ for $eval(q_0, a)$].
2. $eval_M(T_1 \odot_a \cdots \odot_a T_n) = \delta(eval_M(T_1), \ldots, eval_M(T_n), a)$ for $a \in \Sigma$.

We say that $T \in \Sigma^{**}$ is *accepted* iff $eval_M(T) \in F$.

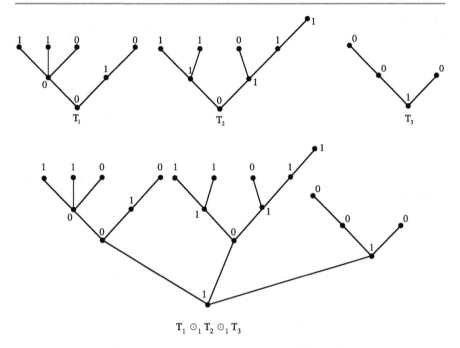

Fig. 12.12   Gluing labeled trees

Example 12.6.1   Specify M as follows. Let $\Sigma = \{0, 1\}$ so that Σ^{**} is the collection of all rooted, labeled binary trees. Let $S = \{q_0\}$ and $F = \{q_1\}$. We specify δ by the following table.

States	Symbol	δ(states, symbol)
q_0	0	q_0
q_0	1	q_1
q_1	0	q_1
q_1	1	q_1
q_0, q_0	0	q_0
q_0, q_0	1	q_1
q_0, q_1	0	q_0
q_0, q_1	1	q_1
q_1, q_0	0	q_0
q_1, q_0	1	q_0
q_1, q_1	0	q_0
q_1, q_1	1	q_1

In Fig. 12.13, we have marked the stages of computing $eval(T)$ for the given tree. Note that $eval(T) = q_0$ and hence $T \notin L(M)$ for the given tree automaton M.

Fig. 12.13 Computation of a tree evaluation

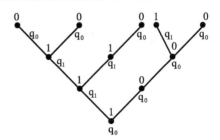

If we were to represent the evaluation of Fig. 12.13 via $\odot$, we would have the following:

$$T = \big[(0 \odot_1 0) \odot_1 (0 \odot_1)\big] \odot_1 \big[(1 \odot_0 0) \odot_0\big],$$

$$eval(T) = \delta\big(eval\big((0 \odot_1 0) \odot_1 (0 \odot_1)\big), eval\big((1 \odot_0 0) \odot_0\big), 1\big)$$

$$= \delta\big(\delta\big(eval(0 \odot_1 0), eval(0 \odot 1), 1\big), \delta\big(eval(1 \odot_0 0), 0\big), 1\big)$$

$$= \delta\big(\delta\big(\delta\big(eval(0), eval(0), 1\big), \delta\big(eval(0), 1\big), 1\big),$$

$$\delta\big(\delta\big(eval(1), eval(0), 0\big), 0\big), 1\big)$$

$$= \delta\big(\delta\big(\delta\big(\delta(q_0, 0), \delta(q_0, 0), 1\big), \delta\big(\delta(q_0, 0), 1\big), 1\big),$$

$$\delta\big(\delta\big(\delta(q_0, 1), \delta(q_0, 0), 0\big), 0\big), 1\big)$$

$$= \delta\big(\delta\big(\delta(q_0, q_0, 1), \delta(q_0, 1), 1\big), \delta\big(\delta(q_1, q_0, 0), 0\big), 1\big)$$

$$= \delta\big(\delta(q_1, q_1, 1), \delta(q_0, 0), 1\big)$$

$$= \delta(q_1, q_0, 1)$$

$$= q_0.$$

We remark that there is no problem with defining *nondeterministic finite tree automata* by generalizing δ to a multifunction as we did in Sect. 12.3. We invite the reader to check that the details of Rabin and Scott's theorem (Theorem 12.3.1) go through exactly as before since the subset/Thompson construction will work in the same way (Exercises 12.6.3 and 12.6.4).

We do encounter a little more difficulty when we hit the result that says "finite-state languages are regular" since we do not yet have an analog of the concept of regularity for trees. We need a syntactic definition of languages accepted by tree automata. As we will see, there are two such notions.

For our first syntactic definition, we will view finite-state tree languages as a subset of the set of *context-free languages*. (See, e.g. Hopcroft and Ullman [401].) For such purposes, it is easiest to think in terms of the algebraic approach given in Example 12.6.1. It is easiest to think in terms of *productions* which involve "replacements," as we see below. For this purpose, we augment the set of trees with some new *leaf* symbols. Let R be any set of new symbols disjoint from Σ. We will say a tree T is a tree *over* Σ_R if T is the result of taking a tree over Σ and (possibly) replacing some of the *leaves* by members of R. We will let Σ_R^{**} denote the collection of trees over Σ_R.

Definition 12.6.3 (Tree grammar) A tree grammar G over Σ is a quadruple $\langle \Sigma, S, R, P \rangle$ where
  (i) R is a set of symbols with $R \cap \Sigma = \emptyset$ called *nonterminal* symbols,
 (ii) $S \in R$ is a distinguished symbol called the *initial* symbol, and
(iii) P is a set of *productions* of the form

$$N \to Q \quad \text{with } N \in R \text{ and } Q \in \Sigma_R^{**}.$$

We will sometimes call Σ the set of *terminal* symbols.

For $T_1, T_2 \in \Sigma_R^{**}$, we will write $T_1 \to T_2$ if there exist words X_i and Y_i for $i \in \{1, 2\}$, a *leaf* $N \in R$ of T_1, and a production $N \to Q$ such that $T_1 = X_1 N Y_1$ and $T_2 = X_2 Q Y_2$. Clearly, we can extend the notion of $\to$ inductively as we did with $\vdash$ and, hence, for $T \in \Sigma_R^{**}$, we say that $T \to^* T'$ iff there are $T_1, \ldots, T_n$ such that

$$T \to T_1 \to \cdots \to T_n \to T'.$$

We refer to such a string as a *derivation* of T' from T. Finally, we say that $T \in L(G)$ if
1. $T \in \Sigma^{**}$ (that is, T has no remaining nonterminal symbols), and
2. there is a derivation of T obtainable from S using the productions of G.

Example 12.6.2 Here is a simple example. Let $\Sigma = \{0, 1, 2\}$. The productions are $S \to N_1$ together with the following.
1. $N_1 \to 0$, $N_2 \to 1$,
2. $N_1 \to (N_1 \odot_0)$,
3. $N_1 \to (N_1 \odot_0 N_2 \odot_0 N_3)$,
4. $N_2 \to ((N_2 \odot_0 2 \odot_0 1) \odot_1 ((0 \odot_0) \odot_1) \odot_1 1)$, and
5. $N_3 \to (1 \odot_2)$.

In Fig. 12.14, we represent the productions above graphically, and represent graphically the derivation of the tree T with algebraic derivation below.

$S \to N_1$

$\quad \to (N_1 \odot_0)$ (by 2)

$\quad \to ((N_1 \odot_0 N_2 \odot_0 N_3) \odot_0)$ (by 3)

$\quad \to ((0 \odot_0 N_2 \odot_0 N_3) \odot_0)$ (by 1)

$\quad \to ((0 \odot_0 ((N_2 \odot_0 2 \odot_0 1) \odot_1 ((0 \odot_0) \odot_1) \odot_1 1) \odot_0 N_3) \odot_0)$ (by 4)

$\quad \to ((0 \odot_0 ((1 \odot_0 2 \odot_0 1) \odot_1 ((0 \odot_0) \odot_1) \odot_1 1) \odot_0 N_3) \odot_0)$ (by 1)

$\quad \to ((0 \odot_0 ((1 \odot_0 2 \odot_0 1) \odot_1 ((0 \odot_0) \odot_1) \odot_1 1) \odot_0 (1 \odot_2)) \odot_0)$ (by 5).

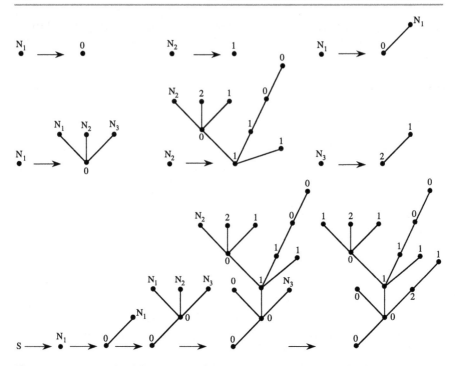

Fig. 12.14 An example of productions and a derivation based on these productions

Lemma 12.6.1 (Grammar normalization lemma) *Suppose that there is a tree grammar G accepting L. Then, we can compute an equivalent grammar G' from G with
the property that if $N \to T$ is a production of G', then the height of T is ≤ 1. Furthermore, we can ensure that G has no productions of the form $N_1 \to N_2$ with both
N_1 and N_2 in $R(G')$. We will call such G' normalized.*

Proof First, note that all productions of the form $N_1 \to N_2$ with both N_1 and N_2
in R can be deleted, provided that we add to P all productions of the form $N_1 \to
Q$ with $Q \in \Sigma_R^{**} - R$ such that, for some $N_3 \in R$, $N_1 \to^* N_3$ and P contains a
production of the form $N_3 \to Q$.

Now, suppose that $N \to T$ is a production of G with the height of $T \geq 2$. Then
the root of T is of the form $G_1 \odot_a \cdots \odot_a G_m$ for some $a \in \Sigma$ and $G_i \in \Sigma_R^{**}$.
We augment R by adding new nonterminal symbols $M_1, \ldots, M_m$, and add new
productions of the form

$$N \to M_1 \odot_a \cdots \odot_a M_m$$

and

$$M_i \to G_i.$$

It is evident that the grammar obtained is equivalent to G, and in this way, we can
normalize G.   □

We can now state our first generalization of Kleene's syntactic characterization of finite-state languages. It is difficult to credit the following theorem to any author or set of authors. We refer the reader to the Historical Notes at the end of the section.

Theorem 12.6.1   *A tree language L is finite state iff there is a tree grammar G with* $L = L(G)$.

Proof Let $L = L(G)$ with G a tree grammar. By the Grammar Normalization Lemma, we can suppose that G is normalized. For each $N \in R$, let $root(N)$ denote the set of labels of roots of trees Q with $N \to Q$, a production of G. If $a \in \Sigma$, let $root(a) = \{a\}$. Finally, if $C \in \Sigma \cup R$, we let $entry(C)$, the entry state of C, be $\{q_0\}$ if $C \in \Sigma$, and $entry(C) = \{q_C\}$ otherwise.

For each $N \in R$, we will create a state q_N. Now consider a production $N \to Q$. If $Q = b \in \Sigma$, then we will create a transition $\delta(q_0, b) = q_N$. If Q has height 1, then Q is of the form $b_1 \odot_a \cdots \odot_a b_m$ for some $a \in \Sigma$ and $b_1, \ldots, b_m \in \Sigma \cup R$. We will associate new states $q_i' = q_{b_i, N, Q}'$ for $1 \le i \le m$, and transitions of the form

$$\delta\big(entry(b_i), root(b_i)\big) = q_i',$$

and

$$\delta\big(q_1', \ldots, q_m', a\big) = q_N.$$

The point here is that the leaves of Q are either labeled with some symbol $M \in R$ or they are in Σ. In the latter case, the only possible entry state that the automaton can have is q_0. On the other hand, if the leaf is labeled $M \in R$, then it can only be replaced by another tree in Σ_R^{**} via some production of the form $M \to Q'$. Now we know that the entry state to Q generated by such a production will be the exit state of Q', which will be q_M. The symbol that corresponds to the production $M \to Q'$ will be the root symbol of Q'.

Conversely, suppose that M is a tree automaton accepting L. We need to construct a tree grammar G for L. For each state q_i of M, we will have a nonterminal symbol Q_i. Corresponding to a transition of the form $\delta(q_{i_1}, \ldots, q_{i_m}, a) = q_k$, we will have a production of the form

$$Q_k \to Q_{i_1} \odot_a \cdots \odot_a Q_{i_m}.$$

We also have productions of the form $Q_0 \to \lambda$ and $S \to Q_d$ for any accept state q_d. The reader should establish by induction that $L = L(G)$ (Exercise 12.6.7).   □

It is also possible to get another characterization of finite-state tree languages via regular tree expressions. We need a definition.

Definition 12.6.4 (Product of tree languages)
1. Let $L, W \in \Sigma^{**}$. For $a \in \Sigma$, the *a-product of L and W* is the language $L \cdot_a W$ obtained by taking trees $T \in W$ and replacing every occurrence of a as a leaf of T by a tree T' of L. (Different leaves can have different T'. If no occurrence of a occurs as a leaf of T, then this action is taken to be vacuous.)

2. We denote by L^{*a} the *a-star* of L,

$$L^{*a} = \bigcup_{n \geq 1} \underbrace{L \cdot_a \cdots \cdot_a L}_{n \text{ times}}.$$

Definition 12.6.5 (Regular tree expression) A *regular (tree) expression* is defined by induction via the following:
1. Ø is a regular expression.
2. If $a \in \Sigma$, then a is a regular expression.
3. If X and Y are regular expressions and $a \in \Sigma$, then $X \cdot_a Y$ is a regular expression.
4. If X is a regular expression and $a \in \Sigma$, then X^{*a} is a regular expression.
5. If $m \geq 1$, $a \in \Sigma$, and $X_1, \ldots, X_m$ are regular expressions, then so is $X_1 \odot_a$ $\cdots \odot_a X_m$.
6. Nothing else.

Of course, we associate a language $L(\alpha)$ with a regular expression α. This is defined in the obvious way: $L(\emptyset) = \emptyset, L(a) = \{a\}, L(X \cdot_a Y) = L(X) \cdot_a L(Y), L(X^{*a}) = [L(X)]^{*a}$, and, finally,

$$L(X_1 \odot_a \cdots \odot_a X_m) = \{T_1 \odot_a \cdots \odot_a T_m : T_i \in L(X_i)\}.$$

Naturally, we say that L is regular iff $L = L(\alpha)$ for some regular expression α.

The true analog of Kleene's theorem (Theorem 12.4.1) is provided by the following.

Theorem 12.6.2 *For Σ^{**}, a language is finite state iff it is regular*

Proof First, suppose that L is finite state and is accepted by the tree automaton $M = \langle K, \Sigma, \delta, S, F \rangle$. We model our proof on Theorem 12.4.1. But we use a trick. Our induction will involve trees in Σ_K^{**}; that is, the induction will involve trees that are normal except that they may have leaves labeled by states from M. The interpretation of the transition is that δ can only act on a state q if it is state q: $\delta(q, q) = q$.

Let $V \subseteq K$. Let $M(V, j, k)$ denote the language that contains all trees in Σ_V^{**} which take a set of states as entry states and transform them via δ to state j without using any state with index $\geq k$.

We prove by induction on k that $M(V, j, k)$ is regular for all V. Notice that this is enough since $V = \emptyset$ means that $\Sigma_K^{**} = \Sigma^{**}$. This is obvious for $k = 1$. For the induction step, a tree T is in $M(V, j, k + 1)$ iff either $T \in M(V, j, k)$, or
1. subtrees of T are taken to state q_k, then
2. perhaps to q_k many times, and finally
3. to q_j from q_k without involving q_i for $i \geq k$.

As with Theorem 12.4.1, such considerations give the regular expression α corresponding to

$$M(V, j, k) \cup M(V, k, k+1) \cdot_{q_k} M\big(V \cup \{q_k\}, k, k+1\big)^{*k} \cdot_{q_k} M(V, j, k).$$

It is easy to argue that $L = L(\alpha)$ (Exercise 12.6.8).

The converse is straightforward and works by building a nondeterministic finite tree automaton corresponding to a regular expression α by induction on complexity of α (Exercise 12.6.9). □

Finally, we turn to an analog of the Myhill–Nerode theorem.

Definition 12.6.6 ($\sim_L$ for trees) Let $x \notin \Sigma$ be a new symbol. For labeled trees T_1, T_2, we will say that $T_1 \sim_L T_2$ iff for all trees $T \in \Sigma^{**}_{\{x\}}$ which have *exactly* one leaf labeled x (and all other nodes in Σ^{**}), we have

$$T_1 \cdot_x T \in L \quad \text{iff} \quad T_2 \cdot_x T \in L.$$

[That is, for all trees T, if we concatenate the root of T_1 to the leaf of T labeled x, the result is in L iff the same holds for T_2. The analogy with the classical $\sim_L$ is obvious.]

In a similar fashion, we may define an equivalence relation $\equiv$ to be a right congruence over Σ^{**} iff for all $T_1, T_2 \in \Sigma^{**}$, $T_1 \equiv T_2$ iff for all $T \in \Sigma^{**}_{\{x\}}$ with exactly one leaf labeled x,

$$T_1 \cdot_x T \equiv T_2 \cdot_x T.$$

Theorem 12.6.3 (The Myhill–Nerode theorem for tree automata) *The following are equivalent*:
  (i) *L is finite state over Σ^{**}.*
 (ii) *L is a union of equivalence classes of some right congruence R of finite index over Σ^{**}.*
(iii) *$\sim_L$ has finite index over Σ^{**}.*

Proof (i) → (ii) and (ii) → (iii) are virtually identical to the arguments for standard languages. For instance, consider (i) → (ii). Let M be finite state accepting L. As with Theorem 12.5.1, we declare that two trees T_1 and T_2 are R-equivalent iff M takes them to the same state (i.e., from q_0 on all the leaves). Then R is a right congruence and has finite index as M has a finite number of states. Hence, (i) → (ii).

We need a little work for (iii) → (i) since δ is now more complex. Suppose that L is the union of equivalence classes of $\sim_L$ which has finite index. Again we can let the states of the machine M we construct be the congruence classes of $\sim_L$. First, we observe that for any trees $T_1 \sim_L T_2$, and $a \in \Sigma$, for $d < f$, the fan-in bound of Σ^{**}, and trees $Q_1, Q_2, \ldots, Q_d \in \Sigma^{**}$, we have

$$T_1 \odot_a Q_1 \odot_a \cdots \odot_a Q_d \sim_L T_2 \odot_a Q_1 \odot_a \cdots \odot_a Q_d.$$

This follows from the fact that $T_1 \odot_a B = T_1 \cdot_x x \odot_a B$ for any $B \in \Sigma^{**}$. Replacing the Q_i one at a time, the observation generalizes to establish the following fact. For any trees $T_1, \ldots, T_m$ and $T'_1, \ldots, T'_m$ with $m \leq f$, if for all i, $T_i \sim_L T'_i$, then for all $a \in \Sigma$

$$T_1 \odot_a \cdots \odot_a T_m \sim_L T'_1 \odot_a \cdots \odot_a T'_m.$$

Now define

$$\delta\big([T_1], \ldots, [T_m], a\big) = [T_1 \odot_a \cdots \odot_a T_m].$$

From the fact above, it follows that if we specify δ in this way, then δ is well defined. The accept states are again $\{[T] : T \in L\}$, and again, as in Theorem 12.5.1, the initial state is $[\lambda]$.   □

Notice that again we will get an implicit minimization algorithm from the Myhill–Nerode theorem for Trees, and a Method of Test Sets (Exercises 12.6.10 and 12.6.11).

12.6.1 Historical Notes

Tree automata were introduced by many authors. The works Doner [228], Thatcher and Wright [640], Mezei and Wright [530] are some of the relevant papers, but we refer the reader to the works Giécseg and Steinby [345, pp. 121–124], Kozen [463] for more detailed discussions of the evolution of this subject. Courcelle and Engelfriet [178] (for instance) give an abstract but more uniform treatment of the material in these sections using notions from abstract algebra.

12.6.2 Exercises

Exercise 12.6.1 1. Define the notion of a root-to-leaf tree automaton.
2. Prove that there are languages accepted by root-to-leaf nondeterministic automata that are not accepted by any root-to-leaf deterministic automaton.

Exercise 12.6.2 Prove that any finite state $L \subseteq \Sigma^{**}$ can be accepted in linear time.

Exercise 12.6.3 Formulate the notion of a (leaf to root) *nondeterministic finite tree automaton*. Carefully verify the details of the analog of Rabin and Scott's theorem (Theorem 12.3.1) and the Thompson construction go through.

Exercise 12.6.4 Formulate the notion of a *nondeterministic finite tree automaton with λ moves*, and check that the analog of Theorem 12.3.2 holds (i.e., prove that λ moves give no additional computational power).

Exercise 12.6.5 Construct a normalized tree grammar corresponding to the grammar of Example 12.6.2.

Exercise 12.6.6 Specialize the notion of grammar to Σ^*; that is, define a notion of regular grammar for strings, and prove that a language $L \subseteq \Sigma^*$ is finite state iff L is accepted by a regular grammar.

Exercise 12.6.7 Complete the details of the proof of Theorem 12.6.1. [That is, complete the proof that $A(G)$ accepts $L(G)$ for a tree grammar G; and prove that the grammar constructed from G accepts $L(M)$.]

Exercise 12.6.8 Carefully argue that α corresponding to the expression $M(V, j, k)$ $\cup M(V, k, k + 1) \cdot_{q_k} M(V \cup \{q_k\}, k, k + 1)^{*k} \cdot_{q_k} M(V, j, k)$ of Theorem 12.6.2 accepts $M(V, j, k + 1)$.

Exercise 12.6.9 Prove that if α is a regular expression over Σ^{**}, then there is an automaton M accepting $L(\alpha)$.
(Hint: Model the proof on Theorem 12.4.1, where the corresponding result is proven for Σ^* by the inductive nondeterministic definition of M.)

Exercise 12.6.10 Formulate an algorithm based on the proof of the Myhill–Nerode theorem for Trees, which minimizes the number of states of an automaton accepting a finite-state language L over Σ^{**}. (Refer to Algorithm 12.5.1.)

Exercise 12.6.11 Formulate a Method of Test Sets for Σ^{**}.

12.7   Parse Trees for Graphs of Bounded Treewidth and an Analogue of the Myhill–Nerode Theorem

The principal tools we use to connect the theory of tree automata with graphs of bounded treewidth are *parse trees*. The idea is that graphs of bounded treewidth can be generated by a set of simple operations, in the same way that one might parse a grammatical expression. The parse trees that we obtain will be used as the input for the relevant tree automata. In turn, the Myhill–Nerode theorem for trees will be used to generate an analog for "t-boundaried" graphs.

Definition 12.7.1 (t-Boundaried graphs) A t-*boundaried* graph is a graph G with t distinguished vertices being given the labels $\{1, \dots, t\}$. The vertices possessing the labels $\{1, \dots, t\}$ are called the *boundary*, and we let $\partial(G)$ denote the set of boundary vertices of G.[5]

Sometimes, such as in the proof of Courcelle's theorem in Sect. 13, we will write a t-boundaried graph as $G = (V, E, B, f)$, where V and E are, as usual, the vertices and edges of G, and where $\partial(G) = B$ with f the labeling function: f is a bijection from $B \mapsto \{1, \dots, t\}$.

---

[5]For simplicity and with no loss of generality, we will insist that all t-boundaried graphs have labels $\{1, \dots, t\}$ and hence have at least t vertices.

As with trees, the basic operation we will perform on t-boundaried graphs will be gluing.

Definition 12.7.2 (Gluing by $\oplus$) Let G_1, and G_2 be t-boundaried graphs. We denote by $G_1 \oplus G_2$ the t-boundaried graph obtained by taking the disjoint union of G_1 and G_2 and identifying each vertex of $\partial(G_1)$ with the vertex of $\partial(G_2)$ with the same label; that is, we glue them together on the boundaries.

Operators on the class of t-boundaried graphs allow us to generate graphs of bounded treewidth. Again, notice the persistent analogy with $\odot$ of the previous sections.

Definition 12.7.3 (t-Boundaried composition operator $\otimes$, Abrahamson and Fellows [5]) An *n-ary t-boundaried composition operator* is a t-boundaried graph $G_\otimes = \langle V, E \rangle$, together with n injective mappings $f_i : \{1, \dots, t\} \mapsto V$.

Suppose that $G_1, \dots, G_n$ are n t-boundaried graphs and that $\otimes$ is a composition operator according to Definition 12.7.3. Then, we will denote by

$$G_1 \otimes \cdots \otimes G_n$$

the t-boundaried graph obtained by gluing each of the G_i to $G_\otimes$ as indicated by f_i, deleting G_i's labels afterward; that is, for each v in $\partial(G_i)$, if v has label ℓ, then we will identify v with the vertex $f_i(\ell)$ of $G_\otimes$ and regard G_i as unlabeled once we have completed all such identifications. In Fig. 12.15, we illustrate this process. The notation $i^{(n)}$ in this figure denotes the attachment point for the vertex labeled i in the nth argument of $\otimes$. Thus, for instance, labeling a vertex v by $1, 2', 3''$ indicates that we should attach to v the vertex labeled 2 of G_1 and the vertex labeled 3 of G_2; and furthermore v would finish up in the boundary of the composite graph with label 1. Notice that, as Fig. 12.15 illustrates, $\oplus$ is a special example of a composition operator.

We will use special families of composition operators for parsing graphs of fixed treewidth. Before giving a general recipe for parsing treewidth k graphs, we will demonstrate the idea by examining treewidth 2. We will use the (composition) parsing operators from Fig. 12.16. We need $\oplus$ and four other operators, $\emptyset$ a *nullary* operator that simply produces two vertices, together with two binary operators $\otimes_a$ and $\otimes_b$.

It is not difficult to establish that a graph G has treewidth 2 iff G can be parsed using the given parse operators (Exercise 12.7.1). We say that these operators form a *complete set* of parsing operators for treewidth 2 graphs. As we will now see, all treewidth t graphs can be parsed by a simple set of operators. For simplicity, we will prove that all treewidth t graphs can be parsed using a set of *binary*, *unary*, and *nullary* operators on the class of $t + 1$ boundaried graphs. In fact, it is possible to prove that all treewidth t graphs can be parsed by a finite set of operators on the class of t-boundaried graphs but these are more complicated (Exercise 12.7.3).

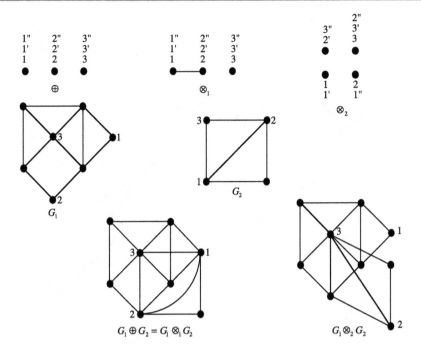

Fig. 12.15 Composition operators and their effects

The collection of size $t + 1$ parsing operators consists of the following set of six operators:

$$\{\emptyset, \oplus, \gamma, i, u, e\}.$$

Here,

1. $\emptyset$ is again the nullary operator[6] which creates boundary vertices labeled $1, \ldots, t + 1$.
2. γ is unary and denotes a cyclic shift of the boundary [that is, γ moves label j to the vertex with label $j + 1$ (mod $t + 1$)].
3. i is a unary operator and denotes an *interchange*: i simply puts label 1 on the vertex currently with label 2, and label 2 on the vertex with label 1.
4. e is the unary operator that adds an edge between the vertices labeled 1 and 2.
5. u is the unary operator that adds a new vertex and labels it 1. (We call this pushing.)

---

[6] Arguably, we could use $\emptyset_{t+1}$ to denote this, but keeping to the spirit of keeping the notation simple, we will not do this and presume that the reader can recognize when the $\emptyset$ denotes the nullary operator, and when it denotes the empty set. We will also similarly suppress the subscript $t + 1$ from the other operators.

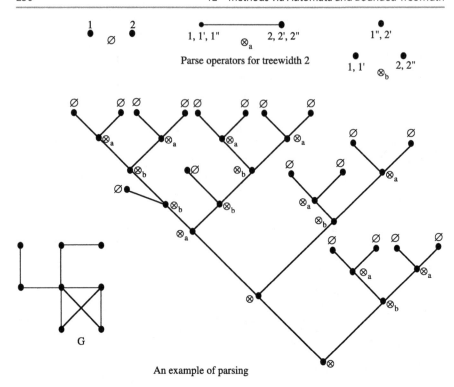

Parse operators for treewidth 2

An example of parsing

Fig. 12.16 Parsing graphs of treewidth 2

Notice that the only operators that add vertices are u and $\oplus$. The only operator that adds an edge is e. Notice that the combination of γ and i enables us to permute the boundary at will, since the symmetric group can be generated by a cycle and a transposition. We remark that the following result is implicit in Abrahamson and Fellows [4], but similar results have been noted by several authors such as Kloks in his Ph.D. thesis.

Theorem 12.7.1 (Parsing theorem)
(i) *There is a linear time algorithm which converts a tree decomposition of width t into a parse tree using the size $t + 1$ parsing operators.*[7]
(ii) *Conversely, if G is any graph generated from a parse tree T using the size $t + 1$ parsing operators, then G has treewidth $\leq t$.*

Proof (i) Let G be a graph with a tree decomposition $\mathcal{T} = \{X_i : i \in T\}$ of width t. We need to construct a parse tree from $\mathcal{T}$. This is done in several steps.
    Step 1. We first perform a preliminary normalization of $\mathcal{T}$.

---

[7]It is important that the reader understand that this theorem only gives a parse tree for the graph *up to isomorphism*. This point is important in some of the work to follow.

1. First, we make the tree $\mathcal{T}$ binary. This is accomplished by duplicating the nodes at which the branch occurs and interpolating as children of the branch node. Thus, if a branch node η had fanout $m > 2$ we would duplicate the node η, $m - 2$ times, eliminating the branches one at a time (see Fig. 12.17).
2. Next, we make all the sets in the tree decomposition uniform of size $t + 1$ by adding new vertices to the node if necessary. This can be accomplished by using vertices from the node to a child to build up the number to $t + 1$. Notice that this action cannot affect the interpolation property (see Fig. 12.17). (This is easily verified by, say, induction.)
3. Now, we ensure that every node has identical children. This property can be ensured by duplicating branch nodes below the branch node.
4. Finally, we ensure that all sets differ by at most one. Thus we will ensure that if $\mathcal{T}' = \{X_i' : i \in T'\}$ is the normalized version of $\mathcal{T}$, then for all edges $uv \in T'$,

$$|X_u \cap X_v| \geq t.$$

This property can be ensured by adding interpolating nodes between X_i and X_j which differ by more than 1 (see Fig. 12.17).

Thus, henceforth, we will presuppose that $\mathcal{T}$ is normalized according to (i)–(iv) of step 1.

Step 2. Now, for each edge $e = uv$ of G, choose a vertex $v(e)$ of $\mathcal{T}$ where e is covered.

Step 3. For each leaf of $\mathcal{T}$, label the vertices with $\{1, \ldots, t + 1\}$ in any order. Now, starting with the leaves of $\mathcal{T}$ and proceeding to parents, generate a parse tree for G via $\mathcal{T}$, as follows:

1. First, locate the edge to be covered. The inductive hypothesis will be that this corresponds to labeled vertices.
2. Using γ and i, rearrange the labels so that 1 and 2 fall on the vertices u and v corresponding to the edge to be added.
3. Now, use e to add the edge between u and v.

There are now two subcases, depending on whether the parent is a branch node or a nonbranch node.

(a) If the parent is a branch node, then use γ and i to align the boundaries so that corresponding vertices have the same labels. (Here is where we use the fact that children of branches are identical.) Branches will be represented by $\oplus$.
(b) If the parent is not a branch node, then without loss of generality, the parent differs from the child by exactly one new vertex (v). First, rearrange the labels on the vertices so that 1 is the label of the vertex to be deleted. Now use u to add a new vertex (labeled 1). This will be the vertex to represent v. Part of this process is represented in Fig. 12.18.

It is clear that in this way we may parse G. Moreover, the size of the parse tree and the time needed to construct is $O(|G|)$.

Part (ii) of the theorem is left to the reader (Exercise 12.7.4). $\square$

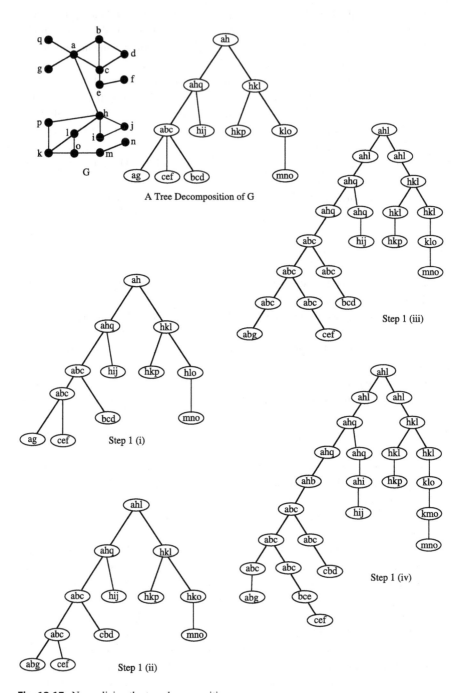

Fig. 12.17 Normalizing the tree decomposition

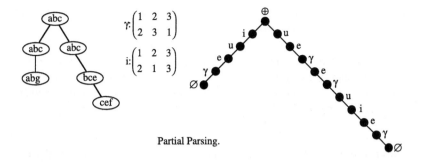

Partial Parsing.

Fig. 12.18 Partial parsing of the normalized decomposition

The way is now clear. To design algorithms for the recognition of whether a bounded treewidth graph obeys a given property, we translate the problem into the arena of tree automata via parse trees. In the arena of finite labeled trees, we have many standard tools at our disposal, and we would hope to be able to use automata theory on parse trees as a *standard design tool* for problems of bounded treewidth. The culmination of this idea is Courcelle's theorem, which we treat in Sect. 13.

There are two universes of discourse for the discussion to follow. For the definition below, we will have fixed a (finite) set of parse operators.

Definition 12.7.4 (Small and large universes)
(1) We let $\mathcal{U}_t^{\text{large}}$ denote the *large* universe, which consists of *all* t-boundaried graphs.
(2) We let $\mathcal{U}_t^{\text{small}}$ denote the *small* universe of t-boundaried graphs obtained from parse trees; that is, for any parse operator $\otimes$ with n arguments, if $H_1, \dots, H_n \in \mathcal{U}_t^{\text{small}}$, then $H_1 \otimes \cdots \otimes H_n \in \mathcal{U}_t^{\text{small}}$.

Completeness Convention   Definition 12.7.4(2) depends upon the particular choice of parse operators, but we will assume that the set we have chosen forms a complete set for treewidth t graphs. Hence by completeness, the small universe is the universe of t-boundaried treewidth $t - 1$ graphs.

Naturally, we would expect that the automata-theoretic properties of the parse trees will have natural reflections in the underlying graphs. Before we give one important example, we will need to note one crucial property of the parsing operators we will use. This property is the following.

The Parsing Replacement Property   For every operator $\otimes$ with arguments $H_1, \dots, H_n \in \mathcal{U}_t^{\text{small}}$ and any i, there is a graph $G \in \mathcal{U}_t^{\text{small}}$ such that

$$H_1 \otimes \cdots \otimes H_n \cong H_i \oplus G \text{ as } \textit{unlabeled} \text{ graphs.}$$

It is important here, and in the following, that the isomorphism $\cong$ has $H_1 \oplus \cdots \oplus H_n$ isomorphic to $H_i \oplus G$, but as graphs where the t-boundaries are forgotten.

Proposition 12.7.1 *The standard size $t+1$ parse operators of Theorem 12.7.1 have the parsing replacement property.*

Proof Since we drop the boundaries for $\cong$, there is nothing to prove for γ, i, and $\emptyset$ since in all cases we can use $\emptyset$ for G. For e, note that $e(H)$ equals $H \oplus e(\emptyset)$. There is also nothing to prove for $\oplus$. Finally, for u, we need to note that $u(H)$ is isomorphic to, as an unlabeled graph, $H \oplus u(\emptyset)$. $\square$

> **Convention** We will henceforth assume that we have a complete set of parse operators for treewidth t graphs that additionally have the parsing replacement property.

Following ideas of Abrahamson and Fellows [5], we can now prove a graph-theoretic analog of the Myhill–Nerode theorem.

Definition 12.7.5 (Canonical and right congruences for graphs) Let $\mathcal{U}$ be a universe of t-boundaried graphs. (For our purposes, $\mathcal{U}$ will either be $\mathcal{U}_t^{\text{large}}$ or $\mathcal{U}_t^{\text{small}}$.)
1. Let $G_1, G_2 \in \mathcal{U}$. Let F be a family of graphs in $\mathcal{U}$. Then, we say that $G_1 \sim_F G_2$ (relative to $\mathcal{U}$) iff for all $H \in \mathcal{U}$,

$$G_1 \oplus H \in F \quad \text{iff} \quad G_2 \oplus H \in F.$$

2. More generally, we will say that $\equiv$ is a *right congruence* over $\mathcal{U}$ iff for all $G_1, G_2 \in \mathcal{U}$,

$$G_1 \equiv G_2 \quad \text{iff} \quad \text{for all } H \in \mathcal{U}, G_1 \oplus H \equiv G_2 \oplus H.$$

Notice that $\sim_F$ is a right congruence since $\oplus$ is associative.

Definition 12.7.6 (*t-Finite state, t-cutset regular*) Let F be a family of graphs. Let F_t denote the restriction of F to $\mathcal{U}_t^{\text{small}}$ ($= F \cap \mathcal{U}_t^{\text{small}}$).
1. We will say that F is *t-finite state*[8] iff the collection of parse trees corresponding to F_t is finite state.
2. We will say that F is *finite state* iff F is t-finite state for all t.

Theorem 12.7.2 (Analog of Myhill–Nerode for treewidth t graphs) *Let F be a family of graphs. Then the following are equivalent:*
  (i) *F is t-finite state.*
  (ii) *F_t is a union of equivalence classes of a right congruence of finite index over $\mathcal{U}_t^{\text{small}}$.*
  (iii) *$\sim_{F_t}$ has finite index over $\mathcal{U}_t^{\text{small}}$.*

---

[8]In the literature, "t-finite state" is often referred to as *t-cutset regular*.

Remark The reader should carefully note that in Theorem 12.7.2, (i) expresses a property *in the universe of trees*, whereas (ii) and (iii) express properties *in the universe of graphs*. A large part of the proof is devoted to moving from one arena to another.

Proof We use the following notation: If T is a parse tree of a t-boundaried graph, then $G(T)$ denotes the corresponding graph. Similarly, if G is a treewidth t graph, let $T(G)$ denote any standard parse tree for G, obtained by, say, the proof of the Parsing Theorem, Theorem 12.7.1.

(i) $\rightarrow$ (ii) [and (iii)]. Suppose that F is t-finite state. Then F_t generates a finite-state collection $L = L(F_t)$ of parse trees relative to Σ^{**}. By the Myhill–Nerode theorem for Tree Automata, Theorem 12.6.3, $\sim_L$ is a right congruence of finite index over Σ^{**}. Recall that $\sim_L$ being a right congruence over Σ^{**} means that for all trees $T_1, T_2 \in \Sigma^{**}$, and $T \in \Sigma_x^{**}$ with exactly one (leaf) node labeled x, $T_1 \sim_L T_2$ means that

$$T_1 \cdot_x T \in L \quad \text{iff} \quad T_2 \cdot_x T \in L.$$

We claim that (iii), and hence (ii), holds. Suppose that $G_1, G_2, \ldots$ are an infinite family of graphs in $\mathcal{U}_t^{\text{small}}$ which are inequivalent with respect to $\sim_{F_t}$. Then, for all $i \neq j$, there exist witnesses $H_{i,j} \in \mathcal{U}_t^{\text{small}}$ such that

$$G_i \oplus H_{i,j} \in F_t \quad \text{iff} \quad G_j \oplus H_{i,j} \notin F_t.$$

Thus, for all $i \neq j$,

$$T(G_i) \cdot_x \left[x \oplus T(H_{i,j}) \right] \in L \quad \text{iff} \quad T(G_j) \cdot_x \left[x \oplus T(H_{i,j}) \right] \notin L,$$

since $G(T(G_k) \cdot_x [x \oplus T(H_{i,j})]) = G_k \oplus H_{i,j}$. But then the family $\{T(G_i) : i \in \mathbb{N}\}$ is an infinite set of trees inequivalent with respect to $\sim_L$, contradicting the Myhill–Nerode theorem for Trees. Hence, $\sim_{F_t}$ has finite index after all.

(ii) $\rightarrow$ (iii). This uses the same argument as (ii) $\rightarrow$ (iii) of Theorems 12.5.1 and 12.6.3, arguing that $\sim_{F_t}$ is the coarsest such right congruence.

(iii) $\rightarrow$ (i). Suppose that $\sim_{F_t}$ has finite index. We need to argue that F is t-finite state. To do this, by the Myhill–Nerode theorem for trees, we need to prove that $\sim_L$ has finite index. Here is where we need the Parsing Replacement Property. Suppose that $T_1, T_2, \ldots$ are trees in $\Sigma_{\{x\}}^{**}$ with $T_i \not\sim_L T_j$ for $i \neq j$. Note that without loss of generality, we must have $G(T_i) \in \mathcal{U}_t^{\text{small}}$ for all i. [It is only possible for $G(T_i) \notin \mathcal{U}_t^{\text{small}}$ for *one* i, since all trees that do not represent t-boundaried graphs are equivalent.] Then, for all $i \neq j$, there is a tree $T_{i,j}$ with x on exactly one leaf such that

$$T_i \cdot_x T_{i,j} \in L \quad \text{iff} \quad T_j \cdot_x T_{i,j} \notin L.$$

Let $T_{i,j}'$ denote the tree obtained by replacing x with $\emptyset$. Note that $G(T_{i,j}') = G(\emptyset \cdot_x T_{i,j}) \in \mathcal{U}_t^{\text{small}}$. We claim that there is a tree $Q_{i,j}$, with $G(Q_{i,j}) \in \mathcal{U}_t^{\text{small}}$ such that

$$G(T_k \cdot_x T_{i,j}) \cong G(T_k \oplus Q_{i,j}), \quad \text{for } k \in \{i, j\}.$$

To see that this holds, we argue inductively on the length ℓ of the path in $T_{i,j}$ from the root to x. If $\ell = 1$, then $T_k \cdot_x T_{i,j}$ is the same as $H_1 \otimes \cdots \otimes T_k \otimes \cdots \otimes H_m$ for some operator $\otimes$, and trees H_w with $G(H_w) \in \mathcal{U}_t^{\text{small}}$. By the Parsing Replacement Property, we have

$$G(H_1) \otimes \cdots \otimes G(T_k) \otimes \cdots \otimes G(H_m) \cong G(T_k) \oplus R_{i,j},$$

for some $R_{i,j} \in \mathcal{U}_t^{\text{small}}$. But then we can take $Q_{i,j} = T(R_{i,j})$.

So suppose that $\ell > 1$. The idea is that we will work our way up the path π from the root to x in $T_{i,j}$. Let W_k be the subtree of $T_k \cdot_x T_{i,j}$ containing T_k, generated by the child of the root of $T_k \cdot_x T_{i,j}$ (see Fig. 12.19).

Note that $G(W_k) \in \mathcal{U}_t^{\text{small}}$. We may argue, as in the case $\ell = 1$, that there are $H_1, \ldots, H_m$ with

$$T_k \cdot_x T_{i,j} = H_1 \otimes \cdots \otimes W_k \otimes \cdots \otimes H_m$$

and with $G(H_b) \in \mathcal{U}_t^{\text{small}}$ for all b. Then, by the Parsing Replacement Property, there is a graph $R_{i,j}^1 \in \mathcal{U}_t^{\text{small}}$ with

$$G(T_k \cdot_x T_{i,j}) = G(H_1 \otimes \cdots \otimes W_k \otimes \cdots \otimes H_m)$$
$$= G(H_1) \otimes \cdots \otimes G(W_k) \otimes \cdots \otimes G(H_m) \cong G(W_k) \oplus R_{i,j}^1.$$

Now, we can argue in exactly the same fashion with W_k in place of $T_k \cdot_x T_{i,j}$; that is, consider the subtree W_k' of W_k above the child of the root of W_k containing T_k. Again, there must be $H_1', \ldots, H_{m'}'$ with $G(H_b') \in \mathcal{U}_t^{\text{small}}$ for all b and an operator $\otimes'$, such that

$$W_k = H_1' \otimes' \cdots \otimes' W_k' \otimes' \cdots \otimes' H_{m'}'.$$

The reasoning above will yield a graph $R_{i,j}^2$ such that

$$G(W_k) \cong G\left(W_k'\right) \oplus R_{i,j}^2.$$

But then

$$G(T_k \cdot_x T_{i,j}) \cong G\left(\left(W_k' \oplus R_{i,j}^2\right) \oplus R_{i,j}^1\right) \cong G\left(W_k' \oplus R_{i,j}'\right) \quad \text{where } R_{i,j}' = R_{i,}^1 \oplus R_{i,j}^2.$$

Continuing in this fashion, we work our way up the path from the root to x in $T_{i,j}$ to eventually get a graph $R_{i,j}^* \in \mathcal{U}_t^{\text{small}}$ such that

$$G(T_k \cdot_x T_{i,j}) \cong G(T_k) \oplus R_{i,j}^*.$$

Hence, $G(T_k) : k = 1, 2, \ldots$ are inequivalent (as witnessed by $R_{i,j}^*$) with respect to $\sim_{F_t}$. This contradicts the fact that $\sim_{F_t}$ has finite index. $\square$

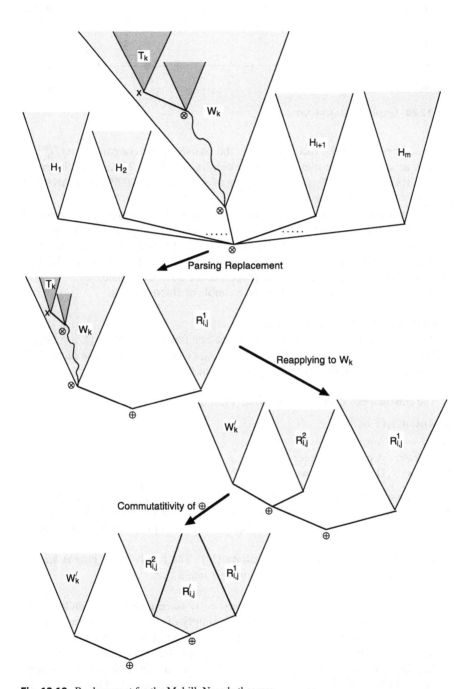

Fig. 12.19 Replacement for the Myhill–Nerode theorem

Fig. 12.20 Graphs for BANDWIDTH

Again, because of the fact that $\sim_{F_t}$ is the coarsest right congruence on $\mathcal{U}_t^{\text{small}}$, there is an implicit state minimization process in the above. Notice that the proof also shows that *if $T_1 \cdot_x T_2$ is the parse tree of a graph G in $\mathcal{U}_t^{\text{small}}$, then there is a parse tree of G of the form $T_1 \oplus T_2'$.*

Abrahamson and Fellows (also) call the property of $\sim_{F_t}$ a right congruence of finite index on $\mathcal{U}_t^{\text{small}}$ *t-cutset regularity*. We will call this property *t-finite index*.

Just as we can use the Myhill–Nerode theorems for trees or formal languages to prove that certain languages are not finite state, we may use Theorem 12.7.2 to establish that various graph-theoretic properties are not t-finite state.

As an illustration, we will look at the example of BANDWIDTH.

Definition 12.7.7 (Bandwidth, layout)
1. A *layout* of a graph $G = (V, E)$ is a one-to-one function $f : V \rightarrow \{1, \ldots, |V|\}$.
2. The *bandwidth* of layout f of G is the $\max_{uv \in E} |f(u) - f(v)|$. The *bandwidth* of G is the minimum bandwidth of a layout of G.

The classical problem stemming from VLSI design is

BANDWIDTH
Input: A graph G.
Parameter: A positive integer k.
Question: Does G have bandwidth $\leq k$?

As we will later see, not only is the nonparameterized version of BANDWIDTH NP-complete, but, in fact, we believe the parameterized version above to be fixed-parameter intractable. Certainly, as we will now see, it is provably not finite state.

Theorem 12.7.3 (Abrahamson and Fellows [5]) *The family F_k of graphs having bandwidth at most k is not t-finite state if $t \geq 1$ and $k \geq 2$.*

Proof It suffices to prove the theorem for $t = 1$, since additional boundary vertices can easily be added. Our proof consists simply of exhibiting an infinite set of representatives of distinct equivalence classes of the equivalence relation $\sim_{F_k}$. The representatives are $(X_i, i \geq 1)$, where X_i is the graph illustrated in Fig. 12.20.

The single boundary vertex of X_i is b_i. The stars at either end of X_i strongly constrain the layouts of bandwidth at most k. They must look like Fig. 12.20, up to permutation of the layouts of the stars. Notice that each X_i is a tree and hence has treewidth 1.

Let Y_i be a path of $2i + 1$ vertices, whose single boundary vertex is the centroid of
the path. Suppose $n \geq m$. The reader can easily check that $X_m \not\sim_{F_k} X_n$ by observing
the following:

(1) $X_n \oplus Y_{kn-n} \in F_k$;
(2) $X_m \oplus Y_{kn-n} \notin F_k$.

The layout needed to verify assertion (1) is constructed by placing $k - 1$ vertices
of Y_{kn-n} between each pair (b_i, b_{i+1}), for $i = 0, \ldots, 2n - 1$. There is not enough
room in X_m to do the same thing, so some edge will have to be stretched to length
more than k to accommodate Y_{kn-n} within the layout for X_m.

Thus, by Theorem 12.7.2, BANDWIDTH for $k \geq 2$ is not t-finite state. $\square$

Theorem 12.7.3 leaves only the case $k = 1$ undetermined. A graph G has a layout
of bandwidth 1 if and only if G is a disjoint union of paths. So each equivalence class
is determined by where the boundary vertices lie on paths: whether they are isolated,
at the end of a path, or somewhere in the interior of a path. Hence, for t-boundaried
graphs, there are 3^t equivalence classes.

Method of Test Sets for Treewidth t Graphs The proof above encapsulates the
Method of Test Sets for Graphs. We need to follow the steps below.

Step 1: Identify a finite set S of *tests* T to be performed on a t-boundaried graph X.
(For example, a test might consist of a t-boundaried graph Y for which we determine whether $X \oplus Y$ is in F.)

Step 2: Argue that the equivalence relation $X \sim Y$ defined as: $X \sim Y$ if and only if
$S_X = S_Y$ where $S_X = \{T \in S : X$ passes the test $T\}$ has the properties (1) $X \sim Y$
and $X \in F$ implies $Y \in F$, and (2) $\sim$ is a right congruence.

We conclude this section with a couple of examples: HAMILTONICITY and
CUTWIDTH. We begin with HAMILTONICITY. Recall that a graph is called Hamiltonian iff it contains a simple cycle through all of the vertices. Determining if a graph
is Hamiltonian is a famous NP-complete problem. The first positive illustration of
the method of test sets for graphs is to prove that HAMILTONICITY is solvable in
time $O(n)$ for graphs of bounded treewidth.

Theorem 12.7.4

(i) HAMILTONICITY *has t-finite index.*
(ii) *Consequently, for any t,* HAMILTONICITY *is* FPT *via an $O(n)$ algorithm for
graphs of treewidth $\leq t$.*

Proof (ii) follows from (i) via Bodlaender's theorem, Theorem 10.4.1. So, we need
only prove (i). This is a classic example of the application of the method of test sets.

Let $\mathcal{H}$ denote the collection of Hamiltonian graphs of treewidth $\leq t$. We need
to establish that for all t, $\sim_{\mathcal{H}}$ has finite index on $\mathcal{U}_t^{\text{small}}$. Actually, we establish
that $\sim_{\mathcal{H}}$ has finite index on $\mathcal{U}_t^{\text{large}}$. Now, the idea is to examine how *information
can flow across a t-boundary*. Remember, we are considering $\sim_{\mathcal{H}}$ and hence looking at graphs of the form $G_i \oplus K$ for $i \in \{1, 2\}$. There are only so many ways

that a graph K can influence whether or not $G_i \oplus K$ is Hamiltonian. Take, for instance, $t = 4$. Then, G_i has four boundary vertices with labels $\{1, 2, 3, 4\}$, say, $V = \{v_1, v_2, v_3, v_4\}$. Now, any Hamilton path in $G_i \oplus K$ must cover the vertices of V. You might, for instance, start in G, pass through v_2 into K back through v_4 into G, then through v_1 but remain in G, and then through v_3 but remain in G. There are also the ones that *must* travel, for instance, from v_i to v_j which can be represented by adding two edges $v_i x$ and $x v_j$ with a new vertex x, in the test. Then to have a Hamilton cycle we would need to include x. Running through possibilities like this gives a canonical set of possibilities, which forms the collection of test sets $\mathcal{T}$.

One easily sees that if

$$G_1 \oplus T \in \mathcal{H} \quad \text{iff} \quad G_2 \oplus T \in \mathcal{H},$$

for all T in the collection of test sets $\mathcal{T}$, then $G_1 \sim_{\mathcal{H}} G_2$ (Exercise 12.7.6). The same reasoning will extend for any t since the number of test sets is finite for each t. Since $\sim_{\mathcal{H}}$ has finite index, for each t, HAMILTONICITY is finite state.   □

A more intricate example is provided by CUTWIDTH. Here we need a looser definition of a graph layout. A *generalized layout* (or *g-layout* for short) of a graph $G = (V, E)$ is an injective function $l : V \to \mathbb{R}$.

Definition 12.7.8 (Cutwidth)
1. Let f be a g-layout of a graph $G = \langle V, E \rangle$. If $\alpha \in \mathbb{R}$, the *value of the cut at* α is the number of edges $uv \in E$ with $f(u) < \alpha$ and $f(v) > \alpha$.
2. The *cutwidth* of a g-layout f of G is the maximum of the value of the cut at α, over all $\alpha \in \mathbb{R}$. The cutwidth of G is the minimum of the cutwidths of all possible g-layouts of G.

CUTWIDTH
Input: A graph G.
Parameter: A positive integer k.
Question: Does G have cutwidth $\leq k$?

Theorem 12.7.5 (Abrahamson and Fellows [5]) *For every k and for every t, the problem of deciding whether a graph has a layout of cutwidth at most k is t-finite state.*

Proof Fix k and t, and let F denote the family of graphs of cutwidth at most k. The argument here will centrally use Theorem 12.7.2, so what we must do is show that F has finite index over $\mathcal{U}_t^{\text{small}}$.

Let $w : \mathbb{R} \to N$ be a non-negative integer-valued function which we will refer to as a *weighting* of the real line $\mathbb{R}$. We will consider the cutwidth of a graph relative to a weighting of $\mathbb{R}$ in the following way.

If l is a g-layout of a graph $G = (V, E)$, the *value of a cut at* $\alpha \in \mathbb{R}$ *with respect to* w is the sum of $w(\alpha)$ and the number of edges uv of G with $l(u) < \alpha$ and $l(v) > \alpha$. The *cutwidth of a g-layout* l *of G with respect to* w is the maximum value of a cut at any point $\alpha \in \mathbb{R}$. The *cutwidth of G with respect to* w is the minimum cutwidth of a g-layout of G with respect to w.

A *test* $T = (\pi, S)$ of size n consists of the following:

(1) A map $\pi : \{1, \ldots, t\} \to \{1, \ldots, n\}$.

(2) A sequence S of non-negative integers $S = (S(0), \ldots, S(n))$.

To a test T of size n we associate a weighting w_T which consecutively assigns to the intervals of $\mathbb{R}$: $(0, 1), (1, 2), \ldots, (n, n+1)$, the values $S(0), \ldots, S(n)$. Precisely, $w_T(\alpha) = S(i)$ if $i < \alpha < i+1$ and $w_T(\alpha) = 0$ otherwise.

We say that a t-boundaried graph $G = (V, E, B, f)$ *passes* a test $T = (\pi, S)$ of size n if there is a g-layout l of G of cutwidth at most k relative to w_T that further satisfies the following conditions:

1. For all $v \in V, 0 < l(v) < n + 1$.
2. If u is a boundary vertex of G with label j, $f(u) = j$, then $l(u) = \pi(j)$. Thus, the boundary vertices are laid out in the order and position described by π.
3. If u is not a boundary vertex, $u \in V - B$, then $l(u) \notin N$. Thus, non-boundary vertices of G are assigned layout positions in the interiors of the weighted intervals of $\mathbb{R}$.

Let G and H be t-boundaried graphs and define $G \sim H$ if and only if G and H pass exactly the same set of tests. We will establish the following two claims that by Theorem 12.7.2 are sufficient to establish our theorem.

Claim 1 *There is a finite set of reduced tests such that G and H agree on all tests if and only if they agree on the set of reduced tests. (From this it follows that $\sim$ has finite index.)*

Claim 2 *If $G \sim H$, then $G \sim_F H$. (From this it follows that $\sim_F$ has finite index.)*

We first argue Claim 2 by contraposition. Suppose Z is a t-boundaried graph such that $G \oplus Z \in F$ and $H \oplus Z \notin F$. Let l be a witnessing g-layout of $G \oplus Z$. From l can be described a test T_l that is passed by G but failed by H. The test T_l records the interval weights and boundary positions given by l restricted to Z. Therefore $G \nsim H$, and our argument is completed.

Finally, we turn to the most important Claim 1. We define an equivalence relation on tests, with $T \sim T'$ if and only if every t-boundaried graph that passes T passes T' and *vice versa*. We will argue that there are a finite number of equivalence classes of tests.

For a test $T = (\pi, S)$, we refer to a subsequence S' of S as a *load pattern* of T in the following circumstances:

- S' is the subsequence of weights assigned to intervals in w_T before the minimum image value of π, or
- S' is the subsequence of weights assigned to intervals in w_T after the maximum image value of π, or

- S' is the sequence of integer weights *between two consecutive images* of $\{1, \ldots, t\}$ under the map π of the test T. Each test $T = (\pi, S)$ can be viewed as π, together with $t + 1$ load patterns S_i that factor S.

For a concrete example, with $t = 3$, suppose the test T consist of the data:
(1) $\pi(1) = 6$, $\pi(2) = 3$ and $\pi(3) = 9$.
(2) The sequence of weights $S = (4, 2, 1, 5, 2, 6, 4, 1, 2, 5, 3, 6, 1)$.
Then S factors into the sequence of π-images (denoted $[-]$) and load patterns

$$(4, 2, 1)[\pi(2)](5, 2, 6)[\pi(1)](4, 7, 2)[\pi(3)](5, 3, 6, 1).$$

We will use the following notation. The symbols s, s', and s_i denote sequences of integers taken from the set $J = \{0, \ldots, k\}$. Letters such as a, b, and c will be used to denote single particular values in J. Two load patterns s and s' are termed *equivalent* if $T \sim T'$ for any test T for which s is a load pattern, where T' is the test obtained by replacing s with s'. Write $s \geq s'$ if every t-boundaried graph which passes a test T for which s is a load pattern also passes the test T' where the load pattern s has been replaced by s'. Thus, for example, we have $s = (5, 4, 1, 3, 2, 7) \geq (5, 1, 1, 1, 1, 7) \geq (5, 1, 7) = s'$ and $(5, 1, 7) \geq (5, 5, 1, 7, 7, 7) \geq (5, 4, 1, 3, 2, 7)$, and therefore $s \sim s'$, since decreasing the weight of an interval only makes finding a cutwidth k layout easier, and consecutively repeating weights (or deleting such repetitions) makes no difference.

We have the following reduction rules for load patterns:
(R1) If $s = s_1 abc s_2$ with $a \leq b \leq c$ (or $a \geq b \geq c$) then $s \sim s_1 ac s_2$.
(R2) If $s = s_1 as_2 bs_3$ where each element of s_2 is greater than or equal to the maximum of a and b, and $c = \max(s_2)$, then $s \sim s_1 acbs_3$.
(R3) If $s = s_1 as_2 bs_3$ where each element of s_2 is at most the minimum of a and b, and $c = \min(s_2)$, then $s \sim s_1 acbs_3$.

(R1) is shown by the sequence of implications: $s_1 abc s_2 \geq s_1 aac s_2 \geq s_1 ac s_2 \geq s_1 accs_2 \geq s_1 abc s_2$. To see (R2) let $m = \max(a, b)$, and suppose c is in the $i + 1$ position of s_2, and that $|s_2| = i + j + 1 = q$. We have the sequence of implications: $s_1 as_2 bs_3 \geq s_1 am^i cm^j bs_3 \geq s_1 a^{i+1} cb^{j+1} s_3 \geq s_1 acbs_3 \geq s_1 ac^q bs_3 \geq s_1 as_2 bs_3$. The argument for (R3) is similar.

Note that since we are concerned with cutwidth k where k is fixed, every load pattern is trivially equivalent to one where the largest integer weight occurring is at most k. Define a load pattern s to be *reduced* if it contains only integer values in the range $\{0, \ldots, k\}$ and none of the reduction rules (R1)–(R3) can be applied to s. Define a test T to be *reduced* if each of the $t + 1$ load patterns of T is reduced.

We next argue that a reduced load pattern s over $\{0, \ldots, k\}$ has length at most $2k + 2$. Otherwise, s must contain some value a at least three times. Consider the factorization $s = s_1 as_2 as_3 as_4$ where a does not occur in s_2 or s_3. Neither of the subsequences s_2 or s_3 can be empty, or s is not reduced. Let b be the rightmost integer in s_2 and let c be the leftmost integer in s_3. If $b < a < c$ or $b > a > c$, then (R1) can be applied. So both of b and c must be greater than a, or both must be less than a. Suppose the latter. (The other case is handled similarly.)

Let d denote the rightmost occurrence in the subsequence as_2 of a value greater than or equal to a. Let e denote the leftmost occurrence in the subsequence s_3a of a value greater than or equal to a. There are at least three integers in s properly between d and e, and each intervening value is at most a. By (R3) s is reducible.

It follows that there are finitely many reduced tests, completing the proof.   □

Since a graph of cutwidth at most t has pathwidth bounded by $2t$, by Bodlaender's theorem, we have the following corollary.

Corollary 12.7.1 CUTWIDTH *is* FPT *with a linear time algorithm for every fixed k.*

12.7.1 Historical Notes

The methods of the present section were introduced by Abrahamson and Fellows [4] from which grew [5]. The proof of HAMILTONICITY being FPT for bounded treewidth t graphs is taken from [247] where it appeared for the first time, based on Xiu Yan Liu's thesis.

12.7.2 Exercises

Exercise 12.7.1 Prove that a graph has treewidth 2 iff it can be parsed by the operators of Fig. 12.16.

Exercise 12.7.2 Construct the parse tree corresponding to the graph of Fig. 12.17.

Exercise 12.7.3 Prove that the class of treewidth t graphs with at least $t + 1$ vertices can be parsed by $\oplus, \emptyset, \gamma$, and i together with the operators generated by taking the collection of all subgraphs of K_{t+1} with $t + 1$ vertices with the labels $\{1', \ldots, t'\}$ and $\{1, \ldots, t\}$ are put on the vertices in all possible ways.
(Hint: Use the partial t-tree characterization, and induction: argue that any partial t-tree P embedded in a t-tree, E can be so parsed so that any given copy of K_t in E will have at the end the labels $\{1, \ldots, t\}$. In the induction, take the target K_t and consider the cutsets.)

Exercise 12.7.4 Prove Theorem 12.7.1(ii) to establish that the six size $t + 1$ operators form a complete set for graphs of treewidth t.

Exercise 12.7.5 Construct a formula which computes the number of test sets for HAMILTONICITY.

Exercise 12.7.6 Generate the complete collection $\mathcal{T}$ of tests for HAMILTONICITY above. Prove that $G_1 \sim_{\mathcal{H}} G_2$ iff for all T in this test set $\mathcal{T}$,

$$G_1 \oplus T \in \mathcal{H} \quad \text{iff} \quad G_2 \oplus T \in \mathcal{H}.$$

12.8   Summary

This chapter has the basic properties of automata, tree automata and methods for running algorithms on graphs of bounded treewidth. Parse trees are thought of as being formal languages processed by tree automata.

Courcelle's Theorem

13

13.1 The Basic Theorem

Courcelle's Theorem is a logic-based meta-theorem for establishing (in conjunction with Bodlaender's Theorem) that various graph-theoretic properties are decidable in linear FPT time, when the parameter is input graph treewidth. Similar results were obtained independently by Borie, Parker and Tovey [108]. Courcelle's Theorem has the form:

> *If the property of interest is expressible in MS_2 logic, then, parameterizing by the treewidth of the input, it can be determined in linear FPT time whether the graph has the property.*

The result can also be viewed in the stronger form:

> *Given a graph G and a formula ϕ in MS_2 logic describing a property of interest, and parameterizing by the combination of* $\mathrm{tw}(G)$ *(the treewidth of G) and the size of the formula ϕ, it can be determined in time $f(\mathrm{tw}(G), |\phi|)n^{O(1)}$ whether G has the property of interest.*

The key player is the fragment of second-order logic in the language of graphs called the *monadic second-order logic of graphs.*

The syntax of the second-order monadic logic of graphs includes the logical connectives $\wedge$, $\vee$, $\neg$, variables for vertices, edges, sets of vertices, and sets of edges, the quantifiers $\forall$, $\exists$ that can be applied to these variables, and the five binary relations:

1. $u \in U$, where u is a vertex variable and U is a vertex set variable.
2. $d \in D$, where d is an edge variable and D is an edge-set variable.
3. $\mathrm{inc}(d, u)$, where d is an edge variable, u is a vertex variable, and the interpretation is that the edge d is incident on the vertex u.
4. $\mathrm{adj}(u, v)$, where u and v are vertex variables and the interpretation is that u and v are adjacent vertices.
5. Equality for vertices, edges, sets of vertices and sets of edges.

We will use lowercase letters for vertices or edge variables and uppercase letters for variables representing sets of edges or sets of vertices.

Example 13.1.1 (Hamiltonicity) A graph G is Hamiltonian if and only if it has a spanning cycle. the edges of G can be partitioned into two sets *red* and *blue* such that:

R.G. Downey, M.R. Fellows, *Fundamentals of Parameterized Complexity*,
Texts in Computer Science, DOI 10.1007/978-1-4471-5559-1_13,
© Springer-Verlag London 2013

1. Each vertex has exactly two incident red edges, and
2. The subgraph induced by the red edges is a connected spanning subgraph of G.

In developing the MS_2 formula that expresses the property of Hamiltonicity, we will represent the set of red edges by the variable R and the set of blue edges by the variable B.

$$\text{Hamiltonian} \equiv \exists R \exists B \forall u \forall v \big[\text{part}(R, B) \land \deg(u, R) = 2 \land \text{span}(u, v, R)$$
$$\land \, \forall x \forall y \exists W (\text{con}(u, v, W, R))\big],$$

where span, deg and part are described below:

$$\text{part}(R, B) \equiv \forall e \big[(e \in R \lor e \in B) \land \neg(e \in R \land e \in B)\big],$$
$$\deg(u, R) = 2 \equiv \big[\exists e_1, e_2 (e_1 \neq e_2 \land \text{inc}(e_1, u) \land \text{inc}(e_2, u) \land e_1 \in R \land e_2 \in R)\big]$$
$$\land \, \neg \big[\exists e_1, e_2, e_3 (e_1 \neq e_2 \neq e_3 \land (\text{inc}(e_i, u) \land e_i \in R \text{ for } i \in \{1, 2, 3\}))\big],$$
$$\text{span}(u, v, R) \equiv \forall V, W \big[(\text{part}(V, W) \land u \in V \land v \in W)$$
$$\rightarrow (\exists e, x, y (\text{inc}(e, x) \land \text{inc}(e, y) \land x \in V \land y \in W \land e \in R))\big],$$
$$\text{conn}(x, y, W, R) \equiv \forall V_1, V_2 \subseteq V \big([V_1 \cup V_2 = V \land V_1 \cap V_2 = \emptyset \land x \in V_1 \land y \in V_2]$$
$$\rightarrow \exists r \in R \exists p \exists q (p \in V_1 \land q \in V_2 \land \text{inc}(r, p) \land \text{inc}(r, q))\big).$$

Actually, this language is sometimes referred to in the literature as the *extended* monadic second-order language, and the logic the *extended* monadic second-order logic or MS_2 logic (e.g. Arnborg [43], Seese [612], and Arnborg, Lagergren, and Seese [46]). This is because we are looking at a two-sorted structure with predicates for edges and vertices plus an incidence relation. Another natural language has only vertex symbols and one must use binary relations for edges. Following Courcelle, we call this one-sorted logic the MS_1 logic. Naturally, the *monadic* second-order theories are different. For instance, Hamiltonicity is *not* MS_1 expressible (Exercise 13.5.1). MS_2 also allows for multiple edges whereas MS_1 does not. For the purposes of the present section, we will concentrate on the logic with the richest expressive power MS_2, and briefly discuss the MS_1 logic in Sect. 13.4.

> Hence, whilst we proceed, until the reader is otherwise ordered about, "monadic second-order logic" means "MS_2 logic".

Theorem 13.1.1 *If F is a family of graphs described by a sentence in second-order monadic logic, then $\sim_F$ has finite index in the large universe of t-boundaried graphs.*

In the language of logic, we consider structures that satisfy the relevant formulae. Here the structures are graphs and we say that $G \models \phi$ for a formula ϕ if the interpretation of ϕ in G is true.

TW-ϕ-MODEL CHECKING FOR ϕ

Instance: A graph $G = (V, E)$, and ϕ.
Parameter: $\text{tw}(G) = t + |\phi|$.
Question: $G \models \phi$?

The main theorem is that this problem is linear time FPT.
Later, we will look at a theorem of Seese which is something of a converse.

Proof of Theorem 13.1.1 Let ϕ be the formula that describes the graph property of interest, and let F denote the corresponding family of graphs. Let t be the treewidth of the input graph G, and recall Bodlaender's algorithm. Write $\sim_\phi$ for $\sim_F$. We prove the theorem by structural induction on formulas. We write $\partial(X)$ to denote the boundary of a t-boundaried graph X. By the *interior* of X, we refer to the vertices in $V(X) - \partial(X)$.

To each formula ϕ [possibly with a nonempty set $Free(\phi)$ of free variables] we associate an equivalence relation $\sim_\phi$ on the set of all t-boundaried graphs that are partially equipped with distinguished vertices, edges, sets of vertices, and sets of edges corresponding to the free variables in the set $Free(\phi)$. (The details follow shortly.)

Two partially equipped t-boundaried graphs X and Y are defined to be $\sim_\phi$ related if and only if they have the "same" partial equipment, and for every compatibly equipped t-boundaried graph Z, the formula ϕ is true for $X \oplus Z$ if and only if it is true for $Y \oplus Z$.

The compatibility condition on Z ensures that the partial equipment of Z agrees with that of X and Y on the boundary vertices and edges and further ensures that all of the free variables have interpretations as the distinguished elements and sets in the product graphs $X \oplus Z$ and $Y \oplus Z$. The formalization of what we mean by "partial equipment" is tedious but unproblematic; it can be accomplished as follows.

Let $fr(0), fr(1), Fr(0)$, and $Fr(1)$ denote, respectively, sets of variables for vertices, edges, sets of vertices, and sets of edges. Let $Free$ denote the disjoint union of these four sets. Allow S^2 to denote the set of all 2-element subsets of a set S. Define a *partial equipment signature* σ for $Free$ to be given by the following data.

1. Disjoint subsets $int_0(\sigma)$ and $\partial_0(\sigma)$ of $fr(0)$, and a map $f_\sigma^0 : \partial_0(\sigma) \to \{1, \ldots, t\}$.
2. Disjoint subsets $int_1(\sigma)$ and $\partial_1(\sigma)$ of $fr(1)$, and a map $f_\sigma^1 : \partial_1(\sigma) \to \{1, \ldots, t\}^2$.
3. For each vertex set variable U in $Fr(0)$, a subset $U_\sigma \subseteq \{1, \ldots, t\}$.
4. For each edge-set variable D in $Fr(1)$, a subset $D_\sigma \subseteq \{1, \ldots, t\}^2$.

Notice there will be a difference between *partial* and *full* equipment. Let $Free(\sigma)$ denote the union of the sets $int_0(\sigma), \partial_0(\sigma), int_1(\sigma), \partial_1(\sigma), Fr(0)$, and $Fr(1)$. We recall the notation $X = (V, E, B, f)$ for a t-boundaried graph: V and E are, as usual, the vertices and edges of G, and $\partial(H) = B$ with f the labeling function: f is a bijection from $B \mapsto \{1, \ldots, t\}$. Using this notation, we will say that a t-boundaried graph $X = (V, E, B, f)$ is σ-*partially equipped* if it has distinguished vertices, edges, sets of vertices, and sets of edges corresponding to the variables in $Free(\sigma)$, where the correspondence is compatible with the data that describe ϕ. More precisely, the following conditions must be satisfied by these distinguished elements and sets:[1]

1. If u is a vertex variable, $u \in int_0(\sigma)$, then the distinguished vertex in X corresponding to u must be in the interior of X.
2. If u is a vertex variable, $u \in \partial_0(\sigma)$, then the distinguished vertex in X corresponding to u must be the unique vertex $x \in V$ for which $f(x) = f_\sigma^0(u)$.
3. If d is an edge variable, $d \in int_1(\sigma)$, then the distinguished edge in X corresponding to d must have at least one endpoint in the interior of X.
4. If d is an edge variable, $d \in \partial_1(\sigma)$, then the distinguished edge in X corresponding to d must have as endpoints the pair of vertices x, y of $\partial(X)$ for which $f_\sigma^1(d) = \{f(x), f(y)\}$.
5. If U is a vertex set variable, $U \in Fr(0)$, then the distinguished set of vertices V_U in X corresponding to U must satisfy $f(V_U \cap \partial(X)) = U_\sigma$.
6. If D is an edge-set variable, $D \in Fr(1)$, then the distinguished set of edges E_D in X corresponding to D must satisfy $\{\{f(x), f(y)\} : \{x, y\} \in E_D\} \cap \partial(X)^2 = D_\sigma$.

All of this is just as expected. We say that partial equipment signatures σ and $\bar{\sigma}$ for $Free$ are *complements* with respect to $Free$ if the data describing them is in agreement with respect to the boundary and if together they provide for a complete interpretation of the variables of $Free$. More precisely, $\bar{\sigma}$ must satisfy the next listed (symmetric) conditions with respect to σ. (Note that $\bar{\sigma}$ is completely determined by σ.)

(1) $\partial_0(\bar{\sigma}) = \partial_0(\sigma)$ and $f_{\bar{\sigma}}^0 = f_\sigma^0$.
(2) $int_0(\bar{\sigma}) = fr(0) - \partial_0(\sigma) - int_0(\sigma)$.
(3) $\partial_1(\bar{\sigma}) = \partial_1(\sigma)$ and $f_{\bar{\sigma}}^1 = f_\sigma^1$.
(4) $int_1(\bar{\sigma}) = fr(1) - \partial_1(\sigma) - int_1(\sigma)$.
(5) For each vertex set variable U in $Fr(0)$, $U_{\bar{\sigma}} = U_\sigma$.
(6) For each edge-set variable D in $Fr(1)$, $D_{\bar{\sigma}} = D_\sigma$.

The important thing is that if partial equipment signatures σ and $\bar{\sigma}$ are complements with respect to $Free$ and if X and Z are t-boundaried graphs with X σ-partially equipped and Z $\bar{\sigma}$-partially equipped then every variable in $Free$ has a consistent interpretation in $X \oplus Z$.

---

[1] The reader needs to be aware that our universe has now changed from the universe of (isomorphism types of) graphs to the universe of (isomorphism types of) σ-*partially equipped* graphs. These objects will thus be *tuples* consisting of graphs together with distinguished parts. Isomorphism here consists of the obvious maps that must send graphs to graphs as well as preserving the relevant distinguished sets of edges, vertices, etc. One can think of these distinguished sets of objects as colored portions of the graphs. It is important that the reader understand that many of the statements to follow talk about differing universes of not just t-boundaried graphs but σ-*partially equipped* t-boundaried graphs for various σ.

Definition 13.1.1 If σ is a partial equipment signature for $Free(\phi)$ and X and Y are σ-partially equipped t-boundaried graphs, then we define $X \sim_\phi Y$ if and only if for every $\bar{\sigma}$-partially equipped t-boundaried graph Z: $X \oplus Z \models \phi$ if and only if $Y \oplus Z \models \phi$.

The induction claim may now be stated precisely. In what follows, we will avoid mentioning signatures when they can be deduced from the context.

Claim *For every second-order monadic formula ϕ, with equipment signature σ, $\sim_\phi$ has finite index on the universe of t-boundaried σ-partially equipped graphs.*[2]

Proof We must first show that the claim holds for atomic formulas. Since vertex adjacency is easy to express as the existence of an edge incident on both vertices, it suffices to assume that the only atomic formulas are $d \in D$, $u \in U$, and $\mathrm{inc}(e, u)$. Our consideration breaks up into cases according to the possible partial equipment signatures. All of these are easy and are left entirely to the reader. For example, let $\phi = \mathrm{inc}(e, u)$ and suppose σ is described by $e \in int_1(\sigma)$ and $u \in \partial_0(\sigma)$. The relation $\sim_\phi$ has index 2. For another example, let ϕ be the formula $d \in D$, where d is an edge variable and D is an edge-set variable, and suppose $d \in \partial_1(\sigma)$ and $D_\sigma = \emptyset$. The equivalence relation $\sim_\phi$ then has index 1.

For the induction step, we may suppose (without loss of generality) that ϕ is a formula obtained from simpler formulas in one of the following ways:

(1) $\phi = \neg\phi'$,

(2) $\phi = \phi_1 \wedge \phi_2$,

(3) $\phi = \exists u \phi'$, where u is a vertex variable free in ϕ',

(4) $\phi = \exists d \phi'$, where d is an edge variable free in ϕ',

(5) $\phi = \exists U \phi'$, where U is a vertex set variable free in ϕ',

(6) $\phi = \exists D \phi'$, where D is an edge-set variable free in ϕ'.

We treat each of these six cases separately, giving a complete argument for (1), (2), (3), and (6). An appropriate equivalence relation is defined for (4) and (5) and the reader will have no difficulty adapting the arguments given for the others to these two cases.

Case 1. $\phi = \neg\phi'$.

It is enough to argue that $X \sim_{\phi'} Y$ implies $X \sim_\phi Y$. By contraposition, if $\exists Z$, $X \oplus Z \models \phi$ and $Y \oplus Z \models \neg\phi$ then immediately $X \oplus Z \models \neg\phi'$ and $Y \oplus Z \models \neg\phi \equiv \phi'$. The converse is just as easy, and in fact $\sim_\phi = \sim_{\phi'}$.

Case 2. $\phi = \phi_1 \wedge \phi_2$.

Here we have $Free(\phi) = Free(\phi_1) \cup Free(\phi_2)$. Let σ be a partial equipment signature for $Free(\phi)$. Our strategy is to define a convenient equivalence relation $\sim$ on the σ-equipped t-boundaried graphs, and then show two things: that $\sim$ has finite index, and that it refines $\sim_\phi$.

---

[2]Note that this is in the large universe.

The sets of variables in $Free(\phi_i), i = 1, 2$, are subsets of the sets of variables in $Free(\phi)$, so the partial equipment signature σ for $Free(\phi)$ induces partial equipment signatures σ_1 for $Free(\phi_1)$ and σ_2 for $Free(\phi_2)$ just by forgetting the unneeded equipment. It is with respect to these induced signatures that we define $\sim$ to be $\sim_{\phi_1} \cap \sim_{\phi_2}$. Since it is, by the induction hypothesis, the intersection of equivalence relations of finite index, $\sim$ has finite index.

It remains to argue that $X \sim Y$ implies $X \sim_\phi Y$. By contraposition, if $\exists Z$, $X \oplus Z \models \phi$ and $Y \oplus Z \models \neg\phi$, then either (i) $X \oplus Z \models \phi_1$ and $Y \oplus Z \models \neg\phi_1$, or (ii) $X \oplus Z \models \phi_2$ and $Y \oplus Z \models \neg\phi_2$. So either (i) X and Y are not $\sim_{\phi_1}$ equivalent, or (ii) X and Y are not $\sim_{\phi_2}$ equivalent. In either case, it follows that X and Y are not $\sim$ equivalent.

Case 3. $\phi = \exists u \phi'$, where u is a vertex variable free in ϕ'.

For $i \in \{1, \ldots, t\}$, let X_i denote the partially equipped t-boundaried graph X further equipped with the boundary vertex with label i in correspondence with the variable u.

Let $\alpha(X, Z)$ denote the statement: There is a vertex x in the interior of X such that $X_u \oplus Z \models \phi'$, where X_u is X additionally equipped with the vertex x in correspondence with the variable u.

Define $X \sim_{\exists u} Y$ if and only if for all Z, $\alpha(X, Z) \leftrightarrow \alpha(Y, Z)$.

This is an equivalence relation of finite index, since, by the commutativity of $\oplus$ and the definition of α, the conditions $\alpha(X, Z)$ and $Z \sim_{\phi'} Z'$ together imply $\alpha(X, Z')$. Hence, $\alpha(X, Z)$ depends in the second argument only on the equivalence class of Z.

Define $X \sim Y$ if and only if

(a) $X \sim_{\phi'} Y$,
(b) $X_i \sim_{\phi'} Y_i$ for all $i \in \{1, \ldots, t\}$, and
(c) $X \sim_{\exists u} Y$.

The relation $\sim$ is an equivalence relation of finite index since it is the intersection of finitely many such relations on the $Free(\phi)$-partially equipped t-boundaried graphs. [In the case of (c), this is the method of test sets in action.]

It remains only to argue that $X \sim Y$ implies $X \sim_\phi Y$. If not, then $\exists Z$ with $X \oplus Z \models \phi$ and $Y \oplus Z \models \neg\phi$. Choose an instantiation of u to a vertex v in $X \oplus Z$ making ϕ true.

Case (a). The vertex v is in the interior of Z. Let Z_u denote Z further equipped with the vertex v in correspondence with the variable u. Then, $X \oplus Z_u \models \phi'$ and by (a), $Y \oplus Z_u \models \phi'$ which implies $Y \oplus Z \models \phi$, a contradiction.

Case (b). The vertex v is a boundary vertex of X. Then for some $i \in \{1, \ldots, t\}$, we have $X_i \oplus Z_i \models \phi'$ which by (b) implies $Y_i \oplus Z_i \models \phi'$ and therefore $Y \oplus Z \models \phi$, a contradiction.

Case (c). The vertex v belongs to the interior of X. Then, (c) implies that there is a vertex v' in the interior of Y such that $Y_u \oplus Z \models \phi'$ where Y_u is Y further equipped with the vertex v' in correspondence with the variable u. This implies $Y \oplus Z \models \phi$, a contradiction.

Case 4. $\phi = \exists d \phi'$, where d is an edge variable free in ϕ'.

For each $\{i, j\} \in \{1, \ldots, t\}^2$, let X_{ij} denote X further equipped with the edge ij in correspondence with the variable d.

Let $\beta(X, Z)$ be the statement: There is an edge d' with at least one interior end-point in X such that $X_d \oplus Z \models \phi'$, where X_d denotes X equipped with d' in correspondence with d.

Define $X \sim_{\exists d} Y$ if and only if for all Z, $\beta(X, Z) \leftrightarrow \beta(Y, Z)$. This is an equivalence relation of finite index, by the commutativity of $\oplus$ and the definition of β.

Define $X \sim Y$ if and only if

(a) $X_{ij} \sim_{\phi'} Y_{ij}$ for all $\{i, j\} \in \{1, \ldots, t\}^2$,

(b) $X \sim_{\phi'} Y$,

(c) $X \sim_{\exists d} Y$.

The relation $\sim$ is an equivalence relation of finite index since it is the intersection of finitely many such relations. The verification that $X \sim Y$ implies $X \sim_\phi Y$ is straightforward.

Case 5. $\phi = \exists U \phi'$, where U is a vertex set variable free in ϕ'.

For each subset $S \subseteq \{1, \ldots, t\}$ let $\gamma_S(X, Z)$, defined for Z equipped with $V_U = S$, be the statement: There is a subset U_0 of the interior of X such that $X_U \oplus Z \models \phi'$ where X_U is X equipped with the vertex set $U_0 \cup S$ in correspondence with the variable U.

Define $X \sim_S Y$ if and only if for all Z, $\gamma_S(X, Z) \leftrightarrow \gamma_S(Y, Z)$.

It follows easily from the commutativity of $\oplus$ and the definition of γ_S that $\gamma_S(X, Z)$ and $Z \sim_{\phi'} Z'$ imply $\gamma_S(Y, Z)$, and from this and the induction hypothesis that $\sim_S$ is an equivalence relation of finite index.

Define $X \sim Y$ if and only if for all $S \subseteq \{1, \ldots, t\}$, $X \sim_S Y$.

The relation $\sim$ is an equivalence relation of finite index, since it is the intersection of finitely many such relations. The verification that $X \sim Y$ implies $X \sim_\phi Y$ is straightforward.

Case 6. $\phi = \exists D \phi'$, where D is an edge-set variable.

For each subset $S \subseteq \{1, \ldots, t\}^2$, let $\delta_S(X, Z)$ be the statement: There is a set of edges D_0, each having at least one interior endpoint in X, such that $X_D \oplus Z \models \phi'$, where X_D is X equipped with the edge set $D_0 \cup S$ in correspondence with the variable D.

Define $X \sim_S Y$ if and only if $\forall Z \delta_S(X, Z) \leftrightarrow \delta_S(Y, Z)$. As in the other cases, this is an equivalence relation of finite index, by the induction hypothesis and the fact that $\delta_S(X, Z)$ depends in the second argument only on the equivalence class of Z with respect to $\sim_{\phi'}$.

Define $X \sim Y$ if and only if $\forall S \subseteq \{1, \ldots, t\}^2 X \sim_S Y$.

The relation $\sim$ is an equivalence relation of finite index since it is the intersection of finitely many such relations. It remains only to show that $X \sim Y$ implies $X \sim_\phi Y$. If not, then $\exists Z$ such that (without loss of generality) $X \oplus Z \models \phi$ and $Y \oplus Z \models \neg\phi$. Fix an instantiation of the variable D to a set of edges E_D in $X \oplus Z$, making ϕ true. Let $S = E_D \cap \partial(X)$. Let Z_D denote Z additionally equipped with the set of edges of E_D in Z in correspondence with the variable D. Thus, we have $\delta_S(X, Z_D)$. Since $X \sim Y$, we have also $\delta_S(Y, Z_D)$, which implies $Y \oplus Z \models \phi$, a contradiction. $\square$

This completes the induction for the Claim, and Theorem 13.1.1 is proved. $\square$

13.2   Implementing Courcelle's Theorem

The current state of the art algorithm for implementing Courcelle's Theorem[3] which seems to work pretty well on problems of interest, and on practical graphs is due to Kneis, Langer, and Rossmanith [453], building on earlier work of, for example, Kneis and Langer [451].

The approach uses the *model checking game* from MSO_2, which is a pebble game between two players *verifier* and *falsifier*. As usual, verifier tries to prove the formula holds and falsifier tries to prove that it does not. Verifier moves on existential formulas and $\vee$ whilst falsifier on universal ones and $\wedge$. Of course this game is simply evaluating the formula like $\exists R\phi(R)$ for a set variable R, and this entails checking sets $\hat{R}$ in the structure. The recursive algorithm so generated would naively take time $O((2^n + n)^r)$ for a universe of size n and ϕ of quantifier rank r.

The key idea is to use dynamic programming on structures of treewidth t to make this linear. As with our dynamic programming examples, we work bottom up. When we are evaluating the substructure we have seen so far $U' \preceq U$, it is possible that we can already decide that $U \models \phi$ or $U \models \neg\phi$. For example, in 3-COLORABILITY we might have found that we have locally violated the possibility, or we could be in the situation that we cannot yet decide ϕ, and we need to transfer the bag information upwards. This is all familiar from the examples done in the dynamic programming section.

What Kneis, Langer, and Rossmanith do is to exploit the fact that MSO and FO formulae with bounded quantifier rank have quite limited ability to distinguish substructures. They define a notion of *r-equivalent subgames* for a formulation of equivalence and this allow them to delete redundant equivalent subgames. The key fact is that non-equivalent subgames is bounded by a constant, and this allows for running times linear in the size of the tree decomposition. In practical experiments, this approach works well. The number of table entries and the like is *much smaller* than the worst case and seem to work pretty well on test problems in graphs of size up to about treewidth 8.

The details are, as you might expect, are quite intricate. We refer the reader to [453] for such details, and to Langer, Reidl, Rossmanith, and Sikdar [477] for some evaluation data. We also refer the reader to Langer, Rossmanith, and Sikdar [478] for similar methods applied to other width metrics such as those we meet in Sect. 17.3.5. We expect that this is not the end of the story.

13.3   Seese's Theorem

In a pretty paper [612], Detlef Seese proved a converse to Courcelle's Theorem.

---

[3]There is also "on the fly" work of Courcelle and Durant [177] which use what they call *fly-automata,* and seem to be implementable.

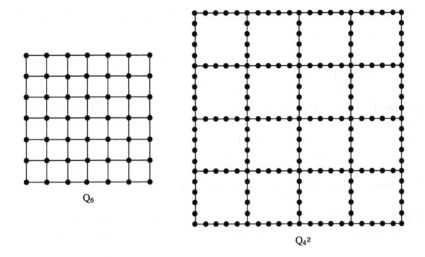

Fig. 13.1 Examples of grids

Theorem 13.3.1 (Seese [612]) *Suppose that $\mathcal{F}$ is any family of graphs with a decidable monadic second-order (MS_2) logic. Then, there is a number n such that for all $G \in \mathcal{F}$, the treewidth of G is less than n.*

Notice the very interesting dichotomy from linear time decidability: the monadic second-order logic of bounded treewidth graphs to the undecidability of the monadic second-order logic for families of graphs failing to have a bound on the treewidth.

In this section, we sketch a proof of Seese's Theorem. It relies on material from Robertson and Seymour and on some classical undecidability results.

We begin with *generalized grids* Q_n and Q_n^m. For $n \geq 2$ and $m \geq 0$, we define these as follows:

$$V(Q_n) = \{(i, j) : 0 \leq i, j \leq n\},$$
$$E(Q_n) = \{\langle (i, j), (i', j') \rangle : |i - i'| + |j - j'| = 1 \text{ and } 0 \leq i, i', j, j' < n\},$$
$$V(Q_n^m) = \{(i, j) : 0 \leq i, j \leq (m + 1)(n - 1)\},$$
$$E(Q_n^m) = \{\langle (i, j), (i' j') \rangle : 0 \leq i, i', j, j' \leq (m + 1)(n - 1)$$
$$\wedge |i - i'| + |j - j'| = 1 \wedge [(i \equiv 0 \ (\mathrm{mod} \ (m + 1)) \wedge i = i')$$
$$\vee (j \equiv 0 \ (\mathrm{mod} \ (m + 1)) \wedge j = j')]\}.$$

Clearly, Q_n is the $n \times n$ grid and Q_n^m is a grid with extra points. For instance, see Fig. 13.1, where we exhibit Q_6 and Q_4^3.

The following is an easy exercise left to the reader (Exercise 13.5.3).

Proposition 13.3.1 *If G is any planar graph, then there is a number n such that* $G \leq_{\text{minor}} Q_n$.[4]

As well as the well-quasi-ordering of graphs, Robertson and Seymour proved the following remarkable result.

Theorem 13.3.2 (Robertson and Seymour [584]) *Suppose that H is a finite planar graph, and C is a class of graphs no member of which has H as a minor. (In standard terminology, we say that C excludes H.) Then there is a number n such that for all* $G \in C$, *the tree width of G is less than n.*

The state of the art here is the following result.

Theorem 13.3.3 (Demaine and Hajiaghayi [214]) *For any fixed graph H, each H-minor-free graph of treewidth w has a* $\Omega(w) \times \Omega(w)$ *grid as a minor.*

It follows that if a class has unbounded treewidth, then it contains graphs containing arbitrarily large grids Q_n as minors.[5] Now in MS_2 for a fixed n, one can express the fact that a graph has Q_n as a minor (Exercise 13.5.4). The point is that this result means that the monadic second-order logic of grids can be interpreted in C if C is a class with unbounded treewidth. But one can code Turing machines or Undecidable Tiling Problems into this logic, the details being spelled out in Seese [611]. It follows that the MS_2 logic of C is undecidable if C contains graphs of arbitrarily large treewidth.

13.4   Notes on MS₁ Theory

The material on Courcelle's Theorem and Seese's Theorem both looked at the MS_2 logic based on the two-sorted language with predicate symbols for edges, vertices, and incidence. As we mentioned earlier, there is a long history concerning the "basic" monadic second-order logic where we have no predicate for edges but need binary relations for these objects. Now, the expressive power changes. For instance, "G is Hamiltonian" is no longer expressible in our logic (Exercise 13.5.1).

Naturally, the methods we have used for Courcelle's MS_2 Theorem still work. What is unclear at the time of writing of this book is whether the analog of Seese's Theorem holds for MS_1 theories.

Seese was able to prove the following theorem.

---

[4]Recall that a graph G is said to be a minor of a graph H iff G is represented in H by disjoint paths; that is, there is an injection $f : V(G) \mapsto V(H)$ such that for all $xy \in E(G)$, there is a path $p(xy)$ from $f(x)$ to $f(y)$ in H such that if $xy \neq x'y'$, then $p(xy)$ is disjoint from $p(x'y')$. We write $G \leq_{\text{minor}} H$. We refer the reader to Definition 17.1.7. We will discuss this concept in some detail in the next chapter.

[5]The proof of this result is beyond the scope of this book, but a related "excluded minor" theorem is treated in Sect. 18.2.

Theorem 13.4.1 (Seese [612]) *If a class of planar graphs has a decidable MS_1 theory, then that class has uniformly bounded treewidth.*

To prove Theorem 13.4.1, Seese needed some intricate arguments to interpret the logic of the Q_n's into a give class of planar graphs of unbounded treewidth. The work build heavily on Seese [611], and Kuratowski's Theorem (Theorem 17.1.2 of Sect. 17.1). The decisive lemma is the following.

Lemma 13.4.1 (Seese [612]) *Let $\mathcal{K}$ be any class of graphs such that for every planar graph H there is a planar $G \in \mathcal{K}$ with $H \leq_{\text{minor}} G$. Then, the monadic second-order (MS_1) theory of $\mathcal{K}$ is undecidable.*

The principle difficulty in the proof is then to prove that MS_1 interpretability occurs. This is quite intricate and heavily relies on planarity to be able to interpret the Q_n's into the class as minors. We refer the reader to [612] for details.

Subsequently, Courcelle and others have extended Theorem 13.4.1 to much wider classes of graphs. In particular, Courcelle and Oum have Seese's MS_2 theorem have extended to MS_1 via the notion of cliquewidth, a width metric with a similar parse language to treewidth, we will meet in Chap. 16. That is, Courcelle and Oum prove the following.

Theorem 13.4.2 (Courcelle and Oum [183]) *If a set of directed or undirected graphs has a decidable MS_1 theory (even with the addition of the "even cardinality" predicate), then it has bounded cliquewidth.*

The main idea in these other results is to replace Q_n by more intricate structures such as the Q_n^m to make up for the lack of expressive power. More can be found in Courcelle [170, 172–174].

13.5 Exercises

Exercise 13.5.1 Prove that Hamiltonicity is not expressible in MS_1. (This is not easy.)

Exercise 13.5.2 Construct a test set for the class of graphs that contain a cycle of length 4.

Exercise 13.5.3 Prove that if G is planar then there is some Q_n with $G \leq_{\text{minor}} Q_n$. (Hint: Use induction.)

Exercise 13.5.4 Prove that $G \leq_{\text{minor}} H$ is MS_2 expressible for a fixed graph G.

Exercise 13.5.5 (Cygan, Fomin, and van Leeuwen [187]) Recall the firefighting problem from Exercise 5.3.6. Show that SAVING k VERTICES from that exercise is FPT on graphs of bounded treewidth.

13.6   Historical Notes

The study of decidability results for monadic second-order theories of classes of combinatorial objects and their connections with automata has a long and interesting history, as can be found in Büchi [113]. Probably the first result is the one of Büchi [113] who proved that a language is regular iff it is definable by some formula in monadic second-order logic (of strings). The famous hallmark result here is the decidability result of Rabin [568], with its simplified proof by Harrington and Gurevich [367].

Detleff Seese was the first to conjecture that classes of graphs with decidable theories were "similar to trees" (Seese [611]). Seese was able to prove that if a class of planar graphs with maximal degree ≤ 3 avoids certain graphs as induced subgraphs, then it has a very nice parse tree. Since Rabin [568] had proven the decidability of the monadic second-order logic of trees, Seese was able to deduce that these theories were decidable. Of course, Rabin's method is not very feasible. Seese continued to pursue his conjecture that "tree-like" graphs were the boundary of decidability culminating in the results of this section.

Courcelle began from a more syntactical viewpoint. Courcelle and Franchi-Zannettacci [179] began to study the sets of dependency graphs generated by attribute grammars. Each such set has bounded treewidth. Being acyclic was an important nontrivial monadic second-order property, and its algorithmic solution suggested generalizations to other logically defined properties. In this and other articles, Courcelle and others were following the finite model logic path of enriching first-order logic with additional predicates (such as "path predicates") and proving various recognizability properties.

The breakthrough was the idea to take the classical connection between Monadic Second-Order Logic and classes of languages and trees and seek a model in "tree-like" graphs. What was needed was a set of parse operations. In the mid-1980s Courcelle was able to prove the principal lemma that the k-theory of a graph $\oplus(G, H)$ is a function of the k-theories of the graphs G and H, where the k-theory of a structure is the (finite) set of monadic second-order formulas of quantifier depth at most k which are valid in the structure and $\oplus$ is an appropriate gluing operation. Actually, this is essentially the same as the Feferman–Vaught Theorem for first-order formulas and Shelah [616] for monadic second-order formulas. See also Makowsky [507] for more on the uses of Feferman–Vaught techniques.

Courcelle's results and their linear time decidability for graphs given by parse trees were presented in a conference [166]. The full paper is Courcelle [167]. Arnborg, Lagergren and Seese extended the basic Courcelle results to a wider class of formulas and to some minimization and maximization problems. Arnborg et al.'s results were presented to ICALP in 1988 and appeared in final form in [46]. The "basic Lemma" follows from a general one given in [171] for relational structures.

Courcelle's results were obtained by Feferman–Vaught style proofs, other proofs such as Arnborg et al.'s using the so-called method of interpretation. For this method, one takes some known decidable language and finds an effective interpretation of the logic or theory in question in the other. In this case we move from bounded treewidth graphs back to trees. This is a longstanding technique in mathematical logic. Early on, and more recently in Courcelle and Engelfriet [178], Courcelle's methods and results were couched in the language of universal algebra.

Abrahamson and Fellows in [4, 5] were able to give (reasonably) straightforward automata-theoretic proofs of Courcelle's Theorems. This is the approach we use here. In fact, both of these papers contain flawed proofs of the Myhill–Nerode Theorem for graphs. These flaws are remedied here as well as a consequential change in the proof that CUTWIDTH is finite state. The results connecting finite-state graph languages and well-quasi-orderings also appear in Abrahamson–Fellows [4, 5].

Seese's Theorem appears in Seese [612].

It is still unknown if there is a full analog of Büchi's Theorem [113] (that regularity is equivalent to monadic second-order definability for formal languages) that holds for graphs. The conjecture of Courcelle is that definability in *Counting Monadic Second-Order Logic* is equivalent to finite state for graphs of bounded treewidth. Here, as we mentioned earlier, to get counting monadic second-order logic, one adds predicates of the form $\mod_{p,q}$ for $p < q$ where $\mod_{p,q}(V)$ has the interpretation true iff $|S| = p \mod q$, where S is the set denoted by the set variable V. Kabanets [426] has verified the conjecture for graphs of bounded pathwidth,[6] and the conjecture has been verified for treewidth ≤ 3 (Courcelle [169] for ≤ 2 and Kaller [427] for width 3 and k-connected graphs of width k). A student of Courcelle announced the proof of the full analogue (Lapoire [481]), but no details have ever emerged, and hence the problem (certainly for treewidth) is now regarded as *open*.

There are many applications of Courcelle's Theorem and Courcelle-like theorems. One recent one was to (treewidth bounded versions of) default logic and autoepistemic logic in [528] by Meier, Schmidt, Thomas, and Vollmer.

As well as the recent implementations, there are other proofs such as Flum and Grohe [312, Chap. 10] and similar results appeared by Arnborg, Lagergren, and Seese in [46] and Borie, Parker, and Tovey in [108]. Also, an alternative game theoretic-proof has appeared recently in [451, 453]. It also has numerous generalizations for other extensions of MSOL (see [168, 175, 181]). Courcelle's Theorem and its extensions can be applied to a very wide family of problems due to the expressibility power of MS_2. As a general principle, "most normal" purely graph theoretical problems in NP can be expressed in MS_2, and hence they are classified in FPT for graphs of bounded width.

However, there are still graph problems that remain parameterized-intractable even when parameterized in terms of the branchwidth, treewidth, or any other width of their inputs. As examples, we mention BANDWIDTH (notably) by Bodlaender, Fellows, and Hallett [87] LIST COLORING, EQUITABLE COLORING, PRECOLORING EXTENSION by Fellows, Fomin, Lokshtanov, Rosamond, Saurabh, Szeider, and Thomassen [286] and BOUNDED LENGTH EDGE-DISJOINT PATHS by Golovach and Thilikos [349].

---

[6]Kabanets [426] is an extended abstract and the full proof only appears in his masters thesis, Kabanets [425].

13.7   Summary

We give a proof of Courcelle's Theorem that MS_2 model checking is linear time FPT for graphs of bounded treewidth. We use the automata approach. We also look at implementations and at Seese's Undecidability Theorem.

More on Width-Metrics: Applications and Local Treewidth

14

14.1 Applications of Width Metrics to Objects Other than Graphs

It might seem that bounded treewidth algorithmics is useful only for problems about graphs—but this is not the case! There are several routes to such generalization:

(1) Other kinds of mathematical objects may admit representations by (annotated) graphs of bounded treewidth, opening up the possibility of applying the algorithmic machinery of bounded treewidth to address decision problems about these objects, parameterized by the width of the representation.

(2) Bounded width tree-decompositions of graphs (where the bags are sets of vertices of a graph, and a bag is always a cutset) may be directly generalized to handle other kinds of objects that can be viewed as consisting of other kinds of elemental objects, and each bag has a property analogous to being a cutset.

(3) The objects can be viewed as associated with graphs that capture important structural information that can be exploited efficiently if these associated graphs have bounded treewidth.

Explorations of these routes to generalization have revealed that bounded treewidth structure (and algorithmic opportunity!) is nearly ubiquitous in real-world datasets over a wide range of mathematical objects and situations. This remains a hot area of current investigation.

Illustrative Examples

Routes (1) and (3) Knots can be represented by *knot diagrams*, which, assuming the projection is in *standard position*, have an obvious underlying structure of a 4-regular multigraph, where each vertex is a crossing. This invites us to visit the most famous open algorithmic problem in knot theory, KNOT TRIVIALITY (not known to be either NP-hard or in P), parameterized by the treewidth of the underlying graph of the knot diagram. If we could directly exploit this underlying structure in order to give an FPT algorithm, that would be a nice example of Route (3). No one presently knows how to do this. But we can go further. If the knot is *oriented* then each crossing is either *left-handed* or *right-handed*, and the (oriented) knot can

R.G. Downey, M.R. Fellows, *Fundamentals of Parameterized Complexity*,
Texts in Computer Science, DOI 10.1007/978-1-4471-5559-1_14,
© Springer-Verlag London 2013

be completely represented by the underlying 4-regular multigraph annotated by two corresponding vertex colors, thus by a digraph with two vertex colors. The FPT bounded treewidth machinery easily generalizes to handle this, with the outcome that a knot diagram of (underlying) bounded treewidth, can be parsed in FPT time into a rooted binary tree of symbols. This is a nice example of Route (1). In the Myhill–Nerode spirit, we can ask if there is a finite-state tree automaton recognizing precisely the parse trees that represent the *Unknot*, for knot diagrams of bounded treewidth.

Route (2) For matroids, there are notions analogous to bounded treewidth, where instead of bags of vertices, there are bags of vectors, and the analog of *being a cutset in a graph* is formulated in terms of algebraic linear independence.

In the remainder of this chapter we survey some of the range of ideas flowing from these routes to generalization that have emerged so far.

14.1.1 Logic

Graphs are a universal data structure, and hence most structures from computer science can be thought of as graphs. For example, if A is a structure in monadic second order logic, then the *Gaifman graph* of the structure is $G(A) = (V, E)$ where $ab \in E$ if and only if $a \neq b$ and there is an n-ary relation R of the signature of A, and a tuple $a_1, \ldots, a_n$ such that $A \models R(a_1, \ldots, a_n)$ such that $a, b \in \{a_1, \ldots, a_n\}$.

We can therefore think of formulas as having an associated underlying treewidth. This turns out to be very productive in the parameterized complexity perspective.

A relational database consists of a finite set of finite relations on a finite set U. Each relation can be viewed as a set of tuples listed in a table for the relation. The size of a database is the length of a description of all the relations by listing the contents of the relational tables. For a finite set U of attributes, each with an associated domain. A *relational schema R* is a subset of U. A *database schema* $D = (R_1, \ldots, R_n)$ is a set of relation schemas. A relation r_i over a relation schema R_i is a set of R_i-tuples, where a R_i-tuple is a mapping from the attributes of R_i to their domains. A database d over the schema D is a set $\{r_1, \ldots, r_n\}$ of finite relations.

There is an extensive literature exploring the computational complexity of queries formulated in the *relational calculus* for relational databases. Formulas of the calculus are built from atomic formulas of the form $x_1 = x_2$ and $R_i(x_1, \ldots, x_m)$, where the x_i are variables or constants of the appropriate domains, using Boolean connectives and quantifiers. A formula ϕ with free variables $y_1, \ldots, y_k$ specifies a query relation

$$r_\phi = \big\{(a_1, \ldots, a_k) : \phi(a_1, \ldots, a_k) \text{ is true in } d\big\}.$$

The following problem is simply stated and is central.

QUERY NON-EMPTINESS

Instance: A database d, and a query ϕ.
Parameter: $|\phi|$.
Question: Is the query relation r_ϕ nonempty?

Given the closeness of the problem to logic, it is not surprising that the general form is very hard, even from a parametric point of view. We will prove that the general problem is complete for the parameterized intractability class $AW[*]$ in Chap. 26. Further, we will see that even the restricted variation of the problem, where we only consider existential conjunctive queries, is hard for $W[1]$.

Structurally parameterized forms of the problem are FPT. If the query is of bounded treewidth then it is FPT (Flum and Grohe [309]). This more or less follows by Courcelle's Theorem. Interestingly, this correlates to Vardi's [655] proof that queries parameterized by the number of variables is FPT.

Whilst we do not have room to discuss this here, if we look beyond relational databases, we can look at LTL documents. Linear temporal logic is used extensively in program specification. Here we state the following FPT result.

Theorem 14.1.1 (Lichtenstein and Pnueli [491]) *Evaluation for* LTL *on the class of all Kripke structures is solvable in time* $O(2^k n)$ *where k is the size of the input and n the size of the instance.*

This result is quite practical, widely implemented and improved using heuristics, and is proven by bounded search trees.

14.1.2 Matroid Treewidth and Branchwidth

The idea that a graph is tree-like if it can be decomposed entirely across small separations can be extended to algebraic structures, and, in particular, matroids using algebraic independence instead of topological separation as the central decomposition criteria.

This programme was initiated by Hliněný and Whittle [397, 398], and is part of a long-term program to generalize the Graph Minors Project of Robertson and Seymour (the originators) to an analogous matroid structure theory (and associated FPT algorithmic methods). The program is surveyed by Geelen, Gerards, and Whittle in [342].

In the setting of matroid theory, *matroid branchwidth* is easier and more natural to define than the matroid analog of *treewidth* (although Hliněný and Whittle [397, 398] showed that you can define a matroid analog of graph treewidth).

In the definition of *branchwidth* for graphs, the key idea in the representation of the separation properties is to make a "data-structure" for the graph, where the edges of the graph (the essential elements of topological connectivity), are in one-to-one correspondence with the leaves of a ternary tree.

In the setting of matroids, we need an analog of the notion of topological separation, and this is provided by the *rank* of a set of vectors, a measure of algebraic linear independence of the set of vectors. The connectivity or width function is $\lambda(A) = r(A) + r(E \setminus A) - r(E) + 1$, where r is the rank function of the matroid. A *separation* can be defined in the same way as for graphs, and this results in a partition of the set E of matroid elements into two subsets A and $B = E \setminus A$.

The branchwidth of a graph and the branchwidth of the corresponding graphic matroid may differ. For instance, the three-edge path graph and the three-edge star have different branchwidths, 2 and 1, respectively, but they both induce the same graphic matroid with branchwidth 1 (Hicks and McMurray [392], Mazoit and Thomassé [520]). Hicks and McMurray and independently, using completely different techniques, Mazoit and Thomassé showed also that for graphs that are not forests, the branchwidth of the graph is equal to the branchwidth of its associated graphic matroid.

Robertson and Seymour conjecture that the matroids representable over any finite field are well-quasi-ordered by matroid minors, analogously to the Robertson–Seymour theorem for graphs. So far, this conjecture has been proven only for the matroids of bounded branchwidth.

It is possible to prove a version of Courcelle's Theorem for matroids of bounded branchwidth. The syntax consists of variables for matroid elements and predicates $e \in F$ where F is a variable for sets of elements, and indep(F) which is true if and only if F is an independent set.

Theorem 14.1.2 (Hliněný [394]) *Let F be a finite field and ϕ a sentence of MMS, matroid monadic second order logic as described above. Suppose that the n element matroid M is given a vector representation over F together with a branch decomposition of width k. Then there is a linear FPT algorithm (in F, ϕ, k) deciding whether $M \models \phi$.*

Of course in this brief section, we can only give a flavor of the results, and we refer the reader to the source papers and recent ones for this thriving subject.

14.1.3 Knots, Links, and Polynomials

Once we realize that many (even, perhaps, most?) mathematical objects support analogs of various width measures of graphs (which are merely the most elemental relational structures) then all kinds of opportunities open up, with respect to efficient algorithms. We firmly believe that only the surface of this opportunity landscape has been scratched.

We will confine ourselves to one more such application. Some of the original applications were to graph polynomials. Graph polynomials are ubiquitous in physics, chemistry, and biology. They are almost always #P-hard. Andrzejak [39], and Noble [548], who first studied the complexity of computing Tutte polynomials for graphs of bounded treewidth. They showed that evaluation of the Tutte polynomials for graphs of bounded treewidth was FPT. Later, Janos Makowsky and his co-authors were the first to realize that the ideas from MS_2 FPT theory and treewidth structuring the actual polynomials could be applied in this area.

In general, a graph polynomial is of the form P: graphs $\to \mathcal{R}[X]$, where $\mathcal{R}[X]$ denoted polynomials over some commutative ring. Typically, the polynomials are of the form

$$p(G) = \sum_{H \subseteq G, H \in \mathcal{H}} c(H) \left(\underbrace{\prod w_1}_{\text{case } 1} \cdots \underbrace{\prod w_d}_{\text{case } d} \right), \quad \text{where,}$$

1. $\mathcal{H}$ is some specified family of subgraphs of G.
2. The products range over elements or subsets of vertices and edges of H where p is some fixed integer.
3. w_j are certain weighting functions with values in $\mathcal{R}$.
4. $c(H)$ is some coefficient depending on H.

The original example is the *characteristic polynomial*

$$\chi(G, \lambda) = \det(\lambda I - A(G)),$$

where $A(G)$ is the adjacency matrix of G. There is a huge industry around graph polynomials, such as the chromatic, Tutte, and Jones polynomials. It is possible to define many of these polynomials using algorithmically efficient formal logics. Some typical results in this area:

Theorem 14.1.3 (Makowsky [511]) *The following problems are* FPT.

STW KAUF
Instance: *A signed diagram $S(L)$ of a link L.*
Parameter: tw($S(L)$).
Task: *Compute the Kaufmann bracket of L.*

ALG-HOMFLY
Instance: *A k-expression $T(L)$ of a link L with n operations.*
Parameter: *A positive integer k.*
Task: *Compute* HOMFLY *using $T(L)$.*

We refer the reader to the following sources for more on these fascinating and deep results: Andrzejak [39], Makowsky [508, 509], Makowsky and Marino [510, 511], and Makowsky, Rotics, Averbouch, and Godlin [512].

14.2 Local Treewidth

14.2.1 Definitions

A generalization of the idea of treewidth to a wide class of other graphs is based upon the growth rate the treewidth of neighborhoods of vertices in graphs. The r-neighborhood $N^r[v]$ in a graph G is $\{u \mid D(u, v) \leq r\}$ where d is the usual graph distance measuring the number of vertices on a geodesic from u to v.

> **Definition 14.2.1** (Frick and Grohe [332]) We say that a *family $\mathcal{F}$ of graphs* has $f(r)$ bounded *local treewidth* if for all $G \in \mathcal{F}$, and all r and $v \in V(G)$,
>
> $$\mathrm{tw}\big(N^r[v]\big) \leq f(r).$$
>
> The family has *bounded local treewidth* if there is some function f such that $\mathcal{F}$ has $f(r)$ bounded local treewidth.

Clearly, any graph family of bounded treewidth has bounded local treewidth. However, there are other examples.

Theorem 14.2.1 *The following classes of graphs are families with bounded local treewidth.*
(1) (Frick and Grohe [332]) *Graphs of bounded degree d.* $f(r) = d^r$.
(2) (Frick and Grohe [332]) *Planar graphs.* $f(r) = 3r$.
(3) (Demaine and Hajiaghayi [210]) *Moreover, if $\mathcal{F}$ is any class which is minor closed,[1] then the local treewidth of the class is bounded by $O(r) = f(r)$.*

Proof The material of Demaine and Hajiaghayi is a bit beyond the scope of this book. We prove the other two statements. 1 is easy since the degree bound means that the possible number of vertices in $N^r(v)$ is $\leq d \sum_{i \in [r]} (d-1)^{i-1} \leq d^r$.

Now for 2. We prove that if G is a planar graph with a spanning tree S of height r then $\mathrm{tw}(G) \leq 3r$, giving the result. We suppose that we have a plane embedding of G. As with elimination orders, we can assume that G is triangulated. We define a tree $T = (T, F)$ by letting T be the set of faces of G, and for $t, t' \in T$, $\{t, t'\} \in F$ iff the boundaries of t and t' share an edge that is not an edge of S. Then T is acyclic. To see this, let $t_1, \ldots, t_s$ be a cycle in T, and let e_i be an edge shared by t_i and t_{i+1}. Choose an interior point a_i on the embedding of e_i and b_i and edge with endpoints a_i and a_{i+1}, with $a_{s+1} = a_1$, and interior in t_i. Then $b_1 \cup \cdots \cup b_s$ defines a polygon B which (by the Jordan Curve Theorem) divides the plane into two regions. Both regions contain vertices of G. Since no edge of S crosses B, this contradicts the fact that S is a spanning tree for G. Thus T is acyclic.

---

[1]See Sect. 17.1 if you need reminding what minors are, though we did meet them in Chap. 13.

It will be shown that T is connected and hence will allow us to define a tree decomposition using it. The bags of the decomposition are as follows. For $t \in T$, let X_t denote the vertices in t together with their ancestors in the spanning tree. If X_t contains a descendant of v it contains v, so the root is in every bag.

First note that every bag has at most $3r + 1$ elements. This follows since G is triangulated, and hence the boundary of each face has at most 3 vertices. They share at least one ancestor and they have at most r ancestors, giving the bound.

If uv is an edge, then $u, v \in X_t$ for any face t containing that edge. So every edge of G is included in some bag.

The most difficult step is the following. We need to prove that for each $v \in V(G)$, $X^{-1}(v) = \{t \in T \mid v \in X_t\}$ is connected in T.

This is proven by induction on the height of v in S. If v is a leaf, then $X^{-1}(v)$ is the set of faces whose boundary contains v. Let $e_1, \ldots, e_m$ be the edges incident with v. Then exactly one of these e_1, say, is in S. Suppose that the plane embedding maps the edges $e_1, \ldots, e_m$ clockwise in order around v. Let t_i be the face whose boundary contains e_i and e_{i+1}. Then the faces $t_1, \ldots, t_m$ form a path in T.

For the induction step, suppose that v is a non-leaf of G. Let $e_1, \ldots, e_m$ be the edges incident with v. Let $e_{m+1} = e_1$ for convenience. Again suppose that the embedding maps the edges in clockwise order around v, and again let t_i be the face whose boundary contains e_i and e_{i+1}. Assume that e_1 is the edge in S that connects v with its parent in S. Let $1 < i_1 < i_2 < \cdots < i_k < m$ be such that $e_{i_1}, \ldots, e_{i_k} \in S$ and let $w_1, \ldots, w_k$ the corresponding children.

Then v is in the bags $X_{t_1}, \ldots, X_{t_m}$ and all bags that contain a child w_i of v. By the induction hypothesis, $X^{-1}(w_j)$ is connected in T for all $j \in [k]$. Moreover, $t_{i_j-1}, t_{i_j} \in X^{-1}(w_j)$. Therefore it is enough to show that for $1 \le i \le m$, there is a path from t_1 to t_i in $X^{-1}(v)$. Let $2 \le i \le m$. If $i \notin \{i_1, \ldots, i_k\}$, then t_{i-1} and t_i are adjacent in T. If $i = i_j$ then t_{i-1} and t_i are in $X^{-1}(w_j)$, and hence there is a path from t_{i-1} to t_i in $X^{-1}(w_j)$. The case that v is the root is entirely similar. This gives the interpolation property. The root v has the property that $X^{-1}(v) = T$, hence T is connected, and thus a tree. Then $\{X_t \mid t \in T\}$ is the desired tree decomposition. $\square$

As remarked by Flum and Grohe [312], the above can be used to give a quick (and dirty) proof that PLANAR DOMINATING SET is FPT. The proof is by applying Courcelle's Theorem after determining if the diameter of a graph is $\le 3k - 1$, and the following proposition, since Theorem 14.2.1(2) would imply that a planar graph with diameter d has a tree decomposition of width $\le 3d$ which can be found in linear time by breadth first search.

Proposition 14.2.1 *A connected planar graph with a dominating set of $\le k$ elements has diameter $\le 3k - 1$.*

Proof Let D be a dominating set in G with $|D| = k$. Let $v, w \in V(G)$ and P a shortest path from v to w. Each vertex $u \in P$ can dominate at most 3 vertices of P since otherwise there would be a shorter path from v to w, and hence $|P| \le 3k$. $\square$

The method of proof is generalized to obtain the meta-theorem we next turn to.

14.2.2 The Frick–Grohe Theorem

The main result of the present chapter is the following.

Theorem 14.2.2 (Frick and Grohe [332]) *First order model checking for families of graphs of local treewidth at most r, is FPT, when parameterized by the local treewidth and the size of the formula. That is, there is a an algorithm, that, given a graph G of local treewidth at most r, and first order formula ϕ decides if $G \models \phi$ in time $O(|G| + f(|\phi|, r)|G|^2)$ for some function f.*

We remark that most of the classical results for planar graphs or ones of bounded degree are special cases of this meta-theorem. The main idea behind the proof of the Frick–Grohe Theorem is that while second order formulas allow us to talk about things like connectivity, *first* order formulas are essentially local. We will stick to graphs, but the following applies to any signature. It is easy to see that for all r there is a first order formula δ_r such that $d(u, v) \leq r$ holds in G iff $G \models \delta_r(u, v)$, and, as usual, we will identify $d(u, v) \leq r$ with $\delta_r(u, v)$.

Definition 14.2.2 A first order formula ϕ is called *r-local* if and only if for all graphs G and $v \in G$,

$$G \models \phi(v) \quad \text{if and only if} \quad G[N^r(v)] \models \phi(v).$$

The following is the fundamental fact about first order formulas.

Theorem 14.2.3 (Gaifman's Locality Theorem—Gaifman [335]) *For each first order sentence ϕ we can effectively compute an equivalent boolean combination of sentences of the form*

$$\exists x_1, \ldots, x_q \left(\bigwedge_{1 \leq i < j \leq q} d(x_i, x_j) > 2r \wedge \bigwedge_{i \in [q]} \psi(x_i) \right),$$

where $r, q \geq 1$ and ψ is r-local.

The proof is not difficult but is technical, and it is basically done by structural induction on the formulas. We will not give it here, but refer the reader to any standard text in finite model theory such as Ebbinghaus and Flum [266] or Libkin [489].

For Frick and Grohe's proof we need the following extension of the concept of an independent set.

Definition 14.2.3 Let $G = (V, E)$, $S \subseteq V$ and $r, q \geq 1$. We say that S is *(q, r)-scattered* if there exists $v_1, \ldots, v_q \in S$ with $d(v_i, v_j) > r$ for all $1 \leq i < j \leq q$.

Lemma 14.2.1 *Let $\mathcal{F}$ be a family of graphs of (effectively) bounded local treewidth. Then there is an algorithm and a computable function g which, given $G \in \mathcal{F}$, $S \subseteq V$ and $r, q \geq 1$, decides if S is (q, r)-scattered in time $g(q, r)|V|$.*

Proof The reader should write out a sentence expressing the definition of being scattered, and from this and Courcelle's Theorem it follows that if we parameterize also by treewidth then the lemma holds. The next part is to show that there is a computable function $f(q, r)$ with the property that for any $T \subseteq V$ with $|T| \leq q$,

$$\text{tw}\big(G[N^{2r}(T)]\big) \leq f(q, r).$$

This follows since, by bounded local treewidth of $G \in \mathcal{F}$ there must be a computable h with $\text{tw}(G[N^{2r}(v)]) \leq h(2r)$ for all $v \in G$. Now define a new graph $(T, \hat{E})$ with $\hat{E} = \{vw \mid v, w \in T \wedge d(v, w) \leq 4r + 1\}$. Since $|T| \leq q$, the diameter of each connected component C of this graph is $\leq q - 2$, and hence for *every* $v \in C$, $N^{2r}(C) \subseteq N^{(4r+1)(q-2)+2r}(v)$. This implies $\text{tw}(G[N^{2r}(T)]) \leq h((4r + 1)(q - 2) + 2r)$.

Now the lemma will follow by application and correctness of the following algorithm.

Algorithm 14.2.1 (Scattered(G, S, q, r))
1. If $S = \emptyset$, reject.
2. Compute a maximal $T \subseteq S$ of vertices of pairwise distance $> r$.
3. If $|T| > q$, accept.
4. Else if S is (q, r)-scattered in $G[N^{2r}(T)]$, accept. Otherwise reject.

Notice that the algorithm computes a maximal T with $d(v, w) > r$ for all distinct $v, w \in T$. It is easy to see that T can be computed by the greedy algorithm. If $|T| > q$, then T and hence S is (q, r)-scattered. If $|T| < q$, note that by maximality, $S \subseteq G[N^{2r}(T)]$. Since the treewidth computations run in linear time, the proof is finished once we prove that S is (q, r) scattered in G iff it is (q, r)-scattered in $G[N^{2r}(T)]$. The point is that every path of length r between two vertices in $G[N^{2r}(T)]$ is already in $G[N^{2r}(T)]$. $\qquad \square$

We are now ready to complete the proof of the Frick–Grohe Theorem.

Proof of Theorem 14.2.2 Let ϕ be first order. By Gaifman's Locality Theorem we will assume that

$$\phi \equiv \exists x_1, \ldots, x_q \bigg(\bigwedge_{1 \leq i < j \leq q} d(x_i, x_j) > 2r \wedge \bigwedge_{i \in [q]} \psi(x_i) \bigg),$$

for $q, r \geq 1$ and ψ r-local. Let $G \in \mathcal{F}$ and note that $q, r \leq |\phi|$. Let $S = \{v \in V(G) \mid G \models \psi(v)\}$. Using r-locality, we see $S = \{v \in V(G) \mid G[N^r(v)] \models \psi(v)\}$. Invoking Courcelle's Theorem, since the treewidth of $G[N^r(v)]$ is effectively bounded in terms of r, we can decide whether $v \in S$ in linear time. Since $\sum_{v \in V} |G[N^r(v)]| \leq |G|^2$, S can be computed in time $O(|G|^2)$. Now we can invoke Lemma 14.2.1, since

$$G \models \phi \quad \text{iff} \quad S \text{ is } (q, r)\text{-scattered}.$$

This completes the proof. $\qquad \square$

14.3   Exercises

Exercise 14.3.1 (Szeider [634]) In Exercise 23.5.12, the reader will show that the following problem is $W[1]$-hard (i.e. likely not FPT) even for 2-CNF formulas.

k-FLIP MAX SAT
Instance: A CNF formula φ and a truth assignment τ from the variables of φ to $\{0, 1\}$.
Parameter: A positive integers k.
Question: Is there in assignment of Hamming distance $\leq k$ of τ satisfying more clauses of φ?

Use the Frick–Grohe Theorem to show that the same problem for q-CNF formulas and where each variable occurs in at most p clauses is FPT for the parameter (p, q).

Exercise 14.3.2 (Cygan, Fomin, and van Leeuwen [187]) Recall the firefighting problem from Exercise 5.3.6. Show that SAVING k VERTICES from that exercise is FPT on families of graphs of bounded local treewidth.

Exercise 14.3.3 (Lampis [476]) Show that the tower of 2's which we get for the standard MS_2 Courcelle's Theorem and also the MS_1 version can be avoided in graphs of bounded VERTEX COVER. To wit, show there is an algorithm, which, give a first order sentence φ with q vertex quantifiers, s set quantifiers and a graph G with vertex cover at most k decides if $G \models \varphi$ in time $2^{2^{O(q+s+k)}}$.
(Hint: The kernel should have size $k + q 2^k 2^s$.)

Exercise 14.3.4 (Ganian [336]) (Compare with the previous exercise.) Recall the notion of TWIN COVER from Exercise 4.11.10.
1. Prove that there is an algorithm, which, give a first order sentence φ with q variables and a graph G with twin cover at most k decides if $G \models \varphi$ in time $2^{O(kq+q \log q)}$.
   (Hint: Dividing the vertices into $k + 2^k$ indistinguishable classes will not work as there might be cliques. However, as there are only q variables available, a first order sentence cannot distinguish between a graph that has q cliques of some type and one with more than q cliques of the same type. Such considerations should yield a kernel of size $\leq k + 2^k(2 + 2q + \cdots + q^2) = O(2^k q^3)$.)
2. Hence deduce that there is an algorithm, which, give a first order sentence φ with q vertex quantifiers, s set quantifiers and a graph G with vertex cover at most k decides if $G \models \varphi$ in time $2^{2^{O(q+s+k)}}$.

14.4   Historical Notes

Complexity theory and computability theory were born of logic. Certainly early work in databases shed much light on the relationship between logical form and complexity. As noted earlier, Vardi [655], who suggested that classical complexity was the wrong notion for databases, since the queries were very small compared to the database. In particular, Vardi pointed out that the input for database-query evaluation consists of two components, query and database. The FPT result for LTL is due to Lichtenstein and Pnueli [491]. Flum and Grohe [312] base their whole approach to parametric complexity on model checking. This enables a unifying approach at the expense of developing the logic. We do not choose to use this approach here, and hence will not develop much of the applications of width metrics to logical formulas. Matroid treewidth is due to Hliněný and Whittle [397, 398]. The analogue of Courcelle's Theorem for representable matroids is due to Hliněný [394]. The material on knots polynomials is due to Makowsky. The definition of, and results concerning, local treewidth is due to Frick and Grohe [332]. The later developments leading to bidimensionality theory are due to Demaine and Hajiaghayi [210, 212, 213]. The Gaifman's Locality Theorem is a classic result in finite model theory and is proven in Gaifman [335]. Finite model theory is a thriving subject and an excellent recent text is Libkin [489].

14.5 Summary

Applications of the ideas of width metrics are discussed in logic, matroids, knot polynomial and other settings. The notion of local treewidth is introduced and the analogue of Courcelle's Theorem for first order formulas and local treewidth is proven.

Depth-First Search and the Plehn–Voigt Theorem

15

15.1 Depth-First Search

Depth-first search is the basis of a number of fundamental polynomial-time complexity results, including the linear-time algorithm to determine whether a graph is planar, due to Hopcroft and Tarjan [400, 636]. Thus it is not surprising that it has proven to be a powerful method in devising FPT algorithms as well.

A depth-first search of a graph $G = (V, E)$, where $|V| = n$ and $|E| = e$, allows us to compute, in time $O(n + e)$, a rooted spanning tree T of G having the useful property that for every edge uv that is *not* an edge of T, either u is an ancestor of v with respect to T, or v is an ancestor of u. This is illustrated in Fig. 15.1.

Such a spanning tree is termed a *depth-first spanning tree* of G, and the edges of the graph that do not belong to the spanning tree are referred to as *back edges*. The following theorem of Mader is occasionally useful in this context.

Theorem 15.1.1 (Mader [503]) *For every fixed graph H, there is a constant c_H, so that if $G = (V, E)$ is any graph with $e \geq c_H n$, then G necessarily contains a subdivision of H.*

The following is one of the simplest possible illustrations of how depth-first search and bounded-treewidth techniques can be combined to yield interesting FPT results. It is a paradigmatic result in this area.

LONG CYCLE
Instance: A graph G and a positive integer k.
Parameter: A positive integer k.
Question: Does G have a cycle of length at least k?

Theorem 15.1.2 (Fellows and Langston [301]) LONG CYCLE *can be solved for a graph G of order n in time $f(k)n$.*

R.G. Downey, M.R. Fellows, *Fundamentals of Parameterized Complexity*,
Texts in Computer Science, DOI 10.1007/978-1-4471-5559-1_15,
© Springer-Verlag London 2013

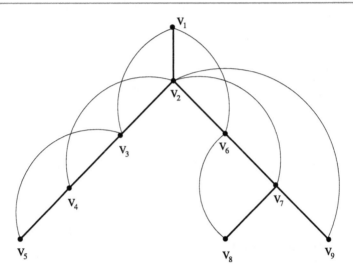

Fig. 15.1 An example of a depth-first spanning tree

Proof We can first apply Mader's Theorem to find if there is an easy reason to an-swer "yes" just by counting to see if there are enough edges in the graph to conclude that G contains a subdivision of the graph $H = C_k$, the cycle of length k. Let $g(k)$ denote this function of the number of edges given by Mader's Theorem. If there are not enough edges in G to conclude that the answer is "yes" then e is bounded by $g(k)n$, and so in time $O(g(k)n)$, we can compute a depth-first spanning tree T of G.

In time $O(n)$, we can examine this depth-first spanning tree to see if any back edge "stretches" a distance greater than or equal to $k-1$, that is, whether the path between its endpoints in the spanning tree has length at least $k-1$. If so, then the back edge together with this path forms a cycle of length at least k, and we are done.

Otherwise, we can read off a tree decomposition of width $k-2$ for G from the spanning tree, and solve the problem in time $O(f(k)n)$ using the techniques of Chap. 13, since the property of having a cycle of length at least k can be expressed in monadic second-order logic. Here the $f(k)$ is determined by Courcelle's Theorem.

The tree that indexes the vertex sets of the tree decomposition can be identified with T. The set S_u associated to a vertex u of T consists of u together with its $k-2$ immediate ancestors (or as many as there are, in the case of vertices close to the root of T). Given that no back edge stretches a distance greater than $k-2$, it is easy to verify that the properties of a tree decomposition are satisfied.   □

It is not really necessary to use Mader's result for the theorem above. We can avoid this by doing a depth-first traversal of G and checking for stretched back edges as we go, progressively computing a tree decomposition or stopping because a long cycle has been found.

Fig. 15.2 A circus graph of size 4

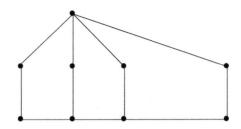

It is interesting to note that the seemingly similar problem, EXACT LONG CYCLE, of determining whether G has a cycle of length *exactly* k cannot be handled by any easy (linear-time) modification of the above algorithm. In fact, this was first shown to be FPT by the very different methods of the next section.

The approach embodied in the above (easy) theorem has been generalized by Bodlaender [71, 72]. The basic idea of the above theorem is to perform a depth-first search of G, and by studying the structure of the resulting depth-first spanning tree T, either identify an interesting substructure (e.g., a long cycle) or output a bounded-width tree decomposition based on T (in the sense that T provides the index structure for the decomposition). In the theorem above, the interesting structure was provided by a "stretched" back edge. By considering the number of back edges that are *around* a vertex v with respect to T (either joining v to a predecessor of v or joining a predecessor to a descendant), a similar (although much more intricate) argument can be carried forward, a result due to Bodlaender [71].

Theorem 15.1.3 (Bodlaender [71]) *Let k and l be constants. There is an algorithm that in time $O(n)$ either:*
(1) *identifies in G a subgraph H that is contractible to the $2 \times k$ grid, or*
(2) *identifies in G a subgraph contractible to the lth circus graph, or*
(3) *outputs a tree decomposition of G of width at most $2(k-1)^2(l-1)+1$.*

An example of a circus graph is shown in Fig. 15.2.
Some applications of this theorem are explored in the exercises.

15.2 Exercises

Exercise 15.2.1 Prove that there is a linear-time algorithm which takes as input a graph G and outputs a depth-first spanning tree for G.
(Hint: Suppose that $G = (V, E)$ is connected. Let $V = \{v_0, \ldots, v_n\}$ list the vertices of G. Construct T in n steps. Let the root of T be v_0, and $V_0 = \{v_0\}$ and $E_0 = \emptyset$. Now in step $i + 1$, let v_j be the largest number with v_j adjacent in G to some $y \notin V_i$. Let $v_{i+1} = y$, $V_{i+1} = V_i \cup \{y\}$ and $E_{i+1} = E_i \cup \{v_j y\}$.)

Exercise 15.2.2 Compute a depth-first spanning tree of the graph of Fig. 15.3, with root v_0.

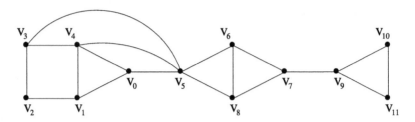

Fig. 15.3 Graph for Exercise 15.2.2

Exercise 15.2.3 Use depth-first search to construct an algorithm which determines if a graph has a vertex whose removal disconnects the graph.

Exercise 15.2.4 Use the theorem of Bodlaender [71] discussed in this section to prove that this problem, of determining whether a connected graph G has a spanning tree with at least k leaves (k-LEAF SPANNING TREE), is solvable in time $f(k)n$. We have already met this problem with various FPT approaches in earlier chapters. The method here was the *original* used to establish parameterized tractability.

Exercise 15.2.5 The *cycle cover number* $\gamma(G)$ is the minimum number of vertices in set $V' \subseteq V$ such that $G - V'$ is acyclic. The FEEDBACK VERTEX SET problem asks, for a graph G and integer parameter k, whether $\gamma(G) \leq k$. Use Exercise 15.2.6 below to show that the FEEDBACK VERTEX SET problem is solvable in time $f(k)n$.

Exercise 15.2.6 Use the theorem of Bodlaender to prove that in time $f(k)n$, we can accomplish one of the following:
1. Find a tree decomposition of width at most $g(k)$.
2. Find $k + 1$ vertex disjoint cycles in G.
3. Find a vertex u with the property that if $V' \subseteq V$ is any k-element feedback vertex set in G, then $u \in V'$.

Exercise 15.2.7 (Downey and Fellows [241]) A theorem of Erdös and Posa [277] shows that if $\gamma(G) \geq C \cdot k \log k$, then G has more than k disjoint cycles. Use this theorem and Exercise 15.2.6 to show that the DISJOINT CYCLES problem, which seeks to determine for a graph G and integer parameter k whether G contains k vertex disjoint cycles, can be solved in time $f(k)n$.

Exercise 15.2.8 For any fixed k, the graphs that do not contain more than k disjoint cycles are well-quasi-ordered by topological containment (e.g. Bollobas [103]). Argue that if $\mathcal{F}$ is a family of graphs that is a topological lower ideal and for which there is a bound on $\gamma(G)$ for $G \in \mathcal{F}$, then $\mathcal{F}$ can be recognized in linear time.

15.3 Bounded-Width Subgraphs, the Plehn–Voigt Theorem, and Induced Subgraphs

A useful result in the toolkit of fixed-parameter tractability is the theorem of Plehn and Voigt that for subgraphs restricted by a fixed bound on treewidth, the SUB-GRAPH ISOMORPHISM problem is FPT. This problem takes two graphs G and H as input, and asks whether H is a subgraph of G. They proved a related result for the INDUCED SUBGRAPH ISOMORPHISM problem when the maximum degree of G is bounded.

An (ordinary) subgraph of a graph G can be viewed as constituted by some subset of the vertices of the G together with some subset of the edges of G that are between these vertices. An *induced* subgraph of G, by contrast, is constituted by some subset of the vertex set of G together with *all* of the edges of G between these vertices.

Formally, a graph $H = (V_H, E_H)$ is a *subgraph* (*induced subgraph*) of a graph $G = (V_G, E_G)$ if there is a $1 : 1$ map $f : V_H \to V_G$ such that $uv \in E_H$ implies $f(u)f(v) \in E_G$ [$uv \in E_H$ if and only if $f(u)f(v) \in E_G$]. A map f satisfying the appropriate condition is termed an *embedding* of H in G (as a subgraph or induced subgraph).

Plehn and Voigt [562] proved the following result by an elegant argument.

Theorem 15.3.1 (Plehn and Voigt [562]) *For graphs H of treewidth at most w, it can be determined in time $f(|H|)|G|^{w+1}$ whether H is a subgraph of G.*

Proof We prove the theorem for the case of trees ($w = 1$) and indicate how this can be generalized. This special case encompasses all of the main ideas and the intuition for the general result, but is much less notationally cumbersome and is free of some tedious details concerning parsing schemes.

Let H be a fixed forest (i.e., a graph of treewidth 1) and suppose that H is rooted (and thus the root can be viewed as the boundary of H, of size $t = 1$). It is easy to establish that a complete set of 1-boundary operators for the parsing of graphs of treewidth at most 1 is provided by the binary $\oplus$ operator and by the unary operators $\otimes_0$ and $\otimes_1$, where $\otimes_0$ adds a new isolated vertex that becomes the new root (i.e., the boundary of the resulting graph), and $\otimes_1$ adds a new vertex that becomes the new root, together with an edge between the old root and the new root. We will assume that H is given by parse tree P over this operator set in the manner of Sect. 12.7. Since $\otimes_0$ and $\otimes_1$ are handled in essentially the same way in the algorithm, we will conveniently assume that H is connected (i.e., a tree) and parsed by the operator subset consisting of $\oplus$ and $\otimes_1$. We will assume also that each leaf of P represents a single edge between two vertices, one of which is the boundary.

Let $k = |H|$ denote the number of vertices of H and let $n = |G|$ be the order of G.

The algorithm proceeds leaf-to-root in the parse tree P. Suppose that the current node in P is labeled $\oplus$. The algorithm will receive from the argument branches, two sets of embeddings (in G) $\mathcal{S}_1$ and $\mathcal{S}_2$ of the two rooted subtrees T_1 and T_2 of H (respectively), whose roots are identified (by the $\oplus$ operator) at this step of the parse. The goal of the algorithm, at this point, is to calculate a set S of embeddings of $T = T_1 \oplus T_2$ sufficient to allow a determination, recursively, when the root of H is reached, of whether there are any embeddings of H in G, by just checking whether the set S is nonempty.

If S_i consisted of *all possible* embeddings of T_i, then S could be calculated by combining pairs of embeddings (of T_1 and T_2, respectively) that are disjoint except for their common root (in $T_1 \oplus T_2$). However, a moment's thought shows that we would then have sets S_i of order n^k. The algorithm is based on a neat trick that allows us to get away with sets S_i of size bounded by only k^k, even though the calculation of S consists of combining interior-disjoint root-compatible pairs as suggested. The trick is originally due to Monien [534] and depends on the data structure of *representation systems* that we next describe.

For each vertex u of the parse tree P, let P_u denote the subparse tree rooted at u and let T_u denote the rooted subtree of H parsed by P_u. Let T_u' denote the interior vertices of T_u, and let r_u denote the root vertex of T_u. Let $k_1(u) = |T_u'|$ and let $k_2(u) = k - k_1(u) - 1$. Thus, $k_2(u)$ is the number of vertices of H that are not in the subtree T_u of H.

Definition 15.3.1 (Representative system) Let u be a vertex of the parse tree P, and let v be a vertex of G. A *representation system $S_u(v)$ for P_u at v* consists of the following information, subject to various compatibility conditions:
1. A rooted *representation tree* $\Gamma_u(v)$ of depth at most $k_2(u)$, with each vertex that is not a leaf having $k_1(u)$ children.
2. To each node α of $\Gamma_u(v)$ is associated a subset of the vertices of G, $A_\alpha \subseteq V_G$, which we will term the *avoidance set* for the node α.
3. To each node α of $\Gamma_u(v)$ is associated either:
   (a) an embedding h_α of T_u in $G - A_\alpha$ that maps the root of T_u to v, $h_\alpha(r_u) = v$,
   or
   (b) the label λ.
4. Each edge of $\Gamma_u(v)$ is labeled with a vertex of G.

The compatibility conditions of Definition 15.3.1 are as follows:
1. If α is the root of $\Gamma_u(v)$, then $A_\alpha = \emptyset$.
2. If α is a node of $\Gamma_u(v)$ and the set of vertices of G occurring as labels on the path from the root of $\Gamma_u(v)$ to α is the set A, then the associated avoidance set is $A_\alpha = A$.
3. If λ is associated to a node α of $\Gamma_u(v)$ then
   • α is a leaf, and
   • it is not possible to embed T_u in $G - A_\alpha$ so that r_u is mapped to v.
4. If an embedding h_α is associated to a node α of $\Gamma_u(v)$ and if J is the set of image vertices of $h_\alpha(T_u')$, then the vertices in J each occur once as a label on an edge in $\Gamma_u(v)$ between α and a child of α. For a vertex u of the parse tree P, a *representation system S_u* is defined to be a collection of representation systems $S_u(v)$, one for each vertex $v \in V_G$. We will refer to the $S_u(v)$ as *subsystems* of S_u.

Intuitively, one can think about the situation addressed by a representation system in the following way. Imagine that we wish to run the algorithm sketched above, but instead of keeping *all possible* embeddings of T_i in G, we decide to keep just one (or more precisely, at most one) for each possible vertex $u \in V_G$ to which we could conceivably map the root of T_i. (This will not work, but let us continue until this is clearer.)

In processing the current node u of the parse tree P that we assume is labeled with $\oplus$, we could see if there is a vertex $v \in V_G$ such that the (single) embeddings h_1 in $S_1(v)$ (for T_1) and h_2 in $S_2(v)$ (for T_2) are compatible; i.e., the interior vertices are mapped disjointly into G, so that h_1 and h_2 can be combined into an embedding h of $T_1 \oplus T_2$. This might "usually" be the case, since the tree H that we are trying to embed is fixed (as the parameter) and therefore "small". If we are unlucky, we might want to inquire if we can have an embedding that is *like* h_1, except that there is just one vertex that we now need to avoid in order to have h_1 and h_2 with disjoint images of the interiors T_1' and T_2' (so that they can be combined). It is the job of the representation systems to supply these "variations" on the (one) embedding that is associated to the root of a representation tree, if they exist. Note that the branching factor in a representation tree is the number of vertices in the interior of the subtree that is embedded. This is the number of different "single vertex errors" that might occur and need to be repaired by a variation. All of this discussion is just motivational; we now describe the algorithm.

Algorithm 15.3.1 For each vertex u of the parse tree P, compute (from leaf-to-root) a representation system S_u. Answer "yes" if and only if for the root r of P, one of the subsystems $S_r(v)$ is nonempty in the sense that the root of the representation tree $\Gamma_r(v)$ is not labeled by λ. To complete the description, we must describe how the computation is performed at each vertex u of P. There are three cases: (1) u is a leaf, (2) u is labeled $\oplus$, and (3) u is labeled $\otimes_1$.

(1) *u is a leaf.*

In this case, u represents a single edge, one endpoint of which is the boundary. For a vertex $v \in V_G$, the representation tree $\Gamma_u(v)$ will have a branching factor of 1, since there is a single internal vertex of T_u. The representation system $S_u(v)$ is easily computed in time $O(n)$. The main task is checking, at most $k_2(u) = k - 2$ times, whether there is any edge in G that avoids a set of at most $k - 2$ vertices.

(2) *u is labeled $\oplus$.*

Let u' and u'' denote the children of u in the parse tree P. We assume that we have already computed representation systems $S_{u'}$ and $S_{u''}$. For each vertex $v \in V_G$, we compute $S_u(v)$ by starting with the root of the representation tree $S_u(v)$ and proceeding toward the leaves.

Let α be a node of $\Gamma_u(v)$ for which we wish to compute a label. To do this, try all possible pairs of embeddings (h', h'') where h' is an embedding of $T_{u'}$ that occurs as a label in $\Gamma_{u'}(v)$, and h'' is an embedding of $T_{u''}$ that occurs as a label in $\Gamma_{u''}(v)$. Given such a pair, check:

1. whether $h'(T_{u'}') \cap h''(T_{u''}') = \emptyset$, i.e., the embeddings of the interiors of the sub-trees joined by the $\oplus$ operator are disjoint, and

2. whether $A_\alpha \cap (h'(T_{u'}) \cup h''(T_{u''})) = \emptyset$, i.e., the embedding of T_u obtained by combining h' and h'' avoids A_α.

If both of these conditions are met, then label α with the combined embedding. Otherwise, label α with the "impossibility" label λ.

(3) *u is labeled* $\otimes_1$.

Let u' denote the child of u in the parse tree P. We assume that a representation system $\mathcal{S}_{u'}$ has already been computed. Let $v \in V_G$ and let α be a node of $\Gamma_u(v)$ for which we wish to compute a label. For every edge $vw \in E_G$, see if there is an embedding h used as a label in $\Gamma_{u'}(w)$ that meets the conditions:

1. $v \notin h(T'_{u'})$;
2. $A_\alpha \cap (\{v\} \cup h(T_{u'})) = \emptyset$.

If both of these conditions are met for some edge uw, then label α with the "extension" of h that sends r_u to v and otherwise coincides with h on $T_{u'}$. Otherwise, label α with λ.

The operator $\otimes_0$ would be handled similarly by considering all $w \neq v$ (i.e., not necessarily adjacent vertices in G).

We now argue that the algorithm is correct. Note that by compatibility condition (3) in the definition of a representation system (Definition 15.3.1), the only issue is whether the above procedure correctly computes representation systems. We argue by structural induction on the parse tree P. The representation systems computed for the leaves of P are trivially valid. The crucial point is that the computation procedure will find a valid embedding to associate to a node α if one exists. We separately consider the cases of parse tree nodes labeled $\oplus$ and $\otimes_1$.

For the case of $\oplus$, suppose there is some embedding h of $T_u = T_{u'} \oplus T_{u''}$ that maps r_u to v and is disjoint from A_α. By the induction hypothesis, there must be some h' in $\Gamma_{u'}(v)$ that avoids $A' = A_\alpha \cup h(T_{u''})$, noting that the height of $\Gamma_{u'}(v)$ is $k_2(u') \geq |A_\alpha| + |T'_{u''}|$. Similarly, there must be some h'' in $\Gamma_{u''}(v)$ that avoids $A'' = A_\alpha \cup h(T_{u'})$. In order to locate such an h' that avoids A' in $\Gamma_{u'}(v)$, start at the root of the representation tree and follow an edge to a child that is labeled by a vertex $a \in A'$ (if none exists, then we are done). This puts a into the avoidance sets of this subtree, by the compatibility conditions (2) and (4). Such a descent needs to be made at most $|A'|$ times until a suitable h' is located; similarly for h''.

For the case of $\otimes_1$, correctness is obvious.

The running time of the algorithm can be roughly bounded as follows. The size of the parse tree P can be bounded by $C \cdot |H|$ for a small constant C. For each vertex u of P, we must compute a representation system $\mathcal{S}_u$ that consists of n representation systems $\mathcal{S}_u(v)$, one for each vertex $v \in V_G$. To compute $\mathcal{S}_u(v)$ takes time $C(k^k)^2$ in the case that the vertex u of the parse tree P represents the $\oplus$ operator and requires time Cnk^k in the case that u represents the $\otimes_1$ operator. The total runtime is therefore at most $Ck^{2k+1}n^2$.

We next describe how the above algorithm can be generalized for $w \geq 2$.

We make use of the fact that there is a complete set of parsing operators for graphs of treewidth at most w over the set of graphs of boundary size w (where one of these operators $\otimes$ has arity w, and the graph of $\otimes$ has order $w + 1$). The details concerning this parsing scheme are not too important.

The modifications to the algorithm described above for $w = 1$ are straightforward. First of all, each subtree of the parse tree P for H now represents a graph with boundary size w. So, instead of having one representation system $\mathcal{S}_u(v)$ for each vertex v of G that might be the boundary of an embedding in G of T_u (the subgraph of H parsed by P_u, where u is a vertex of P), we need to maintain a representation system $\mathcal{S}_u(B)$ for each w-element subset $B \subseteq V_G$ that might be the boundary of the subgraph of H parsed by P_u in this generalized situation. The internal vertices T_u' of the subgraph of H parsed by P_u are defined in the obvious way, as is the definition of a representation system.

Because the size of the graph of $\otimes$ is $w + 1$ (and there seems to be no way to avoid this for treewidth w while maintaining boundary size w), we have to consider all subsets of V_G of size $w + 1$ in computing the representation systems $\mathcal{S}_u(B)$ when u represents the operator $\otimes$. Consequently, the running time of the generalized algorithm is $O(n^{w+1})$. $\qquad\square$

Generalization to graphs that are edge- and vertex-colored with a fixed finite set of colors, can be straightforward. In Plehn and Voigt [562] the extension to *minimally weighted* subgraphs is also proved.

The technique of representation systems can be applied to the INDUCED SUBGRAPH ISOMORPHISM problem when the maximum degree of G is bounded.

Theorem 15.3.2 (Plehn and Voigt [562]) *For graphs H of treewidth at most w, it can be determined in time $g(|H|, d)|G|^{w+1}$ whether H is an induced subgraph of G for graphs G of maximum degree d.*

Note that the above theorem is "best possible" in the sense that we would not expect to be able to get rid of the bound on the maximum degree of G, since determining whether there is an induced subgraph consisting of k isolated vertices (a graph of width $w = 0$) is equivalent to determining whether G has a k-element independent set, a problem we do not expect to be able to solve in linear time for fixed k. (In fact, we see that this problem is one of the "basic hardness" results in Part II.)

15.4 Exercises

Exercise 15.4.1 Give a parsing scheme for graphs of treewidth at most w that fulfills the requirements of the proof in this section. What is the minimum number of operators in such a scheme?

Exercise 15.4.2 Use the methods of this section to show that the problem EXACT LONG CYCLE, of determining whether a graph G has a cycle of length exactly equal to k, can be solved in time $f(k)n^3$.

Exercise 15.4.3 Use the 3-edge-colored version of a theorem of this section to show that the parameterized 3-DIMENSIONAL MATCHING (SHORT 3-DIMENSIONAL MATCHING) problem is FPT. This problem takes as input a set $M \subseteq W \times X \times Y$, where W, X, and Y are disjoint sets. The question is whether there is a subset $M' \subseteq M$ of size k such that no two elements of M' agree in any coordinate. Can this trick be extended to r-dimensional matching?

15.5   Historical Remarks

The Plehn–Voigt Theorem can be found in Plehn and Voigt [562]. The FPT results of this section are either due to Fellows and Langston [301], Bodlaender [71], or from Downey and Fellows [247]. The Fellows–Langston material is an important antecedent of the basic papers on parameterized complexity. The reader should consult Langston [479] and Downey [237] for full accounts of this history.

15.6   Summary

We show how to use depth-first search as a parameterized algorithm design tool, particularly using the Plehn–Voigt Theorem.

Other Width Metrics

16.1 Branchwidth

We have seen by Thomas' Lemma that there are always tree decompositions that portray the separations in a graph. In this section, we will look at a notion called branchwidth. The idea behind the notion of branchwidth is to have a decomposition, which is not a tree decomposition, but instead a road map to *separations in the graph*. We say an (unrooted) tree is *ternary* or *sub-cubic* if every internal node has degree 3.

> **Definition 16.1.1** (Robertson and Seymour)
> 1. A *branch decomposition* of $G = (V, E)$ is a pair (T, λ) consisting of a ternary tree and a bijection $\lambda : E(G) \to L(T)$, where $L(T)$ denotes the leaves of T.
> 2. For each edge $e \in E(T)$, the *width* of e is the (vertex) connectivity of the induced separation. That is, e separates $T - e$ into two components X and $E - X$, and we can calculate the separation number of $G[X]$ and $G[E - X]$. We denote the e-induced partition by $(\tau^{-1}(L_e^1), \tau^{-1}(L_e^2))$ of $E(G)$ where, for $i = 1, 2$, L_e^i contains the leaves of the connected components of $T \setminus e$.
> 3. The *branchwidth* of the decomposition is the maximum of all such widths for $e \in T$. The branchwidth of G, bw(G), is the minimum of such widths taken over all branch decompositions.

An important concept associated with a branch decomposition is the following. The *middle set* or *guts function* mid : $E(T) \to 2^{V(G)}$ of a branch decomposition maps every edge e of T to the set (called *middle set*) containing every vertex of G that is incident *both* to some edge in $\lambda^{-1}(L_e^1)$ and to some edge in $\tau^{-1}(L_e^2)$. Clearly the branchwidth of a particular decomposition is the maximum size of a guts.

Having bounded branchwidth imposes strong structure on a graph. For instance, as the reader will discover in Exercise 16.8.1, for a graph G, bw$(G) \leq 1$ iff G is a forest of stars and G has branchwidth ≤ 2 iff it has treewidth ≤ 2. This suggests that

R.G. Downey, M.R. Fellows, *Fundamentals of Parameterized Complexity*,
Texts in Computer Science, DOI 10.1007/978-1-4471-5559-1_16,
© Springer-Verlag London 2013

perhaps branchwidth and treewidth are related. In fact, branchwidth is a $\frac{3}{2}$ approximation for treewidth. We will let tw(G) denote the treewidth of G.

Theorem 16.1.1 (Robertson and Seymour [586]) *Suppose that* bw(G) > 1. *Then* bw(G) ≤ tw(G) + 1 ≤ $\lfloor\frac{3}{2}$bw(G)$\rfloor$.

Proof (Hliněný, Oum, Seese, and Gottlob [396]) Take a branch decomposition (T, λ) of G of width $b =$ bw(G). Let v be an internal vertex of the decomposition and since it has degree 3, we have the *guts* of the decomposition generated by v, W_v^i for $i \in \{1, 2, 3\}$, the three separations. Any vertex of G must occur in at least two of the W_v^i, and hence the cardinality of $B_v = W_v^1 \cup W_v^2 \cup W_v^3$ is at most $\lfloor\frac{3}{2}$bw(G)$\rfloor$. Now form a tree decomposition of G with the same underlying tree T and bags B_v. This gives a decomposition of treewidth $\lfloor\frac{3}{2}$bw(G)$\rfloor - 1$.

Conversely, start with a tree decomposition of width t, and use the method of Theorem 12.7.1 to make the underlying tree ternary, and having one leaf for each edge of G. Then the width is at most $t + 1$ by the interpolation property of the definition of treewidth. $\qquad\qquad\qquad\qquad\qquad\qquad\qquad\qquad\qquad\qquad\qquad\qquad\qquad\quad$ □

16.2   Basic Properties of Branchwidth

In the next section we will see how to run dynamic programming on branch decompositions. The methods of course are only of relevance if branchwidth is algorithmically tractable. The analog of Bodlaender's Theorem is the following.

Theorem 16.2.1 (Bodlaender and Thilikos [100]) k-BRANCHWIDTH *is linear time strongly* FPT.

The algorithm is significantly more complex than Bodlaender's algorithm for treewidth and seems far from practical. The general unparameterized problem is, NP-complete by Seymour and Thomas [615]. However, they prove surprisingly (and technically difficult) result that for planar graphs the problem can be solved in polynomial time.

Theorem 16.2.2 (Seymour and Thomas [615]) *Let G be planar. Then there is an* $O((|V(G)| + |E(G)|)^2)$ *algorithm which decides if G has branchwidth k. That is,* PLANAR k-BRANCHWIDTH *is in* P.

The reason this result is especially interesting is that the following question is *open*.

Question 16.2.1 Is PLANAR k-TREEWIDTH $\in$ P?

Remarkably, Theorem 16.2.2 is actually implementable and practical. It is a bit too involved to be included in this book. Hicks [390, 391] describes some computational experiments and implementations.

There has been a bit of work classifying graphs of small branchwidth.

16.3 Dynamic Programming and Branchwidth

As with treewidth, we can do model checking for graphs of bounded branchwidth quickly. The general problem is defined as follows.

BW-φ-MODEL CHECKING FOR φ

Instance: A graph $G = (V, E)$.
Parameter: $\mathrm{bw}(G) + |\varphi|$.
Question: $G \models \varphi$?

We begin with an example, bw-WEIGHTED INDEPENDENT SET. Let $G = (V, E)$ be given and suppose that $\mathrm{bw}(G) = b$, and $|V(G)| = n$. Let each $v_i \in V$ have weight w_i. The algorithm to follow will use dynamic programming by processing T in post-order from the leaves to the root, by analogy with treewidth, on an enhanced *rooted* branch decomposition as defined below. BW-WEIGHTED INDEPENDENT SET is solved by processing the guts $\mathrm{mid}(e)$ noting that an optimal independent set intersects some subset of $\mathrm{mid}(e)$. The set $\mathrm{mid}(e)$ has size $\leq b$, and hence this gives us at most 2^b many subsets to consider. We turn to some details.

Suppose that G is a graph and let (T, λ) be a branch decomposition of it of width at most k. For applying dynamic programming on (T, λ) we consider the tree T to be *rooted* at one of its leaves. Let v_r be this leaf and let e_r be the edge of T that contains it. Also, we slightly enhance the definition of a branch decomposition so that no edge of G is assigned to v_r and thus $\mathrm{mid}(e_r) = \emptyset$ (for this, we take any edge of the branch decomposition, subdivide it and then connect the subdivision vertex with a new (root) leaf t_r, using the edge e_r). The edges of T can be oriented towards the root e_r, and for each edge $e \in E(T)$ we denote by E_e the edges of G that are mapped to leaves of T that are descendants of e. We also set $G_e = G[E_e]$ and we denote by $L(T)$ the edges of T that are incident to leaves of T that are different from v_r. Given an edge e heading at a non-leaf vertex v, we denote by $e_L, e_R \in E(T)$ the two edges (left, right) with tail v.

Let T_e be the subtree of T rooted at e. Let G_e be the subgraph induced by all leaves of T_e. For $U \subset V$ let $w(U) = \sum_{v_i \in U} w_i$. For each independent $U \subset \mathrm{mid}(e)$, let $w_e(U)$ denote the maximum weight of an independent set I in G_e with $I \cap \mathrm{mid}(e) = U$, so that $w(I) = w_e(U)$. If U is not independent set $w_e(U) = \infty$, since it cannot be part of a independent set at all. There are therefore $2^{\mathrm{mid}(e)}$ many possible subproblems associated with each edge $e \in E(T)$. Since T has $O(E)$ many edges, the maximum weight is calculated by taking the maximum over all subproblems associated with the root r.

As with the treewidth case, the information needed to compute $w_e(U)$ has already been computed for the left and right subtrees, and we only need to determine the maximum weight independent sets I_{e_L} for G_{e_L} and I_{e_R} for G_{e_R} with the constraints that $I_{e_L} \cap \mathrm{mid}(e) = U \cap \mathrm{mid}(e_L)$ and $I_{e_L} \cap \mathrm{mid}(e) = U \cap \mathrm{mid}(e_R)$ and $I_{e_L} \cap \mathrm{mid}(e_R) = I_{e_R} \cap \mathrm{mid}(e_L)$.

We can calculate the value of $w_e(U)$ as $w(U)$ plus the maximum of
1. $w_{e_L}(U_{e_L}) - w(U_{e_L} \cap U) + w_{e_R}(U_{e_R} \cap U) - w(U_{e_L} \cap U_{e_R} \setminus U)$,
2. $U_{e_L} \cap \mathrm{mid}(e) = U \cap \mathrm{mid}(e_L)$,

3. $U_{e_R} \cap \mathrm{mid}(e) = U \cap \mathrm{mid}(e_R)$,

4. $U_{e_R} \cap \mathrm{mid}(e_L) = a_{e_L} U \cap \mathrm{mid}(e_R)$.

If we compute all $2^{\mathrm{mid}(e)}$ U associated with e and the values of $w_e(U)$ then we can calculate the above. This takes time $O(2^{|\mathrm{mid}(e_L)|}2^{|\mathrm{mid}(e_R)|})$. The total time on any edge e will be at most $O(8^b)$.

We will discuss improving this in Sect. 16.4 below.

Our second example is a dynamic programming algorithm for the following problem:

BW-3-COLORING

Instance: A graph $G = (V, E)$.
Parameter: $k = \mathrm{bw}(G)$.
Question: Does G have a proper 3-coloring?

We will consider 3-coloring functions of the type $\chi : S \rightarrow \{1, 2, 3\}$ and, for $S' \subseteq S$, we use the notation $\chi\,|_{S'} = \{(a, q) \in \chi \mid a \in S'\}$ and $\chi(S') = \{\chi(q) \mid q \in S'\}$.

Given a rooted branch decomposition (T, λ) an edge $e \in V(T)$, we use the notation $\mathcal{X}_e$ for all functions $\chi : \mathrm{mid}(e) \rightarrow \{1, 2, 3\}$ and the notation $\bar{\mathcal{X}}_e$ for all proper 3-colorings of G_e. We define

$$\alpha_e = \{\chi \in \mathcal{X}_e \mid \exists \bar{\chi} \in \bar{\mathcal{X}}_e : \chi\,|_{\mathrm{mid}(e)} = \bar{\chi}\}.$$

The set α_e stores the restrictions in $\mathrm{mid}(e)$ of all proper 3-colorings of G_e. Notice that for each $e \in E(T)$, $|\alpha_e| \leq 3^{\mathrm{mid}(e)} \leq 3^k$ and observe that G has a 3-coloring iff $\alpha_{e_r} \neq \emptyset$ (if $\alpha_e \neq \emptyset$, then it contains the empty function). For each $e \in E(T)$ we can compute $\mathcal{R}_e$ by using the following dynamic programming formula:

$$\alpha_e = \begin{cases} \{\chi \in \mathcal{X}_e \mid |\chi(e)| = 2\} & \text{if } e \in L(T), \\[4pt] \{\chi \in \mathcal{X}_e \mid \exists \chi_L \in \mathcal{X}_{e_L}, \exists \chi_R \in \mathcal{X}_{e_R} : \\[2pt] \quad \chi_L\,|_{\mathrm{mid}(e_L) \cap \mathrm{mid}(e_R)} = \chi_R\,|_{\mathrm{mid}(e_L) \cap \mathrm{mid}(e_R)} \text{ and} \\[2pt] \quad (\chi_L \cup \chi_R)\,|_{\mathrm{mid}(e)} = \chi\} & \text{if } e \notin L(T). \end{cases}$$

Clearly, this simple algorithm proves that BW-3-COLORING belongs to $2^{O(k)} \cdot n$-FPT. A straightforward extension implies that BW-q-COLORING belongs to $q^{O(k)} \cdot n$-FPT.

In both of the above examples, we associate to each edge $e \in E(T)$ some *characteristic structure*, that, in case of BW-WEIGHTED INDEPENDENT SET and BW-3-COLORING, is $\mathcal{R}_e$ and α_e, respectively. This structure is designed so that its value for e_r is able to determine the answer to the problem. Then, it remains to give this structure for the leafs of T and then provide a recursive procedure to compute bottom-up all characteristic structures from the leaves to the root. This dynamic programming machinery has been used many times in parameterized algorithm design and for much more complicated types of problems. In this direction, the algorithmic challenge is to reduce as much as possible the information that is managed in the characteristic structure associated to each edge of T. Usually, for simple problems as those examined above, where the structure encodes subsets (or a bounded number of subsets) of mid(e), it is easy to achieve a single-exponential parametric dependence. Typical examples of such problems are DOMINATING SET, MAX CUT and INDEPENDENT SET, parameterized by treewidth/branchwidth, where the challenge is to reduce as much as possible the constant hidden in the O-notation of their $2^{O(k)}$-parameter dependence.[1] Apart from tailor-made improvements for specific problems such as TW-DOMINATING SET and TW-VERTEX COVER (see e.g. Alber and Niedermeier [26], Alber, Dorn, and Niedermeier [23] and Betzler, Niedermeier, and Uhlmann [64]), substantial progress in this direction has been achieved using the Fast Matrix Multiplication technique, introduced by Dorn in [229] and the results of Rooij, Bodlaender, and Rossmanith in [654], where they used the Generalized Subset Convolution Technique (introduced by of Björklund, Husfeldt, Kaski, and Koivisto in [66]).

Recently, some lower bounds on the parameterized complexity of problems parameterized by treewidth were given by Lokshtanov, Marx, and Saurabh in [494]. According to [494], unless SAT is solvable in $O^*((2 - \delta)^n)$ steps, BW-INDEPENDENT-SET does not belong to $(2 - \varepsilon)^k \cdot n^{O(1)}$-FPT, BW-MAX CUT does not belong to $(3 - \varepsilon)^k \cdot n^{O(1)}$-FPT, BW-DOMINATING-SET does not belong to $(3 - \varepsilon)^k \cdot n^{O(1)}$-FPT, BW-ODD CYCLE TRANSVERSAL does not belong to $(3 - \varepsilon)^k \cdot n^{O(1)}$-FPT, and BW-$q$-COLORING does not belong to $(q - \varepsilon)^k \cdot n^{O(1)}$-FPT. The assumption that SAT $\notin O^*((2 - \delta)^n)$-TIME, is known as the STRONG EXPONENTIAL TIME HYPOTHESIS (SETH) and was introduced by Impagliazzo and Paturi in [410]. We will look at the methods used to prove these results in Chap. 29.

For more complicated problems, where the characteristic structure encodes pairings, partitions, or packings of mid(e) the parametric dependence of the known FPT algorithms is of the form $2^{O(k \log k)} \cdot n^{O(1)}$ or worse. Usually, these are problems involving some global constraint such as connectivity on the certificate of their solution. Recent complexity results of Lokshtanov, Marx, and Saurabh [496] show that for problems such as the DISJOINT PATHS PROBLEM no $2^{o(k \log k)} \cdot n^{O(1)}$-algorithm exists unless ETH fails. An approach recently introduced by Cygan, Nederlof, M. Pilipczuk, M. Pilipczuk, van Rooij, and Wojtaszczyk in [190], solves many problems of this type in $2^{\text{tw}(G)} \cdot n^{O(n)}$ steps by randomized Monte Carlo algorithms. This includes problems such as HAMILTONIAN PATH, FEEDBACK VERTEX SET and CONNECTED DOMINATING SET. For planar graphs it is possible to design $O^*(2^{O(\text{bw}(G))})$ dynamic programming algorithms using the *Catalan structures* method introduced by Dorn, Penninkx, Bodlaender, and Fomin in [235]. This technique uses a special type of branch decomposition called a *sphere cut decomposition* introduced in Seymour and Thomas [615]. In such decompositions, the vertices of mid(e) are arranged according to a closed curve on the surface where the input graph is embedded. In case the characteristic structure encodes non-crossing pairings, its size is proportional to the kth Catalan number,

---

[1]As here we care about the exact parameter dependence the constants may vary depending on whether we parameterize by treewidth or branchwidth.

which is single-exponential in $k = \mathrm{bw}(G)$. This fact yields the $O^*(2^{O(\mathrm{bw}(G))})$ time bound to the corresponding dynamic programming algorithms. The same technique was extended by Dorn, Fomin, and Thilikos in [232] for bounded genus graphs and in [233] for every graph class that excludes some graph as a minor. Finally, Rué, Sau, and Thilikos in [596] extended this technique to even further.

16.4   Fast Matrix Multiplication and Dynamic Programming

It is beyond the scope of this book to describe all of the methods that have been developed for improving the running times of bounded branchwidth algorithms. We next explore briefly one method that can be applied relatively generally, and will use the BW-WEIGHTED INDEPENDENT SET to illustrate its use. The method was introduced by Dorn in [229]. This is an example of algorithm engineering in the same spirit as Hüffner [404] this time using matrices for the data structures to speed things up.

Following Dorn [229], we first beginning with *ordered tables*. This method of making the data structures smaller goes back to Bodlaender and his students (at least). The idea is to store all sets $U \subset \mathrm{mid}(e)$ in an ordering such that the time used per edge is reduced to $O(2^{1.5b})$.

Two facts to recall about the guts: (1) any vertex v must be in at least two of the three guts of incident edges e, f, g, and (2) a vertex must be in all guts along a path from one leaf to another. Fix a parent edge e and then we will have four sets.

1. *Intersection vertices* $I := \mathrm{mid}(e) \cap \mathrm{mid}(e_L) \cap \mathrm{mid}(e_R)$.
2. *Forget vertices* $F := (\mathrm{mid}(e_R) \cap \mathrm{mid}(e_R)) \setminus I$.
3. *Symmetric difference vertices* $L := (\mathrm{mid}(e) \cap \mathrm{mid}(e_L)) \setminus I$ and R similarly.

Since $\mathrm{mid}(e) \subseteq \mathrm{mid}(e_R) \cap \mathrm{mid}(e_L)$, $w(U) \subseteq w(U_{e_R} \cup U_{e_L})$. Therefore $w_e(U)$ can be rewritten as the maximum of

1. $w_{e_L}(U_{e_L}) + w_{e_R}(U_{e_R}) - w(U_{e_R} \cap U_{e_L})$,
2. $U_{e_L} \cap (I \cup L) = U \cap (I \cup L)$,
3. $U_{e_R} \cap (I \cup R) = U \cap (I \cup R)$,
4. $U_{e_L} \cap (I \cup F) = U_{e_R} \cap (I \cup F)$.

For each edge $table_e$ is labeled with a sequence of vertices of $\mathrm{mid}(e)$ represented by concatenations of sequences of $\{L, R, I, F\}$. Then for parent edge e we would get labels $I * L * R$ for the $table_e$, $I * L * L$ for the $table_{e_L}$, and $I * R * L$ for the $table_{e_R}$. The $table_{e_L}$ contains all of the sets U_{e_L} with value w_{e_L} (and similarly for the $table_{e_R}$).

To compute $w_e(U)$ for each of the $2^{|I|+|L|+|R|}$ entries of $table_e$, we need only consider the $2^{|F|}$ many sets U_{e_R} and U_{e_L} subject to the constraints 1–4 above. Since $\mathrm{mid}(e) \cup \mathrm{mid}(e_L) \cup \mathrm{mid}(e_R) \subseteq I \cup L \cup R \cup F$, we see that $|I|+|L|+|F|+|R| \leq \frac{3}{2}b$. This gives the desired bound.

16.4.1 Improving Things Using Matrix Multiplication

Two $n \times n$ matrices can be multiplied in time n^q for $q \leq 2.3727$ by a generalization of the Coppersmith–Winograd algorithm [165] due to Williams [665]. If we have a pair consisting of an $n \times p$ matrix B and a $p \times n$ matrix C then for $p \leq n$, use the Williams algorithm, and for $p > n$ split the matrices into $\frac{p}{n}$ many $n \times n$ matrices $B_1, \ldots, B_{\frac{p}{n}}$, and similarly C, and assemble the subproducts $A_g = B_g C_g$ giving an $O(\frac{p}{n} n^{2.3727} + \frac{p}{n} n^2)$ upper bound.

The *distance product* of two $n \times n$ matrices which we will denote by $A \otimes B$ is the $n \times n$ matrix with $a_{ij} = \min\{b_{ik} + c_{ik} \mid 1 \leq k \leq n\}$. Yuval [672] shows that if the elements of the matrices are in the integer interval $[-m, m]$ then there is an algorithm to compute this product in time $O(m \cdot n^{2.3727})$. Similarly we can get a bound for the $n \times p$, $p \times n$ case of $O((pm)n^{2.3727})$. This can be applied to boolean matrix distance multiplication, where the entries only have values in $\{0, 1\}$, and where $a_{ij} = \bigvee_{1 \leq k \leq n}(b_{ik} \wedge c_{kj})$, setting $a_{ij} = 1$ if and only if $a_{ij} > 0$.

All these running time-bounds improve if faster algorithms are found for matrix multiplication.

The key idea is to replace the tables above with matrices and use (distance) matrix multiplication. We can reformulate the constraints for $w_e(U)$ above as follows:

1. $U \cap I = U_{e_R} = U_{e_L} \cap I$;
2. $U_{e_L} \cap L = U \cap L$;
3. $U_{e_R} \cap R = U \cap R$;
4. $U_{e_L} \cap L = U_{e_R} \cap F$.

With the new first constraint, we note that every independent set $S_e \subset G_e$ is determined by the independent sets S_{e_R}, S_{e_L} such that all three sets intersect some subset $U_I \subseteq I$. Dorn's idea is to not compute $w_e(U)$ separately for each subset U, but to simultaneously calculate for each subset $U_I \subseteq I$ the values $w_e(U)$ for *all* $U \subseteq \mathrm{mid}(e)$ subject to the constraint that $U \cap I = U_I$. For each U the values $w_e(U)$ are stored in a matrix A. Each row is labeled with a subset $U_L \subseteq L$ and each column with $U_R \subseteq R$. The entry determined by U_L, U_R is filled with $W_e(U)$ for U subject to $U \cap L = U_L$, $U \cap R = U_R$, and $U \cap I = U_I$.

The matrix A can be computed by the distance product of the two matrices B and C assigned already to the children edges e_L and e_R. For e_L, a row of B is labeled with $U_L \subset L$ and column with $U_F \subset F$, appointing an entry subject to the constraints $U_{e_L} \cap L = U_L$, $U_{e_L} \cap F = U_F$ and $U_{e_L} \cap I = U_I$. Similarly we fill the matrix C for e_R with all the values for independent sets U_{e_R} with $U_{e_R} \cap I = U_I$. Now we label a row with $U_F \subseteq F$ and a column $U_R \subseteq R$ storing value $w_{e_R}(U_{e_R})$ for U_{e_R} subject to the constraints $U_{e_R} \cap F = U_F$ and $U_{e_R} \cap R = U_R$. (The value $-\infty$ is assigned if at least one of the sets is not an independent set in G.)

The next lemma is central to the argument that the matrix multiplication approach works.

To apply the lemma below, we assume that the rows and columns of A, B, C are ordered so that two equal subsets are in the same position. U_L must be in the same position in the corresponding rows of A and B, U_R in the corresponding columns of A and C. U_F must be in the same position in the column of B corresponding to the column of C. Finally, change the signs in each entry of B and C since we are dealing with a maximization problem. Set all entries with value ∞ to be $\sum_{v \in V(G)} w_v + 1$.

Lemma 16.4.1 *Given an independent $U_I \subseteq I$, then for all independent $U \subseteq$ mid(e), $U_{e_L} \subseteq$ mid(e_L) and $U_{e_R} \subseteq$ mid(e_R) subject to the constraints $U \cap I = U_{e_L} \cap I = U_{e_R} \cap I$, if we let the matrices B and C have entries $w_{e_L}(U_{e_L})$ and $w_{e_R}(U_{e_R})$, then the entries of $w_e(U)$ of A are computed by $B \otimes C$.*

Proof Since U_{e_L} and U_{e_R} only intersect in U_I and U_F, we will substitute the entry $w_{e_R}(U_{e_R})$ in C for $w_{e_R} - |U_I| - |U_F|$. In the last formulation of the computation of $w_e(U)$, we have the term $w(U_{e_L} \cap U_{e_R})$. Thus we can calculate $w_e(U) = \min\{-w_{e_L}(U_{e_L}) - (w_{e_R}(U_{e_R}) - |U_I| - |U_F|)\}$ where
1. $U \cap I = U_{e_L} \cap I = U_{e_R} \cap I = U_I$;
2. $U_{e_L}\mathcal{L} = U\mathcal{L} = U_{e_L}$;
3. $U_{e_R} \cap R = U \cap R = U_{e_R}$;
4. $U_{e_L} \cap F = U_{e_R} \cap F = U_F$.
The worst case has $|L| = |R|$. Hence we can assume $|U_{e_L}| = |U_{e_R}|$. It follows that A is square. By the above, every value of $w_e(U)$ can be calculated by the distance product of B and C, i.e. taking the minimum over all sums of entries in row U_{e_L} in B and column U_{e_R} in C. $\square$

Theorem 16.4.1 (Dorn [229]) *Dynamic programming for* BW-WEIGHTED INDE-PENDENT SET *on weights $O(m) = n^{O(1)}$ takes time $O(m \cdot 2^{\frac{2.2327}{2}b})$, where b is the branchwidth.*

Proof For each U_I we prove that dynamic programming takes time $O(m2^{1.2327|L|}2^{|F|}2^{|I|})$. We need time $O(2^{|I|})$ for subsets $U_I \subseteq I$. Since $2^{|F|} \geq 2^{|L|}$ we can use the calculation of the running time for $n \times p$, $p \times n$ multiplication giving time $O(m\frac{2^{|F|}}{2^{|L|}}2^{2.2327|L|})$. If $2^{|F|} < 2^{|L|}$, then the running time behavior is more or less the same. By definition of the sets, we see that
1. $|I| + |L| + |R| \leq b$ as mid$(e) = I \cup L \cup R$;
2. $|I| + |L| + |F| \leq b$ as mid$(e_F) = I \cup L \cup F$;
3. $|I| + |R| + |F| \leq b$ as mid$(e_R) = I \cup R \cup F$;
4. $|I| + |L| + |R| + |F| \leq \frac{3}{2}b$, as mid$(e) \cup$ mid$(e_L) \cup$ mid$(e_R) = I \cup L \cup R \cup F$.
Maximizing the objective function $O(m \cdot 2^{1.1327|L|}2^{|F|}2^{|I|})$ gives the worst case running time $O(m \cdot 2^{\frac{2.2327}{2}b})$. $\square$

There is nothing special about BW-WEIGHTED INDEPENDENT SET here. In [229], Dorn observes that this methodology can be used for a large number of problems, and describes sharp algorithms for BW-DOMINATING SET, BW-PERFECT CODE, BW-PERFECT DOMINATING SET, BW-PLANAR DOMINATING SET, BW-PLANAR STEINER TREE and many other graph problems parameterized by branchwidth.

16.5 Cliquewidth and Rankwidth

It is natural to note that many problems seem to be easy for graphs of bounded treewidth, and also for highly dense graphs. One structural parameter that seems to give a common generalization is that of cliquewidth. This is best introduced by considering the way in which bounded cliquewidth structure is *parsed*. We have the following operators on k-colored graphs.

1. $\emptyset_i$: create a vertex colored i;
2. join(i, j): join all vertices of color i to those of color j;
3. ($i \mapsto j$): recolor all vertices i to color j;
4. $\sqcup$: take the disjoint union of G_1 and G_2.

> **Definition 16.5.1** (Courcelle [171], Courcelle and Olariu [182]) The smallest number of colors needed to construct G (i.e. as a colored graph classically isomorphic to G) is called the *cliquewidth* of G, cw(G).

Cliques are easy to construct with a small number of colors. Consider the sequence $\emptyset_1, \emptyset_2, \text{join}(1, 2), 2 \mapsto 1, \emptyset_2, \text{join}(1, 2), 2 \mapsto 1, \ldots$ will construct a clique of arbitrarily large size and hence we see.

Proposition 16.5.1 Cliques have cliquewidth 2.

Proposition 16.5.2 Suppose that G has bounded treewidth, then it also has bounded cliquewidth.

Proof We can generate graphs of bounded treewidth using parse operations as per Chap. 12, and here we will for convenience think of them as working on t-boundaried graphs with operations $c(i)$, create a vertex of color i, $\oplus$, push(i), and $j(i, j)$ to join the boundary vertex of color i to that of color j. Begin with the case of pathwidth. Add one new color $t + 1$. We emulate $c(i)$ by $\emptyset_i$ for $1 \le i \le t$, $j(i, j)$ by join(i, j), push(i) by $i \mapsto t + 1$ followed by $\emptyset_i$. The case of treewidth is similar. One way is to add some new colors $c'(i)$ for $1 \le i \le t$. The first guess for $\oplus$ is to color G_1 with the $c(i)$, G_2 with $c'(i)$ (and $c(t + 1)$) and invoke join(i, i') followed by $i' \to i$. However, this has the problem that we get extra edges instead of seamless gluing. The idea is to stop the path decomposition of G_2 earlier. For simplicity, assume that the treewidth is 2. Then assuming that G_2 is nontrivial, it can be decomposed into $Y \oplus G'_2$ where Y is one of a number of simple graphs. And this can be regarded as being as edges from vertices $\{v, u, w\}$ (in G'_2) to the boundary $\{x, y, z\}$ colored, respectively, $c(1')$, $c(2')$, and $c(3')$. For the moment we will ignore the edges in the vertex set $\{v, u, w\}$, and the edges could vary from the $K_{3,3}$ down to a single edge like vy. Each of these can be emulated by cliquewidth operations beginning from G'_2. For example, to construct $G_1 \oplus G_2$ where $Y = K_{3,3}$, we apply join(i, i') to G_1 and G'_2 for all i, i', and then $i' \mapsto 4$. For the case of a single edge vy who are colored 1 and $2'$ respectively, we fist apply $1' \mapsto 4, 3' \mapsto 4$, then join$(1, 2')$ and finally $2' \mapsto 4$. Any other configuration can be so emulated. Finally, if there were edges between, say, x and y in Y, after we have done the above we would then apply join$(1, 2)$ if there is not already an edge there. Armed with this and induction, it is possible to show that graphs of bounded treewidth have bounded cliquewidth.   □

How does Courcelle's Theorem fare for cliquewidth? Can we do dynamic programming?

Using results of [180], Courcelle, Makowsky, and Rotics extended Theorem 13.1.1 for the more powerful parameter of rankwidth (see also [478]). The definition of rankwidth is quite similar to the one of branchwidth.

Definition 16.5.2 (Oum and Seymour [551])

1. A *rank decomposition* of a graph G is a pair (T, τ) with τ is a bijection from $V(G)$ to the set of leaves of T. For each edge e of T, the partition $(V_e^1, V_e^2) = (\tau^{-1}(L_e^1), \tau^{-1}(L_e^1))$ is a partition of $V(G)$ which, in turn, defines a $|V_e^1| \times |V_e^2|$ matrix M_e over the field $GF(2)$ with entries $a_{x,y}$ for $(x, y) \in V_e^1 \times V_e^2$ where $a_{x,y} = 1$ if $\{x, y\} \in E(G)$, otherwise $a_{x,y} = 0$.
2. The *rank-order* $\rho(e)$ of e is defined as the row rank of M_e. Similarly to the definition of branchwidth, the *rankwidth* of (T, τ) is equal to $\max_{e \in E(T)} |\rho(e)|$ and the *rankwidth* of G, rw(G), is the minimum width over all rank decompositions of G.

Rankwidth is a more general parameter than branchwidth in the sense that there exists a function f such that bw$(G) \leq f(\text{rw}(G))$ for every graph G, as proved by Oum and Seymour in [551].

Theorem 16.5.1 (Oum and Seymour in [551]) *The rankwidth of G is less or equal to the cliquewidth of G which, in turn is bounded by $2^{1+\text{rw}(G)} - 1$.*

Consider now the following parameterized meta-problem.

RW-φ-MODEL CHECKING FOR φ

Instance: A graph G.
Parameter: rw$(G) + |\varphi|$.
Question: $G \models \varphi$?

Given the fact that we can define these objects by parse operators it is not surprising that *some* form of Courcelle's Theorem will hold by using automata and the like. The analog of Courcelle's Theorem is the following, *but the reader should note that the logic is the MS$_1$ theory, that is, no variables for edge sets.*

Theorem 16.5.2 (Courcelle, Makowsky, and Rotics [180]) *If φ is an MS_1 formula, then* RW-φ-MODEL CHECKING FOR *φ belongs to $O(f(k, |\varphi|) \cdot n^3)$-FPT and the same is therefore true for cliquewidth.*

At this stage we recall that, as we stated in Theorem 13.4.2, due to Courcelle and Oum, the partial converse of Theorem 16.5.2 above: *If a set of directed or undirected graphs has a decidable MS_1 theory (even with the addition of the "even cardinality" predicate), then it has bounded cliquewidth.*

Space considerations preclude us from giving proofs of the statements above.

The original statement of Theorem 16.5.2 used cliquewidth, instead of rankwidth. However, we avoid the definition of cliquewidth and we use rankwidth instead, in order to maintain uniformity with the definition of branchwidth. We remark, however, that cliquewidth has two advantages over rankwidth. First it has a simple description in terms of simple parsing operators. This tends to make concrete applications easier. Second it can be formulated to work for directed graphs. We refer the reader to Courcelle–Engelfriet [178, Chap. 6]. However, some caution is needed as we now discuss.

We remark that in some sense Theorem 16.5.2 is less algorithmically useful than Courcelle's Theorem. As with treewidth, cliquewidth is NP-complete.

Theorem 16.5.3 (Fellows, Fellows, Rosamond, Rotics, Szeider [303]) *Cliquewidth is NP-complete. Moreover, unless $P = NP$, for each $1 > \varepsilon > 0$ there is no poly algorithm which, given the input of G of n vertices outputs a clique decomposition (in terms of the parse operators) of size $cw(G) + n^\varepsilon$.*

The proof of this result is quite difficult, and has several unusual features. From the point of view of parameterized complexity, the relevant result for treewidth is that it is FPT by Bodlaender's Theorem. However, at the time of writing, it is an open question whether parameterized cliquewidth is FPT. Evidence so far would seem to suggest that it is likely $W[1]$-hard. On the other hand Hliněný and Oum [395] have shown that RANKWIDTH is FPT. On nice method which often allows us to show that certain graph classes have bounded cliquewidth is provided by the following theorem.

Theorem 16.5.4 (Courcelle [171]) *A set of graphs has bounded cliquewidth if and only if it is MS_1 interpretable into the class of binary trees.*

It is interesting to observe that Theorem 16.5.2 is not expected to hold for MS_2 formulas. Indeed, Fomin, Golovach, Lokshtanov, and Saurabh, proved in [314] that there are problems, such as GRAPH COLORING, EDGE DOMINATING SET, and HAMILTONIAN CYCLE that can be expressed in MS_2 but not in MS_1, that are $W[1]$-hard for graphs of bounded cliquewidth (or rankwidth).

16.6   Directed Graphs

It has turned out to be a difficult problem to formulate an analog of treewidth for directed graphs—an obvious quest, given the great successes of treewidth in the structure theory and algorithmics of undirected graphs.

There have been various tilts at this windmill, none with great success. We survey a few of these.

For a DAG $D = (V, A)$ we write $v < v'$ if there is a directed path from v to v'. Of course, $v \leq v'$ if either $v = v'$ or $v < v'$. For $Z \subseteq V$, we say that S is Z-*normal* if there is no directed path in $V \setminus Z$ with first and last vertices in S and using a vertex of $V \setminus (S \cup Z)$.

Definition 16.6.1 (Johnson Robertson, Seymour, and Thomas [422]) An *aboreal decomposition* of a digraph D, is a triple (T, F, W) where T is a directed tree, $F = \{X_e \mid e \in E(R)\}$, and $W = \{W_v \mid t \in V(T)\}$ satisfying the below.
1. W is a partition of $V(G)$ into nonempty sets.
2. For $t = v_1 v_2 \in E(T)$, the set $\bigcup \{W_t \mid t \in V(T) \wedge v \geq v_1\}$ is X_e-normal.
3. The *width* is the least w such that for all $v \in E(T)$, $|W_t \mid t \in e X_e| \leq w$.

The notion has not been pursued much as it has some problems and some errors found particularly by Isolde Adler. We refer the reader to Adler [16], Obdržálek [549], and Johnson et al. [421].

Another attempt was the notion of DAG-width, and uses the idea of a DAG instead of a tree structure for the for the decomposition. This echoes recent work by Grohe on logical aspects of and canonization of excluding minors in undirected graphs in papers such as [360].

Definition 16.6.2 (Obdržálek [549], Berwanger, Dawar, Hunter, and Kreutzer [61]) A DAG-*decomposition* of a digraph $D = (V, A)$ is a pair (R, F) where R is a DAG and $F = \{X_d \mid d \in R\}$ is a family of subsets of $V(G)$ satisfying
1. $V = \bigcup_{v \in V(R)} X_v$.
2. It $dd' \in E(R)$, then for each $uv \in A$ such that $v \in X_{\geq d'} \setminus X_d$ we have $w \in X_{\geq d'}$.
3. If d' lies on the path from d to d'' in R, then $X_d \cap X_{d''} \subseteq X_{d'}$.

As usual the *width* of the decomposition is $\max\{|X_d| \mid d \in R\}$. It is not difficult to prove that acyclic digraphs have DAG-width 1 (Exercise 16.8.7). This notion can be characterized by a directed version of the "cops and robbers game" from [422].

There are other notions, such as ones based on perfect elimination orderings, as proposed by Hunter and Kreutzer [407] *via* a notion called *Kelly width*, whose roots go back to Rose and Tarjan [593]. We will not dwell further on directed treewidth, but we hope that the reader is inspired to solve the problems here.

16.7   *d*-Degeneracy

Another natural notion of graph structure that has been repeatedly rediscovered is that of (in modern terminology) the *degeneracy* of the graph: the extent to which it can be completely decomposed by plucking vertices of "small" degree.

> **Definition 16.7.1** For $d \in \mathbb{N}^+$ we say that a graph $G = (V, E)$ is d-*degenerate* or d-*inductive* if there exits a linear ordering of the vertices $v_1, \ldots, v_n$ which we will refer to as a *layout*, such that for all i,
>
> $$\left| \{ j \mid j > i \wedge v_i v_j \in E \} \right| \leq d.$$

This is an often rediscovered concept. The *degeneracy number* $d(G)$, is the smallest d such that G is d-degenerate. This concept has been rediscovered in bioinformatics where it is known as the *d-core number* (Bader and Hogue [49]) and in computer science by Kirousis and Thilikos [446] who called it *linkage*. It is more or less implicit in Szekeres and Wilf [635], and certainly it is to be found in 1966 in Erdös and Hajnal [275].

A *d-core* of G is a maximal connected graph where no vertices of degree $< d$. The following is an easy lemma, whose proof is left to the reader in Exercise 16.8.8.

Lemma 16.7.1 *The largest d for which G has a d-core is the degeneracy of G.*

This gives an easy algorithm to determine the degeneracy of G. Simply remove vertices in order of degree, starting with those of least degree. That is,

Lemma 16.7.2 *A layout with least d-degeneracy can be calculated in linear time.*

Since Euler's Formula says that every planar graph has a vertex of degree 5, from this the following observation follows.

Proposition 16.7.1 Every planar graph is 5-degenerate.

Using perfect elimination orders we also see the following.

Proposition 16.7.2 If G has treewidth d then it is d-degenerate.

Degeneracy is a reasonable heuristic for graphs arising in nature for treewidth as studied by Fouhy [329]. Big grids have large treewidth, so the converse of Proposition 16.7.2 does not hold. What is useful and interesting is that we can devise algorithms for d-degenerate graphs generalizing those for, e.g., planar graphs, as in the following example.

Theorem 16.7.1 (Alon and Gutner [34]) k-DOMINATING SET *is* $O(k^{O(dk)}|G|)$
FPT *for d-degenerate graphs.*

We begin with a lemma which uses the fact that d-degenerate graphs are sparse
in the sense that they have fewer than dn edges and hence their average degree is
$\leq 2d$. As with the first (correct) algorithm we met in Chap. 3 one idea is to use the
annotated version of the problem where we use a black and white partition of the
vertex set.

ANNOTATED DOMINATING SET
Instance: A graph $G = (V = B \sqcup W, E)$.
Parameter: A positive integer k.
Question: Is there a choice of k vertices $V' \subseteq V$ such that for every vertex $u \in B$,
                there is a vertex $u' \in N[u] \cap V'$?

The reader should recall that in Chap. 3, Alber, Fan, Fellows, Fernau, Niedermeier, Rosamond, and Stege [24] showed that PLANAR DOMINATING SET is
$O(8^k|G|)$-FPT, by using the annotated version, simple reduction rules, but the verification was too complex to include. The argument for d-degenerate graphs is much
simpler and more widely applicable, but the running time is worse.

Lemma 16.7.3 *Let* $G = (B \sqcup W, E)$ *be d-degenerated, and suppose that* $|B| >
(4d + 2)k$. *Then there are at most* $(4d + 2)k$ *vertices that dominate at least* $\frac{|B|}{k}$
vertices of B.

Proof Let $R = \{v \in V \mid |(N[v] \cup \{v\}) \cap B| \geq \frac{|B|}{k}\}$. Suppose that $|R| > (4d + 2)k$.
The induced subgraph $G[R \cup B]$ has at most $|R| + |B|$ vertices and at least
$\frac{|R|}{2}(\frac{|B|}{k} - 1)$ edges and hence the average degree of $G[R \cup B]$ is at least $\frac{|R|(|B|-k)}{k(|R|+|B|)} \geq
\frac{\min\{|R|,|B|\}}{2k} - 1 \geq 2d$. This contradicts the fact that $G[R \cup B]$ is d-degenerate. $\square$

Proof of Theorem 16.7.1 If there is a dominating set of size k then we can split B
into k disjoint pieces so that each cell of the partition has a vertex that dominates it.
If $|B| \leq (4d+2)k$ then there are at most $k^{(4d+2)k}$ ways to partition B into k cells. For
each such partition, we can check in time $O(kdn)$ whether each piece is dominated
by a vertex. On the other hand, if $|B| > (4d + 2)k$, by Lemma 16.7.3, we can let
$R = \{v \in V \mid |(N[v] \cup \{v\}) \cap B| \geq \frac{|B|}{k}\}$, as in the proof of the lemma, and know that
$|R| \leq (4d + 2)k$. Then we can begin a bounded search tree strategy by forking on
R: That is, for each $v \in R$, creating a new graph G_v, marking all the elements of
$N_G(v)$ as white, and removing v from G, putting v into the potential dominating
set, and making $k := k - 1$. The search tree can grow to size at most $(4d + 2)^k k!$
before (possibly) reaching the previous case. The time bound follows. $\square$

We remark that results on d-degenerate graphs can often be extended to graphs
that, for $i \geq j > 1$, omit $K_{i,j}$ as a subgraph. The reader is referred to, for instance,
Philip, Raman, and Sikdar [559, 560], for example.

16.8 Exercises

Exercise 16.8.1 (Robertson and Seymour [586]) Prove the following.
1. $\mathrm{bw}(G) \leq 1$ iff G is a forest of stars.
2. $\mathrm{bw}(G) \leq 2$ iff G is series parallel; and consequently G has branchwidth ≤ 2 iff it has treewidth ≤ 2.

Exercise 16.8.2 Give an example to show that the bounds of Theorem 16.1.1 are sharp.

Exercise 16.8.3 Give the details of the proof of Theorem 16.1.1 second part. Namely, start with a tree decomposition of width t, and use the method of Theorem 12.7.1 to make the underlying tree ternary, and having one leaf for each edge of G. Then the width is at most $t + 1$ by the interpolation property of the definition of treewidth.

Exercise 16.8.4 Use dynamic programming to show that the following is FPT.

BW-3-VERTEX COVER
Instance: A graph $G = (V, E)$ and an integer ℓ.
Parameter: $k = \mathrm{bw}(G)$.
Question: Does G have a size ℓ vertex cover if $\mathrm{bw}(G) \leq k$.

Exercise 16.8.5 Show that the graphs with no induced path of length 4 are precisely those of cliquewidth 2.

Exercise 16.8.6 (Golumbic and Rotics [351]) Prove that the $n \times n$ grid with n^2 vertices has cliquewidth $n + 1$.

Exercise 16.8.7 Prove that acyclic digraphs have DAG-width 1.

Exercise 16.8.8 Prove that the largest d for which G has a d-core is the degeneracy of G.

Exercise 16.8.9 (Philip, Raman, and Sikdar [559, 560])
(1) Prove that the problem INDEPENDENT DOMINATING SET has a polynomial kernel for d-degenerate graphs. (Specifically, one with $O((d + 1)k^{d+1})$ vertices.)
(2) Prove that if G is d-degenerate then for some $i \geq j > 1$, G omits $K_{i,j}$ as a subgraph.
(3) Show that INDEPENDENT DOMINATING SET and DOMINATING SET have polynomial kernels for graphs that do not have the complete bipartite graph $K_{i,j}$ for $i \geq j > 1$ as subgraphs. (This generalizes (1) and the work on DOMINATING SET above.)

Exercise 16.8.10 (Alon and Gutner [34])   Show that for $s \le 5$ the problem of finding out if a d-degenerate graph G has an *induced* cycle[2] of length s is $O(|G|)$ FPT for d-degenerate graphs.

16.9   Historical Notes

For a good survey of graph width metrics as of 2008, we refer the reader to Hliněný, Oum, Seese, and Gottlob [396]. Space considerations have precluded us from including the extensive work on (varieties of) *hypergraph treewidth* that have many connections to *constraint satisfaction* and exciting applications in artificial intelligence, and for many practical problems. See [396], Kolaitis and Vardi [456] (for how this related to database queries), and commercial algorithms like TCLUSTER and HINDGE, e.g. Gyssens and Paredaens [373]. The subject is rapidly evolving.

16.10   Summary

We have introduced the reader to a number of graph width metrics including CLIQUEWIDTH, BRANCHWIDTH, RANKWIDTH, DIRECTED TREEWIDTH, DEGENERACY and indicated how algorithms can be run on such decompositions.

---

[2]Alon and Gutner show that a linear time algorithm for the problem for cycles of size 6 reduces to the classical problem for finding triangles in general graphs in time $O(|V|^2)$, a well known longstanding question.

Part IV
Exotic Meta-techniques

Well-Quasi-Orderings and the Robertson–Seymour Theorems

17.1 Basics of Well-Quasi-Orderings Theory

A *quasi-ordering* on a set S is a reflexive transitive relation on S. We will usually represent a quasi-ordering as $\leq_S$ or simply $\leq$ when the underlying set S is clear. Let $\langle S, \leq \rangle$ be a quasi-ordered set. We will write $x < y$ if $x \leq y$ and $y \nleq x$, $x \equiv y$ if $x \leq y$ and $y \leq x$, and, finally, $x|y$ if $x \nleq y$ and $y \nleq x$. Note that if $\langle S, \leq \rangle$ is a quasi-ordered set, then $S/\equiv$ is partially ordered by the quasi-order induced by $\leq$.[1]

Definition 17.1.1 (Ideal and filter) Let $\langle S, \leq \rangle$ be a quasi-ordered set. Let S' be a subset of S.

1. We say that S' is a *filter* if it is closed under $\leq$ upward; that is, if $x \in S'$ and $y \geq x$, then $y \in S'$. The *filter generated by* S' is the set $F(S') = \{y \in S : \exists x \in S'(x \leq y)\}$.[2]
2. We say that S' is a *(lower) ideal* if S' is $\leq$ closed downward; that is, if $x \in S'$ and $y \leq x$, then $y \in S'$. The *ideal generated by* S' is the set $I(S') = \{y \in S : \exists x \in S'(x \geq y)\}$.
3. Finally, if S' is a filter (an ideal) that can be generated by a finite subset of S', then we say that S' is *finitely generated*.

We will need to distinguish certain types of sequences of elements to further define a finitely generated quasi-ordered set.

Definition 17.1.2 (Good, bad, chain, etc. sequences) Let $\langle S, \leq \rangle$ be a quasi-ordered set. Let $A = \{a_0, a_1, \ldots\}$ be a sequence of elements of S. Then, we say the following.

---

[1] Recall that a *partial ordering* is a quasi-ordering that is also antisymmetric.

[2] Sometimes, filters are called *upper ideals*.

R.G. Downey, M.R. Fellows, *Fundamentals of Parameterized Complexity*,
Texts in Computer Science, DOI 10.1007/978-1-4471-5559-1_17,
© Springer-Verlag London 2013

1. A is *good* if there is some $i < j$ with $a_i \le a_j$.
2. A is *bad* if it is not good.
3. A is an ascending *chain* if for all $i < j$, $a_i \le a_j$.
4. A is an *antichain* if for all $i \ne j$ $a_i | a_j$.
5. $\langle S, \le \rangle$ is *Noetherian* if S contains no infinite (strictly) descending sequences (i.e., there is no sequence $b_0 > b_1 > b_2 > \cdots$).
6. $\langle S, \le \rangle$ has the *finite basis property* if for all subsets $S' \subseteq S$, $F(S')$ is finitely generated.

Theorem 17.1.1 (Folklore, after Higman [393]) *Let $\langle S, \le \rangle$ be a quasi-ordered set. The following are equivalent:*

(i) *$\langle S, \le \rangle$ has no bad sequences.*
(ii) *Every infinite sequence in S contains an infinite chain.*
(iii) *$\langle S, \le \rangle$ is Noetherian and S contains no infinite antichain.*
(iv) *$\langle S, \le \rangle$ has the finite basis property.*

Proof (i) $\to$ (ii) and (ii) $\to$ (iii) are easy (see Exercise 17.3.1).

(iii) $\to$ (iv). Suppose that $\langle S, \le \rangle$ is Noetherian and S contains no infinite antichain. Let F be a filter of S with no finite basis. We generate a sequence from F in stages as follows. Let f_0 be any element from F. As F is not finitely generated, there is some $f_1 \in F$ with $f_1 \notin F(\{f_0\})$. For step $i + 1$, having chosen $\{f_0, \ldots, f_i\}$, we find some $f_{i+1} \in F$ with $f_{i+1} \notin F(\{f_0, \ldots f_i\})$. Since F is not finitely generated, this process will not terminate. Notice that for all $i < j$, it cannot be that $f_i \le f_j$ since, in particular, $f_j \notin F(\{f_i\})$. For $i < j$, color the pairs f_i, f_j either *red* if $f_i > f_j$ or color the pairs *blue* if $f_i | f_j$. (Every pair is colored either red or blue by the construction of the sequence.) By Ramsey's Theorem,[3] there is an infinite homogeneous set; that is there is a sequence $f_{i_0} f_{i_1} \ldots$ and a color $\chi \in \{red, blue\}$, such that for all $i_k < i_t$, the pair f_{i_k}, f_{i_t} is colored χ. If χ is red, then we have an infinite descending sequence. If χ is blue, we have an infinite antichain. In either case, we contradict (iii), and hence we get (iv).

(iv) $\to$ (i) Suppose that $\langle S, \le \rangle$ has the finite basis property. Let $B = \{b_0, b_1, \ldots\}$ be a sequence of elements of S. Let $B' \subseteq B$ be a finite basis for $F(B)$. Let $m = \max\{i : b_i \in B'\}$. Choose $b_j \in B - B'$ with $j > m$. Then, for some $b_i \in B'$, we must have $b_i \le b_j$ since $b_j \in F(B')$, and $i < j$ by construction. Hence, B is good. $\square$

The several equivalent characterizations above lead to an important class of quasi-orderings.

---

[3]Ramsey's Theorem (Ramsey [572]) states that if B is an infinite set and k is a positive integer, then if we color the subsets of B of size k with colors chosen from $\{1, \ldots, m\}$, then there is an infinite subset $B' \subseteq B$ such that all the size k subsets of B' *have the same color*. B' is referred to as a *homogeneous* subset.

Definition 17.1.3 (Well-quasi-ordering) Let $\langle S, \leq \rangle$ be aquasi-ordering. If $\langle S, \leq \rangle$ satisfies any of the characterizations of Theorem 17.1.1, then we say that $\langle S, \leq \rangle$ is a *well-quasi-ordering* (WQO).

The reader might well wonder what this abstract pure mathematics has to do with algorithmic considerations. The key clues are the finite basis characterizations of a WQO. The answer is provided by the WQO principle below.

Well-Quasi-Ordering Principle Suppose that $\langle S, \leq \rangle$ is a WQO. Suppose further that for any x, the parameterized problem ABOVE(x) below is polynomial-time computable. Then, for any filter F of $\langle S, \leq \rangle$, the decision problem "Is $y \in F$?" is answered by whether or not it is polynomial-time computable.

ABOVE(x)

Input: $y \in S$.
Parameter: $x \in S$.
Question: Is $x \leq y$?

The proof of the Well-Quasi-Ordering Principle is clear: If F is a filter, then F has a finite basis $\{b_1, \ldots, b_n\}$. Then, to decide if $y \in F$, we need only ask "$\exists i \leq n (b_i \leq_S y)$?" Since for each fixed i, ABOVE(b_i) is in P, the question "$\exists i \leq n (b_i \leq_S y)$?" is in P too.

Actually, the Well-Quasi-Ordering Principle is often stated in a dual form for ideals. If F is a filter of $\langle S, \leq \rangle$, then $S - F$ is an ideal, and vice versa.

Definition 17.1.4 (Obstruction set and obstruction principle)
1. Let $\langle S, \leq \rangle$ be a quasi-ordering. Let I be an ideal of $\langle S, \leq \rangle$. We say that a set $O \subseteq S$ forms an *obstruction set* for I if

$$x \in I \quad \text{iff} \quad \forall y \in O(y \nleq x).$$

That is, O is an obstruction set for I if I is the complement of $F(O)$.
2. Suppose that ABOVE(x) is in P for all $x \in S$, and that I has a finite obstruction set. Then, membership of I is in P too. Hence, we have the following Obstruction Principle: *If $\langle S, \leq \rangle$ is a WQO, then every ideal of I has a finite obstruction set, and furthermore if ABOVE(x) is in P for all $x \in S$, every ideal has a P-time membership test.*

Fig. 17.1 Kuratowski's
obstructions for planarity

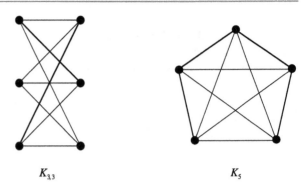

$K_{3,3}$ K_5

So here is our new engine for demonstrating that problems are in P. Prove that the problem is characterized by a finitely generated filter or ideal with a finite obstruction set in a quasi-ordering with ABOVE(q) in P. Even better is to demonstrate that the ordering is a WQO.

"Are there interesting examples of such quasi-orderings?" is one question that often arises. The answer is: "Absolutely". The next few demonstrations that certain quasi-orderings are well-quasi-orderings will take us to some of the deepest theorems of mathematics.

The best known example of an obstruction set is provided by topological ordering, with valuable graph-theoretic finite basis results.

Definition 17.1.5 (Topological ordering) A *homeomorphic or topological embedding* of a graph $G_1 = (V_1, E_1)$ in a graph $G_2 = (V_2, E_2)$ is an injection from vertices V_1 to V_2 with the property that the edges E_1 are mapped to disjoint paths of G_2. (These disjoint paths in G_2 represent possible *subdivisions* of the edges of G_1.) The set of homeomorphic embeddings between graphs gives a partial order, called the *topological order*. We write $G_1 \leq_{\text{top}} G_2$.

The reader should refer to Fig. 17.2. Although topological ordering is not a WQO, there are a number of important finite basis results. The most famous, and the archetype of all graph-theoretic finite basis results, is Kuratowski's Theorem[4] below.

Theorem 17.1.2 (Kuratowski's Theorem, Kuratowski [473]) *The graphs $K_{3,3}$ and K_5 of Fig. 17.1 form an obstruction set for the ideal of planar graphs in the topological ordering.*

---

[4]Apparently, this result was independently discovered by Pontryagin. We refer the reader to Kennedy, Quintas, and Syslo [437].

Proof We give this proof for completeness. We follow the argument of Dirac and Schuster [224] in the style of Harary [378]. We invite the reader to check that if the graph has embedded copies of either K_5 or $K_{3,3}$, then it is nonplanar (Exercise 17.3.4). Now, suppose that G is a nonplanar graph with the minimal number of edges with no embedded copies of either K_5 or $K_{3,3}$. Note that G must have no cutpoints and must have minimal degree ≥ 3. Let $e = uv$ be an edge of G and note that $G' = G - \{e\}$ is planar.

Claim *G' has a cycle containing e.*

To establish the claim, assume that it fails. Then, there must be a cutpoint w of G' lying on each u, v path. Form G'' by adding to G' edges wu and wv if they are not already in G'. Define a *block* to be a maximal component with no cutpoints. Note that in the graph G'', u, and v still lie in differing blocks B_u and B_v. Let B be either B_u or B_v. Note that the minimality of G implies that either B is planar or it contains an embedded copy of either K_5 or $K_{3,3}$. However, in the latter case, if we can embed K_5 or $K_{3,3}$ into G'', then we can also do so in G, by replacing, the edge wu by the path from w to u in G. Therefore, both B_u and B_v are planar. Take a plane embedding of B_u and B_v, where the edges wu and wv occur on the boundary. Now we can remove wu and wv, and put e back in the exterior to make a plane embedding of G, a contradiction. This establishes the claim.

Now, embed G' so that the cycle C containing u and v has the maximal number of regions interior to C possible. Also, we order the edges of the cycle in a cyclic fashion. Let $C[u, v]$ denote the path from u to v in C and similarly for $C[v, u]$. Let $C(x, y)$ denote the subpath resulting from $C[x, y]$ by removing x and y. We define the exterior of C to be the subgraph of G' induced by vertices lying outside of C. We define an *outer* piece of C to be the subgraph of G' induced by all edges incident to least one vertex of a component of G' exterior to C, or by an edge exterior to C meeting two vertices of C. We similarly define interior vertices, components, and pieces.

Definition 17.1.6 (*u, v-separating*) We call an outer piece or inner piece u, v-*separating* if it meets both $C(u, v)$ and $C(v, u)$.

Notice that since there are no cutvertices, each outer piece must meet C in at least two vertices. Moreover, no outer piece can meet $C(u, v)$ or $C(v, u)$ in more than one vertex. To see this, suppose that the outer piece R meets $C(u, v)$, say, in more than one vertex. Suppose that it meets $C(u, v)$ at x and y. Then, we could replace C by a cycle that is the same as C until it reaches x, then travels via R to y, and then copies C. The resulting C' would also be a cycle but would have more internal regions than C. Therefore, no outer piece can meet $C(u, v)$ or $C(v, u)$ in more than one vertex. Similarly, no outer piece can meet u or v. It follows that *every outer piece is u, v-separating and meets C in exactly two vertices $\neq u, v$.* Also note that since we cannot add the edge $e = uv$ and have a planar graph, there must be at least one u, v-separating outer piece.

Claim *There is a u, v-separating outer piece meeting $C(u, v)$ at u' and $C(v, u)$ at v', so that there is an inner piece, which is both u, v-separating and u', v'-separating.*

To see that the claim holds, suppose that it fails. Consider the u, v-separating inner piece I_1 nearest to u, in the sense of encountering inner pieces while moving along C from u. Continuing from u, we can thus order inner pieces $I_1, I_2, \ldots$.

Let f and l be the first and last points of I_1 meeting $C(u, v)$, and f' and l' the first and last points of I_1 meeting $C(v, u)$. Notice that every outer piece has both of its common vertices with C, on either $C[l', f]$ or $C[l, f']$. [Otherwise, there would be an outer piece, which meets $C(u, v)$ at u' and $C(v, u)$ at v', and which is both u, v-separating and u', v'-separating.] Therefore, we can draw an edge E joining l' to f in the exterior of C. It follows that we can transfer I_1 out of the interior of C in a plane manner. Thus, *all* the remaining u, v-separating inner pieces can be transferred outside of C in a plane manner. But then we can add the edge $e = uv$ to G' and the resulting graph will be plane, a contradiction. This contradiction establishes the claim.

To conclude the proof, let H be an inner piece as given by the claim. (So, H is both u, v-separating and u', v'-separating.) Let w_0, w_1 and w'_0, w'_1 be the vertices at which H meets, respectively, $C(u, v)$, $C(v, u)$ and $C(u', v')$, $C(v', u')$.

There are now four cases to consider, depending upon the relative positions of the vertices.

Case 1. *One of the vertices w'_0 or w'_1 is on $C(u, v)$.*

In this case, we can contract w_0 to w'_0 and w_1 to w'_1 (and hence regard $w_0 = w'_0$ and $w_1 = w'_1$) so that G embeds $K_{3,3}$ via the vertex sets $\{u, v, v'\}$ and $\{w_0, w_1, u'\}$.

The other three cases are as follows:

Case 2. *Both of the vertices w'_0 and w'_1 are on either $C(u, v)$ or on $C(v, u)$.*
Case 3. $w'_0 = v$ *and* $w'_1 \neq u$.
Case 4. $w'_0 = v$ *and* $w'_1 = u$.

We leave these cases to the reader (Exercise 17.3.5). (K_5 can only arise in the final case.) ☐

Naturally, we would like to know if the following ordering has ABOVE(P) FPT.

TOPOLOGICAL CONTAINMENT
Instance: A graph $G = (V, E)$.
Parameter: A graph H.
Question: Is H topologically contained in G?

We remark that TOPOLOGICAL CONTAINMENT can be solved in $O(n^{O(|E|)})$ time by brute force together with the k-DISJOINT PATHS algorithm of Robertson and Seymour. Nevertheless, for the example of $H \in \{K_5, K_{3,3}\}$, $H \leq_{\text{top}} G$ is decidable in linear time. Hence, PLANARITY is $O(|G|)$. In 1989, the authors conjectured [247] that TOPOLOGICAL CONTAINMENT is, in fact, FPT.

This conjecture was recently verified using a deep analysis of the whole Robertson–Seymour machinery.

Fig. 17.2 Contractions and a topological embedding

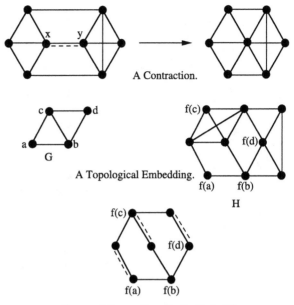

A Contraction.

A Topological Embedding.

Degree 2 Contractions for the Embedding.

Theorem 17.1.3 (Grohe, Kawarabayashi, Marx, Wollan [359]) TOPOLOGICAL CONTAINMENT *is* $O(|H|^3)$-*fixed-parameter tractable.*

It turns out that $\leq_{\text{top}}$ is not the most amenable ordering for graphs, both algebraically and algorithmically. For instance, it is not a WQO (see Exercise 17.3.6). An equivalent formulation of $\leq_{\text{top}}$ is the following. $G \leq_{\text{top}} H$ iff G can be obtained from H by a sequence of the following two operations.
1. (Deletions) Deleting vertices or edges.
2. (Degree 2 contractions) The *contraction* of an edge xy in a graph W is obtained by identifying x with y. A contraction has degree 2 iff one of x or y has degree 2.
   The *minor ordering* is a generalization of $\leq_{\text{top}}$ and is obtained by exactly the same operations, *except that arbitrary edge contractions are allowed.*

Definition 17.1.7 (Minor ordering) We say that G is a minor ordering of H if G can be obtained from H by a sequence of deletions and contractions. We write $G \leq_{\text{minor}} H$.

An example of the minor ordering obtained by contractions and topological embedding is given in Fig. 17.2.
The way to think of $\leq_{\text{minor}}$ is to think of $G \leq_{\text{minor}} H$ as taking $|G|$ many collections C_i of connected vertices of H, and coalescing each collection C_i to a single vertex, then H being topologically embeddable into the new coalesced graph, so that the edges of G become disjoint paths from coalesced vertices. Sometimes this

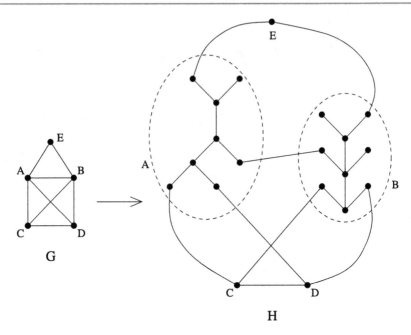

Fig. 17.3 The folio definition

is called the "folio" definition of the minor ordering. Figure 17.3 demonstrates this idea.

Kuratowski's Theorem, Theorem 17.1.2, was about subgraph obstructions to planarity. But an inspection of the *proof* of Kuratowski's famous theorem, also gives the following theorem about graph minor ordering.

Theorem 17.1.4 (Kuratowski's Theorem (II) [473]) $K_{3,3}$ *and* K_5 *are an obstruction set for the ideal of planar graphs under* $\leq_{\mathrm{minor}}$.

Kuratowski's Theorem (II) was so beautiful that it inspired many generalizations and conjectures.

Wagner [660] conjectured the following about the minor ordering of finite graphs.

Wagner's Conjecture *Finite graphs are well-quasi-ordered by the minor ordering.*

Notice that graphs of genus $\leq g$ for a fixed g form an ideal in the minor ordering. Hence, a consequence of Wagner's Conjecture is a Kuratowski Theorem for surfaces. *Graphs of genus* $\leq g$ *have a finite obstruction set.*

One of the triumphs of 20th century mathematics is the following great theorem of Neal Robertson and Paul Seymour.

Theorem 17.1.5 (Graph Minor Theorem [589]) *Wagner's Conjecture holds: Finite graphs are well-quasi-ordered by the minor ordering.*

The proof of Theorem 17.1.5 is spread over 23 articles. To prove Theorem 17.1.5, Robertson and Seymour needed to invent a whole new graph structure theory. A small part of this theory was the invention of the metric of treewidth. In the previous chapters, we have already seen the importance of treewidth.

Robertson and Seymour also demonstrated the startling algorithmic consequences of Theorem 17.1.5 by proving the following.

Theorem 17.1.6 (Robertson and Seymour [588]) MINOR ORDER TEST *below is* $O(n^3)$ *and hence* FPT.

MINOR ORDER TEST
Instance: Graphs $G = (V, E)$ *and* $H = (V', E')$.
Parameter: A graph H.
Question: Is $H \leq_{\text{minor}} G$?

Thus, by the Obstruction Principle, *any* ideal in the minor ordering will not only be a language in P but will have an $O(n^3)$ membership test. The following example was quoted in Chap. 2.

GRAPH GENUS
Instance: A graph $G = (V, E)$.
Parameter: A positive integer k.
Question: Does G have genus k?

The reader should recall that GRAPH GENUS asks whether G be embedded with no edges crossing on a surface with k handles.

We should emphasize that the Graph Minor Theorem is *intrinsically nonconstructive*. By this, we mean that there is no algorithm which, when given an index for a minor closed ideal $\mathcal{I}$, will compute an obstruction set for $\mathcal{I}$. The following is an easy consequence of a well-known theorem from computability theory called Rice's Theorem (Rice [579]). Rice's Theorem asserts, roughly, that no nontrivial property given only by machine description is decidable. Since we will soon improve the Fellows–Langston Theorem below we will not digress to discuss this at length.

Theorem 17.1.7 (Fellows and Langston [297, 299]) *There is no algorithm A which, when given an index e for an algorithm for a minor closed ideal $\mathcal{I}$, outputs A(e), an obstruction set for $\mathcal{I}$. (It is not important what A does on those e which are not indices for minor closed ideals.)*

We are naturally led to investigate what sorts of further information about an ideal might allow the obstruction set for the ideal to be systematically computed. There are a number of apparently difficult unresolved problems in this general area. For example, it is unknown whether the obstruction set for an arbitrary union of ideals $\mathcal{F} = \mathcal{F}_1 \cup \mathcal{F}_2$ can be computed from the two corresponding obstruction sets $\mathcal{O}_1$ and $\mathcal{O}_2$, although this can be accomplished if at least one of these obstruction sets includes a tree (Cattell, Dinneen, Downey, Fellows, and Langston [132]).

The following theorem says that even a monadic second order description of an ideal is not enough to compute obstructions.

Theorem 17.1.8 (Courcelle, Downey and Fellows [176])   *There is no effective procedure to compute the obstruction set for a minor ideal $\mathcal{F}$ from a monadic second order (MS_2) description of $\mathcal{F}$.*

This should be contrasted with general positive results concerning the computation of obstruction sets. Fellows and Langston proved in [301] that $\mathcal{O}$ can be effectively computed if we have access to three pieces of information:
  (i) A decision algorithm for $\mathcal{F}$.
 (ii) A bound B on the maximum treewidth (or pathwidth) of the graphs in the set $\mathcal{O}$ of $\mathcal{F}$ obstructions.
(iii) A decision algorithm for a finite index congruence for $\mathcal{F}$.

Perhaps surprisingly, this algorithm [301] has been implemented. Nontrivial, previously unknown obstruction sets for interesting ideals have also been successfully computed by Cattell and Dinneen [131] and Cattell, Dinneen, and Fellows [133].

Since (i) and (iii) can be effectively derived from an MS_2 description of $\mathcal{F}$ by Courcelle's Theorem, Theorem 17.1.8 emphasizes that (ii) is essential in the earlier positive result of [301].

We turn to the proof of Theorem 17.1.8. We will use the following lemma, which is a consequence of Courcelle's Theorem (because of the definability of $H \leq_{\mathrm{minor}}$) concerning the description in MS_2 logic of generated filters in the minor order.

Lemma 17.1.1   *Given an MS_2 formula φ, we can effectively produce an MS_2 formula φ' such that $\mathcal{F}(\varphi')$ is the filter generated in the minor order by $\mathcal{F}(\varphi)$.*

Proof sketch A graph G has a graph H as a minor if and only if it is possible to identify a set of disjoint connected subgraphs G_v of G indexed by the vertex set of H, such that if u and v are adjacent vertices of H, then the corresponding subgraphs G_u and G_v of G are "adjacent" in the sense that there are vertices $x \in G_u$ and $y \in G_v$ where $xy \in E(G)$. The formula φ' can be constructed by expressing in M2O the statements:
(1) There exists a set of edges that forms a forest in G.
(2) The exists a set V_0 of roots for the trees of the forest of (1), with one root for each tree.
(3) There exists a set E_0 of edges between the trees of the forest, in the sense of "adjacency" described above.

The formula φ' consists of this preface, with φ thereafter modified by significant restrictions and substitutions:

(4) Quantification is restricted to V_0 and E_0. (These are in some sense the "virtual" vertices and edges of the minor H of G that the preface asserts to exist. We are now concerned with expressing that this H is a model of φ.)

(5) Incidence and adjacency terms in φ are replaced by statements concerning suitable paths in the forest of (1). (In other words, incidence and adjacency statements about the virtual vertices and edges, V_0 and E_0, are interpreted with the means provided by (1), (2), and (3).) $\square$

The following proposition is a corollary of a theorem of Trakhtenbrot [648] on the undecidability of the first order logic of graphs. We remark that Seese has shown that undecidability still holds even for planar graphs [612].

Lemma 17.1.2 (Trakhtenbrot [648], Seese [612]) *Given an MS_2 formula φ, there is no algorithm to decide if there is a finite graph G such that $G \models \varphi$.*

Proof of Theorem 17.1.8 We argue that if there were such a procedure, then we could solve the problem of determining whether a formula φ of MS_2 has a finite model, contradicting Proposition 1. Let φ' denote the formula computed from φ by the effective procedure of Lemma 17.1.1. Let $\psi = \neg\varphi'$. We note the following.

- If no finite graph is a model of φ, then the set of models of φ' is empty, and every finite graph is a model of ψ. Thus ψ describes an ideal for which the set of obstructions is the empty set.

- If φ has a finite model, then φ' describes a nontrivial filter and ψ describes a nontrivial ideal complementary to φ' for which the set of obstructions is nonempty.

By computing the obstruction set for the ideal described by ψ we can therefore determine, on the basis of whether this obstruction set is empty or nonempty, whether any finite graph is a model of φ. $\square$

The Graph Minor Theorem thus provides an important philosophical challenge to the notion that polynomial time is a "good" measure of *practical tractability*. Here, we can show that a practical problem is in P, but we have no way of even knowing the polynomial-time algorithm solving the problem. The Graph Minor Theorem only yields *nonuniform fixed-parameter tractability*. Furthermore, the user news is even worse. Even when we can compute the obstruction sets, the *constants* in Theorem 17.1.6 are gigantic (around 2^{500}), and evidence suggests that the sizes of the obstruction sets is very large. At the time of the writing of this book, an exact obstruction set for embedding graphs on the torus is unknown. These facts caused a prominent computer scientist to remark, echoing the famous words of Kronecker,

This is not computer science, it is mathematical curiosity.

We do not agree. We believe that the Graph Minor Theorem is of immense value to computer science. At the very least, an application of the Graph Minor Theorem to yield a $O(n^3)$ algorithm for some natural problem surely demonstrates *potential* practical tractability. (Sometimes, this happens in problems previously not even known to be effectively solvable.) The claim that P is a good measure of tractability is supported by the fact that it has good closure properties—and "Practical Algorithm Builders" believe that

> If a natural problem is in P, then it has a low degree, small constant algorithm demonstrating its membership in P.

If this heuristic principle is correct, then an application of the Graph Minor Theorem to prove a natural problem is in P is an important first step in the gradual discovery of a truly practical algorithm for the problem. For instance, inspired by the Robertson and Seymour result, Mohar [533] has recently demonstrated a constructive and feasible linear time algorithm for GRAPH GENUS.

On the other hand, if this heuristic principle is *not* correct, then the Graph Minor Theorem must surely provide a good vehicle to search for counterexamples.

Practical algorithm builders will, however, notice two *reasons* for the non-computability and impracticality of the Graph Minor Theorem alone. The first reason stems from its great generality as a decision procedure that assumes a grasp of practical factors; yet still lacking is a program to develop techniques to extract truly practical algorithms for classes of natural problems. Practical problems usually help us by offering a lot more combinatorial information than merely a decision procedure. Therefore, the second reason is the impractical insistence that we need to work from only a machine description for membership of the problem class at hand. For example, if we are only provided with a machine description for a class of natural languages, then, in general, there is no process which effectively yields automata. We usually work with *much more* than merely machine descriptions of natural languages. Indeed, in the case of formal languages, we usually have a regular expression describing the language. We believe that similar realistic considerations will apply to natural applications of the Graph Minor Theorem. For instance, we will often have some nice syntactic description of the problem class.

We will soon look at some applications and effectuations of the Graph Minor Theorem in Chap. 19. Later in that chapter, we will look at other WQO theorems such as Robertson and Seymour's result that the "immersion" order is a WQO (Sect. 18.5).

In the next chapter, we give a brief sketch of the Graph Minor Theorem, but a complete proof of the Graph Minor Theorem is beyond the scope of this book.

We now finish the present section by proving the archetype of all WQO results for graphs, Kruskal's Theorem for trees. Kruskal's Theorem and its consequences are very important for what follows.

Theorem 17.1.9 (Kruskal's Theorem [471]) *The class of rooted finite trees are well-quasi-ordered by $\leq_{\text{top}}$ and by $\leq_{\text{minor}}$. [The reader should note that for rooted trees T_1, T_2, if $T_1 \leq_{\text{top}} T_2$ (via the topological embedding) while f and v and w are vertices of branches of T_1 meeting at z, then it can only be that this happens iff $f(v)$ and $f(w)$ are vertices of branches of T_2 meeting at $f(z)$ in T_2.]*

Proof We work with $\leq_{top}$, since $\leq_{minor}$ is coarser. We can consider trees as 1-trees and hence as 1-boundaried graphs. Trees can be parsed by the operators $\oplus$ and 1, where 1 is the operator which adds a vertex, makes it a boundary vertex, and joins an edge from the old boundary vertex to the new one (Exercise 17.3.3).

Suppose that $T_0, T_1, \ldots$ is a minimal bad sequence of trees; that is we suppose that we have chosen T_0 to be the smallest tree (by cardinality) to begin a bad sequence, and from the bad sequences of trees beginning with T_0, we have chosen T_1 to be smallest, etc. Let P_i be a corresponding parse tree for T_i. Either there are infinitely many P_i with root $\oplus$ or almost all P_i have root 1. Both cases are easy.

Consider the case where $P_{i_0}, P_{i_1}, \ldots$ all have root $\oplus$. Now, for a parse tree P, let $T(P)$ denote the tree obtained from P. Let Q_i and R_i denote the parse trees obtained by deleting the roots from P_i, and consider the collection

$$\mathcal{C} = \bigcup_j \left(T(Q_{i_j}) \cup T(R_{i_j}) \right).$$

We claim that $\langle \mathcal{C}, \leq_{top} \rangle$ is a WQO. If this is not the case, then there is a bad sequence $B_0, B_1, \ldots$ from $\mathcal{C}$. For some least k, $B_0 \in \{Q_k, R_k\}$. Then, the sequence $T_0, T_1, \ldots, T_{k-1}, B_0, B_1, \ldots$ is bad, contradicting the minimality of $T_0, T_1, \ldots$. Hence, $\langle \mathcal{C}, \leq_{top} \rangle$ is a WQO. It is easy to prove that $\mathcal{C} \times \mathcal{C}$ is well-quasi-ordered by the product ordering [i.e., $(W_1, V_1) \leq_{top} (W_2, V_2)$ iff $W_1 \leq_{top} W_2 \wedge V_1 \leq_{top} V_2$]. But then there are $m < n$ with

$$T(Q_{i_m}) \leq_{top} T(Q_{i_n}) \wedge T(R_{i_m}) \leq_{top} T(R_{i_n}).$$

Clearly, this implies that $T_{i_m} \leq_{top} T_{i_n}$.

The root 1 case is similar but easier. $\square$

Extend the definition of topological ordering to labeled topological ordering. Let $\langle Q, \leq_Q \rangle$ be a quasi-ordered set. Say that a Q-labeled tree T_1 topologically embeds into a labeled tree T_2 via f, by asking that f takes edges to disjoint paths, and furthermore, if v has label q and $f(v)$ has label q', then $q \leq_Q q'$. The WQO proof above easily extends to the following for Q-labeled trees.

Corollary 17.1.1 (Kruskal's Theorem for labeled trees [471]) *If $\langle Q, \leq \rangle$ is a WQO, then the collection of Q-labeled trees is a WQO under labeled topological embedding.*

The main modification in the proof is that one argues from an infinite sequence of roots $\{q_{i_0}, q_{i_1}, \ldots\}$ with the property $q_{i_0} \leq_Q q_{i_1} \leq_Q \cdots$ —roots that are guaranteed to exist by the fact that Q is a WQO. We remark that standard proofs of Kruskal's Theorem do not use parse trees. Rather, they use Higman's lemma, which is of importance in its own right. (See Higman's lemma below. Also see Exercise 17.3.9.)

Let $\langle S, \leq \rangle$ be any quasi-ordered set. One can define the *induced* product ordering on S, the set of finite sequences of elements of S via $\langle s_1, \ldots, s_n \rangle \leq \langle s'_1, \ldots, s'_m \rangle$ iff there is a strictly increasing map $j : \{1, \ldots, n\} \rightarrow \{1, \ldots, m\}$ such that for all $i \leq n$,

$$s_i \leq s'_{j(i)}.$$

For instance, $\langle 1, 2, 1, 4 \rangle \leq \langle 0, 2, 2, 0, 1, 6, 1 \rangle$, but $\langle 1, 2, 1 \rangle \nleq \langle 0, 1, 1, 2, 0, 0 \rangle$.

Lemma 17.1.3 (Higman's Lemma [393]) *Suppose that $\langle S, \leq \rangle$ is a WQO. Then, the induced ordering is a well-quasi-ordering of S.*

Proof We can give an easy proof of this lemma by using the labeled version of Kruskal's Theorem. We can consider a sequence $\langle a_0, a_1, \ldots a_n \rangle$ as a tree with root labeled a_0, immediate successor $a_1, \ldots, a_n$. So a bad S sequence would correspond to a sequence of trees, bad with respect to labeled topological ordering. $\square$

Before we leave our section on basic WQO theory, we mention two further WQO results of theorems for *induced* and *chain minor* ordering whose proofs are beyond the scope of the present book. The *induced* minor ordering is the same as the normal minor ordering except that only vertex deletions and contractions are allowed.

Theorem 17.1.10 (Thomas [642]) *Finite graphs of treewidth ≤ 2 are well-quasi-ordered by the induced minor ordering.*

Note that the bound 2 is sharp in Theorem 17.1.10 (Exercise 17.3.12).

Finally, we turn to a WQO result in a different finite arena. Let P and Q be partially ordered sets (posets). We say that P is a *chain minor* of Q, $P \ll Q$ iff there exists a partial mapping $f : Q \mapsto P$ such that for each chain C_P of P, there is a chain C_Q of Q such that $f \upharpoonright C_Q$ is an isomorphism of chains from C_Q to C_P.

Theorem 17.1.11 (Gustedt [369]) *Finite posets are well-quasi-ordered by $\ll$.*

Gustedt noted the following.

Theorem 17.1.12 (Gustedt [369])
1. *Let P and Q be finite posets with $c = |P|$. (Assume that $c > 3$.) Then there is a constant l depending only upon P and an algorithm A which decides if $P \ll Q$ running in time $O(c^2|Q|^c + l)$.*
2. *Consequently, if $\mathcal{F}$ is any class of finite posets closed under the chain minor ordering, then $\mathcal{F}$ has a polynomial-time recognition algorithm.*

Proof The proof is left as an exercise for the reader. $\square$

Of course, Theorem 17.1.12 opens the question: Is this chain minor problem FPT[5]?

CHAIN MINOR ORDERING

Input: A finite poset Q.
Parameter: A finite poset P.
Question: Is P a chain minor of Q?

17.2 Connections with Automata Theory and Boundaried Graphs

Well-quasi-orderings have deep connections with automata theory and graphs of bounded treewidth. In this section, we examine some of these connections. We begin with formal languages.

Definition 17.2.1 (Canonical partial order) Let $L \subseteq \Sigma^*$. We define the *canonical partial ordering induced by L*, $\leq_L$, via

$$x \leq_L y \quad \text{iff} \quad \forall u, v \in \Sigma^*, \ uyv \in L \text{ implies } uxv \in L.$$

Lemma 17.2.1 (Abrahamson and Fellows [4, 5]) *Suppose that $\leq_L$ is a well-quasi-ordering of Σ^*. Then, $\leq_{L^c}$ is a well-quasi-ordering of Σ^*, where L^c denotes $\Sigma^* - L$, the complement of L.*

Proof Suppose not. Then, there is a bad L^c sequence $\mathcal{B} = x_0, x_1, \ldots$. Now, by Ramsey's Theorem, $\mathcal{B}$ either has an infinite $\leq_L$ ascending sequence (which it cannot, since it is bad), an infinite $\leq_L$ descending sequence, or an infinite $\leq_L$ antichain. (For Ramsey's Theorem, see Theorem 17.1.1.)

Consider that $\mathcal{B}$ has an infinite $\leq_L$ antichain. Call the antichain $\mathcal{A} = a_0, a_1, \ldots$. Then, for each $i \neq j$, there exists p, q, r, s such that

$$ra_i p \in L^c \wedge ra_j p \notin L^c \quad \text{and} \quad sa_i q \notin L^c \wedge sa_j q \in L^c.$$

But this means that

$$ra_i p \notin L \wedge ra_j p \in L \quad \text{and} \quad sa_i q \in L \wedge sa_j q \notin L.$$

Hence, sequence $\mathcal{A}$ is an L-antichain, too, contradicting the fact that $\leq_L$ is a WQO.

Therefore, $\mathcal{B}$ has an infinite descending sequence. Renaming, we have $x_0 \geq_{L^c} x_1 \geq_{L^c} x_2 \geq_{L^c} \cdots$. This corresponds to an L-ascending sequence, $x_0 \leq_L x_1 \leq_L \cdots$. Since $\mathcal{B}$ is bad, there exists a sequence of pairs of witness strings $\mathcal{W} = (w_{1,2}, v_{1,2}), (w_{2,3}, v_{2,3}), \ldots$ such that for all i,

$$w_{i,i+1} x_{i+1} v_{i,i+1} \in L^c \quad \text{but} \quad w_{i,i+1} x_i v_{i,i+1} \notin L^c.$$

---

[5]CHAIN MINOR ORDERING was recently shown to be FPT by Blasiok and Kaminski [68].

Now, since $\leq_L$ is a WQO, $\leq_L \times \leq_L$ is a WQO of $\Sigma^* \times \Sigma^*$. Thus, $\mathcal{W}$ contains an infinite $\leq_L \times \leq_L$ ascending sequence. And, there exist indices $i < j - 1$ and $w, w', v, v' \in \Sigma^*$ with (i)–(iii) below holding:

(i) $x_i \leq_L x_{i+1} \leq_L x_j \leq_L x_{j+1}$.

(ii) $wx_iv \notin L \wedge wx_{i+1}v \in L$ and $w'x_jv' \notin L \wedge w'x_{j+1}v' \in L$.

(iii) $w \leq_L w'$ and $v \leq_L v'$.

But this is a contradiction:

$$w'x_{j+1}v' \in L \rightarrow wx_{j+1}v' \in L \quad \left(\text{since } w \leq_L w'\right)$$
$$\rightarrow wx_{j+1}v \in L \quad \left(\text{since } v \leq_L v'\right)$$
$$\rightarrow wx_iv \in L \quad \left(\text{since } x_i \leq_L x_{j+1}\right). \qquad \square$$

Lemma 17.2.1 has the following consequence for formal languages that neatly ties together the concepts of well-quasi-ordering and being finite state. This theorem is a well-quasi-ordering analogue of the Myhill–Nerode Theorem.

Theorem 17.2.1 (Abrahamson and Fellows [4, 5]) *A formal language L is finite state iff $\leq_L$ is a well-quasi-ordering.*

Proof Suppose that $\leq_L$ is a WQO. Recall from Definition 12.5.2 that $x \approx_L y$ means that for all $u, v \in L$,

$$uxv \in L \quad \text{iff} \quad uyv \in L.$$

Hence, $x \approx_L y$ iff $x \leq_L y \wedge x \leq_{L^c} y$. By Lemma 17.2.1, if $\leq_L$ is a WQO, then so is L^c. Hence, $\leq_L \times \leq_{L^c}$ is a WQO, and hence $\approx_L$ has finite index. By Exercise 12.5.1, this implies that $\sim_L$ has finite index. By the Myhill–Nerode Theorem, L is finite state.

Conversely, if L is finite state, then $\sim_L$ has finite index by the Myhill–Nerode Theorem. Hence, $\approx_L$ has finite index. If $\leq_L$ is not a WQO, then there is an L-bad sequence $x_0, x_1, \ldots$. Clearly, this L-bad sequence generates infinitely many $\approx_L$ equivalence classes. $\qquad \square$

There is a graph-theoretic analog of Theorem 17.2.1.

Definition 17.2.2 (Canonical partial order for graphs) Let L be a family of t-boundaried graphs in $\mathcal{U}^t_{\text{small}}$. We write $G_1 \leq_L G_2$ iff for all $Z \in \mathcal{U}^t_{\text{small}}$,

$$G_2 \oplus Z \in L \quad \text{implies} \quad G_2 \oplus Z \in L.$$

Theorem 17.2.2 (Abrahamson and Fellows [5]) *A family F of graphs is t-finite state iff $\leq_F$ is a well-quasi-ordering of $\mathcal{U}^t_{\text{small}}$.*

Proof The proof is a simple graph-focused analog of Theorem 17.2.1. We begin with a similar graph-focused analog of Lemma 17.2.1.

Lemma 17.2.2 (Abrahamson and Fellows [5]) *If F is a family of graphs for which $\leq_F$ is a well-quasi-order on U_{large}^t (U_{small}^t), then $\leq_{F^c}$ is a well-quasi-order on U_{large}^t (U_{small}^t), where F^c denotes the complement of F.*

Proof As in Lemma 17.2.1, if the statement is false, then we may take $(X_1, X_2, \ldots)$ to be a bad $\leq_{F^c}$ sequence, which is $\leq_F$ ascending. For $i < j$ let $Z_{i,j}$ be a choice of evidence for the badness of the sequence: $X_i \oplus Z_{i,j} \in F$ and $X_j \oplus Z_{i,j} \in F^c$. Consider the sequence of t-boundaried graphs $(Z_{1,2}, Z_{3,4}, Z_{5,6}, \ldots)$. Since $\leq_F$ is a well-quasi-order, we must obtain the situation (re-indexing for convenience): $X_1 \leq_F X_2 \leq_F X_3 \leq_F X_4$, $X_1 \oplus Z_{1,2} \in F$, $X_2 \oplus Z_{1,2} \notin F$, $X_3 \oplus Z_{3,4} \in F$, $X_4 \oplus Z_{3,4} \notin F$, and $Z_{1,2} \leq_F Z_{3,4}$. But this is a contradiction, since $X_3 \oplus Z_{3,4} \in F$ implies $X_3 \oplus Z_{1,2} \in F$, and this implies $X_2 \oplus Z_{1,2} \in F$. □

Proof of Theorem 17.2.2, completed Now, note that for t-boundaried graphs X and Y, $X \sim_F Y$ if and only if $X \leq_F Y$ and $X \leq_{F^c} Y$, by definition. If $\leq_F$ is a well-quasi-order, then by Lemma 17.2.2, $\leq_{F^c}$ is a well-quasi-order. The Cartesian product of $\leq_F$ and $\leq_{F^c}$ is then a well-quasi-order, and, therefore, $\sim_F$ has finite index. The result then follows by the Myhill–Nerode Theorem for t-boundaried graphs. □

As we will soon see in this section, $\leq_L$ and finite obstructions can be used to provide smooth proofs for algorithms on t-boundaried graphs and formal languages. Before we look at one such application, we would like to point out that $\leq_L$ is also strongly related to *arbitrary* well-quasi-orderings of Σ^* (U_{small}^t).

Definition 17.2.3 (Well-behaved quasi-order)
1. A quasi-order $\leq$ on Σ^* is *well-behaved* if it is (1) Noetherian, and (2) a congruence with respect to concatenation; that is, $x \leq x'$ and $y \leq y'$ imply $xx' \leq yy'$.
2. Similarly, a quasi-order $\leq$ on U_{small}^t is *well-behaved* if (1) $\leq$ is Noetherian and (2) $\leq$ is a congruence with respect to $\oplus$; that is, $X \leq Y$ and $X' \leq Y'$ imply $X \oplus X' \leq Y \oplus Y'$.

Theorem 17.2.3 (Abrahamson and Fellows [5]) *Let $\mathcal{R} \in \{\Sigma^*, U_{\text{small}}^t\}$. A well-behaved quasi-order $\leq$ on $\mathcal{R}$ is a WQO if and only if $\leq_L$ is a WQO for every lower ideal L of $\langle \mathcal{R}, \leq \rangle$.*

Lemma 17.2.3 (Abrahamson and Fellows [5]) *Let $\leq$ be a Noetherian quasi-order on U_{small}^t (on Σ^*.) If every lower ideal of $\leq$ is finite state, then $\leq$ is a well-partial-order.*

Proof If $\leq$ is not a well-quasi-order, then there is an infinite antichain A. Each (possibly infinite) subset S of A determines a distinct lower ideal $I_S = \{x : x \leq a \in S\}$. If S and T are subsets of A with $S \neq T$, then $I_S \neq I_T$. Thus, A yields a collection of uncountably many distinct lower ideals. Since there are only countably many finite state automata, there must be a lower ideal I_S that is not finite state. □

Proof of Theorem 17.2.3 Suppose that $\mathcal{R} = \Sigma^*$. By Lemma 17.2.3 and Theorem 17.2.1, it is enough to argue that if L is a lower ideal of a well-behaved quasi-order $\leq$ on Σ^*, then $\leq_L$ is a well-quasi-order. If not, then there is a bad $\leq_L$ sequence $(x_1, x_2, \ldots)$. But then, since $\leq$ is a WQO, there are indices $i < j$ such that $x_i \leq x_j$. Since it is not the case that $x_i \leq_L x_j$, there are words $u, v \in \Sigma^*$ such that $u x_i v \in L$ and $u x_j v \notin L$. This contradicts that L is a lower ideal, since $x_i \leq x_j$ implies $x_i v \leq x_j v$, which implies $u x_i v \leq u x_j v$.

The case that $\mathcal{R} = \mathcal{U}^t_{\text{small}}$ is entirely similar and is left to the reader (Exercise 17.3.13). $\qquad\square$

17.3   Exercises

Exercise 17.3.1 Prove that (i) implies (ii), and (ii) implies (iii) in Theorem 17.1.1.

Exercise 17.3.2 Prove that if G is a minor of H then $\text{tw}(G) \leq \text{tw}(H)$.

Exercise 17.3.3 Prove that we can parse trees with the operators $\oplus$ and 1; that is, prove that we can consider trees as 1-boundaried graphs which can be parsed by the operators $\oplus$ and 1, where 1 is the operator which adds a vertex, makes it a boundary vertex, and joins an edge from the old boundary vertex to the new one. (Use induction.)

Exercise 17.3.4 Prove that if the graph has embedded copies of either K_5 or $K_{3,3}$, then it is nonplanar.

Exercise 17.3.5 Verify that one can embed K_5 or $K_{3,3}$ in the remaining cases of Theorem 17.1.2.

Exercise 17.3.6 Prove that $\leq_{\text{top}}$ is not a WQO.
(Hint. Consider n circles each consisting of four points $\{x_i, y_i, z_i, q_i\}$ with edges $\{x_i y_i, y_i z_i, z_i, q_i, q_i x_i\}$, joined so that $q_i = x_{i+1}$.)

Exercise 17.3.7 Give a direct proof of Higman's lemma.
(Hint: Suppose that $\langle S, \leq \rangle$ is a WQO. Then, we need to prove that the induced ordering is a well-quasi-ordering of S. Suppose that S is not well-quasi-ordered by the induced ordering $\leq$. Let $a_0, a_1, \ldots$ be a minimal bad sequence for S; that is, we suppose that $a_0, a_1, \ldots$ is a bad sequence of finite sequences of members of S, chosen so that:
1. a_0 is the sequence of minimal length which begins any bad sequence.
2. Among bad sequences beginning with a_0, we choose a_1 to have minimal length, etc.
Note that if $a_0, a_1, \ldots a_{i-1}, b_i, b_{i+1}, \ldots$ is a bad sequence, then $|a_i| \leq |b_i|$. Let q_i denote the final term of a_i. Since $\langle S, \leq \rangle$ is a WQO, there is an infinite collection, $q_{i_0} \leq q_{i_1} \leq \cdots$.

Now, let $\{a'_i : i \in \mathbb{N}\}$ denote the collection obtained from $\{a_i : i \in \mathbb{N}\}$ by deleting q_i. Then, by the minimality condition, the sequence $a_0, a_1, \ldots, a_{i_0-1}, a'_{i_0}, a'_{i_1}, \ldots$ is good. Therefore, there exist $m < n$ with $a'_{i_m} \leq a'_{i_n}$. Since $q_{i_m} \leq q_{i_n}$, we have $a_{i_m} \leq a_{i_n}$ by definition, contradicting the badness of the sequence $a_0, a_1, \ldots$.)

Exercise 17.3.8 Show that finite trees are not well quasi-ordered by the *subgraph* relation.

Exercise 17.3.9 Without using parse trees, give a direct proof of Kruskal's Theorem via Higman's lemma.
(Hint: (Nash-Williams [542]) Let $\mathcal{T} = T_1, T_2, \ldots$ be a bad sequence, minimal by size. Consider the sequence of forests

$$\left(T_1^1, \ldots, T_1^{n_1}\right), \left(T_2^1, \ldots, T_2^{n_2}\right), \ldots$$

obtained by removing the root from each T_i. Use minimality of $\mathcal{T}$ to prove that $\bigcup_{i,j} T_j^i$ is well-quasi-ordered by topological ordering. Now, prove that $\mathcal{T}$ is not bad by using Higman's lemma.)

Exercise 17.3.10 Let $\langle S, \leq \rangle$ be a quasi-order. Let S^ω denote the collection $\{(x_0, x_1, \ldots) : x_i \in S\}$ of countably infinite sequences of members of S. Then, we can extend $\leq$ to S^ω in the obvious way: $(x_0, x_1, \ldots) \leq (y_0, y_1, \ldots)$ iff there is an increasing (1–1) function $f : \mathbb{N} \mapsto \mathbb{N}$ such that for all $i \in \mathbb{N}$, $x_i \leq y_{f(i)}$.
   Give an example of a WQO $\langle S, \leq \rangle$ with S^ω not well-quasi-ordered by the induced ordering.
(Hint: (Rado)[6] Consider $\mathcal{R} = \{(i, j) : i < j < \omega\}$, partially ordered by

$$(i, j) \leq (i', j') \quad \text{iff} \quad j \leq i' \text{ or } i = i' < j \leq j'.)$$

Exercise 17.3.11 (Gustedt [369])   Prove Theorem 17.1.12(i).
(Hint: Prove that if $P \ll Q$, then there is a poset P' with $P \ll P'$, P' an induced subordering of Q, and $|P'| \leq |Q|$.)

Exercise 17.3.12 Prove that there is an infinite bad sequence under the induced minor ordering.

Exercise 17.3.13 Prove the claim at the end of the proof of Theorem 17.2.3.

17.4   Historical Notes

The origin of quasi-order is lost in time. Maybe it is due to Euler's work in the mid-18th century. Well-quasi-order is a frequently re-discovered concept, as carefully discussed in Kruskal [472]. The first attributed example was due to Rado [570], who showed that transfinite sequences of length ω^2 over a finite alphabet are WQO'd, and gave "Rado's counterexample" in 1954. Higman's Lemma on induced WQO'g dates to 1952; see Higman [393]. Topological graph theory (and graph theory) originated from Euler [279] with the Königsberg Bridge Problem that he described in 1741.

---

[6]Rado actually proved that this example is canonical in the sense that if Q is a WQO such that Q^ω is *not* a WQO, then the quasi-ordering $\langle S, \leq \rangle$ of this hint will embed into Q.

Kuratowski's Theorem characterizing planar graphs by excluded graphs appeared in 1930 [473], though it was independently found by Pontryagin. The concept of a minor originated in the early 20th century, and was certainly known by K. Wagner [660] who posed his famous conjecture in 1937. The Robertson–Seymour Theorems are proven in a long series of papers mostly in the *Journal of Combinatorial Mathematics, B*. The theorem showing that MS_2 descriptions of ideals are not enough to give effective ideal membership is due to Courcelle, Downey, and Fellows [176]. Kruskal's Theorem is from 1960 [471]. It was later extensively extended to all trees (not just finite ones) using a *better quasi-ordering* by Nash–Williams [542, 543], who also clarified the minimal bad sequence argument. The connections with automata are due to Abrahamson and Fellows [4, 5].

17.5 Summary

We introduce the main ideas from the theory of WQOs. We prove basic results like Higman's Lemma, and introduce the concept of bad sequences. We prove Kruskal's Theorem on the WQO of trees. We look at the connections between automata theory and WQO theory.

The Graph Minor Theorem

18

18.1 Introduction

It is still not possible to give a short proof of the Graph Minor Theorem, but the last 15 years have seen significant simplifications to the proof, although the overall structure remains the same. One key idea which has really become *central* is the notion of *excluding a graph as a minor*. Thus, if we want to establish some sequence is not bad, then assuming it is bad, if H_1 is the first member of the sequence, all of the remaining members exclude H_1 as a minor. Robertson and Seymour's deep insight is that exclusion of this has profound structural implications on the structure of the remaining graphs both topologically and algorithmically. Moreover, as we see, recent work has also shown that this idea has even more significant *algorithmic* implications than was first expected (at least in theory). This includes the work on protrusions we will briefly describe, and work such as the Grohe–Marx project [360]. We will discuss this in Sect. 19.4.

The ultimate structure of the proof is to give an structure theorem for graphs excluding an arbitrary graph in terms of a "treelike" decomposition of "almost surfaces" with parameters about certain objects like "vortices" and "tangles". Then *structurally* the argument is more or less along the lines of those of Kruskal's Theorem, but simply much more complex.

To begin with, recall how we proved Kruskal's Theorem, Theorem 17.1.9 in Chap. 17.1. We took a minimal bad sequence of boundaried finite trees. The critical case was where the last parsing operation in the formation of the trees was $\oplus$. Then we decomposed the collection into infinite bad sequences generated by the left and right sides of the $\oplus$, say. We used minimality to argue that we could get a good sequence from the right side, than a corresponding one from the left side and glue since the ordering was boundaried topological ordering on trees.

The proof of the Graph Minor Theorem runs along somewhat similar lines. It works by induction on the genus of the excluded graph, and relies on structural properties of the sequences generated by the possible sequences exclude the graph.

R.G. Downey, M.R. Fellows, *Fundamentals of Parameterized Complexity*,
Texts in Computer Science, DOI 10.1007/978-1-4471-5559-1_18,
© Springer-Verlag London 2013

To illustrate the ideas, before we do the sketch, we will prove an excluded minor theorem (excluding a forest) and give the sketch of the fact that graphs of bounded treewidth (those that exclude grids) are well-quasi-ordered under the minor ordering. The results are of independent interest.

18.2   Excluding a Forest

We can give a proof of a related result of Bienstock, Robertson, Seymour and Thomas via an elegant proof of Diestel [220] to give the reader some idea of this style of result. More comments are given in the Historical Notes.

Theorem 18.2.1 (Bienstock, Robertson, Seymour and Thomas [65]) *Let F be a forest, and C a class of graphs excluding F. Then, the pathwidth of the members of C is bounded $< |F| - 1$.*

Proof (Diestel [220]) The proof is inductive and relies upon the following definitions. For $X \subseteq V(G)$, let $G[X]$ denote the subgraph induced by X and attach(X) the set of vertices that attach to $G - X$, namely those vertices in X which have neighbors in $G - X$ (Bienstock, Robertson, Seymour and Thomas [65]). For each $n > 0$ and graph G, let $\mathcal{B}_n$ denote the unique minimal subset of the power set of $V(G)$ such that $\emptyset \in \mathcal{B}_n$ and such that for all $X \in \mathcal{B}_n$,

$$X \subseteq Y \subseteq V(G) \quad \text{and} \quad \left|\text{attach}(X)\right| + |Y - X| \leq n \quad \text{implies} \quad Y \in \mathcal{B}_n.$$

Notice that a set X is in $\mathcal{B}_n$ iff there is a sequence of subsets of $V(G)$ $\emptyset = X_0 \subseteq X_1 \subseteq \cdots \subseteq X_s = X$ such that for all $r < s$, $|\text{attach}(X_r)| + |X_{r+1} - X_r| \leq n$.[1]

Lemma 18.2.1 *If $V(G)$ is $\mathcal{B}_n$, then G has pathwidth $< n$.*

Proof If $V(G)$ is in $\mathcal{B}_n$ then there is a sequence $\emptyset = X_0 \subseteq X_1 \subseteq \cdots \subseteq X_s = V(G)$. Now, consider $W_r = \text{attach}(X_{r-1}) \cup (X_r - X_{r-1})$.   □

Diestel's simplification of the earlier proof uses the following elegant observation.

Lemma 18.2.2 (Diestel [220]) *Suppose that Y is in $\mathcal{B}_n$ and $Z \subseteq Y$. Further suppose that there is a set $\{P_z : z \in \text{attach}(Z)\}$ of disjoint paths such that each P_z starts in Z, has no other vertex in Z, and ends in attach(Y). The Z is in $\mathcal{B}_n$.*

---

[1] The reader should note that if $W_1, \ldots, W_d$ are the bags in order of a path decomposition of G, then the sets $W_1 \cup \cdots \cup W_r$ are in $\mathcal{B}_n$. The reader should compare this with the *running intersection property* of Exercise 10.2.2.

Proof Given the hypotheses, there exist $\emptyset = Y_0 \subseteq Y_1 \subseteq \cdots \subseteq Y_s = Y$ with $|\mathrm{attach}(Y_r)| + |Y_{r+1} - Y_r| \leq n$ for all $r < s$. Let $Z_r = Y_r \cap Z$. The claim is that $|\mathrm{attach}(Z_r)| + |Z_{r+1} - Z_r| \leq n$, giving the lemma. Notice that $Z_{r+1} - Z_r \subseteq Y_{r+1} - Y_r$. Therefore, we need only prove that $|\mathrm{attach}(Z_r)| \leq |\mathrm{attach}(Y_r)|$.

So suppose that $z \in \mathrm{attach}(Z_r) - \mathrm{attach}(Y_r)$. It follows that z has a neighbor in $Y_r - Z_r = Y_r - Z$, and hence $z \in \mathrm{attach}(Z)$. By the hypotheses of the theorem, we have a path P_z from Z_r, and hence Y_r to $\mathrm{attach}(Y)$, and $\mathrm{attach}(Y_r)$ separates these two sets in G. It follows that P_z has a vertex $y \neq z$ in $\mathrm{attach}(Y_r)$. Since z is the only vertex of P_z in Z, it must be that $y \in \mathrm{attach}(Y_r) - \mathrm{attach}(Z_r)$. Since the paths are disjoint, for each such z, there is a distinct y, it must be that $|\mathrm{attach}(Z_r) - \mathrm{attach}(Y_r)| \leq |\mathrm{attach}(Y_r) - \mathrm{attach}(Z_r)|$, and hence $|\mathrm{attach}(Z_r)| \leq |\mathrm{attach}(Y_r)|$. □

Returning to the proof of Theorem 18.2.1, we assume that F is given and is a tree. Suppose that G has pathwidth $\geq n = |F| - 1$. Using topological ordering, say, assume that $V(F) = \{v_1, \ldots, v_n\}$ such that for $i \leq n$, $F[v_1, \ldots, v_i]$ is a subtree of F (i.e., connected) and hence exactly one v_i for $i < n$ is adjacent to v_{i+1}.

For $0 \leq j \leq n$, we shall define a set $\mathcal{S}^i = \{S_0^i, \ldots, S_i^i\}$ of disjoint subgraphs of G so that $S_j^k \subseteq S_j^t$ for all $j \leq k \leq t$, so that for $d > 0$, S_d^i is connected, and such that (i)–(iv) below hold. We define $X^i = V(\bigcup\{S_j^i : j \leq i\})$.

1. Whenever $1 \leq j < k \leq i$ and $v_j v_k \in E(F)$, G contains an edge between S_j^i and S_k^i, so that $F[v_1, \ldots, v_v]$ is a minor of $G[S_1^i \cup \cdots \cup S_i^i]$.
2. $|\mathrm{attach}(X^i)| = i$ and $|V(S_j^i) \cap \mathrm{attach}(X^i)| = 1$ for $1 \leq j \leq i$.
3. X^i is in $\mathcal{B}_n$.
4. $|\mathrm{attach}(X)| > i$ for all $X \in \mathcal{B}_n$ with X^i a proper subset of X.

First, let S_0^0 be the inclusion maximal subgraph of G with $V(S_0^0)$ in $\mathcal{B}_n$, and $|\mathrm{attach}(S_0^0)| = 0$. Clearly, (i)–(iv) hold for S_0^0. For an induction, assume that $\mathcal{S}^i$ has been defined for some $i \leq n$.

Case 1. $i = 0$. Then, $G - S_0^0$ is nonempty since we can apply Lemma 18.2.1 to the fact that $V(S_0^0) \in \mathcal{B}_n$, yet $V(G) \notin \mathcal{B}_n$. Thus, in this case, we let x be any vertex of $G - S_0^0$. Let $X = V(S_0^0) \cup \{x\}$.

Case 2. $i > 0$. In this case, find the unique $j < i$ such that $v_j v_i$ is an edge of F, and let $x \in G - X^i$ be a neighbor of the unique vertex in $V(S_j^i) \cap \mathrm{attach}(X^i)$. We let $X = X^i \cup \{x\}$.

If $i = n$, then (i) and the choice of x together imply that F is a minor of $G[X]$, giving (i)–(iv). Thus, assume that $i < n$. Notice that by (ii) and (iii), we have $X \in \mathcal{B}_n$. Also, it follows that $|\mathrm{attach}(X)| > i$ by (iv). In turn, this means that $\mathrm{attach}(X) = \mathrm{attach}(X^i) \cup \{x\}$ and $|\mathrm{attach}(X)| = i + 1$. Let Y be the inclusion maximal member of $\mathcal{B}_n$ with both $X \subseteq Y$ and $\mathrm{attach}(Y)| = i + 1$. (This set is the later X^{i+1}.)

Now, we apply Menger's Theorem. Menger's Theorem shows that $G[Y]$ contains a set P of disjoint paths from X to $\mathrm{attach}(Y)$ and a separator D between these sets, each path having exactly one member of D. Now, let Z denote the union of D and all the vertex sets of components of $G - D$ not meeting X. Then, X^i is a proper subset of X which is a subset of Z. Furthermore, by choice of D and definition of attach(Y), we see that $Z \subseteq Y$. Observe that attach(Z) $= D$. Consequently, by Lemma 18.2.2, Z is in $\mathcal{B}_n$. By (iv), $|P| = |D| = |\text{attach}(Z)| > i$. By definition of attach(X), each of the paths in P meet attach(X). We conclude that $i < |P| \leq |\text{attach}(X)| = i + 1$, and hence the paths in P give a perfect matching between attach(X) and attach(Y).

We are now ready to define $\mathcal{S}^{i+1}$. Let $S_0^{i+1} = S_0^i \cup G[Y - (X \cup V(\cup P))]$. For $1 \leq j \leq i$, let x_j be the unique vertex of S_j^i in attach(X^i) and the path P_j in P containing it. Define S_j^{i+1} as the union of S_j^i and the final segment of P_j starting at x_j. Finally, we let S_{i+1}^{i+1} denote the final segment from x of the path in P containing x. Then, $X^{i+1} = V(\bigcup \mathcal{S}^{i+1}) = Y$. We note that $\mathcal{S}^{i+1}$ satisfies (i) by choice of x, (ii), and (iii) since $X^{i=1} = Y$, and, finally, (iv) by choice of Y and the fact that $X^i \subseteq Y = X^{i+1}$, together with (iv) for i.

The theorem now follows by Lemma 18.2.1 applied to the definition of X in the case $i = n$. $\square$

18.3   Thomas' Lemma

The next step in the proof of the Graph Minor Theorem is to consider the effect of excluding more complex graphs than forests. We stated in Chap. 13 the following result. Theorem 13.3.2:

- *Suppose that H is a finite planar graph, and C is a class of graphs excluding H as a minor. Then there is a number n such that for all $G \in C$, the tree width of G is less than n.*

Thus, if we consider the effect of excluding graphs of the lowest genus, then we get bounded treewidth. Following the Kruskal roadmap, we could then consider an infinite minimal bad sequence of graphs of treewidth $\leq t$. Again these can be represented as boundaried graphs, and we could imaging a proof where, as with the $\oplus$ case of Kruskal's Theorem, we decompose the given tree decompositions of the graphs at some smallest separation which occurs infinitely often.

Then, we would like to argue, again analogously with the $\oplus$ case, or, as with Higman's Lemma, that we could get an infinite good left side, and a corresponding infinite good right side. The *last step* in any proof is then to argue that for some such infinite collection, H_i, K_i we get a method to lift this to the original bad sequence. The point here is that if we have say H_i present as a minor in H_j, and K_i in K_j, where H_i and K_i are glued at a separation on a boundary, *why* should we be able to contract H_j and K_j to $\hat{H}_j, \hat{K}_j$, say, so that the gluing of H_i and K_i will be present in the gluing of $\hat{H}_j$ and $\hat{K}_j$? What is needed is a method of arguing that the tree decompositions of H_j and K_j have a property that allows us to contract boundaries along disjoint paths. This "linked decomposition" notion was the decisive new idea needed in the proof of the WQO of graphs of bounded treewidth.

The first proof of the well-quasi-ordering of graphs of bounded treewidth was in the original (unpublished) version of Robertson and Seymour [585]. It was a long and complex argument showing that graphs of treewidth t had "linked" decompositions of treewidth $f(t)$ for some function f. Subsequently, the argument was considerably simplified and extended by Robin Thomas [642, 643] and Kriz and Thomas [470]. The Thomas approach was based on an important theorem (Thomas' lemma) which said that graphs of treewidth t had some very well-behaved tree decompositions also of treewidth t. Thomas' lemma was actually used in the final published version of Robertson and Seymour's Graph Minors IV [585] since it simplified Robertson and Seymour's proof so much. The remainder of section is devoted to giving Thomas' proof of the relevant lemma.

The reader should recall Definition 10.1.1 of a tree decomposition as a tree labeled with subsets with each edge included in some bag and the interpolation property. For subsets X and Y of the vertices of G, let $c(X, Y)$ denote the *separation number* of X and Y. That is the size of the smallest set (cutset) A of vertices of G such that each path from X to Y uses a member of A. In [247], we used the terminology "Thomas decomposition" for the below, but we have decided to align ourselves with the new terminology of lean and linked decompositions.

> **Definition 18.3.1** A tree decomposition $\{V_i : i \in \mathcal{T}\}$ of a graph G is called
> (1) a *linked decomposition* if additionally for any k and $i, j \in \mathcal{T}$, either G contains k disjoint $V_i - V_j$ paths, or there is a bag t between x and y with $|V_t| < k$;
> (2) a *lean decomposition relative to* $\mathcal{T}$ if given any $i, j \in \mathcal{T}$ and $Z_i \subseteq V_i$, $Z_j \subseteq V_j$, both of cardinality k, either G contains k disjoint paths from Z_i to Z_j, or there is an edge tt' in $\mathcal{T}$ between x and y, such that $|V_t \cap V_{t'}| < k$.

If we have a lean decomposition, by Lemma 18.3.1 below, then for any subsets $V_i' \subseteq V_i$ and $V_j' \subseteq V_j$ of bags V_i and V_j, if $c(V_i', V_j') < \min\{|V_i'|, |V_j'|\}$, then there is a bag V_q between V_i and V_j in $\mathcal{T}$, with $|V_q| \leq c(V_i', V_j')$. Actually, another consequence of Lemma 18.3.1 is that the conclusion of this statement can be replaced by "$|V_q| = c(V_i', V_j')$". The proof of this easy lemma is left to the reader (Exercise 18.7.1).

Lemma 18.3.1 (Folklore, e.g. Thomas [643]) *Suppose that $\{V_i : i \in \mathcal{T}\}$ is a tree decomposition and P is a path in G which meets bags V_i and V_j in G. Then P meets any bag V_q between V_i and V_j in $\mathcal{T}$.*

It is easy to make a lean decomposition linked without increasing the width by subdividing edges tt' and adding a bag $V_s = V_t \cap V_{t'}$. As pointed out by Bellenbaum and Diestel, for technical reasons, lean decompositions are easier to use than linked ones. The method of the proof below involves separations and surgery. Given a tree decomposition $\{V_t \mid t \in \mathcal{T}\}$, and a separation X the method involves amalgamation of components in of the subgraphs $G[C \cup X]$ for components C in $G \setminus X$, as discussed in Lemma 11.2.1. More generally if C is a *union* of components of $G \setminus X$, set $H = G[c \cup X]$, and let $s \in \mathcal{T}$. For each $x \in X$, pick a *home* node $t_x \in \mathcal{T}$, with $x \in V_{t_x}$, and define for all $t \in \mathcal{T}$,

$$W_t = \left(V_t \cap V(H)\right) \cup \{x \in X \mid t \in t_x T s\},$$

where $t_x T s$ denotes the part of $\mathcal{T}$ between t_x and s inclusive. Then $\{W_t \mid t \in \mathcal{T}\}$ is a tree decomposition of H.

Lemma 18.3.2 (Bellenbaum and Diestel [59]) *If $G - C$ contains a set $\{P_x \mid x \in X\}$ of disjoint $X - V_s$ paths with $x \in P_x$ for all x, then $|W_t| \leq |V_t|$ for all $t \in \mathcal{T}$.*

Proof For each $x \in X$ with $x \in W_t \setminus V_t$ we have $t \in t_x T s$, and hence V_t separates V_{t_x} from V_s by Lemma 18.3.1. Hence V_t contains some other vertex of P_x. That vertex does not lie in W_t as $W_t \setminus X \subseteq V(C)$ while $P_x \subseteq G - C$.   □

Theorem 18.3.1 (Thomas' Lemma [642, 643]) *A treewidth k graph has a treewidth k lean decomposition.*

Thomas' lemma can be viewed as an extension for graphs of bounded treewidth of Menger's Theorem from Chap. 35.

Proof (Bellenbaum and Diestel [59]) Let $|G| = n$. Define the *fatness* as the n-tuple $(a_0, \ldots, a_{n-1})$ where a_i is the number of bags with exactly $n - h$ many vertices. Choose the tree decomposition $\mathcal{T}$ of G of width $\mathrm{tw}(G)$ of lexicographically minimal fatness. The claim is that $\mathcal{T}$ is lean. The technique of proof is to show that it is not lean then we can do some surgery and make it less fat.

Thus suppose that $\{V_t \mid t \in \mathcal{T}\}$ is not lean. Then there is a quadruple (i, j, Z_i, Z_j) violating 1 above, and we can choose i, j to be as close as possible. Among the $Z_i - Z_j$ separators X of smallest size in G choose the one with $\sum_{x \in X} d_x$ as small as possible, where

$$d_x = \min\{d_T(t, t_x) \mid t \in iTj \land x \in V_{t_x}\}.$$

Let C_1 be the union of the components of $G \setminus X$ that meet Z_i, and C_2 the union of the other components of $G \setminus X$. Let $H_q = G[C_q \cup X]$ for $q \in \{i, j\}$. By Menger's Theorem, there is a set $\{P_x \mid x \in X\}$ of disjoint $Z_i - Z_j$ paths in G such that $x \in P_x$ for all x. Each P_x is the union of two paths $P_x^q \subseteq G \setminus C_q$ meeting exactly at x. We will use Lemma 18.3.2 applied to these paths.

For each $x \in X$, choose t_x at minimum distance from iTj in T with $x \in V_{t_x}$. Let T^i, T_j be disjoint copied of $\mathcal{T}$. Let t^q denote the copy of t in T^q. Put $\mathcal{T}' = T^i \cup T^j \cup j^i i^j$, and let $\{W_t^q \mid t \in T\}$ be the tree decomposition of H_q obtained by Lemma 18.3.2 applied to $s = j$ if $q = i$ and $s = i$ if $q = j$. Rewriting $\mathcal{T}$ as T^q and W_t^q as W_{t^q} we combine the tree decompositions to obtain a tree decomposition $\{W_t \mid t \in \mathcal{T}'\}$ of G. This decomposition satisfies the interpolation property as $V(H_i \cap H_j) = X \subseteq W_{ji} \cap W_{ij}$.

Thus to complete the proof we need only show that $\{W_t \mid t \in \mathcal{T}'\}$ has less fatness than the original decomposition. We claim

(1) For all $t \in \mathcal{T}$, for all $q \in \{i, j\}$, $|W_t^q| = |V_t| \to W_t^{q'} \subseteq X$ where $q' = j$ if $q = i$ and $q' = i$ if $q = j$.

(2) For some $t \in iTj$, $|W_t^i|, |W_t^j| < V_t$.

This suffices since $|W_t^q| \leq |V_t|$ by Lemma 18.3.2 and assuming that $\{V_t \mid t \in \mathcal{T}\}$ fails to satisfy the leanness property, e have $|X| < k \leq |V_t|$ for all $t \in iTj$. Assuming (1) and (2), by 1, for each $h > |X|$, the number of parts of cardinality h is no greater in $\mathcal{T}'$ than in $\mathcal{T}$ whilst 2 implies that for some such h this number has gone down.

To establish (1), let $t \in \mathcal{T}$ be given and (e.g.) assume that $|W_t^i| = |V_t|$. We show that $W_t^j \subseteq X$. If not then $V_t \cap C_j \neq \emptyset$ and for each $v \in C_j$ some $x \in X$ was included in W_t^i when it was constructed from V_t. Let Y be the collection of such x, to wit $Y = W_t^i \setminus V_t = \{x \in X \mid t \in iTj\} \setminus V_t$. Let $X' = (X \setminus Y) \cup (V_t \cap C_j)$. We have seen $|W_t^i| = |V_t|$ implies $|Y| = |V_t \cap V(C_j)|$, and hence $|X'| = |X| < k$.

The claim is that we could have chosen X' in the place of X for our supposedly minimal counterexample. To see this, we show that X' separates both Z_i and V_t from Z_j in G. Any path P in $G \setminus X'$ from either Z_i or V_t to Z_j has last vertex $y \in Y$. Then $yP \subseteq C_j$ and contains a $V_{t_y} - V_{t_j}$ path. since $t \in t_y Tj$, since $Y = W_t^i \setminus V_t = \{x \in X \mid t \in iTj\} \setminus V_t$, yP meets V_t by Lemma 18.3.2. As $y \notin V_t$, it must do so in C_j and hence in X'. It $t \in iTj$ (and $i \neq t$) then this implies that for any k-set $Z \subseteq V_t$, the quadruple (i, j, Z, Z_j) violates the leanness condition, contradicting the choice of i, j. Thus if $t \neq i$ then $t \notin iTj$. The proof is completed by showing that $d_{x'} < d_y$ for all $x' \in X' \setminus X$, and $y \in Y = X \setminus X'$. It will then follow that we could have chose X' in place of X for the counterexample (i, j, Z_i, Z_j).

Let x' and y be given. As $Y = W_t^i \setminus V_t = \{x \in X \mid t \in iTj\} \setminus V_t$, t separates t_y from j in $\mathcal{T}$. If $t \in iTj$ with $t \neq i$, then t must separate t_y from the whole path in $\mathcal{T}$. If d is the distance of t from iTj in $\mathcal{T}$, we see that $d_{x'} \leq d < d_y$ as $x' \in V_t$ and the definition of t_y.

For the proof of (2), that for some $t \in iTj$, $|W_t^i|, |W_t^j| < V_t$, it suffices to find some $t \in iTj$ such that V_t meets both C_i and C_j. If there is no such t, then since V_i meets C_i and V_j meets C_j, we see that iTj has an edge tt' with $V_t \subseteq H_i$ and $V_{t'} \subseteq H_j$. Then $V_t \cap V_{t'} \subseteq X$ and satisfies the leanness condition, a contradiction. $\square$

18.4 The Graph Minor Theorem

We sketch the proof of the Graph Minor Theorem. First we consider the case where we exclude a planar graph, that is, of bounded treewidth. Now we follow the method sketched in the last section. We have run the Kruskal argument on lean linked decompositions, and roughly speaking, break the graphs on some minimal boundary which occurs infinitely often, and then argue that the leanness allows the separate minor embeddings we get for the left and right sequences can be glued by "Mengering" the boundaries to the separations.

The general case will follow similar lines. The first thing that is needed to be proven is that if we exclude a general graph H rather than just a planar one, we get a well-behaved "bounded surfacewidth (mythical)" structure. It suffices to consider $H = K_n$ for some n and then what is needed is that for each such n, there is a finite set of surfaces S such that each graph excluding H is *almost* embeddable on some member of S. Almost embeddable means that there is at most a finite number of pieces (the number bounded by parameters depending on H), such that, modulo them, the graph becomes embedded on one of the surfaces.

To give an example of the type of structural result obtained by Robertson and Seymour, we need some definitions. A graph J is called a clique join of graphs of G and H if $G \cap H$ is a clique and J is a subgraph of $G \cup H$. (Think here of $\oplus$.) Now, let P_n be a path with vertices $t_1, \ldots, t_n$ and C_n the circuit on the same vertices. Given an integer r, an r-ring with perimeter $t-1, \ldots, t_n$ is a graph R on the vertices $t_1, \ldots, t_n$ such that there is a collection of bags $X = \{X_{t_i} : i = 1, \ldots, n\}$ for which (P_n, X) is a path decomposition of R or width $r-1$, and for each i, $t_i \in X_{t_i}$. Assume that we are give a hypergraph where each edge is incident with at least two edges. If a hyperedge is incident with exactly two vertices, we call it an edge, and otherwise we will say it is a hyperedge. We regard a hypergraph as embedded on a surface Σ if the following all hold. Each vertex is a point, each edge is a simple closed curve, and each hyperedge is a closed disk with its incident vertices on its boundary and no other intersections occur.

Given a surface Σ and integer r, a (Σ, r)-outgrowth is a pair (G, H), where G is a graph, H is a hypergraph embedded on Σ, such that for each hyperedge h of H, if $(h_1, \ldots h_n)$ is a cyclic ordering of the vertices of h induced by the embedding in Σ, then there is an r-ring R_h with perimeter $h_1, \ldots, h_n$ such that G is the graph induced by the edges of $H \cup \bigcup_{h \text{ a hyperedge of } H} R_h$. Note that $V(G) = V(H)$. If Θ is a set of surfaces, then a graph G is a (Θ, r)-outgrowth if there is a $\Sigma \in \Theta$ and an embedded hypergraph such that (G, H) is a (Σ, r)-outgrowth. [We ask that $(\emptyset, \emptyset)$, the pair of empty graphs, are a (Σ, r)-outgrowth for any (Σ, r).] Finally, given an integer k, we say that M is a $\leq k$-vertex extension of G, if there is a set S of at most k vertices in M such that $G = M - S$.

The following is an example of one of the main technical results used to prove the Graph Minor Theorem as we moved to surfaces.

Theorem 18.4.1 (Robertson and Seymour [589]) *Let G be a graph and $\Theta(G)$ the set of surfaces into which G cannot be embedded. Then, there are numbers $r(G)$ and $w(G)$ such that every graph with no G-minor can be constructed by clique joins, starting from $\leq w(G)$ vertex extensions of $(\Theta(G), r(G))$-outgrowths.*

Note that in the case of G a planar graph, then $\Theta(G) = \emptyset$, and the theorem says that graphs excluding G can be obtained from graphs with at most $w(G)$-vertices by clique joins. That is, Theorem 18.4.1 generalized to surfaces gives the result that graphs excluding a planar graph must have bounded treewidth. We remark that Theorem 18.4.1 has already found other applications such as in Ding, Oporowski, Sanders, and Vertigan [221] where it was used to verify a conjecture of Thomas by proving the following theorem.

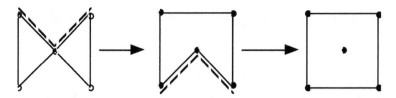

Fig. 18.1 The immersion ordering

Theorem 18.4.2 (Ding et al. [221]) *For every graph G, there is an integer k such that every graph excluding G has a two coloring of either its vertices or its edges where each color induces a graph of treewidth at most k.*

Anyway, the point of all of this is that we only need to consider infinite bad sequences almost embeddable on a fixed surface. The case of planar graphs is genus 0, and hence we work by induction on the genus. So if we have a bad sequence H_i for $i \in \mathbb{N}$, then if we assume that the surface is of genus > 0, then each H_i has an embedding which meets some circle C_i on the surface which does not bound a disk in the surface and in no more than a bounded number of vertices and no edges of H_i. The idea is then to cut along C_i, cap the holes and obtain two new surfaces of lower genus. Structural analysis allows one to be able to lift embeddings, in the same way that leanness allowed us to contract along boundaries.

The details are more or less along these lines, but are complicated. A full proof of the case for bounded treewidth is found in Downey and Fellows [247], but we decided to omit it for space considerations.

18.5   The Immersion and Topological Orders

In the same series of articles where they proved the Graph Minor Theorem, Robertson and Seymour verified another famous WQO conjecture made by Nash–Williams [542].

Definition 18.5.1 (Immersion ordering, lifting) Let G and H be graphs. We say that $G \leq_{\text{immersion}} H$ if we can obtain G from H by applications of the operations *deletion* and *edge lifting.*

Here, if G is a graph with a vertex v having edges wv and qv, $w \neq q$, then *lifting* through $\{w, v, q\}$ is the result of removing the edges wv and qv and adding an edge wq. The reader is referred to Fig. 18.1.

Theorem 18.5.1 (Graph Immersion Theorem, Robertson and Seymour [583]) *Finite graphs are well-quasi-ordered by $\leq_{\text{immersion}}$.*

Until very recently, the algorithmic applicability of the Graph Immersion Theorem relies on the Obstruction Principle, together with the following result.

Theorem 18.5.2 (Fellows and Langston [298]) *For every fixed graph H, the parameterized problem which takes as input a graph G and asks if $H \leq_{\text{immersion}} G$ is solvable in polynomial time.*

Proof Letting k denote the number of edges in H, we replace $G = \langle V, E \rangle$ with $G' = \langle V', E' \rangle$, where $|V'| = k|V| + |E|$ and $|E'| = 2k|E|$. Each vertex in V is replaced in G' with k vertices. Each edge e in E is replaced in G' with a vertex and $2k$ edges connecting this vertex to all of the vertices that replace e's endpoints. We can now apply the disjoint-connecting paths algorithm of [588], since it follows that $H \leq_i G$ if and only if there exists an injection from the vertices of H to the vertices of G' such that each vertex of H is mapped to some vertex in G' that replaces a distinct vertex from G and such that G' contains a set of k vertex-disjoint paths, each one connecting the images of the endpoints of a distinct edge in H. □

Notice that the time bound generated by the proof of Theorem 18.5.2 is $O(n^{h+3})$, where h is the order of the largest graph in the obstruction set. In [247], we conjectured that the following is FPT.

IMMERSION ORDER TEST
Instance: A graph $G = (V, E)$ and a graph $H = (V', E')$.
Parameter: A graph H.
Question: Is $H \leq_{\text{immersion}} G$?

The matter was put to rest in a wonderful paper [359].

Theorem 18.5.3 (Grohe, Kawarabayashi, Marx, and Wollan [359]) TOPOLOGICAL ORDERING *and* IMMERSION ORDERING *are* $O(n^3)$-*FPT.*

18.6   The Summary

Thus we see that the following are FPT: immersion, minor, and topological orderings. The proofs are beyond the scope of the present book. However, we will attempt a sketch.

The proofs are all similar and split into several cases. The first case for testing $H \leq G$ is if G has bounded treewidth. Then the route is through the MS_2 theory. The next case is that G has no large clique minor. This case is quite technical and runs through something called the unique linkage theorem which enables the proof of the existence of a "large flat wall"; then using something called the irrelevant vertex technique we will examine in the chapter on bidimensionality. In the case of topological embedding, the method is similar, and both now work through Kawarabayashi and Wollan [436], which enables the algorithmic part of the minor theorem without the complex machinery of the later Robertson–Seymour papers. If there is a large clique minor, then in the case of the minor ordering, the proof is relatively easy. However, the case of topological ordering, the proof is complex and involves finding well-behaved folios and separators, and analysis of the situations of whether there are many high degree vertices or few. The immersion case is a corollary of the topological ordering case by converting it into a topological embedding via subdivision.

18.7 Exercises

Exercise 18.7.1 Prove Lemma 18.3.1.

Exercise 18.7.2 Construct a linked decomposition for the width-3 graph of Fig. 10.2.

Exercise 18.7.3 (Dinneen [222]) Let $\gamma(G) \leq p(|G|)$ map graphs to integers, where p is some bounding polynomial. Assume $\mathcal{F}_k = \{G \mid \gamma(G) \leq k\}$ is a minor-order lower ideal. Also let $p_\gamma(G, k)$ denote the corresponding graph problem that determines if $\gamma(G) \leq k$ where both G and k are part of the input.

Show that if $p_\gamma(G, k)$ is NP-complete then $f_k = |\mathcal{O}(\mathcal{F}_k)|$ must be superpolynomial else the polynomial-time hierarchy $\mathcal{PH}$ collapses to Σ_3^P.
(Hint. Use the Yap Theorem, Theorem 1.2.1.)

Exercise 18.7.4 Give an upper bound on the smallest $m \times m$ grid in which the complete graph K_n is immersed.

Exercise 18.7.5 Prove that acyclic graphs are well-quasi-ordered by immersions.

Exercise 18.7.6 Prove that IMMERSION ORDER TEST is linear fixed-parameter tractable for planar graphs.

Exercise 18.7.7 Recall that a *plane graph* is a planar graph G together with an embedding f of G in the plane. Define a natural notion of the immersion of one plane graph in another and investigate whether plane graphs are well-quasi-ordered by plane immersions.

Exercise 18.7.8 (Govindan, Langston and Plaut [352]) Use the fact that the immersion order is a WQO to prove that each slice of the following problem can be solved in polynomial time.

MINIMUM DEGREE GRAPH PARTITION
Instance: A graph $G = (V, E)$ and positive integers k and d.
Parameter: k, d.
Question: Can V be partitioned into disjoint subsets $V_1, \ldots, V_m$ so that, for $1 \leq i \leq m$, $|V_i| \leq k$ and at most d edges have exactly one endpoint in V_i?

18.8   Historical Notes

The material of Sect. 18.2 is drawn from Diestel [220]. The first "excluding forest" result was due to Robertson and Seymour [581], who demonstrated that for each forest F, there exists an integer m such that each class of graphs that exclude F has pathwidth below m. The argument was complex and long. Subsequently, the argument was sharpened to the bound $m = |F| - 1$ of this section and considerably simplified by Bienstock, Robertson, Seymour and Thomas [65]. Nevertheless, the argument still relied on a minimax theorem and still remained relatively involved. The short proof of the present section is due to Diestel [220], who eliminated the need for the minimax theorem. We remark that the bound $|F| - 1$ is sharp because K_{n-1} has pathwidth $n - 2$ but has no forest minor on n vertices. Furthermore, if F is not a forest, then Theorem 18.2.1 fails in general. For instance, trees can have arbitrarily large pathwidth, but cannot contain cycles as minors.

As we remarked earlier, there are many beautiful "excluded minor" theorems, and the idea of excluding graphs remains a cornerstone of the proof of the Graph Minor Theorem. A noteworthy one is that of Archdeacon [41] who exhibited a 35-element obstruction being embeddable upon the projective plane. Robertson and Seymour [584] proved that if a class of graphs excludes a planar graph, then it must have bounded treewidth, and the treewidth bound can be theoretically calculated by the methods of Fellows and Langston [299, 300]. However, there are very few obstruction sets that have been explicitly calculated due to the complexity of the computations currently involved. For instance, the obstruction set for embeddability upon the torus is, at present, open. Finally, we remark that there are many other "forbidden subobject" results for other orderings. We met one in Leizhen Cai's Theorem of Sect. 4.9. Others are beyond the scope of the book.

The Robertson–Seymour Theorems are proved in a large sequence of papers called collectively *Graph Minors*. Some can be found in the bibliography. The proofs are very long and difficult, and in some sense this is provably necessary since Friedman, Robertson, and Seymour [333] have shown that the Graph Minor Theorem is not provable in a logical system that encompasses most of classical mathematics. (Π_1^1 comprehension, see Simpson [619].)

18.9   Summary

We give a broad sketch of the ideas of the Graph Minor Theorem. We discuss excluding a forest and more general consequences of excluding a graph as a minor. We prove the Thomas Lemma on the existence of lean and linked decompositions. We discuss the related immersion and topological orderings.

Applications of the Obstruction Principle and WQOs

19

19.1 Methods for Tractability

It is important to point out that the minor and immersion WQO's do not exhaust
the interesting possibilities for general methods based on the Obstruction Principle.
These orders are defined in terms of sequences of local operations on graphs by
which one graph can be derived from another. Given the set of allowed operations,
the order is defined by $G \geq H$ if H can be derived from G by a sequence of these
operations. The nature of this definition immediately yields transitivity. There are
obvious opportunities for generalizing by considering various sets of local operations on various kinds of combinatorial objects. Many of these avenues have not
been explored, and some present difficult mathematical challenges.

The following are some examples of simple local operations on graphs:

(E) Delete an edge.

(V) Delete a vertex.

(C) Contract an edge.

(T) Contract an edge incident on a loopless vertex of degree 2. (Also called *removing a subdivision.*)

(L) Lift an edge. Two edges incident on a vertex v are "unplugged" from v and
made into a single edge that now bypasses v.

(F) Fracture a vertex. Replace a vertex v with two new vertices v_1 and v_2 and
partition the edges incident on v into two classes, making these incident, respectively, on v_1 and v_2.

(I) Identify two vertices.

(S) Simplify. Delete an edge of which there is more than one copy, or remove a
loop.

The minor order is generated by the set of operations $\{E, V, C\}$ and the immersion order is generated by the operations $\{E, V, L\}$. The topological order is
generated by $\{E, V, T\}$ and the *induced minor* order, by $\{V, C\}$. As a matter of historical interest, the minor order was quite obscure *until* the work of Robertson and
Seymour.

R.G. Downey, M.R. Fellows, *Fundamentals of Parameterized Complexity*,
Texts in Computer Science, DOI 10.1007/978-1-4471-5559-1_19,
© Springer-Verlag London 2013

Fig. 19.1 Induced minor obstructions for interval graphs

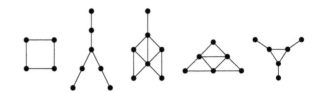

Most graph properties are preserved by some set of local operations that can be used to define a quasi-order for which the property defines an ideal, thus potentially allowing an algorithmic approach based on the Obstruction Principle. We illustrate this with two brief examples.

Example 19.1.1 (Cutwidth) We recall from Sect. 12.7 the cutwidth problem.

CUTWIDTH

Instance: A graph $G = (V, E)$.

Parameter: k.

Question: Can the vertices of G be linearly ordered by a $1 : 1$ function $f :$
              $V \rightarrow \{1, \ldots, |V|\}$ so that $\forall i \in \{1, \ldots, |V| - 1\}$: $|\{uv : f(u) \leq i$
              and $f(v) > i\}| \leq k$?

For each fixed k, the family of graphs that are yes instances is closed under the set of operations $\{E, V, T, L, F, S\}$. Since this includes the set of operations that generate the immersion order, we immediately know that the problem can be solved in polynomial time for every fixed k. It is not hard to show that graphs of cutwidth at most k have pathwidth bounded by a function of k. By Bodlaender's theorem and the fact that immersions are expressible in monadic second-order logic (and using Courcelle's theorem), we can conclude that CUTWIDTH is linear time FPT.

In the immersion order there are 15 obstructions to cutwidth $k = 2$, whereas in the "improved" order native to the problem there are 5 (Makedon and Sudborough [506]).

Example 19.1.2 (Interval graphs) Interval graphs are defined in Sect. 4.9. Like many families of graphs that are defined by the existence of representations based on intersections (in the case of interval graphs, these are intersections of line segments; in the case of chordal graphs, these are intersections of subtrees of a tree), the family of interval graphs is closed under the set of operations $\{V, C\}$ and is therefore a lower ideal in the induced minor order. Even though induced minors are not a WQO in general, interval graphs have a finite obstruction set in this order consisting of the five graphs shown in Fig. 19.1.

We can assume that many of the quasi-orders that are generated by natural sets of closure operations for interesting graph properties are not WQOs in general, but there are two ways in which they may still be useful from the point of view of the Obstruction Principle:

1. Particular lower ideals might be characterized by finite obstruction sets, as in the example of interval graphs in the induced minor order.
2. The order of interest may be a WQO on important restricted classes of graphs. For example, the topological order is a WQO for trees and for graphs of maximum degree 3, the induced minor order is a WQO for series–parallel graphs, and graphs of bounded *vertex cover number* are WQO in the induced subgraph order, generated by $\{V\}$. The important question to ask about an interesting order is: *Where* is this a WQO?

Many of the FPT applications of the Graph Minor Theorem and other WQO results can be discussed in terms of general parameterized operators that take as an argument one or more lower ideals and from these describe a new lower ideal. Rather than give a formal definition of an operator on ideals, it is best to illustrate with some examples that cover a number of familiar computational problems.

Example 19.1.3 (Within k vertices of $\mathcal{F}$) Let $\mathcal{F}$ be an ideal in the minor order and k a positive integer. Say that a graph $G = (V, E)$ is *within k vertices of $\mathcal{F}$* if there is a set $V' \subseteq V$ of at most k vertices such that $G - V' \in \mathcal{F}$. It can be easily verified that for every k, the family $\mathcal{W}_k(\mathcal{F})$ of graphs that are within k vertices of $\mathcal{F}$ is a minor order ideal.

By taking the argument ideal $\mathcal{F}$ to be the family of graphs with no edges, we can conclude that is FPT. By taking $\mathcal{F}$ to be the family of acyclic graphs (i.e., forests), we can conclude that FEEDBACK VERTEX SET is FPT. Graphs that are within k vertices of planar are also of interest; for $k = 1$, these are termed *apex* graphs. The obstruction set for apex graphs is currently unknown.

Example 19.1.4 (k-star augmentation) Let $\mathcal{F}$ be an ideal in the minor order and k a positive integer. Say that a graph $G = (V, E)$ is *k-star augmentable to $\mathcal{F}$* if there is a partition of the vertices of G into k classes $V_1, \ldots, V_k$ so that the graph G' obtained from G by adding k new vertices $v_1, \ldots, v_k$ with (for $i = 1, \ldots, k$) v_i connected to each vertex $u \in V_i$ belongs to $\mathcal{F}$. It is not hard to show that for each positive integer k, the family $\mathcal{S}_k(\mathcal{F})$ of graphs that are k-star augmentable to $\mathcal{F}$ is a minor ideal.

By taking $\mathcal{F}$ to be the family of planar graphs, we can conclude that PLANAR FACE COVER below is FPT.

PLANAR FACE COVER
Instance: A graph $G = (V, E)$.
Parameter: A positive integer k.
Question: Can G be embedded in the plane so that there are k faces which cover all vertices?

This problem has been studied by several authors and has some applications in VLSI layout. For $k = 1$, we obtain the family of outerplanar graphs.

The notion of a *covering space* of a topological space is fundamental in some areas of algebra, topology, and complex analysis. Since graphs can be considered as topological 1-complexes, the notion is available in graph theory, and covering spaces

Fig. 19.2 A 2-fold planar
cover of K_5

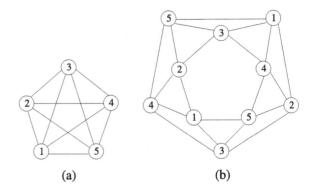

(a)   (b)

of graphs have been considered in the context of parallel and distributed computing
by, for instance, Angluin [40], Abello, Fellows, and Stillwell [1], and Fellows [284].

To illustrate the notion of a covering space, consider the problem of the Flatlanders attempting to build a 5-processor parallel machine with the complete graph
connection topology. By Theorem 17.1.2, this is impossible. However, they can
build the machine shown in Fig. 19.2.

With this, the Flatlanders can emulate a K_5 machine by running two identical
copies of each of the five processes. One can see by examining the figure that each
copy of each of the five processes has a copy of each of the other processes in its
immediate neighborhood. Another way to say this is that the nonplanar graph K_5
has a 2-fold planar cover. A beautiful unsolved problem is the conjecture of Negami
that a graph has a finite planar cover if and only if it embeds in the projective plane.
For many years, this conjecture has been in the following curious state: if one can
show that the single graph on seven vertices $K_{1,2,2,2}$ does *not* have a finite planar
cover, then Negami's conjecture is true! Short of a final answer, it would be nice to
know if it is effectively decidable.

Generalizing from the above discussion, we have the following parameterized
operator on minor ideals.

Example 19.1.5 (*k*-fold $\mathcal{F}$-coverability) Let $\mathcal{F}$ be a minor ideal and k a positive
integer. It is not hard to show that for every k, the family $C_k(\mathcal{F})$ of graphs that have
a k-fold cover in $\mathcal{F}$ is a minor ideal.

WQOs and the Obstruction Principle can be used directly, as in the above examples, to prove FPT and polynomial-time complexity bounds, and they can also
be employed indirectly. We next describe two general methods for indirect applications.

Definition 19.1.1 (RS poset) Let $(S, \leq)$ denote a partially ordered set. We will say
that $(S, \leq)$ is an *RS-poset* if:
(1) $(S, \leq)$ is a WQO.
(2) The order testing problem is FPT.

Theorem 19.1.1 (Method A, Fellows and Langston [297]) *If $(S, \leq)$ is an RS-poset and T is a subset of S such that*
(1) $\mathcal{F} \subseteq T$, *and*
(2) $x \in \mathcal{F}$, $y \leq x$, *and* $y \in T$ *together imply* $y \in \mathcal{F}$,
then there is a polynomial-time decision algorithm to determine for input $x \in T$ whether $x \in \mathcal{F}$.

Proof Let M denote the (finite) set of minimal elements in $T - \mathcal{F}$. Hypothesis (2) implies that if x is in T, then $x \in \mathcal{F}$ if and only if it is not the case that $z \leq x$ for some $z \in M$. Since $(S, \leq)$ is RS and M is finite, this can be determined in polynomial time.   □

Theorem 19.1.2 (Method B, Fellows and Langston [297]) *If $(S, \leq)$ is an RS-poset and if there is a polynomial-time computable map $t : D \to S$ such that for $\mathcal{F} \subseteq D$,*
(1) $t(\mathcal{F})$ *is closed under* $\leq$, *and*
(2) *either* $t^{-1}(x) \cap \mathcal{F} = \emptyset$ *or* $t^{-1}(x) \subseteq \mathcal{F}$ *for each* $x \in t(D)$,
then there is a polynomial-time decision algorithm to determine for input $z \in D$ whether $z \in \mathcal{F}$.

Proof First, compute $t(z)$. There is a finite set M of minimal elements in the complement of the image of $\mathcal{F}$ because $(S, \leq)$ is a WQO. By hypothesis (1), $t(z)$ is in the image of $\mathcal{F}$ if and only if it is not the case that $y \leq t(z)$ for some $y \in M$. This can be determined in polynomial time by performing an order test for each element of M. By hypothesis (2), $z \in \mathcal{F}$ if and only if $t(z)$ belong to the image of $\mathcal{F}$.   □

We give two examples of how these indirect methods can be applied.

Example 19.1.6 (Compact routing schemes) A fundamental problem in parallel processing networks is that messages between processors must somehow be routed through the network. The classical solution is to provide each processor v with a routing table T_v that specifies, for each possible destination x, the neighbor $u \in N[v]$ to which a message having the destination x should be passed along. In very large networks, this classical routing scheme becomes impractical and inefficient. *Interval routing* is one method that has been proposed to deal with this situation (see, for instance, Santoro and Khatib [600], Frederickson and Janardan [330] and Bodlaender, van Leeuwen, Tan, and Thilikos [102] and that has been used in hardware such as the INMOS T9000 Transputer. In a k-interval routing scheme, each neighbor of a vertex is assigned k intervals in a circular ordering of the processors. (The details are not important for this illustration of Method A, and can be found in Bodlaender, van Leeuwen, Tan, and Thilikos [102].)

What graphs admit a k-interval routing scheme that is *optimal*, i.e., as good as any classical scheme? It should be obvious (regardless of the details of interval routing) that this family of graphs cannot be closed in the minor order *in general*, since routing schemes are not defined for disconnected graphs!

It is shown in [102] that Method A can be applied to this problem when T is taken to be the family of connected graphs and $\mathcal{F}$ is the family of graphs that *do* admit an optimal k-interval routing scheme, all this with respect to the minor order. It can be shown that if G is a connected graph that admits an optimal k-interval routing scheme and if H is a *connected* minor of G, then H admits an optimal k-interval routing scheme. By Method A, this means that the problem of recognizing graphs that admit optimal k-interval routing schemes is FPT for parameter k.

Example 19.1.7 (Gate matrix layout) This is a problem that has practical applications in VLSI layout, for a variety of layout styles, and that has appeared in the literature in a number of combinatorially equivalent forms, being known also as "PLAs with multiple folding" and "Weinberger arrays". The GATE MATRIX LAYOUT problem was first defined in the form we consider by Lopez and Law [498]. The problem is defined as follows.

GATE MATRIX LAYOUT[1]
Instance: A boolean matrix M and a positive integer k.
Parameter: k.
Question: Is there a permutation of the columns of M so that if in each row we
               change to $*$ every 0 lying between the row's leftmost and rightmost 1,
               then no column contains more than k 1's and $*$'s?

To apply technique B, we map a matrix M to the graph $f(M)$ for which the vertex set consists of the rows of the matrix, with two vertices joined by an edge if there is some column in which both corresponding rows have entry 1. The map is many-to-1, so that condition hypothesis (b) is clearly necessary. The proof that this mapping works was first given in [297], and can safely be left as an exercise (Exercise 19.5.7).

In addition to providing an important paradigm for the exact solution of problems, the Obstruction Principle also offers the possibility of new kinds of heuristic and approximation algorithms. For example, although Kuratowski's theorem gives us two obstructions that characterize the property of being planar, almost all of the information is provided by just one of these. In fact, it can be shown that any non-planar graph of minimum degree 3 is either *isomorphic* to K_5 or has a $K_{3,3}$ minor. Another example of this phenomena is the GATE MATRIX LAYOUT problem. For a layout cost of $k = 2$ (pathwidth 1), there are two obstructions, for $k = 3$ (pathwidth 2), there are 110 obstructions, and for $k = 4$, there are at least 6×10^7 obstructions in the minor order. For VLSI applications, it seems that the single obstruction K_4 provides almost all of the useful information for $k = 3$. It is conceivable that finite approximate obstruction sets might be a useful algorithmic approach even for some lower ideals of non-WQO's.

---

[1] We remark that the problem is equivalent to GRAPH PATHWIDTH.

19.2   Effectivizations of Obstruction-Based Methods

The techniques of WQOs provided by the deep results of Robertson and Seymour have posed a number of profound philosophical challenges to some of the prevailing habits of algorithm design and complexity analysis. For example, consider the problem defined as follows.

GENUS k-COVER

Instance: A graph G.
Parameter: k.
Question: Is there a finite graph H of genus at most k that is a cover of G? [H is
              a *cover* of G if there is a projection map $p : V(H) \to V(G)$ such that
              (1) p is onto and (2) if $p(u) = x$ and $xy \in E(G)$, then there is a unique
              vertex $v \in V(H)$ with $uv \in E(H)$ and $p(v) = y$.]

This is nonuniformly fixed-parameter tractable by the Graph Minor Theorem, and this is the only way we currently know how to prove that the problem is decidable. It represents well the situation we are in when no other information is available.

We are confronted by two different and basic kinds of non-constructivity:

1. We know that for each k, there is an $O(n^3)$ algorithm for the kth slice of the problem based on the Obstruction Principle—but we have no idea what this algorithm is, since we only know that a finite obstruction set for the kth slice exists, and have no means of identifying it. Let us call this the problem of *algorithm non-constructivity*.

2. Second, even if we knew the obstruction set, we would have a very peculiar algorithm. It would correctly answer the yes/no question, "Does a finite cover H of genus at most k exist?" but would give us no clue about how to efficiently find such an H nor guarantee that this is even possible. Let us call this the problem of *solution non-constructivity*.

In this section, we look at two general techniques for dealing with these constructivity issues. These are sufficient to deal with almost all of the algorithmic applications of WQOs that have been considered. The first and most general technique is based on *self-reduction* algorithms. The second technique is a systematic means of computing obstruction sets in favorable circumstances.

19.2.1 Effectivization by Self-reduction

The natural situation of most computations is most fully modeled in terms of relations. Let $\Pi \subseteq \Sigma^* \times \Sigma^*$ be a binary relation. Typically, we are thinking of something like the following.

Example 19.2.1 (The HAMILTON CIRCUIT Relation)   The relation Π_{HC} consists of all the pairs (G, σ), where G is a graph and σ is an ordering of the vertices of G that constitutes a Hamilton circuit[2] in G.

---

[2]Defined in Sect. 12.7, a *Hamilton circuit* is "a simple cycle through all the vertices".

Definition 19.2.1 (Polynomially balanced) A relation Π is *polynomially balanced* if there is a polynomial q such that $\forall (x, y) \in \Pi$: $|x| \leq q(|y|)$ and $|y| \leq q(|x|)$.

Given such a relation Π, we can describe three basic computational problems:
1. *The Checking Problem* is that of determining, for input (x, y), whether $(x, y) \in \Pi$.
2. *The Domain Decision Problem* is that of determining, for input x, whether there exists a y such that $(x, y) \in \Pi$.
3. *The Search Problem* is the problem of computing a search function for Π, where such a function

$$f : \Sigma^* \to \Sigma^* \cup \{F\}$$

satisfies:
(a) $f(x) = y$ implies $(x, y) \in \Pi$, and
(b) $f(x) = F$ implies that there is no y with $(x, y) \in \Pi$.

A standard characterizations of NP is that it consists of those languages $L \subseteq \Sigma^*$ that are the domain of a polynomially balanced relation for which the checking problem can be solved in polynomial time.

In many ways, it is more natural to discuss the objectives of computation in terms of search problems. The original article by Levin [488] that introduced the notion of NP-hardness independently of Cook was formulated in terms of relations and search problems. The relational point of view is capable of supporting interesting systematic development, especially since most known reductions between computational problems show how to translate between solutions as much as between decision instances. (See, for instance, Abrahamson, Fellows, Langston, and Moret [7].)

Given that our natural concern in computer applications is almost always with search problems, it should be somewhat surprising that the theory of computational complexity has been developed almost entirely in terms of decision problems. The justification that is usually given is that there is no substantial loss of generality in the restriction of the basic theory to decision problems, since there is almost always an efficient reduction of a search problem to its naturally associated decision problem. WQOs may be a source of new reasons to question this assumption.

In any case, the reduction of search to decision is generally termed *self-reducibility*, and this is the sense in which we use the word here.[3]

The following is an example of a "typical" polynomial-time self-reduction of a Π-search problem to the domain decision problem for Π.

Example 19.2.2 (A self-reduction for HAMILTON CIRCUIT) Using an algorithm for the domain decision from for Π_{HC}, we can compute a search function for Π_{HC} as follows. First, use the domain decision "oracle" to determine if the decision output should be "no" in the case that G does not have a Hamilton circuit. If the oracle

---

[3]We remark that there are various formalizations of the basic idea that have been made purely within the framework of decision problem complexity theory that are not equivalent, in fact as different as apples and oranges, to a relational formulation of the issue (Schnorr [604]).

answers "yes"—G is Hamiltonian—then remove any edge uv from G to obtain G' and use the domain decision oracle to determine if G' is Hamiltonian. There are two possible cases:

1. G' is Hamiltonian. In this case, uv is "unnecessary" and we can consider it permanently removed.
2. G' is not Hamiltonian. This implies that every Hamilton circuit in G traverses uv and we can mentally mark uv as "never to be removed".

We use the domain decision oracle in this way to progressively simplify G, each round of consultation with the oracle resulting either in the removal of an edge or the identification of an essential edge. After $|E| - |V|$ consultations, the only edges remaining are those of a Hamilton circuit.

Finding an efficient self-reduction algorithm is the obvious way to deal with the problem of solution non-constructivity that arises in applications of WQOs. Problem parameters may interact non-trivially with self-reduction. A self-reduction algorithm for CUTWIDTH that uses $O(n)$ calls to a decision oracle for $2k$-CUTWIDTH in order to solve the search problem for k-CUTWIDTH is described in Fellows and Langston [302]. The important point is that to efficiently solve the search problem for a given parameter value, we might be able to make effective use of a decision algorithm for a *different* value of the parameter.

Surprisingly, self-reduction *also* gives us a general way to deal with the problem of algorithm non-constructivity. The technique that we next describe will allow us to write down, for almost all known WQO applications, a correct algorithm to compute a search function in polynomial time—but without ever identifying the complete obstruction set, and without knowing the polynomial that bounds the running time of the algorithm!

Definition 19.2.2 (Uniformly enumerable) A quasi-order $(R, \leq)$ is *uniformly enumerable* if there is a recursive enumeration $(r_0, r_1, r_2, \ldots)$ of R with the property that $r_i \leq r_j$ implies $i \leq j$.

In the minor and immersion orders, for example, a natural uniform enumeration is to generate all finite graphs based on a monotonically nondecreasing sequence of number of vertices, with graphs having the same number of vertices generated based on a monotonically nondecreasing sequence of number of edges, with ties (graphs with the same number of vertices and the same number of edges) broken arbitrarily.

Definition 19.2.3 (Uniform self-reduction) Under a uniform enumeration of the elements of domain(Π), we say that a self-reduction algorithm for Π is *uniform* if, on input r_j, the oracle for domain(Π) is consulted concerning r_i only for $i \leq j$.

Definition 19.2.4 (Honest) An oracle algorithm is *honest* if, on inputs of size n, its oracle is consulted only concerning instances of size $O(n)$.

Definition 19.2.5 (Robust) An oracle algorithm A with overhead bounded by $T(n)$ is *robust* [with respect to $T(n)$] if A is guaranteed to halt within $T(n)$ steps for any oracle language.

Theorem 19.2.1 (Fellows and Langston [301]) *Let $F = \text{domain}(\Pi)$ be an ideal in a uniformly enumerable WQO, and suppose the following are known:*
(1) an algorithm that solves the checking problem for Π in $O(T_1(n))$ time,
(2) order tests that require $O(T_2(n))$ time,
(3) a uniform self-reduction algorithm (its time bound is immaterial),
(4) an honest robust self-reduction algorithm that requires $O(T_3(n))$ overhead.
Then an algorithm requiring $O(\max\{T_1(n), T_2(n) \cdot T_3(n)\})$ time is known that solves the search problem for Π.

Proof Let I denote an arbitrary input instance and let K denote the known elements of the (finite) obstruction set for F. (Initially, K can be empty.) We treat K as if it were the correct obstruction set, that is, as if we *did* know a correct decision algorithm. Since the elements of K will always be obstructions, if we find that $H \leq I$ for some $H \in K$, then our algorithm correctly reports "no" and halts. Otherwise, after checking each element of K to confirm that it is not less than or equal to I, we attempt to self-reduce (under the assumption that K is the complete and correct obstruction set—or at least "good enough" for the input under consideration) and check the solution obtained, if one is produced. If the solution is correct, then our algorithm correctly reports "yes", outputs the solution, and halts.

In general, however, it may turn out that our check of the solution reveals that we have self-reduced to a nonsolution. It might also turn out that our attempt to produce a solution by self-reduction, using our currently knowledge of K, goes into an infinite loop or otherwise "takes too long", that is, violates our robustness bound. This can only mean that there is at least one obstruction $H \notin K$. In this event, we proceed by generating the elements of R in uniform order until we find a new obstruction (an element that properly contains no other obstruction but that cannot be uniformly self-reduced to a solution). When such an obstruction is encountered, we need only augment K with it and start over on I.

Let C_1 denote the cardinality of the correct obstruction set and let C_2 denote the largest number of vertices in any of its elements. Then, for some suitably chosen function f, the total time spent by this algorithm is bounded by $O(C_1(T_1(n) + T_2(n) \cdot T_3(n)) + f(C_2))$. □

In the statement of Theorem 19.2.1, it is possible to replace the hypothesis that a uniform self-reduction is known with the alternate hypothesis that a domain decision algorithm is known.

Although, at first blush, any attempt to implement the algorithm used in the proof of Theorem 19.2.1 appears to be completely out of the question, closer inspection reveals that it may in fact provide the basis for novel approaches in algorithm design. Note that this scheme can be viewed as a "learning algorithm" that gradually accumulates a useful subset of the obstruction set, and invokes its exhaustive "learning component" only when forced to do so by input it cannot handle with this subset. Furthermore, some obstructions are already known or are easy to identify for most problems, so that we need not start with the empty set. We give an example of an application.

Example 19.2.3 (PLANAR DIAMETER IMPROVEMENT) Given a planar graph G and a positive integer parameter k, the question is whether there is a planar super-graph G' of G of diameter at most k (Exercise 19.5.4). It is not hard to see that for each k there is a bound on the treewidth of the yes instances, so this problem is linear time FPT, because for each fixed k the yes instances are an ideal in the minor order. For this problem (and others like it), there is no obvious way to express the property in MS$_2$ logic, since the natural definition quantifies over sets of edges that are *not* in the graph. In the absence of any further information, we know only that a linear time FPT algorithm *exists* for this problem and have no idea what it is. The problem is easy to self-reduce, however, just by progressively adding edges to the graph. By Theorem 19.2.1, we can therefore write down an FPT algorithm that runs in time $f(k)n^3$ for some unknown function $f(k)$.

19.2.2 Effectivization by Obstruction Set Computation

The second general technique of constructively producing the FPT and polynomial-time algorithms that are proven to exist by applications of WQO theorems is based on computing the relevant obstruction sets.

We have seen that knowing only a decision algorithm for a minor order lower ideal $\mathcal{F}$ is not enough to allow us to compute the obstruction set for $\mathcal{F}$. The next theorem is in some sense complementary in that it establishes not-too-difficult conditions under which we *can* compute the obstruction set for $\mathcal{F}$.

Theorem 19.2.2 (Fellows and Langston [300]) *Suppose that $\mathcal{F}$ is an ideal in the minor order of finite graphs and suppose that we have the following three pieces of information about $\mathcal{F}$:*
1. *An algorithm to decide membership in $\mathcal{F}$ (of any time complexity).*
2. *A bound B on the maximum treewidth of the obstructions for $\mathcal{F}$.*
3. *For $t = 1, \ldots, B + 1$, a decision algorithm for a finite index right congruence $\sim$ on t-boundaried graphs that refines the canonical congruence for $\mathcal{F}$ on the small universe $\mathcal{U}_t^{\mathrm{small}}$.*
   Then, we can effectively compute the obstruction set $\mathcal{O}$ for $\mathcal{F}$.

Proof The algorithm is outlined as follows. For $t = 1, \ldots, B + 1$, we generate in a systematic way, using the standard set of parsing operators (see Theorem 5.72), the t-boundaried graphs of $\mathcal{U}_t^{\mathrm{small}}$ until a certain *stop signal* is detected. At this point, for a given t, we will have generated a finite set of graphs $\mathcal{G}_t$. Of particular interest among these are the graphs $\mathcal{M}_t \subseteq \mathcal{G}_t$ that are minimal with respect to a certain partial order $\leq$ on t-boundaried graphs. We will prove that $\mathcal{O}$ is a subset of

$$\mathcal{M} = \bigcup_{t=1}^{B+1} \mathcal{M}_t$$

considering the graphs of $\mathcal{M}$ with the boundaries forgotten.

There are three items to be clarified:

(1) how the graphs of the small universe are generated,
(2) the *search ordering* $\leq$,
(3) the nature of the stop signal for width t.

(1) *The order of generation of* $\mathcal{U}_t^{\text{small}}$.

Suppose X is a t-boundaried graph, $X \in \mathcal{U}_t^{\text{small}}$. By the *size* of X, we will refer to the number of nodes in a smallest possible parse tree for X over the standard set of parsing operators given by the Parsing Theorem (Theorem 12.7.1). For a given t, we generate the t-boundaried graphs of $\mathcal{U}_t^{\text{small}}$ in order of increasing size. By the jth generation, we refer to all of those graphs of size j in this process.

(2) *The search ordering* $\leq$.

To define $\leq$, we first extend the minor ordering of ordinary graphs to t-boundaried graphs in the natural way by holding the boundary fixed. In other words, the boundaried minor order is defined by the same local operations as the minor order, except that we are not allowed to delete boundary vertices or to contract edges between two boundary vertices. This can be shown to be a WQO on $\mathcal{U}_t^{\text{large}}$ by using the Graph Minor Theorem for edge-colored graphs (see Exercise 19.5.6). Let $\leq_m$ denote the minor order on ordinary graphs and let $\leq_{\partial m}$ denote the boundaried minor order.

For $X, Y \in \mathcal{U}_t^{\text{small}}$, define $X \leq Y$ if and only if $X \leq_{\partial m} Y$ and $X \sim Y$. This is a WQO since there are only finitely many equivalence classes of $\sim$ on $\mathcal{U}_t^{\text{small}}$.

(3) *The stop signal*.

The graphs of $\mathcal{U}_t^{\text{small}}$ are generated by size, one generation at a time (where the jth generation consists of all those of size j). We say that there is *nothing new at time j* if none of the t-boundaried graphs of the jth generation are minimal with respect to the search order $\leq$.

A *stop signal is detected at time $2j$* if there is nothing new at time i for $i = j, \ldots, 2j - 1$.

We have now completely described the algorithm. For $t = 1, \ldots, B + 1$ we generate the t-boundaried graphs in the manner described until a stop signal is detected. We form the set $\mathcal{M}$ and output the list of elements of $\mathcal{M}$ (with boundaries forgotten) that are obstructions for $\mathcal{F}$. Note that having a decision algorithm for $\mathcal{F}$ is sufficient to determine if any particular graph H is an obstruction, just by checking that $H \notin \mathcal{F}$ while each minor of H is in $\mathcal{F}$. This same procedure and the decision algorithm for $\sim$ allow us to compute whether it is time to stop.

The correctness of the algorithm is established by the following claims.

Claim 1 *For each value of t a stop signal is eventually detected.*

This follows immediately from the fact that $\leq$ is a WQO on $\mathcal{U}_t^{\text{small}}$ and, therefore, there are only a finite number of minimal elements.

Claim 2 *Suppose that for a given t a stop signal is detected at time $2j$. Then, no obstruction for $\mathcal{F}$ that can be parsed with the t-boundaried set of operators has size greater than $2j$.*

If T is rooted tree, then by a *rooted subtree T'* of T we mean a subtree that is generated by some vertex r of T (the root of T'), together with all of the vertices descended from r in T. For t-boundaried graphs X and Y, we say that X is a *prefix* of Y if, in a parse tree T for Y, X is parsed by a rooted subtree T' of T. To denote that X is a prefix of Y, we write $X \prec Y$.

Now suppose that T is a parse tree of minimum size for a counterexample H to Claim 2. Since all of the operators in the standard set are either binary or unary, there must be a prefix H' of H of size at least j. Since there is nothing new during the times when H' would have been generated, Claim 2 follows from Claim 3.

Claim 3 *A prefix of a graph that is minimal with respect to $\leq$ must also be minimal.*

If X is a prefix of Y and X is not minimal, then $X \geq_{\partial m} X'$ with $X \neq X'$ and $X \sim X'$. Since $\sim$ is a right congruence, $Y \sim Y'$, where Y' is obtained from Y by substituting a parse tree for X' for the subtree that parses X in a parse tree for Y. Using the Parsing Replacement Property (see Observation 12.7.1) and since X' is a proper boundaried minor of X, Y' is a proper boundaried minor of Y. This implies that Y is not minimal with respect to $\leq$.

Claim 4 *If $X \in \mathcal{O}$, then for some $t \leq B + 1$, $X \in \mathcal{M}_t$.*

Since the treewidth of X is at most B, X can be parsed by the standard set of t-boundaried operators for some $t \leq B + 1$ by the Parsing Theorem, Theorem 12.7.1. It remains to argue that X (with the boundary supplied by this parse) is $\leq$ minimal. But this is obvious, since any proper minor is in $\mathcal{F}$ and since $\sim$ refines the canonical $\mathcal{F}$-congruence. $\square$

The above theorem really tightens the connections between finite index congruences, WQOs, finite obstruction sets, and finite-state recognizability that we have seen emerging from a variety of different angles. We have the following notable corollary.

Corollary 19.2.1 *If $\mathcal{F}$ is a minor ideal of bounded treewidth, then the following sets of information about $\mathcal{F}$ are effectively interchangeable:*
(1) *A decision algorithm for $\mathcal{F}$ and (for all t) a finite index right congruence on $\mathcal{U}_t^{\text{small}}$ that refines the canonical congruence for $\mathcal{F}$.*
(2) *For all t, a finite automaton that recognizes the graphs of $\mathcal{F}$ from parse trees.*
(3) *The obstruction set for $\mathcal{F}$.*

Proof The information (2) immediately gives us (1). Starting from (1), (2) can be computed by Theorem 12.7.2 and the subsequent discussion. From the obstruction set (3), we can compute the information (2) by the methods of Chap. 12. Theorem 19.2.2 completes the picture.   □

Jens Lagergren [475] has shown that the use of the Graph Minor Theorem in proving that the algorithm of Theorem 19.2.2 terminates can be replaced by an explicit calculation of a "stopping time" computable from the index of the congruence $\sim$. More complex methods using second-order stopping signals can be found in Cattell, Dinneen, Downey, Fellows, and Langston [132].

Perhaps surprisingly, Theorem 19.2.2 can be implemented and a number of previously unknown obstruction sets have been mechanically computed (Cattell and Dinneen [131] Cattell, Dinneen, and Fellows [133]). The "Holy Grail" of such efforts would be a computation of the obstruction set for a torus embedding, which probably contains about 2000 graphs.

Theorem 19.2.2 can be adapted to other partial orders, including those such as the topological order, that are not WQO's. It can be shown in this case that the (adapted) algorithm will correctly terminate if and only if the ideal $\mathcal{F}$ has a finite obstruction set—thus providing a potentially interesting way to mechanically prove the *existence* of a finite basis for particular ideals.

We conjecture that the obstruction set for an arbitrary ideal $\mathcal{F}$ can be effectively computed from just the following information:

1. A decision algorithm for $\mathcal{F}$ membership.
2. For all t, a decision algorithm for a finite index right congruence on $\mathcal{U}_t^{\text{large}}$ that refines the canonical $\mathcal{F}$-congruence on the large universe.

19.3   The Irrelevant Vertex Technique

One of the most important contributions of the Graph Minors project to parameterized algorithm design was the proof, in [588], that the following problem is in $f(k) \cdot n^3$-FPT (a faster, $f(k) \cdot n^2$ step algorithm appeared recently in [433]).

k-DISJOINT PATHS PROBLEM

Instance: A graph G with k pairs of terminals $T = \{(s_1, t_1), \ldots, (s_k, t_k)\} \in V(G)^{2(k)}$.

Parameter: A positive integer k.

Question: Are there k pairwise vertex disjoint paths $P_1, \ldots, P_k$ in G such that P_i has endpoints s_i and t_i?

Whilst the algorithm *in detail* is too complex for us, we will briefly comment on the technique since it gives rise to an area of quite active research. Following this description, we give some pointers as to further reading.

The algorithm for the above problem is based on an idea known as the *irrelevant vertex technique* and revealed strong links between structural graph theory and parameterized algorithms. In general the idea is described as follows.

Let (G, T) be an input of the k-DISJOINT PATHS. We say that a vertex $v \in V(G)$ is *solution-irrelevant* when (G, T) is a YES-instance if only if $(G \setminus v, T)$ is a YES-instance. If the input graph G violates some structural condition, then it is possible to find a vertex v that is solution-irrelevant. One then repeatedly removes such vertices until the structural condition is met which means that the graph has been simplified and a solution is easier to be found.

The structural conditions used in the algorithm from Robertson and Seymour [588] are two:

(i) G excludes a clique, whose size depends on k, as a minor and

(ii) G has treewidth bounded by some function of k.

In [588] Robertson and Seymour proved, for some specific function $h : \mathbb{N} \to \mathbb{N}$, that if the input graph G contains some $K_{h(k)}$ as a minor, then G contains some solution-irrelevant vertex that can be found in $h(k) \cdot n^2$ steps. This permits us to assume that every input of the problem is $K_{h(k)}$-minor free, thus enforcing structural condition (i). Of course, such graphs may still be complicated enough and do not necessarily meet condition (ii). Using a series of (highly non-trivial) structural results [587, 590], Robertson and Seymour proved that there is some function $g : \mathbb{N} \to \mathbb{N}$ such that every $K_{h(k)}$-minor free graph with treewidth at least $g(k)$ contains a solution-irrelevant vertex that can be found in $g(k) \cdot n^2$ steps. This enforces structural property (ii) and then k-DISJOINT PATHS can be solved in $f(k) \cdot n$ steps, using Courcelle's theorem (Theorem 13.1.1).

In fact, the above idea was used in [588] to solve a more general problem that contains both k-disjoint Paths and H-MINOR CHECKING. The only "drawback" of this algorithm is that its parametric dependence, i.e., the function f is immense. Towards lowering the contribution of f, better combinatorial bounds where provided by Kawarabayashi and Wollan in [436]. Also, for planar graphs a better upper bound was given by Adler, Kolliopoulos, Krause, Lokshtanov, Saurabh, and Thilikos [19]. The irrelevant vertex technique has been used extensively in parameterized algorithm design. For a sample of results that use this technique, see Dawar, Grohe, and Kreutzer [199, 200], Golovach, Kamiński, Paulusma, and Thilikos [348], Kawarabayashi and Kobayashi [434, 454], Kawarabayashi and Reed [435], and Grohe, Kawarabayashi, Marx, and Wollan [359], for instance. These applications are beyond the scope of this book.

19.4 Protrusions

An algorithmic method for efficient kernelization for parameterized problems in FPT that is similar in spirit to the irrelevant vertex technique, was introduced by Bodlaender, Fomin, Lokshtanov, Penninkx, Saurabh, and Thilikos [88]. This is the technique of *protrusion-based reduction*. The method works by iteratively finding structures of small treewidth with small boundaries, and replacing them. The approach is particularly applicable to various classes of sparse graphs.

An important result in the area of kernelization is by Alber et al. [25]. They obtained a linear-sized kernel for the DOMINATING SET problem on planar graphs, where the parameter is the size of the dominating set. This work triggered an explosion of papers on kernelization, and in particular, on kernelization for FPT problems on planar and other classes of sparse graphs. Combining the ideas of Alber et al. with *problem specific* data reduction rules, linear kernelization algorithms were obtained for a wide variety of parameterized problems on planar graphs (parameterized by the solution size), including CONNECTED VERTEX COVER, INDUCED MATCHING and FEEDBACK VERTEX SET. In 2009 Bodlaender et al. [88] obtained powerful meta-kernelization algorithms that eliminated the need for the design of problem specific reduction rules by providing an automated process that generates them. They show that all problems that have a "distance property" and are expressible in a certain kind of logic or "behave like a regular language" admit a polynomial kernel on graphs of bounded genus. In what follows we give a short description of these meta-theorems.

19.4.1 Informal Description

The notion of "protrusions and finite integer index" is central to recent meta-kernelization theorems. In the context of problems on graphs, there are three central ideas that form the basis of all protrusion-based reduction rules, and meta-algorithms for kernelization:

- The description of an equivalence relation that classifies all instances of a given parameterized problem in an useful manner.
- General methods for determining, for a given problem, whether the corresponding equivalence relation has *finite integer index* (a key property, defined below).
- General criteria for the existence of suitable large subgraphs that can be surgically replaced with smaller subgraphs so that the result is algorithmically equivalent to the original—and efficient algorithms to find both these large subgraphs and the replacement smaller subgraphs.

One of the critical aspects of this development is in coming up with the right definition for describing the circumstances in which a subgraph may be replaced. This is captured by the notion of a protrusion.

An *r-protrusion* in a graph G is a subgraph $H = (V_H, E_H)$ such that the number of vertices in H that have neighbors in $G \setminus H$ is at most r and the treewidth of H is at most r. The *size* of the protrusion is the number of vertices in it, that is, $|V_H|$. The vertices in H that have neighbors in $G \setminus H$ comprise the *boundary* of H. Informally, H may be thought of as a part of the graph that is separated from the "rest of the graph" by a small-sized separator, and everything about H may be understood in terms of the graph induced by H itself and the limited interaction it has with $G \setminus H$ via its boundary vertices. If the size of the protrusion is large, we may want to replace it with another graph X that is much smaller but whose behavior with respect to $G \setminus H$ is identical to H in the context of the problem that we are studying. Specifically, we would like that the solution to the problem in question does not change after we have made the replacement (or changes in a controlled manner that can be tracked as we make these replacements).

This motivates us to define an equivalence relation that captures the essence of what we hope to do in terms the replacement. We would like to declare H equivalent to X if the size of the solution of G and $(G \setminus H) \cup^* X$ changes in an algorithmically controlled manner, where $\cup^*$ is some notion of a replacement operation that we have not defined precisely yet. Notice, however, that a natural notion of replacement would leave the boundary vertices intact and perform a cut-and-paste on the rest of H. This is precisely what the protrusion-based reduction rules do. Combined with some combinatorial properties of graphs this results in polynomial, and in most cases, linear kernels for a wide variety of problems.

19.4.2 Overview of the Protrusion Based Meta-kernelization Results

Given a graph $G = (V, E)$, we define $\mathbf{B}_G^r(S)$ to be the set of all vertices of G whose distance from some vertex in S is at most r. Let $\mathcal{G}$ be the family of planar graphs and let integer $k > 0$ be a parameter. We say that a parameterized problem $\Pi \subseteq \mathcal{G} \times \mathbb{N}$ is *compact* if there exist an integer r such that for all $(G = (V, E), k) \in \Pi$, there is a set $S \subseteq V$ such that $|S| \leq r \cdot k$, $\mathbf{B}_G^r(S) = V$ and $k \leq |V|^r$. Similarly, Π is *quasi-compact* if there exists an integer r such that for every $(G, k) \in \Pi$, there is a set $S \subseteq V$ such that $|S| \leq r \cdot k$, $\mathrm{tw}(G \setminus \mathbf{B}_G^r(S)) \leq r$ and $k \leq |V|^r$ where $\mathrm{tw}(G)$ denotes the treewidth of G. Notice that if a problem is compact then it is also quasi-compact. For ease of presentation the definitions of *compact* and *quasi-compact* are more restrictive here than in the paper [88].

The following theorem from [88] yields linear kernels for a variety of problems on planar graphs. To this end they utilize the notion of *finite integer index*. This notion originated in the work of Bodlaender and de Fluiter [81] and is similar to the concept of *finite state*. The reader should recall the definition of t-boundaried graphs and the gluing operation $\oplus$ from Sect. 12.7.

For a parameterized problem Π, defined on graphs in $\mathcal{G}$, and two t-boundaried graphs G_1 and G_2, we say that $G_1 \equiv_\Pi G_2$ if there exists a constant c such that for all t-boundaried graphs G_3 and for all k we have:

- $G_1 \oplus G_3 \in \mathcal{G}$ if and only if $G_2 \oplus G_3 \in \mathcal{G}$, and
- $(G_1 \oplus G_3, k) \in \Pi$ if and only if $(G_2 \oplus G_3, k + c) \in \Pi$.

Note that for every t, the relation $\equiv_\Pi$ on t-boundaried graphs is, by the definition, trivially an equivalence relation. A problem Π has *finite integer index* (FII), if and only if for every t, $\equiv_\Pi$ is of finite index, that is, has a finite number of equivalence classes. Compact problems that have FII include DOMINATING SET and CONNECTED VERTEX COVER while FEEDBACK VERTEX SET has FII and is quasi-compact, but not compact. We are now in position to state the theorem.

Theorem 19.4.1 *Let $\Pi \subseteq \mathcal{G} \times \mathbb{N}$ be a quasi-compact parameterized problem, and suppose that the problem has the FII property. Then Π admits a linear kernelization algorithm.*

19.4.3 Overview of the Methods

We give an outline of the main ideas used to prove Theorem 19.4.1. For a problem Π and an instance $(G = (V, E), k)$ the kernelization algorithm repeatedly identifies a part of the graph to reduce and replaces this part by a smaller equivalent part. Since each such step decreases the number of vertices in the graph the process stops after at most $|V|$ iterations. In particular, the algorithm identifies a *constant size separator* S that cuts off a *large* chunk of the graph of *constant treewidth*. This chunk is then considered as a $|S|$-boundaried graph $G' = (V', E')$ with boundary S. Let G^* be the other side of the separator, that is, $G' \oplus G^* = G$. Since Π has FII there exists a finite set $\mathcal{S}$ of $|S|$-boundaried graphs such that $\mathcal{S} \subseteq \mathcal{G}$ and for any $|S|$-boundaried graph G_1 there exists a $G_2 \in \mathcal{S}$ such that $G_2 \equiv_\Pi G_1$. The definition of "large chunk" is that G' should be larger than the largest graph in $\mathcal{S}$. Hence we can find a $|S|$-boundaried graph $G_2 \in \mathcal{S}$ and a constant c such that $(G, k) = (G' \oplus G^*, k) \in \Pi$ if and only if $(G_2 \oplus G^*, k - c) \in \Pi$. The reduction is just to change (G, k) into $(G_2 \oplus G^*, k - c)$. Given G' we can identify G_2 in time linear in $|V'|$ by using the fact that G' has constant treewidth and that all graphs in $\mathcal{S}$ have constant size.

We now proceed to analyze the size of any reduced yes-instance of Π. We show that if Π is compact (not quasi-compact), then the size of a reduced yes-instance (G, k) must be at most $O(k)$. Since $(G = (V, E), k) \in \Pi$ and Π is compact, there is an $O(k)$ sized set $S' \subseteq V$ such that $\mathbf{B}^r_G(S') = V$ for some constant r depending only on Π. One can show that if such a set S' exists there must exist another $O(k)$ sized set S such that the connected components of $G[V \setminus S]$ can be grouped into $O(k)$ chunks as described in the paragraph above. If any of these chunks have more vertices than the largest graph in $\mathcal{S}$ we could have performed the reduction. This implies that any reduced yes-instance has size at most ck for some fixed constant c. Hence if a reduced instance is larger than ck the kernelization algorithm returns.

To prove Theorem 19.4.1 even when Π is quasi-compact, they show that the set of reduced instances of a quasi-compact problem is in fact compact! Observe that it is the set of reduced instances that becomes compact and *not* Π itself. The main idea is that if $G = (V, E)$ has a set $S \subseteq V$ such that the treewidth of $G[V \setminus \mathbf{B}^r_G(S)]$ is constant and there exists a vertex v which is far away from S, then we can find a large subgraph to reduce. The argument utilizes deep results from Robertson and Seymour's Graph Minors project.

The parameterized versions of many basic optimization problems, where the parameter is the solution size, have finite integer index, including DOMINATING SET, (CONNECTED) r-DOMINATING SET, q-THRESHOLD DOMINATING SET, EFFICIENT DOMINATING SET, VERTEX COVER, CONNECTED r-DOMINATING SET, CONNECTED VERTEX COVER, MINIMUM-VERTEX FEEDBACK EDGE SET, MINIMUM MAXIMAL MATCHING, CONNECTED DOMINATING SET, FEEDBACK VERTEX SET, EDGE DOMINATING SET, CLIQUE TRANSVERSAL, INDEPENDENT SET, r-SCATTERED SET, MIN LEAF SPANNING TREE, INDUCED MATCHING, TRIANGLE PACKING, CYCLE PACKING, MAXIMUM FULL-DEGREE SPANNING TREE, and many others [88, 202].

There are problems like INDEPENDENT DOMINATING SET, LONGEST PATH, LONGEST CYCLE, MAXIMUM CUT, MINIMUM COVERING BY CLIQUES, INDE-PENDENT DOMINATING SET, and MINIMUM LEAF OUT-BRANCHING and various edge packing problems which are known not to have FII [202]. It was shown in [88] that compact problems expressible in an extension of Monadic Second Order Logic, namely Counting Monadic Second Order Logic, have polynomial kernels on planar graphs. This implies polynomial kernels for INDEPENDENT DOMINAT-ING SET, MINIMUM LEAF OUT-BRANCHING, and some edge packing problems on planar graphs. The results from [88] hold not only for planar graphs but for graphs of bounded genus. It was shown in [324] that if instead of quasi-compactness, we request another combinatorial property, bidimensionality with certain separability properties, then an analogue of Theorem 19.4.1 can be obtained for much more general graph classes, such as the families of graphs that exclude a fixed apex graph as a minor.

19.5 Exercises

Exercise 19.5.1 Verify that for each fixed k, the family of graphs having cutwidth at most k is closed under the operation of fracturing a vertex.

Exercise 19.5.2 Prove that the minor ordering and the topological ordering coincide on the family of graphs of maximum degree 3.

Exercise 19.5.3 The complete graph K_5 has an interesting "Ramsey theoretic" property: No matter how the edges are labeled with positive integers, there is a cycle whose label sum is 0 mod 3. Generalizing this, consider the following problem:

UNIT CYCLE AVOIDANCE
Instance: A graph G and an Abelian group A.
Parameter: A.
Question: Can the edges of G be labeled by elements of A so that no cycle in G has label sum equal to the identity element of A?

Prove that UNIT CYCLE AVOIDANCE is nonuniformly FPT if A is restricted to be a group of the form $(Z_2)^m$, or if G is restricted to graphs of maximum degree 3. Is this problem in NP? (See Arkin [42] and Barefoot, Clark, Douthett, Entringer [56].)

Exercise 19.5.4 (M.R. Fellows, J. Kratochvíl, M. Middendorf and F. Pfeiffer [296]) Show that the following problem is linear FPT.

INDUCED MINOR TESTING FOR PLANAR GRAPHS
Instance: Planar graphs G and H.
Parameter: H.
Question: Is $G \geq H$ in the induced minor order?

(Hint: First, use Theorem 13.3.2 to show that for each fixed planar graph H there is a constant C_H such that if the answer is "yes", then the treewidth of G is at most C_H.)

Exercise 19.5.5 (Dejter and Fellows (unpublished))   Show that the following problems are linear FPT.

PLANAR DIAMETER IMPROVEMENT
Instance: A planar graph G and a positive integer k.
Parameter: k.
Question: Is there a planar graph G' with $G \subseteq G'$, such that the diameter of G' is at most k? (The *diameter* of a graph is the maximum distance between a pair of vertices in the graph.)

PLANAR DOMINATION IMPROVEMENT NUMBER
Instance: A planar graph G and a positive integer k.
Parameter: k.
Question: Is there a planar graph G' with $G \subseteq G'$, such that G' has a k-element dominating set?

Generalizing from these two examples, define an operator that takes as arguments a minor order ideal $\mathcal{I}$ and an ideal $\mathcal{J}$ in the order generated by the set of local operations $\{C, S\}$ and produces a minor ideal.

Exercise 19.5.6 Robertson and Seymour have proven even stronger results concerning graph minors than we have discussed. They proved that finitely edge-colored, directed graphs are an RS-poset in the minor order extended to these "annotated" graphs. (The local operations V, E, and C are obviously still well-defined.) Use this result and Method A to prove that k-boundaried graphs are an RS-poset under k-boundaried minors, where the minor order is extended to boundaried graphs by requiring that the boundary is held fixed while applying the operations that define the minor order.

Exercise 19.5.7 Verify that Method B applies as indicated for the GATE MATRIX LAYOUT problem. Prove that the image under the map f of the "yes" instances for k-GML consists of precisely the graphs of pathwidth at most $k - 1$.

Exercise 19.5.8 Find an FPT self-reduction algorithm for the MINIMUM DEGREE GRAPH PARTITION problem (see Exercise 18.7.8).

Exercise 19.5.9 Find a self-reduction for the HAMILTON CIRCUIT problem that makes $O(n \log n)$ calls to the decision problem oracle.

Exercise 19.5.10 Suppose that we have a linear FPT "black box" decision algorithm for the parameterized problem:

LONGEST CYCLE
Instance: A graph G and a positive integer k.
Parameter: k.
Question: Does G have a cycle of length at least k?

By means of an efficient self-reduction algorithm, show that we can solve the corresponding search problem (i.e., find a cycle of length $\geq k$ when one exists) in time $f(k)n^2 \log n$.

Exercise 19.5.11 It is possible to formalize search problems for bounded pathwidth computations as follows (this can be generalized to bounded treewidth). A *balanced string relation* is a set R of pairs of strings of equal length: $R \subseteq \Sigma^* \times \Gamma^*$ where $(x, y) \in R$ implies $|x| = |y|$. Because the paired strings of R have equal length, we can equivalently view (x, y) as a string over the alphabet Δ, where $\Delta = \Sigma \times \Gamma$. The search problem for R is: Given $x \in \Sigma^*$, compute $y \in \Gamma^*$ such that $(x, y) \in R$ (if such a y exists). In Bodlaender, Evans, and Fellows [86] it is shown that this search problem can be solved by a two-way finite-state computation (R is *finite-state searchable*) if R, when viewed as a language over the alphabet of symbol-pairs Δ, is finite-state recognizable. Fix t and let Σ be the standard set of parsing operators for pathwidth t. Let Γ be a set of color symbols that can be used to indicate a 3-coloring of a graph by pairing a color symbol to a parse symbol that introduces a new vertex. Prove that this R is finite-state recognizable and therefore finite-state searchable.

19.6 Historical Remarks

Most of the early results we present are due to Fellows and Langston [299, 300]. It is worth remarking that these papers and earlier grants based on the ideas were the grandparents of parameterized complexity, as discussed by Langston in [479]. The irrelevant vertex technique was developed by Robertson and Seymour [588]. It has found extensive applications in structural graph theory based on minor exclusion, and WQO theory. Bodlaender, Fomin, Lokshtanov, Penninkx, Saurabh, and Thilikos [88] were the first to use protrusion techniques (or rather graph reduction techniques) to obtain kernels, but the idea of using graph replacement for algorithms has been around for long time. The idea of graph replacement for algorithms dates back to Fellows and Langston [300]. Arnborg, Courcelle, Proskurowski, and Seese [45] essentially showed that effective protrusion reduction procedures exist for many problems on graphs of bounded treewidth. Using this, Arnborg et al. [45] obtained a linear time algorithm for MSO expressible problems on graphs of bounded treewidth. Bodlaender and Fluiter [80, 81, 202] generalized these ideas in several ways—in particular, they applied it to some optimization problems. It is also important to mention the work of Bodlaender and Hagerup [89], who used the concept of graph reduction to obtain parallel algorithms for MSO expressible problems on bounded treewidth graphs.

This protrusion material is still in a rapid state of development. A detailed treatment can be found in the upcoming book of Fomin, Lokshtanov, and Saurabh [323]. For more on computing obstruction sets and excluded minors, we refer the reader to Adler [17] and Adler, Grohe, and Kreutzer [18].

19.7   Summary

We introduce methods to apply the Graph Minor Theorem and demonstrate theoretical tractability.

Part V
Hardness Theory

Reductions

20

20.1 Introduction

In the first four Parts of the book, we looked at methods for establishing fixed-parameter tractability. Part V is devoted to problems where we can prove, or at least have good evidence to believe, that no such methods exist. For a few problems, we can actually *prove* intractability. However, as with classical intractability indicators, such as NP and PSPACE hardness, in the parameterized complexity framework we must also generally rely on a completeness program based on parameterized problem reductions.

As in Chap. 1, the basic idea behind virtually all completeness results is a *reduction*. A reduction of a language L to a language L' (in whatever complexity-theoretic framework) is a procedure that computes an instance of L' from an instance of L, having the crucial property that if there were an efficient algorithm to decide L', then this could be used in conjunction with the translation procedure, yielding an efficient algorithm to decide L. If we have reasons to believe that L does not admit an efficient recognition algorithm, then a reduction allows us to conclude that likely also L' does not admit an efficient recognition algorithm (for whatever notion of efficiency the framework is concerned with). In parameterized complexity, the notion of efficiency we are concerned with is fixed-parameter tractability.

The main use of reductions is thus to transfer pessimism! And further, if we can also find a reduction of L' to L, equivalence classes of problems with respect to the efficiency notion that the complexity-theoretic framework is focused on. It is a remarkable "empirical" fact of both the classical complexity framework (focused on polynomial time) and the parameterized complexity framework (focused on FPT) that most natural problems belong to only a handful of equivalence classes. This suggests that the equivalence class structure is the primary phenomena, and the nature of our reasons for pessimism relatively secondary—whether that pessimism is referenced to machine models, or simply collective frustration.

Notions of reduction that accomplish the above tasks usually come in several varieties. The best known classical examples are *polynomial-time m-reductions* ($\leq_m^P$) and *polynomial-time T-reductions* ($\leq_T^P$). For languages L and L', we recall that

R.G. Downey, M.R. Fellows, *Fundamentals of Parameterized Complexity*,
Texts in Computer Science, DOI 10.1007/978-1-4471-5559-1_20,
© Springer-Verlag London 2013

$L \leq_T^P L'$ if and only if there is a polynomial-time Oracle Turing Machine Φ such that for all $x \in \Sigma^*$, $x \in L$ iff $\Phi^{L'}(x) = 1$. We also recall that $L \leq_m^P L'$ iff there is a polynomial-time computable function $f : \Sigma^* \to \Sigma^*$, such that for all $x \in \Sigma^*$, $x \in L$ iff $f(x) \in L'$. Karp [431] demonstrated the importance of polynomial-time m-reductions by constructing transformations between many natural combinatorial problems in a wide variety of diverse areas, thereby revealing that the NP-completeness phenomenon is widespread. We recall, from Chap. 1, the following polynomial-time m-reduction.

Example 20.1.1 (A polynomial-time m-reduction)   Polynomial-time m-reductions include the following classical problems; the first considers a formula of conjunctive normal form (CNF).

CNF SATISFIABILITY
Input: A formula X of propositional logic in conjunctive normal form (i.e., a collection of clauses).
Question: Is X satisfiable? That is, is there an assignment of the variables that makes X true?
3SAT
Input: A formula X of propositional logic in conjunctive normal form with maximum clause size 3.
Question: Is X satisfiable?

The classical result of Karp [431] is that CNF SATISFIABILITY $\leq_m^P$ 3SAT and hence CNF SATISFIABILITY $\equiv_m^P$ 3SAT. The polynomial-time problem reduction (also termed a *polynomial-time transformation*) is the following. Let X be an instance of CNF SATISFIABILITY. We compute in polynomial time $f(X)$ an instance of 3SAT such that X is satisfiable iff $f(X)$ is satisfiable. Without loss of generality, we may assume that each clause of X has ≥ 3 literals. Let $U = \{u_1, \ldots, u_m\}$ be a clause of X. We replace this clause by the following clauses:

$$U' = \{u_1, u_2, z_1\}, \{\overline{z_1}, u_3, z_2\}, \{\overline{z_2}, u_4, z_3\}, \ldots, \{\overline{z}_{m-3}, u_{m-1}, u_m\},$$

where $\{z_1, \ldots, z_{m-3}\}$ are new variables, chosen for this clause. Let $f(X)$ be the result obtained from X by performing this local replacement for each clause of X. Notice that the size of $f(X)$ is bounded by $3mq$, where m is the size of the largest clause and q is the number of clauses. Also, notice that any satisfying assignment for the collection U' must make U true, and hence X is satisfiable iff $f(X)$ is satisfiable.

We recall from Chap. 1 and Sect. 5.1 that sometimes we need to use many queries to achieve some kinds of reduction/kernelization.

Example 20.1.2 (A polynomial-time T-reduction)   Nice examples of this reduction are often provided by showing that a *search* problem reduces to a *decision* problem.

Although HAMILTON PATH is not a decision problem, we are often interested in the associated search problem: If G has a Hamilton path, *find* one or else return that G has no Hamilton path. Of course, if we can do the search problem in polynomial time, we can do the decision problem. The algorithm that reduces the search problem to the decision one is the following. Given an oracle Ω for the decision problem and a graph G, first use Ω to see if G has a Hamilton path. If the answer is no, then output that G has no Hamilton path. If Ω tells us that G does have a Hamilton path, we will build a Hamilton path H in stages. Initially, let $H = \emptyset$. Let v_1 be a vertex of G and e an edge of G containing v_1. Let G' be the result of removing e from G. Now, ask Ω if G' has a Hamilton path. If the answer is yes, continue the algorithm with G' as the new version of G. We must reach a stage where G' has no Hamilton path; then we know that e must lie on *every* Hamilton path of the current G. We add e to our current approximation to the Hamilton path, and replace G by G'' which is the result of removing from G v_1 and all the edges emanating from v_1. For the next step, we also replace v_1 by v_m, where e is the edge $\langle v_1, v_m \rangle$.

The key difference between the T- and m-reductions above, is that: in the m-reduction we compute $f(x)$ and then ask a single question of Ω, namely if $f(x)$ is in L'; whereas in the T-reduction, *many* questions need to be asked of the oracle Ω to get the result. Also the T-reduction is *adaptive*, in the sense that answers from earlier questions affect the next question. Even the two examples above suggest many natural questions. For instance, is the solution of the search problem from the decision problem *intrinsically sequential*? It has been shown that if we can solve the search problem from the decision problem without the use of sequential questions then many unlikely collapses occur for complexity theory.

20.2   Parameterized Reductions

Reductions allow us to partially order languages in terms of their computational complexity. Two languages, L and L', are taken to have the same complexity from the point of view of the given reduction iff they are in the same *degree*; that is, there is a reduction from L to L' and vice versa. We will need to create reductions which express the fact that two languages have the same *parameterized* complexity. What is needed are reductions that ensure that if L' is computable in time $f(k)n^c$ "by the slice", then there is a parameterized algorithm for L running in time $g(k)n^{c'}$ "by the slice". After a moment's thought, we realize that we can achieve such reductions only if we allow each slice of L to reduce to a finite number of slices of L'. The easiest way is to reduce, in parameterized polynomial time, the kth slice of L to the k'th slice of L'. This leads us to the *working definition* of a parameterized problem reduction. From the point of view of our hardness results, Definition 20.2.1 below, is the most important definition in the book.

Definition 20.2.1 (Basic working definition of a parameterized reduction for combinatorial reductions)   We say that L reduces to L' by a *standard* parameterized m-reduction if there are functions $k \mapsto k'$ and $k \mapsto k''$ from $\mathbb{N}$ to $\mathbb{N}$, and a function $\langle x, k \rangle \mapsto x'$ from $\Sigma^* \times \mathbb{N}$ to Σ^*, such that

$$\langle x, k \rangle \mapsto x' \text{ is computable in time } k'' |x|^c, \quad \text{and} \quad \langle x, k \rangle \in L \text{ iff } \langle x', k' \rangle \in L'.$$

Example 20.2.1   The following is a simple example of a standard parameterized reduction. It comes from an observation about a classical reduction by Karp. Define a formula (of propositional logic) to be *monotone* if it contains no negation symbols, and *antimonotone* if it contains no negation symbols except for the literals, which are *all* negated. Consider the following problem:

WEIGHTED CNF SATISFIABILITY
Input: A propositional formula X in conjunctive normal form.
Parameter: A positive integer k.
Question: Does X have a satisfying assignment of Hamming weight k? (An assignment has *weight k* iff it has exactly k variables set to be *true*.)

WEIGHTED CNF SATISFIABILITY generates a parameterized language, which is a natural parameterized version of CNF SATISFIABILITY. We can similarly define a parameterized version of 3SAT, WEIGHTED 3SAT, by considering weight k solutions of problems in 3CNF form. Finally, we can define WEIGHTED MONOTONE CNF SATISFIABILITY and WEIGHTED ANTIMONOTONE CNF SATISFIABILITY by considering only monotone and antimonotone formulas.

We need another problem to consider weighted functions.

WEIGHTED BINARY INTEGER PROGRAMMING
Instance: A binary matrix A and a binary vector **b**.
Parameter: A positive integer k.
Question: Does $A \cdot \mathbf{x} \geq \mathbf{b}$ have a binary solution of weight k?

We have the following:

Theorem 20.2.1   WEIGHTED MONOTONE CNF SATISFIABILITY *reduces to* WEIGHTED BINARY INTEGER PROGRAMMING *via a standard reduction.*

Proof Let (X, k) be an instance of MONOTONE WEIGHTED CNF SAT. Let $C_1, \ldots, C_p$ list the clauses of X and $x_1, \ldots, x_m$ list the variables. Let A be the matrix $\{a_{i,j} : i = 1, \ldots p, j = 1, \ldots, m\}$ with $a_{i,j} = 1$ if x_j is present in C_i, and $a_{i,j} = 0$ otherwise. Let **b** be the vector with 1 in the jth position for $j = 1, \ldots, p$. It is easy to see that $A \cdot \mathbf{x} \geq \mathbf{b}$ has a solution of weight k iff X has a satisfying assignment of weight k (and the reasoning is reversible).   □

Actually, the situation in Theorem 20.2.1 is very unusual. It is almost *never* the case that a classical reduction turns out to be *also* a parameterized reduction for some natural parameterized versions of the problems. Classical reductions almost never seem to carry enough structure. This lack of structure in the classical reductions causes us to lose control of the parameter. For instance, consider the situation of Example 20.1.1 where CNF SATISFIABILITY is classically reduced to 3SAT. There, we replace a clause with m literals by $m - 2$ clauses, each consisting of 3 literals. For each clause, we introduced $m - 3$ new variables. Now, suppose the original formula has a weight k satisfying assignment that makes exactly one variable u_j in clause U of X true. Then, to make $f(X)$ true, we would need to make $z_1, \ldots, z_{j-2}$ all *true*. So the weight of the satisfying assignment of $f(X)$ would not only depend upon k but would depend on j, and hence the clause size of X. So this reduction from CNF SATISFIABILITY to 3SAT does *not* give a *parameterized* reduction from WEIGHTED CNF SATISFIABILITY to WEIGHTED 3SAT. It is not known if there is any such parameterized reduction. We will soon see that in fact, there is some evidence that there really is no such reduction, and thus, from a parameterized point of view, CNF SATISFIABILITY and 3SAT are *not* the same. If some NP-complete problems are not fixed-parameter tractable, then it must be that some classical reductions do not lift to parameterized ones, since we have seen that some NP-complete problems are fixed-parameter tractable.

Although Definition 20.2.1 is satisfactory as a working definition, for the general situation we need to be more precise. In Definition 2.2.1, we have seen that there are three basic flavors of fixed-parameter tractability, depending on the level of uniformity. Naturally, the three types of tractability will relativize to three flavors of reducibility. We get the following technical definitions.

Remark 20.2.1   However, readers only interested in the concrete hardness results of Part V can ignore the definitions to follow and take Definition 20.2.1 as the actual definition.

Definition 20.2.2 (Uniform fixed-parameter reducibility)   Let A and B be parameterized problems. We say that A is *uniformly fixed-parameter-reducible* to B if there is an oracle procedure Φ, a constant α, and an arbitrary function $f : \mathbb{N} \mapsto \mathbb{N}$ such that

1. the running time of $\Phi(B; \langle x, k \rangle)$ is at most $f(k)|x|^\alpha$,
2. on input $\langle x, k \rangle$, Φ only asks oracle questions of $B^{(f(k))}$, where

$$B^{(f(k))} = \bigcup_{j \le f(k)} B_j = \big\{ \langle x, j \rangle : j \le f(k) \ \& \ \langle x, j \rangle \in B \big\},$$

3. $\Phi(B) = A$.

If A is uniformly P-reducible to B, we write $A \leq_T^u B$. Where appropriate we may say that $A \leq_T^u B$ *via* f. If the reduction is many : 1 (an *m-reduction*), we will write $A \leq_m^u B$.

Definition 20.2.3 (Strongly uniform reducibility) Let A and B be parameterized problems. We say that A is *strongly uniformly fixed-parameter-reducible* to B if $A \leq_T^u B$ via f where f is recursive. We write $A \leq_T^s B$ in this case.

Definition 20.2.4 (Nonuniform reducibility) Let A and B be parameterized problems. We say that A is *nonuniformly fixed-parameter-reducible* to B if there is a constant α, a function $f : \mathbb{N} \mapsto \mathbb{N}$, and a collection of procedures $\{\Phi_k : k \in \mathbb{N}\}$ such that $\Phi_k(B^{(f(k))}) = A_k$ for each $k \in \mathbb{N}$, and the running time of Φ_k is $f(k)|x|^{\alpha}$. Here, we write $A \leq_T^n B$.

These are good definitions, since whenever $A \leq B$ with $\leq$ any of the reducibilities, if B is fixed-parameter tractable, so too is A. Note that the above definitions also allow our previous ideas of fixed-parameter tractability to be more specific. Now, nonuniformly fixed-parameter tractability corresponds to being $\leq_T^n \emptyset$. *We will henceforth write* FPT($\leq$) *as the fixed-parameter-tractable class corresponding to the reducibility* $\leq$.

If $A \leq B$ via a standard reduction in the sense of Definition 20.2.1, then $A \leq_m^u B$. Furthermore, in most cases, we will look at $A \leq_m^s B$ since the function $k \mapsto k'$ will be computable. The standard definition works for natural languages, since they all have the following property: for all k, L_k is "directly coded into" (or "trivially represented in") L_{k+1}. For instance, for any formula φ in variables $x_1, \ldots, x_n$, φ has a weight k satisfying assignment iff $\varphi \wedge x_{n+1}$ has a weight $k + 1$ satisfying assignment. Thus, in a "natural" language L if $L' \leq L$, then L' reduces to a single slice of L because of this upward encoding of the slices.

20.3   Exercises

Exercise 20.3.1   Prove that INDEPENDENT SET $\leq_m^s$ MAXIMAL IRREDUNDANT SET via a standard reduction. These problems are defined as follows. Bodlaender and Kratsch have proven the apparently stronger result that CNF SATISFIABILITY $\leq_m^s$ MAXIMAL IRREDUNDANT SET, and Cesati [136] proved that MAXIMAL IRREDUNDANT SET$\leq_m^s$CNF SATISFIABILITY, and hence MAXIMAL IRREDUNDANT SET$\equiv_m^s$ CNF SATISFIABILITY. In Theorem 22.1.1, we will look at Cesati's techniques as they are of independent interest. The Bodlaender–Kratsch–Cesati results are apparently stronger because, as we will soon show, INDEPENDENT SET $\equiv_m^s$ WEIGHTED 3SAT, and we believe that WEIGHTED 3SAT $<_m^s$ WEIGHTED CNF SATISFIABILITY.)

(k-)INDEPENDENT SET

Input: A graph G.
Parameter: A positive integer k.
Question: Does G have a set of k vectors $\{x_1, \ldots, x_k\}$ such that all $i \neq j$, x_i is not adjacent to x_j?

(k-)MAXIMAL IRREDUNDANT SET

Input: A graph G.
Parameter: A positive integer k.
Question: Does G have a set X of k vertices such that for each member x of X, there is a $y \in V(G)$ such that either $y = x$ or (x, y) is an edge, and y is not adjacent to nor a member of S, and, furthermore, S is maximal with this property?

(Hint: Given a graph G and a positive integer k, construct a graph H and a k' so that G has a k-element independent set iff H has a maximal irredundant set of size k'. H is constructed as follows. Construct k blocks consisting of $k + 1$ columns, each consisting of $n = |G|$ points. So column i in block j consists of n points $x(j, i, t)$ for $t = 1, \ldots, n$. A block represents a choice for a vertex in the independent set. Now, within a block, join all vertices except those with the same last coordinate. (We refer to this as a *row*.) The first column is not connected to anything else and is called the *dummy column*. Now, for each block i pick a column in each block $i' \neq i$ to identify with block i, and similarly identify a corresponding column i'' in block i. Do this in such a way that for each pair (i, j), there is exactly one column $c_i(i, j)$ identified with exactly one column $c_j(i, j)$ in block j, and these columns are used for no other identifications, for any other pair except (i, j). Let (q, r) be some such pair of columns. Now, connect each $x(i, q, t)$ to $x(j, r, s)$ iff t and s are adjacent vertices in G.

Now, let $k' = k(k + 1)$. First, if G has an independent set of size k, let $\{t_1, \ldots, t_{k-1}\}$ be this independent set. The relevant irredundant set for H is obtained by taking the union of vertices of the t_ith row of block i, for $i = 1, \ldots, k$. They are each their own private neighbors, as the corresponding set in G is independent, and it is maximal by the fact that the blocks are almost cliques. Conversely, take S to be a size k' maximal irredundant set in H. Prove that it must be of the form above and, hence, corresponds to an independent set in G of size k. There are two cases depending on whether some block B has more than k vertices of S. Remember that a maximal irredundant set must also be a dominating set.)

Exercise 20.3.2 Prove that WEIGHTED ANTIMONOTONE 2SAT (see Example 20.2.1) reduces to COLORED AUTOMORPHISM (below) by a standard reduction.

COLORED AUTOMORPHISM

Instance: A 2-colored (bipartite) graph G.
Parameter: A positive integer k.
Question: Is there an automorphism preserving colors moving exactly k blue vertices?

(Hint: Let X be an instance of WEIGHTED ANTIMONOTONE 2-CNFSAT with parameter k. Let $C_1, \ldots, C_m$ be the clauses of X and $u_1, \ldots, u_n$ the variables which will all appear in the clauses as negated since X is antimonotone. The graph $G(X)$ is constructed as follows:

For each u_i, we have two blue nodes $\overline{u(i)}$ and $u(i)$. For each clause C_j, we have three red nodes $C(j, k) : k = 1, 2, 3$. For each i, append a rigid tree T_i to each $u(i)$ and $\overline{u(i)}$ so that additionally for all $i \neq j$, T_i is not isomorphic to T_j. All the nodes of the T_i are red. (This will force $u(i)$ to go to $\overline{u(i)}$ if it moves at all.) Finally, we need some further edges. Let $C_j = \{\overline{u_p}, \overline{u_q}\}$. Add the edges $\overline{u(p)}C(j, 1)$, $\overline{u(q)}C(j, 1)$, $\overline{u(p)}C(j, 2)$, $u(q)C(j, 2)$, and $u(p)C(j, 3)$, $\overline{u(q)}C(j, 3)$.

Prove that X has a weight k satisfying assignment iff $G(X)$ has an automorphism that moves exactly $2k$ blue vertices.)

Exercise 20.3.3 (The advice view) In Exercise 2.3.4, we saw that an alternative view of parameterized tractability is provided by considering the FPT to be the class of parameterized problems that are solvable in time $c \cdot n^\alpha$ by an oracle Turing machine Φ with a finite piece of advice $f(k)$. We can define $A \leq_{\text{advice}} B$ via $f : \mathbb{N} \mapsto \Sigma^*$ if there exists an oracle Turing machine Φ running in time n^α such that

$$\langle x, k \rangle \in A \quad \text{iff} \quad \Phi^{B(f(k)) \oplus f(k)}(\langle x, k \rangle) = 1.$$

Prove that $A \leq_T^n B$ iff $A \leq_{\text{advice}} B$.

Exercise 20.3.4 1. Prove that there exists a language $L \in \text{FPT}(\leq_m^u)$ such that $L \notin \text{FPT}(\leq_T^s)$.

2. Prove that there exists a computable language $L' \in \text{FPT}(\leq_m^n)$ with $L' \notin \text{FPT}(\leq_T^u)$.

(Hint: (i) Construct a language $L = \bigcup_s L_s$ in stages to diagonalize against all possible $\leq_T^s$ reductions.)

20.4 Historical Notes

The concepts of reducibilities and computability were developed in the 1930s in the setting of classical recursion theory by many authors, notably Post, Kleene, and Church. Of course, the idea of a reduction is in some sense much older. Consider the reduction that $a \leq b$ iff a is, say, algebraic over b. Recognition of the difficulty of NP-complete problems has a very interesting history, and the reader is referred to the illuminating article of Hartmanis [380]. The groundbreaking articles here are Karp [431], Cook [163], and Levin [488]. The definitions of reducibilities we have used were developed by the authors in Downey–Fellows [240, 241, 243]. We give fuller historical remarks on parameterized complexity at the end of the next chapter, after reviewing basic classes of intractable problems.

20.5 Summary

We introduced the central notions of parameterized problem reductions.

The Basic Class $W[1]$ and an Analog of Cook's Theorem

21

21.1 Introduction

In the last chapter, we looked at *one* of the ingredients of classical hardness results: the notion of a reducibility. The *other* basic ingredient is the identification of classes of intractable problems. In the case of NP-completeness, recall that a language L is a member of the class NP iff there is a polynomial-time relation R and a polynomial p such that

$$x \in L \quad \text{iff} \quad \exists y\big[|y| \le p(|x|) \wedge R(x, y) \text{ holds}\big].$$

For instance, if L is the collection of satisfiable formulas, then a formula x is satisfiable iff there is some satisfying assignment y of the variables. Recall that L is NP-complete iff $L \in$ NP, and for any language $L' \in$ NP, $L' \le_m^p L$. The hidden beauty and power of these definitions comes from the profound discovery that there are literally thousands of important and natural NP-complete problems. If *any* of them were solvable in polynomial time, then *all* of them would be. Moreover, other than a few exceptions, if a natural problem is in NP and is not in P, then the problem seems to always be NP-complete.

Again, following this methodology, we will need to identify a class of problems that we feel are fixed-parameter intractable. We would then need to show that this class corresponds to a wide collection of problems. For this class, if any of the problems were fixed-parameter tractable, then all of the problems in the class would be fixed-parameter tractable.

An important element of the classical theories of, for instance, NP- and PSPACE-completeness is the identification of a basic generic problem. The classical problem identified by Cook–Levin is the following:

TURING MACHINE ACCEPTANCE

Input: A nondeterministic Turing machine M, a string x, and a number n (in unary).

Question: Is there a computation path of M accepting $\langle x \rangle$ in $\le n$ steps?

R.G. Downey, M.R. Fellows, *Fundamentals of Parameterized Complexity*,
Texts in Computer Science, DOI 10.1007/978-1-4471-5559-1_21,
© Springer-Verlag London 2013

It is clear that TURING MACHINE ACCEPTANCE is NP-complete, since $\exists y(|y| \leq p(|x|) \wedge R(x, y)$ holds) can be modeled by a nondeterministic Turing machine M that guesses y and checks if $R(x, y)$ holds. Thus, $x \in L$ iff $\langle M, x, 1^{p(|x|)} \rangle$ is in TURING MACHINE ACCEPTANCE. So, assuming the polynomial-time intractability of TURING MACHINE ACCEPTANCE, we now have a way of demonstrating polynomial-time intractability of another problem X simply by showing that TURING MACHINE ACCEPTANCE $\leq_T^p X$. Based upon the NP-completeness of TURING MACHINE ACCEPTANCE, strong intuitive and philosophical arguments can be proposed for the thesis that $P \neq NP$. Many NP-complete combinatorial problems such as INTEGER PROGRAMMING, say, have some sort of natural algebraic structure. On the face of it, one might believe that with sufficiently clever algebraic insight, we might just be able to exploit this algebraic structure to prove that INTEGER PROGRAMMING is in P and hence is theoretically tractable. But INTEGER PROGRAMMING is NP-complete, so if we can prove it to be in P, we *must* also have TURING MACHINE ACCEPTANCE in P. However, in the case of TURING MACHINE ACCEPTANCE, the point is that a nondeterministic Turing machine is such an opaque and generic object that it simply does not seem reasonable that we should be able to decide in polynomial time whether a given Turing machine on a given input has some accepting path. Cook's theorem therefore gives powerful evidence that $P \neq NP$.

21.2   Short Turing Machine Acceptance

In the present chapter, we will prove our own analog of Cook's theorem by proving that a number of combinatorial problems are of the same fixed-parameter complexity as a generic problem about nondeterministic Turing machines. The key underlying problem is the following.

SHORT TURING MACHINE ACCEPTANCE
Input: A nondeterministic Turing machine M and a string x.
Parameter: A positive integer k.
Question: Does M have a computation path accepting x in $\leq k$ steps?

It seems to us that if one accepts the philosophical argument that TURING MACHINE ACCEPTANCE is intractable, then the same reasoning would suggest that SHORT TURING MACHINE ACCEPTANCE is fixed-parameter intractable. We will soon establish that there are a large number of problems of the same fixed-parameter complexity as SHORT TURING MACHINE ACCEPTANCE. We believe that the existence of such a large number of problems of the same fixed-parameter complexity as SHORT TURING MACHINE ACCEPTANCE gives further weight to our thesis that SHORT TURING MACHINE ACCEPTANCE is not fixed-parameter tractable.

The main theorem of this chapter is an analog of Cook's theorem. We will first establish some preliminary results. We will need some definitions. It is convenient to consider a 3CNF formula as a (boolean) *circuit*. Thus, a 3CNF formula is considered as a circuit consisting of one input (of unbounded fanout) for each variable, possibly inverters below the variable, and structurally a large *and* of small *or*'s (of

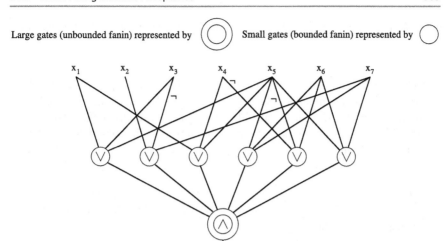

A 3CNF Formula is a large and of small or's.

$$\varphi = (x_1 + x_3 + x_5)(x_2 + \neg x_3 + x_7)(x_1 + x_4 + x_5)(x_5 + x_6 + x_7)(\neg x_4 + \neg x_5 + x_6)(x_5 + x_6 + x_7)$$

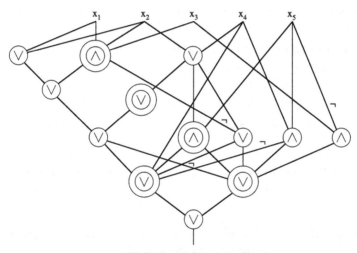

A Weft 2 Depth 5 Decision Circuit.

Fig. 21.1 Examples of circuits

size 3) with a single output line. The reader should refer to Fig. 21.1. We can similarly consider a 4CNF formula to be a large *and* of small *or*'s, where "small" is defined to be 4. More generally, it is convenient to consider the model of a *decision circuit*. This is a circuit consisting of large and small gates with a single output line, and no restriction on the fanout of gates. For such a circuit, the *depth* is the maximum number of gates on any path from the input variables to the output line, and the *weft* is the "large gate depth". More formally, we define the weft of a circuit as follows.

Definition 21.2.1 (Weft) Let C be a decision circuit. The *weft* of C is defined to be the maximum number of large gates on any path from the input variables to the output line. (A gate is called large if its fanin exceeds some preagreed bound. The reader should note that *none* of our results depend upon what the bound actually is.)

In Fig. 21.1, we have an example of a weft 2 depth 5 decision circuit.

Let $\mathcal{F} = \{C_1, \ldots, C_n, \ldots\}$ be a family of decision circuits. Associated with $\mathcal{F}$ is a basic parameterized language

$$L_{\mathcal{F}} = \{\langle C_i, k \rangle : C_i \text{ has a weight } k \text{ satisfying assignment}\}.$$

For instance, if $\mathcal{F}$ is the family of boolean circuits corresponding to propositional formulas in 3CNF form, then $L_{\mathcal{F}}$ corresponds to WEIGHTED 3CNF SATISFIA-BILITY. A generalization of the class of circuits corresponding to 3CNF formulas is provided by the following.

WEIGHTED WEFT t DEPTH h CIRCUIT SATISFIABILITY (WCS(t, h))

Input: A weft t depth h decision circuit C.
Parameter: A positive integer k.
Question: Does C have a weight k satisfying assignment?

Note that WEIGHTED 3CNF SATISFIABILITY is subsumed by WEFT 1 DEPTH 2 DECISION CIRCUIT SATISFIABILITY.

Notation We will denote by $L_{\mathcal{F}(t,h)}$ the parameterized language associated with the family of weft t depth h decision circuits.

Definition 21.2.2 (Basic hardness class) We define a language L to be in the class $W[t]$ iff L is fixed-parameter reducible to $L_{\mathcal{F}(t,h)}$ for some h.

We remark that we have not specified the type of parameterized reduction in Definition 21.2.2. As we said earlier, in all natural instances one may take it to be the one from Definition 20.2.1. As we noted earlier, in all cases of hardness results of which we are aware, the reduction is a standard one and hence a $\leq_m^s$ reduction. (Note the parallel with the classical NP-completeness results where all known hardness results turn out to be Karp reductions.)

Theorem 21.2.1 (Analog of Cook's Theorem (I)) *The following are complete for* $W[1]$:

(i) WEIGHTED n-SATISFIABILITY *for any fixed* $n \geq 2$.
(ii) SHORT TURING MACHINE ACCEPTANCE.

The remainder of the chapter is devoted to proving Theorem 21.2.1. The proof is quite technical and involves a number of intermediate steps. These steps are of independent interest and establish a number of other $W[1]$ completeness results. We begin with some preliminary *normalization* results about weft 1 circuits.

By definition, a parameterized problem $L \in W[1]$ reduces to $L_{\mathcal{F}(1,h)}$ for the family $\mathcal{F}(1, h)$ of weft 1 circuits of depth bounded by h, for some h. The following argument shows that we may assume the circuits in F to have depth 2 and a particularly simple form, consisting of a single output *and* gate which receives arguments from *or* gates having fanin bounded by a constant h'. Thus, each such circuit is isomorphically represented by a boolean expression in conjunctive normal form having clauses with at most h' literals. We will say that a family of circuits having this form is *normalized*. With this in mind, we have the following definition.

Definition 21.2.3   The family of parameterized problems $W[1, s]$ is defined to be those parameterized problems in $W[1]$ reducible to $L_{\mathcal{F}(s)}$ for the family $\mathcal{F}(s)$ of depth 2 weft 1 normalized circuits, with the *or* gates on level 1 having fanin bounded by s.

Lemma 21.2.1   *Let $\mathcal{F}$ be a family of weft 1 circuits of depth bounded by a constant h. Then, $L_{\mathcal{F}}$ is reducible (by a standard reduction) to $L_{\mathcal{F}(s)}$ for $s = 2^h + 1$, and hence $L_{\mathcal{F}} \in W[1, s]$.*

Proof   Let $C \in \mathcal{F}$ and let k be a positive integer. We describe how to produce a circuit $C' \in \mathcal{F}(s)$ and an integer k' such that C accepts a weight k input if and only if C' accepts an input of weight k'.

Step 1. The reduction to tree circuits.

The first step is to transform C into an equivalent weft 1 *tree circuit* C' of depth at most h. In a tree circuit, every logic gate has fanout one, and thus the circuit can be viewed as equivalent to a boolean formula. The transformation can be accomplished by replicating the portion of the circuit above a gate as many times as the fanout of the gate, beginning with the top level of logic gates, and proceeding downward level by level. The creation of C' from C may require time $O(|C|^{O(h)})$ and involve a similar blowup in the size of the circuit. This is permitted since h is a fixed constant independent of k and $|C|$.

Step 2. Moving the *not* gates to the top of the circuit.

Let C denote the circuit we receive from the previous step (we will use this notational convention throughout the proof). Transform C into an equivalent circuit C' by commuting the *not* gates to the top, using DeMorgan's laws. This may increase the size of the circuit by at most a constant factor. The tree circuit C' thus consists (from the top) of the input nodes, with *not* gates on some of the lines fanning out from the inputs. In counting levels, we consider all of this as level 0.

Step 3. A preliminary depth 4 normalization.

The goal of this step is to produce a tree circuit C' of depth 4 that corresponds to a boolean expression E in the following form. (For convenience, we use product notation to denote logical $\land$ and sum notation to denote logical $\lor$.)

$$E = \prod_{i=1}^{m} \sum_{j=1}^{m_i} E_{ij},$$

where

1. m is bounded by a function of h,
2. for all i, m_i is bounded by a function of h,
3. for all i, j, E_{ij} is either:

$$E_{ij} = \prod_{k=1}^{m_{ij}} \sum_{l=1}^{m_{ijk}} x[i, j, k, l]$$

or

$$E_{ij} = \sum_{k=1}^{m_{ij}} \prod_{l=1}^{m_{ijk}} x[i, j, k, l],$$

where the $x[i, j, k, l]$ are literals (i.e., input boolean variables or their negations) and for all i, j, k, m_{ijk} is bounded by a function of h. The family of circuits corresponding to these expressions has weft 1, with the large gates corresponding to the E_{ij}. (In particular, the m_{ij} are *not* bounded by a function of h.)

To achieve this form, let g denote a large gate in C. An input to g is computed by a subcircuit of depth bounded by h consisting only of small gates, and so is a function of at most 2^h literals. This subcircuit can thus be replaced, at constant cost, by either a product-of-sums expression (if g is a large $\land$ gate) or a sum-of-products expression (if g is a large $\lor$ gate). In the first case, the product of these replacements over all inputs to g yields the subexpression E_{ij} corresponding to g. In the second case, the sum of these replacements yields the corresponding E_{ij}.

The output of C is a function of the outputs of at most 2^h large gates. This function can be expressed as a product-of-sums expression of size at most 2^{2^h}. At the cost of possibly duplicating some of the large gate subcircuits at most 2^{2^h} times, we can achieve the desired normal form with the bounds $m \le 2^{2^h}$, $m_i \le 2^h$, and $m_{ijk} \le 2^h$.

Step 4. Employing additional nondeterminism.

Let C denote the normalized depth 4 circuit received from Step 3 and corresponding to the boolean expression E described above. For convenience, assume that the E_{ij} for $j = 1, \ldots, m_i'$ are sums-of-products and the E_{ij} for $j = m_i' + 1, \ldots, m_i$ are products-of-sums. Let $V_0 = \{x_1, \ldots, x_n\}$ denote the variables of E.

In this step, we produce an expression E' in product-of-sums form with the size of the sums bounded by $2^h + 1$ that has a satisfying truth assignment of weight

$$k' = 2k + k\left(1 + 2^h\right)2^{2^h} + m + \sum_{i=1}^{m} m_i'$$

if and only if C has a satisfying truth assignment of weight k. The main idea is to use additional (bounded weight) nondeterminism to guess both (1) a weight k input x for C and (2) additional information that will allow us to check that $C(x) = 1$ with a $W[1, s]$ circuit, $s = 2^h + 1$.

The set V of variables of E' is $V = V_0 \cup V_1 \cup V_2 \cup V_3$, where

$$V_1 = \left\{x[i, j]: 1 \le i \le n, 0 \le j \le \left(1 + 2^h\right)2^{2^h}\right\},$$

$$V_2 = \left\{u[i, j]: 1 \le i \le m, 1 \le j \le m_i\right\},$$

$$V_3 = \left\{w[i, j, k]: 1 \le i \le m, 1 \le j \le m_i', 0 \le k \le m_{ij}\right\}.$$

The expression E' is a product of subexpressions $E' = E_1 \wedge \cdots \wedge E_8$ described by

$$E_1 = \prod_{i=1}^{n} \left(\neg x_i + x[i, 0]\right)\left(\neg x[i, 0] + x_i\right),$$

$$E_2 = \prod_{i=1}^{n} \prod_{j=0}^{2^{2^h}+1} \left(\neg x[i, j] + x[i, j + 1 \ (\mathrm{mod}\ r)]\right), \quad r = 1 + \left(1 + 2^h\right)2^{2^h},$$

$$E_3 = \prod_{i=1}^{m} \sum_{j=1}^{m_i} u[i, j],$$

$$E_4 = \prod_{i=1}^{m} \prod_{j=1}^{m_i-1} \prod_{j'=j+1}^{m_i} \left(\neg u[i, j] + \neg u[i, j']\right),$$

$$E_5 = \prod_{i=1}^{m} \prod_{j=1}^{m_i'} \prod_{k=0}^{m_{ij}-1} \prod_{k'=k+1}^{m_{ij}} \left(\neg w[i, j, k] + \neg w[i, j, k']\right),$$

$$E_6 = \prod_{i=1}^{m} \prod_{j=1}^{m_i'} \left(\neg u[i, j] + \neg w[i, j, 0]\right),$$

$$E_7 = \prod_{i=1}^{m} \prod_{j=1}^{m_i'} \prod_{k=1}^{m_{ij}} \prod_{l=1}^{m_{ijk}} \left(\neg w[i, j, k] + x[i, j, k, l]\right),$$

$$E_8 = \prod_{i=1}^{m} \prod_{j=m_i'+1}^{m_i} \prod_{k=1}^{m_{ij}} \left(\neg u[i,j] + \sum_{l=1}^{m_{ijk}} x[i,j,k,l] \right).$$

To see that Step 4 works correctly, suppose τ is a weight k truth assignment to V_0 that satisfies E. We describe how to extend τ to weight k' truth assignment τ' to the variables V that satisfies E' as follows:

1. For each i such that $\tau(x_i) = 1$ and for $j = 0, \ldots, (1 + 2^h)2^{2^h}$, set $\tau'(x[i,j]) = 1$.
2. For each $i = 1, \ldots, m$, choose an index j_i such that E_{i,j_i} evaluates to 1 (this is possible, since τ satisfies E) and set $\tau'(u[i,j_i]) = 1$.
3. If, in (2), E_{i,j_i} is a sum-of-products, then choose an index k_i such that

$$\prod_{l=1}^{m_{i,j,k_i}} x[i,j,k,l]$$

   evaluates to 1, and correspondingly set $\tau'(w[i,j,k_i]) = 1$.
4. For $i = 1, \ldots, m$ and $j = 1, \ldots, m_i'$ such that $j \neq j_i$, set $\tau'(w[i,j,0]) = 1$.

It is straightforward to check that the above described weight k' extension τ' satisfies E'.

Conversely, suppose υ' is a weight k' truth assignment to the variables of V that satisfies E'. We argue that the restriction υ of υ' to V_0 is a weight k truth assignment that satisfies E.

Claim 1 υ sets at most k variables of V_0 to 1.

If this were not so, then the clauses in E_1 and E_2 would together force at least $(k+1)(2+(1+2^h)2^{2^h})$ variables to be 1 in order for υ' to satisfy E', a contradiction as this is more than k'.

Claim 2 υ sets at least k variables of V_0 to 1.

The clauses of E_4 ensure that υ' sets at most m variables of V_2 to 1. The clauses of E_5 ensure that υ' sets at most $\sum_{i=1}^{m} m_i'$ variables of V_3 to 1. If Claim 2 were false, then for υ' to have weight k', there must be more than k indices j for which some variable $x[i,j]$ of V_1 has the value 1, a contradiction in consideration of E_1 and E_2.

The clauses of E_3 and the arguments above show that υ' necessarily has the following restricted form:

1. Exactly k variables of V_0 are set to 1.
2. For each of the k in (1), the corresponding $(1+2^h)2^{2^h} + 1$ variables of V_1 are set to 1.
3. For each $i = 1, \ldots, m$, there is exactly one j_i for which $u[i,j_i] \in V_2$ is set to 1.
4. For each $i = 1, \ldots, m$ and $j = 1, \ldots, m_i'$, there is exactly one k_i for which $w[i,j,k_i] \in V_3$ is set to 1.

To argue that υ satisfies E, it suffices to argue that υ satisfies every E_{i,j_i} for $i = 1, \ldots, m$.

The clauses of E_6 ensure that if $\upsilon'(u[i, j]) = 1$, then $k_i \neq 0$. This being the case, the clauses of E_7 force the literals in a subexpression of E_{i,j_i} to evaluate in a way that shows E_{i,j_i} to evaluate to 1. The clauses of E_8 enforce that E_{i,j_i} evaluates to 1 for $j_i > m'_i$. $\qquad\qquad\qquad\qquad\qquad\qquad\qquad\qquad\qquad\qquad\qquad\qquad\qquad\square$

Thus, we have the following stratification of $W[1]$ that will be useful to our arguments:

$$W[1] = \bigcup_{s=1}^{\infty} W[1, s].$$

Our next preliminary result is to show that $W[1]$ collapses to $W[1, 2]$. Recall that a circuit C is termed *monotone* if it does not have any *not* gates. Equivalently, C corresponds to boolean expressions having only positive literals. If we restrict the definition of $W[t]$ and $W[1, s]$ to families of monotone circuits, we obtain the family of parameterized problems MONOTONE $W[t]$ and MONOTONE $W[1, s]$.

Recall also, we defined a circuit to be ANTIMONOTONE if all the input variables were negated and the circuit has no other inverters. Thus, in an antimonotone circuit, each fanout line from an input node goes to a *not* gate (and in the remainder of the circuit, there are *no* other *not* gates). The restriction to families of antimonotone circuits yields the classes of parameterized problems ANTIMONOTONE $W[t]$ and ANTIMONOTONE $W[1, s]$.

Theorem 21.2.2 $W[1, s] =$ ANTIMONOTONE $W[1, s]$ *for all* $s \geq 2$.

Proof The plan of our argument is to identify a problem (s-Red/BLUE NON-BLOCKER) that belongs to antimonotone $W[1, s]$, and then show that the problem is hard for $W[1, s]$. RED/BLUE NONBLOCKER is the following parameterized problem.

RED/BLUE NONBLOCKER
Input: A graph $G = (V, E)$ where V is partitioned into two color classes $V = V_{\text{red}} \cup V_{\text{blue}}$.
Parameter: A positive integer k.
Question: Is there is a set of red vertices $V' \subseteq V_{\text{red}}$ of cardinality k such that every blue vertex has at least one neighbor that does not belong to V'.

The *closed neighborhood* of a vertex $u \in V$ is the set of vertices $N[u] = \{x : x \in V \text{ and } x = u \text{ or } xu \in E\}$.

It is easy to see that the restriction of RED/BLUE NONBLOCKER to graphs G with blue vertices of maximum degree s belongs to antimonotone $W[1, s]$ since the product-of-sums boolean expression

$$\prod_{u \in V_{\text{blue}}} \sum_{x_i \in N[u] \cap V_{\text{red}}} \neg x_i$$

has a weight k truth assignment if and only if G has size k nonblocking set.

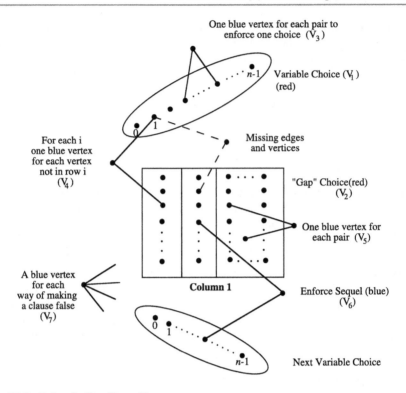

One blue vertex for each pair to
enforce one choice (V_3)

Variable Choice (V_1)
(red)

For each i
one blue vertex
for each vertex
not in row i
(V_4)

Missing edges
and vertices

"Gap" Choice(red)
(V_2)

One blue vertex for
each pair (V_5)

A blue vertex
for each
way of making
a clause false
(V_7)

Column 1

Enforce Sequel (blue)
(V_6)

Next Variable Choice

Fig. 21.2 Gadget for RED/BLUE NONBLOCKER

Such an expression corresponds directly to a circuit meeting the defining conditions for antimonotone $W[1, s]$. We will refer to the restriction of RED/BLUE
NONBLOCKER to graphs with blue vertices of maximum degree bounded by s as
s-RED/BLUE NONBLOCKER. We next argue that s-RED/BLUE NONBLOCKER is
complete for $W[1, s]$.

By Lemma 32.4.1, we can assume that we are given a normalized boolean expression. Thus, let X be a boolean expression in conjunctive normal form with clauses
of size bounded by s. Suppose X consists of m clauses $C_1, \ldots, C_m$ over the set of
n variables $x_0, \ldots, x_{n-1}$. We show how to produce in polynomial time by local replacement, a graph $G = (V_{\text{red}}, V_{\text{blue}}, E)$ that has a nonblocking set of size $2k$ if and
only if X is satisfied by a truth assignment of weight k.

Before we give any details, we give a brief overview of the construction, whose
component design is outlined in Fig. 21.2. There are $2k$ "red" components arranged
in a line. These are alternatively grouped as blocks from $V_{\text{red}} = V_1 \cup V_2$ ($V_1 \cap V_2 =$
$\emptyset$), with a block of vertices from V_1 and then V_2 to be precisely described below. The
idea is that V_1 blocks should represent a positive choice (corresponding to a literal
being true) and the V_2 blocks corresponding to the "gap" until the next positive
choice. We think of the V_1 blocks as $A(0), \ldots, A(k - 1)$. [With a V_2 block between
successive $A(i)$, $A(i + 1)$ blocks, the last group following $A(k)$.] We will ensure

that for each pair in a block, there will be a blue vertex connected to the pair and nowhere else (these are the sets V_3 and V_5). This device ensures that at most one red vertex from each block can be chosen, and since we must choose $2k$, this ensures that we choose *exactly* one red vertex from each block. The reader should think of the V_2 blocks as arranged in k columns. Now, if i is chosen from a V_1 block, we will ensure that column i gets to select the next gap. To ensure this we connect a blue degree 2 vertex to i and each vertex not in the ith column of the next V_2 block. Of course, this means that if i is chosen, since these blue vertices must have an unchosen red neighbor, we must choose from the ith column. The final part of the component design is to enforce consistency in the next V_1 block; that is, if we choose i and have a gap choice in the next V_2 block, column i, of j, then the next chosen variable should be $i + j + 1$. Again, we can enforce this by using many degree 2 blue vertices to block any other choice. (These are the V_6 vertices.) Also, we keep the k choices for V_1 blocks in ascending order $1 \leq c_1 \leq \cdots \leq c_k \leq n$ with c_i in $A(i)$ by the use of extra blue enforcers, the blue vertices V_8. The idea here is that we ensure that for each j in $A(i)$ and each $j' \leq j$ in $A(q)$ with $(q > i)$, there is a blue vertex v in V_8 adjacent to both j and j'. This ensures that if j is chosen in block $A(i)$, then we cannot choose any $j' \leq j$ in any subsequent V_1 block. The last part of the construction is to force consistency with the clauses. We do this as follows. For each way a nonblocking set can correspond to making a clause false, we make a blue vertex and join it up to the s relevant vertices. This ensures that they cannot *all* be chosen. (This is the point of the V_7 vertices.) We now turn to the formal details.

The red vertex set V_{red} of G is the union of the following sets of vertices:

$$V_1 = \{a[r_1, r_2] : 0 \leq r_1 \leq k - 1, 0 \leq r_2 \leq n - 1\},$$
$$V_2 = \{b[r_1, r_2, r_3] : 0 \leq r_1 \leq k - 1, 0 \leq r_2 \leq n - 1, 1 \leq r_3 \leq n - k + 1\}.$$

The blue vertex set V_{blue} of G is the union of the following sets of vertices:

$$V_3 = \{c[r_1, r_2, r_2'] : 0 \leq r_1 \leq k - 1, 0 \leq r_2 < r_2' \leq n - 1\},$$
$$V_4 = \{d[r_1, r_2, r_2', r_3, r_3'] : 0 \leq r_1 \leq k - 1, 0 \leq r_2, r_2' \leq n - 1, 0 \leq r_3,$$
$$r_3' \leq n - 1 \text{ and either } r_2 \neq r_2' \text{ or } r_3 \neq r_3'\},$$
$$V_5 = \{e[r_1, r_2, r_2', r_3] : 0 \leq r_1 \leq k - 1, 0 \leq r_2, r_2' \leq n - 1, r_2 \neq r_2',$$
$$1 \leq r_3 \leq n - k + 1\},$$
$$V_6 = \{f[r_1, r_1', r_2, r_3] : 0 \leq r_1, r_1' \leq k - 1, 0 \leq r_2 \leq n - 1,$$
$$1 \leq r_3 \leq n - k + 1, r_1' \neq r_2 + r_3\},$$
$$V_7 = \{g[j, j'] : 1 \leq j \leq m, 1 \leq j' \leq m_j\},$$
$$V_8 = \{h[r_1, r_1', j, j'] : 0 \leq r_1 < r_1' \leq k - 1 \text{ and } j \geq j'\}.$$

In the description of V_7, the integers m_j are bounded by a polynomial in n and k whose degree is a function of s, which will be described below. Note that since s is a fixed constant independent of k, this is allowed by our definition of reduction for parameterized problems.

For convenience, we distinguish the following sets of vertices:

$$A(r_1) = \{a[r_1, r_2] : 0 \le r_2 \le n - 1\},$$

$$B(r_1) = \{b[r_1, r_2, r_3] : 0 \le r_2 \le n - 1, 1 \le r_3 \le n - k + 1\},$$

$$B(r_1, r_2) = \{b[r_1, r_2, r_3] : 1 \le r_3 \le n - k + 1\}.$$

The edge set E of G is the union of the following sets of edges. In these descriptions, we implicitly quantify over all possible indices for the vertex sets $V_1, \ldots, V_8$.

$$E_1 = \{a[r_1, q]c[r_1, r_2, r_2'] : q = r_2 \text{ or } q = r_2'\},$$

$$E_2 = \{b[r_1, q_2, q_3]d[r_1, r_2, r_2', r_3, r_3'] : \text{either } (q_2 = r_2 \text{ and } q_3 = r_3)$$

$$\text{or } (q_2 = r_2' \text{ and } q_3 = r_3')\},$$

$$E_3 = \{a[r_1, r_2]e[r_1, r_2, q, q']\},$$

$$E_4 = \{b[r_1, q, q']e[r_1, r_2, q, q']\},$$

$$E_5 = \{b[r_1, r_2, r_3]f[r_1, r_1', r_2, r_3]\},$$

$$E_6 = \{a[r_1 + 1 \bmod n, r_1']f[r_1, r_1', r_2, r_3]\},$$

$$E_7 = \{a[r_1, j]h[r_1, r_1', j, j']\},$$

$$E_8 = \{a[r_1', j']h[r_1, r_1', j, j']\}.$$

We say that a red vertex $a[r_1, r_2]$ *represents the possibility* that the boolean variable x_{r_2} may evaluate to *true* (corresponding to the possibility that $a[r_1, r_2]$ may belong to a $2k$-element nonblocking set V' in G). Because of the vertices V_8 and the edge sets E_7 and E_8, any k nonblocking elements one from each V_1 block, $\{a[r_j, r_{i_j}] : j = 0, \ldots, k - 1\}$, $i_0 < i_1 < \cdots < i_k \le k - 1$. We say that a red vertex $b[r_1, r_2, r_3]$ *represents the possibility* that the boolean variables $x_{r_2+1}, \ldots, x_{r_2+r_3-1}$ (with indices reduced mod n) may evaluate to *false*.

Suppose C is a clause of X having s literals. There are $O(n^{2s})$ distinct ways of choosing, for each literal $l \in C$, a single vertex representative of the possibility that $l = x_i$ may evaluate to *false*, in the case that l is a positive literal or in the case that l is a negative literal $l = \neg x_i$, a representative of the possibility that x_i may evaluate to *true*. For each clause C_j of X, $j = 1, \ldots, m$, let $R(j, 1), R(j, 2), \ldots, R(j, m_j)$ be an enumeration of the distinct possibilities for such a set of representatives. We have the additional sets of edges for the clause components of G:

$$E_9 = \{a[r_1, r_2]g[j, j'] : a[r_1, r_2] \in R(j, j')\},$$

$$E_{10} = \{b[r_1, r_2, r_3]g[j, j'] : b[r_1, r_2, r_3] \in R(j, j')\}.$$

Suppose X has a satisfying truth assignment τ of weight k, with variables $x_{i_0}, x_{i_1}, \ldots, x_{i_{k-1}}$ assigned the value *true*. Suppose $i_0 < i_1 < \cdots < i_{k-1}$. Let $d_r = i_{r+1 \pmod{k}} - i_r \pmod{n}$ for $r = 0, \ldots, k-1$. It is straightforward to verify that the set of $2k$ vertices

$$N = \{a[r, i_r] : 0 \leq r \leq k-1\} \cup \{b[r, i_r, d_r] : 0 \leq r \leq k-1\}$$

is a nonblocking set in G.

Conversely, suppose N is a $2k$-element nonblocking set in G. It is straightforward to check that a truth assignment for X of weight k is described by setting those variables *true* for which a vertex representative of this possibility belongs to N, and by setting all other variables to *false*.

Note that the edges of the sets E_1 (E_2) which connect pairs of distinct vertices of $A(r_1)$ [$B(r_1)$] to blue vertices of degree 2 enforce that any $2k$-element nonblocking set must contain exactly one vertex from each of the sets $A(0), B(0), A(1), B(1), \ldots, A(k-1), B(k-1)$. The edges of E_3 and E_4 enforce (again, by connections to blue vertices of degree 2) that if a representative of the possibility that x_i evaluates to *true* is selected for a nonblocking set from $A(r_1)$, then a vertex in the ith row of $B(r_1)$ must be selected as well, representing (consistently) the interval of variables set false (by increasing index because of the E_7 and E_8 edges) until the "next" variable selected to be *true*. The edges of E_5 and E_6 ensure consistency between the selection in $A(r_1)$ and the selection in $A(r_1 + 1 \bmod n)$. The edges of E_9 and E_{10} ensure that a consistent selection can be nonblocking if and only if it does not happen that there is a set of representatives for a clause witnessing that every literal in the clause evaluates to *false*. (There is a blue vertex for every such possible set of representatives.) $\square$

Theorem 21.2.2 provides the starting point for demonstrating the following dramatic collapse of the $W[1]$ stratification.

Theorem 21.2.3 $W[1] = W[1, 2]$.

Proof It suffices to argue that for all $s \geq 2$, ANTIMONOTONE $W[1, s] = W[1, 2]$. The argument here consists of another change of variables. Let C be an antimonotone $W[1, s]$ circuit for which we wish to determine whether a weight k input vector is accepted. We show how to produce a circuit C' corresponding to an expression in conjunctive normal form with clause size 2 that accepts an input vector of weight

$$k' = k2^k + \sum_{i=2}^{s} \binom{k}{i}$$

if and only if C accepts an input vector of weight k. (The circuit C' will, in general, not be antimonotone, but this is immaterial by Theorem 21.2.2. Actually, in the proof of Theorem 30.12.1, another reduction is given, which is polynomial in both k for a fixed s.)

Let $x[j]$ for $j = 1, \ldots, n$ be the input variables to C. The idea is to create new variables representing all possible sets of at most s and at least two of the variables $x[j]$. Let $A_1, \ldots, A_p$ be an enumeration of all such subsets of the input variables $x[j]$ to C. The inputs to each *or* gate g in C (all negated, since C is antimonotone) are precisely the elements of some A_i. The new input corresponding to A_i represents that all of the variables whose negations are inputs to the gate g have the value *true*. Thus, in the construction of C', the *or* gate g is replaced by the negation of the corresponding new "collective" input variable.

We introduce new input variables of the following kinds:
1. One new input variable $v[i]$ for each set A_i for $i = 1, \ldots, p$ to be used as above.
2. For each $x[j]$, we introduce 2^k copies $x[j, 0], x[j, 1], x[j, 2], \ldots, x[j, 2^k - 1]$.

In addition to replacing the *or* gates of C as described above, we add to the circuit additional *or* gates of fanin 2 that provide an enforcement mechanism for the change of variables. The necessary requirements can be easily expressed in conjunctive normal form with clause size 2, and thus can be incorporated into a $W[1, 2]$ circuit.

We require the following implications concerning the new variables:
1. The $n \cdot 2^k$ implications, for $j = 1, \ldots, n$ and $r = 0, \ldots, 2^k - 1$,

$$x[j, r] \Rightarrow x[j, r + 1 \; (\mathrm{mod} \; 2^k)].$$

2. For each containment $A_i \subseteq A_{i'}$, the implication

$$v[i'] \Rightarrow v[i].$$

3. For each membership $x[j] \in A_i$, the implication

$$v[i] \Rightarrow x[j, 0].$$

It may be seen that this transformation may increase the size of the circuit by a linear factor exponential in k. We make the following argument for the correctness of the transformation.

If C accepts a weight k input vector, then setting the corresponding copies $x[i, j]$ among the new input variables accordingly, together with appropriate settings for the new "collective" variables $v[i]$, yields a vector of weight k' that is accepted by C'.

For the other direction, suppose C' accepts a vector of weight k'. Because of the implications in (1) above, exactly k sets of copies of inputs to C must have value 1 in the accepted input vector (since there are 2^k copies in each set). Because of the implications described in (2) and (3) above, the variables $v[i]$ must have values in the accepted input vector compatible with the values of the sets of copies. By the construction of C', this implies there is a weight k input vector accepted by C. $\square$

We have now done most of the work required to show that the following well-known problems are complete for $W[1]$:

INDEPENDENT SET
Instance: A graph $G = (V, E)$.

Parameter: A positive integer k.
Question: Is there a set $V' \subseteq V$ of cardinality k, such that for $\forall u, v \in V'$, $uv \notin E$?

CLIQUE

Instance: A graph $G = (V, E)$.
Parameter: A positive integer k.
Question: Is there a set of k vertices $V' \subseteq V$ that forms a complete subgraph of G (that is, a clique of size k)?

Theorem 21.2.4
 (i) INDEPENDENT SET *is complete for* $W[1]$.
(ii) CLIQUE *is complete for* $W[1]$.

Proof It is easy to observe that INDEPENDENT SET belongs to $W[1]$. By Theorems 21.2.2 and 32.4.4, it is enough to argue that INDEPENDENT SET is hard for ANTIMONOTONE $W[1, 2]$. Given a boolean expression X in conjunctive normal form (product of sums) with clause size 2 and all literals negated, we may form a graph G_X with one vertex for each variable of X and having an edge between each pair of vertices corresponding to variables in a clause. The graph G_X has an independent set of size k if and only if X has a weight k truth assignment.

(ii) This follows immediately by considering the complement of a given graph. The complement has an independent set of size k if and only if the graph has a clique of size k.   □

Finally, we are in a position to establish our first analog of Cook's theorem, which we restate for the reader's convenience. In fact, we can now state an extended form.

Theorem 21.2.5 (Parameterized Cook's Theorem (I)) *The following are complete for* $W[1]$:
1. (Downey, Fellows, Liming Cai and J. Chen) SHORT TURING MACHINE ACCEPTANCE.
2. WEIGHTED 2-SATISFIABILITY.
3. s-RED/BLUE NONBLOCKER *for* $s \geq 2$.
4. INDEPENDENT SET.
5. CLIQUE.

Proof We need only prove that SHORT TURING MACHINE ACCEPTANCE is $W[1]$-complete to finish the result. To show hardness for $W[1]$, we reduce from CLIQUE. $W[1]$-hardness will then follow by Theorem 32.4.5(ii). Let $G = (V, E)$ be a graph for which we wish to determine whether it contains a k-clique. We have shown how to construct a nondeterministic Turing machine M that can reach an accept state in $k' = f(k)$ moves if and only if G contains a k-clique. The Turing machine M is designed so that any accepting computation consists of two phases. In the first phase, M nondeterministically writes k symbols representing vertices of G in the first k tape squares. (There are enough symbols so that each vertex of G is represented by a symbol.) The second phase consists of making $\binom{k}{2}$ scans of the k tape squares, each scan devoted to checking, for a pair of positions i and j, that the vertices represented by the symbols in these positions are adjacent in G. Each such pass can be accomplished by employing $O(|V|)$ states in M dedicated to the ijth scan.

In order to show membership in $W[1]$, it suffices to show how the SHORT TURING MACHINE ACCEPTANCE problem for a Turing machine $M = (\Sigma, Q, q_0, \delta, F)$ and positive integer k can be translated into one about whether a circuit C accepts a weight k' input vector, where C has depth bounded by some t (independent of k and the Turing machine M) and has only a single large (output) *and* gate, with all other gates small. We arrange the circuit so that the k' inputs to be chosen to be set to 1 in a weight k' input vector represent the various data: (1) the ith transition of M, for $i = 1, \ldots, k$, (2) the head position at time i, (3) the state of M at time i, and (4) the symbol in square j at time i for $1 \leq i, j \leq k$. Thus, we may take $k' = k^2 + 3k$. In order to force exactly one input to be set equal to 1 among a pool of input variables (for representing one of the above choices), we can add to the circuit for each such pool of input variables, and for each pair of variables x and y in the pool, a small "not both" circuit representing $(\neg x \vee \neg y)$. It might seem that we must also enforce (e.g., with a large *or* gate) the condition "at least one variable in each such pool is set true"—but this is actually unnecessary, since in the presence of the "not both" conditions on each pair of input variables in each pool, an accepted weight k' input vector *must have* exactly one variable set true in each of the k' pools. Let n denote the total number of input variables in this construction. We have $n = O(k\delta + k^2 + k|Q| + k^2|\Sigma|)$ in any case.

The remainder of the circuit encodes various checks on the consistency of the above choices. These consistency checks conjunctively determine whether the choices represent an accepting k-step computation by M, much as in the proof of Cook's theorem. These consistency checks can be implemented so that each involves only a bounded number b of the input variables. For example, we will want to enforce that if five variables are set true indicating particular values of the tape head position at time $i + 1$ and the head position, state, scanned symbol, and machine transition at time i, then the values are consistent with δ. Thus, $O(n^5)$ small "checking" circuits of bounded depth are sufficient to make these consistency checks; in general, we will have $O(n^b)$ "checking" circuits for consistency checks involving b values. All of the small "not both" and "checking" circuits feed into the single large output *and* gate of C. The formal description of all this is laborious but straightforward.   □

We remark that Theorem 21.2.1 depends crucially on there being no bound on the size of the Turing machine alphabets in the definition of the problem. If we restrict SHORT TURING MACHINE ACCEPTANCE to Turing machines with $|\Sigma|$ bounded by some constant b, then the number of configurations is bounded by $b^k |Q| k$ and the problem becomes fixed-parameter tractable (see Exercise 21.3.7). Later we will see that the alphabet bounded versions of the results will provide a possible base for lower bound arguments on the size of kernels. This idea, due to Hermelin, Kratsch, Soltys, Wahlström, and Wu [388] is discussed in Sect. 30.12.

We also remark that there are other approaches to defining the class $W[1]$. Notably, Flum and Grohe [309, 312] use Fagin type second order model checking (as per Theorem 9.4.1 and Courcelle's Theorem) as a basis for the complexity classes. Their basic model is a random access machine, and for classes of structures you would be looking at Fagin definability problems (cf. Theorem 9.4.1) of the form:

φ-MODEL CHECKING PARAMETERIZED BY k

Instance: A graph $G = (V, E)$[1] and a first order formula φ with free relation
           variable X of arity s.
Parameter: φ and k.
Question: Is there $S \subseteq V^s$ with $|S| = k$ such that $G \models \varphi(k)$?

This extends to classes of formulas Φ in place of φ so that Φ-MODEL CHECK-ING for k is the collection of languages in φ-MODEL CHECKING for k for some $\varphi \in \Phi$. Then if $t \geq 1$, and $\Phi = \Pi_t$, Flum and Grohe *define*

$$W[t] = \big\{ L \mid L \leq_{\text{fpt}} \{\widehat{L} \mid \widehat{L} \text{ is an instance of } \Phi\text{-MODEL CHECKING for } k\} \big\}.$$

This definition evidently makes some of the completeness proofs easier since logic is close to SAT, but the machine characterizations remain more or less as difficult.

The following is a related hierarchy that we will meet in Exercise 26.5.3, called the A-hierarchy.

φ-MODEL CHECKING PARAMETERIZED BY $|\varphi|$

Instance: A graph $G = (V, E)$.
Parameter: $|\varphi|$.
Question: $G \models \varphi$?

In fact, the whole book Flum and Grohe [312] makes logic the central player of the complexity theory in the spirit of the area of descriptive complexity and finite model theory. This tradition goes back to earlier studies on relational databases.

We assume the reader is familiar with the basic ideas of relational database theory. A relational database is essentially just a finite set of finite relations on a finite set U. Each relation can be viewed as a set of tuples listed in a table for the relation. The size of a database is the length of the description of all the relations by listing the contents of the relational tables.

Formally, there is a finite set U of attributes, each with an associated domain. A *relational schema* R is a subset of U. A *database schema* $D = (R_1, \dots, R_n)$ is a set of relation schemas. A relation r_i over a relation schema R_i is a set of R_i-tuples, where a R_i-tuple is a mapping from the attributes of R_i to their domains. A database d over the schema D is a set $\{r_1, \dots, r_n\}$ of finite relations.

We consider here the complexity of queries formulated in the *relational calculus*. Formulas of the calculus are built from atomic formulas of the form $x_1 = x_2$ and $R_i(x_1, \dots, x_m)$, where the x_i are variables or constants of the appropriate do-

---

[1]Of course, here we are using graphs for simplicity, but really this would be a τ-*structure* $\mathcal{A}$ for some signature τ of the formulae φ. Graphs are universal anyway.

mains, using Boolean connectives and quantifiers. A formula φ with free variables $y_1, \ldots, y_k$ specifies a query relation

$$r_\varphi = \{(a_1, \ldots, a_k) : \varphi(a_1, \ldots, a_k) \text{ is true in } d\}.$$

A formula will be said to be in *normal form* if all quantifiers are up front and the body of the formula is in conjunctive normal form. For example, the query

$$\varphi = \exists a \forall b \forall c \big(R_1(a, b, d) \vee \neg R_2(a, c) \vee \neg R_1(a, c, e)\big) \wedge \big(\neg R_1(a, d, e) \vee R_2(d, e)\big)$$

is in normal form.

Various problems may be considered concerning the complexity of computations associated with a query φ. The following problem is simply stated, and a number of other problems can be easily reduced to it. (For example, the problem of determining whether a target tuple x belongs to the relation defined by the query is easily reduced to the non-emptiness problem.)

QUERY NON-EMPTINESS

Instance: A database d, and a query φ.
Parameter: $|\varphi|$.
Question: Is the query relation r_φ nonempty?

Later this problem will be shown the be hard for a class called $AW[*]$ in Chap. 26 in Theorem 26.4.1.

In this section we consider the complexity of the restriction of the CONJUNCTIVE QUERY NON-EMPTINESS problem to conjunctive queries. A query φ is termed *monotone* if: (1) if it has only existential quantifications, and (2) only conjunctions.

Theorem 21.2.6 (Papadimitriou and Yannakakis [554], Downey, Fellows, and Taylor [258]) CONJUNCTIVE QUERY NON-EMPTINESS *is* $W[1]$-*hard.*

Proof A reduction from CLIQUE is straightforward and can be sketched as follows. Let $G = (V, E)$ be the graph for which we wish to determine whether it contains a k-clique. Let V be the base set of the database d that has binary relation R recording the edges of the graph. Order V arbitrarily (denoted $<$), and let

$$R = \{(u, v) : u < v \text{ and } uv \in E\}.$$

The database d also has a "padding" unary relation T:

$$T = \{(u) : u \in V\}.$$

The query to determine whether there is a k-clique is just

$$\varphi = \exists x_1 \exists x_2 \ldots \exists x_k R(x_1, x_2) \wedge R(x_1, x_3) \wedge \cdots \wedge R(x_{k-1}, x_k) \wedge T(x).$$

Note that the size of φ is a function only of k.

The only free variable in φ is x. If there is a k-clique in G, then the unary relation R_φ defined by φ is

$$R_\varphi = \bigl\{(v) : v \in V\bigr\}$$

and if there is no k-clique in G then $R_\varphi = \emptyset$. The ordering of V used in defining R implicitly forces a successful instantiation of the existentially quantified variables to represent k distinct vertices of G. Note that φ is monotone. □

The expression of the CLIQUE problem as a database query problem is really quite straightforward. One can easily work out a similar representation for the DOM-INATING SET problem.

Later, in Chap. 26, we prove that MONOTONE QUERY NONEMPTINESS is and hence CONJUNCTIVE QUERY NONEMPTINESS is $W[1]$-complete. As mentioned above, the general QUERY NONEMPTINESS is hard for a class we call $AW[*]$. These results are due to Downey, Fellows, and Taylor improving earlier results of Papadimitriou and Yannakakis.

21.3 Exercises

Exercise 21.3.1 Show that if we consider RED/BLUE NONBLOCKER where all degrees are bounded by s, then it is FPT.

Exercise 21.3.2 (Fellows, Hermelin, Rosamond, Vialette [291]) Using a reduction from CLIQUE prove that the following is $W[1]$-complete.

k-MULTICOLORED CLIQUE

Input: A graph G with a vertex coloring $\chi : V(G) \to [k]$.
Parameter: A positive integer k.
Question: Does G have a colorful clique of k vertices? That is, is there a set $\{v_1, \ldots, v_k\}$ forming a clique with $\chi(v_j) \neq \chi(v_i)$ if $i \neq j$?

Exercise 21.3.3 (Cygan, Fomin and van Leeuwen [187]) Reducing CLIQUE, prove that the problem SAVING k VERTICES from Exercise 5.3.6 is $W[1]$-hard even for bipartite graphs.

Exercise 21.3.4 (Fellows, Hermelin, Rosamond, Vialette [291]) Recall that an interval graph is one of bounded pathwidth, and is one where the vertices of the graph can be represented as subintervals of a fixed interval, with xy an edge of G implying $I_x \cap I_y \neq \emptyset$, where I_v is the interval representing v. A t-interval graph is the intersection graph of a family of t intervals. Multiple interval graphs (i.e. t-interval for some t) have numerous applications to scheduling and to resource allocation (e.g. Bar-Yehuda, Halldórsson, Naor, Shachnai, and Shapira [55]). We have seen that interval graphs usually have FPT solutions to most NP-complete graph problems. The situation for multiple interval graphs is different. Prove

1. For any fixed t, CLIQUE is FPT for t-interval graphs.

   (Hint: Prove that if G is a t-interval graph with no k-clique, then G has a vertex of degree $\leq 2tk$.)

2. Prove that INDEPENDENT SET is $W[1]$-complete for 2-interval graphs.[2]

   (Hint: This is not easy. Reduce k-MULTICOLORED CLIQUE from the previous exercise.)

Exercise 21.3.5 (Fellows, Hermelin, Müller, and Rosamond [290]) In the definition of $W[1]$ replace the gates by majority gates, which output "yes "if at least half the inputs are "yes". This defines a class $\widehat{W}[1]$. Prove that $W[1] = \widehat{W}[1]$.

Exercise 21.3.6 (Jiang and Zhang [418]) Prove that INDEPENDENT DOMINATING SET is $W[1]$-hard for 2-interval graphs.[3]

Exercise 21.3.7 Prove that the problem RESTRICTED SHORT TURING MACHINE ACCEPTANCE is strongly FPT. This is the restriction of SHORT TURING MACHINE ACCEPTANCE to machines with Σ (the number of states) bounded by some fixed constant b.

Exercise 21.3.8 Prove that the INDEPENDENT SET is FPT when restricted to:
1. graphs with fixed maximum degree m,
2. planar graphs.

Exercise 21.3.9 (Some $W[1]$-complete problems related to algebraic structures.) Prove that the following problems are $W[1]$ complete.
  (i) SEMIGROUP EMBEDDING

   Input: A semigroup $S = (S, \cdot)$.
   Parameter: A semigroup $G = (G, \times)$.
   Question: Is $(G, \times)$ isomorphic to a subsemigroup of $(S, \cdot)$?
 (ii) SEMILATTICE EMBEDDING

   Input: An (upper) semilattice $(X, \leq_X)$.
   Parameter: A semilattice $(Y, \leq_Y)$.
   Question: Is $(Y, \leq_Y)$ a subsemilattice of $(X, \leq_X)$?
(iii) BIPARTITE GRAPH EMBEDDING

   Input: A bipartite graph G.
   Parameter: A bipartite graph H.
   Question: Can H be embedded into G?
(Hint. (i) First, we need to observe that SEMIGROUP EMBEDDING is in $W[1]$. Let $G = \{g_1, \ldots, g_k\}$ be the parameter semigroup, $G = (G, \times)$ with the appropriate multiplication table $M(G)$, and let S be the given semigroup $S = (S, \cdot)$ with $S = \{x_1, \ldots, x_n\}$. The weft 1 circuit consists of kn input variables labeled $\{(x_i, g_j) := 1, \ldots, n, j = 1, \ldots k\}$. The circuit needs a small *or* gate for each j,

---

[2]Note that the dual graph of a 2-interval graph is not necessarily a 2-interval graph so there is no contradiction from 1.

[3]The definition of multiple interval graph in the previous exercise. For more hardness results for multiple interval graphs we refer the reader to Jiang [417] and Jiang and Zhang [418].

j, and p for $j \neq p$ to express that if (x_i, g_j), then $\neg(x_i, g_p)$. This needs $\leq k^2 n$ small or gates. We also need $\leq k^3 n^2$ many small *or* and *and* gates to express the fact that the tables are compatible; that is, we express the implication "if (s_i, g_j) and (s_p, g_q), then (s_t, g_m)" where $s_i \cdot s_p = s_t$ and $g_j \times g_q = g_m$ is read off from, respectively, $M(G)$ and $M(S)$. All of these gates are then joined by a large *and* gate. Note that if the circuit has a weight k accepting input, then exactly one g_j can be chosen for an S_i, and hence G_j is a subsemigroup of S_i. (This technique works for many embedding results, and, in particular works for (ii) and (iii).)

As with all of (i) to (iii), the proof that they are $W[1]$-hard uses reductions from the literature, and, in particular Power [563] and Booth [105]. We reduce CLIQUE to SEMIGROUP EMBEDDING. Given a graph $G = (V, E)$ and parameter k, we use Booth's semigroup defined as $S_G = V \cup E \cup \{0\}$, where

$$
\begin{aligned}
y \cdot x = x \cdot y &= x && \text{if } x = y \\
&= y && \text{if } x \in y \text{ and } x \in V \text{ and } y \in E \\
&= \{x, y\} && \text{if } \{x, y\} \in E \\
&= 0 && \text{otherwise.}
\end{aligned}
$$

One must observe that if G contains a k clique with $\varphi : K_k \to G$ then $\widetilde{\varphi} :_{K_k} \to S_G$ defined via $0 \to 0$ and $x \to \varphi(x)$ for $x \in V \cup E$ induces an embedding of S_{K_k} into S_G. Similarly, if $\varphi :_{K_k} \to S_G$ and $k \geq 3$, then since 0_{S_G} is the only element y of S_G with at least 4 elements x and $x \times y = y$ we must have φ takes 0 to 0. φ maps edges to edges since if y is an edge then there exist at least 3 elements with $x \cdot y = y$ yet if y is a vertex there is exactly one such x, and $\varphi(x) \neq 0$ by the facts that $0 \to 0$ and φ is 1–1. By the homomorphism property, and definition of φ, φ takes vertices to vertices and hence induces an injection from K_k to G. So G contains a k-clique.

Cases (ii) and (iii) are rather similar, but we give some details since Power [563] is not published. For SEMILATTICE EMBEDDING, on a derived Booth semigroup S_H if we define $\leq_H$ on S_H via $a \leq_H b$ iff $a \cdot b$, then $(S_H, \leq_H)$ is a lower semilattice (the meet of a and b is $a \cdot b$ and since all elements of S_G are idempotent, this follows by some standard semigroup theory e.g. Howie [402]). It is straightforward to show that S_{K_k} is embeddable into S_G if $(S_{K_k}, \leq_{K_k})$ is a sublattice of $(S_G, \leq_G)$.

Finally we consider (iii). To see that BIPARTITE GRAPH EMBEDDING is $W[1]$-complete, the Hasse Diagram of $(S_{K_k}, \leq_{K_k})$ and $(S_G, \leq_G)$ are both bipartite graphs and one can simply argue that the former is embeddable in the latter iff the first poset is embeddable in the latter.)

Exercise 21.3.10 Prove that MONOTONE $W[1] \subseteq$ FTP.
(Hint: First prove that if C is a $W[1]$ circuit of depth d then C has $\leq 2^d$ many large gates. (That is, assuming that "small" means fanin 2.) Now, normalize the circuit a large *And* of mostly small *or*'s but with possibly $\leq 2^d$ large *or*'s. Now argue by the method of search trees. See Corollary 24.1.1.)

Exercise 21.3.11 Using the classical proof of Cook's Theorem, Theorem 1.1.1, as a model carefully complete the details of the proof of Theorem 21.2.1(i).

Exercise 21.3.12 Prove that the following problem IRREDUNDANT SET is in $W[1]$. (Downey, Fellows, and Raman [254] have proven that IRREDUNDANT SET is $W[1]$-hard.) The problem MAXIMAL IRREDUNDANT SET (see Exercise 20.3.1) is known to be $W[2]$-hard.)

IRREDUNDANT SET
Instance: A graph $G = (V, E)$, a positive integer k.
Parameter: A positive integer k.

Question: Is there a set $V' \subseteq V$ of cardinality k having the property that each vertex $u \in V'$ has a private neighbor? (A *private neighbor* of a vertex $u \in V'$ is a vertex u' (possibly $u' = u$) with the property that for every vertex $v \in V'$, $u \neq v$, $u' \notin N[v]$.)

(Hint: Let $G = (V, E)$ be a graph for which we wish to determine if G has a k-element irredundant set. We construct the circuit C corresponding to the following boolean expression E. The variables of E are:

$$\{p[i, x, y] : 1 \leq i \leq k, x, y \in V\} \cup \{c[i, x] : 1 \leq i \leq k, x \in V\}.$$

The expression E has the clauses:
(1) $(\neg p[i, x, y] + \neg p[i, x', y'])$ for $1 \leq i \leq k$ and either $x \neq x'$ or $y \neq y'$, $x, x', y, y' \in V$.
(2) $(\neg c[i, x] + \neg c[i, x'])$ for $1 \leq i \leq k$ and $x \neq x'$, $x, x' \in V$.
(3) $(c[i, x] + \neg p[i, x, y])$ for $1 \leq i \leq k$ and $x, y \in V$.
(4) $(\neg c[j, u] + p[i, x, y])$ for $1 \leq i, j \leq k$, $i \neq j$ and $u \in N[y]$.
    Now argue that the circuit C accepts a weight $2k$ input vector if and only if G has a k-element irredundant set.)

Exercise 21.3.13 (Downey, Estivill-Castro, Fellows, Prieto, and Rosamond [238]) Prove that the following problems are $W[1]$-hard.
1. GRAPH k-CUT
    Input: A graph $G = (V, E)$, an edge-weighting function $w : E \to \mathbb{N}$, and positive integers k, m.
    Parameter: k.
    Question: Is there a set $C \subseteq E$ with $\sum_{e \in C} w(e) \leq m$, such that the graph $G' = G - C$ obtained from G by removing the edges of the *cut* C has at least k connected components?
2. CUTTING k VERTICES FROM A GRAPH
    Input: A graph $G = (V, E)$, an edge-weighting function $w : E \to \mathbb{N}$, and positive integers k, m.
    Parameter: k.
    Question: Is there a set of k vertices $V' \subseteq V$ such that
    $$\sum_{uv \in E : u \in V', v \in V - V'} w(uv) \leq m?$$

(Hint: They are both similar. For both reduce from CLIQUE. For 1, let $G = (V, E)$ be the graph on n vertices for which we wish to determine if G has a k-clique. We construct a graph G' as follows: (1) Start with $n + 2$ disjoint cliques of size n^4. We will refer to these as the *top clique* C_{top}, the *bottom clique* C_{bottom} and the n *vertex cliques* C_v, $v \in V$. (2) For each $v \in V$, connect C_v to C_{bottom} by a matching of size e $n^2 + n - \deg(v)$. (3) For each $v \in V$, connect C_v to C_{top} by a matching of size $\binom{k}{2}$. (4) For each edge $uv \in E$, make a single edge from C_u to C_v. Each edge weight is 1; take $m = k(n^2 + n) + (k - 1)\binom{k}{2}$; take $k' = k + 1$.
    Now prove that G has a k-clique if and only if (G', k', m) (with edge weights all 1) is a yes-instance for the GRAPH k-CUT problem. The second problem is similar.)

Exercise 21.3.14 (Cai [118], Marx [513]) Fix integers $\alpha, c \geq 1$, with $d \geq \alpha^c$. Prove that the problem of CLIQUE restricted to d-regular graphs is $W[1]$-complete.
(Hint: Reduce CLIQUE.)

Exercise 21.3.15 (Cai [118])   Show that the following problem is $W[1]$-hard on regular graphs.

k-INDUCED MATCHING

Input: A graph $G = (V, E)$.
Parameter: A positive integer k.
Question: Does G have k edges $M \subseteq E$ with $G[V(M)]$ mutually non-adjacent?

(Hint: Use the previous problem under the guise of INDEPENDENT SET.)

21.4   Historical Notes

A number of authors have looked at (machine dependent) complexity classes within P. These include the Kintala–Fischer β-hierarchy [444], and the work of Buss and Goldsmith [115]. These and similar investigations were mainly technical structural investigations and looked at problems for a *single k* rather than examining the asymptotic behavior of the parameterized problem as $k \to \infty$. This forces one to look at reductions such as "quasi-linear time reducibility" and the like (see, e.g., Naik, Regan and Sivakumar [540] and Gurevich-Shelah [368]). We will discuss some connections further in Part VI, Sect. 29.4. All of these investigations could be looked at as "miniaturizations" of classical complexity.

As best the authors can tell, the first author to note that the asymptotic behavior of fixed-parameter languages might be of interest was Ken Regan in [577]. Regan did so in some comments where he suggested that it might be important that DOMINATING SET might require time $\Omega(n^{k+1})$. The first article to attempt to study the asymptotic complexity of parameterized problems and the issue of n^{α} vs. $n^{f(k)}$ with $f(k) \to \infty$ was by Abrahamson, Ellis, Fellows and Mata [6], with some of those results recast in Buss-Goldsmith [115]. In that article, the objects were P-checkable, P-indexed relations, called (polynomial-time) *generator tester* pairs. For each k, one needed to be able to generate a P-time list of potential candidates for solutions that were easy to test. The actual definition is very involved, but the following example gives the flavor. If we consider the problem VERTEX COVER, then for an instance G and a parameter k, the potential witnesses would be pairs consisting of G and a vertex cover V with the index of $V \le j = \binom{n}{0} + \cdots + \binom{n}{k}$. The ideas only seemed to apply to relations in NP. Although there is a notion of parameterized tractability in [6], it is roughly equivalent to our notion of nonuniform fixed-parameter tractability and hence suffers from the problem that the tractable classes can be noncomputable. The real problems with that article are that (1) the notion of reducibility, which is defined on relations rather than parameterized languages, is rather unnatural and very unwieldy and (2) the notion of intractability is that of (essentially) being P-complete (or "dual P-complete") under *logspace* reducibility "by the slice"; that is, in [6], problems are "PGT-complete" (in their notation) only when for each k they are more or less P-complete. Of course, this means that the [6] ideas cannot address things such as DOMINATING SET and INDEPENDENT SET, nor apparently anything in the $W[t]$ classes of the next chapters. As we will see in Chap. 26 the [6] results can be easily placed in our setup, as they give $W[P]$-completeness results.

In this context, we see our major contributions as being the identification of the correct notions of reducibility, identifying the correct setting for the study of parameterized intractability, and, finally, identifying some "good" problems with which to measure hardness. The number of problems that have now been identified as $W[1]$-hard, and hence apparently intractable, would seem to support our claims.

All of the results of this chapter are taken from Downey–Fellows [244], except for the reduction from CLIQUE to SHORT TURING MACHINE ACCEPTANCE and the proof that SHORT TURING MACHINE ACCEPTANCE is in $W[1]$. These last two results are due to Cai, Chen, Downey, Fellows [124]. The reduction for RED/BLUE NONBLOCKER from Downey–Fellows [244] contained a flaw that was spotted by Alexander Vardy. This flaw is corrected here.

Further historical remarks can be gleaned from the account of Downey [237] which discusses the birth and early years of the area.

21.5 Summary

We introduce the basic class $W[1]$, and prove some basic completeness results.

Some Other $W[1]$ Hardness Results

22

22.1 PERFECT CODE and the Turing Way to W-Membership

PERFECT CODE

Instance: A graph $G = (V, E)$.

Parameter: A positive integer k.

Question: Does G have a k-element perfect code? A *perfect code* is a set of vertices $V' \subseteq V$ with the property that for each vertex $v \in V$ there is precisely one vertex in $N[v] \cap V'$.

We will begin by proving the following. The proof of hardness is relatively straightforward, but the proof of membership of $W[1]$ involves a new idea due to Cesati of using generic simulation by a Turing machine which seems a rather flexible technique. (See also the exercises.)

Theorem 22.1.1

1. (Downey and Fellows [244]) PERFECT CODE *is $W[1]$-hard.*
2. (Cesati [135]) PERFECT CODE *is in $W[1]$.*

Proof We first show that PERFECT CODE is $W[1]$-hard. We reduce from INDEPENDENT SET. Let $G = (V, E)$ be a graph. We show how to produce a graph $H = (V', E')$ that has a perfect code of size $k' = \binom{k}{2} + k + 1$ if and only if G has a k-element independent set. The vertex set V' of H is the union of the sets of vertices:

$$V_1 = \{a[s] : 0 \le s \le 2\},$$

$$V_2 = \{b[i] : 1 \le i \le k\},$$

$$V_3 = \{c[i] : 1 \le i \le k\},$$

$$V_4 = \{d[i, u] : 1 \le i \le k, u \in V\},$$

$$V_5 = \{e[i, j, u] : 1 \le i < j \le k, u \in V\},$$

R.G. Downey, M.R. Fellows, *Fundamentals of Parameterized Complexity*,
Texts in Computer Science, DOI 10.1007/978-1-4471-5559-1_22,
© Springer-Verlag London 2013

$$V_6 = \{f[i, j, u, v] : 1 \leq i < j \leq k,\ u, v \in V\}.$$

The edge set E' of H is the union of the sets of edges:

$$E_1 = \{a[0]a[i] : i = 1, 2\},$$

$$E_2 = \{a[0]b[i] : 1 \leq i \leq k\},$$

$$E_3 = \{b[i]c[i] : 1 \leq i \leq k\},$$

$$E_4 = \{c[i]d[i, u] : 1 \leq i \leq k, u \in V\},$$

$$E_5 = \{d[i, u]d[i, v] : 1 \leq i \leq k, u, v \in V\},$$

$$E_6 = \{d[i, u]e[i, j, u] : 1 \leq i < j \leq k, u \in V\},$$

$$E_7 = \{d[j, v]e[i, j, u] : 1 \leq i < j \leq k, v \in N[u]\},$$

$$E_8 = \{e[i, j, x]f[i, j, u, v] : 1 \leq i < j \leq k, x \neq u, x \notin N[v]\},$$

$$E_9 = \{f[i, j, u, v]f[i, j, x, y] : 1 \leq i < j \leq k, u \neq x \text{ or } v \neq y\}.$$

An overview of this construction and a (partial) example are given in Figs. 22.1 and 22.2.

Suppose C is a perfect code of size k' in H. Since $a[1]$ and $a[2]$ are pendant vertices attached to $a[0]$, neither vertex belongs to C because both cannot belong to C, and if only one belongs to C, then C fails to be a dominating set. It follows that $a[0] \in C$. This implies that none of the vertices in V_2 and V_3 belong to C (V_3 would kill V_2), and it implies also that exactly one vertex in each of the cliques formed by the edges of E_5 belongs to C (to cover V_3). Note that each of these k cliques has n vertices indexed by V, the vertex set of G (this is the *selection* gadget). Let I be the set of vertices of G corresponding to the elements of C in these cliques. We argue that I is an independent set of order k in G.

Suppose $u, v \in I$ and that $uv \in E$. Then there are indices $i < j$ between 1 and k such that (without loss of generality) $d[i, u] \in C$ and $d[j, v] \in C$. By the definition of E_6 and E_7 each of these vertices is adjacent to $e[i, j, u]$, which contradicts that C is a perfect code in H. Thus I is an independent set in G.

Conversely, we argue that if $J = \{u_1, \ldots, u_k\}$ is a k-element independent set in G, then H has a perfect code C_J of size k'. We may take C_J to be the following set of vertices:

$$C_J = \{a[0]\} \cup \{d[i, u_i] : 1 \leq i \leq k\} \cup \{f[i, j, u_i, u_j] : 1 \leq i, j \leq k\}.$$

That C_J is a perfect code can be verified directly from the definition of H.

Now we show membership using Cesati's method. The idea is to make a parametric reduction from PERFECT CODE to SHORT NTM ACCEPTANCE. Let $G = (V, E)$ have n vertices and let the parameter be k. We construct a NTM $M = (\Sigma, Q, \Delta)$ with alphabet Σ including $\{\Box\} \cup \{\sigma_v, s_i \mid v \in V, 1 \leq i \leq n\}$. The state set Q include $\{q_A, q_R\} \cup \{q_i \mid i \in [0, k]\} \cup \{q_j^l, q_j^r \mid j \in [1, n]\}$. (Here q_A s the accept state and q_R

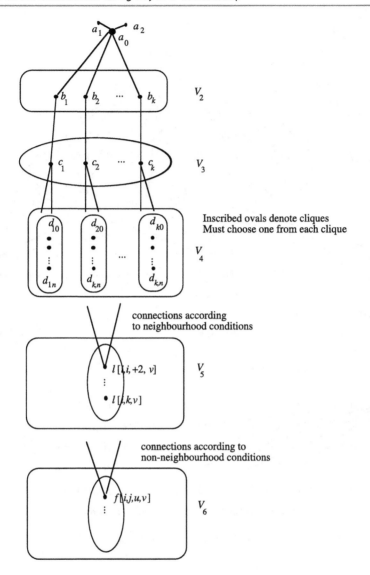

Inscribed ovals denote cliques
Must choose one from each clique

connections according
to neighbourhood conditions

connections according to
non-neighbourhood conditions

Fig. 22.1   Overview of INDEPENDENT SET $\leq_m^s$ PERFECT CODE

the reject one.) The Turing machine M starts in state q_0 and all tape cells have the
blank symbol $\square$. It then operates in three phases (we use the notation $+1$ for move
right, -1 for move left and 0 for do not move).

   Phase 1. *Guess k vertices.* The Turing machine nondeterministically chooses k
vertices of G via the transitions in Δ below.

- $\langle \square, q_i, \sigma_v, q_{i+1}, +1 \rangle$, for all $i \in [0, k-2]$, $v \in V$ (guess $k-1$ vertices moving
  right), and $\langle \square, q_{k-1}, \sigma_v, q_k, 0 \rangle$ (guess the last vertex and do not move).

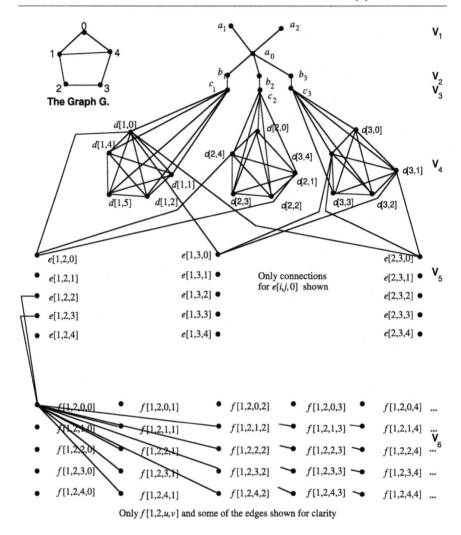

Fig. 22.2 Example of INDEPENDENT SET $\leq_m^s$ PERFECT CODE

Phase 2. Check that the selected vertices are perfect. The Turing machine scans
the guessed vertices and rejects if it finds two vertices that violate at least one of the
conditions below.
1. For each pair of guessed vertices x, y, x, and y are different.
2. For each pair of guessed vertices x, y, x, and y are not adjacent.
3. For each pair of guessed vertices x, y, there is no $z \in V$ adjacent to both of them.
In this phase each symbol σ_v is replaced by a the symbol s_m where m represents the
size of the neighborhood $N[v]$. We add the following transitions.
- $\langle \sigma_v, q_k, \sigma_v, q_k^l, -1 \rangle$ for each $v \in V$ (enter a loop for checking the vertex v under
  the head)

- $\langle \sigma_w, q_k^l, \sigma_w, \widehat{q}, -1 \rangle$ where $\widehat{q}$ is determined by the graph as follows:
  1. $\widehat{q} = r_R$ if $v = w$.
  2. $\widehat{q} = q_R$ if $vw \in E$.
  3. $\widehat{q} = q_R$ if $\exists z \in V(vz \in E \wedge wz \in E)$.
  4. $\widehat{q} = q_L$ otherwise.

  The interpretation here is that we move to the left if the vertex w under the head satisfies the conditions with respect to v, and reject otherwise. Also for all $v, w \in V$ with $v \neq w$, we need $\langle \square, q_v^l, \square, q_v^r, +1 \rangle$ and $\langle \sigma_w, q_v^r, \sigma_w, q_w^r, +1 \rangle$. (At the end of the symbols, go back to the right up to v.)
- $\langle \sigma_v, q_v^r, s_m, q_k, -1 \rangle$ for all $v \in V$ where $m = |N[v]|$. (Replace v with the symbol denoting the size of its closed neighborhood, then move left and enter the state q_k, thus restarting the loop with another vertex.)
- $\langle \square, q_k, \square, q_k, +1 \rangle$ (no more symbols to check, end the phase).

Phase 3. Taking the sum. The tape now can construct exactly k symbols of $\{s_1, \ldots, s_n\}$. Each of them represents the neighborhood size of a guessed vertex. The Turing machine accepts iff the sum of the sizes equals n. The check in phase 2 guarantee that no vertex in G belongs to mode than one such neighborhood. Thus the sum cannot be $> n$, and hence $= n$ iff the guess vertices form a k element perfect code.

- $\langle s_i, q_k, s_i, q_i^s, +1 \rangle$, for $i = 1, \ldots, n$ (initialize the state counter, and move to the right).
- For all $i, j \in \{1, \ldots, n\}$ with $i + j \leq n$ and $t = i + j$, $\langle s_j, q_i^s, s_j q_t^s, +1 \rangle$ (add the size under the head to the internal state counter and move right).
- For all $j \in \{1, \ldots, n-1\}$, $\langle \square, q_j^s, \square, q_R, 0 \rangle$ and $\langle \square, q_n^s, \square, q_A, o \rangle$ (at the end if in state q_n accept, else reject).

This Turing machine uses $\frac{5}{2}n^2 + \frac{k+7}{2}n + 1$ many instructions, and has size polynomial in $|G|$. It accepts in $k^2 + 4k + 2$ many steps iff G has a size k perfect code. $\square$

We remark that McCartin [523] has given a circuit version of the membership of PERFECT CODE. However, the method of Cesati has been used for other membership proofs, and is quite flexible. (See Exercises 23.5.8 and 25.2.5.)

In the exercises we will also look at problems coming from algebra and number theory. These include the following problems.

SIZED SUBSET SUM

Input: A set $X = \{x_1, \ldots, x_n\}$ of integers and a positive integer S.
Parameter: A positive integer k.
Question: Do there exist k elements of X that sum to S?

PERMUTATION GROUP FACTORIZATION

Instance: A set A of permutations $A \subseteq S_n$, and $x \in S_n$.
Parameter: A positive integer k.
Question: Does x have a factorization of length k over A?

22.2   The VC Dimension

All of these areas are very good arenas for the study of parameterized complexity since one is often concerned with small parameters but large objects. We begin by looking at a problem arising from logic, or more properly, computational learning theory. The *Vapnik–Chervonenkis dimension* is the central parameter in computational learning theory both in the PAC model and in the Angluin-type exact learning model. It has proved to be especially important in the determination of lower bounds in computational learning theory. We give this reduction since also it is definitely a parameterized one and not a polynomial one. That is exponential in the parameter. This behavior is somehow necessary as evidenced by its location (level 3) in the E-hierarchy of Flum, Grohe, and Weyer [313], a hierarchy not treated in the current book, that is, concerned with the constants in FPT algorithms. The interested reader should refer to Flum, Grohe and Weyer [313].

Definition 22.2.1 (Vapnik–Chervonenkis dimension)   The *Vapnik–Chervonenkis dimension* (VC dimension) of a family of subsets F of a base set X is the maximum cardinality of a set $S \subseteq X$ such that for every subset $S' \subseteq S$, $\exists Y \in F$ such that $S \cap Y = S'$. In general, we say that such a subset S' of S is *generated* in S by F. The VC dimension of F is thus the largest cardinality of $S \subseteq X$ such that every subset of S is generated by F.

We can naturally associate a (parameterized) language with the concept of the VC dimension.

VAPNIK–CHERVONENKIS DIMENSION
Instance: A family of subsets F of a base set X.
Parameter: A positive integer k.
Question: Is the VC dimension of F at least k?

The VC dimension provides a very nice concrete example of the situation where problems with very different classical complexities can have the same parameterized ones. The VC dimension of a family of sets $\mathcal{F}$ over a base set X of cardinality n can be shown to be at most $\log n$. Consequently, the above problem is unlikely to be NP-complete [553]. It is in fact LOGNP-complete and hence can only be NP-complete in the unlikely event that $NP \subseteq DTIME(n^{\log n})$. However, the following theorem shows that VAPNIK–CHERVONENKIS DIMENSION is $W[1]$-complete.[1] *In the following theorem, both membership and hardness for the parameterized complexity class $W[1]$ involve reductions that are exponential in the parameter k.*

---

[1] Notice that parameterized complexity analysis thus provides an indirect method of demonstrating *practical* intractability of problems either not likely to be NP-complete, or problems whose complexity is currently unclassified. One example of this is the proof by Bodlaender, Fellows and Hallett [87] that PRECEDENCE CONSTRAINED SCHEDULING is $W[1]$-hard and hence will not have practical algorithms for reasonable k. This is one of the open problems from Garey and Johnson [337]. A really nice candidate for such analysis is GRAPH ISOMORPHISM.

Theorem 22.2.1 (Downey and Fellows [245])   VAPNIK–CHERVONENKIS DI-
MENSION *is complete for* $W[1]$.

Proof (i) *Membership of* $W[1]$ (Downey and Fellows [245]).
   Let $\mathcal{F} \subseteq 2^U$ denote the family of sets under consideration. Suppose $\mathcal{F} = \{X_j : 1 \leq j \leq m\}$ and $U = \{1, \dots, n\}$. It suffices to show that in time $f(k) \cdot (mn)^\alpha$ we can produce a product-of-sums boolean expression E in which the clauses have size bounded by some constant, and such that E has a satisfying truth assignment of weight $k' = g(k)$ if and only if the VC dimension of $\mathcal{F}$ is at least k.
   The set of variables for E is $V = V_1 \cup V_2$ where:

$$V_1 = \{a[i, j] : 1 \leq i \leq 2^k, \ 1 \leq j \leq m\},$$

$$V_2 = \{b[r, s] : 1 \leq r \leq k, \ 1 \leq s \leq n\}.$$

   The variables of V_1 serve to indicate which sets in $\mathcal{F}$ witness the shattering of the k-element subset of U indicated by the variables of V_2. Let γ be a fixed $1 : 1$ correspondence between the integers i in the range $1 \leq i \leq 2^k$ and the length k 0–1 vectors, and write $\gamma_i(l) \in \{0, 1\}$ to indicate the value of the lth component of the vector associated to i.
   We will take $k' = k + 2^k$, and $E = E_1 \cdot E_2$ as follows.
   E_1 is a product of small clauses that enforces the conditions:
(1) For each index value of i (r) in the definition of V_1 (V_2), at most one variable of V_1 (V_2) is set to *true*.
(2) If $b[i, j]$ and $b[i', j']$ are set to *true*, with $i < i'$, then $j < j'$.
This can be accomplished with clauses of size 2.
   Note that any satisfying truth assignment to E_1 of weight $k + 2^k$ must set *exactly* one variable of V_1 (V_2) *true* for each index value of i (r).
   E_2 is the product of clauses expressing the implications: $a[i, j] \Rightarrow \neg b[r, s]$ for all *incompatible* pairs of indices (i, j) and (r, s), where such a pair is defined to be *incompatible* if and only if either (1) $\gamma_i(r) = 1$ and $s \notin X_j$, or (2) $\gamma_i(r) = 0$ and $s \in X_j$.
   The verification that this works correctly is straightforward.
   (ii) $W[1]$-*Hardness* (Downey, Evans and Fellows [239]).
   We use a reduction from CLIQUE. Given a graph $G = (V, E)$ and a positive integer k we describe how to compute a family of sets F over a base set X, so that F has VC dimension k if and only if G has a k-clique. The cardinality of the family F that we will describe is $O(k^2 n^2 + 2^k)$, and the size of the base set X is kn, where n is the order of G. For convenience we will assume that $V = \{1, \dots, n\}$. We write $[m]$ to denote the set $\{1, \dots, m\}$.
   The base set X is simply

$$X = \{(u, i) : u \in V, \ i \in [k]\}.$$

The family F consists of four subfamilies, $F = F_0 \cup F_1 \cup F_2 \cup F_3$ which are described as follows. (These correspond, roughly, to the cardinality of the sets in the subfamilies.)

$$F_0 = \{\emptyset\},$$

$$F_1 = \{\{(u, i)\} : u \in V, \ i \in [k]\},$$

$$F_2 = \{\{(u, i), (v, j)\} : uv \in E, \ i, j \in [k]\},$$

$$F_3 = \{\{(u, i) : i \in S\} : S \subseteq [k], \ \#(S) \geq 3\}.$$

To see this that this construction works, let C be the clique in G and let f be any $1 : 1$ map from C to $\{1, \ldots, k\}$. Consider the set $S \subseteq X$ of cardinality k:

$$S = \{(u, f(u)) : u \in C\}.$$

If $S' \subseteq S$ has cardinality at least 3, then it is generated by the corresponding set in F_3. It is straightforward to verify that subsets of S of cardinality smaller than 3 are generated by $F_0 \cup F_1 \cup F_2$.

Conversely, suppose S is a k-element subset of X, every subset of which is generated by F. For each subset $S' \subseteq S$ choose a witness $W \in F$ with $W \cap S = S'$. If S' has cardinality at least 3, then its witness must be chosen from F_3. But this implies that every set in F_3 must serve as a witness for some $S' \subseteq S$ of cardinality at least 3.

The witnesses for sets $S' \subseteq S$ of cardinality 2 must therefore belong to F_2. We cannot have both (u, i) and (u, j) in S for $i \neq j$, else there is no witness possible for the 2-element set consisting of these (by the definition of F_2). Consequently S must range over k different vertex indices. The fact that there are witnesses for all of the 2-element subsets implies that there is a corresponding k-clique in G.   □

22.3   Logic Problems

Turning from computational learning theory, we look at one other problem from logic and formal language theory. Specifically, we consider the following two problems.

SHORT DERIVATION (FOR UNRESTRICTED GRAMMARS)

Instance: The two pieces of information:
   (1) A grammar $G = (N, \Sigma, \Pi, S)$, where N is a finite set of "nonterminal symbols", Σ is a finite set of "terminal symbols", Π is a finite set of "production rules" of the form $(\alpha \to \beta)$ with $\alpha, \beta \in (N \cup \Sigma)^*$, and $S \in N$ is the "start symbol".
   (2) A word $x \in \Sigma^*$.

Parameter: A positive integer k.

Question: Is there a G-derivation of x of length k? That is, is there a sequence of words $x_0, \ldots, x_k$, with $x_i \in (N \cup \Sigma)^*$ for $i = 0, \ldots, k$, that satisfies the requirements:

(1) $x_0 = S$,

(2) $x_k = x$, and

(3) for each $i = 1, \ldots, k$ there is a production rule $\pi_i = (\beta \to \gamma) \in \Pi$ such that $x_{i-1} = \alpha \beta \delta$ and $x_i = \alpha \gamma \delta$?

We note in passing that if the above problem is restricted to grammars having production rules in which the left hand side always consists of a single nonterminal symbol (termed a *context-free grammar*), then it is fixed-parameter tractable (Exercise 22.4.1).

SHORT POST CORRESPONDENCE

Instance: A Post system Π and a positive integer k, where a *Post system* consists of a finite alphabet Σ and two sequences $\alpha = (\alpha_1, \ldots, \alpha_n)$ and $\beta = (\beta_1, \ldots, \beta_n)$.

Parameter: k.

Question: Is there a length k solution for Π? That is, is there a sequence of integers $i_1, \ldots, i_k$ (not necessarily distinct) such that $\alpha_{i_1} \cdots \alpha_{i_k} = \beta_{i_1} \cdots \beta_{i_k}$?

Lemma 22.3.1 (Cai, Chen Downey and Fellows [123]) CLIQUE *reduces to* SHORT DERIVATION.

Proof Let $G = (V, E)$ be a simple graph, and k a positive integer. We describe a grammar $\mathcal{G} = (N, \Sigma, \Gamma, S)$, a word $x \in \Sigma^*$, and a positive integer $k' = f(k)$ such that the start symbol S derives x in k' steps if and only if G has a k-clique.

The terminal symbols of the grammar are

$$\Sigma = \{\#\} \cup \{u : u \in V\} \cup \{l\}.$$

The nonterminal symbols are

$$N = N_1 \cup N_2 \cup N_3 \cup N_4$$

where

$$N_1 = \{S, S', M, W, Z\},$$

$$N_2 = \{X_i : 1 \leq i \leq k\},$$

$$N_3 = \{R[i, j] : 1 \leq i \leq k,\ 0 \leq j \leq k - 1\},$$

$$N_4 = \left\{ T[i, u, j] : 1 \leq i \leq k,\ u \in V,\ 1 \leq j \leq \binom{k}{2} \right\}.$$

The start symbol is S.

The target string $x \in \Sigma^*$ is

$$x = \#^{3\binom{k}{2}} l^{k+2}$$

where the exponents indicate symbol repetition.

The set of productions Γ is

$$\Gamma = \Gamma_1 \cup \Gamma_2 \cup \cdots \cup \Gamma_{12}.$$

In the formal description of some of these production rule sets (in particular, Γ_6, Γ_7 and Γ_8) we make reference to the following sets of indices. Let σ denote the sequence of symbols in the set $\mathcal{S} = \{\sigma[i,j] : 1 \le i < j \le k\}$ in increasing lexicographic order. Thus the length of the sequence σ is $\binom{k}{2}$. Let s_r denote the rth symbol of the sequence σ. For $i = 1, \ldots, k$ define

$$J_i = \{r : s_r = \sigma[i,j],\ 1 \le j \le k\},$$

$$J_i' = \{r : s_r = \sigma[j,i],\ 1 \le j \le k\},$$

$$J_i'' = \left\{1, \ldots, \binom{k}{2}\right\} - (J_i \cup J_i').$$

Thus for $k = 4$ we have $\sigma = \sigma[1,2]\sigma[1,3]\sigma[1,4]\sigma[2,3]\sigma[2,4]\sigma[3,4]$, $J_1 = \{1,2,3\}$, $J_1' = \emptyset$, $J_1'' = \{4,5,6\}$, $J_2 = \{4,5\}$, $J_2' = \{1\}$ and $J_2'' = \{2,3,6\}$.

Intuitive overview. Let S denote the starting symbol for the grammar. The initial production rule (the only one that can be applied) is

$$S \to S'R[k,1] \cdots R[1,1]\#X[1]X[2]\#X[1]X[3]\# \cdots \#X[k-1]X[k]Z.$$

There are productions by which the pairs of symbols $X[i]X[j]$ may produce a pair of terminal symbols representing an edge of the graph; this represents in some sense a "guess" about a k-clique in G. Note that there are $\binom{k}{2}$ pairs corresponding to the $\binom{k}{2}$ edges of a k-clique. For such a guess to be consistent, it must be verified that each occurrence of the symbol $X[i]$ produces the same guessed vertex (as the endpoint of a guessed edge). The R and T symbols make this consistency check by "commuting across" the intermediate string. The last index of these symbols functions as a counter that keeps track of which edge of the potential clique is being checked. The $R[i, *]$ ("read") symbols commute until the first occurrence of the ith vertex (produced by $X[i]$) of the potential k-clique is encountered. At this point the symbol $R[i, *]$ is commuted past the edge of this first occurrence, but transformed into a T ("test") symbol that records the identity of the ith vertex (as the second component in the indexing of the symbol). As this symbol continues to commute across the portion of the string corresponding to the guessed edges, there are three possibilities that arise with respect to next guessed edge of the commute: (1) the first endpoint of the guessed edge should correspond to the recorded vertex of the symbol, (2) the second endpoint of the guessed edge should correspond to the recorded vertex of the symbol, or (3) neither endpoint should correspond to the recorded vertex of the symbol. The positions in the sequence of $\binom{k}{2}$ guessed edges corresponding to these three possibilities are indexed by J_i, J_i' and J_i'', respectively, and the corresponding sets of production rules are Γ_6, Γ_7, and Γ_8. The "read and test" nonterminal symbols that implement the consistency checking can only be eliminated from the string by successfully commuting to the Z symbol on the right end of the string. In the final phase of a successful derivation of the target string x the symbol S' must also commute across in order to be eliminated at the right end. As it commutes, it replaces the guessed vertices with the place holding symbol #.

The details of the production rules of the grammar are as follows:

$$\Gamma_1 = \big\{ S \to S'R[k,1] \cdots R[1,1] \# X_1 X_2 \# X_1 X_3 \# \cdots$$
$$\cdots \# X_1 X_k \# X_2 X_3 \# X_2 X_4 \# \cdots \# X_{k-1} X_k Z \big\},$$

$$\Gamma_2 = \{ X_i X_j \to uv : 1 \le i < j \le k, \ uv \in E \},$$

$$\Gamma_3 = \big\{ R[1,1]\#uv \to \#uvT[1,u,2] : uv \in E \big\},$$

$$\Gamma_4 = \big\{ R[i,j]\#uv \to \#uvR[i,j+1] : 2 \le i \le k, \ 1 \le j \le i-2, \ uv \in E \big\},$$

$$\Gamma_5 = \big\{ R[i,j]\#uv \to \#uvT[i,v,j+1] : 2 \le i \le k, \ j = i-1, \ uv \in E \big\},$$

$$\Gamma_6 = \big\{ T[i,u,j]\#uv \to \#uvT[i,u,j+1] : 1 \le i \le k, \ j \in J_i, \ uv \in E \big\},$$

$$\Gamma_7 = \big\{ T[i,v,j]\#uv \to \#uvT[i,v,j+1] : 1 \le i \le k, \ j \in J_i', \ uv \in E \big\},$$

$$\Gamma_8 = \big\{ T[i,u,j]\#xy \to \#xyT[i,u,j+1] : 1 \le i \le k, \ j \in J_i'' \big\},$$

$$\Gamma_9 = \left\{ T\left[i,u,\binom{k}{2}+1\right]Z \to ZM : 1 \le i \le k-1, \ u \in V \right\},$$

$$\Gamma_{10} = \left\{ T\left[k,u,\binom{k}{2}+1\right]Z \to WM \right\},$$

$$\Gamma_{11} = \big\{ S'\#uv \to \#\#\#S' : uv \in E \big\},$$

$$\Gamma_{12} = \big\{ S'WM^k \to l^{k+2} \big\}.$$

It remains only to specify the number of steps for the derivation:

$$k' = 1 + \binom{k}{2} + (k+1)\left[\binom{k}{2}+1\right].$$

Half of the correctness argument for the reduction is straightforward. If G has a k-clique then a derivation of x of length k' can be written down by following the sketch above, noting that all of the necessary means for commuting symbols from left to right are available. For the other half of the argument, it is easy to see that the necessary first step of the derivation (since it is the only one possible) creates a situation where the R symbols (or the T symbols into which they are transformed) must be commuted across the string. The key point is that this is possible only if the guessed edges are consistent in providing evidence of a k-clique in G. □

Corollary 22.3.1 SHORT DERIVATION *is* $W[1]$-*hard.*

Proof By Theorem 21.2.5, CLIQUE is $W[1]$-complete. □

Recall that a *context-sensitive* grammar is one where the productions $\alpha \to \beta$ satisfy the length restriction $|\alpha| \leq |\beta|$ and $\alpha, \beta \neq \varepsilon$. Our argument for Lemma 22.3.1 actually shows that SHORT DERIVATION remains $W[1]$ hard for this special case. We next establish membership in $W[1]$.

Lemma 22.3.2 (Cai, Chen Downey and Fellows [123]) SHORT DERIVATION *is in* $W[1]$.

Proof To show that SHORT DERIVATION belongs to $W[1]$ we will use Theorem 21.2.1. That is, we will describe (at a high level) how to reduce the problem of determining whether a word x can be generated in k steps from a given (unrestricted) grammar $G = (N, \Sigma, \Pi, S)$, to the problem of determining whether a nondeterministic Turing machine can reach a halting configuration in $f(k)$ steps.

The computation of the Turing machine is organized in two phases. The first phase nondeterministically guesses a description of the k-step derivation of x. (This description can be recorded by $O(k^2)$ symbols written in as many tape squares; we describe how this can be accomplished below.) The second phase consists of a deterministic computation that checks whether the guessed derivation of x is valid.

Suppose the sequence $x_0, \ldots, x_k$ corresponds to a k-step derivation of x in the grammar G in the sense that: (1) $x_0 = S$ is the start symbol of G, (2) $x_k = x$, and (3) for each i, $i = 1, \ldots, k$, x_i can be obtained from x_{i-1} by the application of the production rule $\alpha_i \to \beta_i$. For any word $w \in (N \cup \Sigma)^*$ let $w[j]$ denote the jth symbol of w, and for $s \leq t$ let $w[s, t]$ denote the substring of w consisting of the symbols $w[s] \ldots w[t]$. Thus for the production rule $\alpha_r \to \beta_r$ we write $\beta_r[s, t]$ to denote the substring of the yield β_r consisting of the sth through tth symbols.

We employ an alphabet with a distinct symbol for each possible substring $\beta_r[s, t]$ of the yield of a production rule. Note that the number of symbols required is bounded by a polynomial in the size of the description of the grammar G. Let Γ denote this alphabet.

The description of the k-step derivation of x consists of, for $i = 1, \ldots, k$:
(1) A factorization of x_i

$$x_i = \beta_{i_1}[s_{i_1}, t_{i_1}] \cdots \beta_{i_m}[s_{i_m}, t_{i_m}]$$

represented by symbols of Γ.
(2) The production rule $\alpha_i \to \beta_i$ that yields x_i from x_{i-1}.
(3) The substring of symbols of (1) for x_{i-1} that represents the symbols consumed in the application of the production rule identified in (2).
(4) The substring of symbols of (1) for x_i that represents the symbols yielded by the application of the production rule identified in (2).

We can describe appropriate factorizations for the strings $x_0, \ldots, x_k$ in more detail as follows. To each symbol of each of the strings $x_0, \ldots, x_k \in (N \cup \Sigma)^*$ we can associate a *time of production*: the index i of the string x_i in which the symbol first appeared (i.e., was in the yield of the application of the production rule that produced x_i from x_{i-1}), and a *time of consumption*: the index j (if one exists, otherwise say ∞) such that the symbol is consumed by the application of the production rule that yields x_{j+1} from x_j. Say that two symbols in x_i are *equivalent* if: (1) they are adjacent in x_i, and (2) they have the same times of production and consumption. The relevant factorization of x_i is then given by the equivalence classes. Each factor can be expressed by a symbol $\beta_r[s, t] \in \Gamma$. By an easy induction, the factorization for each x_i has a representation over the symbols of Γ of length at most $1 + 2(k - 1)$.

It follows from the definition of the equivalence relation that if x_i is produced from x_{i-1} by the production rule $\alpha_i \to \beta_i$, then α_i is represented by a substring of the symbols of Γ that represent the factorization of x_{i-1} and β_i is represented by a substring of the symbols of Γ that represent the factorization of x_i.

The second computational phase for the Turing machine consists of k deterministic checks, with the ith check verifying that the guessed information of the first phase relevant to the derivation of x_i from x_{i-1} is consistent. It is easy to see that each such check can be accomplished by $g(k)$ moves through a state space of polynomial size, for an appropriately chosen $g(k)$. $\qquad\square$

As a consequence of Corollary 22.3.1 and Lemma 22.3.2 we have proved the following.

Theorem 22.3.1 (Cai, Chen Downey and Fellows [123]) SHORT DERIVATION *is complete for* $W[1]$.

We next consider the complexity of SHORT POST CORRESPONDENCE.

Lemma 22.3.3 (Cai, Chen Downey and Fellows [123]) SHORT POST CORRESPONDENCE *reduces to* SHORT TURING MACHINE ACCEPTANCE.

Proof Given an instance (Π, k) of the SHORT POST CORRESPONDENCE problem, we can easily express the question in terms of whether a particular Turing machine has a k'-step accepting computation, for k' determined by an appropriate function of k, as follows. Consider a machine M that computes in two phases, over an alphabet that includes one symbol for each pair of strings in the Post system, and one symbol for each of the positive integers $1, \ldots, m$ where m is a bound on the total number of symbols (in the strings of the Post system) for any solution. (Clearly $m \le k|\Pi|$.) In the first phase, M writes on $3k$ tape squares indicating (nondeterministically) a *guessed solution* including the information: (1) what Post pair is the ith factor for $i = 1, \ldots, k$, (2) on what symbol positions the ith factors begin (in the two concatenated strings). In the second phase, M conducts a number of "checks", each consisting of: (1) a scan of the *guess*, recording in the resulting state q, e.g., that the ith factor is (x_j, y_j) and begins in the first component in symbol position r, and that the next factor is due to begin in symbol position s. There is a transition out of q in the transition table for M if and only if the information recorded in the state q is "valid", i.e., $s = r + |x_j|$. The number of states required for this check is $O(|\Pi|^3)$. It is not too hard to see that $2k$ checks of this sort are enough to ensure that the guessed starting positions of the factors are consistent. Similarly, it is necessary to make k^2 checks that each guessed (factor + starting position) in the first component is compatible with each (factor + starting position) in the second component. That is, we must ensure that this information does not imply any mismatched symbols in the two solution strings for the Post problem. A successful check will make $f(k)$ moves, where f is an appropriately chosen function of k. □

Theorem 22.3.2 (Cai, Chen Downey and Fellows [123]) SHORT POST CORRE-SPONDENCE *is complete for* $W[1]$.

Proof Lemma 22.3.3 shows that the problem belongs to $W[1]$ via the analogue of Cook's Theorem 21.2.5. To show that it is hard for $W[1]$ we compose the reduction of Lemma 22.3.1 (from CLIQUE) with the reduction of Theorem 4.2 of Davis [197] (attributed to an unpublished manuscript of Floyd). That is, we demonstrate that SHORT DERIVATION in k steps for reduces to SHORT POST CORRESPONDENCE.

Actually, Floyd's reduction applies to what are called "Semi-Thue" processes. A semi-Thue process Π consists of a finite alphabet $A = \{a_1, \ldots, a_n\}$ together with a set of productions of the form $g \to \overline{g}$. We will write $u_1 \Rightarrow_\Pi u_2$ to mean that there is a production $g \to \overline{g}$ and strings a and b such that $u_1 = agb$ and $u_2 = a\overline{g}b$. Similarly we will, as usual, denote the transitive closure by $\Rightarrow_\Pi^*$. The parametric problem at hand is

SHORT SEMI-THUE PROCESS

Input: A semi-Thue process Π, and words w, v.
Parameter: A positive integer k.
Question: Does $w \Rightarrow_\Pi^* v$ in $\leq k$ steps?

In general, Floyd's reduction below does *not* constitute a *parameterized* reduction of SHORT DERIVATION to SHORT POST CORRESPONDENCE. However, the image of the reduction from CLIQUE described by Lemma 22.3.1 consists of instances of the SHORT DERIVATION problem where: (1) the grammar is context sensitive and is, indeed a semi-Thue process, and (2) the word to be derived in k steps has length bounded by a function of k. Thus to complete our proof we need only describe a parameterized reduction from SHORT SEMI-THUE PROCESS to SHORT POST CORRESPONDENCE. As we see below Floyd's reduction does the trick. Here is Floyd's according to Davis [197].

Let Π be a semi-Thue process on the alphabet $A = \{a_1, \ldots, a_n\}$, and let u, v be words on A. We shall construct a Post correspondence problem P on the alphabet

$$B = \{a_1, \ldots, a_n, a_1', \ldots, a_n', [,], *, *'\},$$

consisting of $2n + 4$ symbols. For any word w on A we write w' for the word obtained from w by placing $'$ after each symbol of w.

Let the productions of Π be $g_i \to \overline{g}_i$, $i = 1, 2, \ldots, k$. We shall assume that included among these productions are the n "identity" productions $a_i \to a_i$, $i = 1, 2, \ldots, n$. This last assumption does not restrict generality (i.e. the class of pairs $\langle r, s \rangle$ for which $r \Rightarrow_\Pi^* s$ is in no way changed by including the identity productions). However, we may now state that $u \Rightarrow_\Pi^* v$ if and only if we can write

$$u = u_1 \Rightarrow_\Pi u_2 \Rightarrow_\Pi \cdots \Rightarrow_\Pi u_m = v$$

where m is *odd*. The Post correspondence system P on the alphabet B is then to consist of the pairs:

$$\langle [u*, [\rangle, \quad \langle *, *'\rangle, \quad \langle *', *\rangle,$$

$$\left.\begin{array}{c} \langle \overline{g}_j, g'_j \rangle \\ \langle \overline{g}'_j, g_j \rangle \end{array}\right\} j = 1, 2, \ldots, k, \quad \langle], *'v] \rangle.$$

Now, let

$$u = u_1 \Rightarrow_{\Pi} u_2 \Rightarrow_{\Pi} \cdots \Rightarrow_{\Pi} u_m = v$$

where m is odd. Then the word

$$w = \left[u_1 * u'_2 *' u_3 * \cdots * u'_{m-1} *' u_m \right]$$

is a solution of P as is obvious from the two decompositions

$$w = \left[u_1 * | u'_2 *' | u_3 * | \cdots | \right]$$
$$= \left[| u_1 * | u'_2 *' | \cdots | *' u_m \right].$$

To see for example that $u'_2 *'$ corresponds to $u_1 *$ we note that we can write $u_1 = r g_j s$, $u_2 = r \overline{g}_j s$ for some $j = 1, 2, \ldots, k$. Then $u'_2 = r' \overline{g}'_j s'$ and the correspondence is obvious (recalling that the productions $a_1 \to a_1, \ldots, a_n \to a_n$ are present in Π).

Conversely, if $\overline{w}$ is a solution then to avoid mismatches on the extreme left and right (between primed and unprimed symbols), $\overline{w}$ must begin [and end]. Hence, letting w be the portion of $\overline{w}$ up to the first],

$$w = \left[u * \cdots *' v \right],$$

where to begin with we have the correspondences

$$w = \left[u * | \cdots *' v | \right],$$
$$w = \left[| u * \cdots | *' v \right].$$

Hence we must have $u*$ corresponding to some $r'*'$ and $*'v$ to some $*s'$, where $u \Rightarrow_{\Pi}^* r$ and $s \Rightarrow_{\Pi}^* v$. Continuing this procedure, we see that $u \Rightarrow_{\Pi}^* v$. This completes the construction and one easily sees that this is indeed a parameterized polynomial time reduction from SHORT SEMI-THUE PROCESS to SHORT POST CORRESPONDENCE. □

Corollary 22.3.2 SHORT SEMI-THUE PROCESS *is $W[1]$ complete*.

As is well known there is an intimate connection between logical decision problems and tiling problems. We consider the complexity of tiling small regions; the problem is defined as follows (see Garey and Johnson [337]).

SQUARE TILING

Instance: Set C of *colors*, collection $T \subseteq C^4$ of *tiles* (where $\langle a, b, c, d \rangle$ denotes a tile whose top, right, bottom, and left sides are colored a, b, c, d, respectively), and a positive integer k.

Parameter: k.

Question: Is there a tiling of a $k \times k$ square using the tiles in T, i.e., an assignment f of a tile $A(i, j) \in T$ to each ordered pair i, j, $1 \le i \le k$, $1 \le j \le k$, such that (1) if $f(i, j) = \langle a, b, c, d \rangle$ and $f(i + 1, j) = \langle a', b', c', d' \rangle$, then $a' = c$, and (2) if $f(i, j + 1) = \langle a', b', c', d' \rangle$ then $b = d'$?

Theorem 22.3.3 (Cai, Chen Downey and Fellows [123]) SQUARE TILING *is complete for* $W[1]$.

Proof Membership in $W[1]$ is straightforward again using SHORT TURING MACHINE ACCEPTANCE and Theorem 21.2.5. Given an instance of the tiling problem as defined above we can create a Turing machine M that reaches a halting configuration in k' steps (where k' is an appropriate function of k) as follows. In the first phase of computation, M writes onto k^2 tape squares nondeterministically a choice of tiles. In the second phase of computation M makes $2k(k - 1)$ passes, each pass dedicated to checking the compatibility of the tiles chosen for two adjacent positions in the $k \times k$ square. In any given pass, the first recorded tile of the pair is "remembered" as state information.

To show that the problem is hard for $W[1]$ we reduce from the $W[1]$-complete problem CLIQUE in two steps. The most important step is the reduction of CLIQUE to the problem of tiling a $k_1 \times k_2$ rectangular region. The second and easier step is to reduce the rectangular tiling problem to SQUARE TILING. We address the easier step first. Suppose $k_1 \le k_2$.

Let T denote the set of tiles over the set of colors C. We describe how to construct a set of tiles T' over a set of colors C' such that a $k_1 \times k_2$ region can be tiled from T if and only if a $k' = k_2$ square region can be tiled from T'.

Suppose $z \notin C$. We may take $C' = C_1 \cup C_2$ where

$$C_1 = C \times \{0, \ldots, k_2\},$$

$$C_2 = \{z\} \times \{0, \ldots, k_2\}.$$

The set of tiles T' is the union $T' = T_1 \cup T_2 \cup T_3$ where

$$T_1 = \{\langle (a, i), (b, j + 1), (c, i + 1), (d, j) \rangle : 0 \le i \le k_1 - 1, 0 \le j \le k_2 - 1,$$
$$\langle a, b, c, d \rangle \in T\},$$

$$T_2 = \{\langle (a, k_1), (z, j + 1), (z, k_1 + 1), (z, j) \rangle : 0 \le j \le k_2 - 1, a \in C\},$$

$$T_3 = \{\langle (z, i), (z, j + 1), (z, i + 1), (z, j) \rangle : k_1 + 1 \le i \le k_2 - 1,$$
$$0 \le j \le k_2 - 1\}.$$

The basic idea is that the tiles of T_2 and T_3 provide for additional rows of padding to fill out the $k_2 \times k_2$ square. The verification that a $k_1 \times k_2$ rectangular tiling with tiles from T yields a $k_2 \times k_2$ square tiling with tiles from T' is straightforward and left to the reader.

Now suppose there is a $k_2 \times k_2$ square tiling f with tiles from T'. The second components of the C' colors of the tiles in T' forces the tile $A(i, j) = f(i, j)$ to have the form $\langle (x_1, i-1), (x_2, j), (x_3, i), (x_4, j-1) \rangle$, for $1 \le i, j \le k_2$. By the definition of T_1, and forgetting the second components of the colors, f must also describe a tiling of the $k_1 \times k_2$ rectangle in the first k_1 rows with tiles from T.

We next argue that CLIQUE can be reduced to the rectangular tiling problem. Let $G = (V, E)$ and k be an instance of CLIQUE. We will describe a set of colors C and a set of tiles T that can tile a $k_1 \times k_2$ rectangle if and only if G has a k-clique, where $k_1 = k$ and $k_2 = \binom{k}{2}$. We will consider the columns of the $k_1 \times k_2$ rectangle as indexed by J_k in lexicographic order. Let J_k denote the set of $\binom{k}{2}$ ordered pairs (r, s) with $1 \le r < s \le k$. Let J_k^+ denote J_k augmented with the single additional element 0. Consider J_k^+ to be linearly ordered by the lexicographic ordering inherited from J_k together with taking 0 to be the minimal element. For $\alpha \in J_k^+$ let $pre(\alpha)$ denote the immediate predecessor of α in the ordering of J_k^+.

The set of colors is $C = C_1 \cup C_2$ where

$$C_1 = J_k^+ \times V,$$
$$C_2 = \{0, \ldots, k\} \times E.$$

The set of tiles is

$$T = \{\langle (i-1, uv), (pre(\alpha), w), (i, uv), (\alpha, w) \rangle : 1 \le i \le k, \; uv \in E, \; w \in V,$$
$$\alpha = (r, s) \in J_k, \; \text{with } (w = u \text{ if } i = r) \text{ and } (w = v \text{ if } i = s) \}.$$

In discussing tiling it is useful to have the notion of the sequence of colors in a row or column. Given a valid tiling f of the $k_1 \times k_2$ rectangle, write

$$f(i, j) = \langle f_1(i, j), f_2(i, j), f_3(i, j), f_4(i, j) \rangle.$$

That is, f_1 describes the top color assignment, f_2 describes the right side color assignment, etc. Define the *sequence of colors of the ith row* to be the sequence of length $k_1 + 1$:

$$f_4(i, 1), f_2(i, 1) = f_4(i, 2), f_2(i, 2) = f_4(i, 3), \ldots,$$
$$f_2(i, k_1 - 1) = f_4(i, k_1), f_2(i, k_1).$$

Similarly define the sequence of colors of a column.

Claim 1 *If there is a k-clique in G then the $k_1 \times k_2$ rectangle can be tiled from T.*

Proof Let $v_1, \ldots, v_k$ be the vertices of the clique. We describe a tiling by describing the row and column color sequences. In the ith row the sequence of colors is

$$(0, v_i), \big((1, 2), v_i\big), \big((1, 3), v_i\big), \ldots, \big((1, k), v_i\big), \big((2, 3), v_i\big), \ldots, \big((k-1, k), v_i\big).$$

In the column indexed by $\alpha = (r, s)$ the sequence of colors is

$$(0, v_r v_s), (1, v_r v_s), (2, v_r v_s), \ldots, (k, v_r v_s).$$

From this point, it remains only to check that these sequences can be realized by tiles of T, which we leave to the reader. □

Claim 2   *If the $k_1 \times k_2$ rectangle can be tiled from T then G has a k-clique.*

Proof Let f denote the tiling. By forgetting the first components, we may refer to the colors of C_1 as *representing* a vertex, and we may similarly refer to the colors of C_2 as representing an edge. First note that in any valid tiling from T we must have the same vertex represented in all of the colors of a row sequence, simply because every tile has the same vertex represented on its sides. Similarly, the same edge is represented in the sequence of colors of any column. Furthermore, the first components of the sequence of colors in a column must be increasing and is thus forced to be: $0, 1, 2, \ldots, k$. A similar statement holds for the sequence of colors in a row. Write v_i to denote the vertex represented in row i. Let $i < j$ and consider the column indexed by (i, j). It is straightforward to check that the definition of T forces the edge represented by this column to be $v_i v_j$, which implies that the v_i are distinct and pairwise adjacent in G. □

Claims 1 and 2 complete the proof of the theorem. □

22.4   Exercises

Exercise 22.4.1   Prove that SHORT DERIVATION is FPT if the left hand side always consists of a single nonterminal symbol.

Exercise 22.4.2 (Downey and Fellows [244])   Consider the following problem.

WEIGHTED EXACT CNF SATISFIABILITY
Instance: A boolean expression E in conjunctive normal form.
Parameter: A positive integer k.
Question: Is there a truth assignment of weight k to the variables of E that makes exactly one literal in each clause of E *true*?

Prove the following: PERFECT CODE $\equiv^s_m$ WEIGHTED EXACT SATISFIABILITY.

Exercise 22.4.3 (Downey and Fellows [244]) Prove that PERFECT CODE $\leq^s_m$ SIZED SUBSET SUM and hence SIZED SUBSET SUM is hard for $W[1]$. (See Exercise 22.4.2 above.)

SIZED SUBSET SUM

Input: A set $X = \{x_1, \ldots, x_n\}$ of integers and a positive integer S.
Parameter: A positive integer k.
Question: Do there exist k elements of X that sum to S?

(Hint: Let $G = (V, E)$ be a graph for which we wish to determine whether G has a perfect code of size k. Suppose for convenience that the vertex set of the graph $V = \{0, \ldots, n-1\}$. We can easily compute the list of positive integers $L = (x[i, j] : 1 \leq i \leq k, \ 0 \leq j \leq n-1)$ and the positive integer M, where

$$x[i, j] = (k+1)^{n+k-i} + \sum_{u \in N[j]} (k+1)^u,$$

$$M = \sum_{t=0}^{n+k-1} (k+1)^t$$

such that L has a sublist of size k summing to M if and only if G has a k-element perfect code. The correctness of this transformation is easily observed if the numbers of L are represented in base $k+1$, and it is noted that there can be no carries in a sum of k integers from L expressed in this way.)

Exercise 22.4.4 (Downey and Fellows [244]) Consider the problem

SHORT CHEAP TOUR

Instance: A weighted graph and a positive integer S.
Parameter: A positive integer k.
Question: Is there a tour through at least k vertices of cost at most S?

The precise difficulty of this problem is at present open but a variation is hard for $W[1]$. Let SHORT EXACT TOUR be the same as SHORT CHEAP TOUR except that we ask that the tour costs *exactly* S. Prove that EXACT CHEAP TOUR is hard for $W[1]$.
(Hint: Reduce SIZED SUBSET SUM to EXACT CHEAP TOUR.)

Exercise 22.4.5 (Short algebraic factorizations) (Cai, Chen, Downey, and Fellows [123]) Prove that the following problem is hard for $W[1]$ via a reduction from PERFECT CODE above.

PERMUTATION GROUP FACTORIZATION

Instance: A set A of permutations $A \subseteq S_n$, and $x \in S_n$.
Parameter: A positive integer k.
Question: Does x have a factorization of length k over A?

(Hint: Let $G = (V, E)$ be a graph of order n for which we wish to determine whether there is a perfect code of size k. Let $n' = (k+1)n$. We show how to produce an equivalent factorization problem instance for $S_{n'}$.

View n' as divided into n blocks of size $k+1$, with these blocks in $1:1$ correspondence with the vertices of G. Let γ denote a cyclic permutation of the elements of a block. Our set A consists of n permutations, one for each vertex of G. For a vertex u and with $N[u]$ denoting the solid neighborhood of u in G, let a_u denote the element of $S_{n'}$ which acts on the blocks corresponding to $v \in N[u]$ according to γ, and which is the identity map on all other blocks. The permutation x to be factored consists of the permutation γ on each of the n blocks.)

22.5   Historical Notes

Many of the results are taken from the basic papers in parameterized complexity Downney and Fellows [240, 241, 243, 244]. Cesati's Turing Way is from his Ph.D. thesis and papers such as Cesati [135]. The logic problems are due to Cai, Chen Downey and Fellows [123].

22.6   Summary

We prove $W[1]$-hardness and completeness results for a number of problems ranging from graph theory, through learning theory and into logic. We introduce reductions which are not polynomial in the parameter. We introduce Cesati's "Turing way" for membership.

The W-Hierarchy

23

23.1 Introduction

Key to the definitional scheme for the W-hierarchy is the WEIGHTED WEFT t DEPTH h CIRCUIT SATISFIABILITY problem, $WCS(t, h)$ (Definition 21.2.1). This problem takes as input a boolean circuit with large-gate depth t and bounded depth h and asks if it has a weight k satisfying assignment. The parameter is k.

This is a very general problem scheme, and many natural parameterized problems can be "represented" in this format, by a trivial parameterized reduction. For example, the parameterized graph problem INDEPENDENT SET that asks if a graph G has a set of k vertices with no two adjacent, is easily represented by a circuit with:

- one input for each vertex of G,
- for each edge uv of G, an *or* gate that expresses "not u or not v", and
- an *and* gate giving the output of the circuit, which gathers the results of the *or* gates (and thus has unbounded fan-in), which we will call a *large* gate.

Asking if the circuit has a Hamming weight k input accepted (where the 1s in the input indicate the chosen k vertices), is clearly equivalent to what is asked in the parameterized INDEPENDENT SET problem—this is just a "circuit-based" representation. The reader will have no trouble finding a similar representation for the parameterized DOMINATING SET problem, of depth 2.

The W-hierarchy was originally motivated by observations such as above, stimulated by the empirical situation that while the INDEPENDENT SET can be relatively easily reduced to DOMINATING SET, no reduction in the other direction could be found. So what is the difference between these two parameterized problems, both presumably intractable in the parameterized framework?

Considering the "circuit representations" of these problems, one notices that the circuit representation of INDEPENDENT SET involves only one large gate (and depth 2), and the circuit representation of DOMINATING SET also has depth 2, but involves, on each input–output path in the circuit, *two* gates of unbounded fan-in.

R.G. Downey, M.R. Fellows, *Fundamentals of Parameterized Complexity*,
Texts in Computer Science, DOI 10.1007/978-1-4471-5559-1_23,
© Springer-Verlag London 2013

In Chap. 21 we proved that in the case $t = 1$ there is no difference between the parameterized power of weft t circuits and boolean expressions of logical depth t, in the sense that we proved that WEIGHTED 3SAT is $W[1]$-complete.

As noted in Example 20.1.1, *classically* there is no difference in the computational complexity of determining satisfiability of classes of propositional formulas no matter what their form (3CNF, CNF, or polynomial sized). However, *parameterized* reductions tend to be much more *structure preserving* than classical reductions, and certainly most classical reductions, such as the reduction between CNF SATISFIABILITY and 3SAT, are definitely not parameterized reductions. In fact we believe that there is *no* parameterized reduction between WEIGHTED 3SAT and WEIGHTED CNF SATISFIABILITY. We offer some evidence for this claim in the next chapter. It seems to us that the situation is the following. Classical reductions are so coarse that the vast majority of natural problems in NP that are apparently not in P are precisely NP-complete. Remarkably, in parameterized complexity, something similar occurs, in that almost all natural parameterized problems belong to about five complexity-equivalent classes, most prominently, the first three classes of the W-hierarchy.

In this chapter we will define a hierarchy of classes of problems of (presumably) increasing parameterized complexity. This hierarchy is based on the belief that WEIGHTED CNF SATISFIABILITY and WEIGHTED 3SAT are of differing complexity, and the underlying reason for this differing complexity is the difference in logical depth of the formulas. If CNF is thought of as products-of-sums of literals, then we can define a propositional formula to be t-*normalized* as follows.

> **Definition 23.1.1** We say that a propositional formula X is t-*normalized* if X is of the form products-of-sums-of-products ... of literals with t-alternations.

According to Definition 23.1.1, 2-normalized is the same as CNF. Now we believe that for all t, WEIGHTED t-NORMALIZED SATISFIABILITY, that is, satisfiability for t-normalized formulas, is strictly easier than WEIGHTED $t + 1$-NORMALIZED SATISFIABILITY. This suggests a hierarchy based on the complexity of t-normalized formulas. As in Chap. 4, we can consider t-normalized formula as special examples of weft t circuits, namely the weft t boolean circuits. Again we are able to prove that weft is, indeed, the key notion that determines computational power. The reader should recall from Definition 21.2.2 that

$$L \in W[t] \quad \text{iff} \quad L \text{ reduces to WCS}(t, h)$$

Definition 23.1.2 (The W-hierarchy) We term the union of these $W[t]$ classes together with two other classes $W[\text{SAT}] \subseteq W[\text{P}]$, the W-*hierarchy*. Here $W[\text{P}]$ denotes the class obtained by having no restriction on depth, i.e. P-size circuits, and $W[\text{SAT}]$ denotes the restriction to circuits corresponding to boolean formulas of P-size. Hence the W-hierarchy is

$$\text{FPT} \subseteq W[1] \subseteq W[2] \subseteq \cdots \subseteq W[\text{SAT}] \subseteq W[\text{P}].$$

We conjecture that each of the containments is proper.

We will explore $W[\text{SAT}]$ and $W[\text{P}]$ in Chap. 25. The remainder of the present chapter will be devoted to establishing some results about the $W[t]$ classes for $t \geq 2$. The first is a higher level version of Cook's Theorem, the *Normalization Theorem* below. Armed with the normalization theorem, we will then establish some concrete hardness results for $\bigcup_t W[t]$.

23.2 The Normalization Theorem

Theorem 23.2.1 (The Normalization Theorem, Downey and Fellows [241, 243, 244]) *For all* $t \geq 1$, WEIGHTED t-NORMALIZED SATISFIABILITY *is complete for* $W[t]$.

We remark that the Normalization Theorem plays a role in our theory analogous to Cook's Theorem for NP-completeness, in the following sense. Usually membership in a particular $W[t]$ class is easy so the circuit definition is easy to reduce *to*, while the t-normalized formulas provide problems that are easy to reduce *from* to establish hardness. The other view of the Cook(–Levin) theorem is that it connects a generic problem from Turing Machines to SATISFIABILITY and hence SATISFIABILITY is very unlikely to be tractable. This other view of the Cook(–Levin) theorem was realized in Theorem 21.2.1.

23.2.1 The DOMINATING SET Reduction

A very important part of the proof of the Normalization Theorem consists of the "hardwiring" of the gadget in the following key combinatorial reduction (Lemma 23.2.1). (None of the techniques used in the case $t = 1$ of Chap. 21 pertain to $t \geq 2$ *or conversely*.) The natural parameterized version of the DOMINATING SET problem is the following.

DOMINATING SET
Input: A graph G.
Parameter: A positive integer k.

Question: Does G have a dominating set of size k? (A dominating set for G is a set $X \subseteq V(G)$ such that for all $y \in V(G)$, there is an $x \in X$ with $\langle x, y \rangle \in E(G)$.)

Lemma 23.2.1   WEIGHTED CNF SATISFIABILITY $\equiv_m^s$ DOMINATING SET.

Proof Let X be a boolean expression in conjunctive normal form consisting of m clauses $C_1, \ldots, C_m$ over the set of n variables $x_0, \ldots, x_{n-1}$. We show how to produce in polynomial-time by local replacement, a graph $G = (V, E)$ that has a dominating set of size $2k$ if and only if X is satisfied by a truth assignment of weight k.

A diagram of the gadget used in the reduction is given in Fig. 23.1. The idea of the proof is as follows. There are k of the gadgets arranged in a vertical line. Each of the gadgets has 3 main parts. Taken from top to bottom, these are variable selection, gap selection and gap and order enforcement. The variable selection component $A(r)$ is a clique and the gap selection component $B(r)$ consists of n cliques which we call columns. Our first action is to ensure that in *any* dominating set of $2k$ elements, we must pick one vertex from each of these two components. This goal is achieved by the $2k$ sets of $2k + 1$ enforcers, vertices from V_4 and V_5. (The names refer to the sets below.) Take the V_4, for instance. For a fixed r, these $2k + 1$ vertices are connected to all of the variable selection vertices in the component $A(r)$, and nowhere else. Thus if they are to be dominated by a $2k$ dominating set, then we must choose *some* element in the set $A(r)$, and similarly we must choose an element in the set $B(r)$ by virtue of the V_5 enforcers. Since we will need exactly $2k$ (or even $\leq 2k$) dominating elements it follows that we must pick *exactly* one from each of the $A(r)$ and $B(r)$ for $r = 1, \ldots, k$.

As the name suggests these will be picked by the variable selection components, $A(r), r = 0, \ldots, k - 1$. Each of these k components consists of a clique of n vertices labeled $0, \ldots, n - 1$. The intention being that the vertex labeled i represents a choice of variable i being made true in the formula X. Correspondingly in the next $B(r)$ we have columns (cliques) $i = 0, \ldots, n - 1$. The intention is that column i corresponds to the choice of variable i in the preceding $A(r)$. The idea then is the following. We join the vertex $a[r, i]$ corresponding to variable i, in $A(r)$, to all vertices in $B(r)$ *except* those in column i. This means that the choice of i in $A(r)$ will cover all vertices of $B(r)$ except those in this column. It follows that we *must* choose the dominating element from this column and nowhere else. (There are no connections from column to column.) The columns are meant to be the gap selection saying how many 0's there will be till the next positive choice for a variable. We finally need to ensure that (i) if we chose variable i in $A(r)$ and gap j in column i from $B(r)$ then we need to pick $i + j + 1$ in $A(r + 1)$ and (ii) that the selections are in order. This is the role of the gap and order enforcement component which consists of a set of n vertices (in V_6.) For $r < k - 1$, the method is to connect vertex j in column i of $B(r)$ to all of the n vertices $d[r, s]$ *except* to $d[r, i + j + 1]$ *provided that* $i + j + 1 \leq n$. (For $r = k - 1$ we simply connect vertex j in column i of $B(r)$ to all of the n vertices $d[r, s]$ *except* to $d[r, i + j + 1]$, since this will need to "wrap around" to $A(0)$.) The first point of this process is that if we choose and j in column i with $i + j + 1 > n$

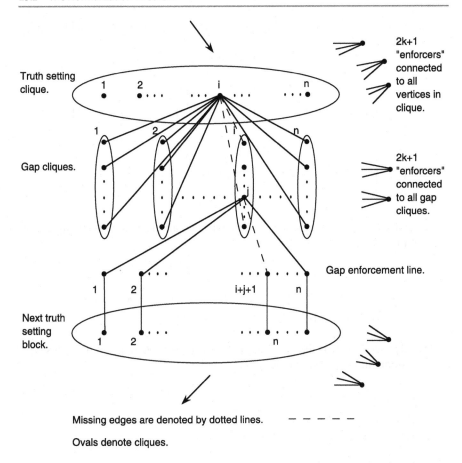

Fig. 23.1 CNFSAT reducing to DOMINATING SET

then *none* of the vertices of the enforcement line are dominated. Since the only other edges from a gap enforcement vertex the single edge to the corresponding vertex in $A(r + 1)$, this would mean that no size $2k$ dominating set is compatible with such a choice. It follows that we must choose some j with $i + j + 1 \leq n$ from column i. The point of this is that if we choose j in column i we will dominate all of the $d[r, s]$ except $d[r, i + j + 1]$. Since we will only connect $d[r, s]$ additionally to $a[r + 1, s]$ and nowhere else, to choose an element of $A[r + 1]$ and still dominate all of the $d[r, s]$ we must actually choose $a[r + 1, i + j + 1]$.

Thus the above provides a selection gadget that chooses k true variables with the gaps representing false ones. We enforce that the selection is consistent with the clauses of X via the clause variables V_3. These are connected in the obvious ways. One connects a choice in $A[r]$ or $B[r]$ corresponding to making a clause C_q true to the vertex c_q. Then if we dominate all the clause vertices too, we must have either chosen in some $A[r]$ a positive occurrence of a variable in C_q or we must have chosen in $B[r]$ a gap corresponding to a negative occurrence of a variable in C_q, and conversely. We now turn to the formal details.

The vertex set V of G is the union of the following sets of vertices:

$$V_1 = \{a[r,s] : 0 \le r \le k-1, 0 \le s \le n-1\},$$
$$V_2 = \{b[r,s,t] : 0 \le r \le k-1, 0 \le s \le n-1, 1 \le t \le n-k+1\},$$
$$V_3 = \{c[j] : 1 \le j \le m\},$$
$$V_4 = \{a'[r,u] : 0 \le r \le k-1, 1 \le u \le 2k+1\},$$
$$V_5 = \{b'[r,u] : 0 \le r \le k-1, 1 \le u \le 2k+1\},$$
$$V_6 = \{d[r,s] : 0 \le r \le k-1, 0 \le s \le n-1\}.$$

For convenience, we introduce the following notation for important subsets of some of the vertex sets above. Let

$$A(r) = \{a[r,s] : 0 \le s \le n-1\},$$
$$B(r) = \{b[r,s,t] : 0 \le s \le n-1, 1 \le t \le n-k+1\},$$
$$B(r,s) = \{b[r,s,t] : 1 \le t \le n-k+1\}.$$

The edge set E of G is the union of the following sets of edges. In these descriptions we implicitly quantify over all possible indices.

$$E_1 = \{c[j]a[r,s] : x_s \in C_j\},$$
$$E_2 = \{a[r,s]a[r,s'] : s \ne s'\},$$
$$E_3 = \{b[r,s,t]b[r,s,t'] : t \ne t'\},$$
$$E_4 = \{a[r,s]b[r,s',t] : s \ne s'\},$$
$$E_5 = \{b[r,s,t]d[r,s'] : s+t+1 \le n \wedge s' \ne s+t\} \cup \{b[k-1,s,t]d[k-1,s'] :$$
$$\qquad s' \ne s+t \ (\mathrm{mod}\ n)\},$$
$$E_6 = \{a[r,s]a'[r,u]\},$$
$$E_7 = \{b[r,s,t]b'[r,u]\},$$
$$E_8 = \{c[j]b[r,s,t] : \exists i \overline{x_i} \in C_j, s < i < s+t\},$$
$$E_9 = \{d[r,s]a[r',s] : r' = r+1 \ \mathrm{mod}\ k\}.$$

Suppose X has a satisfying truth assignment τ of weight k, with variables $x_{i_0}, x_{i_1}, \ldots, x_{i_{k-1}}$ assigned the value *true*. Suppose $i_0 < i_2 < \cdots < i_{k-1}$. Let $d_r = i_{r+1 \ (\mathrm{mod}\ k)} - i_r \ (\mathrm{mod}\ n)$ for $r = 0, \ldots, k-1$. It is straightforward to verify that the set of $2k$ vertices

$$D = \{a[r,i_r] : 0 \le r \le k-1\} \cup \{b[r,i_r,d_r] : 0 \le r \le k-1\}$$

is a dominating set in G.

Conversely, suppose D is a dominating set of $2k$ vertices in G. The closed neighborhoods of the $2k$ vertices $a'[0, 1], \ldots, a'[k-1, 1], b'[0, 1], \ldots, b'[k-1, 1]$ are disjoint, so D must consist of exactly $2k$ vertices, one in each of these closed neighborhoods. Also, none of the vertices of $V_4 \cup V_5$ are in D, since if $a'[r, u] \in D$ then necessarily $a'[r, u'] \in D$ for $1 < u' < 2k + 1$ (otherwise D fails to be dominating), which contradicts that D contains exactly $2k$ vertices. It follows that D contains exactly one vertex from each of the sets $A(r)$ and $B(r)$ for $0 \le r \le k - 1$.

The possibilities for D are further constrained by the edges of E_4, E_5 and E_9. The vertices of D in V_1 represent the variables set to *true* in a satisfying truth assignment for X, and the vertices of D in V_2 represent intervals of variables set to *false*. Since there are k variables to be set to *true* there are, considering the indices of the variables mod n, also k intervals of variables to be set to *false*. Furthermore the set E_5 forces the chosen variables to be chosen so that if $r < r'$ and we choose $a[r, q]$ and $a[r', q']$ then $q < q'$.

The edges of E_4, E_5 and E_9 enforce that the $2k$ vertices in D must represent such a choice consistently. To see how this enforcement works, suppose $a[3, 4] \in D$. This represents that the third of k distinct choices of variables to be given the value *true* is the variable x_4. The edges of E_4 force the unique vertex of D in the set $B(3)$ to belong to the subset $B(3, 4)$. The index of the vertex of D in the subset $B(3, 4)$ represents the difference (mod n) between the indices of the third and fourth choices of a variable to receive the value *true*, and thus the vertex represents a range of variables to receive the value *false*. The edges of E_5 and E_9 enforce that the index t of the vertex of D in the subset $B(3, 4)$ represents the "distance" to the next variable to be set *true*, as it is represented by the unique vertex of D in the set $A(4)$.

It remains only to check that the fact that D is a dominating set ensures that the truth assignment represented by D satisfies X. This follows by the definition of the edge sets E_1 and E_8.   □

Remark 23.2.1 A very important fact about the above proof is the following. For our fixed k the enforcement gadgetry causes us to choose the $2k$ vertices, k from the $A(r)$'s and k from the $B(r)$'s, and there is a 1–1 correspondence between weight k assignments to X and size $2k$ sets that *dominate the graph G' which denotes the gadget, that is, G without the clause connections and vertices*. Define a *weak dominating set* to be a set of $(2k)$ vertices that dominates the gadget of G. The fact we will use in the proof of the normalization theorem is that under the 1–1 correspondence between weight k assignments to X and weak dominating sets, if clauses $C_{i_1}, \ldots, C_{i_p}$ are the subset of the set of clauses satisfied by some weight k assignment, then $c_{i_1}, \ldots, c_{i_p}$ are exactly the clause vertices dominated by the corresponding size $2k$ weak dominating set in G. That is, not only is there a correspondence between weight k satisfying assignments for X and weight $2k$ dominating sets in G, but in fact there is an exact correspondence between *all* weight k assignments together with the clauses they satisfy, and with *all* weak dominating sets and the clause vertices they dominate. This fact is also used in Chap. 29 when we look at parametric miniatures and in Chap. 30 when we look a miniature parametric miniatures.

Corollary 23.2.1

(i) MONOTONE CNF SATISFIABILITY $\equiv_m^s$ CNF SATISFIABILITY.

(ii) WEIGHTED CNF SATISFIABILITY $\leq_m^s$ WEIGHTED BINARY INTEGER PRO-
GRAMMING.

Proof (i) See Exercise 23.5.1. (ii) See Theorem 20.2.1.   □

23.2.2 The Proof of the Normalization Theorem

Let $L \in W[t]$. By Theorem 21.2.1, we can suppose that $t \geq 2$. Let $\mathcal{F}$ be the family of
circuits of depth bounded by h and weft bounded by t to which L reduces. It suffices
to reduce $L_{\mathcal{F}}$ to WEIGHTED t-NORMALIZED SATISFIABILITY. An instance of the
latter problem may be viewed as a pair consisting of a positive integer k and a circuit
having t alternating layers of *And* and *Or* gates corresponding to the t-normalized
expression structure P-o-S-o-P-$\cdots$, and having a single output *And* gate. Thus the
argument essentially shows how to "normalize" the circuits in $\mathcal{F}$.

Let (C, k) be an instance of $L_{\mathcal{F}}$. We show how to determine whether C accepts
a weight k input vector, by consulting an oracle for WEIGHTED t-NORMALIZED
SATISFIABILITY (viewed as a problem about circuits) for finitely many weights k'.
The algorithm for this determination will be uniform in k, and run in time $f(k)n^\alpha$
where n is the size of the circuit C. The exponent α will be a (possibly exponential)
function of h and t. This is permissible, since every circuit in $\mathcal{F}$ observes these
bounds on depth and weft.

Step 1. The Reduction to Tree Circuits The first step is to transform C into a *tree
circuit* C' (or *formula*) of depth and weft bounded by h and t, respectively. In a tree
circuit every logic gate has fanout 1. (The input nodes may have large fanout.) The
transformation is accomplished by replicating the portion of the circuit above a gate
as many times as the fanout of the gate, beginning with the top level of logic gates,
and proceeding downward level by level. (We regard a decision circuit as arranged
with the inputs on top and the output on the bottom.) The creation of C' from C
may require time $O(n^{O(h)})$ and involve a similar blowup in the size of the circuit.
The tree circuit C' accepts a weight k input vector if and only if the original circuit
C accepts a weight k input vector.

Step 2. Moving the *Not* Gates to the Top of the Circuit Let C denote the circuit
we receive from the previous step (we will use this notational convention throughout
the proof). Transform C into an equivalent circuit C' by commuting the *Not* gates
to the top, using DeMorgan's laws. This may increase the size of the circuit by at
most a constant factor. The tree circuit C' thus consists (from the top) of the input
nodes, with *Not* gates on some of the lines fanning out from the inputs. In counting
levels we consider all of this as level 0, and may refer to negated fanout lines from
the input nodes as negated inputs. Next, there are levels consisting only of large and
small *And* and *Or* gates, with a single output gate (which may be of either principal
logical denomination at this point).

Step 3. Homogenizing the Layers   The goal of this step is to reduce to the situation where all of the large gates are at the bottom of the circuit, in alternating layers of large *And* and *Or* gates. To achieve this we work from the bottom up, with the first task being to arrange for the output gate to be large.

Let C denote the circuit received from the previous step. Suppose the output gate z is small. Let $C[z]$ denote the connected component of C including z that is induced by the set of small gates. Thus all gates providing input to $C[z]$ are either large or are input gates of C. Because of the bound h on the depth of C, there are at most 2^h inputs to $C[z]$. The function of these inputs computed by $C[z]$ is equivalent to a product-of-sums expression E_z having at most 2^{2^h} sums, with each sum a product of at most 2^h inputs. Let C' denote the circuit equivalent to C obtained by replacing the small gate output component $C[z]$ with E_z, duplicating subcircuits of C as necessary to provide the inputs to the depth 2 circuit representing E_z. (The "product" gate of E_z is now the output gate of C'.) This entails a blowup in size by a factor bounded by 2^{2^h}. Since h is an absolutely fixed constant (not dependent on n or k) this blowup is "linear" and permitted. Note that E_z and therefore C' are easily computed in a similar amount of time to this size blowup.

Let p denote the output *And* gate of C' (corresponding to the product in E_z). Let $s_1, \ldots, s_m$ denote the *Or* gates of C' corresponding to the sums of E_z. We consider all of these gates to be *small*, since the number of inputs to them does not depend on n or k. (Equivalently, if the gates of these two levels were replaced by binary input gates, we would see that the reduction of C to C' has increased circuit depth from h to 2^h.)

Each *Or* gate s_i of C' has 3 kinds of input lines: those coming from large *And* gates, those coming from large *Or* gates, and those coming from input gates of C'. We will use the same symbol to denote an input line, the subcircuit of C' that computes the value on that line, or the boolean expression corresponding to the subcircuit (since C' is a tree circuit, it is equivalent to a boolean expression). Let these three groups of inputs be denoted, respectively:

$$S_{i,\wedge} = \left\{ s_i[\wedge, j] : j = 1, \ldots, m_{i,\wedge} \right\},$$
$$S_{i,\vee} = \left\{ s_i[\vee, j] : j = 1, \ldots, m_{i,\vee} \right\},$$
$$S_{i,\top} = \left\{ s_i[\top, j] : j = 1, \ldots, m_{i,\top} \right\}$$

and define

$$S_i = S_{i,\wedge} \cup S_{i,\vee} \cup S_{i,\top}.$$

For each line $s_i[\vee, j]$ of C' coming from a large *Or* gate u, let

$$S_{i,\vee,j} = \left\{ s_i[\vee, j, k] : k = 1, \ldots, m_{i,\vee,j} \right\}$$

denote the set of input lines to u in C'. Similarly, for each line $s_i[\wedge, j]$ of C' coming from a large *And* gate v, let

$$S_{i,\wedge,j} = \left\{ s_i[\wedge, j, k] : k = 1, \ldots, m_{i,\wedge,j} \right\}$$

denote the set of input lines to v in C'.

Let

$$k' = \sum_{i=1}^{m} (1 + m_{i,\vee}).$$

The integer k' is the number of *or* gates (counting both large and small gates) that are either part of $C[z]$ or directly supply input to $C[z]$. Note that k' is bounded above by $2^h \cdot 2^{2^h}$.

We describe how to produce a weft t circuit C'' from C' that accepts an input vector of weight $k'' = k + k'$ if and only if C' (and therefore C) accepts an input vector of weight k. The tree circuit C'' will have a large *And* gate giving the output.

Let $x_1, \ldots, x_n$ denote the inputs to C'. The circuit C'' has additional input variables that, for the most part, correspond to the input lines to the *or* gates singled out for attention above. The set V of new input variables is the union of the following groups of variables:

$$V = \left(\bigcup_{i=1}^{m} V_i \right) \cup \left(\bigcup_{i=1}^{m} \bigcup_{j=1}^{m_{i,\vee}} V_{i,j} \right)$$

where

$$V_i = \left\{ v_i[\wedge, j] : 1 \le j \le m_{i,\wedge} \right\} \cup \left\{ v_i[\vee, j] : 1 \le j \le m_{i,\vee} \right\}$$
$$\cup \left\{ v_i[\top, j] : 1 \le j \le m_{i,\top} \right\}$$

and

$$V_{i,j} = \left\{ v_i[\vee, j, k] : 1 \le k \le m_{i,\vee,j} \right\} \cup \left\{ n[i, j] \right\}.$$

The circuit C'' is represented by the boolean expression:

$$C'' = E_1 \cdot E_2 \cdot E_3 \cdot E_4 \cdot E_5 \cdot E_6 \cdot E_7$$

where

$$E_1 = \prod_{i=1}^{m} \left(\sum_{u \in V_i} u \right),$$

$$E_2 = \prod_{i=1}^{m} \prod_{u \neq v, \, u,v \in V_i} (\neg u + \neg v),$$

$$E_3 = \prod_{i=1}^{m} \prod_{j=1}^{m_{i,\vee}} \left(\sum_{u \in V_{i,j}} u \right),$$

$$E_4 = \prod_{i=1}^{m} \prod_{j=1}^{m_{i,\vee}} \prod_{u \neq v}^{u,v \in V_{i,j}} (\neg u + \neg v),$$

$$E_5 = \prod_{i=1}^{m} \prod_{j=1}^{m_{i,\wedge}} \prod_{k=1}^{m_{i,\wedge,j}} \left(s_i[\wedge, j, k] + \neg v_i[\wedge, j] \right),$$

$$E_6 = \prod_{i=1}^{m} \prod_{j=1}^{m_{i,\vee}} \left(\neg v_i[\vee, j] + \neg n[i, j] \right),$$

$$E_7 = \prod_{i=1}^{m} \prod_{j=1}^{m_{i,\vee}} \prod_{k=1}^{m_{i,\vee,j}} \left(s_i[\vee, j, k] + \neg v_i[\vee, j, k] \right).$$

The size of C'' is bounded by $|C'|^2$.

Claim 1 *The circuit C'' has weft t.*

To see this, note first of all that since $t \geq 2$, any input–output path beginning from a new input variable (in V) has at most 2 large gates as the expression for C'' is essentially a product-of-sums. In E_5 and E_7 the sums involve subexpressions of C'; any input–output path from an original input variable (of C') passes through one of these *or* gates. Observe that in C' these subexpressions have weft at most $t - 1$. The sums of E_5 and E_7 are small, so these paths further encounter only the bottommost large *And* gate.

Claim 2 *The circuit C'' accepts an input vector of weight k'' if and only if C' accepts an input vector of weight k.*

First note that any input vector of weight k'' accepted by C'' must set exactly one variable in each of the sets of variables V_i (for $i = 1, \ldots, m$) and $V_{i,j}$ (for $i = 1, \ldots, m$ and $j = 1, \ldots, m_{i,\vee}$) to 1 and all of the others in the set to 0 in order to satisfy $E_1 \cdot E_2 \cdot E_3 \cdot E_4$. It follows that any such accepted input must set exactly k of the original variables of C' to 1, by the definition of k''.

The role of the (new) variables set to 1 in the sets of variables that represent inputs to *or* gates is to indicate an accepting computation of C' on the weight k input of old variables. The expressions $E_5, \ldots, E_7$ enforce the correctness of this representation in C'' of the computation of C'.

The expression E_5 ensures that if the new variable $v_i[\wedge, j]$ is set to 1, indicating that the subexpression $s_i[\wedge, j]$ of C' evaluates to 1, then every argument $s_i[\wedge, j, k]$ must evaluate to 1. (Note that subexpressions $s_i[\wedge, j, k]$ appear in C'' while the subexpressions $s_i[\wedge, j]$ do not. The computations performed in C' by the latter are simply represented by the values of the input variables in V.) The role of the variables $n[i, j]$ is to represent that "none of the inputs" to the *or* gate has the value 1. The expression E_6 enforces that if this situation is represented, then the output of the gate is not represented as having the value 1. The expression E_7 ensures that if the new variable $v_i[\vee, j, k]$ has the value 1, indicating that the subexpression $s_i[\vee, j, k]$ of C' evaluates to 1, then this subexpression must in fact evaluate to 1.

By the above, we may now assume that the circuit with which we are working has a large output gate (which may be of either denomination). Renaming for convenience, let C denote the circuit with which we are working, under this assumption.

If g and g' are gates in C of the same logical character ($\wedge$ or $\vee$) with the output of g going to g', then they can be consolidated into a single gate without increasing weft if g is small and g' is large. We term this a *permitted contraction*. Note that if g is large and g' is small then contraction may not preserve weft. We will assume that permitted contractions are performed whenever possible, interleaved with the following two operations.

(1) *Replacement of bottommost small gate components.*

As at the beginning of Step 3, let $C_1, \ldots, C_m$ denote the bottommost connected components of C induced by the set of small gates, and having at least one large-gate input. Since the output gate of C is large, each C_i gives output to a large gate g_i. If g_i is an *And* gate, then C_i should be replaced with product-of-sums circuitry equivalent to C_i. If g_i is an *Or* gate, then C_i should be replaced with equivalent sum-of-products circuitry. Note that in either case this immediately creates the opportunity for a permitted contraction. As per the discussion at the beginning of Step 3, this replacement circuitry is small, and this operation may increase the size of the circuit by a factor of 2^{2^h}. This step will be repeated at most h times, as we are working from the bottom up in transforming C.

(2) *Commuting small gates upwards.*

After (1), and after the permitted contractions, the bottommost small gate components are each represented in the modified circuit C' by a single small gate h_i giving output to g_i. Without loss of generality, all of the arguments to h_i may be considered to come from large gates. (The only other possibility is that an input argument to h_i may be from the input level of the circuit—but there is no increase in weft in treating this as an honorary large gate for convenience.) Suppose that g_i is an *And* gate, and that h_i is an *or* gate (the other possibility is handled dually).

There are three possible cases:

 (i) All of the arguments to h_i are from *Or* gates.

 (ii) All of the arguments to h_i are from *And* gates.

(iii) The arguments to h_i include both large *Or* gates and large *And* gates.

In case (i), we may consolidate h_i and all of the gates giving input to h_i into a single large *Or* gate without increasing weft.

In case (ii), we replace the small $\vee$ (h_i) of large $\wedge$'s with the equivalent (by distribution) large $\wedge$ of small $\vee$'s. Since h_i may have 2^h inputs, this may entail a blowup in the size of the circuit from n to n^{2^h}. This does not increase weft, and creates the opportunity for a permitted contraction.

In case (iii), we similarly replace h_i and its argument gates with circuitry representing a product-of-sums of the inputs to the arguments of h_i. The difference is that in this case, the replacement is a large $\wedge$ of large (rather than small) $\vee$ gates. Weft is preserved when we take advantage of the contraction now permitted between the large $\wedge$ gate and g_i.

We may achieve our purpose in this step by repeating the cycle of (1) and (2). At most h repetitions are required. The total blowup in the size of the circuit in this step is crudely bounded by $n^{2^{h^2}}$.

Step 4. Removing a Bottom Most *Or* Gate By a Turing reduction, we can determine whether a tree circuit giving output from an *Or* gate accepts a weight k input vector, by simply making the same determination for each of the input branches (subformulas) to the gate.

In order to accomplish this step by a many : 1 reduction, we do the following. Let b be the number of branches of the circuit C with bottommost *Or* gate that we receive at the beginning of this step. We modify C by creating new inputs $x[1], \ldots, x[b]$. The purpose of these input variables is to indicate which branch of C accepts a weight k input vector. Let $C_1, \ldots, C_b$ be the branches of C, so that C is represented by the expression $C_1 + \cdots + C_b$. The modified circuit C' is represented by the expression

$$C' = \left(x[1] + \cdots + x[b] \right) \cdot \prod_{1 \le i < j \le b} \left(\neg x[i] + \neg x[j] \right) \cdot \prod_{1 \le i \le b} \left(C_i + \neg x[i] \right).$$

The first two product terms of the above expression ensure that exactly one of the new variables must have value 1 in an accepted input vector. The modified circuit C' accepts a weight $k + 1$ input vector if and only if C accepts a weight k input vector. For weft at least two, the transformation is weft-preserving, and yields a circuit C' with bottommost *And* gate, but possibly with *not* gates at the lower levels. Thus it may be necessary to repeat steps 2 and 3 to obtain a homogenized circuit with bottommost *And* gate.

Step 5. Organizing the Small Gates The tree circuit C received from the previous step has the properties: (i) the output gate is an *And* gate, (ii) from the bottom, the circuit consists of layers which alternately consist of only *And* gates or only *Or* gates, for up to t layers, and (iii) above this, there are branches B of height $h' = h - t$ consisting only of small gates. Since a small gate branch B has bounded depth, it has at most $2^{h'}$ gates, and thus in constant time (since h is fixed), we can find either (1) an equivalent sum-of-products circuit with which to replace B, as required by Case 1 of Step 6 below, or (2) an equivalent product-of-sums circuit, as required by Case 2 of Step 6.

In this step, all such small gate branches B of C are replaced in this way, appropriately for the relevant case of Step 6. In Case 1, the depth 2 sum-of-products circuits replacing the small gate branches B have a bottommost *or* gate g_B of fanin at most $2^{2^{h'}}$, and the *and* gates feeding into g_B have fanin at most $2^{h'}$, so the weft of the circuit has been preserved by this transformation, which may increase the size of C by the constant factor $2^{2^{h'}}$. The topmost level of large gates (to which the branches B are attached in C) consists of *Or* gates, in Case 1, so that the gates g_B can be merged into this topmost level. Merging is performed similarly in Case 2, where the replacing circuits are products-of-sums, and the topmost level of large gates consists of *And* gates. For the next step we consider two cases, depending on whether the topmost level of large gates consists of *Or* gates or *And* gates. (Essentially, this corresponds to whether weft t is even or odd.)

Step 6. A Monotone Change of Variables. (Two Cases) In this step (in both cases) we employ a "change of variables" based on the combinatorial reduction of Lemma 23.2.1. The goal is to obtain an equivalent circuit that has the property that either all the inputs are MONOTONE (Case 1) (i.e. no inverters in the circuit), or all the inputs are negated with no other inverters in the circuit, that is, ANTIMONO-TONE. (Actually in this case we will have *some* of the inputs positive but these will only be *enforcers* as we will see. So we should call this case NEARLY ANTIMONO-TONE, Case 2.) The point of this step becomes apparent in the next step when we use the special character of the circuit thus constructed to enable us to eliminate the small gates.

Consider the reduction of Lemma 23.2.1. This reduction consists of two parts. The first is the ring of selection gadgets which allow variable choice, gap choice, and then gap enforcement and the second part is consistency obtained by clause wiring. The idea is to "hard wire" the selection and consistency parts of the construction into the circuit. The point being that we can replace positive instances of variable fanout in the original circuit by outputs corresponding to choice of that variable in the positive selection component. We can replace negative fanout in the original circuit by the appropriate sets of gap variables. Finally we can wire in the fact that we need a dominating set and other enforcements by using the facts that we will only look at a weight $2k$ input, and an *And* of *Or*'s, which will not add to the *weft* of the circuit. We argue more precisely below, and also prove in two parts that the whole process can be accomplished without increasing *weft*, given that the weft is ≥ 2.

Suppose the inputs to the circuit C received at the beginning of this step are $x[1], \ldots, x[n]$, and suppose that the output gate of C is an *And* gate. Let Y denote the boolean expression having $2n$ clauses, with each clause consisting of a single literal, and with one clause for each of the $2n$ literals of the n input variables. The reduction of Lemma 23.2.1 allows us to translate Y into a monotone formula via dominating set, thus capturing monotonically all the relevant input settings. Let G_Y be the graph constructed for this expression as in the proof of Lemma 23.2.1. Note that only part of G_Y will actually be wired into C.

Keeping this in mind, and using the variable (vertex) set obtained from G_Y, the change of variables is implemented for C as follows. (1) Create a new input for each vertex of G_Y that is not a clause vertex. (2) Replace each positive input fanout of $x[i]$ in C with an *Or* gate having k new input variable arguments corresponding to the vertices to which the clause vertex for the clause $(x[i])$ of Y is adjacent in G_Y. (3) Replace each negated fanout line of $x[i]$ with an *Or* gate having $O(n^2)$ new input variable arguments corresponding to the vertices to which the clause vertex for the clause $(\neg x[i])$ of Y is adjacent in G_Y. (4) Merge with the output *And* gate of C a new circuit branch corresponding to the product-of-sums expression, where the product is taken over all non-clause vertices of G_Y, and the sum for a vertex u is the sum of the new inputs corresponding to the non-clause vertices in $N[u]$ (this is the dominating set and other enforcements.).

The modified circuit C' obtained in this way accepts a weight $2k$ input vector if and only if the original circuit C accepts a weight k input vector. The proof of this is essentially the same as for Lemma 23.2.1. (This is the whole point of the remark following the proof of Lemma 23.2.1.) If all of the *not* gates of C are at the top, then the circuit C' will be MONOTONE. However, to see that this change of variables can be employed to obtain a monotone or nearly antimonotone circuit *without* increasing weft, we must consider two cases.

Case 1. The Topmost Large-Gate Level Consists of Or Gates Let C denote the circuit obtained from Step 5 and perform a change of variables as described above. The sequence of transformations of C for this step is shown schematically in Fig. 23.2.

The result is a circuit C' with no *not* gates. The input weight we are now concerned with is $2k$, and the construction of C' from C may involve quadratic blowup.

Next, we move the small *and* gates on the second level upward past the *Or* gates introduced by the change of variables, and then merge the *Or* gates down to the topmost large layer (of *Or* gates).

Case 2. The Topmost Large-Gate Level Consists of And Gates Here we use a similar argument, beginning with a trick. Below each gate of the topmost large-gate level (of *And* gates), a double negation is introduced (equivalently). One of the *not* gates is moved to the top of the circuit (by DeMorgan's identities). This is followed by a change of variables based on Lemma 23.2.1, as in Case 1. The second level *and* gates are commuted upwards, and the *Or* gates of the second and third levels are merged, as in Case 1. So now the circuit has no negated inputs and no inverters except the residual ones below the top layer of *Or* gates. Finally, these remaining *not* gates are commuted to the top. Note that this means that all fanouts are negated except the ones to the enforcement *Or* gate added during Step 4.

We are now in position for the last step.

Step 7. Eliminating the Remaining Small Gates If we regard the inputs to C as variables, this step consists of another "change of variables". Let k be the relevant weight parameter value supplied by the last transformation. In this step we will produce a circuit C' corresponding directly to a t-normalized boolean expression

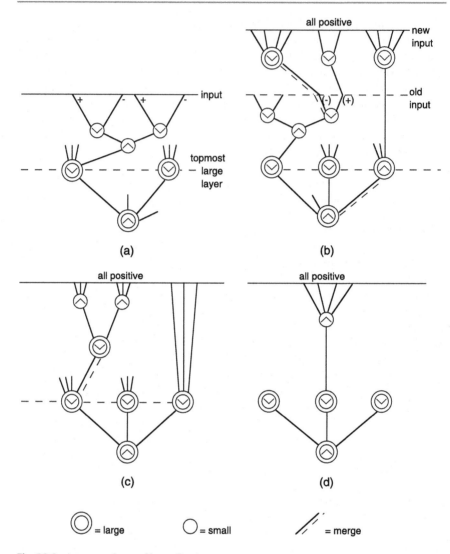

Fig. 23.2 A topmost layer of large *Or* gates

(that is, consisting only of *t* alternating layers of *And* and *Or* gates) such that *C* accepts a weight *k* input vector if and only if *C'* accepts a vector of weight $k' = k \cdot 2^{k+1} + 2^k$.

Suppose that *C* has *m* remaining small gates. In Case 1, these are *and* gates, and the inputs are all positive. In Case 2, these are *or* gates, and the inputs to these gates are all negated. For $i = 1, \ldots, m$ we define the sets A_i of the inputs to *C* to be the sets of input variables to these small gates. The central idea for this step is to create new inputs representing the sets A_i of inputs to *C*.

For example, suppose (Case 1) that the output of the small *and* gate g_i in C is the boolean product $(abcd)$ of the inputs a, b, c, d to C. Thus $A_i = \{a, b, c, d\}$. The gate g_i can be eliminated by replacing it with an input line from a new variable $v[i]$ which represents the predicate $a = b = c = d = 1$. (This representation, of course, will need to be enforced by additional circuit structure.) Similarly (Case 2) if g_i computes the value $(\bar{a} + \bar{b} + \bar{c} + \bar{d})$ then g_i can be replaced by a negated input line from $v[i]$.

Let $x[j]$ for $j = 1, \ldots, s$ be the input variables to C. We introduce new input variables of the following kinds:

(1) One new variable $v[i]$ for each set A_i for $i = 1, \ldots, m$ to be used as indicated above.

(2) For each $x[j]$ we introduce 2^{k+1} copies $x[j, 0], x[j, 1], x[j, 2], \ldots, x[j, 2^{k+1} - 1]$.

(3) "Padding" consisting of 2^k meaningless variables (inputs not supplying output to any gates) $z[1], \ldots, z[2^k]$.

We add to the circuit an enforcement mechanism for the change of variables. The necessary requirements can be easily expressed in P-o-S form, and thus can be incorporated into the bottom two levels of the circuit as additional *Or* gates attached to the bottommost (output) *And* gate of the circuit.

We require the following implications concerning the new variables:

(1) The $s \cdot 2^{k+1}$ implications, for $j = 1, \ldots, s$ and $r = 0, \ldots, 2^{k+1} - 1$,

$$x[j, r] \Rightarrow x\left[j, r + 1 \pmod{2^{k+1}}\right].$$

(2) For each containment $A_i \subseteq A_{i'}$, the implication

$$v[i'] \Rightarrow v[i].$$

(3) For each membership $x[j] \in A_i$, the implication

$$v[i] \Rightarrow x[j, 0].$$

(4) For $i = 1, \ldots, m$ the implication

$$\left(\prod_{x[j] \in A_i} x[j, 0]\right) \Rightarrow v[i].$$

It may be seen that this transformation may increase the size of the circuit by a linear factor exponential in k. We make the following argument for the correctness of the transformation.

If C accepts a weight k input vector, then setting the corresponding copies $x[i, j]$ among the new input variables accordingly, together with appropriate settings for the new "collective" variables $v[i]$ yields a vector of weight between $k \cdot 2^{k+1}$ and $k \cdot 2^{k+1} + 2^k$ that is accepted by C'. The reason the weight of this corresponding vector may fall short of $k' = k \cdot 2^{k+1} + 2^k$ is that not all of the subsets of the k input variables to C having value 1 may occur among the sets A_i. An accepted vector of weight exactly k' can be obtained by employing some of the "padding" input variables $z[i]$ to C'.

Note that the seemingly simpler strategy of creating a new input variable for each set of at most k inputs to C would not serve our purposes, since it would involve increasing the size n of the circuit to possibly n^k. (We are limited in our computational resources for the reduction to $f(k)n^\alpha$. The constant α can be an arbitrary function of the depth and weft bounds h and t, but not k.)

For the other direction, suppose C' accepts a vector of weight k'. Because of the implications in (1) above, exactly k sets of copies of inputs to C must have value 1 in the accepted input vector. Because of the implications (2)–(4), the variables $v[i]$ must have values in the accepted input vector compatible with the values of the sets of copies. By the construction of C', this implies there is a weight k input vector accepted by C. This concludes the proof of the Normalization Theorem.

Corollary 23.2.2 *The following are complete for* $W[2]$:
(1) (Downey and Fellows [241, 243]) DOMINATING SET;
(2) (Downey and Fellows [241, 243]) WEIGHTED MONOTONE CNF SAT;
(3) (Downey and Fellows [241, 243]) WEIGHTED BINARY INTEGER PROGRAMMING;
(4) (Paz and Moran [557] and Wareham) *The problem* SET COVER *below.*

SET COVER
    Instance: A collection of nonempty sets $\{S_i \mid S_i \in C\}$ and an integer k.
    Question: Is there a subcollection of k of the sets $\{S_{i_1}, \ldots, S_{i_k}\}$ such that
$$\bigcup_{j=1}^{k} S_{i_j} \supseteq \bigcup \{S_i \mid S_i \in C\}?$$

(5) *The problem* HITTING SET *below.*

HITTING SET
    Instance: A finite family of sets $S = S_1, \ldots, S_n$ comprised of elements from $U = \{u_1, \ldots, u_m\}$, an integer k.
    Parameter: A positive integer k.
    Question: Is there a subset $T \subseteq U$ of size k such that for all $S_i \in S$, $S_i \cap T \neq \emptyset$?

Proof (1), (2) and (3) all follows by the proof of the Normalization Theorem. (4) follows as the reduction in Lemma 6.4.1 is parameterized. (5) is Exercise 23.5.6. □

23.3 Monotone and Antimonotone Formulas

The proof of the Normalization Theorem allows us to establish some technical results on monotone and antimonotone formulas.

Theorem 23.3.1 (Downey and Fellows [241, 243])

(i) *For t > 0, and t even,*
  (a) MONOTONE $W[t] = W[t]$.
  (b) WEIGHTED MONOTONE t-NORMALIZED SATISFIABILITY *is* $W[t]$-*complete.*
  (c) WEIGHTED MONOTONE $t + 1$-NORMALIZED SATISFIABILITY *is in* $W[t]$.
(ii) *For t > 0, and t odd,*
  (a) ANTIMONOTONE $W[t] = W[t]$.
  (b) WEIGHTED ANTIMONOTONE t-NORMALIZED SATISFIABILITY *is* $W[t]$-*complete.*
  (c) *If, additionally,* $t \geq 3$, WEIGHTED ANTIMONOTONE $t + 1$-NORMALIZED SATISFIABILITY *is in* $W[t]$.

Proof (i) Let t be even. The proof of (a) comes from the analysis of Step 6, Case 1. For (b) we can apply Step 6, Case 1, to a t-normalized formula. The result is a t-normalized monotone formula. For (c), the result follows by the transformations of Step 7, applied to a monotone $t + 1$- normalized formula.

(ii) (a) Again the proof will consist of hardwiring a gadget. This time the gadget is the one concerning the problem RED/BLUE NONBLOCKER from the proof of Theorem 21.2.5. The case $t = 1$ was dealt with in Theorem 21.2.1. Let C be a circuit of weft t for t odd, $t \geq 3$. By The Normalization Theorem, we may assume that C is represented by a boolean expression E_0 that is in (alternating) product-of-sums-of-products ... form (for t alternations). The first level of the circuit below the inputs consists of *And* gates (since t is odd).

Suppose the inputs to C are $x_1, \ldots, x_n$. Let X_1 be the boolean expression with single-literal clauses $X_1 = (x_1)(x_2) \cdots (x_n)$ and let G be the graph constructed from X_1 by the reduction in the proof of Theorem 21.2.1. Let $y_1, \ldots, y_z$ be new variables, one for each red vertex in G.

Let E_1 be the boolean expression

$$E_1 = \prod_{u \in (V_{\text{blue}} - V_7)} \sum_{y_i \in N[u]} \neg y_i$$

and let C_1 be a circuit realizing E_1.

We modify C in the following ways:
(1) Each positive fanout from an input x_i to C is replaced by an *And* gate receiving negated inputs from all of the new input variables y_j for which the corresponding red vertices of G represent the possibility that x_i evaluates to *false*.
(2) Each negated fanout from an input x_i to C is replaced by an *And* gate receiving negated inputs from all of the new input variables y_j for which the corresponding red vertices of G represent the possibility that x_i evaluates to *true*.
(3) The circuit C_1 is conjunctively combined with C at the bottommost (output) *And* gate.

The circuit C' obtained in this way accepts a weight $2k$ input vector if and only if C accepts a weight k input vector. The argument for correctness is essentially the same as for Theorem 21.2.1. The circuit C' has weft t after the *And* gates, replacing the former inputs, are coalesced with the *And* gates of the topmost large-gate level (this is feasible, since t is odd). All of the input fanout lines of C' are negated. (b) and (c) are similar.   □

Note that (i)(c) and (ii)(b) suggest the following structural collapses for all t:
- Monotone $W[2t + 1] = W[2t]$.
- Antimonotone $W[2t + 2] = W[2t + 1]$.

As we will see in Chap. 24, these collapses do indeed occur and hence we finish up with the following picture of the W-Hierarchy.

FPT $=$ Monotone $W[1]$

      $\subseteq W[1] =$ Antimonotone $W[1] =$ Antimonotone $W[2]$

      $\subseteq W[2] =$ Monotone $W[2] =$ Monotone $W[3]$

      $\subseteq \cdots$

      $\subseteq W[\text{SAT}] \subseteq W[\text{P}]$.

We have already seen that there are some natural graph theoretical problems such as Dominating Set which live naturally at the $W[2]$. The restriction of dominating set to tournaments (Tournament Dominating Set) is also complete for $W[2]$ as we will see in Exercise 23.5.3. (A *tournament* is a directed graph $G = (V, A)$ such that for each pair of vertices $u, v \in V$, exactly one of uv or vu is in A.) The fact that the parameterized version of Dominating Set and Tournament Dominating Set are of the same complexity provides another illustration of the fact that problems with unparameterized versions of apparently quite different complexities can have the same parameterized complexity. In this case Papadimitriou and Yannakakis [553] have shown that Tournament Dominating Set is complete for the class LogSNP whereas Dominating Set NP-complete and hence Tournament Dominating Set is of the same classical polynomial time degree as Dominating Set only in the unlikely event that NP $\subseteq$ Dtime$(n^{\log n})$.

23.4 The W^*-Hierarchy and Generalized Normalization to $W^*[t]$

Downey, Fellows, and Taylor (in [258]) studied the following parametric problem in relational database queries: It is well known that for a fixed relational database query ϕ of in m free variables, it can be determined in time polynomial in the size n of the database whether there exists an m-tuple x that belongs to the relation defined by the query. For the best known algorithms, however, the exponent of the polynomial is proportional to the size of the query. Downey, Fellows, and Taylor studied the data complexity of this problem parameterized by the size $k = |\phi|$ of the query, and answer a question recently raised by Yannakakis in STOC 1995. The main results from [258] show: (1) the general problem is complete for the parametric complexity class $AW[*]$ (of Chap. 25) and (2) when restricted to monotone queries, the problem is complete for the fundamental parametric complexity class $W[1]$. We will prove these results in Chap. 26 in Theorem 26.4.1.

An important consequence of the proof of (2) was a significantly improved characterization of the parameterized complexity class $W[1]$. Our basic definition was for families of circuits that satisfy: (1) the depth of the circuits is bounded by a constant c, (2) on any input–output path there is at most one gate having unbounded fan-in (termed a *large* gate), with all other gates having fan-in bounded by c (that is, *small* gates). Downey, Fellows, and Taylor proved that the definition can be broadened by allowing circuits of depth bounded by an arbitrary function $f(k)$.

Definition 23.4.1 (Downey, Fellows, and Taylor [258]) A parameterized language L belongs to $W^*[t]$ if L reduces to $L_{F^*(t,h)}$ for the family $F^*(t,h)$ of mixed type decision circuits of weft at most t, and depth at most $h(k)$, for some arbitrary function h.[1]

23.4.1 W^* Normalization

In this section we sketch the proof of an important lemma concerning $W^*[t]$ that says essentially that the circuits can be put in a normal form that corresponds to a certain kind of Boolean formula. The normal form is defined recursively as follows.

Definition 23.4.2 A Boolean expression E is in 1-alternating normal form with respect to n and k if E is either an n-product of k-sums or an n-sum of k-products. A Boolean expression E is in t-alternating normal form with respect to n and k, for $t \geq 2$, if E either is of the form:

$$E = \prod_{i=1}^{n} \sum_{j=1}^{k} E_{ij}$$

or of the form:

$$E = \sum_{i=1}^{n} \prod_{j=1}^{k} E_{ij}$$

---

[1]It is also possible to consider families of circuits where *small* gates may have fan-in $f(k)$, then we can equivalently substitute a faster-growing function h.

where each sub expression E_{ij} is $(t-1)$-alternating normal with respect to n and k.

Note that the above definition can be equivalently phrased in terms of circuits in a natural way. Define a tree circuit C to be in t-*alternating normal form with respect to n and k* if it has $2t$ layers of gates, with the fan-in alternating between n and k starting with fan-in n for the output gate.

Corresponding to this form we have the following parameterized problem.

t-ALTERNATING WEIGHTED SATISFIABILITY
Instance: A Boolean expression E in t-alternating normal form with respect to n and k.
Parameter: A positive integer k.
Question: Is there a weight k truth assignment that satisfies E?

Theorem 23.4.1 (W^*-normalization—Downey, Fellows, and Taylor [258])
t-ALTERNATING WEIGHTED SATISFIABILITY *is complete for* $W^*[t]$ *for all* $t \geq 1$.

Proof This is similar to the proof of the regular Normalization Theorem, and hence we sketch it. Assume that the given circuit C has an output large gate, since an output small gate subcircuit can be removed by employing additional nondeterminism in the manner used in the proof of Normalization. The idea is basically to progressively analyze and "copy" the circuit C into the required form, working progressively downward from the input level. In the first step, we locate the topmost large gates. Let g denote such a gate, and let $a_1, \ldots, a_r$ denote the inputs to g. Suppose g is a $\wedge$ gate. Each a_i is computed by a subcircuit consisting only of small gates, and therefore there the number of inputs to C on which a_i depends is bounded by a function of k. Thus, a_i can be re-expressed (at a cost that is bounded by a function of k) as a product-of-sums of total size bounded by a function of k. Thus g can be replaced by a product-of-sums "copy-gate" g', where there are at most $f(k)n$ sums (for some function f) and each sum has size bounded by some function $g(k)$. Make a copy of g' for each fanout line of g.

For the case where g is a large $\vee$ gate, we replace each argument a_i with an equivalent sum-of-products, to obtain (for g') an $f(k)n$-sum of $g(k)$-products.

We next identify the topmost layer of large gates *below* the inputs to C and the copy gates created so far. Note that each copy gate has fanout 1. We repeat the above recipe, creating two more layers of copy gates ($f'(k)n$-products of $g'(k)$-sums and $f'(k)$-sums of $g'(k)$-products, for new functions f' and g'). Eventually the entire circuit C is replaced by the copy gates, and the resulting copy gate circuit has the required form.   □

23.4.2 An Improved Characterization of $W[1]$

We are now ready to prove the following.

Theorem 23.4.2 (Downey, Fellows, and Taylor [258]) $W^*[1] = W[1]$.

Proof By the W^*-Normalization Theorem, we just have to show that 1-ALTERNAT-
ING WEIGHTED SATISFIABILITY is in $W[1]$. The only issue is for expressions that
are n-products of k-sums, since the case of n-sums of k-products is in FPT. Let E
denote the Boolean expression. We view E as consisting of a set C of clauses

$$C = \{C_1, \ldots, C_n\}$$

where each clause is a set of k literals. Let X denote the set of variables of E,

$$X = \{x_1, \ldots, x_m\}.$$

Our argument consists of a description of how to produce a Boolean expression
E' over a set of variables X' that is a product of sums of size bounded by a constant,
such that E' is satisfiable by a weight $k' = k + 2^{k+1}$ truth assignment if and only if
E is satisfiable by a weight k truth assignment. By Theorem 21.2.5, this is sufficient
to conclude the theorem.

Let $X_i \subseteq X$ denote the set of variables occurring in the clause C_i. Let X_i^+ denote
the variables that occur positively in C_i and let X_i^- denote the set of variables that
occur negated in C_i. (Below we will make the trivial assumption that $X_i^+ \cap X_i^- = \emptyset$.)

Suppose $X_i^- = \{x_1, x_3, x_5, x_8\}$ The sequence (family) of increasingly larger sets
beginning with the empty set, and progressively including the variable of least index
among those remaining, we will call the ith *trail* $\mathcal{T}_i$. Thus for our example X_i^-, we
would have

$$\mathcal{T}_i = \big\{\emptyset, \{x_1\}, \{x_1, x_3\}, \{x_1, x_3, x_5\}, \{x_1, x_3, x_5, x_8\}\big\}.$$

Let $\mathcal{T}$ be the union of the trails

$$\mathcal{T} = \bigcup_{i=1}^{n} \mathcal{T}_i.$$

For $T \in \mathcal{T}$ the set X_T of *successors* of T is the set of variables $x = x_j$ such that:
(1) $x_j \notin T$,
(2) for all $x_i \in T$, $j > i$, and
(3) $T' = T \cup \{x_j\} \in \mathcal{T}$.
Thus if x is a successor of T, then it will have the largest index in T'. Note also that
the size of $\mathcal{T}$ is bounded by a quadratic polynomial in the size of E.

If $T \in \mathcal{T}$ then since T is a set of indexed variables, T has an ordering inherited
from the indices. We will write $T(s)$ to denote the sth variable of T in this ordering.

For each $T \in \mathcal{T}$ define C_T to be the set of clauses for which:
(1) the set of negated variables in the clause is contained in T, and
(2) the set of positive variables of the clause is disjoint from T.
Formally:

$$C_T = \big\{C_i \in C : X_i^- \subseteq T \text{ and } X_i^+ \cap T = \emptyset\big\}.$$

For example, $C_\emptyset$ is the set of monotone clauses of E.

For each set of clauses $\mathcal{C}_T$ for $T \in \mathcal{T}$ we compute a rooted labeled tree Ω_T as follows. Each node of the tree is labeled by a pair $(\mathcal{A}, V)$ where $\mathcal{A}$ is a subset of $\mathcal{C}_T$, and V is a set of variables. The root of Ω_T is labeled by the pair $(\mathcal{C}_T, \emptyset)$. The tree is generated by beginning at the root and progressively expanding nodes, creating children in the following way.

In order for a node labeled $(\mathcal{A}, V)$ to be expandable, there are two conditions that must be met:

(1) $\mathcal{A}$ must be nonempty, and
(2) $|V| \le k$.

The expansion of a node labeled $(\mathcal{A}, V)$ is achieved as follows. Suppose that the least indexed clause in $\mathcal{A}$ is C_j. One child is created for each variable x_s in X_j^+, that is, for each positive variable of the clause C_j. The child node is labeled $(\mathcal{A}', V')$ where

$$V' = V \cup \{x_s\}$$

and

$$\mathcal{A}' = \{C_i \in \mathcal{A} : X_i^+ \cap V' = \emptyset\}.$$

The tree Ω_T is created by beginning with the root node, and expanding nodes of the tree until no more expansion is possible. Note that by condition (2), the height of Ω_T is at most k, and since the clauses of E have size k, the branching factor of Ω_T is also bounded by k. Consequently there are at most k^k leaves.

A leaf of Ω_T is *interesting* if it is labeled by a pair $(\mathcal{A}, V)$ with $\mathcal{A} = \emptyset$. Let $\mathcal{L}_T$ denote the family of sets of variables V that occur in the labels of interesting leaves of Ω_T:

$$\mathcal{L}_T = \{V : (\emptyset, V) \text{ is the label of an interesting leaf of } \Omega_T\}.$$

Let $\mathcal{L}$ denote the union of the $\mathcal{L}_T$:

$$\mathcal{L} = \bigcup_{T \in \mathcal{T}} \mathcal{L}_T.$$

The usefulness of the sets $\mathcal{L}_T$ is illustrated by the special case of $T = \emptyset$. In this case $\mathcal{C}_T$ is the set of monotone clauses of E, that is, the clauses having only positive literals. The following claim is easy to verify from the construction of $\Omega_\emptyset$.

Claim 1 *E is satisfiable by a weight k truth assignment only if $\exists L \in \mathcal{L}_\emptyset$ such that all of the variables in L are assigned the value true.*

For an arbitrary $T \in \mathcal{T}$, the set of clauses $\mathcal{C}_T$ must be satisfied "positively" in any satisfying truth assignment for E that assigns all of the variables of T to be *true*. By the construction of Ω_T the following claim holds.

Claim 2 *The clauses of $\mathcal{C}_T$ are positively satisfied by a weight k truth assignment that sets all of the variables of T to true, only if $\exists L \in \mathcal{L}_T$ such that all of the variables in L are assigned the value true.*

We next describe the set of variables X', which consists of three principal parts:

$$X' = X'[1] \cup X'[2] \cup X'[3].$$

For convenience, let $[k]$ denote the set $\{1, 2, \ldots, k\}$ and let $2^{[k]}$ (as usual) denote the set of all subsets of $[k]$. If $J \in 2^{[k]}$ then J is naturally ordered. We will write $J(s)$ to denote the sth element of J.

In the definition of $X'[2]$ and $X'[3]$ we employ a "dummy" index $\perp$. In order for our indexing scheme to work properly, we define

$$\mathcal{L}_\perp = \{\perp\}.$$

The constituent sets of variables are:

$$X'[1] = \{a[i, x] : 1 \le i \le k, x \in X\},$$
$$X'[2] = \{b[J, T] : J \in 2^{[k]}, T \in (\mathcal{T} \cup \{\perp\})\},$$
$$X'[3] = \{c[J, T, L, J'] : J, J' \in 2^{[k]}, T \in (\mathcal{T} \cup \{\perp\}), L \in \mathcal{L}_T\}.$$

The expression E' is the product of the following boolean expressions. To simplify the notation, the indices in the products implicitly range over the various important sets used to define the variables of X'. In particular: i and i' range over $[k]$; J, J' and J'' range over $2^{[k]}$; x and x' range over X; T and T' range over $\mathcal{T}$ (but not $\perp$); L and L' range over $\mathcal{L}$.

$$E'[1] = \prod_{i} \prod_{x \ne x'} (\neg a[i, x] \vee \neg a[i, x']),$$

$$E'[2] = \prod_{i \ne i'} \prod_{x} (\neg a[i, x] \vee \neg a[i', x]),$$

$$E'[3] = \prod_{i < i'} \prod_{1 \le j < j' \le m} (\neg a[i, j'] \vee \neg a[i', j]),$$

$$E'[4] = \prod_{J} \prod_{T \ne T'} (\neg b[J, T] \vee \neg b[J, T']),$$

$$E'[5] = \prod_{J, T : |J| \ne |T|} (\neg b[J, T]),$$

$$E'[6] = b[\emptyset, \emptyset],$$

$$E'[7] = \prod_{J, T} \prod_{i > \max(J)} \prod_{x \in X_T} ((b[J, T] \wedge a[i, x]) \to b[J \cup \{i\}, T \cup \{x\}]),$$

$$E'[8] = \prod_{J,T:|J|=|T|} \prod_{s=1}^{|J|} \big(b[J,T] \to a[J(s),T(s)]\big),$$

$$E'[9] = \prod_{J''} \prod_{(T,L,J)\neq(T',L',J')} \big(\neg c[J'',T,L,J] \vee \neg c[J'',T',L',J']\big),$$

$$E'[10] = \prod_{J,T,L\ J':|J'|\neq|L|} \prod \big(\neg c[J,T,L,J']\big),$$

$$E'[11] = \prod_{J,J'\ T\neq T'\ L} \prod \prod \big(\neg b[J,T] \vee \neg c[J,T',L,J']\big),$$

$$E'[12] = \prod_{J,T,J'} \big(\neg b[J,T] \vee \neg c[J,\bot,\bot,J']\big),$$

$$E'[13] = \prod_{J,T,L,J':|L|=|J'|} \prod_{s=1}^{|L|} \big(c[J,T,L,J'] \to a[J'(s),L(s)]\big).$$

The variables of X' and the subexpressions of E' all have intuitive semantic interpretations that we next discuss, in order to clarify the argument.

Note that the variables of $X'[1]$ can be group into k *choice blocks* by the first index, each block having $|X|$ variables. Thus the variables of $X'[1]$ represent the k choices of a variable of X to set to *true* in a weight k truth assignment for E.

The subexpression $E'[1]$ enforces that in each block, at most one choice is indicated in any satisfying truth assignment for E'. The subexpression $E'[2]$ enforces that the indicated choices are distinct, and $E'[3]$ enforces that the choices of variables of X to be set *true* are made in increasing order according to the index of the variables.

The variables of $X'[2]$ pair sets of choice blocks with sets of variables in X. For example, the variable $b[\{2,5,7\},\{x_3,x_4,x_6\}]$ has the intended meaning: "x_3 is chosen in block 2, x_4 in block 5, and x_6 in block 7".

The subexpression $E'[4]$ enforces that for any subset of the k choice blocks, at most one corresponding "macro" set of variables is indicated. Note that it is not possible to supply all possible macros of size at most k since there are too many. Only those that occur in a certain way are available (or needed) as second indices for the variables of X'_2. When a macro corresponding to the choices made in a set of blocks is "not available", the set of blocks can be paired with the dummy index $\bot$. The subexpression $E'[5]$ enforces that in the pairing of a set of blocks with a set of variables, the sets have the same size, as required by the intended meaning.

The subexpression $E'[6]$ is essentially an artifact having to do with the availability of the dummy index. We require that the empty set of blocks is paired with the empty set of variables. Together with the subexpression $E'[7]$, this forces the variables of $X'[2]$ to do their intended job (the formal argument for this is made below). The subexpression $E'[8]$ forces any pairings asserted by the boolean variables of $X'[2]$ to be represented in the truth assignment for $X'[1]$.

The variables of $X'[3]$ each represent a fairly complicated predicate that is perhaps best illustrated by example. Suppose $J = \{2, 3, 5\}$ so that J represents the set of choice blocks indexed 2, 3 and 5. By some variable of $X'[2]$ this is paired with (if it is available as an index, suppose this is so) the set of variables (of X) chosen in these blocks. Suppose that this set of variables is $T = \{x_4, x_6, x_8\}$. There are then various clauses of E that are forced to be positively satisfied in any truth assignment that makes the variables of T *true*, and this situation is analyzed in the tree Ω_T. An index $L \in \mathcal{L}_T$ indicates a choice of further variables to set *true* in order to satisfy these clauses.

In summary, the variable $c[J, T, L, J']$ represents the assertion: "The variables of L are chosen to be *true* in the blocks indicated by J' in order to solve the positive-satisfaction problem created by the variables of T being set *true* in the blocks indicated by J."

The subexpression $E'[9]$ enforces that there is at most one such assertion made for any set of blocks J (by a satisfying truth assignment). $E'[10]$ ensures that the right number of blocks are pointed to in the solution of the positive-satisfaction problem. $E'[11]$ and $E'[12]$ enforce that the solution indicated for the positive-satisfaction problem created by T must be a solution computed in Ω_T. $E'[13]$ forces the indicated solution to be "implemented" in the variables of $X'[1]$.

We next argue more formally for the correctness of the reduction. Suppose τ is a weight k truth assignment that satisfies E. Let K denote the set of variables set to *true* and let $K(i)$ denote the ith variable of K (ordered by the indices of the variables in K).

For $J \subseteq [k]$, let $K(J) = \{x : \exists j \in J, K(j) = x\}$.

For $T \in \mathcal{T}$, let $L(T, K)$ denote the lexicographically smallest element of $S = \{K' \subseteq K : K' \in \mathcal{L}_T\}$.

Claim 3 *$L(T, K)$ is well-defined.*

Proof It is enough to argue that the set S is nonempty. By Claim 2, any truth assignment for X that satisfies E must be a superset of at least one element of $\mathcal{L}_T$, otherwise a clause of $\mathcal{C}_T$ would not be satisfied. $\qquad\square$

Note that $L(T, K)$ is necessarily disjoint from T.

Let τ' be the weight k' truth assignment to the variables of X' that assigns the following variables the value *true*:

(1) $\{a[i, K(i)] : 1 \le i \le k\}$;
(2) $\{b[J, K(J)] : J \in 2^{[k]}, K(J) \in \mathcal{T}\}$;
(3) $\{b[J, \perp] : J \in 2^{[k]}, K(J) \notin \mathcal{T}\}$;
(4) $\{c[J, K(J), L(K(J), K), J'] : J \in 2^{[k]}, K(J) \in \mathcal{T}, K(J') = L(K(J), K)\}$;
(5) $\{c[J, \perp, \perp, \emptyset] : J \in 2^{[k]}, K(J) \notin \mathcal{T}\}$.

Claim 4 *If τ satisfies E then τ' satisfies E'.*

Proof To see that this is so, we just need to check that each of the subexpressions of E' is satisfied. The only case that is nontrivial is $E'[7]$. If one of the implications of this subexpression were false, then we would have a situation where τ' has assigned the value *true* to all possible true assertions represented by the variables of $X'[2]$ of the form: "the variables of T are chosen (in ascending order) in the blocks indicated by J", and where for one of these true assertions represented by $b[J,T]$, and for a further choice of a variable x set *true* by τ' in a block i (of larger index than any block in J), with x a successor of T, τ' has assigned the value *false* to $b[J \cup \{i\}, T \cup \{x\}]$. Since $T \cup \{x\} \in \mathcal{T}$, this is a contradiction, since this variable represents a statement that is clearly true.   □

That completes (the easier) half of the argument for the correctness of the reduction.

Now suppose that υ' is a weight k' truth assignment to the variables of X' that satisfies E'. We argue that necessarily υ has a certain restricted form, from which we can then describe a corresponding weight k truth assignment to the variables of X that satisfies E.

Claim 5 *The truth assignment υ must assign exactly one variable of $X'[1]$ to be true in each choice block of $X'[1]$.*

Proof Since $E'[1]$ is satisfied, there can at most one variable in each block assigned *true*. The satisfaction of $E'[4]$ and $E'[9]$ imply that no more than 2^{k+1} variables are assigned *true* in $X'[2] \cup X'[3]$. This forces k to be assigned *true* in $X'[1]$ and the claim follows.   □

Let υ denote the truth assignment for the variables of X that is indicated in the natural manner (as per Claim 5) by υ'. Our goal is to show that the truth assignment υ satisfies E. Part of our argument is based on the details of how υ' satisfies E'.

Claim 6 *Necessarily υ' assigns true to all possible variables of $X'[2]$ that represent true statements (about υ) of the form: "the variables of T are chosen (in ascending order) in the blocks indicated by J".*

Proof For any $T \in \mathcal{T}$ the proof is a simple induction along the trail leading to T, based on the satisfaction of $E'[6]$ (the base step), and $E'[7]$.   □

Claim 7 *The truth assignment υ satisfies E.*

Proof Suppose on the contrary that some clause C_i is not satisfied. Then all of the variables in C_i^- must be assigned the value *true* by υ. Also, $C_i^- \in \mathcal{T}$. By Claim 6, there is some J such that the variable $b[J, C_i^-]$ is assigned the value *true* by υ'. Since υ' satisfies the subexpressions $E'[10]$, $E'[11]$ and $E'[12]$, it must assign *true* to some variable $c[J, C_i^-, L, J']$ with $|L| = |J'|$ and $L \in \mathcal{L}_{C_i^-}$. Since υ' satisfies

$E'[13]$, υ assigns the value *true* to the variables of L. Since $C_i \in \mathcal{C}_{C_i^-}$, the clause C_i must be satisfied by υ, a contradiction.

That completes the proof of the theorem. $\qquad\qquad\qquad\qquad\qquad\qquad$ □

We remark that a consequence, for instance, of the results of the previous subsection is that the k-WEIGHTED SATISFIABILITY problem for boolean expressions that are products of k-sums is complete for $W[1]$. Downey, Fellows, and Taylor also proved a general normalization lemma for $W^*[t]$ for all t.

Although the Downey, Fellows, Taylor result fell short of the desired "$W[t] = W^*[t]$", which is still conjectural, in a subsequent paper, Downey and Fellows [246] proved that $W[2] = W^*[2]$. This proof is also quite complex and is omitted for space considerations. There has not been too much work on this hierarchy as the methods seem difficult, or perhaps the area has moved more towards the exciting recent results aligning lower and upper bounds we meet in Chap. 30. One notable exception is the work of Chen, Flum, and Grohe, who prove the following.

Theorem 23.4.3 (Yijia Chen, Flum, and Grohe [154]) $\quad W[t] \subseteq W^*[t] \subseteq W[t+2]$
for all $t \geq 1$.

Again these results would take us a bit far to include. Chen, Flum, and Grohe also provided some complete hypergraph related to DOMINATING SET problems for higher levels such as $W^*[3]$. We refer the reader to [154] for more details.

The significance of all of these results come from the fact that the generalized normalization lemmata allow for "easier" membership arguments. For example, MONOTONE QUERY NON-EMPTINESS is easily shown to be $W[1]$-hard as we saw in Theorem 21.2.6, but, for example, Papadimitriou and Yannakakis could only show membership in $W[\text{SAT}]$, and the placement of this problem in the W-hierarchy was quite problematic until the Downey, Fellows, and Taylor results.

The coincidences for the cases $t = 1, 2$ above rely on argument which are quite *ad hoc* and it seems that the general $W[t] = W^*[t]$? question is hard.

23.5 Exercises

Exercise 23.5.1 Prove Corollary 23.2.1(ii). That is prove that DOMINATING SET is in MONOTONE $W[2]$.
(Hint. The reduction is straightforward. Let $\{x_1, \ldots, x_n\}$ denote the set of vertices of the given graph G, and we will interpret them as input variables for our circuit $C(G)$. Have a layer of Or gates directly below the variables. These are also one per input variable and we will label them $g_1, \ldots, g_n$. Make x_i and input to the Or gate g_j precisely in the case that (x_i, x_j) is an edge of G. Now to complete the circuit, we have one large And gate with inputs from each of the Or gates. It is easy to see that satisfying assignments correspond directly to dominating sets and conversely.)

Exercise 23.5.2 The reduction of Theorem 23.2.1 proves also that the problem of determining if a CNF formula has a satisfying assignment of weight $\leq k$ is $W[2]$ complete. Prove, however, that determining if a $W[1]$ circuit has a weight $\leq k$ satisfying assignment is in FPT.

Exercise 23.5.3 Recall that a tournament is a directed graph $G = (V, A)$ such that for all $x, y \in V$, exactly one of xy or yx is in A. Prove that TOURNAMENT DOMINATING SET is $W[2]$-complete (Downey and Fellows [245]).

(Hint. To show that it is hard for $W[2]$, reduce from DOMINATING SET. Let $G = (V, E)$ be an undirected graph for which we wish to determine whether G has a dominating set of size k. We describe how a tournament T can be constructed that has a dominating set of size $k + 1$ if and only if G has a dominating set of size k. The size of T is $O(2^k \cdot n)$ where n is the number of vertices of G, and T can be constructed in time polynomial in n and 2^k.

The vertex set of the tournament T is partitioned into three sets: V_A, V_B and V_C. The vertices in the set V_A are in $1 : 1$ correspondence with the vertices of G, and we write $V_A = \{a[u] : u \in V(G)\}$. The set of vertices V_B of T corresponds to m copies of the vertex set of G and we may write this as $V_B = \{b[i, u] : 1 \leq i \leq m, u \in V(G)\}$. (The appropriate cardinality for m is determined below.) V_C consists of just a single vertex which we will denote c.

The construction of T must ensure that for every pair of vertices x, y of T, one of the arcs xy or yx is present. Let T_0 be any tournament on n vertices (we will use T_0 as "filler"). Include arcs in T to make a copy of T_0 between the vertices of each of the n-element sets V_A and $V_B(i) = \{b[i, u] : u \in V(G)\}$ for $i = 1, \ldots, m$.

Construct a tournament T_1 on $m = O(2^{k+1})$ vertices which has no dominating set of size $k + 1$. Consider that the vertex set of T_1 is $V(T_1) = \{1, \ldots, m\}$. For each arc ij in T_1 include in T an arc from each vertex of $V_B(i)$ to each vertex of $V_B(j)$.

The adjacency structure of G is represented in T in the following way: for each vertex $u \in V(G)$ include arcs from the vertex $a[u]$ to the vertices $b[i, v]$ for every $v \in N_G[u]$ and for each i, $1 \leq i \leq m$, and from every other vertex in V_B include an arc *to* $a[u]$. Thus the neighborhood structure of G is represented in arcs from V_A to V_B, and otherwise there are arcs from V_B to V_A.

Finally, there are arcs in T from c to every vertex in V_A and from every vertex in V_B to c.

Now argue that G has a k element dominating set iff T has a $k + 1$ element dominating set.)

Exercise 23.5.4 (Bodlaender and Kratsch) Using a reduction from DOMINATING SET, prove that the following is $W[2]$-complete.

DOMINATING CLIQUE

Instance: A graph $G = (V, E)$, a positive integer k.
Parameter: A positive integer k.
Question: Is there a k element dominating set that also forms a clique?

Exercise 23.5.5 (Downey and Fellows [246]) Prove that the following is $W[2]$-complete.

THRESHOLD DOMINATING SET

Instance: A graph $G = (V, E)$, integers k, r.
Parameter: Positive integers k and r.
Question: Is there a set $V' \subseteq V$ of at most k vertices such that for every vertex u,
                       $N[u]$ contains at least r elements of V'?

(Hint. One direction reduce DOMINATING SET. For membership use the fact that $W[2] = W^*[2]$.)

Exercise 23.5.6 (Wareham)   Prove that the following is $W[2]$-complete

HITTING SET

Instance: A finite family of sets $S = S_1, \ldots, S_n$ comprised of elements from $U = \{u_1, \ldots, u_m\}$, an integer k.
Parameter: A positive integer k.
Question: Is there a subset $T \subseteq U$ of size k such that for all $S_i \in S$, $S_i \cap T \neq \emptyset$?

Exercise 23.5.7 (Fellows, Hermelin, Müller, and Rosamond [290]) As in Exercise 21.3.5 definition of $W[t]$ replace the gates by majority gates, which output "yes" if at least half the inputs are "yes". This defines a class $\hat{W}[1]$. Prove that $W[1] \subseteq \hat{W}[1]$.

Exercise 23.5.8 (Cesati and Di Ianni [137], Cesati [136])   Prove that the following is $W[2]$-complete.

SHORT MULTI-TAPE NTM

Instance: A multi-tape NTM M and a word w.
Parameter: Positive integers k and m.
Question: Is there a computation of M that reaches the accept state on input w it at most k steps?

(Hint: For hardness use a reduction from DOMINATING SET. For membership of $W[2]$ use the classical emulation.)

Exercise 23.5.9 (Bodlaender and de Fluiter [79], Cesati [136])   Show that the following problem is $W[2]$ complete.

STEINER TREE

Instance: A graph $G = (V, E)$, a set S of at most k vertices in V, an integer m.
Parameter: Positive integers k and m.
Question: Is there a set of vertices $T \subseteq V - S$ such that $|T| \leq m$ and $G[S \cup T]$ is connected?

(Hint: For hardness use a reduction from DOMINATING SET. For membership of $W[2]$ use Cesati's method of Turing machines.)

Exercise 23.5.10 (Monoid Factorization-Cai, Chen, Downey, and Fellows [124]) Prove that the following problem is hard for $W[2]$ using a reduction from DOMINATING SET.

H_n FACTORIZATION

Instance: A set A of self-maps on $[n]$, and a self-map h.
Parameter: A positive integer k.
Question: Is there a factorization of h of length k over A?

(Hint: Let $G = (V, E)$ be a graph for which we are to determine whether there is a k-element dominating set. Let n be the order of G. As in Exercise 22.4.5, we construct a set A of self-maps on $[n']$ where we view $[n']$ as consisting of n blocks. Here we have $n' = 2n$ (the blocks have size 2). Let α denote the self-map on $\{1, 2\}$ that maps both elements to 2. For each vertex u of G we construct a map a_u that consists of α in each block corresponding to a vertex $v \in N[u]$, and that is the identity map on all other blocks. The self-map h to be factored consists of α in each block. Note that $\alpha = \alpha^i$ for any number of compositions of the block map α.)

Exercise 23.5.11 (Chen, Flum, and Grohe see e.g. [312]) Show that the problem below is $W[2]$-complete.

KERNEL
Instance: A graph $G = (V, E)$.
Parameter: A positive integer k.
Question: Does G have a set S of k elements such that no two vertices of S are adjacent and for each vertex $x \in V \setminus S$ there is a vertex $y \in S$ with $xy \in E$.

Exercise 23.5.12 (Szeider [634]) Show that the following problem is $W[1]$-hard even for 2-CNF formulas.

k-FLIP MAX SAT
Instance: A CNF formula φ and a truth assignment τ from the variables of φ to $\{0, 1\}$.
Parameter: A positive integers k.
Question: Is there in assignment of Hamming distance $\leq k$ of τ satisfying more clauses of φ?

(Hint: Reduce INDEPENDENT SET.)

Exercise 23.5.13 (Szeider [634])
1. Show that the following problem is $W[2]$-hard.

   k-FLIP CNF SAT
   Instance: A CNF formula φ and a truth assignment τ from the variables of φ to $\{0, 1\}$.
   Parameter: A positive integers k.
   Question: Is there in assignment of Hamming distance $\leq k$ of τ satisfying φ?

   (Hint: Reduce HITTING SET.)
2. Show that for any fixed q, k-FLIP q-CNF SAT is FPT.
   (Hint: Use the method of bounded search trees.)

Exercise 23.5.14 (Chen and Meng [152]) The following problem comes from applications in the search for genetically targeted drugs, searching for a sequence of genes close to bad ones (S_b, the target) and far from good genes (S_g, to avoid side effects).

DISTINGUISHED SUBSTRING SELECTION

Instance: A pair (x, k) where x is a tuple (n, S_b, S_g, d_b, d_g) with integers
n, d_b, d_g; $d_b \leq d_g$, $S_b = \{b_1, \ldots, b_{n_b}\}$, $|b_i| \geq n$ and $S_g = \{g_1, \ldots, g_{n_g}\}$,
with $|g_j| = n$.

Parameter: A positive integers $k = d_b + d_g$.

Problem: Determine if there is a strings s of length n with Hamming distance
$D_H(s, b_i) \leq d_b$ and $D_H(s, g_j) \leq d_g$ for all $i \in [n_b]$ and $j \in [n_g]$.

Prove that DISTINGUISHED SUBSTRING SELECTION is $W[2]$-hard.
(Hint: Reduce DOMINATING SET.)

23.6   Historical Notes

The W-hierarchy is due to Downey and Fellows [243]. The W^*-hierarchy is due to
Downey, Fellows, and Taylor [258]. Downey and Fellows [246] proved that $W[2] = W^*[2]$. Yijia Chen, Flum, and Grohe [154]] proved that $W[t] \subseteq W^*[t] \subseteq W[t+2]$
for all $t \geq 1$.

23.7   Summary

In this chapter we characterize the W and W^*-hierarchies, using the Normalization
Theorem. We also prove $W[1] = W^*[1]$. This enables easier membership proofs.

The Monotone and Antimonotone Collapse Theorems: MONOTONE $W[2t+1] = W[2t]$ and ANTIMONOTONE $W[2t+2] = W[2t+1]$

24

24.1 Results

Theorem 24.1.1 (Monotone collapse theorem—Downey and Fellows [247]) *For all t, MONOTONE $W[2t+1] = W[2t]$.*

Proof By Theorem 23.3.1(i)(a) and the definition of weft we have already shown

$$W[2t] \subseteq \text{MONOTONE } W[2t] \subseteq \text{MONOTONE } W[2t+1].$$

For the reverse inclusion, we must start again "from the beginning" since the Normalization Theorem, Theorem 23.2.1, does not preserve monotonicity.

Let $L \in$ MONOTONE $W[2t+1]$. Let $\mathcal{F}$ denote a family of monotone circuits of depth bounded by h and weft bounded by $2t+1$ such that L reduces to the WEIGHTED CIRCUIT SATISFIABILITY problem $L_{\mathcal{F}}$. We will show that we can reduce this problem to WEIGHTED CIRCUIT SATISFIABILITY for nonmonotone weft $2t$ circuits.

The fact that MONOTONE $W[1]$ is fixed-parameter tractable plays a central role in our argument. (See Corollary 24.1.1 and Exercise 21.3.10.) Our algorithm for this problem is based on the method of bounded search trees discussed in Chap. 3. For the moment, consider the special case of WEIGHTED MONOTONE q-CNF SATISFIABILITY (cf. Exercise 3.8.9). Suppose E is a monotone CNF formula with maximum clause size q for which we wish to determine if there is a weight k satisfying truth assignment. We can easily build a q-branching tree T of height k with a truth assignment of weight k corresponding to each leaf, such that E is weight k satisfiable if and only if one of these leaf assignments does the job. (The key idea is that for a clause to be satisfied, there are q possibilities, each requiring a variable to be set to 1. The branches of a node in the tree T correspond to these possibilities.) Note that T has at most q^k leaves.

Let (C, k) be an instance of $L_{\mathcal{F}}$. Our proof consists of two parts. The first part transforms C in various ways that consist of replacing or replicating parts of C in monotonicity-preserving ways; the inputs variables remain the same, as does the parameter k and the weft of the circuit. We will term this process *semi-normalization*.

R.G. Downey, M.R. Fellows, *Fundamentals of Parameterized Complexity*,
Texts in Computer Science, DOI 10.1007/978-1-4471-5559-1_24,
© Springer-Verlag London 2013

This transformation will be useful in proving the other main theorem of this chapter. In the second part of the reduction, the input variables are changed in a way that decreases the weft, but results in a nonmonotone circuit.

Part 1. Semi-normalization.
Step 1.1: Transformation to a tree circuit.

We first transform C into a tree circuit (equivalently represented as a boolean formula) in the manner of Step 1 of the Normalization Theorem (Theorem 23.2.1).

Step 1.2: Normalization of small-gate components.

Viewing the circuit as a graph, we consider the connected components induced by the small gates. Each such component is necessarily a tree subcircuit that has a single (small) output gate. Suppose $C(g)$ is such a component, with output gate g. Because of the bound h on the depth of C, there are at most 2^h inputs to $C(g)$. The function of these inputs computed by $C(g)$ is equivalent to a product-of-sums expression $E_\wedge(g)$ having at most 2^{2^h} sums, and to a similar sum-of-products expression $E_\vee(g)$. We now consider three cases.

- If the output of g goes to a large $\wedge$ gate g', then replace $C(g)$ in C with $E_\wedge(g)$, replicating subcircuits of C as necessary to provide the inputs to the subcircuit corresponding to $E_\wedge(g)$. The output $\wedge$ gate of the $E_\wedge(g)$ subcircuit can now be coalesced with g'.
- If the output of g goes to a large $\vee$ gate g', then replace $C(g)$ in C with $E_\vee(g)$ similarly, and coalesce the output gate of the replacing subcircuit with g'.
- If g is the output gate of C, then replace $C(g)$ with $E_\wedge(g)$.

If there is a small gate g anywhere in the resulting circuit that gives output to a large gate g' of the same logical character, then g should be coalesced with g'. This coalescing of "small down to large" does not increase the weft of the circuit. (Note that "large down to small" coalescing may increase weft.)

The above transformations result in a tree circuit C' for which every small-gate-connected component (*small* now means fan-in bounded by 2^{2^h}) is a single gate, with the possible exception of the component containing the output gate of C'.

Step 1.3: Elimination of internal small gates.

By an *internal small gate* of C' we mean a small gate g that gives its output to a large gate of C' and the arguments to which are either (i) all large gates or (ii) a mixture of large gates and inputs to C' that includes at least one large gate. (Thus, an internal small gate is a small gate that is *not* either at the top of the circuit with inputs that are entirely inputs to C' or part of small-gate component that includes the output gate of C'.) Case (ii) can be handled in the same way as case (i) by considering the inputs to g that are inputs to the circuit C' as the outputs of ("honorary") large gates having a single input.

Let g' denote a large gate receiving the output of an internal small gate g. There are two cases.

- The gate g' is a large $\vee$ gate. Necessarily, g is a $\wedge$ gate, else g and g' would have been coalesced. Note that all of the large $\wedge$ gates g_i giving input to g can be combined by replacing them with a single large $\wedge$ gate on all of their inputs.

So we may assume that there is at most one large $\wedge$ gate supplying an argument to g. Let g_1 denote this input to g. The remaining inputs $g_2, \ldots, g_s$ to g are large $\vee$ gates. Note that g is small, so $s \leq 2^{2^h}$. For $i = 1, \ldots, s$, let m_i denote the number of inputs to g_i and write $E(i, j)$ to denote the jth input to g_i. The boolean function computed by g can be written

$$E_g = \prod_{j=1}^{m_1} E(1, j) \cdot \prod_{i=2}^{s} \sum_{l=1}^{m_i} E(i, j).$$

By distributing, this can be rewritten

$$E_g = \sum_{l_2=1}^{m_2} \sum_{l_3=1}^{m_3} \cdots \sum_{l_s=1}^{m_s} \left(E(2, l_2) \cdot E(3, l_3) \cdots E(s, l_s) \cdot \prod_{j=1}^{m_1} E(1, j) \right).$$

Note that this is (potentially) a large sum of large products. We replace g according to this distribution, replicating subcircuits to compute the expressions $E(i, j)$ as needed. This may entail a blowup in the size of C' of $O(n^s)$, which is permitted, since s is small. Weft is preserved by coalescing the new gate g (which is now a large $\vee$ gate) with g'.

- The gate g' is a large $\wedge$ gate. Then, necessarily, g is a large $\vee$ gate. In this case, we make the dual replacement.

We now have a monotone tree circuit C'' that is semi-normalized. By trivial modifications such as inserting unary $\wedge$ and $\vee$ gates where appropriate, we may assume that C'' has the following form, where the levels are indexed from 0 to $2t + 4$ and the inputs to the circuit constitute level 0:

(1) Level 1 consists of small gates, each of opposite character from the large gate on level 2 to which it gives its output.

(2) On any input–output path through the circuit, the gates from levels 2 to $2t + 2$ alternate in logical character and are all large. (Note that there are two possible alternation patterns: beginning with $\wedge$ or beginning with $\vee$.)

(3) The lowest 2 levels implement an all-small product-of-sums expression.

Let s denote the bound on small-gate arity, and let $w = 2t + 1$. We can describe C'' by the following canonical expression E. This will be our starting point for the second part of the reduction:

$$E = \prod_{i=1}^{s} \left(\sum_{j=1}^{s_1} E_1(i, j) + \sum_{j=1}^{s_2} E_2(i, j) \right),$$

where for $i = 1, \ldots, s$ and $j = 1, \ldots, s_1$, the subexpressions $E_1(i, j)$ have the form

$$E_1(i, j) = \prod_{i_1=1}^{m_1} \sum_{i_2=1}^{m_2} \cdots \prod_{i_w=1}^{m_w} \sum_{l=1}^{s} v(1, i, j, i_1, \ldots, i_w, l),$$

and for $i = 1, \ldots, s$ and $j = 1, \ldots, s_2$, the subexpressions $E_2(i, j)$ have the form

$$E_2(i, j) = \sum_{i_1=1}^{n_1} \prod_{i_2=1}^{n_2} \cdots \sum_{i_w=1}^{n_w} \prod_{l=1}^{s} v(2, i, j, i_1, \ldots, i_w, l),$$

and the $v(\cdots)$ are input variables of the circuit. Note that the weft of the circuit corresponding to E is $w = 2t + 1$. We will refer to the subexpressions $E_1(i, j)$ as *type 1*, and the subexpressions $E_2(i, j)$ as *type 2*.

Part 2. Decreasing the weft.
Step 2.1: Replacement of the top small gates of type 1 subexpressions.
    We next make use of the fact that MONOTONE q-CNF SATISFIABILITY is fixed-parameter tractable to replace the tops of the type 1 subexpressions. Let g be a gate at level 2 in a subcircuit corresponding to one of the subexpressions $E_1(i, j)$. The output of g is described by a monotone product-of-sums expression having maximum clause size s. This is equivalent to a logical $\vee$ of at most s^k products of size k that can be computed by the FPT algorithm for the bounded clause size satisfiability problem. In this manner, we replace each top large $\wedge$ gate for each subexpression $E_1(i, j)$. Since the new gate g gives its output to a large $\vee$ gate g' at level 3, we can coalesce g and g'. Note that this "almost" decreases the weft relative to these subexpressions; the hitch is that the top gates now have k inputs, and k is not small, although we are making some sort of progress. The type 1 subexpressions now have the form

$$E_1(i, j) = \prod_{i_1=1}^{m_1} \sum_{i_2=1}^{m_2} \cdots \sum_{i_{w-1}=1}^{m_{w-1}} \prod_{i_w=1}^{k} u(1, i, j, i_1, \ldots, i_w).$$

Step 2.2: Elimination of the topmost conjuncts.
    Our strategy here is similar to Step 7 in the proof of the Normalization Theorem. The main idea is to introduce new variables, each of which represents a predicate indicating that every variable in a set of original variables is 1. Let $x(i)$ for $i = 1, \ldots, n$ denote the input variables for the circuit we are now considering (received from Step 2.1). Let $A(i, j, i_1, \ldots, i_{w-1})$ denote the set of variables that are input to the (appropriately indexed) topmost $\wedge$ gate of a type 1 subcircuit:

$$A(1, i, j, i_1, \ldots, i_{w-1}) = \bigcup_{i_w=1}^{k} \{u(1, i, j, i_1, \ldots, i_{w-1}, i_w)\} \subseteq \{x(1), \ldots, x(n)\}.$$

Similarly for the type 2 subcircuits, let

$$A(2, i, j, i_1, \ldots, i_w) = \bigcup_{l=1}^{s} \{v(2, i, j, i_1, \ldots, i_w, l)\}.$$

We will refer to these as the *macro sets* of the variables of E. We may suppose without loss of generality that $k \geq s$, since a topmost $\wedge$ gate with $s > k$ inputs will necessarily evaluate to false. Let $\mathcal{A}$ denote the family of distinct macro sets

$$\mathcal{A} = \bigcup_{i=1}^{s} \bigcup_{j=1}^{s_1} \bigcup_{i_1=1}^{m_1} \cdots \bigcup_{i_{w-1}=1}^{m_{w-1}} \{A(i, j, i_1, \ldots, i_{w-1})\}.$$

Each $A \in \mathcal{A}$ represents the argument set of a topmost conjunct.

Our construction is based on an augmentation $\mathcal{A}'$ of $\mathcal{A}$.

Let A denote an arbitrary finite set. Say that a tower of sets

$$A_1 \subset A_2 \subset \cdots \subset A_s$$

is a *trail* for A if
  (i) $|A_1| = 1$,
 (ii) $A_s = A$, and
(iii) for $i = 1, \ldots, s$, $|A_i - A_{i-1}| = 1$.

For each $A \in \mathcal{A}$, choose (arbitrarily) a trail for A. Note that the length of each such trail will be at most k. Let $\Omega(A)$ denote the family of sets in the trail for A, that is,

$$\Omega(A) = \{A_1, \ldots, A_s\}.$$

We define $\mathcal{A}'$ to be the union of the trails:

$$\mathcal{A}' = \bigcup_{A \in \mathcal{A}} \Omega(A).$$

We now introduce new variables of the following kinds:
(1) One new variable $y(A)$ for each distinct macro set $A \in \mathcal{A}'$.
(2) For each original variable $x(i)$, we introduce $k2^{k+1}$ new variable "copies"

$$x(i, 0), \ldots, x(i, k2^{k+1} - 1).$$

(3) "Padding" consisting of $k2^k$ meaningless variables (inputs not supplying arguments to any gates) $z(1), \ldots, z(k2^k)$.

The change of variables requires some enforcement that can be expressed as a weft 1 subcircuit, detailed below. The expression E for the circuit is transformed to the following expression E' over the set of new variables. We will argue below that E' has a satisfying truth assignment of weight $k' = k^2 2^{k+1} + k2^k$ if and only if E has a satisfying truth assignment of weight k.

$$E' = \prod_{i=1}^{s} \left(\sum_{j=1}^{s_1} E_1'(i, j) + \sum_{j=1}^{s_2} E_2'(i, j) \right) \cdot E_3',$$

where

$$E_1'(i, j) = \prod_{i_1=1}^{m_1} \sum_{i_2=1}^{m_2} \cdots \sum_{i_{w-1}=1}^{m_{w-1}} y(A(1, i, j, i_1, \ldots, i_{w-1})),$$

$$E_2'(i,j) = \sum_{i_1=1}^{n_1} \prod_{i_2=1}^{n_2} \cdots \sum_{i_w=1}^{n_w} y\big(A(2,i,j,i_1,\ldots,i_w)\big),$$

and E_3' is a product of small sums expression recording the following "enforcement" implications concerning the new variables:

(1) For $i = 1, \ldots, n$ and $r = 0, \ldots, k2^{k+1} - 1$,

$$x(i,r) \Rightarrow x\big(i, r+1 \pmod{k2^{k+1}}\big).$$

(2) For each containment $A \subseteq A'$ for $A, A' \in \mathcal{A}'$,

$$y\big(A'\big) \Rightarrow y(A).$$

(3) For each membership $x(i) \in A$ for $A \in \mathcal{A}'$,

$$y(A) \Rightarrow x(i,0).$$

(4) For each $A \in \mathcal{A}$ and for each $A_i \in \Omega(A)$,

$$\big(y(A_i) \wedge x(i)\big) \Rightarrow y(A_{i+1}),$$

where $x(i) = A_{i+1} - A_i$.

E' may be larger than E by a linear factor exponential in k. Correctness for the transformation can be argued as follows.

If E is satisfied by a weight k truth assignment τ, then a weight k' satisfying truth assignment τ' for E' can be had by the following:

- Setting $\tau'(x(i,0)) = \tau'(x(i,1)) = \cdots = \tau'(x(i, k2^{k+1} - 1)) = 1$ for any i for which $\tau(x(i)) = 1$.
- Setting $\tau'(y(A)) = 1$ for any set $A \in \mathcal{A}'$ for which $\tau(x(i)) = 1$ for all $x(i) \in A$.
- Setting $\tau'(z(i)) = 1$ for some number of the padding variables, in order to bring the weight of τ' to exactly k'. (The reason for this is that not all of the subsets of the k variables set to 1 by τ may occur in $\mathcal{A}$.)
- Setting everything else to 0.

Conversely, suppose E' is satisfied by a weight k' truth assignment σ'. Because the implications of (1) are satisfied, exactly k sets of copies of the original variables must have value 1 for σ'. Suppose the indices of these sets of copies are $l_1, \ldots, l_k$ and let $K = \{x(l_1), \ldots, x(l_k)\}$. We will argue that the truth assignment σ that sets the variables of K to 1 and all else to 0 is a satisfying truth assignment for E.

The implications of (4) ensure that $\sigma'(y(A)) = 1$ for any $A \subseteq K$, $A \in \mathcal{A}$.

The implications of (2), (3), and (4) together imply that the settings of the macro variables by σ' must be exactly consistent with the values of the sets of copies of the original variables. By the construction of E', we see that σ is a weight k satisfying truth assignment for E.

We are definitely making progress. Note that E' would now have weft $w - 1 = 2t$ except for the subexpressions of type 2. The main idea of our next step is (roughly) to remove the bottom most large $\vee$ gates of the type 2 subexpressions by introducing additional nondeterminism.

Step 2.3: *More nondeterminism, less weft.*

Let C' now denote the tree circuit corresponding to the expression E' received from Step 2.2. The bottom of the circuit is a small $\wedge$ of small $\vee$ gates. The arguments for these small $\vee$ gates are supplied from large $\wedge$ gates in the case of type 1 subexpressions, and large $\vee$ gates in the case of type 2 subexpressions. Furthermore, all of the small gates are at the bottom of the circuit. We begin by coalescing the small $\vee$ gates with the bottommost large $\vee$ gates of the type 2 subexpressions.

After renaming and reindexing, the current form of the circuit can be described by the weft $2t + 1$ expression

$$F = \prod_{i_1=1}^{s_1} \sum_{i_2=1}^{n_2} \prod_{i_3=1}^{n_3} F(i_1, i_2, i_3),$$

where s_1 is a constant independent of k and n. Let k_F denote the relevant weight parameter; that is, we have completed a reduction to the situation where we are concerned with whether F has a truth assignment of weight k_F, where k_F is purely a function of k.

We next introduce new variables $v(i_1, i_2)$ for $i_1 = 1, \ldots, s_1$ and $i_2 = 1, \ldots, n_2$ to indicate how subexpressions of F are satisfied. Let

$$k'_F = k_F + s_1$$

and consider

$$F' = F'_1 \cdot F'_2 \cdot F'_3,$$

where

$$F'_1 = \prod_{i_1=1}^{s_1} \sum_{i_2=1}^{n_2} v(i_1, i_2),$$

$$F'_2 = \prod_{i_1=1}^{s_1} \prod_{1 \leq j < j' \leq n_2} \left(\neg v(i_1, j) + \neg v(i_1, j')\right),$$

$$F'_3 = \prod_{i_1=1}^{s_1} \prod_{i_2=1}^{n_2} \prod_{i_3=1}^{n_3} \left(\neg v(i_1, i_2) + F(i_1, i_2, i_3)\right).$$

The weft of F' is $2t$. We claim that F' has a truth assignment of weight k'_F if and only if F has a truth assignment of weight k_F.

A truth assignment of weight k'_F that satisfies F' must set exactly s_1 of the new variables $v(i_1, i_2)$ to 1, by the restrictions imposed by F'_1 and F'_2. Consequently, exactly k_F of the original variables of F must also be set to 1, in order for the total weight of the assignment to be k'_F. Moreover, these must be one variable for each of the s_1 branches of F. Since F'_3 is satisfied, F is satisfied by this weight k_F assignment. In the other direction, the conversion of a weight k_F satisfying assignment for

F into a weight k'_F satisfying assignment for F' follows the same correspondence. This completes the theorem. □

Corollary 24.1.1 MONOTONE $W[1]$ *is fixed-parameter tractable.*

Proof (See also Exercise 21.3.10.) In the proof of the theorem we have used the fact that the special case of WEIGHTED MONOTONE q-CNF SATISFIABILITY is fixed-parameter tractable. The steps of the proof through Step 2.1 all preserve both monotonicity and the original input variables, and consist only in rewriting the boolean function computed by the circuit. Thus, it is enough to argue that it is fixed-parameter tractable to determine if an expression in the form resulting from Step 2.1 has a weight k truth assignment. This form is a monotone product-of-sums-of-k-products

$$E = \prod_{i=1}^{s} \sum_{j=1}^{m} \prod_{l=1}^{k} x(i, j, l),$$

where s is a constant independent of k and the size of the expression.

We can simply try all m^s possibilities for choosing branches of E to be satisfied. □

Theorem 24.1.2 (Antimonotone collapse theorem—Downey and Fellows [247]) *For all t,* ANTIMONOTONE $W[2t + 2] = W[2t + 1]$.

Proof Half of the theorem is easy. For the other half, we may make use of the same initial transformations to semi-normalized form as in the first part of the proof of Theorem 24.1.1, since these are based on reorganizations of the circuit computations that have no effect on monotonicity or antimonotonicity. After these transformations (and some trivial replications of parts of the circuit to simplify the indexing), our starting point is a circuit C of size n corresponding to an expression E in the form

$$E = \prod_{i=1}^{s} \left(\sum_{j=1}^{s} E_1(i, j) + \sum_{i=1}^{s} E_2(i, j) \right),$$

where s is a constant independent of k and n, and where the subexpressions have the form

$$E_1(i, j) = \prod_{i_1=1}^{m} \sum_{i_2=1}^{m} \cdots \sum_{i_w=1}^{m} \prod_{l=1}^{s} \neg v(1, i, j, i_1, \ldots, i_w, l),$$

$$E_2(i, j) = \sum_{i_1=1}^{m} \prod_{i_2=1}^{m} \cdots \prod_{i_w=1}^{m} \sum_{l=1}^{s} \neg v(2, i, j, i_1, \ldots, i_w, l),$$

with the $v(\cdots)$ being input variables of the circuit and $w = 2t + 2$. We wish to determine whether there is a weight k input accepted by C. The remaining steps of the transformation parallel the steps of Part 2 of the proof of Theorem 24.1.1, although there are a few differences.

Step 2.1: Replacement of the top small gates of type 1 subexpressions.
    The top three levels of a type 1 subexpression have the form

$$F = \prod_{i=1}^{m} \sum_{j=1}^{m} \prod_{l=1}^{s} \neg x(i, j, l).$$

We have the logical equivalences

$$F = \neg\neg F = \prod_{i=1}^{m} \left(\neg \prod_{j=1}^{m} \sum_{l=1}^{s} x(i, j, l) \right).$$

The top two levels of the above expression are monotone products-of-sums, so as in the proof of Theorem 24.1.1, we can write

$$F = \prod_{i=1}^{m} \left(\neg \sum_{j=1}^{s^k} \prod_{l=1}^{k} y(i, j, l) \right),$$

where the $y(\cdots)$ are appropriate circuit inputs. Commuting the negations to the top of the circuit gives

$$F = \prod_{i=1}^{m} \prod_{j=1}^{s^k} \sum_{l=1}^{k} \neg y(i, j, l).$$

By coalescing the $\wedge$ gates corresponding to the products, we have "almost" decreased the weft of these subexpressions by 1, except that the topmost sums have size k, which is not small.

Step 2.2: Elimination of the topmost disjuncts.
    We now have an antimonotone circuit, the first two levels of which are various products-of-sums of negated inputs, with the sums having size k in the case of the type 1 subexpressions and size s in the case of the type 2 subexpressions. This is the dual of our situation for Step 2.2 of the proof of Theorem 24.1.1.
    Note that if $v(a, b, c)$ is a new variable that represents the predicate $a = b = c = 1$, then we have the logical equivalence

$$(\neg a) \vee (\neg b) \vee (\neg c) = \neg v(a, b, c).$$

This provides the basis for our transformation. A complication arises because we are unable to argue for well-behavedness as we did for Theorem 24.1.1, so that our argument must be modified slightly.
    As in Step 2.2 of the proof of Theorem 24.1.1, let $\mathcal{A}$ denote the family of sets of variables that, negated, are the inputs to the topmost $\vee$ gates. For each $A \in \mathcal{A}$, choose a trail $\Omega(A)$ for A and let $\mathcal{A}'$ be the union of the trails.
    We introduce new variables of the following kinds:

(1) One new variable $y(A)$ for each $A \in \mathcal{A}'$.

(2) For each original variable $x(i)$, the copy variables $x(i, 0), \ldots, x(i, k2^{k+1} - 1)$.

(3) Padding variables $z(1), \ldots, z(k2^k)$.

Each top level disjunct of the negations of the variables in a set $A \in \mathcal{A}$ is replaced by $\neg y(A)$; this is essentially the dual of the substitution made in Step 2.2 of Theorem 24.1.1.

The remainder of the details of this step are left to the reader, as from this point on they are essentially identical to the previous argument.

Step 2.3: More nondeterminism, less weft.

Because of the weft 1 enforcements introduced in Step 2.2, we are no longer dealing with an antimonotone circuit, but the same was true in the proof of Theorem 24.1.1. In fact, we are now in essentially the same position and can apply *the same* transformation as in the earlier argument, with one complication.

The problem is that this works if $t \geq 1$, but not for $t = 0$, since the enforcement expressions for the change of variables have weft 2. As is so often the case, we are forced into making a special argument to fit our conclusions to weft 1.

For the case of $t = 0$, we may take our current starting point to be described by the weft 2 expression

$$F = \prod_{i=1}^{s} \sum_{j=1}^{m} \prod_{l=1}^{m} \lambda(i, j, l),$$

where $\lambda(i, j, l)$ is a literal, that is, either an input variable or the negation of an input variable. Assume that we are interested in whether there is a weight k_F truth assignment for F, and note that s is small. The basic idea of our transformation for this case is the same; we are simply forced to do a little more work to obtain weft 1 enforcement for our change of variables.

Let $x(1), \ldots, x(n)$ denote the variables of F. We introduce new variables $V = V_1 \cup V_2$, where

$$V_1 = \{v(i, j) : 1 \leq i \leq s, \ 1 \leq j \leq m\},$$
$$V_2 = \{r(i, j) : 1 \leq i \leq n, \ 0 \leq j \leq s - 1\}.$$

The transformed expression is $F' = F_1' \cdot F_2' \cdot F_3' \cdot F_4'$, where

$$F_1' = \prod_{i=1}^{s} \prod_{j=1}^{m} \prod_{l=1}^{m} \left(\neg v(i, j) + \lambda(i, j, l) \right),$$

$$F_2' = \prod_{i=1}^{s} \prod_{1 \leq j < j' \leq m} \left(\neg v(i, j) + \neg v(i, j') \right),$$

$$F_3' = \prod_{i=1}^{n} \left(\neg x(i) + r(i, 0) \right) \cdot \left(x(i) + \neg r(i, 0) \right),$$

$$F_4' = \prod_{i=1}^{n} \prod_{j=0}^{s-1} \left(\neg r(i, j) + r\left(i, j+1 \ (\text{mod } s)\right) \right).$$

The transformed weight parameter is

$$k_F' = s + k_F(s + 1).$$

Suppose τ is a weight k_F truth assignment that satisfies F. Then, for each i, $1 \le i \le s$, we can identify an argument with index j to the ith large $\vee$ gate that evaluates to 1. Correspondingly, set $\tau'(v(i, j)) = 1$. Let τ' correspond to τ on $\{x(1), \ldots, x(n)\}$, and for each i such that $\tau(x(i)) = 1$, let $\tau'(r(i, j)) = 1$ for $j = 0, \ldots, s - 1$. Then, τ' is a truth assignment to the variables of F' of weight k_F' and it is easy to check that τ' satisfies F'.

Conversely, suppose σ' is a truth assignment of weight k_F' that satisfies F'. Because of the implications of F_3' and F_4', the number p of original variables $x(i)$ and variables of V_2 set to 1 by σ' must be a multiple of $s + 1$, and if $p = q(s + 1)$, then precisely q of the variables $x(i)$ are set to 1. The definition of k_F' forces $p = k_F$. Thus, the restriction σ of σ' to the variables $x(i)$ constitutes a weight k_F truth assignment for the variables of F. We argue that σ satisfies F. Since σ' has weight $k_F(s + 1)$ on the original variables together with V_2, it must have weight s on V_1. The expression F_2' ensures that for each i, $1 \le i \le s$, there is exactly one j such that $\sigma'(v(i, j)) = 1$. It is easy to see that these $v(i, j)$ indicate how F is satisfied by σ.

Since F' is an expression of weft 1, we are done. $\qquad\qquad \square$

24.2 Historical Notes

All of the results of this section are from Downey and Fellows [247].

24.3 Summary

We prove tight relationships between monotone and antimonotone problems and their classes.

Beyond $W[t]$-Hardness

25

25.1 $W[P]$ and $W[\textsc{Sat}]$

The $W[t]$-classes reflect the intrinsic difficulty of solution checking formulas of bounded logical depth. Naturally, the question arises as to what happens if we have no bound on the depth and simply look at parameterized problems of "polynomial size". When we do this we arrive at classes hard for $\bigcup_t W[t]$. The study of such classes is the theme of the present chapter. Two important classes immediately suggest themselves. These are the classes $W[\textsc{Sat}]$ and $W[P]$ generated by the following problems.

WEIGHTED SATISFIABILITY
Input: A (boolean) formula X.
Parameter: A positive integer k.
Question: Does X have a weight k satisfying assignment?

WEIGHTED CIRCUIT SATISFIABILITY
Input: A decision circuit C.
Parameter: A positive integer k.
Question: Does C have a weight k satisfying assignment?

The classes are then the following.
- $W[\textsc{Sat}]$, the class of problems reducible to WEIGHTED SATISFIABILITY.
- $W[P]$, the class of problems reducible to WEIGHTED CIRCUIT SATISFIABILITY.

The standard translation of Turing machines into circuits shows that WEIGHTED CIRCUIT SATISFIABILITY is the same as the problem of ascertaining if a deterministic Turing machine has a weight k satisfying assignment. In the same way that we have seen that WEIGHTED CNF SATISFIABILITY and WEIGHTED 3SAT seem to have different parameterized complexities, it seems that the containment $W[\textsc{Sat}] \subset W[P]$ is proper. $W[\textsc{Sat}]$ does not seem to be well populated with natural parameterized problems, but $W[P]$ contains many.

We begin with some technical results. Consider the following problem.

R.G. Downey, M.R. Fellows, *Fundamentals of Parameterized Complexity*,
Texts in Computer Science, DOI 10.1007/978-1-4471-5559-1_25,
© Springer-Verlag London 2013

SHORT CIRCUIT SATISFIABILITY

Instance: A decision circuit C with at most n gates and $\leq k \log n$ inputs.
Parameter: A positive integer k.
Question: Is there any setting of the inputs that causes C to output 1?

Lemma 25.1.1 (Abrahamson, Downey, and Fellows [3]) SHORT CIRCUIT SATIS-
FIABILITY *is* $W[P]$-*complete.*

Proof To see that it is $W[P]$-hard, take an instance C of WEIGHTED CIRCUIT SAT-
ISFIABILITY, with parameter k and inputs $x_1, \ldots, x_n$. Let $z_1, \ldots, z_{k \log n}$ be new
variables. Using lexicographic order in polynomial time (independent of k) we can
have a surjection from $Z = \{z_1, \ldots, z_{k \log n}\}$ to the (characteristic function of the)
k-element subsets of $X = \{x_1, \ldots, x_n\}$. Representing this as a circuit with inputs Z
and outputs X we can put this circuit on top of C to obtain a circuit C' such that C
accepts a weight k input if and only if C' accepts some input.

That $W[P]$ contains SHORT CIRCUIT SATISFIABILITY is equally easy, just use
the polynomial embedding of the $k \log n$ into the $k + 1$-element subsets of n for n
sufficiently large. $\square$

We remark that the $k \log n$ trick above is very useful in many arguments. For
the following result we shall employ ideas from Abrahamson, Ellis, Fellows and
Mata [6]. Let φ be a formula of propositional logic (or a circuit). Suppose we wish
to extend a partial truth assignment t of a (circuit or) formula φ to a satisfying
assignment t'. If there is a clause, all but one of whose literals are forced to be false
by t, then t forces the remaining literal to be made *true* and we can extend t with
this setting without losing any information. This may in turn lead to further such
extensions. We repeat this process until we can go no further. If this process yields
a truth assignment t' that is a satisfying assignment for φ all of whose literals are
specified, we say that t causes φ to *unravel*. Following [6], we have the following
problem definition.

SHORT SATISFIABILITY

Input: A formula φ of n variables and a list of at most $k \log n$ variables of φ.
Parameter: A positive integer k.
Question: Is there any setting of the distinguished variables that causes φ to un-
             ravel?

Theorem 25.1.1 SHORT SATISFIABILITY *is* $W[P]$-*complete.*

Proof We reduce from SHORT CIRCUIT SATISFIABILITY. Let $x_1, \ldots, x_m, C$ be an
instance of SHORT CIRCUIT SATISFIABILITY with parameter k and $m \leq k \log n$.
These variables will be the distinguished variables for the input $\varphi = \varphi(C)$. We introduce new variables and define φ by locally replacing gates of C as follows. We
replace a *not* gate $y = \overline{x}$ by the formula $(x \vee y) \wedge (\overline{x} \vee \overline{y})$ and an *and* gate $z = x \wedge y$
by $(x \vee \overline{z}) \wedge (y \vee \overline{z}) \wedge (\overline{x} \vee \overline{y} \vee z)$. The formula generated by the circuit is at most

quadratically larger than the circuit, and clearly a setting of $x_1, \ldots, x_m$ satisfies C if and only if it unravels φ.   □

Remark 25.1.1 Since the writing of Downey and Fellows [247], the method above has been "officially" coined the "$k \log n$ trick" probably originating in Flum–Grohe [312].

The basis for many of our reductions is the $W[P]$-completeness of WEIGHTED MONOTONE CIRCUIT SATISFIABILITY, which is the restriction of the problem WEIGHTED CIRCUIT SATISFIABILITY to circuits with no inverters. This result was first announced in Downey, Fellows, Kapron, Hallett, and Wareham [248]. In [248], Downey et al. analyzed the parameterized tractability of problems stemming from logic and linguistics. A key problem identified by those authors is the following.

MINIMUM AXIOM SET
Input: A finite set S of "sentences", an "implication relation" R consisting of pairs (A, t) where $A \subseteq S$ and $t \in S$, and a positive integer k.
Parameter: A positive integer k.
Question: Is there a set $S_0 \subseteq S$ with $|S_0| \leq k$ and a positive integer n such that if we define S_i, $1 \leq i \leq n$, to consist of exactly those $t \in S$ for which either $t \in S_{i-1}$ or there exists a set $U \subseteq S_{i-1}$ such that $(U, t) \in R$, then $S_n = S$?

The decisive result is the following, announced in [248], but whose full proof only appeared in Abrahamson, Downey, and Fellows [3]

Theorem 25.1.2 (Downey, Fellows, Kapron, Hallett and Wareham [248], Abrahamson, Downey, and Fellows [3]) MINIMUM AXIOM SET *is $W[P]$-complete.*

Lemma 25.1.2 MINIMUM AXIOM SET *belongs to* MONOTONE $W[P]$.

Proof We describe a parameterized polynomial-time transformation of an instance (S, R) of MINIMUM AXIOM SET to a circuit C that accepts a weight k vector if and only if (S, R) has an axiom set of size k. For convenience of description, we view C as a directed acyclic graph for which (1) each vertex is assigned a logic function in the set $\{\wedge, \vee, 1\}$, where 1 denotes the identity function, and (2) some (appropriate) vertices have been designated as inputs.

Suppose $|S| = n$. The vertex set of C is $V(C) = V_1 \cup V_2$ where

$$V_1 = \big\{ t[u, i] : u \in S, \ 0 \leq i \leq n \big\}.$$

Each vertex of V_1 is assigned the identity logic function. The inputs to C are the vertices $\{t[u, 0] : u \in S\}$. We may think of each vertex of V_1 in the circuit C as representing a boolean variable, the meaning of which is, "statement u is true at time i". Thus a weight k input to C represents k statements (the axioms) being true at time 0.

The vertices of V_2 are the union of n^2 sets of vertices

$$V_2 = \bigcup \{V[u, i] : u \in S,\ 1 \le i \le n\}$$

where the vertices of $V[u, i]$ implement a monotone circuit $C(u, i)$ that computes the value of $t[u, i]$ from the value of variables at the $i - 1$ level of V_1, in accordance with the definition of the sets S_i in the description of the MINIMUM AXIOM SET problem. We may describe $C(u, i)$ by the sum-of-products expression

$$C(u, i) = t[u, i - 1] + \sum_{(A, u) \in R} \left(\prod_{v \in A} t[v, i - 1] \right).$$

The correctness of this construction is straightforward, noting that by time n, all of the statements deducible from the axioms will have been deduced.   □

Lemma 25.1.3 MINIMUM AXIOM SET *is hard for* $W[P]$.

Proof Let C be a decision circuit for which we wish to determine whether there is a boolean vector of weight k accepted by C. We describe how C can be transformed in polynomial time into an instance (S, R) of the MINIMUM AXIOM SET problem that has an axiom set of size $2k$ if and only if C accepts a weight k input vector.

We view the circuit C as a directed acyclic graph $C = (V, A)$. We may assume that the circuit C is *leveled* in the sense that the vertices of V are partitioned into m sets $V[1], \ldots, V[t]$ ("levels"), with the inputs constituting level 1, and such that for every arc $uv \in A$, if u is in level i then v is in level $i + 1$. Since C is a general circuit, each vertex of $V - V[1]$ is assigned a logic function from the set $\{\wedge, \vee, \neg\}$. Thus $V - V[1]$ can be expressed as the disjoint union

$$V - V[1] = V_\wedge \cup V_\vee \cup V_\neg.$$

We assume without loss of generality that $\wedge$ and $\vee$ vertices have in-degree two and that $\neg$ vertices have in-degree one. We will assume that C has n inputs $V[1] = \{u_0, u_1, \ldots, u_{n-1}\}$. Let z denote the output vertex of C.

The set of "statements" S for the instance (S, R) of MINIMUM AXIOM SET is the union of three sets, $S = S[1] \cup S[2] \cup S[3]$. Similarly, we describe the deductive structure R as the union of several parts, $R = R[1] \cup R[2] \cup R[3] \cup R[4] \cup R[5]$. The deductive structure $R[1]$ on $S[1]$ has the primary role of simulating the logic gates of the circuit C. The deductive structure $R[2]$ on $S[1]$ and $S[2]$ handles the inputs to the simulation of C. The role of $R[3]$ and $R[4]$ is to serve as an enforcement mechanism for the input simulation. The role of $R[5]$ is in handling the output of the simulation of C.

$$S[1] = \{q[u, i] : u \in V,\ i \in \{0, 1\}\}.$$

The simulation of C provided by the deductive structure on $S[1]$ is straightforward. Note that each vertex u of C is represented by two "statements" $q[u, 0]$ and $q[u, 1]$ in $S[1]$. The deductive structure (S, R) simulates C in the following general way. First, the structure provided by $S[2]$ enforces (in a manner explained below) that "at time 1" exactly one of $q[u, 0]$ and $q[u, 1]$ has been deduced for each input vertex $u \in V_0$ (that is, will belong to the set S_1 in the definition of the MINIMUM AXIOM SET problem), for any $2k$ element axiom set with any hope of success (let us call this "meaningful input" for the moment). Inductively, the simulation of C ensures that for meaningful input, for each vertex $q \in V$, and for $i = 1, \ldots, t$, exactly one of $q[u, 0]$ and $q[u, 1]$ will belong to S_i.

In describing the simulation, we interpret this in the following way: if $q[u, 0] \in S_t$ then the vertex (logic gate) q of C has logic value 0 at time t, and if $q[u, 1] \in S_t$ then q has logic value 1 at time t. It may eventually transpire (at time t, since the circuit is leveled) that for the output vertex z of C, the statement $q[z, 1]$ is deduced. The structure (S, R) is such that if $q[z, 1]$ is deduced (at time t) then "everything" in S can be deduced at time $t + 2$. The details of the circuit simulation are as follows.

For $u \in (V_\wedge \cup V_\vee)$ let u' and u'' denote the input vertices to u, that is, the two vertices for which $u'u$ and $u''u$ are in the set of arcs A describing the circuit. Similarly, for $u \in V_\neg$ let u' denote the single input vertex to u. The implications of $R[1]$ provide the circuit simulation. We have $R[1] = R_\wedge \cup R_\vee \cup R_\neg$ where

$$R_\wedge = \bigcup_{u \in V_\wedge} \{((\{q[u', 1], q[u'', 1]\}, q[u, 1]), (\{q[u', 0]\}, q[u, 0]),$$

$$(\{q[u'', 0]\}, q[u, 0])\},$$

$$R_\vee = \bigcup_{u \in V_\vee} \{((\{q[u', 0], q[u'', 0]\}, q[u, 0]), (\{q[u', 1]\}, q[u, 1]),$$

$$(\{q[u'', 1]\}, q[u, 1])\},$$

$$R_\neg = \bigcup_{u \in V_\neg} \{((\{q[u', 1]\}, q[u, 0]), (\{q[u', 0]\}, q[u, 1])\}.$$

We next describe the structure which simulates the input to the circuit. This construction is based on the ideas of Theorem 23.2.1. We have $S[2] = S[2, 1] \cup S[2, 2] \cup S[2, 3] \cup S[2, 4] \cup S[2, 5]$ where

$$S[2, 1] = \{a[r, s, i] : 0 \le r \le k - 1, \ 0 \le s \le n - 1, \ 1 \le i \le 2\},$$

$$S[2, 2] = \{b[r, s, t, i] : 0 \le r \le k - 1, \ 0 \le s \le n - 1, \ 1 \le t \le n - k + 1,$$

$$1 \le i \le 2\},$$

$$S[2, 3] = \{a'[r, u, i] : 0 \le r \le k - 1, \ 1 \le u \le 2k + 1, \ 1 \le i \le 2\},$$

$$S[2, 4] = \{b'[r, u, i] : 0 \le r \le k - 1, \ 1 \le u \le 2k + 1, \ 1 \le i \le 2\},$$

$$S[2, 5] = \{d[r, s, i] : 0 \le r \le k - 1, \ 0 \le s \le n - 1, \ 1 \le i \le 2\}.$$

Note that the set of statements $S[2]$ can be partitioned into two sets $S[2] = S'[2] \cup S''[2]$ according to the value of the last index of the elements. That is, let $S'[2]$ be those elements of $S[2]$ with last index $i = 1$ and let $S''[2]$ be those elements of $S[2]$

with last index $i = 2$. For $x' \in S'[2]$ let x'' be the element of $S''[2]$ that corresponds to x' by changing the last index from $i = 1$ to $i = 2$. Similarly we define the notation $S'[2, j]$ for $j = 1, \ldots, 5$.

We can now describe the second set of implications

$$R[2] = \{(\{a[r, i, 1] : 0 \leq r \leq k - 1\}, q[u_i, 1]) : 0 \leq i \leq n - 1\}$$
$$\cup \{(\{b[r, s, t, 1]\}, q[u_i, 0]) : s < i < s + t, \ 0 \leq i \leq n - 1\}.$$

We next introduce as an intermediate descriptive device a graph $G = (V, E)$ (the *domination graph*) on the vertex set $V = S'[2]$. For convenience, we introduce notation for the following sets of vertices in this graph.

$$A(r) = \{a[r, s, 1] : 0 \leq s \leq n - 1\},$$
$$B(r) = \{b[r, s, t, 1] : 0 \leq s \leq n - 1, 1 \leq t \leq n - k + 1\},$$
$$B(r, s) = \{b[r, s, t, 1] : 1 \leq t \leq n - k + 1\}.$$

The edge set E of G is the union of the following sets of edges. In these descriptions we implicitly quantify over all possible indices.

$$E_1 = \{a[r, s, 1]a[r, s', 1] : s \neq s'\},$$
$$E_2 = \{b[r, s, t, 1]b[r, s, t', 1] : t \neq t'\},$$
$$E_3 = \{a[r, s, 1]b[r, s', t, 1] : s \neq s'\},$$
$$E_4 = \{b[r, s, t, 1]d[r, s', 1] : s' \neq s + t \ (\mathrm{mod}\ n)\},$$
$$E_5 = \{a[r, s, 1]a'[r, u, 1]\},$$
$$E_6 = \{b[r, s, t, 1]b'[r, u, 1]\},$$
$$E_7 = \{d[r, s, 1]a[r', s, 1] : r' = r + 1 \ (\mathrm{mod}\ n)\}.$$

We use this graph in describing the third set of implications as follows. For a vertex $v \in V$, the *closed neighborhood* $N[v]$ *of* v is defined to be $N[v] = \{w : vw \in E\} \cup \{v\}$.

$$R[3] = \{(N[x'], x'') : x' \in V = S'[2]\}.$$

The third set of statements is

$$S[3] = \{e[i] : 1 \leq i \leq 2k + 1\}.$$

The final sets of implications are

$$R[4] = \{(S''[2] \cup \{q[z, 1]\}, e[i]) : 1 \leq i \leq 2k + 1\},$$
$$R[5] = \{(S[3], s) : s \in S[1] \cup S[2]\}.$$

This completes the description of the instance (S, R) of MINIMUM AXIOM SET. We argue the correctness of the reduction as follows.

We first describe an *interpretation* of a set U of $2k$ vertices in the domination graph, as this plays a key role in the argument. We interpret $a[r, s, 1] \in U$ as the directive, "set the input u_s of C to 1". We interpret $b[r, s, t, 1] \in U$ as the directive, "set the inputs u_i of C to 0, for all $i, s < i < s + t$". If these directives are consistent for the vertices in a set U, call U a *consistent* set of vertices in the domination graph. Say that U is *total* if the directives corresponding to the vertices in U assign each input to C a value. If U is consistent and total, then the *interpretation* τ_U of U is the input to C corresponding to U according to the directives.

Claim 1 *A dominating set D of $2k$ vertices in the domination graph G is consistent and total and the corresponding boolean input vector τ_D to the circuit C has weight k.*

Proof of Claim 1 Suppose D is a dominating set of $2k$ vertices in G. The closed neighborhoods of the $2k$ vertices $a'[0, 1, 1], \ldots, a'[k - 1, 1, 1], b'[0, 1, 1], \ldots, b'[k - 1, 1, 1]$ are disjoint, so D must consist of exactly $2k$ vertices, one in each of these closed neighborhoods. Also, none of the vertices of $V_4 \cup V_5$ are in D, since if $a'[r, u, 1] \in D$ then necessarily $a'[r, u', 1] \in D$ for $1 < u' < 2k + 1$ (otherwise D fails to be dominating), which contradicts that D contains exactly $2k$ vertices. It follows that D contains exactly one vertex from each of the sets $A(r)$ and $B(r)$ for $0 \le r \le k - 1$.

The possibilities for D are further constrained by the edges of E_3, E_4 and E_7. The vertices of D in $S'[2, 1]$ represent the inputs set to 1 in a weight k input vector for the circuit C, and the vertices of D in $S'[2, 2]$ represent "intervals" of inputs set to 0 in a weight k input vector for C. Since there are k inputs to be set to 1 (in a weight k vector), there are, considering the indices of the variables mod n, also k intervals of inputs to be set to 0.

The edges of E_3, E_4 and E_7 enforce that the $2k$ vertices in D must represent such a choice consistently. To see how this enforcement works, suppose $a[3, 4, 1] \in D$. This represents that the third of k distinct choices of inputs to be given the value 1 is the input u_4. The edges of E_3 force the unique vertex of D in the set $B(3)$ to belong to the subset $B(3, 4)$. The index of the vertex of D in the subset $B(3, 4)$ represents the difference (mod n) between the indices of the third and fourth choices of an input to receive the value 1, and thus the vertex represents a range of inputs to receive the value 0. The edges of E_4 and E_7 enforce that the index t of the vertex of D in the subset $B(3, 4)$ represents the "distance" to the next input to receive the value 1, as it is represented by the unique vertex of D in the set $A(4)$.

In this way, a dominating set D of size $2k$ is forced to be both consistent and total, and we may note also that τ_D necessarily has weight k.   □

Claim 2 *If C accepts a weight k input vector, then (S, R) is a yes-instance of* MINIMUM AXIOM SET.

Proof of Claim 2 Let τ denote an input boolean vector of weight k accepted by C that assigns the inputs $V_\tau = \{u_{i_0}, u_{i_1}, \ldots, u_{i_{k-1}}\} \subseteq V[1]$ the value 1. Suppose $i_0 < i_2 < \cdots < i_{k-1}$. Let $d_r = i_{r+1 \,(\text{mod}\,k)} - i_r \pmod{n}$ for $r = 0, \ldots, k - 1$. It is straightforward to verify that the set of $2k$ vertices

$$D = \big\{a[r, i_r, 1] : 0 \le r \le k - 1\big\} \cup \big\{b[r, i_r, d_r] : 0 \le r \le k - 1\big\}$$

is a dominating set in the domination graph G and that $\tau_D = \tau$.

We take the initial set of axioms $S_0 = D$. We will argue that $S_{t+2} = S$. Since D is a dominating set in the graph on $S'[2]$, by the implications in $R[3]$ (based on closed neighborhoods in the domination graph) we may deduce that $S''[2] \subseteq S_1$.

The implications in $R[2]$ yield

$$S[1] \cap S_1 = \big\{q[u, 1] : u \in V_\tau\big\} \cup \big\{q[u, 0] : u \in V[1] - V_\tau\big\}$$

consistent with our description of the circuit simulation (at time 1).

By induction on i, noting the form of the implications in $R[1]$, it is straightforward that the circuit simulation is correct; that is, for $i = 2, \dots, t$ we have

$$S[1] \cap S_i = \big\{q[u, 1] : u \text{ evaluates to 1 in } C(\tau)\big\}$$

$$\cup \big\{q[u, 0] : u \text{ evaluates to 0 in } C(\tau)\big\}.$$

Since C accepts τ, $q[z, 1] \in S_t$, and consequently by the implications in $R[4]$, $S[3] \subset S_{t+1}$. By the implications in $R[5]$ this yields $S \subseteq S_{t+2}$. $\square$

Claim 3 *If $(S, R, 2k)$ is a yes-instance of* MINIMUM AXIOM SET *then C accepts a weight k input vector.*

Proof of Claim 3 Let S_0 be the set of $2k$ axioms. If S_0 does not contain $2k$ statements in $S'[2]$ then by Claim 1 S_0 is not a dominating set in the domination graph, and consequently there is at least one statement $s \in S''[2]$ that can only be deduced if every statement in $S[3]$ is first deduced (or is an axiom). Not every statement in $S[3]$ can be an axiom (since the cardinality of $S[3]$ is $2k + 1$), so there must be some statement $s' \in S[3]$ that is properly deduced, necessarily by an implication in $R[4]$, prior to the deduction of s. An examination of $R[4]$ shows that s must be deduced prior to s', a contradiction. Thus $S_0 \subseteq S'[2]$, and furthermore, S_0 must be a dominating set in G.

By Claim 1, S_0 must be consistent and total, and we may therefore consider the interpretation τ_{S_0}, which we argue is accepted by C. The only possibility for statements in $S'[2] - S_0$ to be deduced is by the implications in $R[5]$, and these can only be applied if $q[z, 1]$ has been deduced. By the inductive argument for the correctness of the circuit simulation, this can only happen if C accepts the weight k input vector τ_{S_0}. $\square$

That completes the proof of the lemma and hence of the theorem. $\square$

Corollary 25.1.1 WEIGHTED MONOTONE CIRCUIT SATISFIABILITY *is* $W[P]$-*complete. That is,* MONOTONE $W[P] = W[P]$.

Proof By Lemma 25.1.2 MINIMUM AXIOM SET is a special case of WEIGHTED MONOTONE CIRCUIT SATISFIABILITY.   □

The reader might wonder if instead of the intricate argument above we could simply use the reasoning of the Normalization Theorem as follows. First move all the inverters to the top of the circuit. Now consider each line as a "gate" with fanin 1 and fanout 1. Hardwire the DOMINATING SET gadget from the Normalization Theorem. (More correctly, the gadget from Lemma 23.2.1.) This would increase the weft by 1 but make the circuit monotone, and since we do not care about controlling weft, only size, this would seem to give an easy construction of a circuit where the original circuit has a weight k satisfying assignment if and only if the new one has a weight $2k$ satisfying assignment.

The problem is that *the action of moving the inverters to the top can increase the size of the circuit exponentially in the height.* This is because of the fact that the circuits are not tree circuits, they are only decision circuits, and gates can thus have unbounded fanout. However, the reasoning here can clearly be applied to tree circuits, and hence we can prove a similar result to Theorem 25.1.1 for $W[\text{SAT}]$. (Theorem 25.1.3(ii) follows by similarly hardwiring the gadget from Theorem 21.2.1.)

Theorem 25.1.3 (Abrahamson, Downey, and Fellows [3]) *The following are complete for $W[\text{SAT}]$.*
 (i) MONOTONE $W[\text{SAT}]$.
 (ii) ANTIMONOTONE $W[\text{SAT}]$.

The basic results above form a basis for many of the reductions for various concrete problems. The next theorem gives some illustrations. More can be found in the Exercises. We will need the following definitions:

DEGREE 3 SUBGRAPH ANNIHILATOR
Instance: A graph G.
Parameter: A positive integer k.
Question: Is there a set V' of at most k vertices of G, such that $G - V'$ has no minimum degree subgraph?

k-LINEAR INEQUALITIES
Instance: A system of linear inequalities.
Parameter: A positive integer k.
Question: Can the system be made consistent over the rationals by deleting at most k of the inequalities?

Theorem 25.1.4 (Abrahamson, Downey, and Fellows [2, 3]) *The following problems are $W[P]$-complete.*
 (i) DEGREE 3 SUBGRAPH ANNIHILATOR
 (ii) k-LINEAR INEQUALITIES

Proof (i) We reduce from WEIGHTED MONOTONE CIRCUIT SATISFIABILITY. Let
C be an instance of WEIGHTED MONOTONE CIRCUIT SATISFIABILITY with parameter k. We can regard all gates to have only two inputs with only quadratic
increase in the size of the circuit. We shall use local replacement. We replace
the *and* and *or* gates by the gadgets given in Fig. 25.1. Formally, we can define the *and* gadget AND(x, y, z) for a gate $x \wedge y$ with output z via the vertices $v(x, j)$ and $v(y, j)$ for $1 \le j \le 7$ together with vertex $q(x, y)$ and "pendant vertices" $l(x, y, p)$ for $p = 1, 2, 3, 4$. The edges are for $r \in \{x, y\}$, $rv(r, 1)$,
$v(r, 1)v(r, 2)$, $v(r, 1)v(r, 3)$, $v(r, 2)v(r, 4)$, $v(r, 3)v(r, 5)$, $v(r, 4)v(r, 5)$ (these are
the "pentagonal" edges of the gadget), $v(r, 2)v(r, 6)$, $v(r, 4)v(r, 6)$, $v(r, 6)v(r, 7)$,
$v(r, 3)v(r, 7)$, $v(r, 5)v(r, 7)$ (the edges connected to the pentagon), $v(r, 6)q(x, y)$,
$v(r, 7)q(x, y)$, $q(x, y)l(x, y, 1)$ (these are the connecting edges), $l(x, y, 1)l(x, y, 2)$,
$l(x, y, 1)l(x, y, 3)$, $(x, y, 2)l(x, y, 3)$, $l(x, y, 2)l(x, y, 4)$, and $l(x, y, 3)l(x, y, 4)$
(these are the edges of the "pendant diamond").

The vertices for the *or* gadget OR(x, y, z) are $w(x, j)$, $w(y, j)$ for $1 \le
j \le 7$ and 3 additional vertices $m(x)$, $m(y)$, $n(x, y)$ and $g(x, y, p)$ for $p =
1, 2, 3, 4$. The vertices $w(r, j)$ are connected among themselves exactly as we
did for the $v(r, j)$ above, as are the pendant $g(x, y, p)$. We also have edges
$w(r, 6)m(r)$, $w(r, 7)m(r)$, $m(r)n(x, y)$, and $n(x, y)z$.

Additionally, we shall need another sort of gadget, a *fanout* gadget which is used
to cope with the fanout occurring at gates and in the construction. This gadget is
also indicated in Fig. 25.1 for fanouts 3 and 4.

Let the variables of C be $x[1], \ldots, x[n]$. For the construction of the graph $G(C)$,
we first replace all the gates of C and their fanouts by the gadgets of Fig. 1. Call
the result $I(C)$. Using $I(G)$, we make another intermediate graph $I'(G)$ as follows. For each $x[i] \in \{x[1], \ldots, x[n]\}$, create a subgraph consisting of the (new)
vertices $v(i, j) : j = 1, \ldots, 4$ and edges $v(i, 1)v(i, 2)$, $v(i, 1)v(i, 3)$, $v(i, 2)v(i, 3)$,
$v(i, 2)v(i, 4)$, and $v(i, 3), v(i, 4)$. Form $2k + 1$ copies of the $I(C)$ except that in
place of the $k + 1$ versions of the input variable $x[i]$ we connect all relevant gates to
the *single* vertex $v(i, 4)$. (Thus if in the original circuit the variable $x[i]$ had fanout
m then in the graph $I'(C)$ $v(i, 4)$ will have degree $m(k + 1) + 2$.) Finally we form
the final graph $I''(C)$ by taking the vertex $u(i)$ representing the output of the ith
version of $I(C)$ used in $I'(C)$, and for each $j = 1, \ldots, n$, join $o(i)$ to $v(j, 1)$ *via*
one of the outputs u_j of a fanout gadget of fanout n. This concludes the construction.

To see that the construction succeeds, we show that C has a weight k accepting
input iff $I''(C)$ has no k element degree 3 subgraph annihilator. First suppose that
C has a weight k accepting input. Let $x[i_1], \ldots, x[i_k]$ be the true variables in some
satisfying assignment. Remove the vertices $v[q, 4]$ for $q \in \{i_1, \ldots, i_k\}$ from $I''(C)$.
We claim that the result has no min degree 3 subgraph. We shall describe a process
that removes edges that are useless to a min degree 3 subgraph and eventually kills
all of $I''(C)$. The reader should note the following guiding principle to our proof.
Given any graph G the following simple process determines if G has a min degree 3
subgraph.

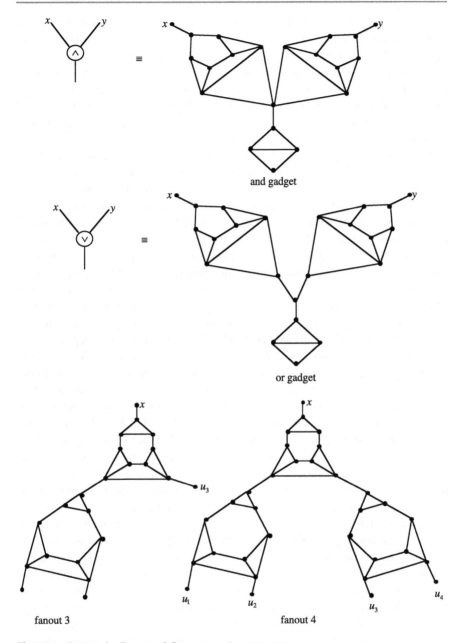

Fig. 25.1 Gadgets for DEGREE 3 SUBGRAPH ANNIHILATOR

1. Repeat until there are no degree ≤ 2 vertices:
   Find any vertex v of degree < 3. Let $G \leftarrow G - v$.
2. G is has a min degree 3 subgraph iff the result of step 1 is nonempty.

In our construction, the interpretation is that "x removed" equates to "x is true". Suppose vertex x of the and gadget $\mathrm{AND}(x, y, z)$ is removed. This will cause the vertex $v(x, 1)$ at the top of the gadget that is joined to x to have degree 2. Thus we may remove $v(x, 1)$ from $I''(C)$ since it cannot be part of a min degree 3 subgraph. In turn this decreases the degree of $v(x, 2)$ and $v(x, 3)$ to 2. Arguing in this way we see that all of the vertices $v(x, j)$ will need to be removed. Note that this causes $l(x, y)$ to have degree 3. Thus if one input of *and* gate is removed then the effect is to remove that half of the gadget.

In the case of an *or* gate we can similarly argue that the removal of either of the inputs will cause the whole gadget to be removed, and the vertices corresponding to the other input and the one corresponding to z to have their degree reduced by 1.

Translating the gadget observations back to C it follows that if a gate is made true then the corresponding gadget can be removed from $I''(C)$. Since the given assignment is one that satisfies C it follows that any vertex u representing the output of a copy of C can be erased. Consequently, using the same reasoning we can erase all the vertices of the fanout gadgets thus dropping the degree of the vertices $v[i, 1]$ by 1. This then allows us to erase the vertices $v[i, j]$ for all i and j and thus erase all of $I''(C)$. Thus $I''(C) - \{v[i_1, 4], \ldots, v[i_k, 4]\}$ has no min degree 3 subgraph.

Conversely, suppose that $I''(C)$ has a collection V' of k vertices such that $G = I''(C) - V'$ has no min degree 3 subgraph. It follows that except for variable components, V' must be disjoint from $k + 1$ of the circuit copies. As the circuit copies are identical, we may suppose without loss of generality that V' has no internal circuit vertices. Now the removal of V' must cause the graph to be erased as above. This means that the removal of V' and the process above must cause *all* of the output nodes u to be removed, or else we can add the 9 nodes below the fanout immediately below u and have a min degree 3 subgraph. Since all the output vertices must be removed by V', and not all can be in V' it follows that the circuit above at least one output vertex u must be removed. (Fanout gadgets can only be removed from above not below.) Since there are $2k + 1$ copies of the circuit, it follows that for one such u, its removal can only be generated by the removal of a set of variables, which we can take to be $\{v[i_1, 4], \ldots, v[i_m, 4]\}$. Now we can argue that since u is removed, if u corresponds to an *and* gate then *both* of its inputs must be removed, and if u is an *or* gate then at least one of its inputs must be removed. Chasing the removed inputs up the circuit, we can argue that eventually a set of vertices must be removed, and we see that this set of $\leq k$ vertices cause the circuit to be erased. The reason these $\leq k$ vertices are able to cause this removal is the following. Apply the removal process to the initial set V'. As the removal process discharges through the circuit the "first" time, only vertices corresponding to variables can cause the removal of u. But if u is not removed on the first sweep then it will never be removed. As above they must correspond to a satisfying assignment for C. Since C is monotone, they can be extended to a weight k satisfying assignment for C. Thus C is satisfiable by a weight k assignment iff $I''(C)$ has a size k degree 3 subgraph annihilator, and hence MIN DEGREE 3 SUBGRAPH ANNIHILATOR is $W[P]$-hard.

(ii) This time we employ the reduction from [6]. Cook observed that if C is a circuit with input variables $x = (x_1, \ldots, x_n)$ and output variables $y = (y_1, \ldots, y_m)$ then there is a LOGSPACE procedure that on input C produces a set L of linear inequalities with the properties below.

(a) The inequalities include variables $X = (X_1, \ldots, X_n)$ and $Y = (Y_1, \ldots, Y_m)$.

(b) For all binary vectors $b = (b_1, \ldots, b_n)$ there is exactly one point that satisfies all the inequalities of L and additionally satisfies $X = b$.

(c) If c is the output of C on input b then the unique assignment that satisfies L and $X = b$ also satisfies $Y = c$.

L is said to simulate C. (The proof of this result is the following: represent *not* gates $a = \bar{b}$ by $a = 1 - b$ and $0 \leq a \leq 1$, and *and* gates $a = b \wedge c$ by $0 \leq a \leq 1$, $a \leq b$, $a \leq c$, and $b + c - 1 \leq a$.) We reduce from WEIGHTED CIRCUIT SATISFIABIL-ITY. Let C be an instance of WEIGHTED CIRCUIT SATISFIABILITY with q input variables $x_1, \ldots, x_q$. Let $X_1, \ldots, X_q$ be rational variables. Consider the inequalities:

$$X_i \leq 0, \text{ and } k + 1 \text{ copies of both } X_1 + \cdots + X_q \geq k \text{ and } 0 \leq X_i \leq 1.$$

To this system we add $k + 1$ copies of the inequalities representing C with $X_1, \ldots, X_q$ also representing the input variables. We also add a variable Z to represent the output variable and add the inequality $Z \geq 1$ $k + 1$ times. Clearly this system is satisfiable by the deletion of k of the inequalities (which must be the $X_i \leq 0$-type) iff C has a weight k satisfying assignment. $\qquad\square$

25.2 Exercises

Exercise 25.2.1 (Abrahamson, Downey, and Fellows [2, 3]) Prove that the following problems are complete for $W[P]$.

(i) t-THRESHOLD STARTING SET

    Instance: A directed graph D.

    Parameter: A positive integer k.

    Question: Is there a set of k vertices of D with the property that it will pebble D under the rule that a vertex can be pebbled iff at least t incoming vertices are pebbled?

(ii) CHAIN REACTION CLOSURE

    Instance: A directed graph D.

    Parameter: k.

    Question: Does there exist a set V' of k vertices of D whose chain reaction closure is D. Here the chain reaction closure of V' is the smallest superset S of V' such that if $u, u' \in S$ and arcs ux, $u'x$ are in D then $x \in S$.

(iii) k-INDUCED SATISFIABILITY

    Input: A formula φ.

    Parameter: A positive integer k.

    Question: Is there a set of k variables, and a truth table assignment to those variables that causes φ to unravel?

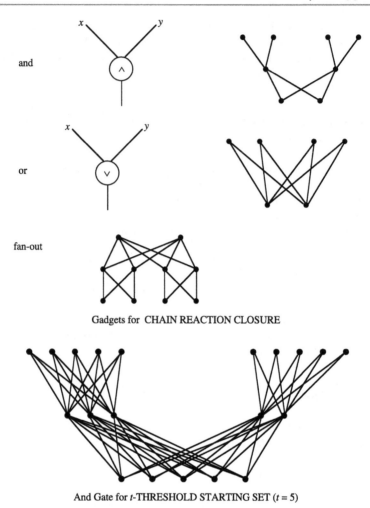

and

or

fan-out

Gadgets for CHAIN REACTION CLOSURE

And Gate for t-THRESHOLD STARTING SET ($t = 5$)

Fig. 25.2 Gadgets for Exercise 25.2.1

(Hint: We look at (i); (ii) and (iii) are similar. Use the technique of Theorem 25.1.4 and reduce from WEIGHTED MONOTONE CIRCUIT SATISFIABILITY. This time we use the *and*, *or* and *fanout* gates of Fig. 25.2, where the arcs go *down*. In the construction, this time we represent the variables by pairs of vertices $x[i]$, $y[i]$. Again we form $I''(C)$ by replacing the gates in the circuit C by gadgets as above and then taking $2k + 1$ copies of the for $i = 1, \ldots, n$. Again each *pair* that represents an output for a circuit fans out to *and* gates above each of the variable pairs. (Or more precisely, to collections of gadgets corresponding to large *and* gates with outputs the variables.) Note, for instance, that it requires both of the input pairs of the *and* gadget to be in the closure before the output pair will be included. Thus for the variable pairs to be included, we need all of the output pairs corresponding to the circuits to be included. Thus as in the previous argument, the original collection of $\leq 2k$ variable pairs included in V' cannot cause any more variable pairs to be included unless they first cause the inclusion of all of the circuit output pairs *first*. But as in the previous argument, this corresponds to a satisfying assignment of C.)

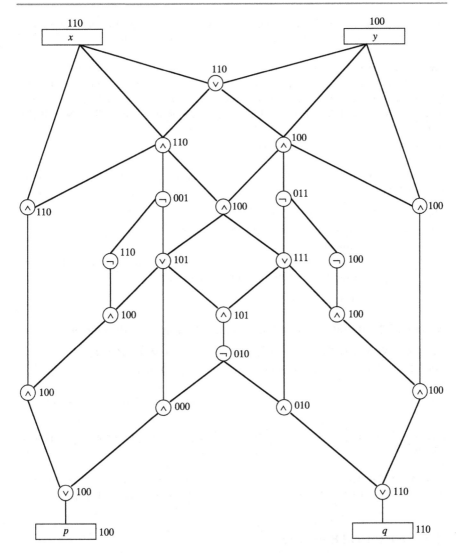

Fig. 25.3 Gadget for WEIGHTED PLANAR CIRCUIT SATISFIABILITY

Exercise 25.2.2 (Abrahamson, Downey, and Fellows [3]) Prove that WEIGHTED PLANAR CIRCUIT SATISFIABILITY, the restriction of WEIGHTED CIRCUIT SATISFIABILITY to planar circuits (i.e. can be drawn on the plane with no edges crossing), is complete for $W[P]$.

(Hint: Reduce WEIGHTED CIRCUIT SATISFIABILITY to the planar version. This involves no more than the use of a crossover gadget along the lines of Lichtenstein [490], or Garey and Johnson [337]. Take an instance φ of WEIGHTED CIRCUIT SATISFIABILITY with parameter k. Suppose a wire from x (a variable *or* gate) is crossing one from y. We replace this crossing by the gadget of Fig. 25.3. On this figure we have listed the values at each gate as a triple corresponding to the settings of x and y. Note that we need only specify the values for a triple by symmetry. Since p agrees with y and q agrees with x we have achieved the desired crossing.)

Exercise 25.2.3 (Abrahamson Downey and Fellows [3]) Prove that WEIGHTED MONOTONE SATISFIABILITY (the restriction of WEIGHTED SATISFIABILITY to monotone formulas) and WEIGHTED ANTIMONOTONE SATISFIABILITY (the restriction of WEIGHTED SATISFIABILITY to antimonotone formulas) are both $W[P]$ complete.

(Hint: Hardwire the gadgets used in the Normalization Theorem.)

Exercise 25.2.4 (Downey and Fellows) Prove that the following problem is complete for $W[P]$.

k-BASED TILING
Instance: A tiling system with distinguished tiles.
Parameter: A positive integer k.
Question: Is there a tiling of the $n \times n$ plane using the tiling system and starting with exactly k distinguished tiles in a line ?

(Hint: Use a generic simulation of a Turing machine.)

Exercise 25.2.5 (Cesati [136])   Show that the following is $W[P]$ complete.

BOUNDED NONDETERMINISM TURING MACHINE COMPUTATION
Instance: A NTM M, and input word w and integer n in unary.
Parameter: k.
Question: Does M accept w in at most n steps and using at most k nondeterministic steps?

(Hint: for hardness reduce from CHAIN REACTION CLOSURE. For membership use a generic reduction.)

Exercise 25.2.6 (Cesati [136]) Using the methods of Exercise 25.2.5, show that MAXIMAL IRREDUNDANT SET of Exercise 20.3.1 is complete for $W[1]$.

25.3   Historical Remarks

Most of the basic results in this section are due to Abrahamson, Downey, and Fellows [3]. $W[P]$-completeness of WEIGHTED MONOTONE CIRCUIT SATISFIABILITY, which is the restriction of the problem WEIGHTED CIRCUIT SATISFIABILITY to circuits with no inverters. This result was first announced in Downey, Fellows, Kapron, Hallett, and Wareham [248]. In [248], Downey et al. analyzed the parameterized tractability of problems arising from logic and linguistics. Many of the proofs presented in this chapter are corrections and adaptations of ideas and results from Abrahamson, Ellis, Fellows and Mata [6]. Some of the problems are from Downey and Fellows [247], with a couple from Cesati [136]. Treatments of the topics in this chapter from the point of view of logic and model checking can be found in Flum–Grohe [312].

25.4 Summary

We introduce the classes $W[P]$ and variations, proving a number of concrete hardness and membership results for this level of the W-hierarchy.

Fixed Parameter Analogues of PSPACE and k-Move Games

26

26.1 Introduction

As we have seen, classical time classes such as NP seem to split into many corresponding parameterized complexity classes. The same seems to occur when we consider parameterized *space*. The following parameterized space class suggests itself naturally.

> **Definition 26.1.1** We say that a parameterized language L is in SLICEWISE PSPACE if there is a procedure Φ, a function f, and a constant α such that for all k,
>
> $$\langle x, k \rangle \in L \quad \text{iff} \quad \Phi(\langle x, k \rangle) \text{ accepts,}$$
> $$\text{and the space bound on } \Phi(\langle x, k \rangle) \text{ is } f(k)|x|^{\alpha}.$$

Cai, Chen, Downey, and Fellows [122] observed that one can similarly define a parameterized class SLICEWISE **C** for any classical complexity class **C**. This was the basis for the "advice view" mentioned in Exercise 2.3.4. Since there is a set in $\text{DSPACE}(|x|^{c+1})$ which in linear time can compute any language in $\text{DSPACE}(|x|^{c})$, it follows that:

Observation *If* $P = \text{PSPACE}$ *then* SLICEWISE PSPACE $= \text{FPT}$.

26.2 k-Move Games

Our main goal in this chapter is to explore an interesting connection between parameterized versions of space complexity and the complexity of k-move games. That there is such a connection is perhaps not surprising, since many classical game

R.G. Downey, M.R. Fellows, *Fundamentals of Parameterized Complexity*,
Texts in Computer Science, DOI 10.1007/978-1-4471-5559-1_26,
© Springer-Verlag London 2013

problems are known to be PSPACE-complete. Typically, such problems ask whether the first player to move has a winning strategy. The reason that the classical problems are PSPACE-complete stems from the following facts:

- QBFSAT is PSPACE-complete, and
- games are naturally in QBF form, since the alternating plays can be thought of as alternating quantifiers.

A natural parameterized version of classical games is whether the first player has a strategy that wins within at most k moves. This in turn leads to natural parameterized versions of QBFSAT. However, it seems that there is no straightforward correspondence between SLICEWISE PSPACE and any parameterized form of QBFSAT.

Classical proofs that QBFSAT is PSPACE-complete use alternations of quantifiers to emulate PSPACE by guessing intermediate steps along the lines of Savitch's Theorem. This does not seem to translate well into the parameterized setting since the intermediate steps in a SLICEWISE PSPACE computation can lose weight control. A parameterized complexity analogue of Savitch's Theorem classical theorem is currently not known.

Natural parameterized versions of *some* hard game problems are fixed-parameter tractable. An example is the ALTERNATING HITTING SET game Garey and Johnson [337], Schaefer [602] restricted to sets of any fixed *size $t \geq 2$*. Considered classically, this problem is PSPACE-complete. More formally, we consider the problem:

RESTRICTED ALTERNATING HITTING SET
Instance: A collection C of subsets of a set B with $|S| \leq k_1$ for all $S \in C$.
Parameter: $\langle k_1, k_2 \rangle$.
Question: Does player I have a winning strategy of at most $\leq k_2$ moves in the following game? Players alternately choose unchosen elements of B until, for each $S \in C$, some member of S has been chosen, that is, every set in C has been *hit*. The player whose choice makes this happen wins.

The reader can prove the following easy result in Exercise 26.5.1.

Theorem 26.2.1 (Abrahamson, Downey, and Fellows [2, 3]) RESTRICTED ALTERNATING HITTING SET *is (strongly) fixed-parameter tractable.*

Some k-move game problems seem likely not to be fixed-parameter tractable. GENERALIZED GEOGRAPHY is a game played on a directed graph G with a distinguished *start vertex* [337, 602]. Players alternate choosing vertices, beginning with the start vertex v_1, in such a way that the chosen vertices form, in sequence, a simple directed path in G. The first player who is unable to choose a vertex loses. Determining whether player 1 has a winning strategy in a GENERALIZED GEOGRAPHY game is PSPACE-complete. In the parameterized problem SHORT GENERALIZED GEOGRAPHY, it is asked whether player 1 has a strategy that wins a given game of GENERALIZED GEOGRAPHY in at most k moves.

In order to address such questions we introduce the classes $AW[P]$, $AW[SAT]$, and $AW[*]$, which plausibly contain problems that are not in FPT. The classes can be viewed as generalizations of the Polynomial Time Hierarchy and QBFSAT. We show that SHORT GENERALIZED GEOGRAPHY is $AW[*]$-complete.

Like $W[SAT]$, $AW[SAT]$ is the closure under parameterized-reductions of a kernel problem of such a general nature that it appears not to be fixed-parameter tractable. This problem is a parameterized version of QBF (QUANTIFIED BOOLEAN FORMULA SATISFIABILITY), defined as follows.

PARAMETERIZED QBFSAT

Instance: A sequence $s_1, \ldots, s_r$ of pairwise disjoint sets of boolean variables, and a boolean formula X involving the variables in $s_1 \cup \cdots \cup s_r$.
Parameters: $r, k_1, \ldots, k_r$.
Question: Is it the case that there exists a size k_1 subset t_1 of s_1 such that for every size k_2 subset t_2 of s_2 there exists a size k_3 subset t_3 of s_3 such that ... (alternating quantifiers) such that when the variables in $t_1 \cup \cdots \cup t_r$ are made true, and all other variables are made false, formula X is true?

We can similarly define PARAMETERIZED QBFCP as with PARAMETERIZED QBFSAT except that quantified object is a circuit rather than a boolean formula.

Definition 26.2.1 $AW[SAT]$ is the set of all problems that *fpt*-reduce to PARAMETERIZED QBFSAT. Similarly $AW[P]$ are those problems *fpt*-reduce to PARAMETERIZED QBFCP.

Clearly, an equivalent formulation of this problem is

Instance: A QBF formula $Q_1 x_1 \ldots Q_n x_n X$.
Parameter: $k = \langle k_1, \ldots, k_n \rangle$.
Question: Is $Q_1^{k_1} x_1 Q_2^{k_2} x_2 \ldots Q_n^{k_n} x_n X$ true? (Here $\exists^{k_i} x$ is interpreted to mean "does there exist a *weight i* x such that ..." and $\forall^{k_i} x$ is interpreted to mean "for all *weight k* x ...".)

Similarly, we can define the problem of PARAMETERIZED QUANTIFIED CIRCUIT SATISFIABILITY (PARAMETERIZED QCSAT), by replacing the X by a circuit with the variables x as the inputs. WEIGHTED CIRCUIT SATISFIABILITY is by definition in PARAMETERIZED Σ_1. This observation leads us to a natural result that is a partial analogue of the classical result that QBFSAT is PSPACE-complete. We need the following problem definition.

COMPACT NTM (COMPUTATION)

Instance: A nondeterministic Turing machine M and a word x.
Parameter: A positive integer k.
Question: Is there an accepting computation of M on input x that visits at most k tape squares?

The problem COMPACT TM (COMPUTATION) is defined identically, except that the Turing machine is deterministic.

Theorem 26.2.2 (Cai, Chen, Downey, and Fellows [122]) COMPACT NTM COMPUTATION *is AW[SAT]-hard.*

Proof Let $X, s_1, \ldots, s_n$ be an instance of $AW[SAT]$ with parameter $k = \langle k_1, \ldots, k_n \rangle$. The reduction is from WEIGHTED MONOTONE CIRCUIT SATISFIABILITY. Let C be a circuit for which we wish to determine whether there is an input vector of weight k accepted by C. We may assume that each logic gate g of C has two inputs. In time polynomial in $|C|$ we can describe a Turing machine M sketched as follows.

M has an alphabet consisting of one letter for each input to C, and the operation of M consists of two phases. In the first phase, M makes k moves nondeterministically, writing down in the first k tape squares k symbols which represent k inputs to C set to 1. In the second phase (and visiting no other tape squares), M checks whether the guess made in the first phase represents a vector accepted by the circuit C.

The key point is that we can structure the transition table of M to accomplish this, with the size of the table polynomial in $|C|$. To do this, we make two states q_{up}^l and q_{down}^l for each connection (or *line*) l of the circuit C. Let g be an *and* gate of C, let l be an output line of g and suppose the input lines to the gate g are l_1 and l_2. We include in the transition table for M transitions from q_{up}^l to $q_{\text{up}}^{l_1}$, from $q_{\text{down}}^{l_1}$ to $q_{\text{up}}^{l_2}$, and from $q_{\text{down}}^{l_2}$ to q_{down}^l. The significance of being a state q_{down}^l is that this represents a value of 1 for the line l as computed by C on the input guessed in the first phase. The state q_{up}^l might be viewed as a state of *query* about the value of l for the circuit C on the input guessed in the first phase. Note that the three transitions described above for the *and* gate g thus enforce that q_{down}^l can be reached only if $q_{\text{down}}^{l_1}$ and $q_{\text{down}}^{l_2}$ can be reached. The appropriate transitions for an *or* gate will differ in the obvious way, i.e., we arrange that the state q_{down}^l can be reached if either of $q_{\text{down}}^{l_1}$ or $q_{\text{down}}^{l_2}$ can be reached.

If l is an input line to the circuit C, then we encode in the state table for M a "check" (involving a scan of the k tape squares) to see if the corresponding input symbol was written during the first phase of computation. The second phase begins in the state $q_{\text{up}}^{l_{\text{out}}}$ where l_{out} is the output line of C, and the only accept state is $q_{\text{down}}^{l_{\text{out}}}$.

Note that the proof above gives a canonical way of going from a proposed collection t_i of true input variables to a Turing acceptance.

Thus to complete the proof we need to say how to introduce layers of quantifiers. Without loss of generality, we can suppose that Q_1 is existential. What we do is break the work tape into n cells of size $k_1, \ldots, k_n$. Our first action is to write a guess for t_1 in the first k_1 squares. We build a recursive algorithm that accepts only if t_1 can be extended to a satisfying pattern according to the quantifier structure. For cell $i \geq 2$, if the quantifier corresponding to k_i is an existential one then on each sweep, in the k_i squares of *cell i* we will write a guess for the variables t_i from s_i we will be assigning true for this sweep. (We shall process the guesses in lexicographic ordering.) If the Q_i corresponding to cell i is a universal quantifier, then for each setting of the cells $1, \ldots, i-1$ we will cycle lexicographically through all the possibilities for t_i, that is, all the k_i element subsets of s_i. As above for each sweep we will get a setting for cells corresponding to k true variables, and we can see if M accepts. Fix a setting of $1, \ldots, t_{n-1}$. Recursively, note that if the last Q_n is universal then cell n will pass back a confirmation of this setting only if it successfully cycles through all possible t_n. Similarly if Q_n is existential cell n can pass a yes for this setting only if it finds a t_n. This process can be continued inductively using at most k counters, and hence we can make a machine that uses at most $2k$ squares and accepts iff $Q_1 x_1 \ldots Q_n x_n C$ is true. $\square$

In earlier papers and in Downey and Fellows [247] we claimed that COMPACT NTM was $AW[P]$-hard. Unfortunately, that proof is flawed, and we do not know how to repair it. We speculate that perhaps COMPACT TM COMPUTATION $\equiv_m^s AW[P]$.

26.3   AW[*]

$AW[P]$ appears to be too large a class for most of our purposes in exploring the parameterized complexity of k-move games. This leads us to concentrate on the class $AW[*]$ defined below.

> **Definition 26.3.1** (PARAMETERIZED QBFSAT$_t$) is the restriction of PARAMETERIZED QBFSAT in which the formula part is t-normalized. $AW[t]$ is the set of all parameterized problems that fp-reduce to PARAMETERIZED QBFSAT$_t$, and $AW[*] = \bigcup_t AW[t]$.

Lemma 26.3.1 (Abrahamson, Downey, and Fellows [3])   SHORT GEOGRAPHY *is in AW[2].*

Proof We use a generic reduction. Let D be a digraph with distinguished vertex v_0 upon which we shall play SHORT GEOGRAPHY with parameter k. Let $\{v_0, \ldots, v_n\}$ list the vertices of D and let E denote the edge set. We shall have variables $\{p_{i,j} : 1 \leq i \leq k \wedge 0 \leq j \leq n\}$. We think of the game as a pebbling game with $p_{i,j}$ denoting that the ith pebble is on vertex v_j. We need clauses as follows:

(1) $p_{1,0}$ [Pebble 1 is on vertex v_0.]

(2) $\bigwedge_{1 \leq i \neq j \leq q, 0 \leq k \leq n}(p_{i,q} \to \overline{p_{j,q}})$ [Only one pebble per vertex.]

(3) $\bigwedge_{v_i v_j \notin E, 1 \leq q \leq k}(p_{q,i} \to (\overline{p_{q+1,j}} \wedge \overline{p_{q-1,j}})$
$\vee (\bigvee_{q' \leq q, q' \text{odd}, 0 \leq t \leq n} p_{q',t} \wedge (\bigwedge_{v_t v_r \in E}(\bigvee_{q'' \leq q'} p_{q'',r}))))$ [If $v_i v_j$ not an edge then for any pebble placed on v_i the preceding pebble and the next pebble must not be on v_j unless player 1 has already won on some vertex v_t pebbled with some $q' \leq q$.]

(4) $\bigwedge_{2\leq j\leq k, 0\leq i\leq n}(p_{j,i} \rightarrow (\bigvee_{v_i v_q \in E} p_{j-1,q})$
$\vee (\bigvee_{q'\leq j, q' \text{odd}, 0\leq t\leq n} p_{q't} \wedge (\bigwedge_{v_t v_r \in E}(\bigvee_{q''\leq q} p_{q'',r}))))$ [If v_i is pebbled by pebble $j \geq 2$ then some vertex adjacent to v_i must be pebbled by pebble $j-1$ unless player 1 has already won on some v_t as in (3).]

(5) $\bigwedge_{1\leq j\leq k-2, j \text{odd}, 0\leq i\leq n}(p_{j,i} \rightarrow (\bigvee(\bigvee_{q'\leq j, q' \text{odd}, 0\leq t\leq n} p_{q',t}$
$\wedge (\bigwedge_{v_t v_r \in E}(\bigvee_{q''\leq q'} p_{q'',r}))) \vee ((\bigwedge_{v_i v_j \in E} \bigvee_{1\leq q\leq j} p_{j,q})$
$\vee (\bigvee_{\{q,r : v_i v_q, v_q v_r \in E\}}(p_{j+1,q} \wedge p_{j+2,r}))))$ [If player 1 pebbles v_i then either she wins at this play, has won at a preceding play, or player 2 pebbles a vertex v_q adjacent to v_i with the next pebble and player 1 pebbles a vertex adjacent to v_q.]

Let $P(p_{i,j} : 1 \leq i \leq k, \ 0 \leq j \leq n)$ denote the conjunction of (1)–(5) expressed in CNF form. Note that this expression has length polynomial in $\langle |D|, k\rangle$. It is by definition in $W[2]$. The expression we then need is the following.

$$\forall y_2 \in \{p_{2,0}, \ldots, p_{2,n}\} \exists y_3 \in \{p_{3,0}, \ldots, p_{3,n}\} \ldots (k \text{ alternations}) P.$$

This is interpreted as making the chosen variable from the set $\{p_{j,0}, \ldots, p_{j,n}\}$ true and the others false. Note that the form of the expression P ensures that the formula is true if and only if player 1 has a winning strategy making at most k moves. □

We prove that PARAMETERIZED QBFSAT$_t$ fp-reduces to SHORT GEOGRAPHY for every t. The reduction is actually from a restricted form of PARAMETERIZED QBFSAT$_t$.

Definition 26.3.2 *Unitary* PARAMETERIZED QBFSAT$_t$ is the restriction of PARAMETERIZED QBFSAT$_t$ in which the parameters $k_1, \ldots, k_r$ are all 1.

Lemma 26.3.2 *For $t > 0$,* UNITARY PARAMETERIZED QBFSAT$_t$ *is AW$[t]$-complete.*

Proof The method is quite simple. Given a quantifier $\exists k$ members of $(S = \{s_1, \ldots, s_n\})(\ldots)$, we can replace it by $2k$ quantifiers

$$\exists x_1 \in S \forall y \in \emptyset \exists x_2 \in S \forall y \in \emptyset \ldots \exists x_k \in S \forall y \in \emptyset \left(\bigwedge_{i\neq j}(x_i \neq x_j) \wedge \cdots\right).$$

(As in Lemma 26.3.1, this formula is interpreted as making the chosen variable from the set S_i true and the others false. We treat universal quantifiers similarly.) Note that the overall parameter is doubled. We also only add an additional large *and* of *or*'s to the circuit. This expression in turn can be put into standard form by replacing the ith occurrence of S by $S_i = \{s_1^i, \ldots, s_n^i\}$ and adding the expression $\bigwedge_{j\in\{1,\ldots,n\}}(s_j \equiv (\bigvee_{i=1}^k s_j^i))$. That is s_j will be false iff all of the k possible representatives are chosen false. This does not increase the weft of the circuit. □

Theorem 26.3.1 (Abrahamson, Downey, and Fellows [2, 3]) SHORT GEOGRAPHY *is AW$[*]$-complete. Hence, AW$[*] = AW[2]$.*

Proof We reduce PARAMETERIZED QBFSAT$_{2t}$ to SHORT GEOGRAPHY for an arbitrary $t > 0$. Let $I = (r, k_1, \ldots, k_r, s_1, \ldots, s_r, X)$ be an instance of PARAMETERIZED QBFSAT$_{2t}$, and assume that r is odd. The leading quantifier is then existential. The reduction uses ideas from Schaefer's polynomial time reduction from QBF to GENERALIZED GEOGRAPHY in [602]. The graph on which the geography game is played has three parts: the choice component, the formula-testing component and the literal-testing component.

The choice component is similar to Shaefer's, and is designed so that player 1 chooses a member of s_1, then player 2 chooses a member of s_2, then player 1 chooses a member of s_3, etc. The choice-testing gadget is given in Fig. 26.1 for successive quantifier pairs $Qt_i\, Qt_{i+1}$. The gadget for Qt_i which asks us to pick *one* member from s_i consists of vertices v_i, w_i and x_j for each $x_j \in s_i$. The edges are $v_i x_j$, and $x_j w_i$ for each j. A total of $3r$ moves are made through this component, and we add two additional edges to ensure that it is player 2's move at the end, where the game enters the formula-testing component.

In the formula-testing component we use player 1's moves to simulate disjunctions, and player 2's moves to simulate conjunctions. A total of $2t$ moves are made through this component. Let y be bottom vertex of the choice component. Let C be a circuit representing X. Reversing the arrows, create a tree representing C with root y. We can assume that the circuit consists of $2t$ layers of, alternately, conjunctions and disjunctions as in [243]. This is why when the arrows are reversed, since y will be representing the output of a conjunction and it will be player 2's turn, player 2 will always be playing a conjunction and player 1 a disjunction. Play ends at a *literal vertex* v, corresponding to a literal in formula X, with the move being made by player 2.

A literal vertex v corresponding to a positive literal x has an edge to the vertex v_x in the choice component t_i that corresponds to x. If v_x was chosen (variable x is true), then player 2 has no move, and player 1 wins. If v_x was not chosen (so x is false), then player 2 moves to v_x and wins.

If literal vertex v corresponds to a negative literal $\bar{x}$, then the literal-testing component has edges (v, u_x) and (u_x, v_x), where u_x is a new vertex and v_x is the vertex corresponding to x in the choice component. Vertex u_x switches the initiative, and causes player 1 to win if v_x was not chosen, and player 2 to win if v_x was chosen.

In Fig. 26.1 we have given an example of this construction for the formula

$$\exists x \in \{x_1, x_2, x_3\} \forall y \in \{y_1, y_2, y_3, y_4\}\big[(x_1 \vee y_1) \wedge (\overline{x_1} \vee \overline{y_2} \vee \overline{y_3})\big].$$

It is easy to see that the parity of the choices allows player 1 to win if and only if the formula is true since all plays begin at v_1. The total number of moves is at most $3r + 2t + 4$. Since t is fixed, the conditions of an fp-reduction are met. □

Remark 26.3.1 The above proof employs a tradeoff between quantification and weft. We will improve upon this result to establish the following in the next chapter.

Theorem 26.3.2 (The Tradeoff Theorem, Downey, Fellows, and Regan [256]) $AW[*] = AW[1]$.

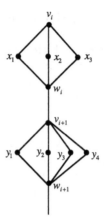

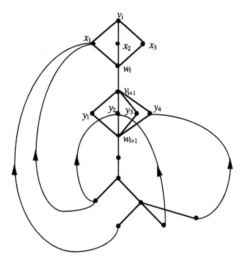

Choice components for Qt_iQt_{i+1} for $t_i = t_{i+1} = 1$, $s_i = \{x_1, x_2, x_3\}$, $s_{i+1} = \{y_1, y_2, y_3, y_4\}$.

Example for $\exists x \in \{x_1, x_2, x_3\} \; \forall y \in \{y_1, y_2, y_3, y_4\} \; ((x_1 \vee y_2) \wedge (\overline{x}_1 \vee \overline{y}_1, \vee \overline{y}_3))$.

[All edges are directed down unless otherwise indicted.]

Fig. 26.1 Gadgets for SHORT GEOGRAPHY

The theorem above suggests that $AW[*]$ is perhaps the natural home of "most" k-move games. Other examples are given in the exercises.

Parameterized space seems a very interesting arena of investigation. For instance, as we mentioned earlier, we do not know if a parameterized version of Savitch's Theorem holds.

Obviously, there are a number of other k-move games that are natural candidates for $AW[*]$-completeness. We mention two. The first is ALTERNATING HITTING SET with no vertex degree restrictions. The other is:

SHORT GENERALIZED HEX (see Even and Tarjan [280])

Instance: A graph G with two distinguished vertices v_1 and v_2.
Parameter: A positive integer k.
Question: Does Player I have a winning strategy in the following game. Player I plays with white pebbles and player II with black ones. Pebbles are to be placed on unpebbled vertices alternately by player I and player II. Player I wins if he can construct a path of white vertices from v_1 to v_2, in $\leq k$ moves.

26.4 Relational Databases

To conclude this chapter, we return again to problems from the theory of databases. We recall from Chap. 21 the following problem.

QUERY NON-EMPTINESS

Instance: A database d, and a query φ.
Parameter: $|\varphi|$.
Question: Is the query relation r_φ nonempty?

In Chap. 21, we proved that the problem is $W[1]$-hard for even simple conjunctive queries. We recall that we considered the complexity of queries formulated in the *relational calculus*. Formulas of the calculus are built from atomic formulas of the form $x_1 = x_2$ and $R_i(x_1, \ldots, x_m)$, where the x_i are variables or constants of the appropriate domains, using Boolean connectives and quantifiers. A formula φ with free variables $y_1, \ldots, y_k$ specifies a query relation

$$r_\varphi = \{(a_1, \ldots, a_k) : \varphi(a_1, \ldots, a_k) \text{ is true in } d\}.$$

A query is in *normal form* if all quantifiers are up front and the body of the formula is in conjunctive normal form. For example, the query

$$\varphi = \exists a \forall b \forall c \big(R_1(a, b, d) \vee \neg R_2(a, c) \vee \neg R_1(a, c, e) \big) \wedge \big(\neg R_1(a, d, e) \vee R_2(d, e) \big)$$

is in normal form. Here we prove the following result about the general problem.

φ-MODEL CHECKING FOR φ

Instance: A graph $G = (V, E)$.
Parameter: $|\varphi|$.
Question: $G \models \varphi$?

This reduced to the problem for relational databases below.

QUERY NON-EMPTINESS

Instance: A database d, and a query φ.
Parameter: $|\varphi|$.
Question: Is the query relation r_φ nonempty?

Here is the theorem promised in Chap. 21.

Theorem 26.4.1 (Downey, Fellows, and Taylor [258])
(1) QUERY NON-EMPTINESS *is AW*[*]-*complete.*
(2) MONOTONE QUERY NON-EMPTINESS *is W*[1]-*complete.*

Proof (1) We will say that circuit C is k-structured if it has its inputs partitioned into k disjoint sets $I_1, \ldots, I_k$, each with an associated logical quantifier $Q_i \in \{\exists, \forall\}$ for $i = 1, \ldots, k$. Another way of expressing the AW-classes is in the language of k-structured circuits. In a manner analogous to the definition of L_F we define

$$
\begin{aligned}
L_F^A = \big\{ & (C, k) : C \text{ is a } k\text{-structured circuit} \\
& Q_1 i_1 \in I_1, Q_2 i_2 \in I_2, \ldots, Q_k i_k \in I_k, \text{ such that} \\
& \text{the weight } k \text{ input } x \text{ with } x_{i_1} = x_{i_2} = \cdots = x_{i_k} = 1 \text{ is accepted by } C \big\}
\end{aligned}
$$

Lemma 26.4.1 QUERY NON-EMPTINESS *is in AW*[1].

Proof At the cost of a blow-up exponential in the size of φ we may assume that φ is in the normal form:

$$\varphi = Q_1 x_1 Q_2 x_2 \ldots Q_r x_r E$$

where the Q_i are logical quantifiers and E is in conjunctive normal form. Using multiplicative notation for convenience:

$$E = \prod_{i=1}^{l} \sum_{j=1}^{m} E_{ij}$$

where each E_{ij} is either a relational atomic formula or its negation. Let $y_1, \ldots, y_s$ denote the free variables of φ, and let U denote the base set over which the database d is defined, $|U| = n$.

We describe an appropriate k-structured circuit C to exhibit membership in AW[4]. Since the AW hierarchy collapses to AW[1], this is sufficient for our purposes. The parameter k is obliged to be purely a function of φ. We may take $k = r + s$, the number of variables (both free and bound) occurring in φ.

The inputs to C are partitioned into k blocks of size n. Let J_i for $i = i, \ldots, r$ denote the blocks corresponding to the r bound variables of φ, and let J_i' for $i = 1, \ldots, s$ denote the blocks corresponding to the free variables of φ. Let

$$J_i = \big\{ v[i, u]; u \in U \big\} \quad \text{for } i = 1, \ldots, r$$

and let

$$J_i' = \big\{ w[i, u] : u \in U \big\} \quad \text{for } i = 1, \ldots, s.$$

To each block J_i for $i = 1, \ldots, r$ we associate the quantifier Q_i. To each block J_i' for $i = 1, \ldots, s$ we associate the existential quantifier.

The circuit C for our reduction corresponds to the logical expression:

$$F = F_1 \cdot F_2 \cdot F_3$$

where

$$F_1 = \prod_{i=1}^{r} \prod_{u,u' \in U, u \neq u'} \left(\neg v[i, u] + \neg v[i, u'] \right),$$

$$F_2 = \prod_{i=1}^{s} \prod_{u,u' \in U, \; u \neq u'} \left(\neg w[i, u] + \neg w[i, u'] \right)$$

and

$$F_3 = \prod_{i=1}^{l} \sum_{j=1}^{m} F_{ij}$$

where the subexpressions F_{ij} correspond to the relational formulas (or negations of these) E_{ij} of E in the following way (which we will explain by example).

Suppose $E_{ij} = R(x_2, x_3, y_5)$. Then

$$F_{ij} = \sum_{(u,u',u'') \in R} \left(v[2, u] \cdot v[3, u'] \cdot w[5, u''] \right).$$

The expression F_{ij} is thus simply a sum-of-products expression that evaluates R in the blocks corresponding to the variables occurring in the atomic relational formula (or its negation, as the case may be).

The reader can easily check that this construction works correctly, as it is little more than a notational translation. The circuit C has depth and therefore weft 4. Since $AW[4] = AW[1]$ by the Tradeoff Theorem, we are done. $\qquad \square$

To complete the proof of (1) we need to show that QUERY NON-EMPTINESS is hard for $AW[*]$.

We reduce from the problem SHORT NODE KAYLES that is shown to be complete for $AW[*]$ in Abrahamson, Downey, and Fellows [3] (see Exercise 26.5.2).

SHORT NODE KAYLES

Instance: A graph G.

Parameter: k.

Question: Does player I have a winning k move strategy in the following game? Players I and II take turns pebbling a vertex not adjacent to any pebbled vertex. The first player with no play loses. Player I plays first.

The database d for our reduction has as a base set the vertex set of the graph, and it has two relations: (1) the binary adjacency relation E of the graph (where we consider also that a vertex is adjacent to itself), and (2) a unary padding relation $T = \{(u) : u \in V\}$.

It is straightforward to concoct a query that will define a nonempty relation if and only if the first player has a winning strategy. For example, the following query will serve when $k = 3$ and is easily generalized to arbitrary k.

$$\varphi = \exists x_1 \big(\forall y_1 \big(E(x_1, y_1) \vee \exists x_2 \big(\forall y_2 \big(E(x_1, y_2) \vee E(x_2, y_2)$$
$$\vee \exists x_3 \big(\forall y_3 \big(E(x_1, y_3) \vee E(x_2, y_3) \vee E(x_3, y_3)\big)\big)\big)\big)\big)\big) \wedge T(x).$$

The only free variable of φ is x. The unary relation defined by φ coincides with T when the first player has a 3-move winning strategy, and is otherwise empty.

Note that the size of φ depends only on k. This concludes the proof of (1).

For (2), we consider the MONOTONE QUERY NON-EMPTINESS problem—the restriction to monotone queries. A query φ is termed *monotone* if: (1) if it has only existential quantifications, and (2) no negations. We need to prove that MONOTONE QUERY NON-EMPTINESS belongs to $W[1]$.

Use the same circuit construction as in the proof of (1). In this case, there are no blocks of input variables (as the circuit is now "ordinary" as opposed to k-structured). The subexpressions F_1 and F_2 enforce, however, that in any weight k input vector accepted by the circuit, there is exactly one input variable in each block with value 1, which is equivalent to an implicit existential quantification, as required for the correctness of the reduction from a monotone query.

An examination of the circuit C shows that on any input-output path there is at most one gate with unbounded fan-in, all other gates having fan-in bounded by a function of k. Thus by Theorem 23.4.2, we have established membership in $W^*[1] = W[1]$. $\qquad\qquad\qquad\qquad\qquad\qquad\qquad\qquad\qquad\qquad\qquad\qquad\qquad\qquad\square$

26.5   Exercises

Exercise 26.5.1 (Abrahamson, Downey, and Fellows [3])   Prove that RESTRICTED ALTERNATING HITTING SET is strongly FPT (Theorem 26.2.1).
(Hint: It is simplest to consider $k_1 = 2$, the analogue of the PSPACE complete problem ALTER-NATING VERTEX COVER. Take an edge (x, y). All vertex covers must include x or y. Try each, generating the tree of possibilities. Terminate a branch and put the cover at the leaf if a branch achieves a vertex cover. This gives a tree with at most $k_1^{k_2} = 2^{k_2}$ leaves (corresponding to possible candidates for vertex covers), at most $k_2 2^{k_2}$ vertices, and all size $\leq k_2$ covers must contain a subset occurring at one of the leaves.

Now we select k_2 additional vertices of G, not occurring at any of the leaves of the tree (we can assume V is large compared to k_2, else the problem is easily done). Consider all possible strategies played on the subgraph induced by these at most $k_2 + k_2 2^{k_2}$ vertices. It is easy to see that player I has a winning strategy in $\leq k_2$ moves in G iff he has one in this set of strategies.)

Exercise 26.5.2 (Abrahamson, Downey, and Fellows [3]) Prove that the following problem is $AW[*]$ complete.

SHORT NODE KAYLES

Instance: A graph G.
Parameter: k.
Question: Does I have a winning k move strategy in the following game? Players pebble a vertex not adjacent to any pebbled vertex. The first player with no play loses. I plays first.

Remark We have stuck with the terminology of Schaefer, although the reader should think of the above as k-move ALTERNATING DOMINATING SET.

(Hint: First we show that the problem is $AW[*]$-hard. In view of Theorem 26.3.1, we only need show that the problem is $AW[2]$-hard. Let $\varphi = \exists x_k \in S_k \forall x_{k-1} \in S_{k-1} \ldots \exists x_1 \in S_1 (C_1 \wedge \ldots C_m)$ be an instance of unitary $AW[2]$. Here, we shall assume that k is odd, and $S_i = \{x_{i,1}, \ldots, x_{i,n_i}\}$ with $S_1 = \{x_{1,1}, x_{1,2}\}$ and $B_1 = x_{1,2} \vee x_{1,2} \vee \overline{x_{1,1}} \vee \overline{x - 1, 2}$. We need the vertex sets

$$V = \bigcup_i S_i, \quad S_0 = \{z_{0,q} : 1 \le q \le m\}, \quad \text{and}$$
$$Y_i = \{y_{i,j} : 0 \le j \le i - 1\} \quad \text{for } 1 \le i \le k.$$

Now we need the edge sets below.

$$D = \{xz_{0,q} : x \text{ occurs in clause } C_q\},$$
$$F = \{xz_{0,q} : y \ne x, x, y \in S_i \text{ and } \overline{y} \text{ occurs in clause } C_q\}, \quad \text{and}$$
$$G = \{y_{i,j}w : w \in \left(\left(\bigcup_{0 \le p < i, p \ne j} S_p \right) \cup \left(\bigcup_{1 \le r < i, r \ne j} Y_r \right) \right)\}.$$

Following ideas of Schaefer, we say that the game is played *legitimately* if the node played at move i is an element of S_{k-i+1}. We claim that if at move i a player does not play legitimately then the other player wins at the next move. Suppose that the first $k - i$ moves have been legitimate. If the player then plays illegitimately, note that he cannot have played any node from $\bigcup_{j \le i+1}(S_{k-j} \cup Y_{k-j})$ as these are already dominated by pebbled vertices. Now if he plays a vertex in $S_j \cup Y_j$ for some $j < i$, then his opponent can win by playing $y_{i,j}$. (Every vertex in $S_j \cup Y_j$ is adjacent to the illegal vertex all the rest are adjacent to either a previously played vertex or $y_{i,j}$.) If the illegitimate play is in Y_i, the only remaining possibility, it must be a $y_{i,j}$, and then the opponent can win by playing a member of S_j if $j > 0$, and either $z_{0,1}$ or $z_{0,2}$ if $j = 0$. This enforcement gadgetry clearly now ensures that player I has a winning strategy of k moves iff φ is true, as the reader can readily check.

To complete the proof, we need to establish membership of $AW[*]$. But this is essentially the same as Theorem 26.3.1.)

Exercise 26.5.3 (The A hierarchy—Flum and Grohe [309]) There are a number of other hierarchies which have been proposed in parameterized complexity.[1] Clas-

---

[1] Some of them, such as the WK-hierarchy will be met in Chap. 30 and the M-hierarchy found in Chap. 29 are quite new and relate to lower bounds and strong complexity hypotheses.

sically, the polynomial time hierarchy can also be based on alternating machines, but in the parameterized setting this apparently gives rise to another hierarchy. Recall that an *alternating Turing machine* is a NTM in whose states are divided into two sets: existential states and universal states. An existential state is accepting if some transition leads to an accepting state and a universal state is accepting if every transition leads to an accepting state. A universal state with no transitions accepts unconditionally; an existential state with no transitions rejects unconditionally. The machine as a whole accepts if the initial state is accepting.

ATM-TIME[t] (Flum and Grohe [309], Y. Chen, Flum and Grohe [153])
Instance: An alternating Turing machine M with at most t alternations.
Parameter: An integer k.
Question: Does M accept the empty string in at most k steps?

Definition 26.5.1 (Flum and Grohe [309], Y. Chen, Flum and Grohe [153])
We define $A[t]$ to be the languages fpt-reducible to ATM-TIME[t].

1. Prove that model checking for Σ_t (i.e. t alternations of quantifiers) is complete for $A[t]$, where, recall

   φ-MODEL CHECKING PARAMETERIZED BY $|\varphi|$
   Instance: A graph $G = (V, E)$ (or a suitable structure).
   Parameter: $|\varphi|$.
   Question: $G \models \varphi$?[2]

2. Prove $A[1] = W[1]$.
3. Prove $W[t] \subseteq A[t]$.
   (Hint: Use the Normalization Theorem.)
4. Show that the problem below is in $A[1]$.

   SUBGRAPH ISOMORPHISM
   Instance: A graph $G = (V, E)$.
   Parameter: A graph $H = (W, F)$.
   Question: Is H isomorphic to a subgraph of G?

5. Prove that the following is in $A[2]$.

---

[2]Flum and Grohe argue that this makes the A-hierarchy a natural analogue of the polynomial time hierarchy. The other possible analogue according to our results would be the $AW[t]$ hierarchy. Space considerations preclude us for a discussion, and we refer the reader to [312].

CLIQUE COVER

Instance: A graph $G = (V, E)$.
Parameter: Integer k.
Question: Is there h, g such that $k = h + g$ and a set of h vertices C such that
              every clique of size g contains at least one vertex of C?

Exercise 26.5.4 (Scott and Stege [608]) Prove that the following problem is $AW[*]$
complete (this is not easy).

PARAMETERIZED CHESS

Instance: An $n \times n$ chessboard position.
Parameter: k.
Question: Does I have a winning k move strategy?

Exercise 26.5.5 (Chen, Flum and Grohe [153])
1. If COMPACT TM $\in W[P]$ then there is an r such that LOGSPACE $\subseteq$ NTIME(n^r).
2. If COMPACT NTM $\in W[P]$ then there is an r such that NLOGSPACE $\subseteq$
   NTIME(n^r).
(Hint: Consider a generic language complete for DSPACE$(k \log n)$.)

Exercise 26.5.6 (Scott and Stege [609])
1. Prove that the following is $AW[*]$-hard (this is not easy).

SEEDED PURSUIT EVASION

Instance: A simple undirected graph $G = (V, E)$.
A set $C \subseteq V$ of starting position for the *cops*.
A starting position $a \in V$, $a \notin C$ for the *robber* subject to the following rules.
   (i) Exactly one cop for each $v \in C$.
   (ii) Cops and robbers alternate at turns.
   (iii) At each stage a cop either moves to an adjacent vertex or stays put. Multiple
         cops can occupy the same vertex at any stage.
   (iv) The robber can move to an adjacent vertex or stay put.
   (v) If the robber moves to a vertex occupied by a cop, the cops win.
   (vi) Both players have complete information.
Parameter: A positive integer $k = |C|$.
Question: Do the cops have a winning strategy?
2. Prove the directed case is also $AW[*]$-hard.

Exercise 26.5.7 (Björklund, Sandberg, Vorobyov [67]) The following problems
concern the parameterized complexity of infinite games. This is important in the
solution to software/hardware problems involving scheduling, routing, etc., and to
LTL and other similar temporal logics; as well as their verification.
   A *game graph* is a tuple $G - (V = V_0 \sqcup V_1, E, v_0)$, with the directed bipartite
graph sink-free. v_0 is the initial vertex. Players 0 and 1, construct an infinite path
$p = v_0, v_1, \ldots$ called a play, where $v_i v_{i+1}$ must be an edge. The *game* is a pair
(G, W) where $W \subseteq 2^V$ is a relation called the *winning condition*. Let $\inf(p)$ denote

the set of vertices visited infinitely often in the play p. We say 0 *wins* iff $W(\inf(p))$ holds.

1. A *parity game* is one where W is a pair $(c, [k])$ where c is a k-coloring of the vertices so $c(v) \in [k]$. 0 wins if the maximum coloring occurring infinitely often is even.

PARAMETERIZED PARITY GAME
Instance: A parity game with k colors.
Parameter: A positive integer k.
Question: Does 0 have a winning strategy?

We remark that, classically, this is a well-known problem in NP $\cap$ co-NP, and not known to be in P.

A *Rabin game* that has the winning condition is a set W of k pairs of vertex sets (I_i, J_i). Player 0 wins p if there is a pair (I_i, J_i) with $\inf(p) \cap I_i \neq \emptyset$ and $\inf(p) \cap J_i = \emptyset$. A *Strett game* is similar except that now we need $\inf(p) \cap I_i \neq \emptyset$ implies $\inf(p) \cap J_i \neq \emptyset$. The parameterized versions are the obvious ones.

Prove that PARAMETERIZED PARITY GAME, PARAMETERIZED RABIN GAME, and PARAMETERIZED STETT GAME are all FPT equivalent.

2. Prove that MINI-PARITY GAMES is FPT, where MINI-PARITY GAMES has G input with k and n in unary and total description size is $k \log n$.

3. Prove that the following is FPT.

GAMES PARAMETERIZED BY WINNING CONDITION NODES SET
Instance: A game where only vertices in W occur in the winning condition.
Parameter: A positive integer $k = |W|$.
Question: Does 0 have a winning strategy?

26.6   Historical Notes

The *AW*-classes were defined by Abrahamson, Downey, and Fellows [3]. Chen, Flum and Grohe [153] used descriptive complexity to analyze these results and pointed out the flaw in the original paper about COMPACT TM COMPUTATION. Downey, Fellows, and Taylor [258] proved the result for query nonemptiness, and introduced the W^*-hierarchy for this reason. This answered questions of Yannakis.

Concerning Problem 26.5.3, alternation was introduced by Chandra and Stockmeyer [630]. It is a central notion of classical complexity. The A-hierarchy was introduced and analyzed by Flum and Grohe in [309]. There are other hierarchies involved in parameterized complexity. One includes the E-hierarchy which, for tractability, required that the running times be $O^*(2^k)$. E-hierarchy was introduced by Flum, Grohe, and Weyer in [313]. These hierarchies are carefully analyzed in Flum and Grohe [312]. Since then parameterized complexity has spawned other hierarchies. One notable one is due to Chen, Huang. Kanj and Xia [143] These authors analyze hardness results using *linear* FPT reduction and use this to create a subhierarchy (which is a proper refinement of the W-hierarchy, assuming the W-hierarchy does not collapse) and prove stronger completeness and hardness results about classical problems like CLIQUE.

26.7 Summary

We defined and explored the *AW*-classes, and their uses in analyzing the parameterized complexity of k-move games. We established a number of hardness and completeness results, and a few fixed-parameter tractability results for these natural parameterized problems. One of the k-move game problems played an important role in understanding the parameterized complexity of relational databases problems parameterized (very naturally) by the size of the database query.

Provable Intractability: The Class XP

27

27.1 Introduction and X Classes

Many classes of parameterized problems are apparently different from FPT, but as with classical complexity theory, we have no proof of this fact. Either distressingly—or intriguingly—these proliferating, but quite intuitively reasonable—mathematical hypotheses about the nature of computational complexity (beginning with the undoubted conjecture that P is not equal to NP), seem to be *almost untouchable* mathematically. We believe there is much to this phenomenon that deserves deep philosophical thought, which to date has not yet surfaced to any significant degree.

Essentially, the entire project of *parameterized complexity* lives in-between the two basic parameterized complexity classes FPT, and XP, with the latter being our main focus in this brief chapter. Almost all of the core research in parameterized complexity begins by identifying a natural parameterized problem that is *trivially* in XP, and then asking whether it might be in FPT, with various outcomes, and when the outcome is (conjecturally) negative, problems may be proved hard for various, presumably intractable, parameterized complexity classes. But the separation conjectures for those parameterized classes *versus* FPT are just as (or even more so) conjectural and apparently, untouchable, as the classical conjecture that P is not equal to NP.

However, in the setting of parameterized complexity, which originated and draws its vitality from the distinction between FPT and XP, we can actually prove (unconjecturally) that these classes *are not equal*. This is in sharp contrast to classical complexity, which lives between P and NP, and that has made essentially no mathematical progress on distinguishing between them.

Our main report in this chapter is that there are natural parameterized problems that are *complete* for XP, and therefore, are unconditionally, unconjecturally, *not* in FTP.

R.G. Downey, M.R. Fellows, *Fundamentals of Parameterized Complexity*,
Texts in Computer Science, DOI 10.1007/978-1-4471-5559-1_27,
© Springer-Verlag London 2013

The class XP consists of those parameterized languages L such that $L_k \in$ P for each k (perhaps with a different algorithm for each k). L is in *uniform* XP if there are computable functions f and g and a single such that for each k, $\Phi_{f(k)}$ accepts L_k and has running time $|x|^{g(k)}$ upon input $\langle x, k \rangle$.

For the purposes of clarity of this chapter, we will only look at XP and remark that all the results hold with no change for uniform XP as well. By standard diagonalization, we have the following basic result:

Proposition 27.1.1 FPT *is a* proper *subset of* XP.

27.2   Pebble Games

There are a number of problems that are complete for XP, which seems to be the parameterized class corresponding to EXP. We briefly look at some such results drawn from the literature.

Definition 27.2.1 (Pebble Game) A *pebble game* is a quadruple (N, R, S, T) consisting of
  (i) N, a finite set of vertices (or nodes).
 (ii) $R \subseteq \{(x, y, z) : x, y, z \in N, x \neq y \neq z\}$, called the set of *rules*.
(iii) $S \subset N$, the start set. $|S|$ is called the *rank* of the game.
 (iv) $T \in N$, called the terminal node of the game.

The pebble game is played by two players I and II, who alternatively move pebbles on the nodes. At the beginning of the game, pebbles are placed on all the start nodes. If $(x, y, z) \in R$ and there are pebbles upon x and y but not on z, then a player can move a pebble from x to z. The winner is the first player to put a pebble on T or can force the other player into a position where he cannot move.

PEBBLE GAME PROBLEM
Input: A pebble game (N, R, S, T).
Parameter: A positive integer k.
Question: If $|S| = k$ determine if player I has a winning strategy.

We reinterpret an old result of Adachi, Iwata, and Kasai [12, 13] to establish the following.

Theorem 27.2.1 (Adachi, Iwata, and Kasai [12, 13])   PEBBLE GAME PROBLEM *is XP-complete.*

In the proof, we begin with proving hardness. Actually, Adachi et al. establish the following sharper characterization for the single-tape Turing machine model.

Theorem 27.2.2 (Adachi, Iwata, and Kasai [12, 13])   *Let* $L \in$ DTIME(n^k). *Then L is* $O(n \log n)$-*reducible* [*using space* $O(n \log n)$ *to the* $(2k + 1)$-*pebble game problem*].

Proof We follow Adachi et al. [12, 13]. The proof uses a generic reduction. Since $L \in \text{DTIME}(n^k)$, there is a single-tape Turing machine M which accepts L within n^k time. We will assume that M has a set of states Q, a set of symbols Γ with initial symbols $\Sigma \subseteq \Gamma$, a blank symbol b, initial state q_0 and final state q_f, and a transition function $\delta : Q \times \Gamma \mapsto Q \times \Gamma \times \{-1, 0, 1\}$. Suppose that $\delta(q, X) = (\widehat{q}, \widehat{X}, j)$. This would have the interpretation that the machine, on reading some symbol X in state q, will either move left ($j = -1$), right ($j = 1$), or remain where it is ($j = 0$), move to some state $\widehat{q}$, and (over-)print the symbol X by symbol $\widehat{X}$. Without loss of generality, we make some assumptions on M.

We use the symbol # as a left endmarker. A *configuration* shown is, as usual, represented by

$$\#X_1 X_2 \ldots X_{i-1}(q X_i) X_{i+1} \ldots X_{n^k},$$

where $X_j \in \Gamma$ for all j ($1 \le j \le n^k$). We treat an element of $Q \times \Gamma$ as an abstract symbol. Hence, $(q X_i)$ is the ith symbol (# is the 0th symbol) of the configuration. We also assume that the head never scans the endmarker. The initial configuration for an input $w = w_1 w_2 \ldots w_n$ is $\#(q_0 w_1) w_2 \ldots w_n b^{n^k - n}$. Whenever M enters the final state q_f, the configuration of M should be of the form

$$\#(q_f b) X_2 X_3 \ldots X_{n^k},$$

where $X_j \in \Gamma$ for all i ($2 \le i \le n^k$). Once M enters q_f, M stays q_f without changing the position of the tape head or the contents of the tape; that is, if M accepts w within time n^k, then M is in q_f at time n^k.

We construct the $(2k + 1)$-pebble game G from M and its input w such that (1) M accepts w within time $|w|^k$ if and only if the first player has a forced win in the $(2k + 1)$-pebble game and (2) the construction is performed within time $O(n \log n)$.

Throughout this proof, we use two global variables t and i for explanation, where $0 \le t \le n^k$ and $1 \le i \le n^k$. The variable t represents the current time of the Turing machine and i represents the position of the symbol of the configuration. We prepare nodes in the pebble game of the form $[qX], [X]$ for every state $q \in Q$ and every tape symbol $X \in \Gamma$ of the Turing machine. We also use nodes of the form $[X_{-1}, X_0, X_1]$, where $X_{-1}, X_0, X_1 \in (Q \times \Gamma) \cup \Gamma \cup \{\#\}$. Let $F(X_{-1}, X_0, X_1)$ be the symbol in $(Q \times \Gamma) \cup \Gamma$ such that if $\alpha X_{-1} X_0 X_1 \beta$ is a configuration of M for some α, β, then one move of M changes the symbol X_0 to $F(X_{-1}, X_0, X_1)$. When $X_{-1} X_0 X_1$ is not part of any configuration, then $F(X_{-1}, X_0, X_1)$ is not defined. We note that $F(X_{-1}, X_0, X_1)$ is uniquely defined and is independent of t and i since M is deterministic.

The game uses $(2k + 1)$ pebbles. One pebble is placed to represent a symbol of a configuration; k pebbles are used for TCOUNTER, a counter for t; and another k pebbles are for ICOUNTER, a counter for i. Each value is represented by k, n-ary counters, and the jth pebble is placed to represent the jth n-ary digit for $1 \le j \le k$.

The pebble game will simulate M backward. At the beginning of the pebble game, a pebble is placed on $[q_f b]$ and $2k$ pebbles are placed in order to represent $t = n^k$ and $i = 1$. This corresponds to $q_f b$ being the first symbol of the configuration

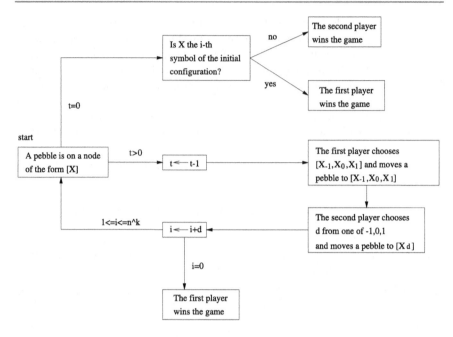

Fig. 27.1 Pebble moves where $X = F(X_{-1}, X_0, X_1)$, $X \in (Q \times \Gamma) \cup \Gamma$, $d \in \{-1, 0, 1\}$. Initially a pebble is on $[q_f b]$ and $t = n^k$, $i = 1$

at time n^k. The rules of the pebble game are constructed in such a way that moves of pebbles are restricted as follows:

(1) Only pebble moves shown in Fig. 27.1 are allowed; that is, both players have to be forced to move pebbles along the edges in Fig. 27.1.

(2) When a pebble is on $[X]$, the first player can move the pebble to a node of the form $[X_{-1}, X_0, X_1]$ such that $X = F(X_{-1}, X_0, X_1)$, and when a pebble is on $[X_{-1}, X_0, X_1]$, the second player can choose d from $\{-1, 0, 1\}$ and can move the pebble to $[X_d]$.

(3) Before and after the renewals of TCOUNTER and ICOUNTER, k pebbles are used to represent the current value of t and another k pebbles are used for i. During these renewal procedures, each player is forced to cooperate with another player to update TCOUNTER and ICOUNTER.

Typical moves in the game are (1) at time $t > 0$ to iterate those pebble moves that are described in the boxes along the cycle in Fig. 27.1 and (2) at time $t = 0$ to check whether X is the ith symbol of the initial configuration to determine which player wins the game. The alternation inherent in the definition of the pebble game allows us to get by without explicitly storing the contents of the tape inside the game. If the first player ever tries to cheat by modifying the initial contents of a tape square illegally, the second player can catch him by forcing a win by staying on this square.

Fig. 27.2 A set of rules, where a, b, and c are nodes and $P(a, c)$ is a condition

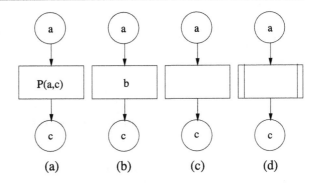

Now we define the rules of the pebble game. For our convenience, let Fig. 27.2(a) denote the set of rules

$$\{(a, b, c)|P(a, c), b \text{ is a node in ICOUNTER}\},$$

where a, b, and c are nodes and $P(a, c)$ is a condition on a and c. Thus, if a pebble is on a, no pebble is on c, and $P(a, c)$ holds, then a player can move the pebble from a to c in his turn since there is some pebble on nodes in ICOUNTER. Let Fig. 27.2(b) denote the rule (a, b, c), and let Fig. 27.2(c) denote the set of rules

$$\{(a, b, c)|b \text{ is in the set of the nodes of the game}\},$$

where the box in the figure is left blank. We let Fig. 27.2(d) denote a predefined set of rules. We use two predefined set of rules, TCOUNTER and ICOUNTER.

The rules of the pebble game are shown in Figs. 27.3 and 27.4. Consider Fig. 27.3. The first player moves a pebble from $[X]$ to $[w1, X]$. For every element $[X]$ in $(Q \times \Gamma) \cup \Gamma$, TCOUNTER uses nodes of the form $[w1, X]$ and $[w2, X]$. Once a pebble is on $[w1, X]$, two players of the game are forced to cooperate for decreasing a value of t by one until the second player puts a pebble on $[w2, X]$ to end the procedure for TCOUNTER. The value of t is represented by the content of TCOUNTER. When $t = 0$, one of the k pebbles of TCOUNTER is placed on a distinguished node [time-zero]. If a pebble is already on [time-zero] and if the first player moves a pebble to $[w1, X]$, then the second player wins the game.

When a pebble is on $[w2, X]$, the first player moves the pebble to $[X_{-1}, X_0, X_1]$ such that $X = F(X_{-1}, X_0, X_1)$. Then, the second player selects one of the three nodes [dec, X_{-1}], $[X_0]$, and [inc, X_1], and moves the pebble from $[X_{-1}, X_0, X_1]$ to the selected node. If a pebble is on a node of the form [dec, Y] or [inc, Y], the two players are forced to operate with ICOUNTER. The contents of ICOUNTER represent the value of i, that is, the symbol position of a configuration. If a pebble is on [dec, Y], then the contents of ICOUNTER are decreased by one, and if the resulting value is zero, then one of the k pebbles of ICOUNTER is moved to a distinguished node [leftend]. If a pebble in on [inc, Y], then the contents of ICOUNTER are increased by one. We do not have to check the right end of the tape, since the value

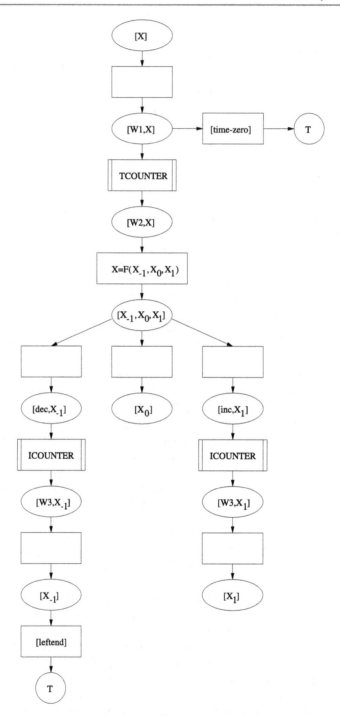

Fig. 27.3 Rules for the pebble game

Fig. 27.4 More rules for the pebble game

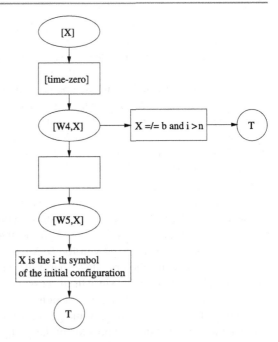

of TCOUNTER would be 0 before the contents of ICOUNTER will be $n^k + 1$. The first player moves a pebble to $[w3, X_1]$ or $[w3, X_{-1}]$ to end the procedure for ICOUNTER. If a pebble is on [leftend], then the first player wins.

Figure 27.4 is a set of rules when $t = 0$. Note that a pebble is on [time-zero] when and only when $t = 0$. Thus, the first player can move a pebble to $[w4, X]$ only when $t = 0$. Assume that the first player places a pebble on $[w4, X]$. If X is the ith symbol if the initial configuration, then the first player wins the game; otherwise the second player wins, where i is the contents of ICOUNTER.

Detailed construction of TCOUNTER, ICOUNTER, and Fig. 27.4 is straightforward and left to the reader. We shall prove the correctness of the construction; that is, M accepts w if and only if the first player has a forced win in pebble game G. We call a position of pebble game G a *disposition* $D(X, t, i)$ of G if

(i) A pebble is on $[X]$, $X \in (Q \times \Gamma) \cup \Gamma$.
(ii) The contents of TCOUNTER is t.
(iii) The contents of ICOUNTER is i.

Note that k pebbles are used for TCOUNTER, and another k pebbles for ICOUNTER.

We now show that X is the ith symbol of the configuration at time t of M if and only if the first player has a forced win from $D(X, t, i)$.

We prove this by induction on t.

Basis. When $t = 0$, it is obvious that the first player has a forced win if and only if X is the ith symbol of the initial configuration. This is illustrated in Figs. 27.1 and 27.4.

Induction step. Assume that the induction hypothesis holds at time $t - 1$. We consider the case at time t $(1 \leq t \leq n^k)$.

Assume that the first player has a forced win from the disposition $D(X, t, i)$ of the game. Thus, there is a way to win the game for the first player even if the second player may select any node to move a pebble. After the first player moves a pebble from $[w2, X]$ to $[X_{-1}, X_0, X_1]$, where $X = F(X_{-1}, X_0, X_1)$, the first player then has a forced win from $D(X_d, t - 1, i + d)$ for all d, $d \in \{-1, 0, 1\}$. By the induction hypothesis, X_d is the $(i + d)$th symbol of the configuration at time $t - 1$, for each d, $d \in \{-1, 0, 1\}$. Thus, X is the ith symbol of the configuration at time t, since $X = F(X_{-1}, X_0, X_1)$.

Conversely, assume that X is the symbol of the configuration at time t $(1 \leq t \leq n^k)$ and position i $(1 \leq t \leq n^k)$. Then, there are symbols X_{-1}, X_0, and X_1 such that $X = F(X_{-1}, X_0, X_1)$, and these symbols occupy positions $i - 1$, i, $i + 1$ of the configuration at time $t - 1$, respectively. By the induction hypothesis, the first player has a forced win for $D(X_d, t - 1, i + d)$ of the game for all d $(d \in \{-1, 0, 1\})$. Thus, the first player has a forced win from $D(X, t, i)$. Note that there is nondeterminism here, since the discussion is made for all d $(d \in \{-1, 0, 1\})$.

This completes the proof of the induction.

We note that at the beginning of the game, pebbles are placed as $D(q_f b, n^k, 1)$, and that the symbol at the first position of the configuration at time n^k is $(q_f b)$ if M accepts w within time n^k. Therefore, we have shown that M accepts w within time n^k if and only if the first player has a forced win in the pebble game G.

Note that the number of the nodes and the rules of G is $O(n)$, and that the construction of G is performed within $O(n \log n)$ time.   □

To complete the proof of Theorem 27.2.1, we need membership of XP.

Lemma 27.2.1 *The* PEG GAME PROBLEM *is in XP.*

Proof To prove Lemma 27.2.1, we need to show that the k-pebble game is in P. Note that since the allowed moves preserve the fact that there are k pebbles placed on the graph, there are at most $\binom{n}{k}$ many configurations. We build a graph G, whose vertices are all possible ways that k pebbles can be placed on the set of nodes of a pebble game. We make a directed edge from configuration u to configuration v whenever there is an allowed move that would transform u in one step to v. There is a single vertex S that represents the starting configuration, and there are $\binom{n-1}{k-1}$ many configurations that would represent a win (that is, with the one remaining pebble on the terminal node T).

The lemma immediately follows from the following general lemma about a game played with one marker. We need a definition. Define the BASIC ONE-PEBBLE GAME to be a directed graph $D = (V, A)$ with a single start vertex S and a set of final vertices F contained in V. Play begins with a single marker on S. A move consists of moving the marker along an arc of D. The first player to move the marker to a vertex of F wins.   □

Lemma 27.2.2 *It can be determined in time $O(|D|)$ whether the first player has a winning strategy for the* BASIC ONE-PEBBLE GAME.

Proof Let D' be the digraph obtained from D in the following way:
(1) Compute the set of winning configurations W (for player 1)—these are the vertices w for which there is an arc from w to a vertex of F. Thus, if the marker is on w and it is player 1's turn, then player 1 can play and win. We assume that there are no out-arcs from vertices in F, since they represent "game over".
(2) Compute the set of vertices F' defined to be the set of all vertices x such that every arc xy leads to vertex y in W.

Then, we define D' to be the directed graph with vertex set $V - F - W$, with s inherited from D, with arcs inherited from D (except delete arcs between F' vertices), and with the set F' of final vertices.

Equivalently, D' is obtained from D by deleting F and W, i.e., deleting arcs between F' vertices. And s is still there if it has not been removed.

Note that player 1 has a winning strategy for D iff player 1 has a winning strategy for D'. (One has a strategy to win the ONE-PEBBLE GAME iff one has a strategy to get the marker moved to a vertex from which every out-arc—your possible moves—leads to a vertex from which one can win.)

The algorithm is just to compute the sequence of BASIC ONE-PEBBLE GAMES $D, D', D'', \ldots$. $\qquad\square$

Corollary 27.2.1 ([13])
(i) *The $(2k + 1)$-pebble game problem requires $\Omega(n^{k-\varepsilon})$ time on* STM *for any $\varepsilon > 0$ and $k > 2$.*
(ii) *Hence, for a fixed k, the* PEBBLE GAME PROBLEM *of rank k requires time $\Omega(n^{\frac{k-1}{4-\varepsilon}})$ for any $\varepsilon > 0$ and $k > 5$ and is hard for* DTIME(n^k).

Proof Exercise 27.4.1. $\qquad\square$

27.3 Other XP-Completeness Results

There are several other XP completeness results which can also be extracted from Adachi, Iwata, and Kasai [12] and [13]. They are proved by the construction of combinatorial reductions from the PEBBLE GAME PROBLEM. We will merely state them here.

The CAT AND MOUSE GAME is a quintuple $G = (X, E, c, M, v)$ with X a set of vertices, E a set of edges, $c \in X$, $M \subseteq X$, and $v \in X$. In the game, player I begins with his token on c and player II begins with tokens on each member of M. Players play alternatively and can move tokens from vertex x to vertex y provided $xy \in E$. Two tokens of player II cannot occupy the same vertex. Player I wins if he can place his token on a vertex with one of player II's token. Player II wins if she can place one of her tokens on vertex v even if it is occupied by player I's token. Player I plays first.

CAT AND MOUSE PROBLEM
Input: A CAT AND MOUSE GAME $G = (X, E, c, M, v)$.
Parameter: A positive integer k.
Question: If $|M| = k$, does player I have a winning strategy?

Let Q be a set of integers. The PEG GAME is a triple $G = (V, k, l)$ with $k, l \in Q$ and $V \subseteq V^l$ such that $(v_1, \ldots, v_l) \subseteq V$ implies that $v_1 + \cdots + v_l = 0$. A play of the PEG GAME is as follows. There are l pegs on the board and k rings. The interpretation of the vector $(v_1, \ldots, v_l) \in V$ is that for each i, if $v_i \geq 0$, then we put v_i rings on the ith peg, and if $v_i \leq 0$, we remove v_i rings from the ith peg. Initially, all k rings are on the first peg. Players play alternatively according to the rules V of the game. The player who wins is the one who places all k rings on the last peg, or who forces the opponent into a position where the opponent has no valid play.

PEG GAME PROBLEM
Input: A PEG GAME $G = (V, k, l)$.
Parameter: A positive integer k.
Question: Does player I have a winning strategy?

27.4   Exercises

Exercise 27.4.1 Prove Corollary 27.2.1.

(Hint: By diagonalization, there exists a language L in $\mathrm{DTIME}(n^k)$ which is not in $\mathrm{DTIME}(n^{k-\varepsilon})$ for any $\varepsilon > 0$. By Theorem 27.2.2, L is $O(n \log n)$-reducible to the $(2k+1)$-pebble game problem. Suppose that the $(2k+1)$-pebble game problem is $n^{k-\varepsilon_1}$ time computable on an STM for some $\varepsilon_1 > 0$. If $n^2 \log^2 n \ll n^{k-\varepsilon_1} \log^{k-\varepsilon_1} n$, then establish that L is $O(n^{k-\varepsilon_1} \log^{k-\varepsilon_1} n)$ time computable on a Turing machine. Otherwise L is $O(n^2 \log^2 n)$ time computable on a Turing machine.)

Exercise 27.4.2 (Chandra and Stockmeyer [629, 630]) Prove that the CAT AND MOUSE GAME PROBLEM is in XP.
(Hint: Use the method of Lemma 27.2.1.)

Exercise 27.4.3 Prove that the PEG GAME PROBLEM is in XP.
(Hint: Use the method of Lemma 27.2.1.)

27.5   Historical Remarks

Diagonalization classically yields languages L in $\mathrm{DTIME}(n^{k+1})$ but not in $\mathrm{DTIME}(n^k)$, and shows that exponential time differs from P. This fact was realized very early. The first examples of "provably difficult" combinatorial problems can be found in Chandra and Stockmeyer [629]. The results of this section are drawn from Adachi, Iwata, and Kasai [12, 13] (as presented first in Downey and Fellows [247]), which gave the first examples of natural combinatorial problems unconditionally provably *not* in FPT. In this regard, we remark that, in some sense, these earlier (unconditional) separation results fit more easily, and profoundly, into the parameterized setting, since XP-completeness seems to be machine independent, whereas the class $\text{DTIME}(n^k)$ is not.

27.6 Summary

We discussed a fundamental (unconditional) separation between FPT and XP, and proved that some specific parameterized problems are in XP and therefore, are *not* in FPT, and this holds *not* on the basis of any yet-unproved mathematical conjecture, in striking contrast to most intractability results in classical complexity.

Another Basis for the W-Hierarchy and the Tradeoff Theorem

28

28.1 Results

The fundamental separation hypothesis upon which our theory of parameterized intractability is based is

$$\text{FPT} \neq W[1].$$

A weaker hypothesis might be that for some t,

$$\text{FPT} \neq W[t]$$

or that

$$\text{FPT} \neq W[P].$$

A more subtle hypothesis is that the W-hierarchy is *proper*, and, in particular, $W[1] \neq W[2]$. In the present chapter, we will look at some evidence that all of these hypotheses may be reasonable.

We begin by concentrating upon the $W[1]$ *versus* $W[2]$ question. The quintessential complete problems here are CLIQUE for $W[1]$ and DOMINATING SET for $W[2]$.

For each k, the language of graphs with a clique of size k is defined by the existential formula

$$\varphi_k := (\exists u_1 \dots \exists u_k) : \bigwedge_{i,j \leq k} E(u_i, u_j),$$

where $E(\cdot, \cdot)$ formalizes the adjacency relation for graphs. By contrast, the language of graphs with a dominating set of size k is known to require two alternating blocks of uniform quantifiers to define in first-order logic, such as by the Σ_2 formula

$$\psi_k := (\exists u_1 \dots \exists u_k) : (\forall v) \bigvee_{i \leq k} \left(v = u_i \vee E(v, u_i) \right).$$

R.G. Downey, M.R. Fellows, *Fundamentals of Parameterized Complexity*,
Texts in Computer Science, DOI 10.1007/978-1-4471-5559-1_28,
© Springer-Verlag London 2013

Both problems are about searching for a set of vertices of size k that satisfy the condition following the ":", but in ψ_k, the condition is more complex because it has the extra universal quantification over all the vertices v of the graph.

To say this another way, once candidate vertices have been assigned to $u_1, \ldots, u_k$, the condition for CLIQUE is entirely "local" in a sense studied for parameterized languages in Paz and Moran [557], whereas that for DOMINATING SET requires a "global" reference to other parts of the graph. Some natural parameterized problems on graphs have conditions that make several alternating first-order quantifications over the vertices or edges of a graph, and are known to belong to $W[t]$ only for higher values of t.

Other natural parameterized problems are concerned with properties that are *not first-order definable at all*, and some of these are complete for $W[P]$ (as per Bodlaender, Downey, Fellows, and Wareham [84]).

Intuitively, the question "Does $W[1] = W[2]$?" asks whether a local check of a fixed-size substructure can do the same work as a global check. The question "For all t, does $W[t] = W[2]$?" asks whether the simple check over vertices v in the $W[2]$-complete DOMINATING SET problem suffices to verify any condition that is definable by circuits of bounded weft t. Similarly, if $W[P] = W[2]$, then fixed-parameter many-one reductions have an enormous power to simplify the checking of properties.

We must be careful not to push this reasoning too far. For instance, VERTEX COVER has a logical definition of form similar to that of the ψ_k above, and yet the problem is fixed-parameter tractable. This is because "logical complexity" of the *reductions* can be more substantial than the logical depth (weft) of the circuits referred to in the definition of the $W[\cdot]$ classes.

Compare the W situation with the classical NP situation. If $L \leq_m^p Q$ and $Q \in \text{NP}$, then L can be put in the NP logical form: for some P-time relation R,

$$x \in L \quad \text{iff} \quad \exists^P y \big(R(x, y) \big).$$

However, if $L' \leq_m^s Q'$ and $Q' \in W[t]$, we have no guarantee that we can actually *express* L' as the weight k' solutions to some collection of weft t circuits. Indeed, we can show that in general this is not true.

The goal of the present chapter is to tie the classes in the W-hierarchy to classes of (parameterized) languages that are much *smaller* than P, and where complexity-based hierarchies are *known* to be proper. Our results bring out the *computational nature* of the W-hierarchy and provide some evidence for a positive answer to the question, "Are all classes in the W-hierarchy distinct?" One way we do this is to prove that the basic defining weft classes (i.e., without the reductions) do, in fact, generate provably distinct classes. We also have a second agenda in that we wish to demonstrate that the notion of weft is, itself, of independent interest even to classical complexity theorists.

Each class $W[t]$ is shown to be definable via existential quantification on the class of parameterized languages recognizable by FPT-sized circuit families (one circuit for each n and k) of constant weft t, analogous to the way NP is defined by existential quantification on P. The circuits we employ are actually AC^0 circuits of depth t except for extra layers of gates of fanin 2, and providing also for parameterization. We call them $G[t]$ circuits.

In this approach, we obtain

$$W[t] = \langle N[t] \rangle$$

where the angle bracketing indicates closure under FPT reductions and where

$$N[t] = \exists \cdot G[t].$$

If we similarly define $N[P] = \exists \cdot G[P]$ (*without FPT closure*) then we obtain the elegant equality $W[P] = N[P]$.

Not only is the $G[t]$-hierarchy proper, but, more interestingly, the $N[t]$-hierarchy is proper. Thus, among the three conceptual building blocks of the W-hierarchy: *parameterized languages, circuit weft*, and FPT-*reductions*, only the last, FPT reductions, could be responsible for any possible collapse of the W-hierarchy.

We explain how these results rule out any "normal" arguments for collapse of the W-hierarchy, and offer this as heuristic evidence that the hierarchy does *not* collapse. We explore analogies with the classical polynomial-time hierarchy, and investigate generalizations of the G-hierarchy based on adding further (alternating) parameterized quantifiers.

This leads to a definition, for each t, of a hierarchy $H[t]$ based on alternating $\forall \cdot$ and $\exists \cdot$ parameterized quantification over $G[t]$. In the notation of Chap. 26, we will obtain $AW[t] = \langle H[t] \rangle$. Improving Theorem 26.3.1, we prove the Tradeoff Theorem, which, unlike the $G[t]$ and $N[t]$ hierarchies, which are unconditionally proper, the $H[t]$ hierarchy (like the $AW[t]$ hierarchy) collapses: for all $t \geq 1$, $H[t] = H[1]$.

Definition 28.1.1 (Parametric connection)   A *parametric connection* is a function $\alpha : (\mathbb{N} \times \mathbb{N}) \to (\mathbb{N} \times \mathbb{N}) : (n, k) \mapsto (n', k')$, a polynomial q, and arbitrary functions $f, g : \mathbb{N} \to \mathbb{N}$ with $n' = f(k)q(n)$ and $k' = g(k)$. A parametric connection is *nice* if it can be computed in time n'' described by another parametric connection.

To economize on notation, we write $n, k, n', k', n'', k'', \ldots$ to indicate that the first four quantities represent one parametric connection, the third through sixth another, and so on. The connection relation is transitive. This notion enables us to define circuit complexity directly for parameterized problems:

Definition 28.1.2 (Parameterized family of circuits) A *parameterized family of circuits* is a bi-indexed family of circuits $\mathcal{F} = \{C_{n,k} : n, k \in \mathbb{N}\}$ such that each $C_{n,k}$ has n inputs and size at most n', where n' is part of a connection with n, k. We say that such a family is FPT-*uniform* if there is an algorithm to produce the circuit $C_{n,k}$ in time $O(n')$.

Definition 28.1.3 $G[t]$ (Uniform $G[t]$) is the class of parameterized languages $L \subseteq \Sigma^* \times \mathbb{N}$ for which there is a parameterized (uniform) family of weft t circuits $\mathcal{F} = \{C_{n,k}\}$ such that for all x and k, with $n = |x|$, $\langle x, k \rangle \in L \Leftrightarrow C_{n,k}(x) = 1$. If there is no restriction on the circuit weft, then we obtain the class of parameterized languages $G[\mathrm{P}]$, and if the circuits are boolean, then we get $G[\mathrm{SAT}]$.

It is not difficult to see that Uniform $G[\mathrm{P}] = \mathrm{FPT}$ (Exercise 28.2.1). Thus, the classes $G[t]$ provide a stratification of the problems that are fixed-parameter tractable. Now we can build upon them in much the same way that NP is definable by bounded existential quantification over P. NP uses a polynomial length bound, whereas our classes $N[t]$ use bounds on Hamming weight.

Definition 28.1.4
(a) For any class $\mathcal{F}$ of parameterized languages, $\exists \cdot \mathcal{F}$ stands for the class of parameterized languages A such that for some $B \in \mathcal{F}$, there are nice parametric connections (n, k, n', k', n'', k''), giving for all $\langle x, k \rangle$, $\langle x, k \rangle \in A \Leftrightarrow (\exists y \in \Sigma^{n'})$ [weight$(y) = k' \wedge (xy, k'') \in B$]. (Here, $n = |x|$, $n' = |y|$, and $n'' = n + n'$.)
(b) For all $t \geq 1$, $N[t]$ stands for $\exists \cdot$ Uniform $G[t]$, and $N[\mathrm{P}]$ stands for $\exists \cdot$ Uniform $G[\mathrm{P}]$.

In a corresponding way, we can define "bounded-weight" versions of the other familiar class operators $\forall$, $\oplus$, and BP. Combining the latter two formally, we see that a language A belongs to $\mathrm{BP} \cdot \oplus \cdot \mathcal{F}$ if there exists $B \in \mathcal{F}$ and nice connections giving for all $\langle x, k \rangle$,

$$\langle x, k \rangle \in A \implies \Pr_{y \in \{0,1\}^{n'}, \text{weight}(y)=k'} \left[\left| z \in \{0, 1\}^{n''} : \text{weight}(z) = k'' \right. \right.$$
$$\left. \wedge \left(xyz, k''' \right) \in B \right| \text{ is odd} \right] > 3/4,$$

whereas $\langle x, k \rangle \notin A \Rightarrow \Pr[\ldots] < 1/4$. If the latter probability is zero (i.e., we have a one-sided error), then we write $A \in \mathrm{RP} \cdot \oplus \cdot G[t]$.

The next theorem provides a powerful characterization of the $W[t]$ classes.

Theorem 28.1.1 (The Replacement Theorem [256]) *For all $t \geq 1$, $W[t] = \langle N[t] \rangle$.*

To see what is interesting about this theorem, consider the special case of $t = 2$ and the $W[2]$-complete parameterized problem DOMINATING SET. The original criterion for showing DOMINATING SET to be in $W[2]$ requires constructing, for each graph G and positive integer k, a weft 2 circuit C_G that accepts a weight k input vector if and only if G has a k-element dominating set. The point is that for each graph G, we construct a *different* circuit, thus perhaps $2^{\binom{n}{2}}$ different circuits for graphs of order n for a fixed value of k. By contrast, to show that DOMINATING SET belongs to the FPT-closure of $N[t]$, we must refer all of the graphs of order n to a *single* circuit $C_{n,k}$. The input to $C_{n,k}$ consists of the concatenation xy of a string x representing G and a string y representing the $k \log n$ bits of nondeterminism. For this particular instance, our proof must devise a bi-indexed family of weft 2 circuits, each circuit $C_{n,k}$ of which is "universal" for the dominating set problem for graphs of order n and for the parameter k. These "universal circuits" will resemble programmable logic arrays.

Proof Assume first that $t \geq 2$ and that t is even. Let L be a parameterized language in $W[t]$. We can assume without loss of generality that the reduction showing membership of L in $W[t]$ maps $\langle x, k \rangle$ to (C_x, k'), where

(i) C_x is a t-normalized circuit.
(ii) C_x has n' inputs.
(iii) C_x has exactly n'' gates on each level other than the input and output levels (achievable by padding).
(iv) k', n', and n'' are described by nice parametric connections.

Let the gates (including inputs) of C_x be described by the set $\{g[s,i] : 0 \leq s \leq t, 1 \leq i \leq n'\}$. Here, the level of the gate is indicated by the first index. Note that on level t, only one gate (the output) is important (the padding is just a notational convenience). We may assume the output gate is $g[t, 1]$.

We consider the following uniform circuit family $\mathcal{F}_L = \{C_{m,k'}\}$, $m = t(n'')^2 + n'$. (To arrange for $\mathcal{F}_L$ to have one circuit for each possible pair of indices, simply pad with nonaccepting empty circuits for index pairs not of the indicated form.)

The circuit $C_{m,k'}$ is described as follows. There are $2t + 1$ levels of gates $L_0, \ldots, L_{2t}$. The inputs to the circuit constitute level 0. The gate sets are described as follows:

$$L_0 = \big\{a_X[s,i,j] : 1 \leq s \leq t, 1 \leq i \leq n'', 1 \leq j \leq n''\big\} \cup \big\{a_Y[i] : 1 \leq i \leq n'\big\},$$

and for $s = 1, \ldots, t$,

$$L_{2s} = \big\{c[2s, i] : 1 \leq i \leq n''\big\},$$
$$L_{2s-1} = \big\{b[2s-1, i, j] : 1 \leq i \leq n'', 1 \leq j \leq n''\big\}.$$

According to our assumption that t is even, we assign the following logic functions to these gates: for $s = 1, \ldots, 2t$, the gates of L_s are $\wedge$ gates if s is congruent to 0 or 1 mod 4; the other levels are $\vee$ gates. The $a_X[*, *, *]$ inputs to the circuit have the role of describing the circuit C_x. The $a_Y[*]$ inputs represent the (nondeterministic) inputs to C_x. The gates on even-indexed levels L_{2s} provide a PLA-type template on which to simulate the circuit C_x. Note that these are large gates of the same logical character as the gates on level s of C_x. The gates on odd-indexed levels L_{2s-1} are small gates whose function is to interpret the description of C_x so that C_x can be simulated.

The $a_X[*, *, *]$ inputs describe C_x in the following way. Set $a_X[s, i, j] = 1$ if and only if, in C_x, the gate $g[s, i]$ takes input from $g[s-1, j]$. Let $\chi(C_x)$ denote the length $t(n'')^2$ 0-1 vector that describes C_x in this way.

The gates in the circuit $C_{m,k'}$ are connected as follows:

(i) For $s = 1, \ldots, t$, the gate $c[2s, i]$ receives input from each of the gates $b[2s-1, j]$ for $j = 1, \ldots, n''$.

(ii) For s odd, $2 \le s \le t$, the gate $b[2s-1, i, j]$ computes the boolean expression $c[2(s-1), j] \wedge a_X[s, i, j]$.

(iii) For s even, $2 \le s \le t$, the gate $b[2s-1, i, j]$ computes the boolean expression $c[2(s-1), j] \vee \neg a_X[s, i, j]$.

(iv) The gate $b[1, i, j]$ computes the boolean expression $a_Y[j] \wedge a_X[1, i, j]$.

Note that since the gates $b[*, *, *]$ are small, the simulation preserves weft. The following claim establishes that the circuit $C_{m,k'}$ works correctly.

Claim 1 *For all* $y \in \Sigma^{n'}$ *of weight* k', $C_{m,k}(\chi(C_x) \cdot y) = 1$ *if and only* $C_x(y) = 1$.

Claim 1 is easily proved by induction on the levels of the circuit simulation.

An essentially identical argument handles t odd, $t \ge 3$. The case of $t = 1$ presents additional difficulties and must be handled as a special case. (The simulation above would result in universal circuits of weft 2.)

It suffices to show a "universal" family of circuits for the $W[1]$-complete problem INDEPENDENT SET. What we want is a weft 1 circuit that takes as input the concatenation of two strings x and y, where x describes a graph of order n and y represents the candidate k-element independent set. We can accomplish this by having the first part of the input $x = (x[1, 2], x[1, 3], \ldots, x[n-1, n])$ represent the adjacencies of G as a 0-1 string of length $\binom{n}{2}$, and letting $y = (y[1], \ldots, y[n])$ (the nondeterministic part of the input) have length n and weight specification k. The circuit can simply represent the boolean expression

$$C = \prod_{1 \le i < j \le n} \left(\neg x[i, j] \vee \neg y[i] \vee \neg y[j] \right).$$

The above arguments show that $W[t] \subseteq N[t]$. For the reverse inclusion, suppose L is a parameterized language in $N[t]$. Then, $\langle x, k \rangle \in L$, $|x| = n$, if and only if $\exists y \in \Sigma^{n'}$ of weight k', such that a nicely produced weft t circuit $C_{n'',k''}$ accepts xy. To exhibit a reduction from L to WEIGHTED CIRCUIT SATISFIABILITY for weft t, we may just take the image of the reduction to be $C_{n'',k''}$ with the first $n'' - n'$ inputs "removed" by being fixed to the value of x. $\qquad \square$

In a similar way, we can prove the following characterization of $W[P]$ (Exercise 28.2.2).

Theorem 28.1.2 (Downey, Fellows, and Regan [256]) $W[P] = \langle N[P] \rangle$.

Let Σ_d^P stand for the class of languages recognized by depth-d unbounded fanin boolean circuits of polynomial size having a single *or* gate at the output, as described in the survey by Boppana and Sipser [106]. Let Π_d^P stand for the complements of these languages, which are recognized by depth-d circuits with an *and* gate at the output. Sipser [620] showed that for all $d \geq 1$, $\Sigma_d^P \neq \Pi_d^P$. It is not surprising that this carries over to the parameterized setting to show that the $G[t]$-hierarchy is proper. We prove that this carries over to the nondeterministic case.

Theorem 28.1.3 (Separation Theorem—Downey, Fellows, and Regan [256]) *For all $t \geq 1$, $N[t] \subset N[t+1]$.*

Proof Suppose $N[t] = N[t+1]$, and let A_0 be a language in Π_{t+1}^{poly}. Define a simple parameterized language A by $A = \{\langle x, k \rangle : x \in A_0\}$. Then, $A \in G[t+1] \subseteq N[t+1]$. By our supposition, $A \in N[t]$. By the definition of $N[t]$, there exists a parameterized language $B \in G[t]$ accepted by a bi-indexed family of circuits $\mathcal{F} = \{C_{n,k}\}$ such that we have for all x and any fixed integer k_0,

$$
\begin{aligned}
x \in A_0 \quad &\Longleftrightarrow \quad (x, k_0) \in A \\
&\Longleftrightarrow \quad \left(\exists y \in \{0,1\}^{n'} \right) \left[\text{weight}(y) = k_0' \wedge \left(xy, k_0'' \right) \in B \right] \\
&\Longleftrightarrow \quad \left(\exists y \in \{0,1\}^{n'} \right) \left[\text{weight}(y) = k_0' \wedge C_{n'', k_0''}(xy) = 1 \right].
\end{aligned}
$$

Here, again, the priming indicates that n and the fixed k_0 are part of nice parametric connections, with $n'' = |xy| = n + n'$.

Using $C_{n'', k_0''}$ as a building block, we can create a circuit $\tilde{C}_{n'', k_0''}$ that evaluates $C_{n'', k_0''}(xy)$ for *all possible* y, with an output $\vee$ gate on all these possibilities. There are $\binom{n'}{k_0'}$ possible y, but this is permitted since k_0' is a constant. The family of circuits constructed from $\mathcal{F}$ in this way over all n show that $A_0 \in \Sigma_{t+1}^{\text{poly}}$, contradicting the fact that Π_d^{poly} is not contained in Σ_d^{poly}, for all d. $\square$

The above theorem does not prove, of course, that the $W[t]$-hierarchy is proper. If we could prove that, then we would have $P \neq NP$. We would need something like $N[t] \not\subseteq G[P]$ for that. What it does show is that any "normal" approach of the kind often employed in the study of the W classes, namely the use of additional (bounded-weight) nondeterminism, will necessarily fail. For example, to show that $W[t+1]$ collapses to $W[t]$, we might hope to design some sort of gadgetry whose operation can be described by a weft t circuit C' that would correctly verify that a circuit C of weft $t+1$ accepts a particular weight k input vector x on the basis of some additional $k' \log n$ bits of nondeterministic information. Collapse would then follow by using C' to process *two* guesses: the input x to C and the "proof" that $C(x) = 1$. Since x has bounded weight and the size of C' can involve a blowup in size of $f(k)n^{g(t)}$ for $|C| = n$ and arbitrary functions f and g, we might well believe that there is some hope for this project. However, if this program were to succeed, then we would, in fact, have shown that $G[t + 1] \subseteq N[t]$. By the following easy but important proposition, in which the transitivity of parametric connections enables us to "coalesce" two like quantifiers into one, we would then have $N[t + 1] \subseteq N[t]$, contradicting the Separation Theorem.

Lemma 28.1.1 *Let $\mathcal{F}$ be any class of parameterized languages. Then, $\exists \cdot \exists \cdot \mathcal{F} = \exists \cdot \mathcal{F}$.*

Although the parameterization of A_0 in the proof of the Separation Theorem is trivial by itself, the manner in which the parameter interacts with the definition of $\exists \cdot$ Uniform $G[t]$ and with the switch between Σ_d and Π_d circuits is noteworthy, and, overall, the information in the theorem seems surprisingly good. It lends support to the conjecture that the $W[t]$-hierarchy is proper.

The classes $N[t]$ are defined by a single bounded-weight existential quantification. It is natural to consider corresponding classes defined by universal and by alternating bounded-weight quantification. In this way, we would expect to obtain analogs of the classes $AW[t]$ of Chap. 26.

Definition 28.1.5 For each $t \geq 1$, define $\Sigma_1[t] = W[t] = \langle \exists \cdot \text{Uniform } G[t] \rangle$. Correspondingly, define $\Pi_1[t] = \langle \forall \cdot \text{Uniform } G[t] \rangle$. For $i \geq 2$, define $\Sigma_i[t] = \langle \exists \cdot \Pi_{i-1}[t] \rangle$ and $\Pi_i[t] = \langle \forall \cdot \Sigma_{i-1}[t] \rangle$. Define $\Sigma_0[t] = \Pi_0[t] = \langle G[t] \rangle = \text{FPT}$. Finally, for each t, define $H[t]$ to be the union of these classes, viz.

$$H[t] = \bigcup_{i=0}^{\infty} \Sigma_i[t] \cup \Pi_i[t].$$

As one would expect, the $\Pi_i[t]$ classes consist of the complements of parameterized languages in the $\Sigma_i[t]$ classes. Moreover, by the methods of Theorem 23.2.1 and induction on i, it follows that the Σ_i-quantified analog of WEIGHTED t-NORMALIZED SATISFIABILITY is complete for $\Sigma_i[t]$. The proof of Theorem 28.1.2 can easily be modified to prove the following (Exercise 28.2.3):

Theorem 28.1.4 $AW[t] = \langle G[t] \rangle$.

The next theorem shows that in contrast to the proper inclusions of the $N[t]$-hierarchy, the $H[t]$-hierarchy collapses to $H[1]$. This result constitutes a significant improvement of Theorem 26.3.1, where it is shown that $AW[*] = AW[2]$. From the following, we have as an immediate corollary that $AW[*] = AW[1]$.

Theorem 28.1.5 (Tradeoff Theorem—Downey, Fellows, and Regan [256]) *For all $t \geq 1$, $H[t] = H[1]$.*

Proof By induction, it suffices to show that $H[t] \subseteq H[t-2]$, for t odd. Let $L \in \Sigma_s[t]$. We argue that $L \in \Sigma_{s+2}[t-2]$. By the above remarks, L is reducible to the Σ_s-quantified version of WEIGHTED t-NORMALIZED SATISFIABILITY. Accordingly, let E be a boolean expression over a set of variables V that is partitioned into sets $V = V_1 \cup \cdots \cup V_s$ with $V_i \cap V_j = \emptyset$ for $1 \leq i < j \leq s$, such that E has the form

$$E = \prod_{i=1}^{m} \sum_{j=1}^{m_i} \prod_{k=1}^{m_{ij}} E[i, j, k],$$

where $E[i, j, k]$ is a literal if $t = 3$, and is otherwise a weft $t-3$ expression that is a large $\vee$ of weft $t-2$ subexpressions. Let $(k_1, \ldots, k_s)$ be a sequence of positive integers. The quantified satisfiability question for E is whether

$\exists$ a weight k_1 truth assignment to the variables of V_1, such that

$\forall$ weight k_2 assignments to the variables of V_2,

$\ldots$,

such that E is satisfied.

Now we describe an expression E' over a set of variables

$$V' = V \cup V_\forall \cup V_\exists,$$

where

$$V_\forall = \{a[i] : 1 \leq i \leq m\}$$

and

$$V_\exists = \{e[i, j] : 1 \leq i \leq m, \ 1 \leq j \leq m_i\},$$

such that the answer to the quantified satisfiability question for E is "yes" if and only if the answer to the quantified satisfiability question for E' is "yes." The latter is defined to hold if and only if

$\exists$ a weight k_1 assignment to V_1, such that

$\forall$ weight k_2 assignments to V_2,

$\ldots$,

$\forall$ weight 1 assignments to $V_\forall$,

$\exists$ a weight 1 assignment to $V_\exists$,

such that E' is satisfied.

The expression E' is described by $E' = E_1' \cdot E_2'$, where the two factors are

$$E_1' = \prod_{i=1}^{m} \prod_{j=1}^{m_i} (e[i, j] \rightarrow a[i])$$

and

$$E'_2 = \prod_{i=1}^{m} \prod_{j=1}^{m_i} \prod_{k=1}^{m_{ij}} (E[i, j, k] \vee \neg e[i, j]).$$

For $t > 3$, since $E[i, j, k]$ is a large logical sum of subexpressions and has weft $t - 3$, the same is true for $(E[i, j, k] \vee \neg e[i, j])$, and therefore E' has weft $t - 2$. If $t = 3$, then E' is a product of sums of size 2 and, thus, has weft 1. The verification that the construction works correctly is straightforward and is left to the reader. ☐

This proof does not tell us whether $\Sigma_s[t]$ is equal to $\Sigma_{s+2}[t - 2]$, and, in general, we do not know exactly how the hierarchies $H[t]$ intercalate for different t.

We can easily follow the analogy with NP and PH for other complexity results and classes. One such class could be generated by the result of Valiant and Vazirani: for randomized reductions, NP is the same detecting unique solutions. We could then look at analogs of this result. Downey *et al.* proved one such result in Downey, Fellows, and Regan [256]. One might then consider the question of whether the W-hierarchy might collapse under *randomized* reductions.

28.2   Exercises

Exercise 28.2.1 Prove that Uniform $G[P] = \text{FPT}$.
(Hint: If a parameterized language L is in Uniform $G[P]$, then membership of $\langle x, k \rangle$ in L, $|x| = n$, can be decided in the right amount of time $O(n')$ by generating the circuit $C_{n,k}$ and evaluating it on input x. The converse also holds by imitating the usual proof that languages in P have polynomial-sized circuits.)

Exercise 28.2.2 Prove that $W[P] = \langle N[P] \rangle$.
(Hint: Consider the defining problem WEIGHTED CIRCUIT SATISFIABILITY and the fact that reductions may be folded into circuits.)

Exercise 28.2.3 Prove $AW[t] = \langle G[t] \rangle$.

Exercise 28.2.4 (Downey and Fellows [243]) As another analog to classical notions, we define the D-hierarchy, $D_p[t]$, as the languages that can be expressed as the intersection of a language in $W[t]$ and one whose complement is in $W[t]$. Prove that the following problem is in $D_p[2]$.

UNIQUE WEIGHTED CNF SATISFIABILITY
Instance: A boolean CNF formula X and an integer k.
Parameter: A positive integer k.
Question: Is there a unique weight k satisfying assignment for X?

(Hint. It suffices to describe how to say that a CNF expression has at least two satisfying expressions in $W[2]$. Let C be the circuit corresponding to X. Take two copies of C. Add $O|X|$ many gates to express the fact that the first copy of C has a satisfying assignment different from the second. Now, for k, choose q and r appropriately and take q copies of the left circuit and r copies of the right. Add new gates to express the fact that the inputs of the left q must all be equal and the fact that the inputs of the right r must all be equal. Now, accept C if the new circuit has a weight $(q+r)k$ accepting input. Then, for the correct choice of (q, r), depending only on k, C has two or more accepting inputs iff the new circuit has one of weight $(q+r)k$.)

28.3 Historical Notes

All of the results of this chapter are taken from Downey, Fellows, and Regan [256].

28.4 Summary

We give a uniform circuit basis for the W-hierarchy. The Tradeoff Theorem is proven.

Part VI
Approximations, Connections, Lower Bounds

The M-Hierarchy, and XP-Optimality

29

29.1 Introduction

This chapter is one of the most important in the book. It shows how the methods used to obtain algorithms we have encountered so far can be shown to be more or less optimal. The authors see this as one of the most important outgrowths of the parameterized complexity story, and the true fruition of the story articulated in, for example, Downey, Fellows, and Stege [257].[1]

In this chapter we introduce the M-hierarchy, which intercalates with the W-hierarchy, to the best of our present knowledge. The overall picture is expressed in the tower of parameterized complexity classes:

$$\text{FPT} = M[0] = W[0] \subseteq M[1] \subseteq W[1] \subseteq M[2] \subseteq W[2] \subseteq \cdots$$

The M-hierarchy provides us with a very fine tool to explore the qualitative optimality of FPT and XP algorithms.

29.2 The W-Hierarchy and SUBEXP Time

The reason for the wide applicability of the M-hierarchy is the fact that there turns out to be a very close connection between the hierarchy and classical complexity. This connection was first noticed by Abrahamson, Cai, Chen, Downey, and Fellows. Consider the following question:

"Would a collapse of the W-hierarchy imply anything unexpected in classical complexity?"

One might hope that $W[P] = \text{FPT}$ *might* imply $P = NP$, which would be very strong evidence that $W[P] \neq \text{FPT}$. Unfortunately, we cannot prove this implication, and in the [247], the authors construct an oracle A relative to which $P \neq NP$ and

---

[1]Known to the authors as "son of feasible" as reported in Downey [237].

R.G. Downey, M.R. Fellows, *Fundamentals of Parameterized Complexity*,
Texts in Computer Science, DOI 10.1007/978-1-4471-5559-1_29,
© Springer-Verlag London 2013

yet $W[P] = \text{FPT}$. This oracle construction would seem to give at least some evidence that the implication does not hold, and in fact, if it does, then it will require nonrelativising techniques to prove it.

In any case, purely on general grounds we believe that all we could hope for is some sort of "*smaller*" collapse than $P = NP$. To express our results, we use the *Limited Nondeterminism* classes introduced by Kintala and Fischer [444]. The computation model is a Turing machine.

Definition 29.2.1

(i) (Kintala and Fischer [444]) Let $\text{NP}[f(n)]$ denote the class of decision problems that are solvable in P-time with an algorithm using only $f(n)$ bits of nondeterminism.

(ii) (Abrahamson, Downey, and Fellows [2, 3]) We define the class SUBEXP-TIME$(f(n))$ to be the set of languages L for which there is a polynomial $p(n)$ such that L is accepted in $\text{DTIME}(p(n)2^{g(n)})$ for some function g in $o(f(n))$ (that is, $\lim_{n \to \infty} f(n)/g(n) = \infty$).

We have the following result connecting parameterized and classical complexity.

Theorem 29.2.1 (Abrahamson, Downey, and Fellows [3]) *For the reducibility $\leq_T^s$, and hence for the standard reducibility, $W[P] = \text{FPT}$ iff for every P-time function f with $f(n) \geq \log n$, there is a recursive function h such that for every $L \in \text{NP}[f(n)]$, $h(L)$ computes machine M which witnesses that $L \in \text{SUBEXPTIME}(f(n))$. [We say that $\text{NP}[f(n)]$ is constructively a subset of $\text{SUBEXPTIME}(f(n))$.]*

Corollary 29.2.1 (Abrahamson, Downey, and Fellows [3]) *For $\leq_T^s$, $W[P] = \text{FPT}$ implies that for all P-time f, $\text{NP}[f(n)] \subseteq \text{SUBEXPTIME}(f(n))$.*

Proof ($\Rightarrow$) Suppose $W[P] = \text{FPT}$. Then WEIGHTED CIRCUIT SATISFIABILITY is solvable by some procedure Φ in time $g(k)p(n)$, where p is a polynomial and g is some function. We can assume that g is strictly increasing and that it grows rapidly, since any function can be bounded above by such a function.

Let $f(n) \geq \log n$ be polynomial-time computable, and let α be a nondeterministic polynomial-time algorithm that uses at most $f(n)$ bits of nondeterminism on inputs of length n. The following algorithm β simulates α, and we will show that β runs in time $q(n)2^{o(f(n))}$ for a polynomial q. For $k < m$, let $E_{m,k}$ be a canonical list of the size k subsets of $\{1, \ldots, m\}$. If s is member of $E_{m,k}$, let $N(s)$ be the index of s in list $E_{m,k}$. List $E_{m,k}$ can be chosen so that $N(s)$ can be computed in polynomial time in m. Let $[N(s)]_d$ be the d least significant bits of the binary representation of $N(s)$.

Algorithm 29.2.1 (Algorithm $\beta(x)$)
$n \leftarrow |x|$.
$k \leftarrow f(n)/\log \log n$.

loop

$m \leftarrow \lceil k2^{f(n)/k} \rceil$.

$C \leftarrow$ a circuit with m inputs, and the following behavior.

On an input vector v of weight k, C considers v to encode a size k subset s_v of $\{1, \ldots, m\}$, and computes $n_v = [N(s_v)]_{f(n)}$.

C then simulates α on input x with guess n_v, and produces the same output as α does with that guess.

Run Φ on input $\langle C, k \rangle$ for at most $p(n)2^{f(n)/\log\log n}$ steps.

If Φ terminates within the allotted time, stop and give the same answer that it gave.

$k \leftarrow k - 1$

end loop

Since $\binom{m}{k} \geq (m/k)^k = [\lceil k2^{f(n)/k} \rceil / k]^k \geq 2^{f(n)}$ at each iteration, all $2^{f(n)}$ binary strings of length $f(n)$ can be passed as guesses to α, when all weight k inputs are tried. It is evident, therefore, that if $\beta(x)$ terminates, then it gives the same answer that $\alpha(x)$ does. We show that $\beta(x)$ terminates within time $q(n)2^{o(f(n))}$.

Variable m is largest for small values of k, so the last loop iteration costs the most. Since Φ terminates in time $g(k)p(n)$, the last iteration is the first one where $g(k) \leq 2^{f(n)/\log\log n}$; that is, in the last iteration, $k = g^{-1}(2^{f(n)/\log\log n})$. [Here, for an increasing function $z(n)$, we define $z^{-1}(n)$ to be the largest m with $z(n) \leq m$.] (Note that the initial value of k is larger than that for sufficiently rapidly growing functions g.) Considering this smallest value of k to be a function $k(n)$ of n, observe that $\lim_{n\to\infty} k(n) = \infty$.

The initial value of k is computable in polynomial time in n. The cost of producing circuit C is polynomial in m and n. But $m(n) = k(n)2^{f(n)/k(n)}$ is in $2^{o(f(n))}$, so any polynomial of m is also $2^{o(f(n))}$. The simulation of Φ is bounded in time by $p(n)2^{o(f(n))}$, so each iteration of the loop uses time at most $q(n)2^{o(f(n))}$ for some polynomial q. Since there are at most $f(n)/\log\log n$ iterations, the entire algorithm runs within time $q(n)2^{o(f(n))}$.

To complete this half of the proof, note that β is computable from α.

($\Leftarrow$) Suppose that NP$[f(n)]$ is constructively a subset of SUBEXPTIME$(f(n))$ for every polynomial time function $f(n) \geq \log n$. Let $f_k(n) = k \log n$.

For each k, let WCS$_k$ be the restriction of WEIGHTED CIRCUIT SATISFIABILITY to inputs of the form (x, k). Then, WCS$_k$ is in NP$[f_k(n)]$. Let α_k be a deterministic algorithm that solves WCS$_k$ in time $p(n)2^{f_k(n)/g(n)} = p(n)n^{k/g(n)}$, where p is a polynomial and $\lim_{n\to\infty} g(n) = \infty$. By the constructive nature of the supposition, α_k is a recursive function of k. So, to solve WEIGHTED CIRCUIT SATISFIABILITY on input (x, k), compute α_k, and run it on input (x, k). Call the resulting algorithm β.

Algorithm β solves WEIGHTED CIRCUIT SATISFIABILITY in time $h(k)p(n) \cdot n^{k/g(n)}$ for some function h. For $k < g(n)$, this time is $h(k)q(n)$ for a polynomial q. For $k \geq g(n)$, the time is bounded by a function of k. So, there is a function h' such that algorithm β runs in time $h'(k)q(n)$. $\square$

Notice that we only explicitly use the fact that we can compute $k(n)$ as a function of n to get the constructivity conclusion in the $\Rightarrow$ part and similarly only use constructivity in the other half to get *strong* uniformity. In both cases, the same proof will show that

Corollary 29.2.2 *For the reducibility* $\leq_T^u$, $W[\mathrm{P}] = \mathrm{FPT}$ *iff for all* P-*time* $f \geq \log n$, $\mathrm{NP}[f(n)] \subseteq \mathrm{SUBEXPTIME}(f(n))$.

The theorems above say nothing about the classes $W[t]$, and, in particular, at least till the next section, the class $W[1]$ which is our current measure of intractability. We will see in the next section that this class and "log"alternating Turing machines are closely related. We remark that all of the results we have seen regarding $W[t]$ collapse seem to be model dependent since the definition of the class $W[t]$ itself comes from restricted logical depth of formulas.

We finish this section with one result, the ideas of the proof of which are recycled and miniaturized later. The material of the next section will use a class $M[1]$ which enables quite tight lower bounds on running times, but relies on a complexity assumption apparently stronger than the "gold standard" $W[1] \neq \mathrm{FPT}$ so the following is of independent importance as it only concerns the W-hierarchy.

Definition 29.2.2 (Abrahamson, Downey, and Fellows [3]) Let C be a class of decision circuit problems. We shall say that C is *nearly polynomial time* if there is an algorithm Φ such that for each $C \in C$, Φ solves $\langle C, k \rangle$ in time $p(n)2^{o(v)}$, where $p(n)$ is a polynomial and v is the number of inputs to C.

Theorem 29.2.2
(a) (Abrahamson, Downey, and Fellows [3]) for t even, and Downey and Fellows [247] for t odd) *For* $t > 0$, $W[t] \subseteq \mathrm{FPT}$ *implies that* $\mathrm{SAT}[t]$ *is nearly polynomial time.*
(b) (Abrahamson, Downey, and Fellows [3]) (i) $W[\mathrm{SAT}] \subseteq \mathrm{FPT}$ *implies* SAT *is nearly polynomial time.*
   (ii) $W[\mathrm{P}] \subseteq \mathrm{FPT}$ *implies* WEIGHTED CIRCUIT SATISFIABILITY *is nearly polynomial time.*

Proof We prove (a), the others follow by the same method (Exercise 29.11.1).
   Case 1. Even weft.
   Suppose that $W[2t]$ is FPT. Then, there is a $f(k)p(n)$ time algorithm for solving WEIGHTED MONOTONE $2t$-NORMALIZED SAT on instances $\langle x, k \rangle$ with $|x| = n$, for some polynomial $p(n)$ and arbitrary recursive function $f(k)$. We assume that $f(k) > k$, and as in the previous proof, we let $f^{-1}(k)$ denote the largest y with $f(y) \leq k$. Furthermore, we can assume that $f(k)$ is computable in time $f(k)$, so that we can easily see if $f(k) \geq y$.
   Now, let X be a member of $\mathrm{SAT}[2t]$, that is, a $2t$-normalized boolean formula, and consider it as a circuit with $2v$ input lines for the variables and their complements. We construct a pair $\langle C, k \rangle$ with C a circuit and k an integer to be chosen later. We ensure that C is a $2t$-normalized monotone circuit and that X is satisfiable iff C has a weight k satisfying assignment. The reader should note that the reduction will *not* be a parameterized one. Let $s = 2^{\lceil v/k \rceil}$. The circuit C has four distinct parts:

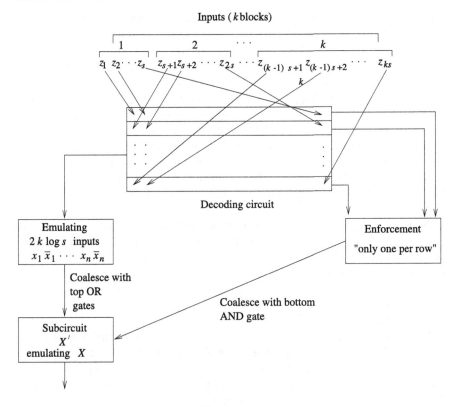

Fig. 29.1 The Even case of Theorem 29.2.2(a)

(1) A decoding circuit E with ks input lines from variables $\{z_1, \ldots, z_{ks}\}$.
(2) A circuit X' emulating the action of X.
(3) An enforcement circuit connected to the ks inputs.
(4) An additional *and* gate taking the outputs of X' and the enforcement circuit. The single output of this *and* gate is the output for the circuit. The reader is referred to Fig. 29.1.

The ks input lines of the encoding circuit can be considered a $k \times s$ matrix. The encoding circuit translates each row of the input into two binary numbers. Each binary number represents the position of the *single* 1 in that row. (The enforcement circuit will be weft 1 and will express the fact that *the input matrix has at most one 1 per row*. Since we will be looking for weight k inputs to C, this will ensure that there is exactly one 1 per row.) Each binary number has length $\log s$ and the collection of possible binary strings of length $\log s$ will be in exact one-to-one correspondence with the position of the 1 on the row. The binary numbers are complements of each other (meaning that whenever one is 1, the other is 0). One corresponds to negated variables and one corresponds to variables. This can all be easily achieved disjunctively using $2\log s$ large *or* gates. For instance, for the positive variables, suppose the binary number is to be represented by $x_1 \ldots x_{\log s}$. We then have one *or* gate corresponding to x_1. This is connected to every second position on the row. Thus, x_1 will be 1 iff the 1 on the row falls on an even square. (Similarly, if we want the negation of x_1 in the other binary string, we use all the odd positions.) For x_2, we have a large *or* attached to positions $4n + 3$ and $4n + 4$ of the row for each n. In general, we use $2^j n + 2^{j-1}n + 1, \ldots, 2^{j+1}n$ for each n for x_j. The idea here is that the $2v$ inputs to X are chopped into k equal pieces (sequentially), each of length $\log s$. We note that each row has length s, so we will indeed be able to create a total of $2k \log s = 2k \times \log 2^{\lceil v/k \rceil} \geq 2v$ lines in this way. It is clear that we can use these lines as the inputs to the circuit X'. Circuit X' is the same as X, except that each literal is replaced by one of the $2v$ output lines. Clearly, the circuit can be massaged to be $2t$-normalized by coalescing the conjunction at the output of X' with the bottom *and*, and the disjunctions of the encoding circuit with those of the upper level of X'.

Now, to complete the proof, for a given input with v variables, we choose $k = f^{-1}(2^{v/\log v})$. Note that $f(k) < o(v)$. This means that the $f(k)$ factor in the solution size is acceptable. The size of C is P-time in X and ks. As k is an unbounded function of v, we see that $s < 2^{o(v)}$. The running time of Φ on C is $f(k)p(|C|)$. This is $2^{o(v)}p(|X|.2^{o(v)})$. But this quantity is $p(|X|).2^{o(v)}$, giving the desired result.

Case 2. Odd weft ≥ 3.

For this case, we use the trick from Step 6, Case 2 of the Normalization Theorem, Theorem 23.2.1. Let $t \geq 1$, and this time, let X be a weft $2t + 1$-normalized circuit and let $A_1, \ldots, A_m$ list the *And* gates at the top layer, below the inputs. For each output line from the A_i, introduce two inverters, and using De Morgan's laws, draw the top inverter past A_i to make it a (large) *Or* gate. Now, apply the process of Case 1 to the resulting equivalent circuit.

Case 3. Weft 1.

This time let X be a weft 1 circuit representing a 3CNF formula. The trouble with the above is that the enforcement gadget is a collection of *large Or* gates. If we use the process of Case 3, we would convert a weft 1 circuit into a weft 2 circuit. To get around this difficulty, we replace the large gates by a collection of small gates using the fact that the clause size is 3.

Again, we will have ks input variables in k blocks of size s. Again we will have the enforcement gadget of Cases 1 and 2. (This was in normalized weft 1.) Again, we will think of positions of the (blocks) rows of the input as encoding input variables or their negations to the circuit X.

However, now we will not explicitly have the circuit X' emulating X directly. Rather, we will have a new circuit or weft 1 taking the ks inputs directly. At the bottom, we will have a large *And* gate A. Suppose that $C = \{x_i, x_j, x_k\}$ is a clause of X. Now, we wish to wire into the circuit all the possibilities that make C *false*. Note that the literals can correspond to at most three blocks of input variables. If all the literals of C come from the same block B, then for each z_m in block B which corresponds to the placement of a 1 in the block corresponding to negating all the literals in clause C, we simply have a line passing through an inverter. This line directly inputs to gate A. Suppose that the literals come from more than one block, and, without loss of generality, from three blocks, B_i, B_j, and B_k. (Here we suppose that x_d is set in block B_d.) Then, for each setting of three input variables z_{d_p} from block B_p, $d_p \in \{i, j, k\}$, one z_{d_p} per block, which make the clause *false*, create a gate expressing the fact that $\overline{z_{d_i} \wedge z_{d_j} \wedge z_{d_k}}$. Run the gate's output directly to gate A. This is a small *or* gate. We create at most $O(|X|^3)$ gates in this way. The circuit we get is satisfiable by a weight k input iff X is satisfiable. The remaining reasoning goes through as before.   □

Since $v < |X|$, the above implies the following corollary:

Corollary 29.2.3 (Downey and Fellows) *If* $W[1] = \text{FPT}$, *then size n*

$$3\text{CNF SATISFIABILITY} \subseteq \text{DTIME}\left(2^{o(n)}\right).$$

It would seem unlikely that $3\text{CNFSAT} \subseteq \text{DTIME}(2^{o(n)})$ if $P \neq NP$, and so this seem to provide some evidence that $W[1] \neq \text{FPT}$.

29.3 The M-Hierarchy

Liming Cai and Jianer Chen [120] realized that the W-hierarchy was related to logarithmically bound versions of the query complexity to a nondeterministic machine. This idea was taken a step further by Liming Cai and David Juedes who realized that what we now call XP-optimality was closely related to something called the *Exponential Time Hypothesis*, and to Downey, Estivill-Castro, Fellows, Prieto, and Rosamond [238] who realized the close connection to *parametric miniatures*.

The key class for these results is one called $M[1]$. There are two routes to the notion. The first is by allowing a logarithmic factor in the solution size of the problem. This approach allows for extensions of the notion of $M[1]$ to higher levels. Consider for example the problem.

$k \log n$ MINI-CIRCSAT

Input: Positive integers k and n in unary, and a Boolean circuit C of total description size n.

Parameter: k.

Question: Is there any input vector x of weight $\leq k \log n$ such that $C(x) = 1$?

Similar problems can be defined based on any of the usual suspects. By controlling the weft of the boolean circuit (and so getting a member of $H[t]$) we get the classes for first defined for $t \geq 2$ by Flum and Grohe [311], and going back to work of Cai and Juedes [126, 127], and as we see below somewhat differently by Downey et al. [238].

Definition 29.3.1 (Flum and Grohe [311]) The class $M[t] =_{\text{def}} k \log n$ MINI-$W[t]$ is the class of problem FPT-reducible to the following core problems.

$k \log n$ MINI-$H[t]$

Input: Positive integers k and n and m, and a weft t Boolean circuit C
                   with $k \log n$ variables and of total description size m.
Parameter: k.
Question: Is there any input vector x such that $C(x) = 1$?

Similarly we can define miniaturizations of any of the standard classes, in particular $M[\text{SAT}]$ and $M[\text{P}]$. It is not difficult to show that this hierarchy interweaves with the W-hierarchy.

Theorem 29.3.1 (Downey et al. [238] plus Flum and Grohe [311])

$$\text{FPT} \subseteq M[1] \subseteq W[1] \subseteq M[2] \subseteq \cdots \subseteq W[\text{SAT}] = M[\text{SAT}] \subseteq W[\text{P}] = M[\text{P}].$$

Proof The proof is a miniaturization of the $k \log n$ trick of Abrahamson, Downey, and Fellows [3] for $W[\text{P}]$. Another example of this is soon seen in Corollary 29.4.1. Of course this trick gives the last two immediately. We leave this proof to the reader in Exercise 29.11.2.   □

Of course we have a new agenda for defining all small classes of classical problems. For example $k \log n$ VERTEX COVER simply asks if there is a vertex cover of a graph G with n vertices of size $k \log n$. For any such problem can easily solve this in time $O(2^{O(k \log n)}) = O(n^k)$ by complete search, and even better using the arsenal of VERTEX COVER algorithms we have. However, this begs the question: is this problem FPT? It would seem not.

The reader might well ask: *Given the large number of hierarchies we have so far, why do we need another one?* The answer comes from the very bottom class, $M[1]$. This has close relationships with approximation and allows for tight lower bounds as we see in the next section where we use an alternate route to the class.

29.4   Parametric Miniatures

It turns out that there is an alternative approach to defining $M[1]$. The other approach is to parameterize the *size* of the problem, i.e. the whole problem itself, and ask for a solution.[2]

---

[2]Many of the key ideas in this were implicit, in some sense, in [126, 127], we are following the route of Downey et al. [238] which contained the definition of MINI-$W[1]$ and of miniatures of NP problems, many combinatorial reductions, especially Theorem 29.4.2, the Turing counting trick, and the inclusion of MINI[1] in $W[1]$. The programme was initiated by Cai and Judes [127].

MINI-CIRCSAT$_t$

Input:	Positive integers k and n in unary, and a weft t Boolean circuit C of total description size at most $k \log n$.
Parameter:	k.
Question:	Is there any input vector x such that $C(x) = 1$?

The price for this simplicity is a seemingly weaker intuitive reference point, codified in the following conjecture.

Conjecture 29.4.1 *There is no algorithm with running time $2^{o(n)}$ that determines, for a weft 1 Boolean circuit C of total description size n, whether there is an input vector x such that $C(x) = 1$.*

Circuits are also unstructured and opaque, and it is hard to imagine radically improving on the brute force approach of trying all possible inputs. In any case, no one knows how to do this. In any case, as we remarked earlier, in the first version of parameterized complexity we used circuits as a base. It was only later we found the reductions to and from SHORT NTM ACCEPTANCE.

Note that n is not the number of inputs to the circuit C, but the total size of a description of C. We will refer to the problem MINI-CIRCSAT$_1$ as CIRCUIT SAT. We will need the following simple algebraic facts whose proof is left to the reader.

Lemma 29.4.1 *The following functions are* FPT:
(a) $2^{o(k \log n)}$;
(b) $(\log n)^k$.

Determining if MINI CIRCUIT SAT is a "yes" instance by trying all possible inputs gives a brute force $O(n^k)$ algorithm. Now we have the crucial foundational lemma, essentially due to Cai and Juedes [127].

Theorem 29.4.1 (Cai and Juedes [126, 127]) MINI-CIRCSAT *is fixed-parameter tractable if and only if the conjecture below fails.*

Exponential Time Hypothesis (ETH—Impagliazzo, Paturi, and Zane [411]) *There is no algorithm with running time $2^{o(n)}$ that determines, for a weft 1 Boolean circuit C of total description size n, whether there is an input vector x such that $C(x) = 1$.*

Proof [3] One direction follows from Lemma 29.4.1(a). In the other direction, suppose we are given a weft 1 Boolean circuit C of size N, and suppose that MINI-CIRCSAT is solvable in FPT time $f(k)n^c$. Take $k = f^{-1}(N)$ and $n = 2^{(N/k)}$. Then, of course,

---

[3]The reader may note that there are in effect three theorems being proven here, depending on the level of uniformity in the notion of $M[1]$. This is kind of important below when we prove lower bounds. For the strongest results, we will be assuming the *nonuniform* version that $M[1] \notin$ nonuniform FPT.

$N = k \log n$. For example, if $f(k) = 2^{2^k}$ then $f^{-1}(N) = \log \log N$. In general, $k = f^{-1}(N)$ will be some slowly growing function of N, and therefore $N/k = o(N)$, and also $cN/k = o(N)$ since c is a constant, and furthermore by trivial algebra $cN/k + \log N = o(N)$. Using the FPT algorithm, we thus have a running time of

$$f\left(f^{-1}(N)\right)\left(2^{N/k}\right)^c = N2^{cN/k} = 2^{cN/k + \log N} = 2^{o(N)}$$

to analyze the circuit C. $\square$

> **Definition 29.4.1** (Downey, Estivill-Castro, Fellows, Prieto, and Rosamond [238]) MINI-$W[1]$ is the collection of problems FTP reducible to MINI-CIRCSAT.

The Exponential Time Hypothesis is usually stated for the 3CNF case, no subexponential time algorithm for n-variable 3SAT.

In a fascinating development of parameterized complexity, we discover that the "parametric miniatures" allow for some elegant and interesting combinatorics, much of which "recycles" some familiar combinatorics from the theory of NP-completeness.

It turns out that *many* (but not all) $k \log n$ *miniatures* of familiar NP-complete problems are MINI[1]-complete. The following are some examples that give us a way to complete our alternative foundational sketch.

MINI-3SAT

Input: Positive integers k and n in unary, and a 3SAT expression E of size at most $k \log n$.

Parameter: A positive integer k.

Question: Is E satisfiable?

One can similarly define MINI-SAT by not insisting on size 3 clauses.

MINI-VERTEX COVER

Input: Positive integers k and n in unary, a graph G of total description size at most $k \log n$, and a positive integer r.

Parameter: k.

Question: Does G have a vertex cover of size at most r?

One can similarly define MINI-INDEPENDENT SET. In fact, note that they are essentially the same problem (because here the parameter is *not* the size of the vertex sets).

Theorem 29.4.2 (Downey et al. [238]) MINI-SAT, MINI-3SAT, MIN-d-COLOR-ABILITY, MINI-VERTEX COVER, *and* MINI-INDEPENDENT SET *are all* MINI[1]-*hard.*

Proof The usual reductions (from NP-completeness theory given in Chap. 1) of CIRCUIT SAT to SAT to 3SAT are all (crucially, for our purposes here) *linear size* reductions. Applied to the miniatures, these are then FPT reductions. The usual reduction ("a triangle for every variable, and a triangle for every clause") of 3SAT to VERTEX COVER is also linear size. □

Our final step in this "easy alternative foundation" for parametric intractability is to reduce MINI-INDEPENDENT SET to the usual parameterized INDEPENDENT SET problem, where the parameter is the size of the independent set.

Theorem 29.4.3 (Downey et al. [238]) INDEPENDENT SET (*parameterized by the size of the independent set*) *is* MINI[1]-*hard.*

Proof Let $G = (V, E)$ be the miniature, for which we wish to determine whether G has an independent set of size r. Here, of course, $|V| \leq k \log n$ and we may regard the vertices of G as organized in k blocks $V_1, \ldots, V_k$ of size $\log n$. We now employ a simple but useful *counting trick* that can be used when reducing miniatures to "normal" parameterized problems. Our reduction is a Turing reduction, with one branch for each possible way of writing r as a sum of k terms, $r = r_1 + \cdots + r_k$, where each r_i is bounded by $\log n$. By Lemma 29.4.1(b) there are FPT-many branches. A branch represents a commitment to choose r_i vertices from block V_i (for each i) to be in the independent set.

We now produce (for a given branch of the Turing reduction) a graph G' that has an independent set of size k if and only if the miniature G has an independent set of size r, distributed as indicated by the commitment made on that branch. The graph G' consists of k cliques, together with some edges between these cliques. The ith clique consists of vertices in $1 : 1$ correspondence with the subsets of V_i of size r_i. An edge connects a vertex x in the ith clique and a vertex y in the jth clique if and only if there is a vertex u in the subset $S_x \subseteq V_i$ represented by x, and a vertex v in the subset $S_y \subseteq V_j$ represented by y, such that $uv \in E$.

Verification that the reduction works correctly is straightforward. □

A consequence of Theorem 29.4.2 is the following, which pre-dated the work of Flum and Grohe [311].

Corollary 29.4.1 (Downey et al. [238]) FPT $\subseteq$ MINI[1] $\subseteq W[1]$.

We conjecture that these containments are *proper.* We now have completed our excursion through an alternative foundational discussion, having arrived (one way or the other) at the point of having k-CLIQUE (trivially FPT equivalent to k-INDEPENDENT SET) available to use as a convenient starting point for parametric intractability demonstrations.

Everything so far seems to be too easy, but that is not necessarily a bad thing. There is a common impression that $W[1]$-hardness demonstrations are typically heroic, and that the foundations of the theory are cryptic and difficult. This only seems really true of the fundamental hierarchy results. However, for the $M[1]$ class, the (easy) results should encourage the use of this as a reasonable basis with easy proofs. It would be interesting to know if, for some unresolved parameterized problems, reductions from parametric miniatures (such as MINI-3SAT, etc.) might provide easier routes for intractability demonstrations.

However, there is one fundamental result which needs some heavy work to establish: not only are the miniatures above $M[1]$-hard, but they are $M[1]$-complete. This fundamental result will allow us to derive some sharp results about running times for problems, assuming ETH. The *proof* of the completeness result needs a difficult theorem called the Sparsification Lemma.

29.5   The Sparsification Lemma

29.5.1 Statement of the Lemma

This section is devoted to the proof of the following theorem. It says that the problem of d-CNF satisfiability is polynomial time Turing reducible to a collection of d-CNF formulas whose size is linear in the number of variables.

Theorem 29.5.1 (Sparsification Lemma of Impagliazzo, Paturi, and Zane [411]) *Let $d \geq 2$. Then there is a computable function $f : \mathbb{N} \to \mathbb{N}$ such that for each k and every instance φ of k-SAT, with n variables, there is a collection $\psi_1, \ldots, \psi_p$ of 2-CNF formulas such that*

1. $\beta = \bigvee_{i=1}^{p} \psi_i \equiv \varphi$.
2. $p \leq 2^{\frac{n}{k}}$.
3. $|\psi_i| \leq f(k) \cdot n$ *for all i.*

Moreover, the algorithm computing the ψ_i runs in time $2^{\frac{n}{k}} |\varphi|^{O(1)}$.

29.5.2 Using the Lemma for $O^*(2^{o(k)})$ Lower Bounds

Armed with the Sparsification Lemma, we can prove the following.

Theorem 29.5.2 (Cai and Juedes [127], Downey et al. [238]) MINI-3CNF SAT *is $M[1]$-complete under parameterized Turing reductions.*

Proof It is quite straightforward to prove that MINI-3CNF SAT $\in M[1]$ (Exercise 29.11.3).

For the hard direction, it suffices to show that $k \log n$ MINI d-CNF SAT is reducible to MINI-3CNF SAT. Let φ be an instance of $k \log n$ MINI d-CNF SAT with size $m, n' = k \log n$ variables so that $k = \lceil \frac{n'}{\log m} \rceil$. We have f depending on d and can construct $\beta = \bigvee_{i=1}^{p} \psi_i$ depending on these parameters given by the Sparsification

Lemma. We note

$$2^{\frac{k\log n}{k}} = \frac{k\log n}{2^{\lceil \frac{k\log n}{\log m}\rceil}} \leq m.$$

We have $p \leq m$, and β computed in time $m^O(1)$. Note that each ψ_i has at most $f(k) \cdot k\log n$ many clauses. Consequently there is a 3-CNF formula ψ_i' with at most $f(k) \cdot d \cdot k\log n$ many variables of length $O(f(k) \cdot k\log n \cdot m)$ such that ψ is satisfiable if and only if ψ' is satisfiable.

Therefore φ is satisfiable if and only if there is an i such that ψ_i' is satisfiable with $1 \leq i \leq p$. Moreover, $k_i' = \lceil \frac{|\mathrm{var}(\psi_i')|}{\log m}\rceil$ where $\mathrm{var}(\psi_i')$ denoted the variables in ψ_i', and hence $k_i' \leq O(\frac{f(k)\cdot d\cdot k\log n}{\log m}) \leq O(f(k) \cdot d \cdot k)$. Therefore we can decide if φ is satisfiable by querying the instances of ψ_i', m for $1 \leq i \leq p$ in MINI 3CNF SAT. $\qquad\square$

Similar methods show the following.

Corollary 29.5.1
1. *The following are all* $M[1]$ *complete under FPT Turing reductions*: MINI-SAT, MINI-3SAT, MIN-d-COLORABILITY, *and* MINI-INDEPENDENT SET.
2. *Hence neither* MINI VERTEX COVER *nor any of these can have a subexponential time algorithm unless the ETH fails.*

The original paper of Cai and Juedes also used similar methods to establish several other lower bound results assuming ETH.

Theorem 29.5.3 (Cai and Juedes [126, 127]) *Assuming ETH there is no* $O^*(2^{o(k)})$ *algorithm for* k-PATH, FEEDBACK VERTEX SET, *and no* $O^*(2^{\sqrt{k}})$ *algorithm for* PLANAR VERTEX COVER, PLANAR INDEPENDENT SET, PLANAR DOMINATING SET, *and* PLANAR RED–BLUE DOMINATING SET.

The proofs are similar and left to the reader in Exercise 29.11.4.

The original Impagliazzo, Paturi, and Zane paper used what they called *SERF* reductions.

Definition 29.5.1 (Impagliazzo, Paturi, and Zane [411]) A SERF Turing reduction (subexponential reduction family) from a parameterized L to $\widehat{L}$ is a Turing reduction A running such that
1. For each $\varepsilon > 0$, and instance $\langle x, k\rangle$ A runs in time $O(2^{\varepsilon k}|x|^c)$ (so it is a parameterized reduction, with controlled constant), and
2. For any query $\langle x', k'\rangle$ $A(\langle x, k\rangle)$ makes to $\widehat{L}$,
   (a) $|x'| = |x|^d$, and
   (b) $k' = \alpha k$.

The methods above show that the following proposition.

Proposition 29.5.1 *If* $L \leq_{\text{SERF}} \widehat{L}$, *then if* $\widehat{L}$ *has a subexponential algorithm so does* L.

There is a definite feeling of deja vu looking at the definition above. In fact Chen and Grohe [155] showed that there is an *isomorphism* between subexponential complexity theory and miniaturized complexity. The proof would take us a little far from the main line of this chapter.

The methods can be used to establish "FPT optimality" lower bounds on problems known to be FPT only "slightly superexponentially". By this we mean that the FPT algorithm is $O^*(f(k))$ where $f(k)$ is of the form $k^{O(k)} = 2^{O(k \log k)}$, such as CLOSEST STRING where we have an FPT algorithm based on a search tree of height k with branching factor k. The following problem provides a key technical starting point.

$k \times k$-CLIQUE

Input: A graph G over the vertex set $[k] \times [k]$.
Parameter: A positive integer k.
Question: Determine if G has a k element clique with exactly one element in each row.

Theorem 29.5.4 (Lokshtanov, Marx, and Saurabh [496]) $k \times k$-CLIQUE *has no* $2^{o(k \log k)}$ *algorithm unless ETH fails.*

Proof Suppose not and we have an algorithm Φ solving $k \times k$-CLIQUE in time $2^{o(k \log k)}$. The method is to show that this would imply 3-coloring on n vertex graphs in time $2^{o(n)}$ which is a contradiction assuming ETH as we have seen. The method, of course, is a reduction.

Let H have n vertices partitioned into groups $X_1, \dots, X_k$ of size $\lceil \frac{n}{k} \rceil$ where k is the least integer with $3^{\frac{n}{k+1}} \leq k$. For each $1 \leq i \leq k$ fix an enumeration of all the 3-colorings of $H[X_i]$, and notice that there are at most $3^{\lceil \frac{n}{k} \rceil} \leq 3^{\frac{n}{k+1}} \leq k$ such colorings. We define a proper 3-coloring c_i of $H[X_i]$ to be *compatible* with c_j of $H[X_j]$ if together they induce a proper 3-coloring of $H[X_i \sqcup X_j]$. That for each $uv \in E[H]$ with $u \in X_i$ and $v \in X_j$, $c_i(u) \neq c_j(v)$. Let G be the graph over $[k] \times [k]$ such that the vertices (i_1, j_1) and (i_2, j_2) are adjacent iff the j_1th proper coloring of X_{i_1} is compatible with the j_2th proper coloring of X_{i_2}.

It is routine to show that using a proper 3-coloring allows us to choose a $k \times k$-clique, and conversely such a clique induces a proper 3-coloring, hence G is a yes instance of $k \times k$-CLIQUE iff H is 3-colorable.

Now use the algorithm Φ for G decides the 3-coloring of H. To do this we need k^2 vertices of G and the polynomial time to construct G. Thus the total running time is $2^{o(k \log k)} k^{O(1)} = 2^{o(n)}$ as $k \log k = O(n)$, and hence we have a $2^{o(n)}$ algorithm for 3-COLORING, a contradiction to ETH. □

Using this artificial problem as a base, Lokshtanov, Marx, and Saurabh proved a number of interesting lower bounds along similar lines. For example they prove the following.

Theorem 29.5.5 (Lokshtanov, Marx, and Saurabh [496]) *Assume ETH.*

1. *There is no* $O^*(2^{o(d \log d)})$ *algorithm for closest string with parameter d for* CLOSEST STRING.
2. *The k-*DISJOINT PATHS PROBLEM *has no* $O^*(2^{o(k \log k)})$ *algorithm.*

Recall that CLOSEST STRING is defined as follows.

CLOSEST STRING

Input: A set of k length ℓ strings $s_1, \ldots, s_k$ over an alphabet Σ and an integer $d \geq 0$.

Parameter: Positive integers d and k.

Question: Find the center string, if any. That is find $s = s_j$ such that $d_H(s, s_i) \leq d$ for all $i = i, \ldots, k$.

We met this problem in Chap. 3 and there in Theorem 3.6.1, Gramm, Niedermeier, and Rossmanith [357] showed that CLOSEST STRING is solvable in time $O((d + 1)^d |G|)$, that is, $2^{O(d \log d)}$.

Similarly the k-DISJOINT PATHS PROBLEM seeks k vertex disjoint paths in a graph. The proofs of these results are too intricate to include in detail, but we will sketch the proof for CLOSEST STRING. We will make use of the intermediate problems:

$k \times k$-PERMUTATION CLIQUE

Input: A graph G over the vertex set $[k] \times [k]$.

Parameter: A positive integer k.

Question: Determine if G has a k element clique with exactly one element in each row and exactly one element from each column.

$2k \times 2k$-BIPARTITE PERMUTATION INDEPENDENT SET

Input: A bipartite graph G over the vertex set $[2k] \times [2k]$ where every edge is between $I_1 = \{(i, j) \mid i, j \leq k\}$, and $I_2 = \{(i, j) \mid i, j \geq k + 1\}$.

Parameter: A positive integer k.

Question: Is there an independent set $\{(1, \rho(1)), \ldots, (2k, \rho(2k))\} \subseteq I_1 \sqcup I_2$ for some permutation ρ of $[2k]$?

$k \times k$-HITTING SET

Input: Sets $S_1, \ldots, S_m \subseteq [k] \times [k]$.

Parameter: A positive integer k.

Question: Determine if there is an S with exactly one element in each row, such that $S \cap S_i \neq \emptyset$ for all $1 \leq i \leq m$.

The crucial idea is to use randomization for the initial reduction and then to derandomize the reduction.

Lemma 29.5.1 (Lokshtanov, Marx, and Saurabh [496]) *If there is a* $2^{o(k \log k)}$ *algorithm for* $k \times k$ PERMUTATION CLIQUE *then there is a randomized* $2^{o(m)}$ *algorithm for m-clause* 3SAT.

Proof We think of $k \times k$ PERMUTATION CLIQUE as selecting solution vertices $(1, \rho(1)), \ldots, (k, \rho(k))$ for some permutation ρ. The idea is as follows: if we apply a random ρ to some instance I of $k \times k$-CLIQUE with solution S, then with nonzero probability the solution S has exactly one vertex from each row and column. Let $c(i, j) : [k] \times [k] \to [k]$ be a random mapping, thought of as a coloring. Let $c'(i, j) = +$ if there is a $j' \neq j$ with $c(i, j) = c(i, j')$ and let $c'(i, j) = c(i, j)$ otherwise. We construct an instance I' of $k \times k$ PERMUTATION CLIQUE as follows. If there is an edge $(i_1, j_1)(i_2, j_2)$ with $c'(i_1, ji_1), c'(i_2, j_2) \neq +$ then add an edge $(i_1, c'(i_1, j_1))(i_2, c'(i_2, j_2))$. Again it is routine to show that if I' has a k-clique for some permutation ρ then for each i there is a unique $\delta(i)$ with $c'(i, \delta(i)) = \rho(i)$. Hence $(1, \delta(1)), \ldots, (k, \delta(k))$ is a clique in I and so if I is a no-instance, so is I'.

If I is a yes instance, we can use Stirling's formula to estimate the probability that the following two events occur.
1. For all $1 \leq i_1 \leq i_2 \leq k$, $c(i_1, \delta(i_1)) \neq c(i_2, \delta(i_2))$.
2. For each $1 \leq i, j \leq k$ with $j \neq \delta(i)$, $c(i\delta(i)) \neq c(i, j)$.

The probability of 1 is $\frac{k!}{k^k} = 2^{-O(k)}$. For 2 we see that it holds for a fixed i if $k - 1$ randomly chosen values are all different from $c(i, \delta(i))$. Thus for a fixed i, this has probability $(1 - \frac{1}{k})^{-(k-1)} \geq e^{-1}$. The probability that 2 holds for every i is at least e^{-k} and since the events 1 and 2 are independent, the probability that both hold is $e^{-O(k)}$. Thus given a yes instance we can produce a yes instance with probability $e^{-O(k)} = 2^{-O(k)}$. Amplifying the probability gives the randomized reduction. $\square$

The next step for which we omit the details is to use one of the efficient hash families met in Chap. 8 to de-randomize the reduction.

Theorem 29.5.6 (Lokshtanov, Marx, and Saurabh [496]) *Assume ETH. There is no* $O^*(2^{o(k \log k)})$ *algorithm for* $k \times k$-PERMUTATION CLIQUE.

The next step in the chain of reductions is the following.

Theorem 29.5.7 (Lokshtanov, Marx, and Saurabh [496]) *Assuming ETH, there is no* $O^* 2^{o(k \log k)}$ *time algorithm for* $2k \times 2k$-BIPARTITE PERMUTATION INDEPENDENT SET.

Proof Let I be an instance of $k \times k$-PERMUTATION INDEPENDENT SET. The for each $1 \leq i \leq k$ and $1 \leq j, j' \leq k$, $j \neq j'$, add the edge between (i, j) and $(i + k, j + k)$ in I'. If there is an edge between (i_1, j_1) and (i_2, j_2) in I add an edge $(i_1, j_1)(i_2 + k, j_2 + k)$ in I'. Then the claim is that I is a yes instance of $k \times k$-PERMUTATION INDEPENDENT SET iff I' is a yes instance of $2k \times 2k$-BIPARTITE PERMUTATION INDEPENDENT SET. This verification is straightforward and left to the reader (Exercise 29.11.7). $\square$

The penultimate step is to show the hardness of hitting.

Theorem 29.5.8 (Lokshtanov, Marx, and Saurabh [496]) *Assuming ETH, there is no $O^*(2^{o(k \log k)})$ time algorithm for $k \times k$-HITTING SET even in the special case, which we denote by $\times k$-HITTING SET when each set contains at most one element from each row.*

Proof Given an instance of $2k \times 2k$-BIPARTITE PERMUTATION INDEPENDENT SET I we construct an instance I' of $k \times k$-HITTING SET^T where we ask instead for exactly one element from each *column*. The universe of I' is $[2k] \times [2k]$. For $1 \leq i \leq k$ let the set S_i contain the first k elements of row i, and if $k < i \leq 2k$, set S_i to be the last k elements of row i. For each edge e of I, construct a set S_e as follows. Because of the way $2k \times 2k$-BIPARTITE PERMUTATION INDEPENDENT SET is defined, we need only consider edges $(i_1, j_1)(i_2, j_2)$ with $i + 1, j_1 \leq k$ and $i_2, j_2 \geq k + 1$. For such edges define $S_e = \{(i_1, j') \mid 1 \leq j' \leq k \wedge j' \neq j\} \cup \{(i_2, j) \mid k + 1 \leq j' \leq 2k \wedge j' \neq j_2\}$. Let $(1, \rho(1)), \ldots, (2k, \rho(2k))$, be a solution to I, then we claim it is also a solution of I'. First ρ is a permutation and hence we get one element from each column. As $\delta(i) \leq k$ iff $i \leq k$ the set S_i is hit for each $1 \leq i \leq 2k$. Suppose that there is an edge $e = (i_1, j_1)(i_2, j_2)$ such that S_e is not hit. Elements $(i_1, \delta(i_1))$ and $(i_2, \delta(i_2))$ have been chosen and we have $1 \leq \delta(i_1) \leq k$ and $k + 1 \leq \delta(i_2) \leq 2k$. Therefore if these two elements do not hit S_e, it must be that $\delta(i_1) = j_1$ and $\delta(i_2) = j_2$. But then the solution for I contains two adjacent vertices, a contradiction.

If $(\rho(1), 1), \ldots, (\rho(2k), 2k)$ is a solution for I'. Because of the sets S_i, the solution contains one element for each row, and hence ρ is a permutation of $[2k]$. $S_1, \ldots, S_k$ must be hit by the k elements in the first k columns because of the way that they are constructed. Thus $\rho(j) \leq k$ if $j \leq k$. Hence $\rho(i) > k$ if $i > k$. Then the claim is that $(\rho(1), 1), \ldots, (\rho(2k), 2k)$ is also a solution for I. To see that it is an independent set, suppose that $(\rho(j_1), j_1)(\rho(j_2), j_2) = e$ is an edge. We may assume that $\rho(j_1) \leq k$ and $\rho(j_2) > k$, and hence $j_1 \leq k$ and $j_2 > k$. Then the solution for I' must hit S_e so that the solution S must either include $(\rho(j_1), j_1)$ or $(\rho(j_2), j_2)$, a contradiction. $\qquad\square$

Finally we are in a position to prove lower bounds for CLOSEST STRING.

Proof of Theorem 29.5.5 We reduce $k \times k$-HITTING SET. Take an instance I with sets $S_1, \ldots, S_m$. The instance I' of CLOSEST STRING has alphabet $\Sigma = [2k + 1]$, $L = k$, and $d = k - 1$. Hence the center string must have one character in common with every string. Instance I' has $(k + 1)m$ input strings $s_{x,y}$ for $1 \leq x \leq m$ and $1 \leq y \leq k + 1$. If S_x contains (i, j) from row i, then the ith character of $s_{x,y}$ is j. If S_x contains no element from row i, then the ith character of $s_{x,y}$ will be $y + k$. Then $s_{x,y}$ described the elements of S_x with a dummy value between $k + 1$ and $2k$ marking the rows disjoint from S_x. Note that the strings $s_{x,1}, \ldots, s_{x,k+1}$ differ only on the choice of dummy values.

The claim is that I has a solution iff I' has. Suppose that $(1, \rho(1)), \ldots, (k, \rho(k))$ is a solution of I. The claim is that we can choose $s = \rho(1) \ldots \rho(k)$ as the center string. If the element $(i, \rho(i))$ of a solution set hits S_x then both s and $s_{x,y}$ have

$\rho(i)$ in position i. For the converse, take a solution s of I'. As there size of the solution is k, there is a $k + 1 \leq y \leq 2k + 1$ not appearing in s If the ith character of s is some $1 \leq p \leq k$, then define $\rho(i) = p$ and otherwise set $\rho(i) = 1$. The claim is $(1, \rho(1)), \ldots, (k, \rho(k))$ is a solution of I. Consider a string $s_{x,y}$ with at least one character in common with s. Suppose that p appears in at position i in both. It is not possible for $p > k$ as character y is the only character larger than k appearing in $s_{x,y}$, but we know y does not appear in s. Hence $1 \leq p \leq k$. Thus we hit S_x.

Finally, note that the size of I' is polynomial in k and m.   □

We feel the ability to prove such tight results on FPT and XP algorithms is quite surprising, and certainly elegant and important.

> In other words, now we have a method of proving that not only are things likely not in P, but the current algorithms for them are asymptotically close to optimal.

29.5.3 No $O^*((1 + \varepsilon)^k)$ Algorithm for VERTEX COVER

The methods of the previous section can be used to investigate limits on how far FPT algorithms can be "improved". For example, problems such as VERTEX COVER have seen steady improvements in the sense of FPT algorithms running in time $O^*(c^k)$ for smaller and smaller c. Can this go on forever? The answer is, "No!" Assuming ETH, some absolute barrier exists:

Theorem 29.5.9 (Impagliazzo, Paturi, and Zane [411]) *Suppose that $M[1] \neq$ FPT. There is an ε such that no strongly FPT parameterized algorithm for parameterized VERTEX COVER can run in time $O^*((1 + \varepsilon)^k)$.*

Proof [4] If VERTEX COVER has arbitrarily small kernels, then the following problem has an algorithm A solving the problem UVCC below in time $f(\ell)2^{\frac{k}{\ell}} n^{f(\ell)}$, for some computable increasing and time constructible f.

UVCC
Input: A graph G and $k, \ell > 0$.
Parameter: A positive integer ℓ.
Question: Determine if G has a k element vertex cover in time in time $O^*(2^{\frac{k}{\ell}})$.

To decide if a graph G with $|V(G)| = n$ and $k > 0$ we want to decide if G has a vertex cover of size k. Kernelize this, using Nemhauser and Trotter to a size $2k$ vertices. Thus we can assume $n = 2k$. Let $\ell = f^{-1}(\log k)$. We apply the algorithm

---

[4]This neat proof is an unpublished one due to Yijia Chen.

A on the instance (G, k, ℓ), and this needs time

$$f(f^{-1}(\log k))2^{\frac{k}{f^{-1}(\log k)}}(2k)^{f(f^{-1}(\log k))} \leq \log k 2^{o(k)} 2^{\log k^2 + \log k} = 2^{o(k)}. \qquad \square$$

We remark that this proof works for any problem which has the same complexity as k-VERTEX COVER using a parameterized reduction $\langle x, k \rangle \mapsto \langle x', k' \rangle$ which is linear time.

Therefore we have the following corollary.

Corollary 29.5.2 *For each of the following, there is an $\varepsilon > 0$ such that there is no strongly FPT $O^*((1+\varepsilon)^k)$.* MINI-SAT, MINI-3SAT, *and* MINI-INDEPENDENT SET.

Notice how we have extended the Abrahamson, Downey, and Fellows [3] material. In Theorem 29.2.2 they showed that $W[1]$-hard problems seem hard. The theorem and corollary above give tight lower bounds for $M[1]$ hard problems. One immediate consequence is that if we prove a language L to be $M[1]$-hard or $W[1]$-hard then there will be no subexponential algorithm. In particular, using the standard parameterization $k = \frac{1}{\varepsilon}$ for a PTAS with approximation ratio ε, you can use this technique to obtain lower bounds on approximation ratios assuming ETH.

Some care is needed for all of this material. For example, since a clique of size m has $O(m^2)$ edges, CLIQUE $\in$ DTIME$(n^{O(\sqrt{n})})$ and hence

Proposition 29.5.2 MINI-CLIQUE $\in$ FPT.

However, this result is sensitive to how we measure size. If we measure size by the number of vertices, this makes a difference.

29.6   The Proof of the Sparsification Lemma

The proof visualizes the problem as an instance of the SET COVER. problem we met in Chap. 6. The k-SET COVER[5] problem. The instance has a universe of n elements $\{x_1, \ldots, x_n\}$ a collection S of subsets $S \subseteq \{x_1, \ldots, x_n\}$, with $|S| \leq k$. It asks for $C \subseteq \{x_1, \ldots, x_n\}$ such that $C \cap S \neq \emptyset$ for all $S \in \mathcal{S}$. We let $\sigma(\mathcal{S})$ denote the collection of all set covers of $\mathcal{S}$. The set cover problem is to find a minimum sized set cover. A *restriction* $\mathcal{T}$ of $\mathcal{S}$ simply has for each $T \in \mathcal{T}$, there is an $S \in \mathcal{S}$ with $T \subseteq S$. If $\mathcal{T}$ is a restriction of $\mathcal{S}$, then $\sigma(\mathcal{T}) \subseteq \sigma(\mathcal{S})$.

Note that k-CNF SAT can be regarded as an instance of k-SET COVER by having a universe of $2n$ literals defining $\mathcal{U}$ to be the family of sets $\{x, \overline{x}\}$, and then note that set covers of size n will necessarily be consistent assignments. Then simply add for each clause the set consisting of the literals of the clause. (This reduction goes back to Karp [431].)

---

[5]To distinguish this from k-HITTING SET.

Now we can restate what we need to prove.

For all $\varepsilon > 0$ and $k > 0$, there is a constant C and ad algorithm which, given an instance $\mathcal{S}$ of k-SET COVER on a universe of size n, produces a list of $t \leq 2^{\varepsilon n}$ restrictions $\mathcal{T}_1, \ldots, \mathcal{T}_t$ of $\mathcal{S}$ with $\sigma(\mathcal{S}) = \bigcup_{i=1}^{t} \sigma(\mathcal{T}_i)$ and such that for each i, $|\mathcal{T}_i| \leq C \cdot n$. Moreover, the algorithm runs in time polynomial in n and $2^{\varepsilon n}$.

To prove this, choose an integer α such that $\frac{\alpha}{\log(4\alpha)} > \frac{4k2^k}{\varepsilon}$. Let $\theta_0 = 2$, $\beta_i = (4\alpha)^{2^{i-1}-1}$, and $\theta_i = \alpha\beta_i$ for $1 \leq i \leq k - 1$. Impagliazzo, Paturi, and Zane then call the collection $S_1, \ldots, S_c$ of sets of size j a *weak sunflower* with *heart* H and *petals* $S_i - H$ if $H = \bigcap_{i=1}^{c} S_i \neq \emptyset$. Note that each petal has the same size $j - |H|$, called the petal size. The size of the sunflower is the number of elements of the collection, and a set of size j is called a j-set. Finally, for a family of sets $\mathcal{S}$, let $\pi(\mathcal{S}) \subseteq \mathcal{S}$ be the family of all sets $S \in \mathcal{S}$ with no $S' \subset S$ and $S' \in \mathcal{S}$. Note that $\pi(\mathcal{S})$ is an antichain.

The proof then starts with the collection $\mathcal{S}$ and applies the recursive algorithm below. The tree describing this algorithm has at each node a family $\mathcal{S}'$ of sets. π is applied to $\mathcal{S}'$ to eliminate supersets. If $\pi(\mathcal{S}')$ contains a sunflower the node branches into two children. One family is created to by adding the petals of the sunflower to $\mathcal{S}'$, and the other adds the heart to $\mathcal{S}'$. Else the node is a leaf.

Algorithm 29.6.1 (Reduce($\mathcal{S}$))
1. $\mathcal{S} := \pi(\mathcal{S})$.
2. For $j = 2$ to k do:
3.   For $i = 1$ to $j - 1$ do:
4.   If there is a weak sunflower $\{S_1, \ldots, S_c\} \in \mathcal{S}$, $|S_d| = j$, of j-sets and petal size i where the number of petals is at least θ and with $|H| =_{\text{def}} \bigcap_{d=1}^{c} S_d = j - i$,
5.   Then Reduce($\mathcal{S} \cup H$); Reduce($\mathcal{S} \cup \{S_1 - H, \ldots, S_c - H\}$); Halt.
6. Return $\mathcal{S}$, Halt.

Note that new sets are added to the collection at a node due to the existence of a sunflower, and existing sets are eliminated by the application of π. A set S' can be eliminated if some proper subset S is in $\mathcal{S}'$, and we say that S eliminates S'.

Suppose that $\mathcal{T}$ is associated with any node except the root and $\mathcal{T}'$ is associated with its parent. If $S \in \mathcal{T}$ eliminated S' in $\mathcal{T}$, then S' must be either the heart or a petal of a sunflower in $\mathcal{T}'$ that caused the branching. After applying π to $\mathcal{T}$, no proper superset of S would ever be associated with the descendants of $\mathcal{T}$. All the sets eliminated by S are eliminated in the next application of π immediately after S is added to the collection.

Lemma 29.6.1 Reduce($\mathcal{S}$) *returns a set of restrictions of* $\mathcal{S}$, $\mathcal{T}_1, \ldots, \mathcal{T}_t$ *with* $\sigma(\mathcal{S}) = \bigcup_{i=1}^{t} \sigma(\mathcal{T}_i)$.

Proof If $S \subset S'$, then $\sigma(\mathcal{S}) = \sigma(\mathcal{S} \setminus S')$. If $S_1, \ldots, S_c \in \mathcal{S}$ form a weak sunflower with heart H, and T is a cover of $\mathcal{S}$, then either $T \cap H \neq \emptyset$, or $T \cap (S_i \setminus H) \neq \emptyset$ for all i. Hence $\sigma(\mathcal{S}) = \sigma(\mathcal{S} \cup H) \cup \sigma(\mathcal{S}) \cup \bigcup_{i=1}^{c} \sigma(\mathcal{S} \cup S_i)$. Now the lemma follows by induction. □

Lemma 29.6.2
1. *For $1 \leq i \leq k$, every family $\mathcal{T}$ output by* Reduce($\mathcal{S}$) *has no more than* $(\theta_{i-1} - 1)n$
   sets of size i.
2. *Consequently, every element produced by* Reduce($\mathcal{S}$) *has no more than Cn sets
   where $C = \sum_{i=1}^{k}(\theta_{i-1} - 1)$. Hence the sparseness property is satisfied.*

Proof If there were more than $(\theta_{i-1} - 1)n$ sets of size i, in $\mathcal{T}$, then there would
beat least one x in at least θ_{i-1} of them. The sets contains x would form a weak
sunflower with petal size at most $i - 1$, and hence the program would branch rather
than halt.   □

Now we get to the hard part of the proof. If t is the number of (i.e. the number
of leaves) families output we need to show that $t \leq 2^{\varepsilon n}$. This is done by showing
that the maximum length of any path is $O(n)$, whereas the maximum number of
nodes created by adding petals rather than the heart is at most $\frac{kn}{\alpha}$. For this proof,
we use the invariant $J_i(\mathcal{T})$ for $1 \leq i \leq k$, which holds for a family $\mathcal{T}$ on any path
as long as a set of size i to one of the predecessor families on the path. Note that
$J_i(\mathcal{T})$ is the condition that no element x appears in more than $2\theta_i - 1$ sets of size
i in $\mathcal{T}$. Because of this invariant, it will be shown that any set added eliminates at
most $2\theta_i - 1$ of size i during the next application of π. As there are at most n sets
of size one in any family, we show by induction that the total number of sets of size
at most i ever added to all the families along any path if at most $\beta_i n$. Since at least
θ_i many petals are added to any family created by adding petals of size i, it follows
at most $\frac{kn}{\alpha}$ families are created due to petals.

Lemma 29.6.3   *For any family $\mathcal{T}$ such that a set of size i has been added to either
it or a predecessor, $J_i(\mathcal{T})$ holds, viz no element x appears in more than $2\theta_{i-1}$ sets
of size i in $\mathcal{T}$.*

Proof We need only consider a family $\mathcal{T}$ created by adding sets of size i to a family $\mathcal{T}'$. The sunflower in $\mathcal{T}'$ creating these sets must be a family of sets of petal size
$j > i$. Furthermore there is no sunflower of i-sets in $\mathcal{T}'$ with θ_{i-1} petals of petal
size $i - 1$, thus no x appears in more than $\theta_{i-1} - 1$ sets of size i in $\mathcal{T}'$. Now $\mathcal{T}$ is
created by adding the petals or the heart. If the petals were added, no x is in more
than $\theta_{i-1} - 1$ petals lest $\mathcal{T}'$ would contain a sunflower of j-sets with θ_{i-1} petals of
size $i - 1$. Thus no x is contained in more than $2\theta_{i-1} - 1$ i-sets of $\mathcal{T}$. If the heart is
added only one i-set is added, and since $\theta_{i-1} \geq 1$, the result follows.   □

Lemma 29.6.4   *At most $2\theta_{i-1}$ of the added i-sets can be eliminated by a single set.*

Proof If S is an i-set added to $\mathcal{T}$, then after the application of π, $\mathcal{T}$ and all of
its successor families would satisfy the invariant by Lemma 29.6.3. Thus for such
families no x appears in more than $2\theta_{i-1} - 1$ sets i-sets. Let S be added to $\mathcal{T}$, and
$x \in S$. All the sets S eliminates contain x during the next application of π. Therefore
S can eliminate at most $2\theta_{i-1} - 1$ i-sets.   □

Lemma 29.6.5 *The total number of sets of size $\leq i$ added to the families along any path is $\leq \beta_i n$.*

Proof This is proven by induction on i. For $i = 1$, $\beta_1 = 1$ and since each newly added element is a new element of the underlying universe, there are at most $n = \beta_1 n$ such sets. Assume for $i - 1$. Then each added i-set is either subsequently eliminated by an $(i - 1)$-set or is in the final collection of sets output by the algorithm. By Lemma 29.6.2, there are at most $\theta_{i-1} n$ i-sets in the final collection. By the induction hypothesis and Lemma 29.6.4, each of at most $\beta_{i-1} n$ of added smaller sets eliminates at most $2\theta_{i-1}$ i-sets. The total number of $\leq i$-sets is $\leq 2\beta_{i-1}\theta_{i-1} n + \theta_{i-1} n \leq 4\beta_{i-1}\theta_{i-1} n = 4\alpha\beta_{i-1}^2 n = \beta_i n$. $\qquad\qquad\square$

Lemma 29.6.6 *Along any path at most $\frac{kn}{\alpha}$ families are created by adding petals.*

Proof By Lemma 29.6.5, there are at most $\beta_i n$ sets of size $\leq i$ ever added to all the families along any path. If a family is created by adding the petals of size i, then at least θ_i such petals are added. Thus there are at most $\frac{\beta_i n}{\theta_i} = \frac{n}{\alpha}$ giving a total of $\frac{kn}{\alpha}$ since we consider all $1 \leq i \leq k$. $\qquad\qquad\square$

The decisive and final lemma is the following, which completes the proof by Lemmas 29.6.1, 29.6.2, 29.6.7 and setting $C = \sum_{i=1}^{k-1} \theta_{i-1}$.

Lemma 29.6.7 *The algorithm has at most $2^{\varepsilon n}$ many paths, and hence outputs at most $2^{\varepsilon n}$ families.*

Proof A path is completely determined by its transcript: its sequence of decisions at the node along the path. Since a set of size at most $k - 1$ is added to the family at each step, there are at most $\beta_{k-1} n$ sets ever added to the families by Lemma 29.6.5. By Lemma 29.6.6, at most $\frac{nk}{\alpha}$ of these families are created using petals. We conclude there are at most $\sum_{r=0}^{\lfloor \beta_{k-1} n \rfloor} \binom{\beta_{k-1} n}{r}$ many paths. This quantity is $\leq 2^{\varepsilon n}$. There are several ways to prove this, and the method of Impagliazzo, Paturi, and Zane is to use the binary entropy function for the probability of p heads in a Bernoulli trial. $H(p) = -p \log p - (1 - p) \log(1 - p)$. Then we note $\binom{m}{pm} \leq 2^{H(p)m}$, and hence the total number of paths is at most

$2^{\beta_{k-1} n H(\frac{k}{\alpha\beta_{k-1}})} \leq 2^{((\frac{2k}{\alpha}) \log(\frac{\alpha\beta_{k-1}}{k}))n}$ as $H(\frac{k}{\alpha\beta_{k-1}}) \leq \frac{2k}{\alpha\beta_{k-1}} \log(\frac{k}{\alpha\beta_{k-1}})$ since $\frac{k}{\alpha\beta_{k-1}} \leq \frac{1}{2}$.

This quantity is $\leq 2^{((\frac{2k}{\alpha}) \log(\alpha(4\alpha)^{2^{k-2}-1}))n} \leq 2^{2k2^{k-2}(\frac{\log(4\alpha)}{\alpha})n} \leq 2^{\varepsilon n}$, the last inequality following by choice of α. The result follows. $\qquad\qquad\square$

29.7 XP-Optimality

The insight of Chen, Chor, Fellows, Huang, Juedes, Kanj, and Xia [139] was that using ETH allowed for very tight bounds for *membership in* XP. What do we mean by this? As with the analogy of P and NP, we have problems like INDEPENDENT

SET which are believed in $W[1] \setminus$ FPT. But in the same way that being in NP $\setminus$ P says only that you do not have a polynomial-time algorithm, being in $W[1] \setminus$ FPT only says that you likely do not have an algorithm to determine if $\langle G, k \rangle \in L$ in time $f(k)n^c$. This does not rule out the possibility that there might be an algorithm running in time $n^{O(\log \log k)}$ for example.

Chen et al. [139] introduced new techniques to show that lower bounds of the form $n^{o(k)}$ can be found for a number of $W[2]$-hard problems (such as DOMINATING SET) assuming FPT $\neq W[1]$, and use the ETH to rule out lower bounds of this type for $W[1]$ hard problems. Later this was improved by Jianer Chen, Huang, Kanj, and Xia [143, 144] to rule out $f(k)n^{o(k)}$ algorithms. We refer to the behavior below as XP-*optimality*, since it says that locating the problem in XP is optimal in terms of its parameterized complexity.

We remark that one view of this result is that classes like $W[1]$ are (useful) artifacts of parameterized complexity, and, in fact, $M[1]$ plays a central role.

Theorem 29.7.1 (Chen et al. [143, 144]) *Assuming ETH there are no $f(k)n^{o(k)}$ algorithms for either* CLIQUE *or* INDEPENDENT SET.

Proof Suppose that there is an $f(k)n^{o(k)}$ algorithm for CLIQUE. Then there is an algorithm in time $f(k)n^{\frac{k}{s(k)}}$ where $s(k)^6$ is monotone nondecreasing and unbounded. Let $f^{-1}(n)$ be the usual "pseudo inverse" of f, namely the largest integer i with $f(i) \leq n$. Then f^{-1} is nondecreasing and unbounded. Now let G be a given graph with n vertices. Let $k = f^{-1}(n)$. As in the proof of Theorem 29.5.4, partition the vertices of G into k equal groups, and as above build a graph where each vertex corresponds to a proper 3-coloring of one of the groups. We then connect two vertices of they are compatible. Clearly a k-clique of H corresponds to a 3-coloring of G, and this can be found in time $f(k)n^{\frac{k}{s(k)}} \leq n(3^{\frac{n}{k}})^{\frac{k}{s(k)}} = n3^{\frac{n}{s(f^{-1}(n))}} = 2^{o(n)}$. This means we can decide if G has a proper 3-coloring in time $2^{o(n)}$, contradicting ETH. The INDEPENDENT SET result follows from the usual reduction. $\square$

29.8 Treewidth Lower Bounds

Let G be a graph of bounded treewidth. Then as the number of vertices of G is an upper bound for $\mathrm{tw}(G)$, we get as corollaries the following.

Corollary 29.8.1 *Assume ETH. Then there is no $O^*(2^{o(\mathrm{tw}(G))})$ algorithm for* INDEPENDENT SET, DOMINATING SET, VERTEX COVER, *etc.*

As we have seen in Chap. 10, natural dynamic programming algorithms yield FPT algorithms for various problems like HAMILTONICITY and naturally give algorithms in time $O^*(2^{O(\mathrm{tw}(G)\log(\mathrm{tw}(G)))})$.

---

[6]Here definitely note that if we do not assume more than the existence of an algorithm, then f may not be computable, and hence we will need the nonuniform version of ETH.

The methodology introduced by Lokshtanov, Marx, and Saurabh can be applied to obtain qualitatively matching lower bounds for such problems.

Theorem 29.8.1 (Cygan, Nederlof, Pilipczuk, Pilipczuk, J. van Rooij, and Wojtaszczyk [190], and Lokshtanov, Marx, and Saurabh [496]) *Assume ETH. There are no $O^*(2^{o(\mathrm{tw}(G)\log(\mathrm{tw}(G)))})$ algorithms for any of the problems* CHROMATIC NUMBER, CYCLE PACKING, DISJOINT PATHS.

It is also possible to get XP-optimality results for treewidth. We conclude this section with one example to give the reader a flavor as to how such results are proven.

Theorem 29.8.2 (Fellows, Fellows, Fomin, Lokshtanov, Rosamond, Saurabh, Szeider, Thomassen [286]) *Assume ETH. Then the following problem cannot be solved in time $O^*(f(\mathrm{tw}(G))n^{o(\mathrm{tw}(G))})$.*

LIST COLORING

Input: A graph G and for each vertex v, a list $L(v)$ of permitted colors.
Parameter: $\mathrm{tw}(G)$.
Question: Does G have a proper coloring with $c(v) \in L(v)$ for all v?

Proof We will prove this using a reduction from MULTICOLORED CLIQUE that the reader has proven XP-optimal in Exercise 29.11.5. Let G be an instance of MULTICOLORED CLIQUE. We construct an instance of LIST COLORING G' that is a yes instance iff the original G contains a k-clique. In the construction below, we make the colors on the lists of vertices of G' have 1–1 correspondence with the vertices of G, and we do not distinguish between the vertex of G and the color it represents in G'. Every vertex v in the instance G of MULTICOLORED CLIQUE has a color in $1, \ldots, k$. Let V_i denote the vertices of color i. For G' construct as follows.

1. Make k vertices $v[i]$ in G' one for each color class of G and $L_{v[i]} = V_i$.
2. For $i \neq j$, there is a degree two vertex in G' adjacent to $v[i]$ and $v[j]$ for each pair x, y of *nonadjacent* vertices in G with x of color i and y of color j. Label this vertex as $v_{i,j}[x, y]$, and make $L_{v_{i,j}[x,y]} = \{x, y\}$.

First note that $\mathrm{tw}(G') \leq k$ since removing the k vertices $v[i]$ from G' makes an edgeless graph. Suppose that G has a multicolored clique, $K = \{c_1, \ldots, c_k\}$ with $c_i \in V_i$. Then color $v[i]$ with c_i. Ever degree 2 vertex in G' has at least one color free on its list (due to the nod-adjacency requirement). Therefore this induces a "yes" instance of LIST COLORING. For the converse direction, suppose that G' is properly list colored. Let $K = \{c_1, \ldots, c_k\}$ denote the set of vertices in G corresponding to the colors with $v[i]$ colored c_i. We claim that for all $i \neq j$ $c_i c_j \in E(G)$. Otherwise there would be a degree 2 vertex adjacent to each of c_i and c_j in G' which is only colorable by one from $\{c_i, c_j\}$, a contradiction. $\square$

We remark that the results above also apply to applications of treewidth such as constraint satisfaction. For example, assuming ETH there is no $O^*(f(\text{tw}(G)) \cdot n^{o(\text{tw}(G))})$ solution to constraint satisfaction on instance I with Gaifman graph G.

29.9 The Strong Exponential Time Hypothesis

The Sparsification Lemma has certain consequences for the complexity of d-CNFSAT as d grows. Let

$$s_d = \inf\{\varepsilon \mid \exists O^*(2^{\varepsilon n}) \text{ algorithm for } n \text{ variable } d\text{-CNFSAT}\}.$$

Clearly $s_d \leq s_{d+1}$. We can define $s_\infty = \lim_{d \to \infty} s_d$.

Theorem 29.9.1 (Impagliazzo and Paturi [410]) *Assume ETH. Then $s_d < s_{d+1}$ infinitely often. Moreover, $s_d \leq s_\infty(1 - \frac{c}{d})$ for some constant $c > 0$.*

The proof of Theorem 29.9.1 uses sparsification plus some ideas from algebraic coding theory. Space considerations preclude us from including it here.

Impagliazzo and Paturi's [410] *Strong Exponential Time Hypothesis* (SETH) is the statement that $s_\infty = 1$. There is less evidence for it, and some authors think that it might not be true. However, if true, it yields some really tight bounds for the running times of algorithms. For example, we have the following.

Theorem 29.9.2 (Cygan, Dell, Lokshtanov, Marx, Nederlof, Okamoto, Paturi, Saurabh, Wahlström [186]) *The following are equivalent to SETH.*

1. *For all $\varepsilon < 1$, there is a d such that n-variable d-CNFSAT is not solvable in time $O(2^{\varepsilon n})$.*
2. *For all $\varepsilon < 1$, there is a d such that HITTING SET over $[n]$ with sets of size bounded by d is not solvable in time $O(2^{\varepsilon n})$.*
3. *For all $\varepsilon < 1$, there is a d such that SET SPLITTING for systems over $[n]$ for sets of size at most d is not solvable in time $O(2^{\varepsilon n})$.*

Theorem 29.9.3 (Lokshtanov, Marx, and Saurabh [494]) *If INDEPENDENT SET can be solved in time $O^*((2 - \varepsilon)^{\text{tw}(G)})$ for some $\varepsilon > 0$, then for some $\delta > 0$ we can solve SAT in time $O^*((2 - \delta)^n)$.*

We give one example of this phenomenon.

Theorem 29.9.4 (Pătraşcu and Ryan Williams [555]) *Assume SETH. Then there is no $O(n^{k-\varepsilon})$ algorithm for DOMINATING SET for any $\varepsilon > 0$.*

Theorem 29.9.4 follows from the following lemma.

Lemma 29.9.1 (Pătrşcu and Williams [555]) *Suppose that there is a $k \geq 3$ and a function f such that k-DOMINATING SET is solvable in $O(n^{f(k)})$ time. Then n-variable m-clause CNFSAT is solvable in time $O((m + k2^{\lceil \frac{n}{k} \rceil})^{f(k)})$.*

Proof To prove the lemma, fix $k \geq 3$. Take the CNF formula φ. Without loss of generality, we will assume k divides n. Partition the set of variables into k parts of size $\frac{n}{k}$. For each part make a list of all the $2^{\frac{n}{k}}$ partial assignments to variables in those parts. Each partial assignment will correspond to a node of a graph $G = G_\varphi$ we build.

Now make each of the k parts into a clique and hence there are now k disjoint cliques each with $2^{\frac{n}{k}}$ vertices and $O(2^{\frac{2n}{k}})$ edges. We add m more vertices, one for each clause, and make an edge from the vertex corresponding to a clause to the partial assignment vertex iff the partial assignment satisfies the clause. To finish the construction, for each clique add an enforcer vertex which has edges only to that clique and to every vertex of the clique.

Suppose that G_φ has k dominating set D. Then by the enforcers no clause node is in the dominating set, and moreover exactly one vertex per clique must be chosen. Therefore the collection of partial assignments correspond to the vertices of D is a satisfying assignment of φ since all the clauses are dominated.

There are $k2^{\frac{n}{k}} + m + k$ many vertices, giving the lemma. $\square$

As pointed out by Pătrşcu and Williams, the above can be viewed in terms of k-SET COVER, where we have a collection of n sets in a universe of size m, and need to find a S of sets covering the family with $|S| = k$. This is done by defining the family as $S_v = N[v] \cup \{v\}$ for each $v \in G$.

Corollary 29.9.1 (Pătrşcu and Williams [555]) *If there is a $k \geq 2$ such that k-SET COVER can be solved in time $O(n^{k-\varepsilon})$ for some $\varepsilon > 0$ and for a family of n sets and universe of size $poly(\log n)$, then SETH fails.*

We finish this section by observing that the theorem above is extremely tight. For a long time the best algorithm for DOMINATING SET as that of trying all possibilities which has time $\Omega(n^{k+1})$. Pătrşcu and Williams gave a slight improvement of this using fast matrix multiplication. Consider the special case of 2-DOMINATING SET. Take the boolean matrix A of a graph G, complement it by flipping 0's and 1's, giving $\overline{A}$, and then observe the following easy lemma.

Lemma 29.9.2 (Pătrşcu and Williams [555]) *G has a 2-DOMINATING SET if and only if for some i, j $B[i, j] = 0$ where $B = \overline{A} \cdot \overline{A}^T$.*

Proof Exercise 29.11.9. $\square$

Recall from Sect. 16.4 that two $n \times n$ matrices can be multiplied in time n^q for $q \leq 2.3727$ by a generalization of the Coppersmith–Winograd algorithm [165] due to Vanessa Williams [665]. The conclusion is that we can solve 2-DOMINATING SET in time $O(n^{2.3727})$. Then to generalize this to k-DOMINATING SET, let $v_1, \ldots, v_n$

list the vertices, and $S_a, \ldots, S_{\binom{n}{\frac{k}{2}}}$ list all the $\frac{k}{2}$-sets of vertices. Then define a $\binom{n}{\frac{k}{2}} \times n$ boolean matrix A_k where $A_k = 0$ iff v_j is dominated by S_i. Then the following lemma holds similarly (Exercise 29.11.9).

Lemma 29.9.3 *The product matrix* $B_k = A_k \times A_k^T$ *is an* $\binom{n}{\frac{k}{2}} \times \binom{n}{\frac{k}{2}}$ *matrix such that* $B_k[i, j] = 0$ *iff* $S_i \cup S_j$ *is a dominating set.*

Theorem 29.9.5 (Pătraşcu and Williams [555]) *For* $k \geq 7$, k-DOMINATING SET *is solvable in time* $O(n^{k+o(k)})$.

Proof Either using Coppersmith [164] or the methods of Sect. 16.4 we can multiply an $n \times n^{0.294}$ matrix by an $n^{0.294} \times n$ matrix in $n^{2+o(1)}$ many operations. The product B_k is basically an $N \times N^{\frac{2}{k}}$ times an $N^{\frac{2}{k}} \times N$ matrix product, where $N = \binom{n}{\frac{k}{2}}$. If $k \geq 7$, $\frac{2}{k} \leq 0.0294$ as hence we can apply [164].   □

29.10 Subexponential Algorithms and Bidimensionality

Our next step is to deal with the classification of parameterized problems in the class SUBEPT, the class of FPT parameterized problems with parameter k solvable in time $O^*(2^{o(k)})$. The name is derived from the class EPT introduced by Flum, Grohe, and Weyer [313]. EPT is the class of parameterized problem solvable in time $O^*(2^{O(k)})$. As with all FPT material, we need to have an ongoing discourse with a problem. Thus, hand in hand with the limiting material above, we have some new positive techniques via a new technique called *bidimensionality theory*. Unfortunately, this is sufficiently beyond the scope of the present book that we will merely give a sketch of the material, following the account of Downey and Thilikos [261].

We present techniques that provide algorithms with subexponential parameter dependence for variants of parameterized problems where the inputs are restricted by some sparsity criterion. The most common restriction for problems on graphs is to consider their *planar* variants where the input graph is embeddable in a sphere without crossings. As mentioned in Theorem 29.5.3, Cai and Jeudes [127], established that k-PLANAR DOMINATING SET does not belong to $2^{o(\sqrt{k})} \cdot n$-FPT, unless $M[1] = $ FPT and the same holds for several other problems on planar graphs such as PLANAR VERTEX COVER, PLANAR INDEPENDENT SET, PLANAR DOMINATING SET, and PLANAR RED/BLUE DOMINATING SET. (Also see [139, 147].) This implies that for these problems, when the sparsity criterion includes planar graphs, the best running time we may expect is $2^{O(\sqrt{k})} \cdot n^{O(1)}$. The first subexponential parameterized algorithm on planar graphs appeared by Alber, Bodlaender, Fernau, and Niedermeier in [22] for DOMINATING SET, INDEPENDENT DOMINATING SET, and FACE COVER. After that, many other problems were classified in $2^{c\sqrt{k}} \cdot n^{O(1)}$-FPT, while there was a considerable effort towards improving the constant c for each one of them. Examples include [21, 208, 217, 305, 307, 325, 428, 450, 462].

Bidimensionality theory was proposed in [209] as a meta-algorithmic framework that describes when such optimal subexponential FPT algorithms are possible. Our first step is to illustrate the idea of bidimensionality for the following problem.

k-PLANAR LONGEST CYCLE
Instance: A planar graph G and a nonnegative integer k.
Parameter: A positive integer k.
Question: Does G contain a cycle of length at least k?

We define the graph parameter **lc** where

$$\mathbf{lc}(G) = \min\{k \mid G \text{ does not contain a cycle of length at least } k\}.$$

Our subexponential algorithm for k-PLANAR LONGEST CYCLE will be based on the Win/win technique whose main combinatorial ingredient is the following result of Gu and Tamaki.

Theorem 29.10.1 (Gu and Tamaki [362]) *Every planar graph G where* $\text{bw}(G) \geq 3k$ *contains a $(k \times k)$-grid as a minor.*

Actually, Theorem 29.10.1 is an improvement of the result of Robertson, Seymour, and Thomas in [591] where the lower bound to branchwidth was originally $4k$ instead of $3k$.

Notice that **lc** is closed under taking of minors; the contraction/removal of an edge in a graph will not cause a bigger cycle to appear. This means that if G is a planar graph and $\mathbf{lc}(G) \leq l^2$, none of the minors of G can contain a cycle of length l^2. As the $(l \times l)$-grid contains a cycle of length l^2, we conclude that G does not contain an $(l \times l)$-grid as a minor. From Theorem 29.10.1, $\text{bw}(G) < 3l$. In other words, $\text{bw}(G) < 3 \cdot \sqrt{\mathbf{lc}(G)}$.

To solve k-PLANAR LONGEST CYCLE we apply the following steps:
(a) We compute an optimal branch decomposition of G.[7]
(b) We check whether $\text{bw}(G) \geq 3\sqrt{k}$. If this is the case then $\mathbf{lc}(G) > k$ and we can safely return that G contains a cycle of length k.
(c) If $\text{bw}(G) < 3\sqrt{k}$, then we solve the following problem by using dynamic programming.

BW-PLANAR LONGEST CYCLE
Instance: A planar graph G and a nonnegative integer l.
Parameter: $\text{bw}(G)$.
Question: Does G contain a cycle of length at least l?

What we have done so far is to reduce the k-PLANAR LONGEST CYCLE to its bounded branchwidth counterpart. For general graphs, dynamic programming for this problem requires $O^*(2^{O(k \log k)})$ steps. However, for planar graphs, one may use

---

[7]According to to [361], this can be done in $O(n^3)$ steps (the result in [361] is an improvement of the algorithm of Seymour and Thomas in [615]).

the technique introduced in Dorn, Penninkx, Bodlaender, and Fomin [235] yielding a $2^{O(k)} \cdot n^{O(1)}$ step dynamic programming. In fact, according to [230], BW-PLANAR LONGEST CYCLE can be solved in $O(2^{2.75 \cdot \text{bw}(G)} \cdot n + n^3)$ steps. We conclude that k-PLANAR LONGEST CYCLE belongs to $O(2^{8.25 \cdot \sqrt{k}} \cdot n + n^3)$-SUBEPT.

Notice that, in order to apply the above machinery, we made the following two observations on the parameter **lc**: (1) it is closed under taking of minors and (2) its value with the $(l \times l)$-grid as input is $\Omega(l^2)$, i.e. a certificate of the solution for the t-PLANAR LONGEST CYCLE problem spreads along *both* dimensions of a square grid (i.e. it virtually spreads *bidimensionally* on the grid). These two observations apply to a wide range of parameters where the same approach can be used for the corresponding problems on planar graphs. Bidimensionality theory was introduced in Demaine, Fomin, Hajiaghayi, and Thilikos [209]. It used. the above observations in order to derive subexponential algorithms for graphs embeddable in surfaces of bounded genus also being used in Demaine, Hajiaghayi, and Thilikos [218]. Later, the same theory was developed for parameters that are closed under *contractions* in Fomin, Golovach, and Thilikos [315] and for classes of graphs excluding specific graphs as a minor in Demaine and Hajiaghayi [214] and Demaine, Fomin, Hajiaghayi, and Thilikos[207].

Consider the following parameterized meta-problem where $\mathcal{C}$ is a class of graphs.

k-PARAMETER CHECKING FOR **p** ON $\mathcal{C}$

Instance: A graph $G \in \mathcal{C}$ and an integer $k \geq 0$.
Parameter: A positive integer k.
Question: **p**$(G) \leq k$?

Briefly, this theory can be summarized by the following meta-algorithmic framework.

Theorem 29.10.2 (Minor bidimensionality—Demaine, Fomin, Hajiaghayi, and Thilikos, Demaine, and Hajiaghayi [209, 214]) *Let $\mathcal{G}$ be an H-minor free graph class. Let also **p** be a graph parameter which satisfies the following conditions:*
  (i) **p** *is minor closed,*
  (ii) *if A_k is the $(k \times k)$-grid, then* **p**$(A_k) = \Omega(k^2)$*, and*
  (iii) *for graphs in $\mathcal{G}$,* **p** *is computable in time $2^{O(\text{bw}(G))} \cdot n^{O(1)}$.*
*Then, k-PARAMETER CHECKING FOR **p** ON $\mathcal{C}$ belongs to $O^*(2^{O(\sqrt{k})})$-SUBEPT.*

Notice that Theorem 29.10.2 can be applied to maximization and minimization problems. We provide a typical example of its application on a maximization problem through k-PATH. Here, the associated parameter is

$$\mathbf{p}(G) = \min\{k \mid G \text{ does not contain a path of length } k\}.$$

It is easy to check that conditions (i) and (ii) are satisfied. Indeed, no bigger path occurs if we remove or contract an edge and the $(k \times k)$-grid has a (Hamiltonian) path of length k^2. Let $\mathcal{G}$ be any H-minor free graph class. Condition (iii) holds for $\mathcal{G}$, because of the results in Dorn, Fomin, and Thilikos [233]. Therefore, k-PATH restricted to graphs in $\mathcal{G}$ belongs to $O^*(2^{O(\sqrt{k})})$-SUBEPT.

Problems for which Theorem 29.10.2 proves SUBEPT-membership for graphs excluding some graph as a minor are k-VERTEX COVER, k-FEEDBACK VERTEX SET, k-ALMOST OUTERPLANAR, k-CYCLE PACKING, k-PATH, k-CYCLE, k-d-DEGREE-BOUNDED CONNECTED SUBGRAPH, k-MINIMUM MAXIMAL MATCHING, and many others.

We say that a graph H is a *contraction* of a graph G if H can be obtained from G after applying a (possibly empty) sequence of edge contractions.[8] A parameter $\mathbf{p}$ is *contraction closed* if $H \leq_c G$ implies that $\mathbf{p}(H) \leq \mathbf{p}(G)$.

Clearly, there are parameters escaping the applicability of Theorem 29.10.2 due to the fact that they are not minor-closed. The most typical example of such a parameter is the dominating set number: take a path P of length $3k$ and connect all its vertices with a new vertex. Clearly, the resulting graph has a dominating set of size 1, while it contains P as a minor (actually it contains it as a subgraph) and every dominating set of P has size at least k. However, the dominating set number is contraction closed and the good news is that there is a counterpart of Theorem 29.10.2 for contraction closed parameters. Before we state this result, we need two more definitions. An *apex* graph is defined to be a graph that becomes planar after the removal of a vertex. A graph class $\mathcal{G}$ is *apex-minor free* if it is H-minor free for some apex graph H. An apex-minor free graph can be seen as having some big enough "flat region", provided that the graph has big treewidth. Graph classes that are apex-minor free are the graph classes of bounded genus and the classes of graphs excluding some single-crossing graph as a minor Demaine, Hajiaghayi, and Thilikos [217].

The graph Γ_k is the graph obtained from the $(k \times k)$-grid by triangulating internal faces of the $(k \times k)$-grid such that all internal vertices become of degree 6, all non-corner external vertices are of degree 4, and one corner is joined by edges with all vertices of the external face (the *corners* are the vertices that in the underlying grid have degree two). For the graph Γ_6, see Fig. 29.2.

Theorem 29.10.3 (Contraction bidimensionality—Fomin, Golovach, and Thilikos [315]) *Let $\mathcal{G}$ be an apex-minor free graph class. Let also $\mathbf{p}$ be a graph parameter which satisfies the following conditions:*

(i) $\mathbf{p}$ *is closed under contractions,*

(ii) $\mathbf{p}(\Gamma_k) = \Omega(k^2)$, *and*

(iii) *for graphs in $\mathcal{G}$, $\mathbf{p}$ is computable in time $2^{O(\mathrm{bw}(G))} \cdot n^{O(1)}$.*

Then, k-PARAMETER CHECKING FOR $\mathbf{p}$ ON $\mathcal{C}$ belongs to $O^*(2^{O(\sqrt{k})})$-SUBEPT.

---

[8]Formally, the contraction relation occurs if in the definition of the minor relation we additionally demand that

(c) for every $\{x, y\} \in E(G)$, either $\varphi(x) = \varphi(y)$ or $\{\varphi(x), \varphi(y)\} \in E(H)$.

Fig. 29.2 The graph Γ_6

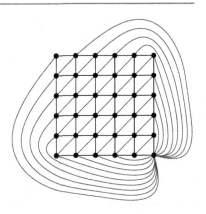

Problems for which Theorem 29.10.3 (but not Theorem 29.10.2) proves SUB-EPT-membership for apex-minor free graph classes are k-DOMINATING SET, $k - r$-DISTANCE DOMINATING SET, k-FACE COVER (for planar graphs), k-EDGE DOMINATING SET, k-CLIQUE-TRANSVERSAL SET, k-CONNECTED DOMINATING SET, and others.

For surveys on the application of Bidimensionality Theory on the design of subexponential parameterized algorithms, see Dorn, Fomin, and Thilikos [234] and Demaine and Hajiaghayi [212]. Finally, further applications of bidimensionality, appeared in Sau and Thilikos [601] Demaine and Hajiaghayi [211, 324] and Fomin, Lokshtanov, Raman, and Saurabh [322], the last exhibiting a strong connection between bidimensionality and the existence of EPTAS's. Finally, we should stress that bidimensionality is not the only technique to derive subexponential parameterized algorithms. Alternative approaches have been proposed in Alon, Lokshtanov, and Saurabh [35], Dorn, Fomin, Lokshtanov, Raman, and Saurabh [231] (for directed graphs), Fomin and Villanger [326], Tazari [639] and Thilikos [641], to name but a few. As usual, space considerations preclude us from including these studies. Moreover, they will be treated in detail in the upcoming book on Kernelization by Fomin, Lokshtanov, and Saurabh [323].

29.11 Exercises

Exercise 29.11.1 (Downey and Fellows [247]) (Abrahamson, Downey, and Fellows [3]) Prove (ii) and (iii) of Theorem 29.2.2.

Exercise 29.11.2 (Flum and Grohe [311]) Prove Theorem 29.3.1 using the methods of the proof of Theorem 25.1.1.

Exercise 29.11.3 Prove that MINI-3CNF SAT $\in M[1]$.
(Hint: Show MINI-3CNF SAT is reducible to $k \log n$ MINI 3-CNF SAT.)

Exercise 29.11.4 (Cai and Juedes [126, 127])   Prove Theorem 29.5.3. That is, assuming ETH there is no $O^*(2^{o(k)})$ algorithm for k-PATH, FEEDBACK VERTEX SET, and no $O^*(2^{\sqrt{k}})$ algorithm for PLANAR VERTEX COVER, PLANAR INDEPENDENT SET, PLANAR DOMINATING SET, and PLANAR RED–BLUE DOMINATING SET.
(Hint: The first ones follow the same route as for VERTEX COVER. For the PLANAR VERTEX COVER one, first note that there is a linear FPT reduction from an instance of VERTEX COVER to one of HAMILTON CYCLE, and Garey, Johnson, and Tarjan [339] established that given an instance φ of classical 3SAT, with n variables and m clauses, we can construct a planar graph G with $O(m^2)$ vertices and edges in polynomial time such that φ is satisfiable iff G has a Hamilton cycle. The other planar ones are similar.)

Exercise 29.11.5 (Lokshtanov, Marx, and Saurabh [495])   Prove that, assuming ETH, there are no $f(k)n^{o(k)}$ algorithms for MULTICOLORED CLIQUE defined below.

MULTICOLORED CLIQUE

Input: A graph G and a proper coloring of G into colors $\{1, \ldots, k\}$.
Parameter: k.
Question: Is there a size k clique with one vertex from each color class?

(Hint: Reduce from INDEPENDENT SET.)

Exercise 29.11.6 (Lokshtanov, Marx, and Saurabh [495])   Prove that, assuming ETH, there are no $f(k)n^{o(k)}$ algorithms for DOMINATING SET.
(Hint: Reduce from MULTICOLORED CLIQUE. Let (G, k) be an instance of that. Construct a new graph G' as follows. For each $i \le k$ let V_i be the vertices of G colored i, and for each i, j let $E_{i,j}$ be the edges in $G[V_i \sqcup V_j]$. Make G' by taking a copy of V_i and make it a clique. For every $i \le k$ add a set S_i of $k + 1$ vertices all adjacent to those in V_i. Then for each $i \ne j$, and for each $u \in V_i$ and $v \in V_j$ with $uv \notin E_{i,j}$ add a new vertex x_{uv} making at adjacent to all vertices in $(V_i \setminus \{u\}) \cup (V_j \setminus \{v\})$.) Argue that G has a k-clique iff G' has a dominating set of size k.)

Exercise 29.11.7   Verify the correctness of reduction in Theorem 29.5.7.

Exercise 29.11.8 (Lokshtanov, Marx, and Saurabh [496])   Assume ETH. Prove that the following two problems have no $O^*(2^{o(k \log k)})$ algorithms.

CONSTRAINED PERMUTATION

Input: Subsets $S_1, \ldots, S_n$ of $[k]$.
Parameter: k.
Question: Is there a permutation ρ of $[k]$ such that for each $i \in [1, n]$, there is a j with $\rho(j), \rho(j + 1) \in S_i$?

DISTORTION

Input: A graph G.
Parameter: k.
Question: Is there an embedding $g : V(G) \to \mathbb{Z}$ such that for all $u, v \in V(G)$, the shortest path distance between u and v, $D(u, v) \le |g(u) - g(v)| \le k \cdot D(u, v)$?

(Hint. These are not easy. For the first reduce $2k \times 2k$-BIPARTITE PERMUTATION INDEPENDENT SET. For the second use the first. For the second, given an instance I of CONSTRAINED PERMUTATION with subsets $S_1, \ldots, S_n$ and representing $[k]$ as $u_1, \ldots, u_k$, we make an instance of DISTORTION with parameter $2k$ as follows. The vertex set will contain for $1 \leq i \leq n$ and $1 \leq j \leq k$ a vertex u_j^i, a vertex s_i for each S_i, two cliques C_a and C_b of size $k+1$ with vertices $c_c^1, \ldots, c_a^{k+1}$ and $c_b^1, \ldots, c_b^{k+1}$, respectively, and finally a path with n edges of vertices $v_1, \ldots, v_{n+1}$. Additional edges are added to this gadget as follows. Add all edges for all vertices in the clique C_a except c_a^1 to v_1, and from all in C_b except c_b^1 to v_{n+1}. For $1 \leq i < n$ and $1 \leq j \leq k$, make u_j^i adjacent to v_i, v_{i+1}, u_j^{i+1}. For $1 \leq j \leq k$ make u_j^n adjacent to v_n, v_{n+1}. The construction of I' is finished by making s_i adjacent to u_j^i if $u_j \in S_i$. Prove that I is a "yes" instance of k-CONSTRAINED PERMUTATION iff I' is a "yes" instance of $2k$-DISTORTION.)

Exercise 29.11.9 Prove Lemmas 29.9.2 and 29.9.3.

29.12 Historical Notes

29.12.1 Prehistory

Ever since the Kintala–Fischer [444] article, authors have looked at limiting the amount of nondeterminism allowed in an oracle computation. We have already seen that our work has impact in these smaller classes. Up to the work of Abrahamson, Fellows, Ellis, and Mata [6], authors had always studied these smaller classes "by themselves". In some ways, this is akin to our studying a single k-slice of a problem rather than looking at the asymptotic behavior as $k \to \infty$. Immediately, one is faced with the problem of finding good reducibilities. The trouble is that we are now looking at "nondeterminism in P", and, hence, the reducibilities will need to be bounded by some polynomial and hence be machine dependent. Typically, these studies fix a model. This model is usually a multitape Turing machine, so that the Linear Speedup Theorem works, and we do not need to worry about multiplicative constants. Such studies into limited nondeterminism in P include Álvarez, Díaz, Torán [38], Buss and Goldsmith [115], Gurevich and Shelah [368], Joseph, Pruim, and Young [423], Geske [344], Regan [577] and others. All of these studies attempt to study the problems by limiting the number of nondeterministic moves available and keeping the complexity of the reductions down. The Kintala–Fischer [444] article is rather technical. As we have seen in a Sect. 29.2, Kintala and Fischer define a class we denote by $\mathrm{NP}[f(n)]$. This class is the collection of languages accepted by nondeterministic (multitape) Turing machines allowing only $f(n)$ c-ary nondeterministic moves. Concentrating on $\mathrm{NP}[\log^k n]$ for $k > 0$, they use a Baker–Gill–Solovay [50] type construction to get an oracle separation of the hierarchy $\mathrm{NP}[\log n] \subseteq \mathrm{NP}[\log^2 n] \subseteq \cdots \subseteq \mathrm{NP}[n]$.

Geske [344] similarly studies the relationship between the classes $\mathrm{DTIME}(n^q)$ and $\mathrm{NTIME}(n^q)$. The classes in P looked at by Buss and Goldsmith [115] are rather more elaborate, and, furthermore, they define reduction procedures to enable them to address completeness issues. Fix the model as a multitape machine. They define

$$P_q = \bigcup_{d \in \mathbb{N}} \text{DTIME}\big[n^q \log^d n\big]$$

and $N^m P_q$ to be the collection of languages accepted by nondeterministic machines running in time $O(n^q \log^d n)$ (for some d), making at most $m \log n$ nondeterministic binary moves in the computation. In the Buss–Goldsmith article, it is shown that the language

$$G_q^m = \big\{ m\#x\#1^j : M \text{ accepts } x \text{ within } j|x|^{q-1} \text{ steps}$$

$$\text{using at most } m \log |x| \text{ many choices}\big\}$$

is complete for $N^m P_q$ under the quasi-linear reductions. Here, a reduction is called *quasi-linear* if it runs in time $O(n \log^d n)$ for some d. (The notion of quasi-linear time also occurs in Geske [344].) The Buss–Goldsmith article clearly demonstrates the problems which arise when one tries to study the types of computational problems that are the basis of this book, by only studying a *single* slice of a parameterized language. One is forced to only consider reductions that are naturally small with respect to the possible amount of nondeterminism in that particular slice. Not only does this fact make the setting rather sensitive to the model chosen, but, as with the Abrahamson et al. [6] article, one then needs the languages themselves to be complete for P subject to the refined reductions, and hence there seems to be no way to address the apparent parametric intractability of 3SAT, DOMINATING SET, CLIQUE, or many other natural problems. Buss and Goldsmith do show that the Abrahamson et al. [6] completeness results and hence some of our W-completeness results can be recast in their setting. To do this, Buss and Goldsmith observe that the reductions of Abrahamson et al. [6] can be made to be quasi-linear in the following sense: For a parameterized set C, it becomes what they call $N^* P_q$-complete if for each k, there is a $k' \geq k$ such that $C_{k'}$ is hard for $N^k P_q$ with respect to quasi-linear reductions. In this setting, for instance, we discover that WEIGHTED SATISFIABIL-ITY is $N^* P_2$-complete. These results can be viewed as an analysis of the complexity of the reductions used in the Abrahamson et al. [6] article, completeness results from the Buss–Goldsmith setting will lift to ones for W.

The article of Regan [577] takes a different tack. Regan noted that the question of the existence of an n^c algorithm to determine if a k-clique existed in a graph G of size n was nontrivial (this seems to be the first place this observation is mentioned). This observation led him to study the structure of the kth slice of what we call parameterized problems. He considers the local rather than the global behavior, concentrating on a fixed k. In [577], he analyzes the structure of such sets by analogy with sparse sets. He studies the density of hard instances and the extent to which an analog of Mahaney's theorem (the existence of sparse NP-complete sets imply P = NP) holds. He leaves open several interesting questions.

29.12.2 History

In the context of parameterized complexity, the first real studies connecting the classes to limited nondeterminism were Cai and Chen [120, 121] and Abrahamson, Downey, and Fellows [3]. Cai and Chen [120, 121] looked at another variation of the Kintala–Fischer-type model which they refer to as the *Guess and Check* model.

Definition 29.12.1 (Guess and Check [120]) Let $s(n)$ be a function and let C be a complexity class. A language L is in the class $GC(s(n), C)$ if there are a language $A \in C$ and an integer $c > 0$ such that for all x, $x \in L$ if and only if $\exists y \in \{0, 1\}^*$, $|y| \leq c \cdot s(|x|)$, and $(x, y) \in A$.

Intuitively, the first component $s(n)$ in the $GC(s(n), C)$ model specifies the length of the guessed string y, which is the amount of nondeterminism allowed to be made in the computation, whereas the second component C specifies the verifying power of the computation.

The classes relevant to the W-hierarchy are those of the form $GC(s(n), \Pi_h^B)$, where $s(n) = \omega(\log n)$, and Π_h^B is the class of languages accepted by $O(\log n)$ time alternating Turing machines that make at most h alternations and always begin with universal states (such alternating Turing machines will be called "Π_h^B-ATMs"). Cai and Chen note that, to our current knowledge, a deterministic polynomial-time computation can only "guess" a string of length at most $\Theta(\log n)$ by exhaustively enumerating all strings of length $\Theta(\log n)$. Moreover, Yao [669] has proven that the class Π_h^B is a proper subclass of the class P. It seems a reasonable conclusion that the model $GC(s(n), \Pi_h^B)$ has a (presumably) stronger guessing ability but (provably) weaker computational power than deterministic polynomial-time computations.

They prove the following result.

Theorem 29.12.1 (Cai and Chen [120, 121]) *For any nondecreasing function $t(n)$ constructible in polynomial time and for any integer $h > 1$, the following language* CSAT(t, h) *is complete for the class $GC(t(n) \log(n), \Pi_h^B)$ under the polynomial-time reduction.*

CSAT(t, h): *the set of all strings of the form $x = \alpha \# w$, where w is an integer not larger than $t(|x|)$, and α is a Π_h-circuit that accepts an input of weight w.*

Using this, they gave the following necessary and sufficient condition for the W-hierarchy to collapse.

Theorem 29.12.2 (Cai and Chen [121]) *For all $h > 1$, $W[h] = $ FPT if and only if there is an unbounded, nondecreasing function $t(n) \leq n$ computable in polynomial time such that the class $GC(t(n) \log(n), \Pi_h^B)$ is a subclass of the class P.*

The material of Sect. 29.2 is due to Abrahamson, Downey, and Fellows [3] and Downey and Fellows [247]. In particular, Downey and Fellows were the first authors to note the connection between collapse of $W[1]$ to FPT and 3SAT being in DTIME$(2^{o(n)})$.

Matters stood there until Impagliazzo, Paturi, and Zane [411] introduced ETH, proved the Sparsification Lemma, and showed the equivalence to ETH to n-variable 3SAT being in DTIME$(2^{o(n)})$. This insight allowed for quantitative lower bounds on other computational problems using (in the [411] paper) SERF reductions. Cai and Juedes [126, 127] realized the connection between ETH and sharp lower bounds for parameterized problems. Downey, Estivill-Castro, Fellows, Prieto, and Rosamond [238] introduced the class $M[1]$ and established the results above. The $M[t]$ classes were introduced by Flum and Grohe [311]. The $O^*(2^{o(k \log k)})$ lower bounds for problems like CLOSEST STRING are due to Lokshtanov, Marx, and Saurabh [496]. Further historical comments on treewidth computations can be found in Chaps. 11 and 16. The proof of the Sparsification Lemma is drawn from [411]. No simple proof for this is result is known. The STRONG EXPONENTIAL TIME HYPOTHESIS (SETH) and was introduced by Impagliazzo and Paturi in [410]. Bidimensionality theory was introduced by Demaine, Fomin, Hajiaghayi, and Thilikos [209]. Contraction bidimensionality was introduced in Fomin, Golovach, and Thilikos [315].

29.13 Summary

We relate NP to parameterized complexity via the ETH, SETH, and prove results on XP-optimality. Namely, we give methods to show that, under reasonable hypotheses, current algorithms for various computational tasks are *optimal* up to an O-factor. Positive techniques aligned to this material obtained using the bidimensionality theory are discussed.

Kernelization Lower Bounds

30

30.1 Introduction

We have seen that FPT is equivalent to polynomial-time (many : 1) kernelization taking a parameterized problem instance (x, k) to a reduced instance (x', k') where the size of the instance $|(x', k')|$ is bounded by a function (perhaps exponential) of the parameter k. We have also seen that for many FPT problems, one can described polynomial-time kernelization algorithms that achieve a reduced instance size of polynomial or even linear size bounds.

In this chapter, we explore the natural questions:

What FPT parameterized problems admit many : 1 polynomial-time kernelization to kernels of size bounded by a polynomial function of the parameter (polynomial kernels)? *Are there plausible lower-bound techniques? Are there plausible lower-bound techniques for* Turing kernelization?

Of course, if $NP = P$, then all parameterized versions of problems in NP have constant size kernels. So we will need some complexity hypothesis in order to establish any lower bounds.

Most kernelization algorithms for FPT problems reduce a problem instance (x, k) to a reduced instance (x', k'), where in addition to the basic requirement that (x, k) is a yes instance if and only if (x', k') is a yes instance, the reduction procedure is such that it is "reversible" in the sense that (informally):

- A solution S for (x, k) can be computed in polynomial time from a solution S' for (x', k').
- A c-approximate solution S for (x, k) can be computed in polynomial time from a c' approximate solution S' for (x', k'), with $c \leq c'$.

This can all be suitably formalized in the notion of an *approximation-preserving* kernelization.

In Chap. 4, we briefly discussed how the hypothesis that $P \neq NP$ allows us to establish lower bounds for approximation-preserving kernelization for the parameterized VERTEX COVER problem, yielding a lower bound for kernel size of $1.36k$. There are other results of this kind in the literature. We do not pursue the topic of

R.G. Downey, M.R. Fellows, *Fundamentals of Parameterized Complexity*,
Texts in Computer Science, DOI 10.1007/978-1-4471-5559-1_30,
© Springer-Verlag London 2013

lower bounds for approximation-preserving kernelization any further, but do reiterate that this is generally a very natural hypothesis as most known kernelization algorithms have this property. As always, much remains to be explored.

For a different sort of lower bound, we have the following argument by Chen, Fernau, Kanj, and Xia.

Theorem 30.1.1 (Chen et al. [140]) *For all $\varepsilon > 0$, PLANAR VERTEX COVER has no kernel consisting of a planar graph of vertex size $(\frac{4}{3} - \varepsilon)k$ unless* NP = P.

Proof Suppose there was such an ε. We argue that then NP = P as follows. Given an instance (G, k) of PLANAR VERTEX COVER, if $k \leq \frac{4}{\frac{4}{3}-\varepsilon+4} \cdot |V|$, run the Nemhauser–Trotter algorithm on G; otherwise, run the size $4k$ kernel (or better) for PLANAR INDEPENDENT SET on the dual problem $(G, |V| - k)$. Let $\hat{G} = (\hat{V}, \hat{E})$ be the reduced instance obtained by this process. If $k \leq \frac{4}{\frac{4}{3}-\varepsilon+4} \cdot |V|$, then

$$|\hat{V}| \leq \frac{4(\frac{4}{3} - \varepsilon)}{\frac{4}{3} - \varepsilon + 4} \cdot |V| < |V|.$$

Else,

$$|\hat{V}| \leq 4\big(|V| - k\big) < 4\left(\frac{\frac{4}{3} - \varepsilon}{\frac{4}{3} - \varepsilon + 4}\right) \cdot |V| < |V|.$$

The fact that $|\hat{V}| < |V|$ yields a polynomial-time algorithm to decide PLANAR VERTEX COVER, which is known to be NP-complete. $\square$

We next explore new generic (rather than *ad hoc*) techniques, which, under the hypothesis that the classical polynomial-time hierarchy does not collapse, give us the means to show that a wide range of natural parameterized problems, although they are FPT, do not admit polynomially size-bounded, polynomial-time many : 1 kernelization.

30.2   A Generic Engine for Lower Bounds of Kernels

In this section we develop the main engine for proving Theorems 30.3.1 and 30.4.1. For this, we develop a generic framework for showing the unlikelihood of polynomial kernels for parameterized problems fulfilling certain criteria. As a first step, we describe the original framework of Bodlaender, Downey, Fellows, and Hermelin [83], and then consider extensions of that framework, such as the notion of *cross-composition*.

Our framework revolves around the notion of *distillation algorithms* for NP-complete problems. We first introduce this notion, and then use an argument of Fortnow and Santhanam [328] to show that a distillation algorithm for any NP-complete problem implies the collapse of the polynomial hierarchy to the third level.

We then define a parameterized analog of a distillation algorithm, which we term a *composition algorithm*. Following this, we show that if a compositional parameterized problem has a polynomial kernel, then its classical counterpart has a distillation algorithm. We begin with the central notion of our framework:[1]

Definition 30.2.1 (Harnik and Naor [379], Bodlaender, Downey, Fellows, and Hermelin [83])   An OR-*distillation algorithm* for a classical problem $L \subseteq \Sigma^*$ is an algorithm that
- receives as input a sequence $(x_1, \ldots, x_t)$, with $x_i \in \Sigma^*$ for each $1 \leq i \leq t$,
- uses time polynomial in $\sum_{i=1}^{t} |x_i|$,
- and outputs a string $y \in \Sigma^*$ with
  1. $y \in L$ iff $x_i \in L$ for some $1 \leq i \leq t$.
  2. $|y|$ is polynomial in $\max_{1 \leq i \leq t} |x_i|$.
We can similarly define AND-distillation by replacing the first condition of the third item by
- $y \in L$ iff $x_i \in L$ for *all* $1 \leq i \leq t$.

We remark that the above is a *self-distillation* in the sense that $y \in L$ iff $x_i \in L$ for some $1 \leq i \leq t$. It was pointed out by Fortnow and Santhanam that the target might be different, in the same way that a kernelization could be into, for example, an annotated kernel. Then a *weak distillation* of L into $\hat{L}$ is identical to the above except we replace this line by $y \in \hat{L}$ iff $x_i \in L$ for some $1 \leq i \leq t$.

Given a sequence of t instances of L, an OR-distillation algorithm gives an output that is equivalent to the sequence of instances, in the sense that a collection with at least one yes instance (i.e. instance belonging to L) is mapped to a yes instance, and a collection with only no instances is mapped to a no instance. Of course, this is easy, as we could simply make $y = \vee x_i$, but that would contradict the size bound on y. The algorithm is allowed to use polynomial time in the total size of all instances. The crux is that its output must have size bounded by a polynomial in the size of the largest of the instances from the sequence, rather than in the total length of the instances in the sequence.

It seems highly implausible that NP-complete problems have distillation algorithms. We give evidence for this in the following lemma, which shows that a distillation algorithm for an NP-complete problem implies a collapse in the polynomial hierarchy.

Lemma 30.2.1 (Fortnow and Santhanam [328])
1. *If any* NP-*complete problem has a distillation algorithm then* co-NP $\subseteq$ NP/poly *and hence* $\Sigma_n^p = \Sigma_3^p$, *for all* $n \geq 3$.

---

[1] In the definition below, we have taken the liberty of re-expressing the Harnik–Naor notions in the manner of [83].

2. *Moreover, if any* NP-*hard problem has a weak distillation algorithm to an* NP-*complete problem then* co-NP $\subseteq$ NP/poly *and hence* $\Sigma_n^p = \Sigma_3^p$, *for all* $n \geq 3$.

Proof We prove 1, 2 being entirely similar. Let L be an NP-complete problem with a distillation algorithm $\mathcal{A}$, and let $\overline{L}$ denote the complement of L. We show that using $\mathcal{A}$, we can design a nondeterministic Turing-machine (NDTM) that, with the help of the polynomial advice, can decide $\overline{L}$ in polynomial time. This will that show co-NP $\subseteq$ NP/poly, and combined with Yap's theorem, Theorem 1.2.1, we get the result.

Let $n \in \mathbb{N}$ be a sufficiently large integer. Denote by $\overline{L}_n$ the subset of strings of length at most n in the complement of L, i.e. $\overline{L}_n = \{x \notin L : |x| \leq n\}$. By its definition, given any $x_1, \dots, x_t \in \overline{L}_n$, the distillation algorithm $\mathcal{A}$ maps $(x_1, \dots, x_t)$ to some $y \in \overline{L}_{n^c}$, where c is some constant independent of t. Any sequence containing a string $x_i \notin \overline{L}_n$ is mapped to a string $y \notin \overline{L}_{n^c}$. The main part of the proof consists in showing that there exists a set $S_n \subseteq \overline{L}_{n^c}$, with $|S_n|$ polynomially bounded in n, such that for any $x \in \Sigma^{\leq n}$ we have the following:

- If $x \in \overline{L}_n$, then there exist strings $x_1, \dots, x_t \in \Sigma^{\leq n}$ with $x_i = x$ for some i, $1 \leq i \leq t$, such that $\mathcal{A}(x_1, \dots, x_t) \in S_n$.
- If $x \notin \overline{L}_n$, then for all strings $x_1, \dots, x_t \in \Sigma^{\leq n}$ with $x_i = x$ for some i, $1 \leq i \leq t$, we have $\mathcal{A}(x_1, \dots, x_t) \notin S_n$.

Given such a set $S_n \subseteq \overline{L}_{n^c}$ as advice, a NDTM M can decide whether a given $x \in \Sigma^{\leq n}$ is in $\overline{L}$ as follows: It first guesses t strings $x_1, \dots, x_t \in \Sigma^{\leq n}$, and checks whether one of them is x. If not, it immediately rejects. Otherwise, it computes $\mathcal{A}(x_1, \dots, x_t)$, and accepts if and only if the output is in S_n. It is immediate that M correctly determines (in the nondeterministic sense) whether $x \in \overline{L}_n$. In the remaining part of the proof, we show that there exists such an advice $S_n \subseteq \overline{L}_{n^c}$ as required above.

We view $\mathcal{A}$ as a function mapping strings from $(\overline{L}_n)^t$ to $\overline{L}_{n^c}$, and say a string $y \in \overline{L}_{n^c}$ *covers* a string $x \in \overline{L}_n$ if there exist $x_1, \dots, x_t \in \Sigma^{\leq n}$ with $x_i = x$ for some i, $1 \leq i \leq t$, and with $\mathcal{A}(x_1, \dots, x_t) = y$. Clearly, our goal is to find polynomial-size subset of $\overline{L}_{n^c}$ which covers all strings in $\overline{L}_n$. By the pigeonhole principle, there is a string $y \in Y$ for which $\mathcal{A}$ maps at least $|(\overline{L}_n)^t|/|\overline{L}_{n^c}| = |\overline{L}_n|^t/|\overline{L}_{n^c}|$ tuples in $(\overline{L}_n)^t$ to. Taking the tth square root, this gives us $|\overline{L}_n|/|\overline{L}_{n^c}|^{1/t}$ distinct strings in $\overline{L}_n$ which are covered by y. Hence, by letting $t = \lg|\overline{L}_{n^c}| = O(n^c)$, this gives us a constant fraction of the strings in $\overline{L}_n$. It follows that we can repeat this process recursively in order to cover all strings in $\overline{L}_n$ with a polynomial number of strings in $\overline{L}_{n^c}$. $\qquad\square$

We remark that the Fortnow–Santhanam argument has a number of strong consequences for the assumption co-NP $\not\subseteq$ NP/poly. For example we have the following theorem of Buhrman and Hitchcock. We will say that a language A has an *and function* iff there is a polynomial-time computable function g such that for all n, $x_1, \dots, x_n \in A$ iff $g(x_1, \dots, x_n) \in A$; and similarly *or function*. For example, $\overline{\text{SAT}}$ has an and function.

Theorem 30.2.1 (Buhrman and Hitchcock [114])

(1) *If S is a set* NP-*hard under Karp reductions, then S must have exponential density unless* co-NP $\subseteq$ NP/poly. *That is, there is some* $\varepsilon > 0$ *such that* $S_{=n} =_{\text{def}} \{x \in S \mid |x| = n\}$ *has size at least* $2^{n^{\varepsilon}}$ *for infinitely many n.*

(2) *Furthermore, if* $\overline{\text{SAT}}$, *or any other* co-NP-*complete language A with an and function, Karp reduces to a set of subexponential density, then* co-NP $\subseteq$ NP/poly.

Proof Clearly (2) implies (1). Suppose that A satisfies the hypotheses of (2). Let $g(x_1, \ldots, x_n)$ be an *and* function for A. Assume $A \leq_m^p S$ via a polynomial-time computable f, so that $x \in A$ iff $f(x) \in S$ for all x. Now suppose that S has subexponential density, so that for all $\varepsilon > 0$,

$$\left| S_{=n} =_{\text{def}} \{x \in S \mid |x| = n\} \right| < 2^{n^{\varepsilon}}.$$

For strings x with $|x| = n$, we will say that $z \in S$ *is an* NP-*proof for* $x \in A$ if and only if there exist $x_1, \ldots, x_n$ such that for all i, $|x_i| = n$, and there exists an i with $x = x_i$ and $f(g(x_1, \ldots, x_n)) = z$.

As the Fortnow–Santhanam proof of Lemma 30.2.1, we argue that there is a string z_1 that is an NP-proof for at least half the strings of $A_{=n}$. Exactly as in the Fortnow–Santhanam proof, we then use recursion on the remaining strings for which z_1 is not an NP proof. The advice is the sequence of z_i's, and showing that $A \in$ NP/poly. Note that if z is an NP-proof for precisely t strings $x \in A$, then

$$\left| \{ (x_1, \ldots, x_n) \mid |x_i| = n \wedge f(g(x_1, \ldots, x_n)) = z \} \right| \leq t^n.$$

Assuming that both f and g run in time n^c, if we let $m_n = n^{2c^2}$, we see

$$\left| f(g(x_1, \ldots, x_n)) = z \right| \leq m_n.$$

Since S has subexponential density, for large enough n, $|S_{\leq m_n}| < 2^n$.

Now let t be the largest integer such that some z_1 is a NP-proof for t elements of length n in A. Again as in the Fortnow–Santhanam proof, we observe that for each tuple $(x_1, \ldots, x_n)$, with $x_i \in A$ for all i, $f(g(x_1, \ldots, x_n))$ maps to some fixed z in $S_{\leq m_n}$. Therefore, $t^n |S_{\leq m_n}| \geq |A_{=n}|^n$. This fact implies

$$t^n 2^n \geq |A_{=n}|.$$

Therefore, $t \geq \frac{|A_{=n}|}{2}$. That is, z_1 is an NP proof for at least half of the elements of A. Now recursing on this process on the remaining elements gives the desired sequence $z_1, \ldots, z_p$ of advice and the argument follows as in the Fortnow–Santhanam proof. □

We remark that in [114], Buhrman and Hitchcock prove a stronger version of the result above.

Theorem 30.2.2 (Buhrman and Hitchcock [114]) *Sets S* NP-*hard under polynomial-time Turing reductions making at most* $n^{1-\varepsilon}$ *queries must have exponential density unless* co-NP $\subseteq$ NP/poly.

We invite the reader to prove this in Exercise 30.13.3.

We next introduce the notion of a composition algorithm for parameterized problems. In some sense, one can view a composition algorithm as the parametric-analogue of a distillation algorithm.

Definition 30.2.2 (Bodlaender, Downey, Fellows, and Hermelin [83]) A *composition algorithm* for a parameterized problem $L \subseteq \Sigma^* \times \mathbb{N}$ is an algorithm that

- receives as input a sequence $((x_1, k), \ldots, (x_t, k))$, with $(x_i, k) \in \Sigma^* \times \mathbb{N}^+$ for each $1 \leq i \leq t$,
- uses time polynomial in $\sum_{i=1}^{t} |x_i| + k$,
- and outputs $(y, k') \in \Sigma^* \times \mathbb{N}^+$ with
  (1) $(y, k') \in L$ iff $(x_i, k) \in L$ for some $1 \leq i \leq t$,
  (2) k' is polynomial in k.

We can similarly define AND-composition.

A composition algorithm outputs an equivalent instance to an input sequence in the same logical sense as for a distillation algorithm. The difference is that here, the *parameter* of the output instance is required to be bounded by a polynomial function of the common parameter k of the input sequence, rather than that the *size* of the output instance is bounded by a polynomial function of the maximum size of the instances in the input sequence.

We call classical problems with distillation algorithms OR, AND-*distillable problems*, and parameterized problems with composition algorithms OR, AND-*compositional problems*. There is a deep connection between distillation and composition, obtained via polynomial kernelization.

Lemma 30.2.2 (Bodlaender, Downey, Fellows, and Hermelin [83]) *Let L be a compositional parameterized problem whose derived classical problem L_c is NP-complete. If L has a polynomial kernel, then L_c is also distillable.*

Proof Let $x_1^c, \ldots, x_t^c \in \Sigma^*$ be instances of L_c, and let $(x_i, k_i) \in \Sigma^* \times \mathbb{N}^+$ denote the instance of L from which x_i^c is derived, for all $1 \leq i \leq t$. Since L_c is NP-complete, there exist two polynomial-time transformations $\Phi : L_c \to \text{SAT}$ and $\Psi : \text{SAT} \to L_c$. We use the composition and polynomial kernelization algorithms of L, along with Φ and Ψ, to obtain a distillation algorithm for L_c. The distillation algorithm proceeds in three steps.

Set $k = \max_{1 \leq i \leq t} k_i$. In the first step, we take the subsequence in $((x_1, k_1), \ldots, (x_t, k_t))$ of instances whose parameter equals ℓ, for each $1 \leq \ell \leq k$. We apply the composition algorithm on each one of these subsequences separately, and obtain a new sequence $((y_1, k_1'), \ldots, (y_r, k_r'))$, where (y_i, k_i'), $1 \leq i \leq r$, is the instance obtained by composing all instances with parameters equaling the ith parameter value in $\{k_1, \ldots, k_t\}$. In the second step, we apply the polynomial kernel on each instance of the sequence $((y_1, k_1'), \ldots, (y_r, k_r'))$, to obtain a new sequence $((z_1, k_1''), \ldots, (z_r, k_r''))$, with (z_i, k_i'') the instance obtained from (y_i, k_i'), for each $1 \le i \le r$. Finally, in the last step, we transform each z_i^c, the instance of L_c derived from (z_i, k_i''), to a Boolean formula $\Phi(z_i^c)$. We output the instance of L_c that Ψ maps the disjunction of these formulas to, that is, the instance $\Psi(\bigvee_{1 \le i \le r} \Phi(z_i^c))$.

We argue that this algorithm distills the sequence $(x_1^c, \ldots, x_t^c)$ in polynomial time, and therefore is a distillation algorithm for L_c. First, by the correctness of the composition and kernelization algorithms of L, and by the correctness of Φ and Ψ, we have

$$\Psi\left(\bigvee_{1 \le i \le r} \Phi(z_i^c)\right) \in L_c \quad \text{iff} \quad \bigvee_{1 \le i \le r} \Phi(z_i^c) \in \text{SAT}$$

$$\text{iff} \quad \exists i, 1 \le i \le r : \Phi(z_i^c) \in \text{SAT}$$

$$\text{iff} \quad \exists i, 1 \le i \le r : z_i^c \in L_c$$

$$\text{iff} \quad \exists i, 1 \le i \le r : (z_i, k_i'') \in L$$

$$\text{iff} \quad \exists i, 1 \le i \le r : (y_i, k_i') \in L$$

$$\text{iff} \quad \exists i, 1 \le i \le t : (x_i, k_i) \in L$$

$$\text{iff} \quad \exists i, 1 \le i \le t : x_i^c \in L_c.$$

Furthermore, as each step in the algorithm runs in polynomial time in the total size of its input, and since the output of each step is the input of the next step, the total running-time of our algorithm is polynomial in $\sum_{i=1}^t |x_i^c|$. To complete the proof, we show that the final output returned by our algorithm is polynomially bounded in $n = \max_{1 \le i \le t} |x_i^c|$.

The first observation is that since each x_i^c is derived from the instance (x_i, k_i), $1 \le i \le t$, we have $r \le k = \max_{1 \le i \le t} k_i \le \max_{1 \le i \le t} |x_i^c| = n$. Therefore, there are at most n instances in the sequence $((y_1, k_1'), \ldots, (y_r, k_r'))$ obtained in the first step of the algorithm. Furthermore, as each (y_i, k_i'), $1 \le i \le r$, is obtained via composition, we know that k_i' is bounded by some polynomial in $\ell \le k \le n$. Hence, since for each $1 \le i \le r$, the instance (z_i, k_i'') is the output of a polynomial kernelization on (y_i, k_i'), we also know that (z_i, k_i'') and z_i^c have size polynomially bounded in n. It follows that $\sum_{i=1}^r |z_i^c|$ is polynomial in n, and since both Φ and Ψ are polynomial time, so is $\Psi(\bigvee_{1 \le i \le r} \Phi(z_i^c))$. $\qquad\square$

30.3   Applications of OR-Composition

In this section we will prove the following result.

Theorem 30.3.1 (Bodlaender, Downey, Fellows, and Hermelin [83]) *Unless* co-NP $\subseteq$ NP/poly, *none of the following* FPT *problems have polynomial kernels:*

- k-PATH, k-CYCLE, k-EXACT CYCLE *and* k-BOUNDED TREEWIDTH SUB-GRAPH.

- k-MINOR ORDER TEST *and* k-IMMERSION ORDER TEST.
- k-PLANAR SUBGRAPH TEST *and* k-PLANAR INDUCED SUBGRAPH TEST.
- k, σ-SHORT NONDETERMINISTIC TURING MACHINE COMPUTATION (*this is defined below*).
- w-VERTEX COVER, w-INDEPENDENT SET *and* w-DOMINATING SET (*these are treewidth bounded versions of the given problems and are defined below*).

Lemma 30.3.1 (Bodlaender, Downey, Fellows, and Hermelin [83]) *Let L be a parameterized graph problem such that for any pair of graphs G_1 and G_2, and any integer $k \in \mathbb{N}$, we have $(G_1, k) \in L \vee (G_2, k) \in L$ iff $(G_1 \cup G_2, k) \in L$, where $G_1 \cup G_2$ is the disjoint union of G_1 and G_2. Then L is compositional.*

Proof Given $(G_1, k), \ldots, (G_t, k)$, take G to be the disjoint union $G_1 \cup \cdots \cup G_t$. The instance (G, k) satisfies all requirements of Definition 30.2.2. $\qquad\square$

As a direct corollary of the lemma above, we find that all of the following FPT problems are compositional: k-PATH, k-CYCLE, k-CHEAP TOUR, and k-EXACT CYCLE. Other examples include k-CONNECTED MINOR ORDER TEST, in FPT by Robertson and Seymour's Graph Minor Theorem, k-PLANAR CONNECTED INDUCED SUBGRAPH TEST, in FPT due to Eppstein [273], and k-MULTIPLE INTERVAL CLIQUE—the problem of deciding whether a given multiple interval graph has a k-clique (Fellows, Hermelin, Rosamond, and Vialette [291]), and the k-PLANAR SUBGRAPH ISOMORPHISM problem (shown FPT by Eppstein [273]).

As an example of a non-graph-theoretic problem which is distillable, consider the parameterized variant of the Levin-Cook generic NP-complete problem—the k, σ-SHORT NONDETERMINISTIC TURING MACHINE COMPUTATION problem. In this problem, we receive as input a nondeterministic Turing machine M with alphabet-size σ, and an integer k, and the goal is to determine in FPT-time, with respect to both k and σ, whether M has a computation path halting on the empty input in at most k steps. This problem can be shown to be in FPT by applying the algorithm which exhaustively checks all global configurations of M as per Cesati and Di Ianni [137], as we discussed in Chap. 22.

Lemma 30.3.2 (Bodlaender, Downey, Fellows, and Hermelin [83]) k-SHORT NONDETERMINISTIC TURING MACHINE COMPUTATION *is compositional.*

Proof Given $(M_1, k), \ldots, (M_t, k)$, we create a new NDTM M, which is the disjoint union of all M_i's, in addition to a new unique initial state which is connected the initial states of all M_i by an ε-edge. (That is, by a nondeterministic transition that does not write anything on the tape, nor moves the head.) Note that M has alphabet-size σ. Letting $k' = 1 + k$, the instance $(M, (k', \sigma))$ satisfies all requirements of Definition 30.2.2. $\qquad\square$

30.3.1 Notes on Turing Kernels

We recall from Chap. 5 the problems

ROOTED k-LEAF OUT BRANCHING
Instance: A directed graph $D = (V, E)$, and a distinguished vertex r of G.
Parameter: An integer k.
Question: Is there an out tree in D with at least k leaves and having r as the root?

ROOTED k-LEAF OUT TREE
Instance: A directed graph $D = (V, E)$ (with root r).
Parameter: An integer k.
Question: Is there a tree in D with at least k leaves (with root r)?

and the unrooted variants k-LEAF OUT BRANCHING, and k-LEAF OUT TREE.

The rooted versions were shown to have quadratic kernels in Corollaries 5.1.3 and 5.1.4. In Sect. 5.1.5 we showed that both k-LEAF OUT BRANCHING and k-LEAF OUT TREE have Turing kernels. We note the following.

Theorem 30.3.2 (Fernau, Fomin, Lokshtanov, Raible, Saurabh, and Villanger [306]) k-LEAF OUT TREE *has no polynomial kernel unless* co-NP $\subseteq$ NP/poly.

Proof Lemma 30.3.1 clearly applies. □

With somewhat more work, we can also show the same for k-LEAF OUT BRANCHING. But the rooted case *does* have a polynomial kernel. The same holds for the problem of CLIQUE in graphs of max degree d, as observed by Hermelin, Kratsch, Soltys, Wahlström, and Wu [388]. This is easily seen to be compositional, but the rooted case has an almost trivial polynomial kernelization. So for both of these problems, which do not admit polynomial many : 1 kernelization (in the unrooted case), we do have polynomial Turing kernelization, just by trying each vertex as the root (and transferring to the rooted case). This suggest that lower bounds for many : 1 kernelization may be too pessimistic from a practical perspective.

The question of whether there is similar machinery for proving the non-existence of Turing kernels (under some reasonable complexity assumption) remains a very interesting question. Currently, the best that has been offered is a completeness program. We discuss this at the end of the chapter, in Sect. 30.12.

30.4 AND Composition

We now turn to extending the framework presented in the previous section so that it captures other (arguably more) important FPT problems not captured by Theorem 30.3.1. In particular, we provide a complete proof for the following result.

Theorem 30.4.1 (Bodlaender, Downey, Fellows, and Hermelin [83]) *Unless NP-complete problems have AND-distillation algorithms, none of the following* FPT *problems have polynomial kernels*:

- k-TREEWIDTH, k-PATHWIDTH, k-CUTWIDTH, *and* k-MODIFIED CUTWIDTH.

For a long time, the situation was as above. There was no "classical" evidence that AND-distillation implied any collapse until a remarkable solution to this conundrum was announced by a graduate student from MIT.

Theorem 30.4.2 (Drucker [263])   *If* NP-*complete problems have* AND-*distillation algorithms, then* co-NP $\subseteq$ NP/poly.

The Fortnow–Santhanam proof was relatively easy, but Drucker's proof of his most general result is complex. In the next section we will give a more or less elementary proof of Drucker's Theorem supplied by Drucker in correspondence. (This proof is also presented in [263]. It sacrifices some strength and generality in favor of a more elementary proof.) The proof of Drucker' most general result uses sophisticated ideas and also has consequences for quantum computation, and is too long to include. The proof presented here uses some probabilistic concepts and results from the theory of interactive proofs; however, aside from that it is self-contained.

The reader can skip the whole section and simply take the result as a black box *for applications*. We only include a proof for completeness. We get the following corollary.

Corollary 30.4.1 *Unless* co-NP $\subseteq$ NP/poly, *none of the following* FPT *problems have polynomial kernels*:

- k-TREEWIDTH, k-PATHWIDTH, k-CUTWIDTH, *and* k-MODIFIED CUTWIDTH.

The main observation we use for establishing these results is that an AND-variant of a composition algorithm for a parameterized problem L, yields a composition algorithm for $\overline{L}$, the complement of L. This observation is useful since a lot of problems have natural AND-compositions rather than OR-compositions. As any FPT problem has a polynomial kernel if and only if its complement also has one, showing that a co-FPT problem is compositional is just as good for our purposes as showing that its complement in FPT is compositional.

Lemma 30.4.1 (Bodlaender, Downey, Fellows, and Hermelin [83]) *Let L be a parameterized graph problem such that for any pair of graphs G_1 and G_2, and any integer $k \in \mathbb{N}$, we have $(G_1, k) \in L \wedge (G_2, k) \in L$ iff $(G_1 \cup G_2, k) \in L$, where $G_1 \cup G_2$ is the disjoint union of G_1 and G_2. Then L is AND-compositional.*

Proof Given $(G_1, k), \ldots, (G_t, k)$, take G to be the disjoint union $G_1 \cup \cdots \cup G_t$. Then $(G, k) \in L$ iff $(G_i, k) \in L$ for all i, $1 \le i \le t$. But then $(G, k) \in \overline{L}$ iff $(G_i, k) \in \overline{L}$ for any i, $1 \le i \le t$. It follows that (G, k) satisfies all requirements of Definition 30.2.2.   □

There are many FPT problems with a natural composition as above. These include "width problems" such as k-TREEWIDTH, k-PATHWIDTH, k-CUTWIDTH, and k-MODIFIED CUTWIDTH that we met in Chap. 10.

One example we will follow is

w-INDEPENDENT SET

Instance: A graph G, a tree-decomposition $\mathcal{T}$ of G, and an independent set I in G.

Parameter: Positive integers w.

Question: Find a maximal independent set in G.

These also include all problems parameterized by treewidth (problems that are solved by treewidth dynamic programming) such as w-VERTEX COVER, w-INDEPENDENT SET and w-DOMINATING SET.

We call the unparameterized variant of the problem above the INDEPENDENT SET WITH TREEWIDTH problem. Clearly, INDEPENDENT SET WITH TREEWIDTH is NP-complete by the straightforward reduction from INDEPENDENT SET which appends a trivial tree-decomposition to the given instance of INDEPENDENT SET.

w-INDEPENDENT SET REFINEMENT

Instance: A graph G, a tree-decomposition $\mathcal{T}$ of G, and an independent set I in G.

Parameter: A positive integer w.

Question: Does G have an independent set of size $|I| + 1$?

The unparameterized variant of the problem above is INDEPENDENT SET REFINEMENT WITH TREEWIDTH. It is easy to see that this problem is NP-complete by the reduction from INDEPENDENT SET WITH TREEWIDTH—Given an instance $(G, \mathcal{T}, k)$, construct the instance G', where G' is the graph obtained by adding k new vertices I to G which are adjacent to all the old vertices, and $\mathcal{T}'$ is the tree-decomposition obtained by adding the set of new vertices I to each node in $\mathcal{T}$.

Lemma 30.4.2 (Bodlaender, Downey, Fellows, and Hermelin [83]) w-INDEPENDENT SET REFINEMENT *is compositional, and furthermore, if* w-INDEPENDENT SET *has a polynomial kernel then so does* w-INDEPENDENT SET REFINEMENT.

Proof To prove the first part of lemma, suppose we are given t instances $(G_1, \mathcal{T}_1, I_1)$, $\ldots, (G_t, \mathcal{T}_t, I_t)$ of w-INDEPENDENT SET REFINEMENT. Consider the algorithm which maps this sequence of instances to $(G, \mathcal{T}, I)$, with G and $\mathcal{T}$ the disjoint unions of all G_i's and $\mathcal{T}_i$'s, respectively, $1 \le i \le t$, and with $I = \bigcup_{i=1}^{t} I_i$. Note that G has an independent set of size $|I| + 1$ if and only if there exists an i, $1 \le i \le t$, such that G_i an independent set of size $|I_i| + 1$. Moreover, as the width of each tree-decomposition $\mathcal{T}_i$ is w, $1 \le i \le t$, $\mathcal{T}$ also has width w.

We next show that a polynomial kernel for w-INDEPENDENT SET implies a polynomial kernel for w-INDEPENDENT SET REFINEMENT. For this, suppose w-INDEPENDENT SET has a polynomial kernel, and consider a given instance

$(G, \mathcal{T}, I)$ of w-INDEPENDENT SET REFINEMENT. Forgetting I, we create an equivalent instance $(G, \mathcal{T})$ of w-INDEPENDENT SET, and apply the polynomial kernelization algorithm on this instance to obtain the instance $(G', \mathcal{T}')$, with $|G'|$ and $|\mathcal{T}'|$ polynomially bounded by the width w of $\mathcal{T}$. We now consider the instance $(G', \mathcal{T}')$ as an instance of the unparameterized INDEPENDENT SET WITH TREEWIDTH problem. Using the reduction discussed above, we transform $(G', \mathcal{T}')$ in polynomial time to an equivalent instance $(G'', \mathcal{T}'', I'')$ of INDEPENDENT SET REFINEMENT. The parameterized instance $(G'', \mathcal{T}'', I'')$ is equivalent to $(G, \mathcal{T}, I)$, and has size polynomial in the width w of $\mathcal{T}$. $\square$

30.5   Proof of Drucker's Theorem

In this section we give Drucker's proof of the AND conjecture. The proof does use ideas from computational complexity theory not to be found elsewhere in the book, but the result is of sufficient importance that we have chosen to include it here.

Again we emphasize: For those interested only in applications, this whole section can be skipped.

We recall from the proof of the Sparsification Lemma that the *binary entropy function* $H(\alpha) : [0, 1] \to [0, 1]$ is defined by

$$H(\alpha) := -\alpha \log_2 \alpha - (1 - \alpha) \log_2(1 - \alpha)$$

on $(0, 1)$, with $H(0) = H(1) := 0$. Drucker's proof uses the following standard result bounding the number of binary strings of low Hamming weight:

Proposition 30.5.1 *For $t \in \mathbb{N}$ and $\alpha \in (0, 0.5)$,*

$$\sum_{0 \le \ell \le \alpha t} \binom{t}{\ell} \le 2^{H(\alpha)t}.$$

30.5.1 Statistical Distance and Distinguishability

The proof uses concepts from the theory of statistical distributions. All distributions in this section take only finitely many values; let $\mathrm{supp}(\mathcal{D})$ be the set of values assumed by $\mathcal{D}$ with nonzero probability, and let $\mathcal{D}(u) := \Pr[\mathcal{D} = u]$.

For a probability distribution $\mathcal{D}$ and $t \ge 1$, $\mathcal{D}^{\otimes t}$ will denote a t-tuple of independent samples from $\mathcal{D}$. Also, $\mathcal{U}_K$ will denote the uniform distribution over a multiset K.

Definition 30.5.1 Define the *statistical distance* of distributions $\mathcal{D}, \mathcal{D}'$ as

$$\left\| \mathcal{D} - \mathcal{D}' \right\|_{\mathrm{stat}} := \frac{1}{2} \sum_{u \in \mathrm{supp}(\mathcal{D}) \cup \mathrm{supp}(\mathcal{D}')} \left| \mathcal{D}(u) - \mathcal{D}'(u) \right|.$$

We use the following easily proved facts.

Proposition 30.5.2 *If X, Y are two random variables over a set S, and $\Delta :=$ $\|X - Y\|_{\text{stat}}$, then there is a subset $T \subseteq S$ such that*

$$\Pr_{x \sim X}[x \in T] \geq \Delta \quad and \quad \Pr_{y \sim Y}[y \notin T] \geq \Delta.$$

Proposition 30.5.3 *If X, Y are random variables over some shared domain S, and $R(\cdot)$ is a function taking inputs from S, then*

$$\big\| R(X) - R(Y) \big\|_{\text{stat}} \leq \|X - Y\|_{\text{stat}}.$$

Theorem 30.5.1 (Sahai and Vadhan [598], Fact 2.3) *Suppose (X_1, X_2, Y_1, Y_2) are distributions on a shared probability space Ω, that X_1 is independent of X_2, and that Y_1 is independent of Y_2. Then,*

$$\big\| (X_1, X_2) - (Y_1, Y_2) \big\|_{\text{stat}} \leq \|X_1 - Y_1\|_{\text{stat}} + \|X_2 - Y_2\|_{\text{stat}}.$$

Our goal will be to prove the following result, which implies and strengthens Theorem 30.4.2:

Theorem 30.5.2 *Let L be any NP-complete language. Suppose there is a deterministic polynomial-time reduction R that takes an arbitrarily long list of input strings $(x^1, \ldots, x^t)$ and outputs a string z, with*

$$z \in L \quad \Longleftrightarrow \quad \bigwedge_{j \in [t]} [x^j \in L].$$

Suppose further that R obeys the output-size bound $|z| \leq (\max_{j \leq t} |x^j|)^{O(1)}$, with the polynomial bound independent of t. Then, $\mathrm{NP} \subseteq \text{co-NP/poly}$.

A fortiori, we establish the following. Suppose there is any second, "target" language L', a pair of polynomially bounded functions $t(n), t'(n) : \mathbb{N} \to \mathbb{N}$ with $t(n) = \omega(1)$ and $t'(n) + 1 \leq t(n)/2$, and a deterministic polynomial-time reduction $R : \{0, 1\}^{t(n) \times n} \to \{0, 1\}^{\leq t'(n)}$, such that

$$R\big(x^1, \ldots, x^{t(n)}\big) \in L' \quad \Longleftrightarrow \quad \bigwedge_{j \in [t(n)]} [x^j \in L].$$

Then $\mathrm{NP} \subseteq \text{co-NP/poly}$.

30.5.2 Proof of Theorem 30.5.2

We consider mappings $R : \{0, 1\}^{t \times n} \to \{0, 1\}^{\leq t'}$, for fixed n, t, t'. For $A \subseteq \{0, 1\}^n$, define the distribution $\mathbf{R}_A := R(\mathcal{U}_A^{\otimes t})$, and for each $a \in \{0, 1\}^n$, define the distribution

$$\mathbf{R}_A[a] := R\big(\mathcal{U}_A^{\otimes(\mathbf{j}-1)}, a, \mathcal{U}_A^{\otimes(t-\mathbf{j})}\big),$$

where $\mathbf{j} \sim \mathcal{U}_{[t]}$.

Define $\beta(a, A)$, the *standout factor of a with respect to A*, as

$$\beta(a, A) := \left\| \mathbf{R}_A[a] - \mathbf{R}_A \right\|_{\text{stat}}. \tag{30.1}$$

Essentially, the proof of Theorem 30.5.2 works as follows: for each $n > 0$, we exhibit a $\operatorname{poly}(n)$-size collection of $\operatorname{poly}(n)$-size sets $A_i \subseteq L_n,$[2] such that every other element $x \in L_n$ will have standout factor $\beta(x, A_i) < 1 - \Omega(1)$ for at least one A_i. On the other hand, each $x \notin L_n$ will have standout factor 1 against each A_i. Thus, if a polynomial-time Verifier challenges a Prover by randomly sampling, either from $\mathbf{R}_{A_i}$ or from $\mathbf{R}_{A_i}[x]$ on each i, then Prover will be able to reliably guess which distribution Verifier sampled from exactly if $x \notin L$. By known results, this leads to the conclusion $L \in \text{co-NP/poly}$.

To implement this idea, we need the following technical lemma.

Lemma 30.5.1 *Let $R : \{0, 1\}^{t \times n} \to \{0, 1\}^{\leq t'}$ be given. Let $A \subseteq \{0, 1\}^n$ be a set of size $M \geq 100t$, and suppose that we sample $a^* \sim \mathcal{U}_A$. Then for sufficiently large t and for $t' < 2(t - 1)$, we have*

$$\mathbb{E}\left[\beta\left(a^*, A \setminus a^*\right)\right] \leq 1 - 10^{-4}. \tag{30.2}$$

We make no attempt to optimize the constants involved in Lemma 30.5.1, striving for an elementary proof and self-contained proof instead. Related results, with information-theoretic proofs, appear in Raz [573] and Klauck, Nayak, Ta-Shma, and Zuckerman [447], and these can be used to derive Lemma 30.5.1. A distinctive aspect of Lemma 30.5.1, however, is that it bounds the statistical distance between the output distribution of R induced by an input to R containing a string a^*, to an output distribution induced by an input distribution to R that does not support a^*. This "apples-to-oranges" comparison is key to the application of Lemma 30.5.1: we will use it to build small ($\operatorname{poly}(n)$-size) subsets of L_n that serve as helpful non-uniform advice to prevent *exponential*-size chunks of L_n from being accepted by Verifier. We will do so in an iterative fashion until all of L_n is "covered" by our advice. This is similar to the incremental approach of Fortnow and Santhanam [328] to defining their advice, in the proof of Lemma 30.2.1.

Proof of Lemma 30.5.1 Suppose for sake of contradiction that $\mathbb{E}[\beta(a^*, A \setminus a^*)] > 1 - 10^{-4}$. Call $a \in A$ "distinctive" if $\beta(a, A \setminus a) \geq 0.99$; then clearly more than a 0.99 fraction of $a \in A$ are distinctive.

For each $a \in A$, let $T = T_a$ be the set given by Proposition 30.5.2, with $X := \mathbf{R}_{A \setminus a}[a]$, $Y := \mathbf{R}_{A \setminus a}$; then for all distinctive $a \in A$, we have

$$\Pr_{z \sim \mathbf{R}_{A \setminus a}[a]} [z \in T_a] \geq 0.99, \qquad \Pr_{z \sim \mathbf{R}_{A \setminus a}} [z \notin T_a] \geq 0.99. \tag{30.3}$$

---

[2] Here, $L_n = L \cap \{0, 1\}^n$.

Let us index A as $A = \{a^1, \ldots, a^M\}$. Define a random R-input $\mathbf{x} = (x^1, \ldots, x^t) \sim \mathcal{U}_A^{\otimes t}$, and for $i \in [M]$ let $\mathrm{Incl}_i(\mathbf{x})$ be the 0/1-valued indicator variable for the event that at least one of the elements x^j is equal to a^i. We also define

$$\mathrm{Corr}_i(\mathbf{x}) := \left[\mathrm{Incl}_i(\mathbf{x}) \Leftrightarrow \left(R(\mathbf{x}) \in T_{a^i}\right)\right] = \neg\left[\mathrm{Incl}_i(\mathbf{x}) \oplus \left(R(\mathbf{x}) \in T_{a^i}\right)\right].$$

The idea is that $R(\mathbf{x}) \in T_{a^i}$ "suggests" that a^i was one of the inputs to R, while $R(\mathbf{x}) \notin T_{a^i}$ suggests a^i was not such an input; $\mathrm{Corr}_i(\mathbf{x})$ checks whether the suggestion given is correct.

From the definitions we see that, if we condition on $[\mathrm{Incl}_i(\mathbf{x}) = 0]$, then $R(\mathbf{x})$ is distributed as $\mathbf{R}_{A \setminus a^i}$. Then the conditional probability that $[\mathrm{Corr}_i(\mathbf{x}) = 1]$ holds is at least 0.99 for any distinctive a^i.

Suppose we instead condition on $[\mathrm{Incl}_i(\mathbf{x}) = 1]$. Then the conditional probability that a^i appears *twice* among the coordinates of $\mathbf{x}$ is, by counting, at most $t/M \leq 0.01$. Thus under this conditioning, $R(\mathbf{x})$ is 0.01-close to the distribution $\mathbf{R}_{A \setminus a^i}[a^i]$, so that $[\mathrm{Corr}_i(\mathbf{x}) = 1]$ holds with probability at least $0.99 - 0.01 = 0.98$, for any distinctive a^i.

We also have $\sum_{i \in [M]} \mathrm{Incl}_i(\mathbf{x}) \geq 0.95t$ with probability at least 0.99 (for large enough t), since $t/M \leq 0.01$. Combining these observation, we find that for large enough t, with probability at least 0.5 the following conditions hold simultaneously:

1. $\sum_{i \in [M]} \mathrm{Incl}_i(\mathbf{x}) \geq 0.95t$;
2. $\sum_{i \in [M]}[\mathrm{Incl}_i(\mathbf{x}) \wedge \mathrm{Corr}_i(\mathbf{x})] \geq 0.9t$;
3. $\sum_{i \in [M]} \mathrm{Corr}_i(\mathbf{x}) \geq 0.9M$.

Let us say that $\mathbf{x}$ is *good* if all of these conditions hold.

Now fix any R-output $z \in \{0, 1\}^{\leq t'}$; we will give an upper bound U on the number of good inputs $\mathbf{x}$ such that $R(\mathbf{x}) = z$. Since every $\mathbf{x}$ maps to a string of length $\leq t'$ under R, we will conclude that

$$2^{t'+1} \geq \frac{0.5|A^{\times t}|}{U} = \frac{0.5M^t}{U}, \tag{30.4}$$

which will give a contradiction, completing the proof.

First, say that for more than $t + 0.1M$ indices $i \in [M]$, we have $z \in T_{a^i}$. Then for any $\mathbf{x}$ such that $R(\mathbf{x}) = z$, there are more than $0.1M$ indices such that $\mathrm{Incl}_i(\mathbf{x}) = 0$ yet $z \in T_{a_i}$. For these indices i, $\mathrm{Corr}_i(\mathbf{x}) = 0$. So $\mathbf{x}$ is not good. Thus in order to have *any* good inputs $\mathbf{x}$ map to it under R, z must satisfy

$$\left|\{i : z \in T_{a^i}\}\right| \leq t + 0.1M. \tag{30.5}$$

Next, suppose that $R(\mathbf{x}) = z$ and that $\mathbf{x} = (x^1, \ldots, x^t)$ contains more than $0.15t$ strings x^j whose value is any element $x^j = a^i \in A$ such that $z \notin T_{a^i}$. If $\mathbf{x}$ is good, then by property 1 of good inputs, from among these components we can find a subcollection of more than $0.1t$ components x^j whose values are pairwise distinct. For each $a^i = x^j$ in this subcollection, it holds that $\mathrm{Incl}_i(\mathbf{x}) = 1$ yet $\mathrm{Corr}_i(\mathbf{x}) = 0$. Thus $\sum_{i \in [M]}[\mathrm{Incl}_i(\mathbf{x}) \wedge \mathrm{Corr}_i(\mathbf{x})] < 0.9t$, so $\mathbf{x}$ is not good—a contradiction. Thus, any good $\mathbf{x}$ for which $R(\mathbf{x}) = z$ can contain at most $0.15t$ components x^j whose value $x^j = a^i$ satisfies $z \notin T_{a^i}$.

Combining this with Eq. (30.5), there is a set $A' \subseteq A$ (depending on z) of size at most $t + 0.1M \leq 0.11M$, such that for any good $\mathbf{x}$ mapping to z under R, at least $0.85t$ components x^j satisfy $x^j \in A'$. This allows us to bound the number of good inputs $\mathbf{x}$ mapping to z under R; any such $\mathbf{x}$ can be specified by:

- a set of at most $0.15t$ "exceptional" indices $j \in [t]$;
- the values of x^j on these exceptional indices;
- the values of x^j on all other indices; these values must lie in A'.

The number of such inputs $\mathbf{x}$ is at most

$$
\sum_{0 \leq t' \leq 0.15t} \binom{t}{t'} M^{t'} (0.11M)^{t-t'} \leq (0.11)^{0.85t} M^t \cdot \sum_{0 \leq t' \leq 0.15t} \binom{t}{t'}
$$

$$
\leq (0.11)^{0.85t} M^t \cdot 2^{H(0.15)t}
$$

$$
< 4^{-t} M^t,
$$

using Fact 30.5.1 and a calculation. Thus we may take as our bound $U := 4^{-t} M^t$, so that by Eq. (30.4),

$$
2^{t'+1} \geq 0.5 \cdot 4^t = 2^{2t-1},
$$

contradicting our assumption that $t' < 2(t-1)$. This proves Lemma 30.5.1. $\qquad \square$

Proof of Theorem 30.5.2 We will use the assumed reduction R for L to construct a two-message, private-coin, *interactive proof system* between a polynomial-time-bounded Verifier and a computationally unbounded Prover to prove that a given string $x \in \{0, 1\}^n$ lies in $\overline{L}$. The proof system will use poly(n) bits of non-uniform advice on length-n inputs; Prover will be able to make Verifier accept with probability 1 if $x \notin L$, and with probability at most $1 - \Omega(1)$ if $x \in L$. It then follows from known results on interactive proof systems and non-uniform derandomization due to Goldwasser and Sipser [347] and Adleman [15] that $\overline{L} \in \text{NP/poly}$, which gives our desired conclusion.

Using the existence of the reduction R and Lemma 30.5.1, we will prove the following:

Claim 30.5.2 *There are multisets* $A_1, \ldots, A_{q(n) \leq \text{poly}(n)} \subseteq L_n$, *each of size at most* $s(n) \leq \text{poly}(n)$, *such that, for all* $x \in \{0, 1\}^n \setminus (\bigcup_{i \in [q(n)]} A_i)$:

(1) *If* $x \in \overline{L}_n$, *then* $\beta(x; A_i) = 1$ *for all* $i \in [q(n)]$;
(2) *If* $x \in L_n$, *there exists an* $i \in [q(n)]$ *for which* $\beta(x; A_i) \leq 1 - 10^{-5}$.

Assume Claim 30.5.2 for the moment; we will use it to prove Theorem 30.5.2. For inputs of length n to our interactive proof system, we let the non-uniform advice be a description of the sets $A_1, \ldots, A_{q(n)}$ given by Claim 30.5.2, along with the value $t(n)$. This advice is of size poly(n), as needed. The proof system proceeds as follows. On input $x \in \{0, 1\}^n$, Verifier first checks if x is in one of the sets A_i. If so, Verifier knows that $x \in L$. Otherwise, Verifier and Prover execute the following procedure in parallel for $i = 1, 2, \ldots, q(n)$:

- Verifier privately flips an unbiased coin $b_i \sim \mathcal{U}_{\{0,1\}}$;
- Verifier privately samples strings $y^{i,1}, \ldots, y^{i,t(n)} \in \{0, 1\}^n$ independently from $\mathcal{U}_{A_i}$;
- If $b_i = 0$ then Verifier sets

$$z = z(i) := R(y^{i,1}, \ldots, y^{i,t(n)});$$

otherwise ($b_i = 1$), Verifier samples $\mathbf{j} = \mathbf{j}(i) \sim \mathcal{U}_{[t(n)]}$ and sets

$$z := R(y^{i,1}, \ldots, y^{i,\mathbf{j}-1}, x, y^{i,\mathbf{j}+1}, \ldots, y^{i,t(n)}).$$

- Verifier sends z to Prover.
- Prover makes a guess $\widetilde{b}_i$ for the value of b_i.

Verifier accepts iff $\widetilde{b}_i = b_i$ for all i.

This protocol is polynomial-time executable by Arthur given our non-uniform advice. Now we analyze the behavior of the protocol, assuming $x \notin \bigcup_i A_i$. First, assume that $x \in \overline{L}_n$. In this case, we have

$$\left\| \mathbf{R}_{A_i}[x] - \mathbf{R}_{A_i} \right\|_{\text{stat}} = 1$$

for each i, by the first property of our sets A_i. Thus, Prover can guess b_i with perfect confidence for each i, causing Verifier to accept with probability 1.

Next, suppose that $x \in L_n$. Then by the second property of our sets, there is an $i^* \in [q(n)]$ such that

$$\left\| \mathbf{R}_{A_{i*}}[x] - \mathbf{R}_{A_{i*}} \right\|_{\text{stat}} \leq 1 - 10^{-5}.$$

This implies that the probability that Prover guesses b_{i*} correctly is at most $1 - 0.5 \cdot 10^{-5}$. Thus Verifier rejects with probability $\Omega(1)$. So our interactive proof has the desired properties. As discussed earlier, this yields $\overline{L} \in \text{NP/poly}$. $\qquad\square$

Proof of Claim 30.5.2 Fixing a value of n, let $(t, t') = (t(n), t'(n))$. Assume that t is large enough to apply Lemma 30.5.1. (Note that then t' satisfies the assumptions of that lemma as well.) Let $M := 100t$. We define a sequence of sets $S_1 \supseteq S_2 \supseteq \cdots \supseteq S_{q(n)+1} = \emptyset$, each contained in L_n, and a sequence of sets $A_1, A_2, \ldots, A_{q(n)}$, with $A_i \subseteq S_i$. Let $S_1 := L_n$. Inductively, having defined S_i, we define A_i, S_{i+1} as follows. If $|S_i| < M$, we let $A_i := S_i$ and $S_{i+1} := \emptyset$, and set $q(n) := i$, ending the construction at this stage. Otherwise (if $|S_i| \geq M$), we let A_i be a uniformly chosen size-$(M - 1)$ subset of S_i. We let

$$S_{i+1} := \left\{ a \in S_i \setminus A_i : \beta(a, A_i) > 1 - 10^{-5} \right\}.$$

The procedure terminates, since $|S_{i+1}| \leq |S_i| - (M - 1)$ whenever $S_{i+1} \neq \emptyset$. Let us check that these sets A_i satisfy conditions (1)–(2) of the claim; we will then argue that $q(n) \leq \text{poly}(n)$ (with high probability over the random construction).

First, suppose $x \in \overline{L}_n \setminus (\bigcup_{i \in [q(n)]} A_i)$. Then with an eye to Eq. (30.1), note that R always outputs an element of $\overline{L'}$ when x is one of the inputs to R. On the other hand, when all inputs to R are drawn from some $A_i \subseteq S_i \subseteq L_n$, R outputs an element of L'. Thus these two cases are perfectly distinguishable, and $\beta(x, A_i) = 1$ for each i, as needed.

Next suppose $x \in L_n \setminus (\bigcup_{i \in [q(n)]} A_i)$. Let $i \in [1, q(n)]$ be the unique index such that $x \in S_i \setminus S_{i+1}$. Then by definition, we must have $\beta(x, A_i) = \|\mathbf{R}_{A_i}[x] - \mathbf{R}_{A_i}\|_{\text{stat}} \leq 1 - 10^{-5}$.

Finally, we argue that $q(n) \leq \text{poly}(n)$ with high probability. Note that when we generate A_i as a uniform set of size $M - 1$, an equivalent way to generate A_i is to first generate a uniform set $\widehat{A}_i \subseteq S_i$ of size M, then select a uniform element a^* of $\widehat{A}_i$ to discard to form A_i.

By Lemma 30.5.1, $\mathbb{E}_{a^*}[\beta(a^*, A_i)] \leq 1 - 10^{-4}$. Then with probability at least 0.9 over our randomness at this stage, a^* satisfies $\beta(a^*, A_i) \leq 1 - 10^{-5}$. But a^* is a uniform element of $S_i \setminus A_i$. Thus,

$$\mathbb{E}\big[|S_{i+1}|\big] \leq 0.1\big(|S_i| - |A_i|\big).$$

So $q(n) = O(n)$ with high probability. This finishes the proof of Claim 30.5.2. $\square$

30.6   Cross-composition

Almost immediately after the papers of Fortnow and Santhanam [328] and Bodlaender, Downey, Fellows, and Hermelin [83], an appropriate notion of problem transformation emerged allows kernelization lower bounds to be studied in the manner that is usual in complexity theory: by means of problem reductions. The key notion of a *polynomial parameter transformation*, or *parameterized polynomial transformation* (terminology has not settled to date—in short, a *PPT-reduction*), was introduced by Bodlaender, Thomassé, and Yeo [101].

Definition 30.6.1 A parameterized many : 1 reduction from L to $\hat{L}$ is *parameterized polynomial* if governed by a polynomial-time computable function h, and a polynomial g, such that $\langle x, k \rangle \in L$ if and only if $\langle h(x, k), g(k) \rangle \in \hat{L}$.

The central result about PPT-reductions is the following.

Corollary 30.6.1 (Bodlaender, Thomassé, and Yeo [101]) *Suppose that L and $\hat{L}$ are parameterized problems and L reduces to $\hat{L}$ by a PPT-reduction. Suppose that the unparameterized version of L is NP-hard and the unparameterized version of $\hat{L}$ is in NP. If $\hat{L}$ has a polynomial kernel, then so does L.*

These ideas rapidly led to the more powerful notion of *cross-composition*.

Definition 30.6.2 (Polynomial-time equivalence relation) An equivalence relation R on Σ^* is called a *polynomial-time equivalence relation* iff

1. There is a poly time algorithm which, given $x, y \in \Sigma^*$, decides if $x R y$ in time $(|x| + |y|)^c$.
2. For any finite set $S \subseteq \Sigma^*$, R partitions the elements of S into at most $(\max_{x \in S} |x|)^{c'}$ many equivalence classes.

> **Definition 30.6.3** (Bodlaender, Jansen, and Kratsch [91]) We say $L \subseteq \Sigma^*$
> *cross-composes* into a parameterized $Q \subseteq \Sigma^* \times \mathbb{N}$ if there is a polynomial-
> time equivalence relation R and an algorithm which given t strings $x_1, \ldots, x_t$
> which are all R-equivalent, computes an instance $\langle x^*, k^* \rangle$ in time polynomial
> in $\sum_{i=1}^{t} |x_i|$ such that
> (1) $\langle x^*, k^* \rangle \in Q$ iff $x_i \in L$ for some i,
> (2) k^* is bounded by $\max_{i=1}^{t} |x_i| + \log t$.

The key differences between composition and cross-composition are that:
(1) The source and target problems can be different.
(2) The input to a cross-composition is a collection of *classical* instances rather than parameterized ones.
(3) The output parameter may depend on the logarithm of the number of input instances.

Theorem 30.6.1 (Bodlaender, Jansen, and Kratsch [91]) *Suppose that $L \subseteq \Sigma^*$ is NP-hard under Karp reductions. If L cross composes into Q and Q has a polynomial kernel, then there is a weak distillation of* SAT *into the unparameterized version of Q and the polynomial-time hierarchy collapses.*

We remark that the above result works for both the OR and AND versions. We remark that the proofs are more or less the same as those in the last section showing that the definitions are specifically tailored to enable this. The point is that if we begin with t distinct instances of SAT, we can transform these into instance of L, pairwise compare them using the cross-composition into groups of equivalent instances, and then apply the kernelization algorithm and convert back to instance of SAT, as we did in the last section. We leave the details to the reader (Exercise 30.13.5).

As an application of Corollary 30.6.1, we show that the problem of k-LEAF OUT BRANCHING does not admit a polynomial kernelization unless the Polynomial Hierarchy collapses. For the proof, we need another concept.

Definition 30.6.4 (Drescher and Vetta [262]) We say that a digraph $D = (V, A_1 \cup A_2)$ is a *willow graph* if $D' = (V, A_1)$ is a directed path $P = p_1 p_2 \ldots p_n$ on all vertices of D and $D'' = (V, A_2)$ is a dag with one root r, such that every arc of A_2 is a backwards arc of P. p_1 is called the *bottom* of the willow graph, p_n the *top*, and P the *stem*. Finally a *nice* willow graph is one where if $p_n p_{n-1}$ and $p_{n-1} p_{n-2}$ are arcs of D, then neither p_{n-1} nor p_{n-2} are incident to any other arcs of A_2 and $D'' = (V, A_2)$ has a p_n-outbranching.

Lemma 30.6.1 (Fernau et al. [306]) *k*-LEAF OUT TREE *on nice willow graphs is NP-complete under Karp reductions.*

Proof Let $F = \{S_1, \ldots, S_m\}$ be an instance of SET COVER over universe U with $|U| = n$, and parameter $b \leq m$ asking if there is a set cover $F' \subseteq F$ with $|F'| \leq b$. Without loss of generality we suppose that $U \subseteq \bigcup S_i$, and $b \leq m - 2$. We build a digraph $D = (V, A_1 \cup A_2)$. V is a root r and vertices s_i corresponding to the S_i $i \leq i \leq m$, e_j, $1 \leq j \leq n$ for the elements of U and two additional vertices p and p'. The arc sets A_2 is as follows. There is an arc from r to each s_i, and $s_i e_j$ iff $e_j \in S_i$. rp and rp' are arcs in A_2. Then set $A_1 = \{e_{i+1} e_i \mid 1 \leq i < n\} \cup \{s_{i+1} s_i \mid 1 \leq i < m\} \cup \{e_1 s_m, s_1 p, pp', p'r\}$. This completes the description of D. The claim is that there is a set cover F' with $|F'| \leq b$ if and only if there is an outbranching with at least $n + m + 2 - b$ leaves.

To see this suppose that such and F' exists. We build a directed tree T rooted at r. For each s_i, and p and p' make the parent be r. For each e_j choose the parent of e_j to be s_i such that $e_j \in S_i$ and for all $i' < i$, $e_j \notin S_{i'}$ or $S_{i'} \notin F'$. The only inner nodes of T are r and vertices representing the set cover, T is an outbranching of D with at least $n + m + 2 - b$ many leaves.

Conversely, let T be an outbranching with at least $n + m + 2 - b$ many leaves. and suppose that T has the most leaves of any outbranching of D. Since D is a nice willow graph with root r, we claim that this implies that this outbranching has root r.

To see this, let D be a nice willow graph as above with top vertex $r = p_n$, and stem $P = p_1 \ldots p_n = r$. Suppose there is a maximum sized outbranching with root p_i and $i < n$. Since D is nice willow, $D' = (V, A_2)$ has a p_n-outbranching T'. Since each arc in A_2 is a back arc of P, we see $T'[\{v_j \mid j \geq i\}]$ is a p_n-outbranching of $D[\{v_j \mid j \geq i\}]$. Then $T'' = (V, \{v_x v_y \in A(T') \mid y \geq i\} \cup \{v_x v_y \in A(T) \mid y < i\})$ is an outbranching of D. If $i = n - 1$ then p_n is not a leaf of T since the only arcs going out of the set $\{p_n, p_{n-1}\}$ start at p_n. Thus in this case, all the leaves of T are leaves of T'' and p_{n-1} is a leaf of $T'' \setminus T$, a contradiction.

Using this fact that the maximum sized outbranching has root r, if there is an arc $e_{i+1} e_i \in A(T)$, let s_j be any vertex with $e_i \in S_j$. Then $T' = (T \setminus \{e_{i+1} e_i\}) \cup \{s_j e_i\}$ is an r-outbranching with as many leaves as T. Hence for any outbranching, we may assume that the parent of e_i is some s_j. Then the inner vertices will define the desired set cover. $\square$

Theorem 30.6.2 (Fernau et al. [306]) *k*-LEAF OUT BRANCHING *has no poly kernel unless co-NP $\subseteq$ NP/poly.*

Proof The proof is to construct a PPT-reduction from *k*-LEAF OUT TREE to *k*-LEAF OUT BRANCHING. Let (D, k) be an instance of *k*-LEAF OUT TREE. For each $v \in V(D)$ make an instance (D, v, k) of ROOTED *k*-LEAF OUT TREE. Clearly (D, k) is a "yes" instance iff for some v, (D, v, k) is a "yes" instance of ROOTED *k*-LEAF OUT TREE. The n instances $(D, v_1, k), \ldots, (D, v_n, k)$, and each has a $O(k^2)$ kernel by Corollary 5.1.4. Lemma 30.6.1 nice willow *k*-LEAF OUT

BRANCHING is NP Karp complete. Hence we can reduce each instance (D, v_i, k) of ROOTED k-LEAF OUT TREE to one (W_i, b_i) of k-LEAF OUT BRANCHING, for a nice willow graph, and this happens in time polynomial in $|D| + |k|$. This makes the reduction for each i a parameterized poly one. Without loss of generality we can simply take $b = b_i$ for all i by subdividing the last arc $b - b_i$ many times if necessary and $b < k^c$. From the instances $(W_1, b), \ldots, (W_n, b)$ build an instance $(D', b + 1)$ as follows. Let r_i and s_i be the top and bottom vertices of W_i, respectively. Take disjoint unions of $(W_1, b), \ldots, (W_n, b)$ and add an arc $r_i s_{i+1}$ for $i < n$ and also add $r_n s_1$.

By the proof of Lemma 30.6.1 if W_i has an outbranching with at least b leaves, it has one rooted at r with at least b leaves. This can be extended to one for D' with at least $b + 1$ leaves by following s_{i+1} from r_i. Conversely, suppose that D' has an outbranching with $b + 1$ many leaves. Let the root be in W_i. For any v in D' outside of W_i, there is only one path from r to v, namely the directed path from r to v. Thus the tree has at most one leaf outside of W_i. Therefore the induced subtree gives a "yes" instance of k-LEAF OUT TREE for W_i. This is clearly a PPT-reduction, so we can apply Corollary 30.6.1. $\qquad\qquad\qquad\qquad\qquad\qquad\qquad\qquad\qquad\qquad\qquad\qquad\quad$ □

For another example of cross-composition we offer the following. Recall the problem.

k-EDGE CLIQUE COVER

Instance: A graph G.
Parameter: A positive integer k.
Question: Can we cover the edges of G with at most k cliques?

We will refer to the unparameterized version as EDGE CLIQUE COVER. In Exercise 30.13.4, the reader showed that this problem has an exponential kernel, a result of Gramm, Guo, Hüffner and Niedermeier [353, 355]. We see that this kernelization is tight.

Theorem 30.6.3 (Cygan, Kratsch, Philipczuk, Wahlström [188])
1. EDGE CLIQUE COVER AND-*cross-composes to* k-EDGE CLIQUE COVER.
2. *Hence there is no polynomial kernel for* k-EDGE CLIQUE COVER *unless* co-NP $\subseteq$ NP/poly.

Proof To define the equivalence relation, declare that (G_1, k_1) and (G_2, k_2) are equivalent if $k_1 = k_2$ and they have the same number of vertices.[3] Thus we can assume we are given an instance (G_i, k) for $i \le t - 1$ of EDGE CLIQUE COVER instances all in the same equivalence class with n vertices each. Without loss of generality we can make $n = 2^{h_n}$ and $t = 2^{h_t}$.

---

[3]Of course, thinking of this as *bitstrings*, then we can put all strings representing malformed instances into one equivalence class, but this is a technicality.

We construct and instance (G^*, k^*) with k^* poly in $n + k + h_t$. Begin by making G^* as the disjoint union of graphs G_i. Now add edges between each pair of vertices from G_i and G_j for $i \neq j$. Let $V(G_i) = V_i = \{v_0, \ldots, v_{n-1}\}$. For each (i, j, r) with $i, j \leq n - 1$ and $r \leq h_t$, add a vertex $w(i, j, r)$ adjacent to exactly one vertex in each V_i, namely adjacent to v_a^b with $b = (i + j \lfloor \frac{b}{2^r} \rfloor)$. Let W denote the set of such $w(i, j, r)$'s. Set $k^* = |W| + k = n^2 h_t + k$.

Now, W is clearly an independent set of nonisolated vertices of G^*. For each $w \in W$, $N_{G^*}[w]$ induces a clique. Therefore, an optimal clique cover of G^* contains the $|W|$ cliques $N_{G^*}[w]$ for $w \in W$. These cliques contain no edges of G_i for any i while covering the other edges of G^*. Thus we need k cliques to cover the remaining edges.

The second part follows as EDGE CLIQUE COVER is a classic NP-complete problem (Garey and Johnson [337], (GT17)). Then we can apply Theorem 30.4.2. $\square$

30.7   Unique Identifiers

Another method of generalizing the lower-bound machinery was discovered by Dom, Lokshtanov, and Saurabh [226, 227]. As with Bodlaender, Thomassé, and Yeo [101], these authors realized that the number of instances could be bounded by 2^{k^c}, else composability was easy. This then allowed them to use what they called *unique identifiers* in their proofs of composability. The other main idea in their arguments was to use colored versions of the problems and then reduce to the (usual) uncolored versions. This is reminiscent of the modern importance of MULTICOLOR CLIQUE in $W[1]$-hardness demonstrations.

The colored versions make the *identifiers* easier to handle in the gadgeteering. The reduction strategies introduced by Dom, Lokshtanov, and Saurabh [226, 227] allowed them to show that a number of problems, apparently "close" to VERTEX COVER, do not have polynomial kernels unless PH collapse occurs.

To illustrate their techniques, we use the following problems.

RED–BLUE DOMINATING SET

Instance: A bipartite graph $B = (U \sqcup W, E)$.
Parameter: Positive integers $|U|$ and k.
Question: Is there $W' \subseteq W$ of size $\leq k$ such that every vertex of U has at least one neighbor in W'?

COLORED RED–BLUE DOMINATING SET

Instance: A bipartite graph $B = (U \sqcup W, E)$, with the vertices of W in colors from $\{1, \ldots, k\}$.
Parameter: Positive integers $|U|$ and k.
Question: Is there $W' \subseteq W$ of size $\leq k$ such that every vertex of U has at least one neighbor in W', and W' has exactly one vertex of each color?

Lemma 30.7.1 (Dom, Lokshtanov, and Saurabh [226, 227])
1. *Unparameterized* COLORED RED–BLUE DOMINATING SET *is NP-complete.*

2. *There is a polynomial parameterized reduction from* COLORED RED–BLUE
DOMINATING SET *to* RED–BLUE DOMINATING SET.
3. COLORED RED–BLUE DOMINATING SET *is solvable in time* $2^{|U|+k} \cdot |U \sqcup W|^{O(1)}$.

Proof 1. We reduce the uncolored version to the colored one. Let $B = (U \sqcup W, E), k$ be an instance of RED–BLUE DOMINATING SET. Define $B' = (U \sqcup W', E'), k, \chi$ with vertex set W' k copies $w_1, \ldots, w_k$ for each $w \in W$, and color via $\chi(w_j) = j$. The edge set

$$E' = \bigcup_{j \in \{1, \ldots, k\}} \{uw_j \mid u \in U \wedge uw \in E(B)\}.$$

2. Take an instance of the colored problem $B = (U \sqcup W, E), k, \chi$ and define an uncolored version by increasing U to U' by adding k additional vertices $u_1, \ldots, u_k$, and the new edge set will additionally have

$$\bigcup_{j \in \{1, \ldots, k\}} \{u_j w \mid w \in W \wedge \chi(w) = j\}.$$

3. Solving the problem in the given time, first reduce using 2 to the uncolored version. Note $|U'| = |U| + k$. Then transform the instance into one $(\mathcal{F}, U, k)$ of SET COVER, with the same universe U, and the elements of the family $\mathcal{F}$ in one to one correspondence with the vertices of W. (This is really the same problem, just considered differently.) SET COVER is easily solvable in time $O(2^{|U|}|U||\mathcal{F}|)$ (for example Fomin, Kratsch, and Woeginger [320] or by the algorithm we met to discuss measure and conquer in Chap. 6, of Fomin, Grandoni, and Kratsch [316]) we get the result. □

Theorem 30.7.1 (Dom, Lokshtanov, and Saurabh [226, 227]) COLORED RED–BLUE DOMINATING SET *is compositional, and hence neither* COLORED RED–BLUE DOMINATING SET *nor* RED–BLUE DOMINATING SET *has a poly kernel unless* co-NP $\subseteq$ NP/poly.

Proof Let $G_i = (U_i \sqcup W_i, E_i), k, \chi_i$ for $1 \le i \le t$ be instances of COLORED RED–BLUE DOMINATING SET, and assume that $|U_i| = p$ for all i, so that we can label $U_i = \{u_1^i, \ldots, u_p^i\}$ and $W_i = \{w_1^i, \ldots, w_{q_i}^p\}$.

To demonstrate the fact that this is compositional, we construct G via sets U and W as follows. First let $\{u_1, \ldots, u_p\} = U$. For each $w_j^i \in W_i$, we color $\chi(w_j^i) = \chi_i(w_j^i)$, so that colors are preserved. Now add the edges $u_{j_1} w_{j_2}^i$ for $u_{j_1} w_{j_2}^i \in E_i$. Note that if G_i corresponds to a yes instance, then G, k, χ is also a yes instance. This follows if a size k subset from W_i dominates all of U_i, then it must dominate U.

The problem is that it might be that all the instances are no instances and yet ("currently") G is a yes instance. To avoid this, Dom, Lokshtanov, and Saurabh modified the reduction via *unique identifiers*. For each graph G_i, we will add a $(p + k)$ bit binary number ID(G_i). The claim is that we may assume that the input has no more than 2^{p+k} many instances. If there are *more*, then we can solve the instances in time $\sum_{1 \le i \le t} 2^{p+k}(p+q_i)^{O(1)} \le t \sum_{1 \le i \le t}(p+q_i)^{O(1)}$, giving a polynomial-time composition algorithm that does the job. For each pair $(a,b) \in [k] \times [k]$ (here choosing $[k] = \{1, \dots, k\}$) with $a \ne b$, add a vertex set $M_{a,b} = \{m_1^{a,b}, \dots, m_{p+k}^{a,b}\}$ to U and for $i \in \{1, \dots, t\}$, $j_1 \in \{1, \dots, q_i\}$ add edges

$$\{m_{j_2}^{a,b} w_{j_1}^i \mid \chi(w_{j_1}^i) = a \land b \in [k] \setminus \{a\} \land \text{the } j_2\text{th bit in ID}(G_i) \text{ is } 1\},$$

$$\{m_{j_2}^{a,b} w_{j_1}^i \mid \chi(w_{j_1}^i) = b \land a \in [k] \setminus \{b\} \land \text{the } j_2\text{th bit in ID}(G_i) \text{ is } 0\}.$$

Note that $|U| = p + k(k-1)(p+k)$, so that the definition of composition applies. It some G_i is a yes instance, then so is G, as the solution lifts. First the vertices u_i are still dominated as before. Additionally, for each color $(a,b) \in [k] \times [k]$ with $a \ne b$, each vertex from $M_{a,b}$ is either connected to all the vertices w from W_i with $\chi(w) = a$, or all the ones with color b. If the solution $W_i' \subset W_i$ contains one element from each color class, each vertex of $M_{a,b}$ is dominated by some vertex of the solution.

Finally, suppose that $W' \subseteq W$ is a solution for G. We argue that W' cannot have vertices originating from differing G_is. Every two distinct vertices of W' must be of distinct colors. Suppose that W' contains $w_{j_1}^{i_1}$ of color a and $w_{j_2}^{i_2}$ of color b originating from G_{i_1} and G_{i_2}, respectively. Crucially, $\text{ID}(G_{i_1}) \ne \text{ID}(G_{i_2})$. Hence, the vertices $m_j^{a,b}$ where the jth bit of $\text{ID}(G_{i_1})$ is 0 and of $\text{ID}(G_{i_2})$ is 1, and $m_j^{b,a}$ where the jth bit of $\text{ID}(G_{i_2})$ is 0 and of $\text{ID}(G_{i_1})$ is 1, are members of $M_{a,b} \cup M_{b,a}$ adjacent to neither $w_{j_1}^{i_1}$ nor $w_{j_2}^{i_2}$. Thus W' does not dominate all of U.   □

Corollary 30.7.1 (Dom, Lokshtanov, and Saurabh [226, 227]) STEINER TREE *parameterized as below has no polynomially bounded polynomial-time kernelization algorithm unless the Polynomial Hierarchy collapses.*

STEINER TREE

Instance: A graph $G = (T \cup N, E)$.
Parameter: Positive integers $|T|$ and k.
Question: Is there $N' \subseteq N$ with $|N'| \le k$ and $G[T \cup N']$ connected?

Proof Exercise 30.13.7.   □

Dom, Lokshtanov, and Saurabh gave the following summary description of their approach.
1. Find a suitable parameterization of the problem.
2. Find a suitable colored version to limit the kinds of solutions.
3. Prove that the unparameterized version is in NP, and show that the colored version is NP-hard.
4. Find a PPT-reduction from the colored to the uncolored versions of the problem.
5. Show that the colored version is FPT solvable in time $2^{k^c} n^{O(1)}$ for a fixed c by an algorithm Φ. (This is to enable the identifiers.)

6. Show that the colored version of the problem is compositional. To do this you need to act as follows.
   - If the number of instances t is greater than 2^{k^c}, then running Φ separately on each of the instances separately, incurs a timecost polynomial in the total input size, and then output a small canonical yes or no instance.
   - If the number of instances is less than 2^{k^c}, associate with each instance an ID, as above, k^c bits per instance. Code as an integer, or as a subset of some set of objects of size polynomial in k.
   - Use the coding power provided by the colors and the ID's to construct the composition.

   Some further applications of this "recipe" are sketched below in the exercises.

30.8 Sharpening Bounds

Dell and van Melkebeek re-examined the kernelization material and established the following lemma.

Theorem 30.8.1 (Dell and van Melkebeek [206]) *Suppose that (L, k) is a parameterized language and $\hat{L}$ is an NP-hard language. Suppose also that there is a polynomial-time many : 1 reduction f from $\mathrm{OR}(\hat{L})$ to L and $d > 0$ such that, given an instance $(x_1, \ldots, x_t)$ of $\mathrm{OR}(\hat{L})$ with $|x_i| \leq s$ for all i, $f((x_1, \ldots, x_t))$ is an instance of L whose parameter $k \leq t^{\frac{1}{d}+o(1)} \cdot \mathrm{poly}(s)$. Then L does not have kernels of size $O(k^{d-\varepsilon})$ for any $\varepsilon > 0$ unless co-NP $\subseteq$ NP/poly.*

Here is one application.

Theorem 30.8.2 (Dell and van Melkebeek [206]) VERTEX COVER *does not have kernels of size[4] $O(k^{2-\varepsilon})$ unless* co-NP $\subseteq$ NP/poly.

Proof (Dell and Marx [205]) The proof uses the following problem as its base.

MULTICOLORED BICLIQUE

Instance: A bipartite graph $B = (U \sqcup W, E)$ and partitions of U and W each into k parts (colors) $U_1, \ldots, U_k$ and $W_1, \ldots, W_k$.

Parameter: A positive integer k.

Question: Does B contain a biclique (i.e. a version of $K_{k,k}$) that has one vertex from each of U_i and W_i for all i?

The problem is NP-complete (Exercise 30.13.12). Let $\hat{L}$ be an instance of MULTICOLORED BICLIQUE. Given and instance of $\mathrm{OR}(\hat{L})$ ($B_1, \ldots, B_t$) we assume that each B_i has the same number k of groups and the same n, and that $\sqrt{t}$ is

---

[4] Here size means overall size, whereas the Buss kernel means vertex set size. A vertex size k kernel can have $O(k^2)$ overall size.

an integer. We will refer to the instance of $OR(\hat{L})$ as $B_{i,j}$ for $1 \le i, j \le \sqrt{t}$, and $V(B_{i,j} = U_{i,j}) \sqcup W_{i,j}$. Now we modify the $B_{i,j}$ to obtain $B'_{i,j}$ so that the $U_{i,j}$ and $W_{i,j}$ become complete k-partite graphs. That is, if two vertices of $U_{i,j}$ or two in $W_{i,j}$ are in different groups make them adjacent. Note that there is a $2k$-clique in the new graph iff there is a correctly partitioned $K_{k,k}$ in the original $B_{i,j}$. Construct G by adding $2\sqrt{t}$ new sets $U^1, \ldots, U^{\sqrt{t}}$ and $W^1, \ldots, W^{\sqrt{t}}$, of kn vertices each. For each $1 \le i \le j \le \sqrt{t}$, copy the graph $B'_{i,j}$ into the new $U^i \sqcup W^j$ by identifying $U_{i,j}$ with U_i and $W_{i,j}$ with W^j. Note that $U_{i,j}$ and $W_{i,j}$ induce the same complete bipartite graph in $B'_{i,j}$ so this copying can be done in such a way that $G[U^i]$ receives the same set of edges when copying $B'_{i,j}$ for any j, and similarly $G[W^i]$, so this is all legal. Thus $G[U^i \sqcup W^j] \cong B'_{i,j}$ for all i, j. Then it is easy to show that G has a $2k$-clique iff one of the $B'_{i,j}$ has a $2k$-clique, and hence one of the $B_{i,j}$ has a correctly colored $K_{k,k}$.

Let $N = 2\sqrt{t}kn$ be the number of vertices in G. Note that $N = t^{\frac{1}{2}} \cdot \text{poly}(s)$ where s is the maximum bitlength of the t instance of the $OR(\hat{L})$ problem. G has a $2k$-clique iff $\overline{G}$ has a vertex cover of size $N - 2k$. Thus $OR(\hat{L})$ can be reduced to an instance of VERTEX COVER with parameter at most $t^{\frac{1}{2}} \cdot \text{poly}(s)$. $\qquad\square$

As we will see in Sect. 30.11, Hermelin and Wu [389] also introduced a method called *weak composition* which allows for sharp polynomial bounds on kernelization. This machinery is based on co-nondeterminism, and we will delay its analysis until after we introduce this idea in Sect. 30.10. In the same paper, Hermelin and Wu also gave methods, similar to the ones described above, to establish sharp *quasipolynomial* kernelization lower bounds. That means bounds of the form $O(2^{O(\log n^c)})$ bounds. These are based on an extension of Yap's Theorem, due to Pavan, Selman, Sengupta, and Viriyam [556]. Their result states that if $NP \subseteq co\text{-}NP/qpoly$ (where qpoly here means an advice string of length bounded by a quasipolynomial function), then the Exponential Hierarchy collapses to the third level.

30.9 The Packing Lemma

One nontrivial product of the work on kernelization lower bounds is the useful lemma below.

Theorem 30.9.1 (The Packing Lemma—Dell and van Melkebeek [206]) *For any integers $p \ge d \ge 2$ and $t > 0$ there exists a p-partite d-uniform hypergraph P on $O(p \cdot \max(p, t^{\frac{1}{d+o(1)}}))$ vertices such that*
1. *The hyperedges partition P into t cliques $K_1, \ldots, K_t$ with p vertices each.*
2. *P contains no cliques on p vertices save the K_i's.*
Furthermore, for any fixed d, P, and the K_i's can be constructed in polynomial time in p and t.

We consider the following two problems.

k–H-MATCHING

Instance: A graph G.
Parameter: A graph H and a positive integer k.
Question: Can I find k vertex disjoint subgraphs isomorphic to H in G?

k–H-FACTOR

Instance: A graph G.
Parameter: A graph H and a positive integer k.
Question: Can I find k vertex disjoint subgraphs isomorphic to H in G covering all the vertices?

This terminology covers a host of problems as discussed in Yuster [671]. Notice that k–H-FACTOR is a special case of k–H-MATCHING in the case that $k = \frac{n}{d}$. Notice that for any fixed H, k–H-FACTOR has a $O(k^2)$ sized kernelization upper bound. This bound is tight.

Theorem 30.9.2 (Dell and Marx [205]) *Let H be a connected graph with $d > 3$ vertices and $\varepsilon > 0$. Then k–H-FACTOR has no kernel of size $O(k^{2-\varepsilon})$ unless co-NP $\subseteq$ NP/poly.*

The proofs of the Packing Lemma and the above theorem are both quite intricate, and they are beyond the scope of this book.

30.10 Co-nondeterminism

30.10.1 Co-nondeterministic Composition, and k-RAMSEY

One final idea in the area of kernelization lower bounds is the use of co-nondeterminism in the proof of lower bounds. We begin with the notion of co-nondeterministic composition. In the following, by $t^{o(1)}$ we mean the class of functions $f(t)$ that are $o(t^\varepsilon)$ for every $\varepsilon > 0$. That is, for every $\varepsilon > 0$, the limit as t goes to infinity, of $f(t)/t^\varepsilon$ is zero.

Definition 30.10.1 (Kratsch [468], also Müller and Yijia Chen (unpublished)) A *co-nondeterministic OR-composition algorithm* for a parameterized problem $L \subseteq \Sigma^* \times \mathbb{N}$ is an algorithm that

- receives as input a sequence $((x_1, k), \ldots, (x_t, k))$, with $(x_i, k) \in \Sigma^* \times \mathbb{N}^+$ for each $1 \leq i \leq t$,
- uses time polynomial in $\sum_{i=1}^{t} |x_i| + k$,
- and outputs $(y, k') \in \Sigma^* \times \mathbb{N}^+$ with
  1. $(x_i, k) \in L$ for some $1 \leq i \leq t$ implies that for $(y, k') \in L$ and all computation paths lead to a "yes" instance.
  2. If all instance (x_i, k) are no instances, then *at least one* computation path leads to the output of a "no" instance.
  3. $k' \leq t^{o(1)} \mathrm{poly}(k)$.

For the proof of the composition theorem below we will need some further concepts.

Definition 30.10.2 (Dell and van Melkebeek [206]) An *oracle communication protocol* for a language L is a communication protocol with two players. Player I is given an input x and runs in time poly$(|x|)$. Player II is computationally unbounded, but is not given any part of x. At the end of the protocol, Player I should be able to decide if $x \in L$. The cost of the protocol is the number of bits of communication from Player I to Player II.

Lemma 30.10.1 (Complementary Witness Lemma—Dell and van Melkebeek [206]) *Let L be a language and $t : \mathbb{N} \to \mathbb{N} \setminus \{0\}$ be poly bounded such that the problem of deciding where at least one out of $t(s)$ inputs of length at most s belongs to L has an oracle communication protocol of cost $O(t(s) \log t(s))$ where the first player can be co-nondeterministic. Then $L \in$ co-NP/poly.*

Theorem 30.10.1 (Kratsch [468], also Müller and Chen) *Suppose that L is a parameterized problem whose unparameterized version is NP-hard. If L has co-nondeterministic OR-composition then L has no poly kernel unless NP $\subseteq$ co-NP/poly.*

Proof Let K be a poly kernel of L of size $h(k) = O(n^c)$. Let C be the co-nondeterministic composition algorithm with parameter bound $t^{o(1)}k^d$. We define a poly bound t bounded by $t(N) := N^{cd+2}$. Using Lemma 30.10.1, we give an oracle communication protocol for L as an unparameterized problem, with the first player co-nondeterministic and with cost at most $O(t(N) \log t(N))$ for t inputs of size at most N.

Fix N, and $t = t(N)$. Let $x_1, \ldots, x_t$ be t instances of size $\le N$ given to the first player and $(x_1, k_1), \ldots, (x_t, k_t)$ the corresponding parameterized instances of L.

Here is the protocol. Player I groups the instances by parameter value giving at most N groups, and applies the co-nondeterminism to each group. In each computation path, this gives $r \le N$ instances $(G', k'_1), \ldots, (G'_r, k'_k)$ and if (G'_i, k'_i) was formed by composing instances with parameter value $\hat{k}$, $k'_i := t^{o(1)}N^d$. Player I applies the assumed poly kernelization to each (G'_i, k'_i), sends the obtained kernels to Player II, who tests membership in L for each. Player II says "yes" if at least one answer is "yes" and "no" otherwise.

Each kernelization has size at most $h(k'_i) = O((k'_i)^c) = O((t^{o(1)}N^d)^c)$. Since $t = N^{cd+1}$, we can therefore bound the cost of sending N kernelized instances as

$$O\left(N\left(\left(t^{o(1)}N^d\right)^c\right)\right) = O\left(N^{cd+1}\left(t^{o(1)}\right)^c\right) = O(t).$$

The remaining part is to verify the correctness of the protocol, using the co-nondeterminism of the composition. If at least one instance is a "yes" instance, the corresponding instance (G'_i, k'_i) will be a "yes" on all computation paths. So the oracle says yes on all computation paths. Otherwise if it is a "no" instance, there must be a computation path on which all N runs of the co-nondeterministic composition return "no". Applying the composition, we therefore create N "no" instances as well. These are sent to the oracle, causing it to be a "no".

Therefore, assuming a poly kernel, we get a communication protocol for deciding OR of t instances of L at cost $O(t)$. By Lemma 30.10.1, this puts L in co-NP/poly and hence NP $\subseteq$ co-NP/poly. $\qquad\square$

One consequence of Theorem 30.10.1 we will use later is the following connection between oracle communication protocols for classical problems and kernels for parameterized problems. For a parameterized problem $L \subseteq \{0, 1\}^* \times \mathbb{N}$, we let $\widetilde{L} := \{x\#1^k : (x, k) \in L\}$ denote the unparameterized version of L.

Lemma 30.10.2 *If $L \subseteq \{0, 1\}^* \times \mathbb{N}$ has a kernel of size $f(k)$, then $\widetilde{L}$ has an oracle communication protocol of cost $f(k)$.*

Co-nondeterminism would seem to have a lot of potential applications. We finish this section showing how to apply co-nondeterminism to the following problem. Later, in Sect. 30.11, we will look at using the technique for sharper lower bounds.

k-RAMSEY

Instance: A graph G.
Parameter: A positive integers k.
Question: Determine if G has *either* an independent set or a clique of size k.

We met Ramsey's Theorem in Chap. 17, at least in the infinite form. It is a beautiful extension of the pigeonhole principle. The finite (2-colored) version of this theorem asserts that for all k there is an $n = R(k)$ so large such that every graph G with $|V(G)| \geq n$ either has an independent set of size k or a clique of size k. The least such n is called the *Ramsey number*. Evidently, the theorem implies that k-RAMSEY is FPT. None of the techniques of the preceding sections seem to apply. For example, if you try to compose disjoint graphs you will get a large independent set, but if you cover them then you get a big clique. Kratsch realized the methods of co-nondeterministic composition will apply. He did so using an auxiliary problem and a combinatorial construction. Here is the auxiliary problem.

k-REFINEMENT RAMSEY

Instance: A graph G with an independent set and a clique of size $k - 1$.
Parameter: A positive integers k.
Question: Determine if G has *either* an independent set or a clique of size k.

The following lemma is easy and left to the reader (Exercise 30.13.19). The fact that k-RAMSEY is NP-complete is a matter of folklore, and surely known to anyone who has thought about it.

Lemma 30.10.3
1. *An instance of k-RAMSEY poly reduces to an equivalent instance of $k + 1$-REFINEMENT RAMSEY*
2. *k-RAMSEY and k-REFINEMENT RAMSEY are NP-complete.*

The combinatorial construction will be to make a *host graph H* ion t' vertices which will take t instances of k-RAMSEY $G_1, \ldots, G_t$ with $t \le t'$, we construct $G' = Embed(H, k; G_1, \ldots, G_t)$. We begin with a *dummy graph* D_c which is the union of a $c - 2$ clique and an independent set of size $c - 1$. Assign each instance G_i with a unique vertex of H. If necessary repeat some instances and make sure that each vertex of H is assigned to an instance. For each copy of an instance G_i assigned to a vertex v in H create a local graph by joining a copy of D_{k-1} to a copy of G_i, joining a copy of $\overline{G_i}$ to another copy of D_{k-1} and forming the disjoint union of these two joins. To get G' we connect all the graphs H_v according to the adjacency of H, fully connecting all H_v to H_w if $vw \in E(H)$.

Lemma 30.10.4 (Kratsch [468]) *Let H be a host graph on t' vertices and $(G_1, k), \ldots, (G_t, k)$ legal instances of k-REFINEMENT RAMSEY. Suppose that every vertex v of H is either contained in a clique of size ℓ or an independent set of size ℓ but H contains neither a clique or independent set of size $\ell + 1$. Then $Embed(H, k; G_1, \ldots, G_t)$ has a clique or an independent of size $\ell(2k - 2)$. Furthermore, H contains a clique or an independent set of size $\ell(2k - 2) + 1$ iff (G_i, k) is a "yes" instance for some $i \in \{1, \ldots, t\}$.*

Proof The H_v have both cliques and independent sets of size $2k - 2$. Furthermore if an instance (G_i, k) is a "yes" instance then H_v contains both an independent set and a clique of size $2k - 1$, in both cases using D_{k-1}.

Suppose that $V' \subseteq V(H)$ is a clique (resp. independent set) of size ℓ in H. Then choose a clique (resp. independent set) of size $2k - 2$ in each H_v. This gives a clique (resp. independent set) size $\ell(2k - 2)$ in $Embed(H, k; G_1, \ldots, G_t)$.

Since every vertex in H is contained in either a clique or an independent set of size ℓ, if some (G_i, k) is a "yes" instance, then we can choose an independent set or clique of size $2k - 2 + 1$ in H_v where v is the vertex to which G_i is assigned. This gives a clique of size $\ell(2k - 2) + 1$ in $Embed(H, k; G_1, \ldots, G_t)$.

If all instances of (G_i, k) are "no" instances, then no clique or independent set in $Embed(H, k; G_1, \ldots, G_t)$ can have more than $2k - 2$ vertices in H_v. No clique or independent set in $Embed(H, k; G_1, \ldots, G_t)$ can contain vertices from more than ℓ different local graphs, $Embed(H, k; G_1, \ldots, G_t)$ cannot contain either a clique or an independent set of size $\ell(2k - 2) + 1$. $\qquad\square$

We need the following combinatorial bound obtained by and easy modification of Erdös classic application of the probabilistic method.

Lemma 30.10.5 (Kratsch [468]) *For each $t \ge 3$, there is an $\ell \in \{1, \ldots, \lceil 8 \log(t) \rceil\}$ with $R(\ell + 1) > R(\ell) + t$.*

Proof If no then $R(\lceil 8 \log(t) \rceil + 1) \le t \cdot \lceil 8 \log(t) \rceil + R(1)$. Erdös [274] classic bound is $R(N) \ge 2^{\frac{N-1}{2}}$. Hence $R(\lceil 8 \log(t) \rceil + 1) \ge 2^{\frac{\lceil 8 \log(t) \rceil}{2}} \ge 2^{4 \log(t)} = t^4$. Hence $t^4 \le t \lceil 8 \log(t) \rceil + R(1)$ which is false for $t \ge 3$ as $R(1) = 1$. $\qquad\square$

The co-nondeterministic algorithm devised by Kratsch is the following.

Algorithm 30.10.1 (Compose)
Input: t instances $(G_1, k), \ldots, (G_t, k)$ of k-REFINEMENT RAMSEY.
Output: "yes" or an instance (G', k') of k-RAMSEY with $k' = O(\log t \cdot k)$.
1. If $k < 3$, simply solve each instance by complete search in time $O(n^4)$.
2. Else, guess $T \in \{1, \ldots, (\lceil 8\log(t)\rceil + 1) \cdot t\}$ and $\ell \in \{1, \ldots, \lceil 8\log(t)\rceil\}$.
3. Guess a host graph H with T vertices.
4. Guess t vertex sets $A_1, \ldots, A_t \in \binom{V(H)}{\ell}$ which are allowed to overlap.
5. Unless all the A_i induce independent sets and cliques and their union has size $\geq t$, return "yes".
6. Let A' denote an arbitrary minimal choice of the sets A_i such their union has size at least t.
7. Let $H' = H[A']$.
8. Let $G' = Embed(H', k; G_1, \ldots, G_t)$
9. Return (G', k') where $k' = \ell(2k - 2) + 1$.

Lemma 30.10.6 *The algorithm Compose is a co-nondeterministic composition for* k-REFINEMENT RAMSEY.

Proof Let t instances $(G_1, k), \ldots, (G_t, k)$ be given and we suppose $k \geq 3$. Assume $t > 3$. Suppose first that at least one input instance is a "yes". The algorithm returns "yes" unless it get to step 6. If it gets to 6 then we need to confirm that the graph (G', k') in step 9 is a "yes". If the host graph used for the embedding contains an independent set or a clique of size $\ell + 1$, then G' contains a clique or independent set of size $(\ell+1)(2k-2) > \ell(2k-2)+1$ because of the local structures, and hence $e(G', k')$ is a yes. There is no clique or independent set of size $\ell + 1$, then since we passed 5, the requirements of Lemma 30.10.4 are satisfied, and again the output is a "yes".

The other case is that all input instances are "no". If we can show that we fins a suitable host graph on one computation path and then Lemma 30.10.4 will ensure that the output is "no". Let ℓ denote the smallest positive integer with $R(\ell + 1) > R(\ell) + t$, and by Lemma 30.10.5, $\ell \leq \lceil 8\log t\rceil \cdot t$. Also note that $R(\ell) \leq (\ell - 1) \cdot t + R(1) \leq \lceil 8\log t\rceil \cdot t$, by the proof, and choice of ℓ. Therefore for some choice $T \in \{1, \ldots, \lceil 8\log t\rceil \cdot t\}$, and $\ell \in \{1, \ldots, \lceil 8\log t\rceil\}$ guessed by Compose we have $T = R(\ell) + t < R(\ell + 1)$. Therefore there is a host H on T vertices containing no clique nor independent set of size $\ell + 1$. At least one computation path will find this H. If $t \leq 3$, $R(3) = 6$ and $R(2) = 2$ guarantee that T and ℓ will be found.

As $T = R(\ell) + t$ there must exist cliques and independent sets A_i each of size ℓ which cover at least t vertices of H, from the definition of the Ramsey numbers. While there are at least $R(\ell)$ uncovered vertices the induced subgraph generated by the uncovered vertices must contain a clique or an independent set of size ℓ. Therefore t sets $A_1, \ldots, A_t$ can be chose in such a way that they cover at least t vertices. Then there is a computation path choosing H and $A_1, \ldots, A_t$.

From step 7, he get H' on at least t vertices containing neither clique nor independent set of size $\ell + 1$. By Lemma 30.10.4, $G' = Embed(H', k; G_1, \ldots, G_t)$ has an independent set or a clique of size $k' = \ell(2k - 2) + 1$ iff at least one G_i contains a clique or independent set of size k. Now $k' = \ell(2k - 2) + 1 \leq t^{o(1)} k^{O(1)}$. $\square$

Corollary 30.10.1 (Kratsch [468]) *Unless* $\text{NP} \subseteq \text{co-NP/poly}$,
(1) k-REFINEMENT RAMSEY *has no poly kernel.*
(2) k-RAMSEY *has no poly kernel.*

Proof (1) is immediate. Let $\hat{G}$ be the poly kernel for an instance G of k-RAMSEY. Then we can apply the reduction of Lemma 30.10.3 to get a poly kernel for $k + 1$-REFINEMENT RAMSEY, a contradiction, giving (2).   □

30.10.2 Co-nondeterministic Cross-composition and Hereditary Properties

It is not surprising that we can combine the ideas of cross-composition and co-nondeterminism, with quite useful applications.

Definition 30.10.3 (Kratsch, Philipczuk, Rai, and Raman [469]) Let $L \subseteq \Sigma^*$ and $Q \subseteq \Sigma^* \times \mathbb{N}$. The we say that L *co-NP cross-composes into* Q if there is a polynomial equivalence relation R and a co-nondeterministic algorithm which, given t strings $x_1, \ldots, x_t$ belonging to the same equivalence class of R, computes on each computation path an instance $(x^*, k^*) \in \Sigma^* \times \mathbb{N}$ in time poly in $\sum_{i=1}^{t} |x_i|$, such that
1. If $x_i \in L$ for some $i \in [t]$, then each computation path returns an instance $(x^*, k^*) \in Q$.
2. If $x_i \notin L$ for all $i \in [t]$ then at least one computation path returns $(x^*, k^*) \notin Q$.
3. k^* is bounded by a polynomial in $\max_{i \in [t]} |x_i| + \log t$ for each output (x^*, k^*).

The same methods used earlier show the following.

Theorem 30.10.2 (Kratsch, Philipczuk, Rai, and Raman [469]) *If $L \subseteq \Sigma^*$ co-NP cross-composes into Q, and Q has a poly kernel, then $L \in \text{NP/poly}$. Furthermore if L is NP-hard, then $\text{NP} \subseteq \text{co-NP/poly}$.*

Those authors also observed that polynomial parameterized transformations also preserve poly kernelizability for this setting. Indeed, they extended this observation to co-nondeterministic reductions as follows.[5]

Definition 30.10.4 If $L, L' \subseteq \Sigma^*$. We say that a co-nondeterministic poly time computable function f is a *co-nondeterministic poly reduction from L to L'* if and only if
(1) whenever $x \in L$ all computation paths for f have $f(x) \in L'$;
(2) if $x \notin L$, then some computation path has $f(x) \notin L'$.

---

[5]This definition is stated in Kratsch, Philipczuk, Rai, and Raman [469], but was certainly known earlier in classical structural complexity theory.

The following is easy, and left to the reader (Exercise 30.13.9).

Lemma 30.10.7 *If $L' \in$ co-NP/poly, and there is a co-nondeterministic poly reduction from L to L', then $L' \in$ co-NP/poly.*

We recall from Chap. 4, Sect. 4.9 the definition of hereditary properties.

Definition 30.10.5 (Hereditary property) If Π is a property of graphs, we say a graph G is a Π graph iff G satisfies Π. We say that a property Π is a *hereditary property*, if given any Π graph G, whenever H is an induced subgraph of G, then H is a Π graph.

In Sect. 4.9, we considered Leizhen Cai's FPT graph modification problems. Those results concerned themselves with induced problems about induced subgraphs and graph modification problems.

Π-INDUCED SUBGRAPH
Instance: A graph $G = (V, E)$.
Parameter: An integer k.
Question: Does there exist an induced Π-subgraph H of G on k vertices?

If Π excludes both large cliques and independent sets, then by Ramsey's Theorem, Π-INDUCED SUBGRAPH is FPT. The general problem otherwise is $W[1]$-hard by Khot and Raman [439].

As with the results on k-RAMSEY from the last subsection, here will again look at iterative versions of the problems. In this case we look at the following.

IMPROVEMENT Π-INDUCED SUBGRAPH
Instance: A graph $G = (V, E)$ and $X \subset V(G)$ with $|X| = k - 1$ and with $G[X] \in \Pi$.
Parameter: An integer k.
Question: Does there exist an induced Π-subgraph H of G on k vertices?

The analogy with the k-RAMSEY is the following theorem.

Theorem 30.10.3 (Kratsch, Philipczuk, Rai, and Raman [469]) *Let Π be any nontrivial hereditary graph class where membership can be tested in deterministic polynomial time. Then if there is a co-NP-cross-composition from IMPROVEMENT Π-INDUCED SUBGRAPH to Π-INDUCED SUBGRAPH, and Π-INDUCED SUBGRAPH has a polynomial kernel, then Π-INDUCED SUBGRAPH $\in$ co-NP/poly and NP $\subseteq$ co-NP/poly.*

Proof By Theorem 30.10.2, we see IMPROVEMENT Π-INDUCED SUBGRAPH $\in$ co-NP/poly. To show that this implies Π-INDUCED SUBGRAPH $\in$ co-NP/poly, and therefore NP $\subseteq$ co-NP/poly, we will use a co-nondeterministic reduction, and invoke Lemma 30.10.7.

Take an instance of Π-INDUCED SUBGRAPH, (G, k). The nondeterministic reduction guesses $k' \in [k]$ and X, a set of $k' - 1$ vertices. In poly time it checks if $G[X] \in \Pi$. If not then a *dummy* "yes" is returned. If yes, (G, X, k') an instance of IMPROVEMENT Π-INDUCED SUBGRAPH is returned.

Clearly if the input is a "yes" instance then all paths yield a "yes" instance of IMPROVEMENT Π-INDUCED SUBGRAPH. On the other hand if the input is a "no" instance, then consider the minimal $k' \in [k]$ which is also a "no" instance. Then the improvement version will be a "no" instance of IMPROVEMENT Π-INDUCED SUBGRAPH. The result follows.   □

The main result of this section is the following broad generalization of Kratsch's results on k-RAMSEY.

Theorem 30.10.4 (Kratsch, Philipczuk, Rai, and Raman [469]) *Unless* NP $\subseteq$ co-NP/poly, *there is no poly kernelization algorithm for any class Π-INDUCED SUBGRAPH for any hereditary property Π that poly recognizable, contains all independent sets and cliques, is closed under embeddings and has the Erdös–Hajnal property*: There exists a $e(\Pi) > 0$ such that every Π-graph either has a clique or independent set of size $|V(G)|^{e(\Pi)}$.

Proof By Theorem 30.10.3, it is enough to prove the following lemma.

Lemma 30.10.8 *If Π is a hereditary class of graphs that is polynomial-time recognizable, contains all independent sets and cliques, is closed under embeddings and has the Erdös–Hajnal property, then* IMPROVEMENT Π-INDUCED SUBGRAPH *co-NP-cross-composes to Π-INDUCED SUBGRAPH.*

Proof of Lemma 30.10.7 As with the proof of the Ramsey result from the last subsection, we will use a suitable host graph. Let e be such that any $G \in \Pi$ has either a clique or independent set of size $|V(G)|^{e(\Pi)}$. Let $e = e(\Pi)$. The following lemma has a similar proof to Lemma 30.10.5.   □

Lemma 30.10.9 *There is a constant $\delta(\Pi)$ depending only on Π such that for any $t > \delta(\Pi)$, there is an $\ell = O(\log^{\frac{1}{e}} t)$, such that $R_\Pi(\ell+1) > R_\pi(\ell)+t$, where $R_\Pi(\ell)$ denotes the Π-Ramsey number which is defined as the smallest number where every graph G of size $R_\Pi(\ell)$ has an induced Π-subgraph with at least ℓ vertices.*

Proof of Lemma 30.10.9 As Π satisfies the Erdös–Hajnal property for $e = e(\Pi)$, we see $R_\Pi(\ell) \geq R_\Pi(\ell^e)$. Hence, and graph having an induced subgraph of size ℓ has a clique or independent set of size ℓ^e. From the previous section, we know the classical Ramsey bound is $R(\ell) \geq 2^{\frac{(\ell-1)}{2}}$. Thus, $R_\Pi(\ell) > t^c$ for $\ell > (2c \log t + 1)^{\frac{1}{e}}$ and any $c > 0$. Thus, $R_\Pi(\ell + 1) - R_\Pi(\ell) = \Omega(t)$ for some $\ell = O(\log^{\frac{1}{e}} t)$.   □

Following the Ramsey construction of the previous subsection, the next step is the construction of a suitable host graph. The construction should be compared with that of Lemma 30.10.4.

Lemma 30.10.10 *There is a nondeterministic algorithm which, given an integer*
$t > \delta(\Pi)$, *in time polynomial in t either answers "fail" or outputs the following.*

1. *An integer* $\ell = O(\log^{\frac{1}{e}} t)$.
2. *A graph H together with a family of sets* $\{A_e \mid x \in V(H) \wedge A_x \subseteq V(H)\}$, *such that* $|V(H)| = t + o(t)$ *and such that for each* $x \in V(H)$,

$$|A_x| = \ell, \quad x \in A_x \quad and \quad H[A_x] \in \Pi.$$

We call this the covering property.
Moreover, for some computation path, H also satisfies the additional property that it contains no induced subgraph of size $\ell + 1$ *belonging to* Π.

Proof of Lemma 30.10.10 Nondeterministically guess $\ell = O(\log^{\frac{1}{e}} t)$ and H_0 with $t^{O(1)}$ vertices. Apply Lemma 30.10.9, there does exist $\ell = (\log^{\frac{1}{e}} t)$ so that $R_\Pi(\ell + 1) > R_\Pi(\ell) + t$. The smallest such choice of ℓ has $R_\Pi(\ell) + t = t^{O(1)}$. Hence there is at least one computation path giving H_0 on $R_\Pi(\ell) + t < R_\Pi(\ell + 1)$ vertices containing no induced Π-subgraph on $\ell + 1$ vertices.

Now we start deleting vertices from H_0. Begin with $S := \emptyset$. Then, whilst $|S| < t$, guess $A \subseteq V(H_0 \setminus S)$ such that $|A| = \ell$ and $H_0[A] \in \Pi$. Then set $S := S \cup A$. At each step if $H_0[A] \in \Pi$ continue, else we output "fail". Since H_0 has $\geq R_\Pi(\ell) + t$ vertices so long as $|S| < t$, we will have $|V(H_0 \setminus S)| \geq R_\Pi(\ell)$, and thus there is always some A with $|A| = \ell$ and $H_0[A] \in \Pi$. Hence there is a computation path finding A's at each step. To complete the construction, when $|S| \geq \ell$, define $H = H_0[S]$. We note that $t \leq |S| < t + \ell = t + o(t)$. □

We now complete the proof of Lemma 30.10.8. Given t instances $I_1, \ldots, I_t$ of IMPROVEMENT Π-INDUCED SUBGRAPH with input size n, as usual with cross-composition we can easily partition into $n^{O(1)}$ equivalence classes consisting of malformed instances, trivial "no" instances where k exceeds the number of vertices, and the remaining instances of equal parameter value k. Thus, without loss of generality, we can assume we have instances $I_j = (G_j, X_j, k)$ with $k \leq n$ and $G_j[X_j] \in \Pi$, and also without loss of generality assume that $t > \delta(\Pi)$.

Again following the k-RAMSEY template, we will begin to apply the algorithm from Lemma 30.10.10 for Π, t. We modify this by asking that if the algorithm would output a "fail" we output a trivial "yes" instance of Π-INDUCED SUBGRAPH.

Otherwise, we use the host graph H and ℓ to construct a graph G' by embedding one graph G_i into each vertex of H. Since $|V(H)| = t + o(t)$ each graph G_j is embedded at least once. That is, let $g : V(G) \to [t]$ be a surjection, and we define $Embed(H; (G_{g(x)})_{x \in V(G)})$ setting $k' = \ell(k - 1) + 1$. The algorithm then returns (G', k') as an instance of Π-INDUCED SUBGRAPH.[6] Note that as $\ell = O(\log^{\frac{1}{e}} t)$, $k' \leq (n + \log t)^{O(1)}$.

---

[6] The reader should compare this process with that of Lemma 30.10.6.

To see that this is all correct, first assume that (G_i, X_i, k) is a "yes" instance of IMPROVEMENT Π-INDUCED SUBGRAPH for some $i \in [t]$. Let $Y \subseteq V(G_i)$, $|Y| = k$, and $G_i[Y] \in \Pi$. Choose $x \in V(G)$ with $g(x) = i$. Note that the construction of H ensures that $x \in A_x$, $A_x \subseteq V(H)$, $|A_x| = \ell$, and $G[A_x] \in \Pi$ (else there is trivially a "yes" output). Then the $k' = \ell(k-1) + 1$ vertices $Y' = \{v(y, u) \mid y \in A_x \wedge u \in Y_y\}$ (where $v(y, u)$ denotes the vertices of the graph replacing y) corresponding to Y and $X_{g(y)}$ for $y \in A_x \setminus \{x\}$, inducing a subgraph isomorphic to $Embed(H[A_x]; Y \cup (X_{g(x)})_{x \in A_x})$. Since $H[A_x] \in \Pi$ and $G_{g(y)}[X_{g(y)}] \in \Pi$ for $y \in A_x$. we see that $G'[Y'] \in \Pi$ and hence we have a yes instance of Π-INDUCED SUBGRAPH for all computation paths.

On the other hand if we assume that no I_i is a yes instance, as with the Ramsey case, we look at the computation path where H and ℓ have the property that H has no induced Π-subgraph of size $\ell + 1$. Then we verify that this path yields a "no" instance of Π-INDUCED SUBGRAPH. Assuming that this is a "yes" instance, let $Y' \subseteq V(G')$, $|Y'| = k'$, and $G[Y'] \in \Pi$. Let $A = \{x \in V(H) \mid \exists u(v(x, u) \in Y')\}$. Then $H[A] \in \Pi$, as it is an induced subgraph of $G'[Y']$. Again, since H has the property that it has no Π-induced subgraph of size $\ell + 1$, $|A| \leq \ell$. As $|Y'| = k' = \ell(k-1) + 1$, there is some $x \in A$ such that $Y = \{u \in V(G_{g(x)}) \mid v(x, u) \in Y'\}$ is of size $\geq k$. Also $G_{g(x)}[Y]$ is an induced subgraph of $G'[Y'] \in \Pi$. Thus Y induces a subgraph of $G_{g(x)}$ of size $\geq k$ in Π, a contradiction. $\square$

30.11 Weak Composition and Sharp Lower Bounds Revisited

In this section we will introduce yet another method of composition; this allowing for sharp lower bounds based on co-nondeterminism and unique identifiers.

Definition 30.11.1 (Weak d-composition-Hermelin and Wu [389]) Let $d \geq 2$ be a constant, and let $L_1, L_2 \subseteq \{0, 1\}^* \times \mathbb{N}$ be two parameterized problems. A *weak d-composition* from L_1 to L_2 is an algorithm $\mathbb{A}$ that on input $(x_1, k), \ldots, (x_t, k) \in \{0, 1\}^* \times \mathbb{N}$, outputs an instance $(y, k') \in \{0, 1\}^* \times \mathbb{N}$ such that:

- $\mathbb{A}$ runs in co-nondeterministic polynomial time with respect to $\sum_i (|x_i| + k)$,
- $(y, k') \in L_2 \Leftrightarrow (x_i, k) \in L_1$ for some i, and
- $k' \leq t^{1/d} k^{O(1)}$.

In "normal" compositions the output parameter is required to be polynomially bounded by the input parameter, while in d-compositions it is also allowed to depend on the number of inputs t.

The basic lemma below follows more or less familiar arguments. We recall that for $L \subseteq \{0, 1\}^* \times \mathbb{N}$, we let $\widetilde{L} := \{x\#1^k : (x, k) \in L\}$ denote the unparameterized version of L. Lemma 30.10.2 stated that *if $L \subseteq \{0, 1\}^* \times \mathbb{N}$ has a kernel of size $f(k)$, then $\widetilde{L}$ has an oracle communication protocol of cost $f(k)$.*

Lemma 30.11.1 (Hermelin and Wu [389]) *Let $d \geq 2$ be a constant, and let $L_1, L_2 \subseteq \{0, 1\}^* \times \mathbb{N}$ be two parameterized problems such that $\widetilde{L_1}$ is NP-hard. Also assume* NP $\not\subseteq$ co-NP/poly. *A weak-d-composition from L_1 to L_2 implies that L_2 has no kernel of size $O(k^{d-\varepsilon})$ for all $\varepsilon > 0$.*

Proof Assume for the sake of contradiction that L_2 has a kernel of size $O(k^{d-\varepsilon})$ for some $\varepsilon > 0$. By Lemma 30.10.2 this implies that $\widetilde{L_2}$ has a communication protocol of cost $O(k^{d-\varepsilon})$. We show that this yields a low cost oracle communication protocol for $\mathrm{ORi}_{n,t(n)}(\widetilde{L_1})$ for some polynomial t. Because $\widetilde{L_1}$ is assumed to be NP-hard, this results in a contradiction to the assumption that NP $\not\subseteq$ co-NP/poly by applying the Complementary Witness Lemma, Lemma 30.10.1.

Consider a sequence of $\widetilde{L_1}$ instances $(\tilde{x}_1, \ldots, \tilde{x}_t)$ with $|\tilde{x}_i| = n$ and $t := t(n)$, where t is some sufficiently large polynomial. Let the corresponding parameterized problem sequence be $((x_1, k_1), \ldots, (x_t, k_t))$. The low cost protocol proceeds as follows:

1. Divide the parameterized problem sequence into *subsequences*, where each subsequence consists of instances with equal parameter values. Clearly there are at most $k := \max_i k_i \leq n$ subsequences.
2. For each subsequence, apply the d-composition from L_1 to L_2. This results in at most n instances of L_2, each with parameter bounded by $k' \leq t^{1/d} k^{O(1)}$.
3. For each instance of L_2, apply the assumed $O(k'^{(d-\varepsilon)})$ protocol to decide it. If one of the composed instances is a YES instance, then *accept*, otherwise *reject*.

It is clear that the protocol has cost $O(n \cdot k'^{(d-\varepsilon)})$, plug in that $k' \leq t^{1/d} k^c$ for some $c > 0$, and write $t = t(n)$. We have:

$$O\left(n \cdot k'^{(d-\varepsilon)}\right) = O\left(n \cdot \left(t^{1/d} k^c\right)^{d-\varepsilon}\right)$$

$$= O\left(n \cdot t^{(1-\varepsilon/d)} k^{c(d-\varepsilon)}\right)$$

$$= O\left(n \cdot t^{(1-\varepsilon/d)} n^{c(d-\varepsilon)}\right) \quad \text{(as } k \leq n\text{)}$$

$$= O\left(n^{1+cd-c\varepsilon} \cdot t^{(1-\varepsilon/d)}\right)$$

$$= O(t) \quad \text{(since } t \text{ is sufficiently large)}$$

$$= O(t \log t).$$

By the Complementary Witness Lemma it follows that $\widetilde{L_1} \in$ co-NP/poly, causing the desired contradiction. $\square$

Hermelin and Wu observed that ppt-reductions are not strong enough to preserve kernel size, and hence introduce the notion of *linear parametric transformations* that facilitate polynomial lower bounds.

Definition 30.11.2 (Linear parametric transformation-Hermelin and Wu [389]) Let L_1 and L_2 be two parameterized problems. We say that L_1 is *linear parameter reducible* to L_2, written $L_1 \leq_{\mathrm{ltp}} L_2$, if there exists a polynomial-time computable function $f : \{0, 1\}^* \times \mathbb{N} \to \{0, 1\}^* \times \mathbb{N}$, such that for all $(x, k) \in \Sigma^* \times \mathbb{N}$, if $(x', k') = f(x, k)$ then:

- $(x, k) \in L_1$ iff $(x', k') \in L_2$, and
- $k' = O(k)$.

The function f is called *linear parameter transformation*.

The following is immediate.

Lemma 30.11.2 *Let L_1 and L_2 be two parameterized problems, and let $d \in \mathbb{N}$ be some constant. If $L_1 \leq_{\text{lpt}} L_2$ and L_2 has a kernel of size $O(k^d)$, then L_2 also has a kernel of size $O(k^d)$.*

Hermelin and Wu have various lower-bound applications. Here we will present a polynomial kernelization lower bound for d-DIMENSIONAL MATCHING using weak composition. Recall that for this problem we are given a set $S \subseteq \mathcal{A} = A_1 \times \cdots \times A_d$ for some collection $A_1, \ldots, A_d$ of pairwise-disjoint sets. The parametric question is whether there is a subset $P \subseteq S$ of size k that are pairwise disjoint. As the reader will recall this was shown to be FPT using either color-coding and greedy localization.

Theorem 30.11.1 (Hermelin and Wu [389]) *Unless NP $\subseteq$ co-NP/poly, d-DIMEN-SIONAL MATCHING has no kernel of size $O(k^{d-3-\varepsilon})$ for any $\varepsilon > 0$.*

Again the proof filters through a colored variant of the problem, and uses unique identifiers in the proof. Here Hermelin and Wu use COLORED d-DIMENSIONAL PERFECT MATCHING. As with the other colorful problems, we have a coloring $col : S \mapsto \{1, \ldots, k\}$, and each dimension A_i, $i \in \{1, \ldots, d\}$, has exactly k elements. The goal is to find a colorful solution, that is, a solution $P \subseteq S$ that sets in P has distinct colors. Note each such solution corresponds to a perfect matching of elements across different dimensions. Hermelin and Wu's d-composition is from COLORED 3-DIMENSIONAL PERFECT MATCHING to $d + 3$-DIMENSIONAL MATCHING. The proof breaks into two steps. The first step we compose to an instance of SET PACK-ING, where the sets can have arbitrary size and are allowed to include more than one element from the same A_i. The reader should recall that this is the following problem.

SET PACKING

Instance: A family of subsets $\mathcal{F} = \{A_i \mid i \in F\}$ of a universe $\mathcal{U}$.

Parameter: An integer k.

Question: Are there k disjoint subsets in $\mathcal{F}$.

In the second step, we will transform the SET PACKING instance to an instance of $d + 3$-DIMENSIONAL MATCHING. First, we will split and pad the sets of high cardinality into many sets of cardinality exactly $(d + 3)$ which include exactly one set of each dimension. Then Hermelin and Wu use an equality gadget and unique identifiers to preserve the correctness of our construction.

The composition algorithm will compose a sequence of t^d instances with parameter k, and its output will be a single instance with parameter bounded by $t \cdot k^{O(1)}$. Again, we assume that $k > d$, since otherwise all instances can be solved in polynomial time by brute force, and a trivial instance of size $O(1)$ can be used as output.

In this step we will compose to the SET PACKING problem. Consider input sequence $\{I_\ell = (S_\ell \subseteq X_\ell \times Y_\ell \times Z_\ell, col_\ell, k) : 1 \le \ell \le t^d\}$ of COLORED 3-DIMENSIONAL PERFECT MATCHING instances, where $X_\ell = \{x_1^\ell, \ldots, x_k^\ell\}$, $Y_\ell = \{y_1^\ell, \ldots, y_k^\ell\}$, and $Z_\ell = \{z_1^\ell, \ldots, z_k^\ell\}$ are pairwise-disjoint sets. We will use $I = (S, k')$ to denote the instance of SET PACKING, which is the output of our composition. We proceed by describing dimensions and sets of S in detail.

- We first create pairwise-disjoint dimensions X, Y, Z of k elements each. Let $X = \{x_1, \ldots, x_k\}$, $Y = \{y_1, \ldots, y_k\}$, and $Z = \{z_1, \ldots, z_k\}$. Then for each set $R \in S_\ell$, we create the same set using elements in X, Y, Z in a way that matches R. That is $(x_a, y_b, z_c) \in S \Leftrightarrow (x_a^\ell, y_b^\ell, z_c^\ell) \in S_\ell$ for all $a, b, c \in \{1, \ldots, k\}$. In the following when we mention $R \in S_\ell$ in I, we mean the set created in S in this manner.
- Create d new dimensions, $P_1, \ldots, P_d$ where each dimension has kt new elements. Every dimension is organized as k layers, with each layer t elements. That is, for $i \in \{1, \ldots, d\}$:

$$P_i = \bigcup_{j \in \{1, \ldots, k\}} \{c_{j,1}^i, \ldots, c_{j,t}^i\}.$$

- For $\ell \in \{1, \ldots, t^d\}$, assign instance I_ℓ a unique d-tuple as its identifier ID_ℓ. This is done by picking an element from $\{1, \ldots, t\}^d$. Let $\text{ID}_\ell(i)$ be the value at index $i \in \{1, \ldots, d\}$.
- For $\ell \in \{1, \ldots, t^d\}$, consider $R \in S_\ell$ of color $j \in \{1, \ldots, k\}$. We extend R by adding elements from $P_1, \ldots, P_d$, such that for $i \in \{1, \ldots, d\}$, $R(P_i) = c_{j,\alpha}^i$ where $\alpha = \text{ID}_\ell(i)$. After this, every set $R \in S_\ell$ has exactly $(d + 3)$ elements.
- Now we construct *gadget sets*. For $r \in \{1, \ldots, t\}$, and $i \in \{1, \ldots, d\}$, construct $W_r^i = \{c_{j,r}^i : 1 \le j \le k\}$. Basically W_r^i occupies every position r of each of the k layers in dimension i. It is straightforward to check that each of the td gadget sets has unbounded size k. Further the elements of W_r^i are from the same dimension P_i.
- To conclude the construction, we set $k' = k + (t - 1)d$.

Lemma 30.11.3 $I \in$ SET PACKING *if and only if* $I_\ell \in$ COLORED 3-DIMENSIONAL PERFECT MATCHING *for some* $\ell \in \{1, \ldots, t^d\}$.

Proof ($\Leftarrow$) Suppose $(S_\ell, k) \in$ COLORED 3-DIMENSIONAL PERFECT MATCHING for some $\ell \in \{1, \ldots, t^d\}$, and let $P_\ell \subseteq S_\ell$ be a solution of size k for (S_ℓ, k). We take P_ℓ and $X = \{W_r^i : r \ne \text{ID}_\ell(i), r \in \{1, \ldots, t\}, i \in \{1, \ldots, d\}\}$ to be our solution for (S, k'). It is easy to check that all sets in X are pairwise disjoint and that $|P_\ell \cup X| = k + (t - 1)d$.

($\Rightarrow$) We show that if I has a pairwise-disjoint solution of size $k + (t - 1)d$, then I_ℓ has a perfect matching for some $\ell \in \{1, \ldots, t^d\}$. For this, fix one $i \in \{1, \ldots, d\}$, and consider the number of sets in the solution for I which are from $\{W_r^i : r \in$

$\{1,\ldots,t\}\}$. Clearly, the solution can contain at most t of them. However, if we pick t of them into solution, then we *cannot* pick any set from $S_1,\ldots,S_{td}$ into the solution due to the disjointness constraint. This means the solution can have at most $td < k + (t-1)d$ pairwise-disjoint sets, which cannot satisfy the number of disjoint sets required by the solution.

Therefore there can be at most $(t-1)$ sets in the solution which are from $\{W_r^i : r \in \{1,\ldots,t\}\}$. This gives us at most $(t-1)d$ pairwise-disjoint sets from td gadget sets. Hence by requirement of the size of the solution, it indicates that we have to pick at least k sets from $S_1,\ldots,S_{td}$. The crucial observation now is that we can pick at most k pairwise-disjoint sets from $S_1,\ldots,S_{td}$, because $|X| = |Y| = |Z| = k$. Therefore the only possibility is that we pick k sets from $S_1,\ldots,S_{td}$, and pick $(t-1)d$ gadget sets, with $(t-1)$ sets from $\{W_r^i : r \in \{1,\ldots,t\}\}$ for each $i \in \{1,\ldots,d\}$.

Now the observation is that, after we fix $(t-1)$ gadget sets for each $i \in \{1,\ldots,d\}$, it leaves, for each dimension, exactly the same position in each of the k layers not used. That is, there exists a set $\mathrm{ID} = \{\alpha_1,\ldots,\alpha_d\} \in \{1,\ldots,t\}^d$ such that c_{j,α_i}^i is not used for $j \in \{1,\ldots,k\}$. By the requirement of disjointness, now the k sets from $S_1,\ldots,S_{td}$ must come from S_ℓ where $\mathrm{ID}_\ell = \mathrm{ID}$. Further for any color $j \in \{1,\ldots,k\}$, c_{j,α_i}^i cannot be matched twice, hence these sets must have distinct colors. This gives $I_\ell \in$ COLORED 3-DIMENSIONAL PERFECT MATCHING, which completes the proof. $\square$

Next we transform the SET PACKING instance $I = (S, k')$ into an equivalent $(d+3)$-DIMENSIONAL MATCHING instance $I^* = (S^*, k^*)$. Note the SET PACK-ING instance derived in the last section fails the requirement of d-DIMENSIONAL MATCHING in two ways. First, gadget sets W_r^i ($i \in \{1,\ldots,d\}, r \in \{1,\ldots,t\}$) are of size k. Second, the elements of W_r^i come from the same dimension P_i. Initially let S^* be S. Fix arbitrary W_r^i, we show how to modify W_r^i so that it is of dimension $(d+3)$ and each of its element comes from different dimension, while preserving the correctness of the construction. Recall that $W_r^i = \{c_{j,r}^i : 1 \le j \le k\}$.

First, we split W_r^i into $W_{1,r}^i, W_{2,r}^i, \ldots, W_{k,r}^i$, where $W_{j,r}^i = \{c_{j,r}^i\}$. In the following, when we mention elements of the form x_*^* ($*$ is the wild-card symbol), it means to extend X with this new element, similar for y_*^* and z_*^*. Now we begin to construct constraints among $\{W_{j,r}^i : j \in \{1,\ldots,k\}\}$. We use two operations, *extend* means extending an existing set with new elements, *add* means adding a new set into S^*. The gadgets are modified as follows: *extend* $W_{1,r}^i$ with $x_{k+1,r}^i$, *add* $U_{1,r}^i = (x_{k+1,r}^i, y_{k+1,r}^i)$, *extend* $W_{2,r}^i$ with $y_{k+1,r}^i, z_{k+2,r}^i$, *add* $U_{2,r}^i = (z_{k+2,r}^i, x_{k+2,r}^i)$, and so on until we extend $W_{k,r}^i$. Now it is clear that the size of all gadget sets is bounded by 3. Further the elements of every gadget set come from different dimensions. Finally to make each gadget set of dimension exactly $(d+3)$, we simply pad each gadget set. To conclude the construction we set $k^* = k + (t-1)kd + (k-1)d$. The following lemma shows that the construction preserves correctness.

Lemma 30.11.4 $(S^*, k^*) \in (d+3)$-DIMENSIONAL MATCHING *iff* $(S, k') \in$ SET PACKING.

Proof ($\Leftarrow$) Assume we found a solution $P \subseteq S$ of $k' = k + (t-1)d$ pairwise-disjoint sets in S. As we have argued, the solution must consist of k sets from some S_ℓ for some $\ell \in \{1, \ldots, t^d\}$ with $\mathrm{ID}_\ell = \{\alpha_1, \ldots, \alpha_d\} \in \{1, \ldots, t\}^d$, and $W_r^i \in P$ for $i \in \{1, \ldots, d\}, r \neq \alpha_i$. Hence for (S^*, k^*), we choose $W_{1,r}^i, \ldots, W_{k,r}^i$ for $r \neq \alpha_i$, and choose $U_{1,r}^i, \ldots, U_{k-1,r}^i$ for $r = \alpha_i$. This gives a solution in S^* of size $k + [(t-1)k + (k-1)] \cdot d = k + (t-1)kd + (k-1)d = k^*$.

($\Rightarrow$) Suppose there is a set of k^* pairwise-disjoint sets in S^*. Fix $i \in \{1, \ldots, d\}$ and $r \in \{1, \ldots t\}$ and consider $W_{1,r}^i, U_{1,r}^i, \ldots, U_{k-1,r}^i, W_{k,r}^i$. The first observation is that, by our construction, we can pick at most k pairwise-disjoint sets from $W_{1,r}^i, U_{1,r}^i, \ldots, U_{k-1,r}^i, W_{k,r}^i$. And if we pick k of them, the k sets must be $W_{1,r}^i, W_{2,r}^i, \ldots, W_{k,r}^i$.

Now if we pick $W_{1,r}^i, W_{2,r}^i, \ldots, W_{k,r}^i$ for every $r \in \{1, \ldots, t\}$, S^* cannot contain any set from $S_1, \ldots, S_{t^d}$. This indicates that the solution size is at most $tkd < k^*$, contradiction. Therefore for every $i \in \{1, \ldots, d\}$, there are at most $k(t-1)+(k-1)$ sets in S^*. Hence S^* contains at most $[k(t-1)+(k-1)] \cdot d = (t-1)kd + (k-1)d$ from gadget sets.

Therefore we have to pick k sets from $S_1, \ldots, S_{t^d}$. Let Q be this family of k sets. Further, after fixing $W_{1,r}^i, \ldots, W_{k,r}^i$ for $(t-1)$ different r's and for every $i \in \{1, \ldots, d\}$. There exists a set $\mathrm{ID} = \{\alpha_1, \ldots, \alpha_d\} \in \{1, \ldots, t\}^d$ such that c_{j,α_i}^i is not matched for $j \in \{1, \ldots, k\}$. Now all sets of Q must come from S_ℓ with $\mathrm{ID}_\ell = \mathrm{ID}$ and of different colors. Choosing Q and $\{W_r^i : r \neq \alpha_i, i \in \{1, \ldots, d\}\}$ gives a solution in (S, k'). This completes the proof. $\square$

We are now in position to complete the proof of Theorem 30.11.1.

Proof of Theorem 30.11.1 Let $d' = d + 3$, and let $t' = t^d = t^{d'-3}$. The composition algorithm presented above composes t' COLORED d-DIMENSIONAL PERFECT MATCHING instances with parameter k to a d'-DIMENSIONAL MATCHING instance with parameter $k^* = O(kt) = O(t'^{1/d'-3}k)$. Thus, our composition is in fact a weak $(d'-3)$-composition from COLORED d-DIMENSIONAL PERFECT MATCHING to d'-DIMENSIONAL MATCHING. Since COLORED d-DIMENSIONAL PERFECT MATCHING is NP-hard, applying Lemma 30.11.1 shows that d-DIMENSIONAL MATCHING has no kernel of size $O(k^{d-3-\varepsilon})$, for any $\varepsilon > 0$, unless co-NP $\subseteq$ NP/poly. $\square$

Hermelin and Wu [389] applied this technology to a number of other problems. For example, the d-SET COVER is the same as SET COVER except that all the underlying sets have cardinality bounded by d. Similarly d-SET COVER, and related here is d-RED–BLUE DOMINATING SET where the vertex degrees are bounded by d. All of these can be shown to have no poly kernels unless co-NP $\subseteq$ NP/poly.

30.12 Miniature Parametric Miniatures

The question arises about what to do with Turing kernels. In this section we will look at recent results which give a completeness programme to give evidence for lower bounds on kernel size for Turing kernels. The idea is to introduce two new hierarchies the MK-hierarchy and WK-hierarchy to base the proofs upon.

Definition 30.12.1 (Hermelin, Kratsch, Soltys, Wahlström, and Wu [388])
1. We define the problem of WEIGHTED t-NORMALIZED SATISFIABILITY$(k \log n)$ $H[t](k \log n)$, as the problem of determining if a t-normalized boolean formula with n variables has a satisfying assignment of weight k, *parameterized by $k \log n$*. That is, a "yes" instance is a pair (Φ, d) where Φ is a t-normalized boolean formula with n variables with a satisfying assignment of weight k, and $d = k \log n$.

   Let WK$[t]$ be the class of problems reducible to $H[t](k \log n)$ under *polynomial-time parameterized reductions*. Similarly define these $k \log n$ bounded analogs of the other W-hierarchy classes, such as WK$[SAT]$ and WK$[P]$.
2. Similarly define MK$[t]$ to be the collection of problem parametrically polynomial reducible to WEFT t d-SAT(n), the satisfiability problem for WEFT t d-SAT formulas for some fixed d and n variables, parameterized by n.

Notice that, for example, a WK$[SAT]$ formula is trivially solvable in time $2^d n^{O(1)}$ and the hierarchies are all in a class called EPT introduced by Flum, Grohe, and Weyer [313]. This is the class of parameterized problem solvable in time $O^*(2^{O(k)})$. That is, all of these problems are in FPT with algorithms with a single exponential dependency on the parameter.

The completeness programme below correlates to a miniaturization of the original W and M hierarchies. Many of the combinatorics in the proofs are lifted from the earlier sections, observing the parameterized polynomial reductions used in those proofs.

Theorem 30.12.1 (Hermelin, Kratsch, Soltys, Wahlström, and Wu [388])
(1) ANTIMONOTONE-WK$[1, 2]$ *the $k \log n$ version of whether there is a weight k satisfying assignment for a* 2-CNF *formula is complete for* WK$[1]$.
(2) MONOTONE WEIGHTED t-NORMALIZED SATISFIABILITY$(k \log n)$, $H[t](k \log n)$ *is complete for* WK$[t]$ *if t is even and $t > 1$*.
(3) ANTIMONOTONE WEIGHTED t-NORMALIZED SATISFIABILITY$(k \log n)$, $H[t](k \log n)$ *is complete for* WK$[t]$ *if t is odd and $t > 1$*.
(4) *n variable d-SAT is* WK$[1]$ *complete for all $d \geq 3$*.

Proof We argue for $t = 1$ the rest are similar. The first reduction is to transform a weighted d-$W[1, d]$ formula of size $k \log n$ into an antimonotone one. This follows immediately by examination of the proof of Theorem 21.2.2 as that gives a parameterized polynomial reduction. Next we reduce such an antimonotone instance (φ, k) to an antimonotone *multicolored* one. Here we ask that the variables come in k colors and we ask also that the solution comes in one of each color.[7] Since the formula is antimonotone, we simply create k copies of the variable set, and denote x_i in the jth one to have color j and be denoted by $x_{i,j}$. Then for each pair $j' \neq j$ add the clause $(\overline{x_{i,j'}} \vee \overline{x_{i,j}})$. Then for each clause $(\overline{x_{i_1}} \vee \cdots \vee \overline{x_{i_d}})$ replace it by the formula $\bigwedge_j (\overline{x_{i_1,j}} \vee \cdots \vee \overline{x_{i_d,j}})$. This simple reduction clearly works with the same parameter.

Next we reduce antimonotone multicolored $W[1, d](k \log n)$ to antimonotone multicolored $W[1, 2](k \log n)$. This can be done by the reduction of Downey and Fellows [244] where the authors reduce $W[1, d]$ to $W[1, 2]$ using a parameterized polynomial-time reduction. Here we will follow the reduction of Hermelin et al. Let φ be the formula on n variables from X with k colors. We create the multicolored φ' as follows. Let the colors of φ' correspond to the d-tuples of colors of φ, $(c_1, \ldots, c_d)$. Thus φ' has $k' = O(k^d)$ many colors. Let $X_C = \{(x_1, \ldots, x_d) \mid x_i \in X \wedge x_i \text{ has color } c_i \text{ for } 1 \leq i \leq d\}$. Remove from X_C each tuple that explicitly falsifies a clause of φ, i.e. iff the φ contains $(\overline{x_1} \vee \cdots \vee \overline{x_d})$ since all the clauses are d-ary. Then we denote the remaining set of tuples as X_C^*. Now for each color tuples C and C' which have at least one common color, enforce consistency by excluding pairs of variable tuples which disagree on some common color using a conjunction of clauses of with exactly two negative literals each. The resulting instance is φ'. A multicolored satisfying assignment of φ' directly corresponds to one of φ. Furthermore, as φ is antimonotone, each clause has been verified in the construction of X_C^*. Thus φ has a weight k multicolored assignment iff φ' has one of weight k'.

Finally there is a simple reduction from the multicolored antimonotone $W[1, 2](k \log n)$ to the uncolored version by adding clauses to emulate the color classes enforcing one from each group.

The relevant reductions given earlier in Chap. 23 are seen to be parameterized polynomial time. They are basically the reduction for DOMINATING SET and for INDEPENDENT SET used locally. (This is explicitly stated in Flum and Grohe [312], Lemma 7.5.) Such observations lead to (2) and (3).

Now for (4). The classic Karp reduction, Theorem 1.1.2, shows that d-SAT(n) is poly reducible to the case of $d = 3$, observing that if all duplicate clauses are removed the total length is n^d, and this is parametrically polynomial as d is constant. □

Now we will characterize the problems which have polynomial kernels as defined by parametric Karp kernelizations using the concepts developed so far. Let NP_{para} denote the collection of parameterized languages whose unparameterized versions are in NP.

---

[7]This trick was first used by Downey, Fellows, and Regan [256]. There the problem was called SEPARATED t-NORMALIZED SATISFIABILITY.

Theorem 30.12.2 (Hermelin, Kratsch, Soltys, Wahlström, and Wu [388]) *Let* PK *denote the collection of parameterized problems with polynomial kernels. Then* $PK \cap NP_{para} = MK[1]$.

Proof Let $L \in PK \cap NP_{para}$. Let K be the poly kernel of L. Since $\hat{L}$, the unparameterized version of L is in NP, there is a Karp reduction from $\hat{L}$ to $3SAT(n)$, and hence there is a parametric polynomial reduction from L to $3SAT(n)$, and so $L \in MK[1]$. Conversely, since an instance of $d\text{-}SAT[n]$ has $O(n^d)$ clauses after removing duplicates, we get a polynomial kernel for each $d \geq 1$. Thus we have the result.   □

Again using the $k \log n$ trick we see the following.

Lemma 30.12.1 $PK \cap NP_{para} = MK[1] \subseteq WK[1]$.

This re-parameterization method allows us to show that a number of problems do not have poly kernels unless collapse occurs. That is because we will see that several problems with this property are $WK[1]$-complete.

The conjecture of Hermelin et al. is the following.

Conjecture 30.12.1 *If a problem is* $WK[1]$-*hard then it does not have a polynomial* Turing *kernelization.*

The conjecture could be disproved by showing that one of the problems with Turing but no Karp kernel is $WK[1]$-hard. Unfortunately, we know precious few of them.

Theorem 30.12.3 (Hermelin, Kratsch, Soltys, Wahlström, and Wu [388]) *The following are* $WK[1]$-*complete, and do not have poly kernels unless* co-NP $\subseteq$ NP/poly.
1. CLIQUE($k \log n$) *and* INDEPENDENT SET($k \log n$).
2. BINARY NDTM HALTING(k), *the k-step halting problem for a binary NTM of size n, and* NTM HALTING($k \log n$).

As remarked in [388], most natural $W[1]$ complete problems re-parameterize under the parameter $k \log n$. Thus, those authors argue for their conjecture as follows. If NTM HALTING($k \log n$) has a polynomial-time Turing kernel, then BINARY NDTM HALTING(k) would reduce k step halting on a size n binary NTM to a polynomial number $f(n)$ of queries about NTM's of size $g(k)$. This does not seem reasonable, if nondeterminism miniaturizes as we think it does. We turn to the proof of the theorem.

Proof First notice that BINARY NDTM HALTING(k) is clearly OR-compositional, and hence has no poly kernel unless co-NP $\subseteq$ NP/poly. The remaining reductions come from inspection of the reductions of Theorem 21.2.5 modulo the $W[1,2] = W[1,s]$ miniaturization above. $\qquad\qquad\square$

30.13 Exercises

Exercise 30.13.1 Prove that PLANAR VERTEX COVER is NP-complete.

Exercise 30.13.2 (Niedermeier [546]) Using the method of Theorem 30.1.1, show that there is no $(2 - \varepsilon)k$ kernel for PLANAR INDEPENDENT SET unless P $=$ NP.

Exercise 30.13.3 Prove Theorem 30.2.2.
(Hint. The method is to be more careful with the proof for the Karp reductions.)

Exercise 30.13.4 (Gramm, Guo, Hüffner and Niedermeier [353, 355]) Prove that EDGE CLIQUE COVER problem has an exponential kernel.

Exercise 30.13.5 Using the methods of Lemma 30.2.2, prove Theorem 30.6.1.

Exercise 30.13.6 (Heggernes and P. van 't Hof and B. Lévêque and D. Lokshtanov and C. Paul [385]) This problem considers contractions to paths and stars.

CONTRACTION TO A PATH
Input: A graph G.
Parameter: A positive integer k.
Question: Can you contract at most k edges so that the resulting graph is a path?

The problem CONTRACTION TO A STAR is the same, except we replace "path" by "star".
1. Prove that if (G, k) is a positive instance of CONTRACTION TO A PATH there exists a path witness structure $W_1, \ldots, W_l$ such that
   (a) the W_i's form a partition of the vertices and each part is connected,
   (b) if $i < j - 1$, there is no edge between vertices of W_i and W_j,
   (c) at most k parts are non-singleton (call these parts the *big parts*),
   (d) and the sum of the size of the big parts is at most $2k$.
2. Prove that for the path problem the following reduction rules is safe.
   If the removal of an edge e separate the graph into components of size at least $k + 2$, contract e.
3. Prove that the path problem has a linear size kernel.
4. For CONTRACTION TO A STAR, prove that if (G, k) is a positive instance there exists star witness structure $W_0, W_1, \ldots, W_l$ with
   (a) the W_i's form a partition of the vertices and each part is connected,
   (b) if $i \neq 0$ and $j \neq 0$, there is no edge between vertices of W_i and W_j,

(c) at most k parts are non-singleton (call these parts the *big parts*),

(d) and the sum of the size of the big parts is at most $2k$.

5. Prove for the star problem that if (G, k) is a positive instance there exists star witness structure in which W_0 is the only non-singleton part.

6. Prove that CONTRACTION TO A STAR has no poly kernel unless co-NP $\subseteq$ NP/poly.

7. Generalize this result for CONTRACTION TO TREES.

(Hint: Use RED–BLUE DOMINATING SET.)

Exercise 30.13.7 Use Theorem 30.7.1 to establish Corollary 30.7.1.

Exercise 30.13.8 (Cygan, Kratsch, Philipczuk, Wahlström [188]) Prove the following.

1. COMPRESSION CLIQUE COVER cross-composes to k-EDGE CLIQUE COVER.

2. Hence there is not poly kernel for k-EDGE CLIQUE COVER unless co-NP $\subseteq$ NP/poly.

COMPRESSION CLIQUE COVER

Instance: A graph G, and a set S of $k + 1$ cliques in G covering all the edges of G.

Parameter: A positive integer k.

Question: Does there exist a size k EDGE CLIQUE COVER?

(Hint: Augment the method of the proof of Theorem 30.6.3.)

Exercise 30.13.9 Prove Lemma 30.10.7.

Exercise 30.13.10 (Cygan, Fomin and van Leeuwen [187]) Recall the firefighting model from Exercise 5.3.6, and the problem shown to be FPT in Exercise 8.8.8, SAVING ALL BUT k VERTICES. Prove that unless co-NP $\subseteq$ NP/poly, SAVING ALL BUT k VERTICES has no polynomial kernel even if the input graph is a tree of maximum degree 4.

Exercise 30.13.11 (Dell and van Melkebeek [206]) Prove Theorem 30.8.1.

Exercise 30.13.12 (Dell and van Melkebeek [206]) Prove that the unparameterized version of MULTICOLORED BICLIQUE is NP-complete.

Exercise 30.13.13 (Bodlaender, Jansen, and Kratsch [91]) Prove that CLIQUE parameterized by the size of a vertex cover has no polynomial kernel unless co-NP $\subseteq$ NP/poly.

(Hint: Again define the equivalence relation so that all bitstrings not encoding valid instances of CLIQUE are equivalent, and two well-formed instances have the same number of vertices and parameters. Given (G_i, ℓ) for $i \in [t]$, construct a single instance (G', Z', ℓ', k') of CLIQUE parameterized by the size of a vertex cover. G' will have vertex cover Z'. Number the vertices of G_i from 1 to n. Construct G' by applying 1–3 below.

1. Create ℓn vertices $v_{i,j}$ with $i \in [\ell]$ and $j \in [n]$. Make an edge $v_{i,j} v_{i',j'}$ if $i' \neq i$ and $j' \neq j$. Let C denote these vertices. Any clique can contain at most ℓ vertices from C.

2. For each $1 \leq p < q \leq n$, (vertices of G_i) create three vertices $w_{p,q}, w_{p,\hat{q}}, w_{\hat{p},q}$, and make $w_{p,q}$ adjacent to all the vertices of C, $w_{p,\hat{q}}$ adjacent to all vertices of C save for $v_{x,j}$ where $j = q$ (and any x), and $w_{\hat{p},q}$ adjacent to all members of C except for $v_{x,j}$ with $j = p$. Also add edges between the $w_{x,y}$ corresponding to distinct pairs from $[n]$ adding in total a set D of $3\binom{n}{2}$ many vertices. Note that any clique can have at most one $w_{x,y}$ for each pair from $[n]$.

3. For each instance with graph G_i make a new vertex u_i and connect to all vertices of C. The adjacency to D is defined as follows. Make u_i adjacent to $w_{p,q}$ if $pq \in E(G_i)$. Otherwise make u_i adjacent to $w_{p,\hat{q}}$ and $w_{\hat{p},q}$. Let B denote this set of t vertices.

Define $\ell' = \ell + 1 + \binom{n}{2}$, and $Z = C \cup D$, with $k' = |Z| = \ell n + 3\binom{n}{2}$. The reader will need to verify that this is a valid cross OR-composition.)

Exercise 30.13.14 (Dom, Lokshtanov, and Saurabh [226, 227]) Show that the following have no poly kernel unless co-NP $\subseteq$ NP/poly.

(Hint: For UNIQUE COVERAGE and SMALL UNIVERSAL HITTING SET, you should consider a colored version and the "cookbook" approach.)

SMALL UNIVERSAL SET COVER

Instance: A family of sets F with universe U and $k \leq |F|$.
Parameter: $k + |U|$.
Question: Is there a subfamily $F' \subseteq F$ with $|F'| \leq k$ such that $\bigcup_{S \in F'} S = U$?

UNIQUE COVERAGE

Instance: A universe U, a family of sets $\mathcal{F}$ over U.
Parameter: A positive integer k.
Question: Find a $\mathcal{F}' \subset \mathcal{F}$ and $S \subseteq U$ with $|S| \geq k$ such that each element of S occurs in exactly one set in $\mathcal{F}'$.

BOUNDED RANK DISJOINT SETS

Instance: A universe U, a family of sets $\mathcal{F}$ all of size at most d over U.
Parameter: Positive integers d and k.
Question: Find a $\mathcal{F}' \subset \mathcal{F}$ with $|\mathcal{F}'| \geq k$ such that each element of for each $S_1 \neq S_2 \in \mathcal{F}', S_1 \cap S_2 = \emptyset$?

BIPARTITE REGULAR PERFECT CODE

Instance: A bipartite $G = (T \sqcup N, E)$ with each vertex in E of the same degree.
Parameter: Positive integers $|T|$ and k.
Question: Is there $N' \subseteq N$ of size at most k, such that every $v \in T$ has exactly one neighbor in N'?

SMALL UNIVERSAL HITTING SET

Instance: A universe U, a family of sets $\mathcal{F}$ with $|U| \leq d$.
Parameter: Positive integers d and k.
Question: Is there $S \subseteq U$ such that every element of $\mathcal{F}$ has nonempty intersection with S?

Exercise 30.13.15 (Dom, Lokshtanov, and Saurabh [226, 227]) Show that the following have no poly kernel unless co-NP $\subseteq$ NP/poly.

SMALL SUBSET SUM

Instance: A set S of n integers and a target t, with each integer at most 2^d.
Parameter: Positive integers d and k.
Question: Is there a subset $S' \subseteq S$ adding up to t?

(Hint: Reduce COLORED RED–BLUE DOMINATING SET to SMALL SUBSET SUM.)

Exercise 30.13.16 (Kirkpatrick and Hell [445])   Prove that k–H-FACTOR is NP-complete even when restricted to $\chi(H)$-partite graphs.

Exercise 30.13.17 (Dom, Lokshtanov, and Saurabh [226])   Show that the following has no poly kernel unless co-NP $\subseteq$ NP/poly.

CONNECTED FEEDBACK VERTEX SET

Instance: A graph $G = (V, E)$ and a set of terminals $T \subset V$.
Parameter: k and T.
Question: Is there a set $S \subset V$ such that $|S| = k$ and $G[S]$ is connected and $G[V \setminus S]$ has no cycles?

Exercise 30.13.18 (Cygan, Philipczuk, Philipczuk, and Wojtaszczyk [191])   Using Problem 30.13.17, show that the following has no poly kernel unless co-NP $\subseteq$ NP/poly.

2-DEGENERATE CONNECTED FEEDBACK VERTEX SET

Instance: A 2-degenerate graph $G = (V, E)$ and a set of terminals $T \subset V$.
Parameter: k and T.
Question: Is there a set $S \subset V$ such that $|S| = k$ and $G[S]$ is connected and $G[V \setminus S]$ has no cycles?

Exercise 30.13.19   Prove Lemma 30.10.3.

Exercise 30.13.20 (Kratsch, Philipczuk, Rai, and Raman [469])   Prove that if Π is a hereditary class such that Π that poly recognizable, closed under disjoint union, and does not contain the biclique $K_{s,s}$ for some $s = s(\Pi)$, then IMPROVEMENT Π-INDUCED SUBGRAPH co-NP-cross-composes to Π-INDUCED SUBGRAPH.

Exercise 30.13.21   Consider the problem of DEGREE d-CLIQUE, which is the k-CLIQUE problem restricted to graphs all of whose valencies are $\leq d$. Prove that this problem is FPT. Use composition to show that it has no poly kernel unless co-NP $\subseteq$ NP/poly. Prove that it has a poly Turing kernel.

Exercise 30.13.22 (Hermelin, Kratsch, Soltys, Wahlström, and Wu [388])   Show that SET COVER(n) and HITTING SET(n) has both WK[1]-complete. Show also the problem of deciding if a graph has k disjoint paths is WK[1]-hard.

30.14 Historical Notes

As pointed out earlier, until quite recently, the only kernelization lower bounds were based on approximation lower bounds, usually via the PCP theorem. An example of this was the $1.36k$ lower bound by Dinur and Safra [223]. The other alternative was using the unique games conjecture, for which there is less compelling evidence that it is true. The lower-bound engine using distillation and composition was due to Bodlaender, Downey, Fellows, and Hermelin [83], and the notion of compression of instances of NP-complete formulas was in the work of Harnik and Naor [379]. A manuscript of preliminary version of [83] had been around for a while, and based on the question of whether there was a connection between classical complexity and distillation, Fortnow and Santhanam [328] proved that a distillation algorithm for any NP-complete problem implies the collapse of the polynomial hierarchy to at least three levels. The AND case remained unconnected to classical considerations until the very recent work of Drucker [263], although a lot of lower bounds had been proven using the hypothesis that AND compression was hard. Whilst *parameterized polynomial-time* reductions had been around since the early papers of Downey and Fellows, it was Bodlaender, Thomassé, and Yeo [101] who introduced them as a tool in kernelization lower bounds. Bodlaender, Jansen, and Kratsch [91] introduced cross-composition. Dell and van Melkebeek [206] established the use of co-nondeterminism and communication protocols as a tool. This was clarified by Kratsch [468], and the attractive work on k-RAMSEY. Hermelin, Kratsch, Soltys, Wahlström, and Wu [388]. The notion of weak composition and its relationship with co-nondeterminism for sharp lower bounds was introduced in Hermelin and Wu [389], who also showed how to get quasipolynomial lower bounds. Hermelin, Kratsch, Soltys, Wahlström, and Wu [388] introduced the new miniaturization methods targeted at Turing kernelization lower bounds. Kernelization lower bounds is a thriving area of research and is an interesting study of the relationship between known algorithms and lower bounds.

30.15 Summary

Several methods of increasing sophistication are introduced to demonstrate that parameterized languages in FPT do not have small kernels. These include *composition, poly parameterized reductions, cross-composition, co-nondeterminism and communication complexity,* and *miniature parametric miniatures.*

Part VII
Further Topics

Parameterized Approximation

31

31.1 Introduction

In this chapter, we look at a parameterized approximation, a relatively new, topic of intense interest. It was introduced independently by three papers presented at the same conference: Downey, Fellows, and McCartin [251], Chen and Grohe and Grüber [156], and Cai and Huang [125]. Arguably it was already present in the work of Alekhnovich and Razborov [27] who proved that WEIGHTED MONOTONE CIRCUIT SATISFIABILITY does not have a 2-FPT approximation algorithm (as described below) unless $W[P] = \mathrm{FPT}$ under randomized parameterized reductions. Alekhnovich and Razborov used this result to show that resolution is not "automatizable" (an important problem in the area of automated proofs and SAT solvers) unless such a collapse occurs.

As we have seen in Chaps. 9 and 30, there are various ways in which parameterized complexity and parameterized computation can interact with approximation algorithms. The interplay between the two fields is surveyed comprehensively in Marx [514]. We consider only the most natural extension of parameterized complexity in this direction. We follow the definitions given in Marx's survey [514].

As we have seen in Chap. 9, an input instance to an NP-optimization problem has associated with it a set of feasible solutions. A cost measure is defined for each of these feasible solutions. The task is to find a feasible solution where the cost measure is as good as possible.

We define an NP-optimization problem as a 4-tuple $(I, sol, cost, goal)$ where

- I is the set of instances.
- For an instance $x \in I$, $sol(x)$ is the set of feasible solutions for x. The length of each $y \in sol(x)$ is polynomially bounded in $|x|$, and it can be decided in time polynomial in $|x|$ whether $y \in sol(x)$ holds for given x and y.
- Given an instance x and a feasible solution y, $cost(x, y)$ is a polynomial-time computable positive integer.
- $goal$ is either max or min.

The cost of an optimum solution for instance x is denoted by $opt(x)$.

$$opt(x) = goal\{cost(x, y') \mid y' \in sol(x)\}.$$

R.G. Downey, M.R. Fellows, *Fundamentals of Parameterized Complexity*,
Texts in Computer Science, DOI 10.1007/978-1-4471-5559-1_31,
© Springer-Verlag London 2013

If y is a solution for instance x then the *performance ratio* of y is defined as

$$R(x, y) = \begin{cases} cost(x, y)/opt(x) & \text{if } goal = \min, \\ opt(x)/cost(x, y) & \text{if } goal = \max. \end{cases}$$

For a real number $c > 1$ (or a function $c : \mathbb{N} \to \mathbb{N}$), we say that an algorithm is a *c-approximation* algorithm if it always produces a solution with performance ratio at most c ($c(x)$). As in Chap. 9, the *standard parameterization* of an optimization problem X, where we define the corresponding parameterized decision problem $X_\le$ as

$X_\le$

Input: An instance x of X.
Parameter: k.
Question: Is $opt(x) \le k$?

We can define $X_\ge$ analogously. As we have seen, many of the standard problems in the parameterized complexity literature are standard parameterizations of optimization problems, for example, k-VERTEX COVER, k-CLIQUE, k-DOMINATING SET, k-INDEPENDENT SET. If such a standard parameterization is fixed-parameter tractable then this means that we have an efficient algorithm for exactly determining the optimum for those instances of the corresponding optimization problem where the optimum is small. An $M[1]$ or $W[1]$-hardness result for a standard parameterization of an optimization problem shows that such an algorithm is unlikely to exist. In this case, we can ask the question: Is it possible to efficiently *approximate* the optimum as long as it is small?

We use the following definition for an *FPT-approximation algorithm* proposed by Chen, Grohe, and Grüber [156].

Definition 31.1.1 (Standard FPT-approximation algorithm)   Let $X = (I, sol, cost, goal)$ be an optimization problem. A *standard FPT-approximation algorithm with performance ratio* $c(k)$[1] for X is an algorithm that, given input (x, k) satisfying

$$\begin{cases} opt(x) \le k & \text{if } goal = \min, \\ opt(x) \ge k & \text{if } goal = \max, \end{cases} \tag{31.1}$$

computes a $y \in sol(x)$ in time $f(k) \cdot |x|^{O(1)}$ such that

$$\begin{cases} cost(x, y) \le k \cdot c(k) & \text{if } goal = \min, \\ cost(x, y) \ge \frac{k}{c(k)} & \text{if } goal = \max. \end{cases} \tag{31.2}$$

For inputs not satisfying (1.1) the output can be arbitrary.

An example of one of these parameterized approximation problems is the following.

BIN PACKING

Input: Given bins $B_1, \ldots, B_k$ with sizes $b_1, \ldots, b_k$ and positive integers $a_1, \ldots, a_n$ (the sizes of the items to be packed).

Parameter: k.

Question: Can pack the a_i into the bins so that for all j, $\sum_{a_i \in B_j} a_i \le b_j$?

A classical result in the theory of NP-completeness is that of *First Fit*, a natural parameterized approximation with $g(k) = 2k$. (See Garey and Johnson [337].) For another example, see Problem 6.6.1. There we let $G = (V, E)$ be a graph and $w :$ $V \to \mathbb{Q}^+$ be a function assigning weights to the vertices. We said there that the weighting function is *degree weighted* if there is a constant c such that $w(v) = c \cdot \deg(v)$. The *optimal vertex cover* C is the one of least overall weight $\sum_{v \in C} w(v)$. It has weight *OPT*. In that exercise, the reader proved that:

1. $w(V)$ is defined to be $\sum_{v \in V} w(v) \le 2 \cdot OPT$.
2. There was a 2-approximation algorithm for MINIMAL WEIGHT VERTEX COVER, assuming arbitrary vertex weights.
3. This results in a 2-approximation solution to SET COVER.

Thus we see that there are natural problems with such multiplicative parameterized approximation algorithms.

One particularly interesting example of such parameterized approximation algorithms comes from computational biology.

k-SUBTREE PRUNE AND REGRAFT

Input: A pair of phylogenetic X-trees, T_1 and T_2 (as per Definition 4.10.1), each an unrooted binary tree with leaves labeled by a common set of species X and (unlabeled) interior nodes having degree exactly three.

Parameter: A positive integer k.

Question: Find at most k *subtree prune and regraft* operations that will transform T_1 into T_2.

Recall that a single *subtree prune and regraft* (SPR) operation on T_1 begins by "pruning" a subtree of T_1 by detaching an edge $e = (u, v)$ in T_1 from one of its (non-leaf) endpoints, say u. The vertex u and its two remaining adjacent edges, (u, w) and (u, x) are then *contracted* into a single edge, (w, x). Let T_u be the phylogenetic tree, previously containing u, obtained from T_1 by this process. We create a new vertex u' by subdividing an edge in T_u. We then create a new tree T' by adding an edge f between u' and v. We say that T' has been obtained from T_u by a single *subtree prune and regraft* (SPR) operation.

A related operation on the phylogenetic tree T_1 is the *tree bisection and reconnection* (TBR) operation. A single *tree bisection and reconnection* (TBR) operation begins by detaching an edge $e = (u, v)$ in T_1 from both of its endpoints, u and v. Contractions are then applied to either one or both of u and v to create two new phylogenetic trees (note that a contraction is necessary only in the case of a non-leaf vertex). Let T_u and T_v be the phylogenetic trees, previously containing, respectively,

u and v, obtained from T_1 by this process. We create new vertices u' and v' by subdividing edges in T_u and T_v, respectively. We then create a new tree T' by adding an edge f between u' and v'. We say that T' has been obtained from T_1 by a single *tree bisection and reconnection* (TBR) operation. Note that the effect of a single TBR operation can be achieved by the application of either one or two SPR operations and that every SPR operation is also a TBR operation.

In Sect. 4.10, we saw that Hallett and McCartin gave an algorithm for the k-MAXIMUM AGREEMENT FOREST problem, which is equivalent to the k-TREE BISECTION AND RECONNECTION problem running in time $O(k^4 \cdot n^5)$. The algorithm employed there uses a kernelization strategy from Allen and Steel [28] as a pre-processing step, followed by a search tree strategy. Neither of the techniques used is applicable in the case of the k-SUBTREE PRUNE AND REGRAFT problem. It had long been conjectured that the k-SUBTREE PRUNE AND REGRAFT problem is fixed-parameter tractable. Proof of the conjecture has been a long-standing open problem in the parameterized complexity community, but it has been recently resolved.

The algorithm given in Sect. 4.10 can easily be adapted to give either a NO answer to the k-TREE BISECTION AND RECONNECTION problem or, otherwise, a minimal set of TBR operations that transforms T_1 into T_2. Given two phylogenetic trees, T_1 and T_2, if there is a set S of at most k SPR operations that transforms T_1 into T_2 then S is also a set of at most k TBR operations transforming T_1 into T_2. Thus, in this case, the Hallett–McCartin algorithm given will return a solution S' consisting of at most k TBR operations. This set S' will translate into a set of at most $2k$ SPR operations transforming T_1 into T_2. If there is no size k set of SPR operations that transforms T_1 into T_2, then the algorithm given in [377] might still return a solution S' consisting of at most k TBR operations and again, this set S' will translate into a set of at most $2k$ SPR operations transforming T_1 into T_2. However, in this case, there is no guarantee of success.

This is currently one of two examples of a problem that is not proven to be FPT but that does have a standard FPT-approximation algorithm, although there are other examples of FPT-approximation algorithms appearing in the literature. The other is the DISJOINT CYCLE PROBLEM, which is known to be $W[1]$-hard, but has a parameterized approximation algorithm. We will discuss this example a bit later.

First, we develop some general theory. The reader may note that another method we might have used to formulate parameterized optimization would have been to parameterize the optimization problem by parameterizing by the optimum. For minimization problems, this is the same. For the time being, we will mean that our FPT algorithms are strongly uniform, in that we have a known algorithm with computable constants.

Theorem 31.1.1 (Chen, Grohe, and Grüber [156]) *Let L be an NP-minimization problem over Σ and assume that c is computable and $kc(k)$ is nondecreasing. Then the following are equivalent.*

(1) *L has a parameterized approximation algorithm with approximation ration c.*

(2) *There is a computable function g, and an algorithm Φ which, on input $x \in \Sigma^*$ computes a solution $y \in opt(x)$ such that $cost(x, y) \leq opt(x)c(opt(x))$ in time $g(opt(x))|x|^d$ for a fixed constant d.*

Proof (1) implies (2). Without loss of generality we assume that we have an algorithm Ψ which always outputs something and is the relevant optimization algorithm for L with ration c, and running in time $f(k)|x|^d$ on input (x, k). Then let Φ simulate Ψ on inputs $(x, 1), (x, 2), \ldots$ until Ψ does not reject for the first time, say on (x, k), and then outputs the output y of Ψ on this instance. Note that since Ψ only rejects if $i < opt(x)$, we conclude $k \leq opt(x)$. The running time of Φ is bounded by $\sum_{i=1}^{opt(x)} f(i)|x|^d$. Moreover, we note $cost(x, y) \leq opt(x) \leq kc(k)$.

For the converse direction, choose Φ satisfying (2). Without loss of generality we can assume g is nondecreasing. Let Ψ be the algorithm that simulates Φ for $g(k)|x|^d$ many steps and outputs the output y of Φ if Φ stops, and rejects else. Then evidently Ψ is a parameterized algorithm. Finally, if $opt(x) \leq k$, then Φ halts in $g(k)|x|^d$ steps and y satisfies $cost(x, y) \leq opt(x)c(opt(x)) \leq kc(k)$. $\square$

For maximization problems there is an analogous (but not directly analogous) result. The proof is analogous and left to the reader (Exercise 31.6.4).

Theorem 31.1.2 (Chen, Grohe, and Grüber [156]) *Let L be an NP-maximization problem over Σ and assume that c is computable and $\frac{k}{c(k)}$ is nondecreasing and unbounded. Suppose that L has a parameterized approximation algorithm with approximation ration c. There is a computable function g, and an algorithm Φ which, on input $x \in \Sigma^*$ computes a solution $y \in opt(x)$ such that $cost(x, y) \geq \frac{opt(x)}{c(opt(x))}$ in time $g(opt(x))|x|^d$ for a fixed constant d.*

The problem with the full analog of Theorem 31.1.1 is the following. Consider a NP-optimization problem where the optimal solution is always large, being $\Omega(|x|)$ for all inputs x. Then an algorithm Φ trivially exists for $c = 1$ as all NP-optimization problems can be solved in exponential time exactly. This does not help for finding a solution of size k for small k in time $f(k)|x|^d$. Chen, Grohe, and Grüber observed that this problem can be overcome in the case of certain self-reducible problems.

Definition 31.1.2 (Chen, Grohe, and Grüber [156]) We say that an NP-maximization problem L is *well-behaved for parameterized approximation with constant q* if
1. Given a nontrivial instance x for L in time polynomial in $|x|$ we can construct another instance x' with $|x'| < |x|$ such that $opt(x) \geq opt(x') \geq opt(x) - q$, and
2. for all x, if x is a valid instance so is x'.

For example MAX-CLIQUE and MAX-DIRECTED VERTEX DISJOINT CYCLES are both well-behaved.

Theorem 31.1.3 (Chen, Grohe, and Grüber [156]) *Let L be an NP-maximization problem over Σ and assume that c is computable and $\frac{k}{c(k)}$ is nondecreasing and unbounded. Suppose that there is a computable function g, and an algorithm Φ which,* *on input* $x \in \Sigma^*$ *computes a solution* $y \in opt(x)$ *such that* $cost(x, y) \geq \frac{opt(x)}{c(opt(x))}$ *in time* $g(opt(x))|x|^d$ *for a fixed constant* d.

Then L *has a parameterized approximation algorithm with approximation ratio* c.

Proof Let Φ be as above, and g nondecreasing. Let $h : \mathbb{N} \to \mathbb{N}$ be a function with $h(k) - k \geq q$ for all k, and without loss, $h(k) = k + q$. We define $\Psi(\langle x, k \rangle)$ as follows. Simulate Φ for $g(h(k))|x|^d$ many steps and if it stops Ψ outputs its output y. Note that Ψ may stop although $k < h(k) < opt(x)$. But then

$$\frac{k}{c(k)} \leq \frac{h(k)}{c(h(k))} \leq \frac{opt(x)}{copt(x)} \leq cost(x, y).$$

Otherwise $h(k) > opt(x)$. Then for input (x, k) with $k \leq opt(x)$ we note that $\frac{k}{c(k)} \leq \frac{opt(x)}{copt(x)} \leq cost(x, y)$. If Φ does not stop, $h(k) < opt(x)$. Now we use the well-behavedness of L to construct an instance (x', k) for L and start the next iteration simulating Φ on (x', k), taking time $g(opt(x))|x'|^d$ many steps, and hence the number of iterations is bounded by $|x|$. Moreover, $h(k) < opt(x)$ implies $k < opt(x)$ at each step, so it is guaranteed that for inputs with $k \leq opt(x)$ a solution y with $cost(x, y) \geq \frac{k}{c(k)}$ is computed.   □

Grohe and Grüber use the theorem above to prove the following.

Theorem 31.1.4 (Grohe and Grüber [358]) DISJOINT CYCLE PROBLEM *has a parameterized approximation algorithm.*

DISJOINT CYCLE PROBLEM
Input: A directed graph D.
Parameter: k.
Question: Find k vertex disjoint cycles in D.

The proof of this result uses the results above together with a lot of graph structure theory. It is beyond the scope of this book. The result is particularly interesting since the authors point out that by an easy modification of the work of Slivkins [623], the problem is $W[1]$-hard to solve exactly. See also Exercise 31.6.9.

31.2   Pathwidth

In this section we look at a classic parametric approximation algorithm. For PATH-WIDTH there is a simple $O(2^k)$-parameterized approximation algorithm which is single exponential. Cattell, Dinneen, and Fellows [134] describe a very simple algorithm, based on "pebbling" the graph using a pool of $O(2^k)$ pebbles, that in linear time (for fixed k) either determines that the pathwidth of a graph is more than k or finds a path decomposition of width at most the number of pebbles actually used.

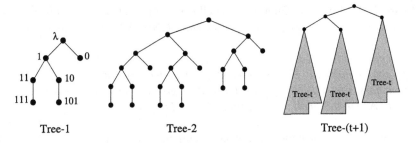

Fig. 31.1 Embedding pathwidth t tree obstructions, tree-t, in binary trees

The main advantages of this algorithm over previous results are (1) the simplicity of the algorithm and (2) the improvement of the hidden constant for a determination that the pathwidth is greater than k. The main disadvantage is in the width of the resulting "approximate" decomposition when the width is less than or equal to k, although it has never been explored on real world data, or heuristically adapted.

It is easy to see that the family of graphs of pathwidth at most t is a lower ideal in the topological (and minor) order. We need the following facts.

Lemma 31.2.1 (Folklore) *If G has pathwidth t and $|V(G)| = n$, then $|E(G)| \leq nt - \frac{(t^2+t)}{2}$.*

Proof Exercise 31.6.1.   □

Let B_h denote the complete binary tree of height h and order $2^h - 1$. Let $h(t)$ be the least value of h such that $B_{h(t)}$ has pathwidth greater than t, and let $f(t)$ be the number of vertices of $B_{h(t)}$. To get a bound for $f(t)$, $B_{h(t)}$ needs to contain at least one obstruction of pathwidth t.

Lemma 31.2.2 (Ellis, Sudborough, and Turner [272]) *All topological tree obstructions of pathwidth t can be recursively generated by the following rules:*
1. *The single edge tree K_2 is the only obstruction for pathwidth 0.*
2. *If T_1, T_2, and T_3 are any three tree obstructions for pathwidth t, then the tree T consisting of a new degree 3 vertex attached to any vertex of T_1, T_2, and T_3 is a tree obstruction for pathwidth $t + 1$.*

Proof Exercise 31.6.2.   □

From Lemma 31.2.2, we see that the orders of the tree obstructions of pathwidth t are precisely $(5 \cdot 3^t - 1)/2$, (e.g., orders 2, 7, 22, and 57 for pathwidth $t = 0, 1, 2$, and 3). We can easily embed at least one of the tree obstructions for pathwidth t, as shown in Fig. 31.1, in the complete binary tree of height $2t + 2$. Thus, the complete binary tree of order $f(t) = 2^{2t+2} - 1$ has pathwidth greater than t.

Using the $f(t)$ bound given above, the main result of the Cattell, Dinneen, and Fellows article [134] now follows:

Theorem 31.2.1 (Cattell, Dinneen, and Fellows [134]) *Let H be an arbitrary undirected graph, and let t be a positive integer. One of the following two statements must hold*:

1. *The pathwidth of H is at most $f(t) - 1$.*
2. *H can be factored*: $H = A \oplus B$, *where A and B are boundaried graphs with boundary size $f(t)$, the pathwidth of A is greater than t and less than $f(t)$. Also, A is in the small universe (boundary is first bag).*

Proof We describe an algorithm that terminates either with a path decomposition of H of width at most $f(t) - 1$, or with a path decomposition of a suitable factor A with the last vertex set of the decomposition consisting of the boundary vertices.

If we find an homeomorphic embedding of the *guest tree* $B_{h(t)}$ in the *host graph* H, then we know that the pathwidth of H is greater than t. During the search for such an embedding, we work with a partial embedding. We refer to the vertices of $B_{h(t)}$ as *tokens*, and call tokens *placed* or *unplaced* according to whether or not they are mapped to vertices of H in the current partial embedding. A vertex v of H is *tokened* if a token maps to v. At most one token can be placed on a vertex of H at any given time. We recursively label the tokens by the following standard rules:

1. The root token is labeled by the empty string λ.
2. The left child token and right child token of a height h parent token $P = b_1 b_2 \ldots b_h$ are labeled $P \cdot 1$ and $P \cdot 0$, respectively.

Let $P[i]$ denote the set of vertices of H that have a token at time step i. The sequence $P[0], P[1], \ldots, P[s]$ will describe a path decomposition either of the entirety of H or of a factor A fulfilling the conditions of Theorem 31.2.1. In the case of outcome (b), the boundary of the factor A is indicated by $P[s]$.

The placement algorithm is described as follows. Initially, consider that every vertex of H is colored *blue*. In the course of the algorithm, a vertex of H has its color changed to *red* when a token is placed on it, and stays red if the token is removed. Only blue vertices can be tokened, and so a vertex can only be tokened once.

Algorithm 31.2.1
    function GrowTokenTree
1   **if** root token λ is not placed on H **then**
        arbitrarily place λ on a blue vertex of H
    endif
2   **while** there is a vertex $u \in H$ with token T and blue neighbor v,
          and token T has an unplaced child $T \cdot b$ **do**
    2.1 place token $T \cdot b$ on v
    endwhile
3   **return** {tokened vertices of H}

    program PathDecompositionOrSmallFatFactor
1    $i \leftarrow 0$
2    $P[i] \leftarrow$ **call** GrowTokenTree

3   **until** $|P[i]| = f(t)$ or H has no blue vertices **repeat**
   3.1 pick a token T with untokened children
   3.2 remove T from H
   3.3 **if** T had one tokened child **then**
          replace all tokens $T \cdot b \cdot S$ with $T \cdot S$
     endif
   3.4 $i \leftarrow i + 1$
   3.5 $P[i] \leftarrow$ **call** GrowTokenTree
   enduntil
   done

Before we prove the correctness of the algorithm, we note some properties:
(1) the root token will need to be placed (step 1 of the GrowTokenTree) at most once for each component of H;
(2) the GrowTokenTree function only returns when $B_{h(t)}$ has been embedded in H or all parent tokens of degree less than 2 have no blue neighbors;
(3) the algorithm will terminate since during each iteration of step 3.2, a tokened red vertex becomes untokened, and this can happen at most n times, where n denotes the order of the host H.

Since tokens are placed only on blue vertices and are removed only from red vertices, it follows that the interpolation property of a path decomposition is satisfied. Suppose the algorithm terminates at time s with all of the vertices colored red. To see that the sequence of vertex sets $P[0], \ldots, P[s]$ represents a path decomposition of H, it remains only to verify that for each edge (u, v) of H, there is a time i with both vertices u and v in $P[i]$. Suppose vertex u is tokened first and untokened before v is tokened. But vertex u can be untokened only if all neighbors, including vertex v, are colored red [see step 3.1 and comment (2) above].

Suppose the algorithm terminates with all tokens placed. The argument above establishes that the subgraph A of H induced by the red vertices with boundary set $P[s]$ has pathwidth at most $f(t)$. To complete the proof, we argue that in this case, the sequence of token placements establishes that A contains a subdivision of $B_{h(t)}$, and hence must have pathwidth greater than t. Since the GrowTokenTree function only attaches pendant tokens to parent tokens, we need to only to observe that the operation in step 3.3 subdivides the edge between T and its parent. $\qquad\square$

Corollary 31.2.1 (Cattell, Dinneen, and Fellows [134]) *Given a graph H of order n and an integer t, there exists a $O(n)$ time algorithm that gives evidence that the pathwidth of H is greater than t or finds a path decomposition of width at most $O(2^t)$.*

Proof We show that program PathDecompositionOrSmallFatFactor runs in linear time. First, if H has more than $t \cdot n$ edges, then the pathwidth of H is greater than t. By the proof of Theorem 31.2.1, the program terminates with either the embedded binary tree as evidence, or a path decomposition of width at most $f(t)$.

Note that the guest tree $B_{h(t)}$ has constant order $f(t)$, and so token operations that do not involve scanning H are constant time. In the function GrowTokenTree, the only nonconstant time operation is the check for blue neighbors in step 2. While scanning the adjacent edges of vertex u, any edge to a red vertex can be removed, in constant time. Edge (u, v) is also removed when step 2.1 is executed. Therefore, across all calls to GrowTokenTree, each edge of H needs to be considered at most once, for a total of $O(n)$ steps. In the program PathDecompositionOrSmallFat-Factor, all steps except for GrowTokenTree are constant time. The total number of iterations through the loop is bounded by n, by the termination argument following the program.   □

The last result for this section shows that we can improve the pathwidth algorithm by restricting the guest tree. This allows us to use the subdivided tree obstructions given in Fig. 31.1.

Corollary 31.2.2 (Cattell, Dinneen, and Fellows [134]) *Any subtree of the binary tree $B_{h(t)}$ that has pathwidth greater than t may be used in the algorithm for Theorem* 31.2.1.

Proof The following simple modifications allow the algorithm to operate with a subtree. The subtree is specified by a set of *flagged* tokens in $B_{h(t)}$. At worst, the algorithm can potentially embed all of $B_{h(t)}$.

In step 2 of GrowTokenTree, the algorithm only looks for a flagged untokened child $T \cdot b$ to place, since unflagged tokens need not be placed. The stopping condition in step 3 of PathDecompositionOrSmallFatFactor is changed to "all flagged tokens of $B_{h(t)}$ are placed or . . . ," so that termination occurs as soon as the subtree has been embedded. The relabeling in step 3.3 can place unflagged tokens on vertices of H, since all the rooted subtrees of a fixed height are not isomorphic. If that happens, we expand our guest tree with those new tokens. (It is easy to see that the new guest is still a tree.) These tokens can then get relabeled by future edge subdivisions that occur above the token in the host tree. Thus, duplication of token labels will not happen, and the $f(t)$ width bound is preserved.   □

Example 31.2.1 Using $K_{1,3}$ as the guest tree of pathwidth 2, the program trace represented in Fig. 31.2 terminates with all vertices colored red (gray) yielding a path decomposition of width 5.

Observe that the Cattell–Dinneen–Fellows pathwidth algorithm provides an easy proof of basically the main result we saw in Sect. 18.2 from Bienstock, Robertson, Seymour, and Thomas [65] (or its earlier variant in Robertson and Seymour [581]) that for any forest F, there is a constant c, such that any graph not containing F as a minor has pathwidth at most c.

Corollary 31.2.3 *Every graph with no minor isomorphic to forest F, where F is a minor of a complete binary tree B, has pathwidth at most $c = |B| - 2$.*

Fig. 31.2 Path
decomposition of width 5

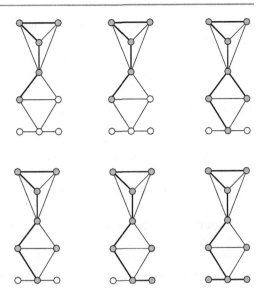

Proof Without loss of generality, we can run our pathwidth algorithm using (as the
guest) any subtree T of B that contains F as a minor. Since at most $|B| - 1$ pebbles
are used when we do *not* find an embedding of T (in any host graph), the resulting
path decomposition has width at most $|B| - 2$.   □

Note that the Cattell–Dinneen–Fellows constant c is identical to the one given
in [65] when $F = B$. This constant is the best possible in that case. We refer to
Sect. 18.2 for more details on excluding forests.

31.3 Treewidth

Earlier in the book, in Sect. 11.2, we looked at heuristics based on safe separators for
treewidth. Here is another notion which allows for a parameterized approximation
algorithm for treewidth.

Definition 31.3.1 We say a separator S of G is *good* for a set W if no com-
ponent of $G \setminus S$ contains more that $\frac{|W|}{2}$ of the vertices of W.

The following observation is easy and left to the reader (Exercise 31.6.7).

Lemma 31.3.1 *If* $\mathrm{tw}(G) \leq k$, *then any* $W \subseteq V(G)$ *has a good separator of size*
$\leq k + 1$.

The basic ideas in the algorithm below are to seek good separators of size $\leq k+1$ for sets W of size $\leq 2k+3$. By Lemma 31.3.1, if no such separator exists, then we have a certificate for $\mathrm{tw}(G) > k$. If we succeed in all recursive steps, then we will succeed in constructing a tree decomposition of width $\leq 3k+3$.

Theorem 31.3.1 (Reed [575])
1. *Algorithm* SAFESEPTREEDECOMP *below either yields a certificate for* $\mathrm{tw}(G) > k$ *or yields a rooted tree decomposition for G of size $\leq 3k+3$.*
2. *The algorithm runs in time $O(|G|^2)$ (specifically, $O(9^k|G|^2)$).*

Algorithm 31.3.1 SAFESEPTREEDECOMP
1. **Input:** A graph G, $W \subseteq V(G)$ with $|W| \leq 2k+3$.
2. **Output:** Either a certificate that $\mathrm{tw}(G) > k$ or a rooted tree decomposition of G of width at most $3k+3$.
3. Find a good separator S for W in G. If none is found output "$\mathrm{tw}(G) > k$" and stop. S generates two vertex sets A and B as it is a separator.
4. Construct recursively a tree decomposition T_A for $G[A]$, W_A, where $W_A = W \cap A$.
5. Construct recursively a tree decomposition T_B for $G[B]$, W_B.
6. Take $T = T_B \cup T_B \cup \{W \cup S\}$, where the roots of T_A and T_B are the only children of $\mathrm{root}(T) = \{W \cup S\}$.

Proof The recursion ensures that the sizes of W_B and W_A are bounded since the definition of good separator ensures that

$$|W_B| \leq |S| + \frac{|W|}{2} \leq k+1 + \frac{2k+3}{2} \leq 2k+3.$$

Therefore the constructed tree decomposition will have width $\leq |W| + |S| - 1 \leq 3k+3$. The output really is a tree decomposition since each edge in G will either be present in the node $W \cup S$, or in one of the recursive trees T_A or T_B. Finally, the interpolation property is obeyed since if a vertex v occurs in $X_{\mathrm{root}(T)}$ and not in G_A it cannot return.

For the running time, good separators exist if and only if W can be partitioned into three parts $W_a \sqcup W_b \sqcup W_c$ each of size at most $k+1$, W_c being part of a separator S of size at most $k+1$. Therefore, in this case, the graph $G \setminus W_c$ has no $k+1-|W|$ vertex disjoint paths between the sets W_a and W_b. Testing this uses standard flow techniques, in time $O(|E(G \setminus W_c)|) = O(n)$.

We exhaustively explore all partitions of W into three parts, there being at most 3^{2k+3} of them, and for each such partition, solve the flow problem. this gives a running time bounded by $O(9^k n^2)$. $\square$

It would be a great achievement to get a practical version of this algorithm. It runs fairly efficiently as a heuristic on reasonable graphs.

31.4   Parameterized Online Algorithms and Incremental Computing

A particularly natural and also underdeveloped direction for parameterized complexity and parameterized approximation comes from the area of *online* algorithms.

Consider BIN PACKING. The algorithm of First Fit has the property that it is online. One packs items into bins and once an item is in a bin it cannot be removed. We are in an *online* situation. This is quite different from the *offline* case where we get to see all the boxes laid out and then can decide how to pack.

Online problems abound in diverse areas of computer science. There are many situations where it is necessary to make decisions without access to future inputs, or where only partial information about the input data is available. Examples include robot motion planning, or robot reconfiguration, in an uncharted environment; visual searching and mapping; resource management for multi-user, or multi-processor, computing systems; and network configuration problems. In short, any programmable system that operates in, or responds to, a dynamic environment can be considered as being "online". Due to advances in hardware design, complex online systems are rapidly becoming feasible.

An online algorithm is basically a response strategy for a sequence of inputs. Formally, we can think of an online problem as being one where the input data is supplied as a sequence of small units, one unit at a time. An online algorithm to solve the problem must respond by producing a sequence of outputs: after seeing t units of input, it must produce the tth unit of output.

For our purposes an online graph would be $G_0 \subset G_1 \subset \cdots$ where I give you the graph one vertex at a time (and its relationship with all the vertices seen so far) and you need to decide what to do with it. That is, you must build a function $f : G_s \to \widehat{G}_s$ where $\widehat{G}_s$ is some kind of *expansion* of G_s before seeing G_{s+1}. The expansion would be a coloring (of G_s), scheduling, packing etc. (There are of course other variations. With a scheduler in a computer, vertices come and go as people log off and on, etc.)

We may not have the mathematical ability to gauge the difficulty of online problems and to measure the performance of online algorithms. The probabilistic approach attempts to analyze the performance of online algorithms on "typical" inputs. It suffers from the criticism that we seldom have a probability distribution to model a typical input.

During the last decade or so this approach has been largely superseded by *competitive analysis*, introduced by Sleator and Tarjan [621]. In the competitive model the input sequence for an online problem is generated by an adversary, and the performance of an online algorithm is compared to the performance of an optimal offline algorithm, that is allowed to know the entire sequence of inputs in advance. Usually we do this by looking at the ratio $\frac{\text{Online}(G)}{\text{Offline}(G)}$ of the best online performance and the best offline one. Competitive analysis gives us the ability to make strong theoretical statements about the performance of online algorithms without making probabilistic assumptions about the input. Nevertheless, practitioners voice reservations about the model, citing its inability, in some important cases, to discern between online algorithms whose performances differ markedly in practice.

31.4.1 Coloring

We all know the famous 4-color theorem that planar graphs can be colored with 4 colors, and trees can be colored with 2. Coloring these objects can be done efficiently in terms of the number of colors because we can "see" the whole graph.

Now suppose that I am giving you the graph one vertex at a time and all I tell you is the relationship of the given vertex with the previously given ones. I will even tell you it is a tree. Can we 2-color it *online*?

So suppose that I give you a single vertex v_1. You color it red. Then I give you one more. Call this one v_2 and I tell you $v_2 v_1$ is an edge. You color it blue. Now I give you a new vertex v_3 and I tell you that it is not connected to v_1 or v_2. You must color it red or blue to keep within the 2 colors. If you say red then I add a v_4 and connect it to v_3 (which is red) and v_1 (which is blue). You then must choose a 3rd color for v_4.

In general, a tree with n vertices can be colored with $\log n$ many colors, and there is no online algorithm that colors it with fewer than $O(\log n)$. It was proved by Sandy Irani [412] that k-degenerate graphs and hence graphs of treewidth k can be colored with $O(k \log n)$ many colors and this is tight (see, e.g., Downey and McCartin [260]) The *performance ratio* comparing the offline and the online performances is $O(\log n)$.

31.4.2 Online Coloring Graphs of Bounded Pathwidth

We have the following result of Kierstead and Trotter.

Theorem 31.4.1 (Kierstead and Trotter [443])
1. *There is an online algorithm that will color a graph of pathwidth k with $3k - 2$ many colors.*
2. *Hence, there is a k, $3k - 2$ parameterized approximation algorithm for this problem.*

Remark 31.4.1 The reader should carefully note how this clearly demonstrates the difference between a *promise* and a *presentation* problem. If we are given G presented as a partial k-path, then clearly we can color it with $k + 1$ many colors. Here we only know that it has pathwidth k (which could be tested of course), and in fact the algorithm either gives a $3k - 2$ coloring or certificate that either the problem is not or pathwidth k or it if not k colorable. Notice that this is also a *heuristic for pathwidth*.

To prove this result we will restate an equivalent formulation in the domain of partial orderings. Recall that a partial ordering $(P, \leq)$ is called *an interval ordering* if P is isomorphic to $(I, \leq)$ where I is a set of intervals of the real line and $x \leq y$ iff the right point of x is left of the left point of y.

Fig. 31.3 The canonical blockage

There is a related notion of interval graph where the vertices are real intervals and xy is an edge if and only if $I_x \cap I_y \neq \emptyset$. The following lemma relates pathwidth to interval orderings.

Lemma 31.4.1 (Folklore) *A graph G has pathwidth k iff G is a subgraph of an interval graph G' of maximal clique size $k + 1$.*

Interval orderings can be characterized by the following theorem.

Theorem 31.4.2 (Fishburn [308])
(i) *Let P be a poset. Then the following are equivalent.*
    (a) *P is an interval ordering.*
    (b) *P has no subordering isomorphic to $2 + 2$ which is the ordering of four elements with $\{a, b, c, d\}$ with $a < b, c < d$ and no other relationships holding (see Fig. 31.3).*
(ii) (Folklore) *G is an interval graph iff G G is the incomparability graph of some interval order P. That is, we join x to y if $x|y$ in P.*

The proofs of the above are not difficult, and we leave them to the reader in Exercise 31.6.10.
    We first prove the following.

Theorem 31.4.3 (Kierstead and Trotter [443]) *Suppose that $(P, \leq)$ is an online interval ordering of width k. Then P can be online covered by $3k - 2$ many chains.*

We need the following lemma. For a poset P, and subsets S, T, we can define $S \leq T$ iff for each $x \in S$ there is some $y \in T$ with $x \leq y$ (similarly $S|T$ etc.).

Lemma 31.4.2 *If P is an interval order and $S, T \subset P$ are maximal antichains the either $S \leq t$ or $T \leq S$.*

Proof The argument is interesting and instructive. It uses induction on k. If $k = 1$ then P is a chain, and there is nothing to prove. Suppose the result for k, and consider $k = 1$. We define B inductively by

$$B = \{p \in P : \text{width}(B^P \cup \{p\}) \leq k\}.$$

Here B^p denotes the amount of B constructed by step p of the online algorithm. Then B is a maximal subordering of P or width k. By the inductive hypothesis the algorithm will have covered B by $3k - 2$ chains. Let $A = P - B$. Now it will suffice to show that A can be covered by three chains.

To see this it is enough to show that every element of A is incomparable with at most two other elements of A. Then the relevant algorithm is the greedy algorithm, which will cover A, as we see elements *not* in B.

Lemma 31.4.3 *The width of A is at most 2.*

Proof To see this, consider 3 elements $q, r, s \in A$. Then there are antichains Q, R, S in P of width k with $q|Q, r|R$ and $s|S$. Applying Lemma 31.4.2, we might as well suppose $Q \leq R \leq S$. Suppose that $r|q$ and $r|s$. Then we prove that $q < s$. Since $q|r$ and width$(P) \leq k + 1$, there is some $r' \in R$ with q and r' comparable. Since $q|Q$, $r' \notin Q$. Since the width of B is $\leq k$, there is some $q' \in Q$ q' and r' comparable. Since $Q \leq R$, there is some $r_0 \in R$ with $q' \leq r_0$. Since eR is an antichain, $q' \leq r'$. Since $q|q'$, $q \leq r'$. Similarly, there exists $r'' \in R$ with $r'' \leq s$. Since P does not have any ordering isomorphic to $\mathbf{2 + 2}$, we can choose $r' = r''$, and hence $q < s$. $\qquad\square$

Now we suppose that r, q, s, t are distinct elements of A with $q|\{r, s, t\}$. Then without loss of generality $r < s < t$ since the width of A is at most 2. Since $s \in A$ there is an antichain $S \subset B$ of length k with $s|S$. Since $s|q$, and width$(P) \leq k + 1$, q is comparable with some element $s' \in S$. If $s' < q$, then $s'|r$ and hence the sub-order $\{s', q, r, s\}$ is isomorphic to $\mathbf{2 + 2}$. Similarly, $q < s'$ implies $s'|t$ and then the subordering $\{q, s', s, t\}$ is isomorphic to $\mathbf{2 + 2}$. Thus there cannot be 4 elements r, q, s, t of A with $q|\{r, s, t\}$. Hence A can be covered by 3 chains. $\qquad\square$

Since the algorithm uses only the information about comparability of various elements, and not their ordering, we have the following corollary.

Corollary 31.4.1 (Kierstead and Trotter [443]) *Suppose that G is an online interval graph of width k. Then G can be online colored by $3k - 2$ many colors.*

It has been shown in [443] that the bound above is *sharp*. The algorithm resembles the simplest of all online algorithms: *first fit*. A first fit algorithm for the problem above would be "color with the first available color". In fact for pathwidth 2 or less, the algorithm above *is* first fit.

Question: Should there be any first fit algorithm for coloring interval graphs online? The answer is yes.

Theorem 31.4.4 (Kierstead [441]) *First fit online colors interval graphs of width k with at most $40k$ colors.*

Kierstead and Qin [442] improved the constant to 25.72. It is unknown what the correct constant is, but the lower bound is 4.4 as shown by Chrobak and Slusarek [160] leaving a rather large gap.

31.5 Parameterized Inapproximability

In this section, we will look at problems that have the property that *any* parameterized approximation algorithm would entail collapse of the $W[1]$-hierarchy. We consider the following problem.

k-APPROXIMATE INDEPENDENT DOMINATING SET
Input: A graph $G = (V, E)$.
Parameter: k and a computable function g.
Output: 'NO', asserting that no independent dominating set $V' \subseteq V$ of size $\geq k$
              for G exists, or an independent dominating set $V' \subseteq V$ for G of size at
              most $g(k)$.

Theorem 31.5.1 (Downey, Fellows, McCartin, and Rosamond [252]) *There is no*
FPT *algorithm for* k-APPROXIMATE INDEPENDENT DOMINATING SET *unless*
$W[2] = \text{FPT}$.

Proof We reduce from DOMINATING SET. Let (G, k) be an instance of this. We construct $G' = (V', E')$ as follows. V' is $S \cup C \cup T$:
1. $S = \{s[r, i] \mid r \in [k] \wedge i \in [g(k) + 1]\}$ (the *sentinel* vertices).
2. $C = \{c[r, u] \mid r \in [k] \wedge u \in V\}$ (the *choice* vertices).
3. $T = \{t[u, i] \mid u \in V \wedge i \in [g(k) + 1]\}$ (the *test* vertices).
E' is the set $E'(1) \cup E'(2) \cup E'(3)$:
1. $E'(1) = \{s[r, i]c[r, u] \mid r \in [k] \wedge i \in [g(k) + 1] \wedge u \in V\}$.
2. $E'(2) = \{c[r, u]c[r, u'] \mid r \in [k] \wedge u, u' \in V\}$.
3. $E'(3) = \{c[r, u]t[v, i] \mid r \in [k], u \in V, v \in N_G[u], i \in [g(k) + 1]\}$.
The idea here is that the $E'(2)$ are k groups of *choice* vertices. They form a clique. To each of these k cliques are *sentinel* vertices S, and the edges $E'(1)$ connect each sentinel vertex to the corresponding choice clique. Note that the sentinel vertices are an independent set. The other independent set is the *test* vertices T, and the edges $E'(3)$ connect the choice vertices to the test vertices according to the structure of G.

First we note that if G has a k-dominating set, $v_1, \ldots, v_k$, then the set $\{c[i, v_i] \mid i \in [k]\}$ is an independent dominating set for G'.

Second, suppose that g' has an independent dominating set D' of size $\leq g(k)$. Then because of the sentinels enforcement, there must be at least one $c[i, u_i]$ per C, and hence exactly one as D' is independent. Let $D = \{u_i \mid c[i, u_i] \in D'\}$. We need to argue that D is a dominating set in G. Note that for all $v \in V$, there is a $t[v, i] \in T$, and by cardinality constraints, at least one is not in D'. This must correspond to a domination in G. $\qquad\square$

A similar nonapproximability result was given by Marx who proved the following. For simplicity, we will say that a minimization problem is FPT *cost approximable* iff there is a computable c for Definition 31.1.1. That is there is no computable function g and FPT algorithm giving a no for one of size k or a solution of size $\leq g(k)$.

Theorem 31.5.2 (Marx [516]) MONOTONE CIRCUIT SATISFIABILITY *is not* FPT *cost approximable unless* FPT $= W[P]$.

Proof Let C be a monotone circuit with n inputs. Note that we can interpret C as a boolean function over subsets of $[n]$. Let $\mathcal{H}$ be a k' perfect family of hash functions from $[n]$ to $[k']$, which the reader will recall from Chap. 8, means that for each k'-element $S \subset [n]$, there is $h \in \mathcal{H}$ with h injective on S. Also from Chap. 8 we know that there is such a family of size $2^{O(k')} \log n$.

Let k be the parameter, and define $k' = g(k)k$. Let $\mathcal{H}$ be the k' perfect family of hash functions from $[n]$ to $[k']$. We define the following function

$$C'(S) = \bigwedge_{h \in \mathcal{H}} \bigvee_{T \in [k'] \mid |T| \leq k} C\big(C \cap h^{-1}(T)\big).$$

Here $h^{-1}(T)$ denotes $\{i \in [n] \mid h(i) \in T\}$. Then we can construct a monotone circuit C' expressing the function $C'(S)$ in time $g(k)|C|^{O(1)}$. The claim is that
(1) C is weight k satisfiable iff C' is;
(2) C not weight k satisfiable implies C' is not weight k' satisfiable.
To see (1), let $S \subset [n]$ be a weight k satisfying assignment of C. As with the proof above, we show that it satisfies C' as well. This is equivalent to showing that the disjunction in C' is true for every $h \in \mathcal{H}$. Let $T = \{h(s) \mid s \in S\}$. The $|T| \leq k$. By definition, $S \subseteq h^{-1}(T)$. Therefore $C(S \cap h^{-1}(T)) = C(S) = 1$. Therefore the disjunction is satisfied by the term corresponding to T.

To see (2), Let S be a weight k' assignment satisfying C'. Let h be a hash function injective on S. We claim that the disjunction corresponding to this is no satisfied. Notice that for each $T \in \{T \in [k'] \mid |T| \leq k\}$, $|S \cap h^{-1}(T)| \leq k$; for every $t \in T$, the function h maps at most one element of S to t. Therefore if the algorithm outputs 1, on input $S \cap h^{-1}(T)$, for some $T \in \{T \in [k'] \mid |T| \leq k\}$, then $S \cap h^{-1}(T)$ is a satisfying assignment of weight at most k in C, a contradiction. $\qquad\square$

Marx points out that the proof above only uses circuits of depth $d + 2$ and weft $w + 2$ on circuits C of weft w and depth d. Thus as a corollary we get the following.

Corollary 31.5.1 (Marx [516]) *If* MONOTONE CIRCUIT SATISFIABILITY *is* FPT *cost approximable for circuits of depth 4 then* FPT $= W[2]$.

Marx also proves that the following antimonotone problem has no FPT approximation unless FPT $= W[1]$.

THRESHOLD SET
Input: A collection $\mathcal{F}$ of subsets of U with a positive integer weight $w(S)$ for each $S \in \mathcal{F}$.

Parameter: k.

Goal: A set $T \subset U$ such that $|T \cap S| \leq w(S)$ for each $S \in \mathcal{F}$ and $|T| \leq k$.

The proof is interesting in that it uses Reed–Solomon codes. Related here are very attractive results by Eickmeyer, Grohe, and Grüber about the approximability of "natural" $W[P]$ problems, meaning those from, for instance, the appendix of Downey and Fellows [247]. They showed that for such *natural* $W[P]$-complete problems have no constant or polylogarithmic FPT approximations unless $W[P] =$ FPT. The point is that Eickmeyer, Grohe, and Grüber proved their inapproximability results by first using MONOTONE CIRCUIT SAT (using, it must be said, very difficult "expander" techniques), and then using *gap preserving* reductions to the other problems they consider. Thus, since Marx's Theorem above improves this result to complete inapproximability for MONOTONE CIRCUIT SAT, we obtain the following.

Theorem 31.5.3 (Marx [516], Eickmeyer, Grohe, and Grüber [268]) Natural $W[P]$-*complete problems have no* FPT *approximations unless* $W[P] =$ FPT. *In particular the following have this property.*
1. MINIMUM CHAIN REACTION CLOSURE.
2. MINIMUM GENERATING SET.
3. MINIMUM LINEAR INEQUALITY DELETION.
4. MINIMUM DEGREE 3 SUBGRAPH ANNIHILATOR.
5. MINIMUM INDUCED SAT.

The proof is to make modify the reductions from Abrahamson, Downey and Fellows [3] so as to make them gap preserving. We invite the reader to do this in Exercise 31.6.11.

31.6 Exercises

Exercise 31.6.1 Prove Lemma 31.2.1; that is, prove that if G has pathwidth t, and n vertices, then $|E(G)| \leq nt - \frac{(t^2 + t)}{2}$.

Exercise 31.6.2 Prove Lemma 31.2.2.
(Hint: Use a case analysis and induction.)

Exercise 31.6.3 Show that the first fit gives a 2-approximation algorithm for BIN PACKING.

Exercise 31.6.4 (Chen, Grohe, and Grüber [156]) Prove Theorem 31.1.2.

Exercise 31.6.5 (Downey, Fellows, McCartin, and Rosamond [252]) The reader might wonder if we can vary the definition of parameterized approximation to consider *additive* rather than *multiplicative* factors. Consider the following.

ADD-APPROX k-INDEPENDENT SET

Input: A graph $G = (V, E)$.
Parameter: k, c.
Output: 'NO', asserting that no independent set $V' \subseteq V$ of size $\geq k$ for G exists, or an independent set $V' \subseteq V$ for G of size at least $k - c$.

Prove that ADD-APPROX k-INDEPENDENT SET is $W[1]$-hard. Similarly, define and prove that ADD-APPROX k-DOMINATING SET is $W[2]$-hard.

Exercise 31.6.6 (Downey and McCartin [260])
1. Prove that first-Fit will use at most $3k + 1$ colors to color any online *tree* $T^<$ of pathwidth k.
2. for each $k \geq 0$, there is an online tree $T^<$, of pathwidth k, such that First-Fit will use $3k + 1$ colors to color $T^<$.

Exercise 31.6.7 Prove Lemma 31.3.1, that is, if $\mathrm{tw}(G) \leq k$, then any $W \subseteq V(G)$ has a good separator of size $\leq k + 1$.

Exercise 31.6.8 (Downey and McCartin [260]) One problem with pathwidth is that a star-like vertex must be present in many bags, arguably detracting from the idea that the graph looks like a path. Downey and McCartin defined a variation of this more appropriate for online algorithms. We say that a path decomposition of width k, in which every vertex of the underlying graph belongs to at most l nodes of the path, has width k and *persistence l*, and say that a graph that admits such a decomposition has *bounded persistence pathwidth*.
    Consider the following problems.

BANDWIDTH

Input: A graph $G = (V, E)$.
Parameter: A positive integer k.
Question: Is there a bijective *linear layout* of V, $f : V \to \{1, 2, \ldots, |V|\}$, such that, for all $(u, v) \in E$, $|f(u) - f(v)| \leq k$?

BOUNDED PERSISTENCE PATHWIDTH

Input: A graph $G = (V, E)$.
Parameter: A pair of positive integers (k, l).
Question: Is there a path decomposition of G of width at most k and persistence at most l?

Show that BANDWIDTH FPT-reduces to BOUNDED PERSISTENCE PATHWIDTH. From this we can conclude that BOUNDED PERSISTENCE PATHWIDTH is $W[t]$-hard for all t since Bodlaender, Fellows, and Hallett [87] proved that BANDWIDTH is $W[t]$ hard for all t.

Exercise 31.6.9 (Slivkins [623]) Using a reduction from CLIQUE, or otherwise, prove that the following is $W[1]$-hard.

DISJOINT PATHS IN A DAG
Input: A directed acyclic graph D.
Parameter: k.
Question: Does D have at least k edge disjoint paths?

Exercise 31.6.10 Prove Theorem 31.4.2.

Exercise 31.6.11 (Eickmeyer, Grohe, Grüber [268]) Show that there is no FPT-approximation algorithm for MINUMUM CHAIN REACTION CLOSURE unless FPT = $W[P]$.
(Hint: Use Marx's Theorem on MONOTONE CIRCUIT SATISFIABILITY and modify the reduction of Exercise 25.2.)

31.7 Historical Notes

The roots of parameterized approximation go back to a question Fellows stated in [247] asking whether DOMINATING SET has a FPT-approximation algorithm with ratio 2. This question remains open. As an explicit subject it began with the three conference papers Downey, Fellows, and McCartin [251], Chen and Grohe and Grüber [156], and Cai and Huang [125]. Clearly the material on pathwidth and treewidth demonstrate that the idea had been around implicitly for a long time; certainly since the work of Robertson and Seymour and Reed [575]. The set-up we use here is drawn from Chen, Grohe, and Grüber [156]. As the reader can see above, earlier work by Kierstead and others shows that, like kernelization, the subject was in the subconscious of others. Grohe and Grüber [358] proved the material on DISJOINT CYCLES, thereby giving the first example of a $W[1]$ hard problem with a FPT-approximation; perhaps the only example at present. The material on online algorithms, an area ripe for parameterized development, particularly in the area of incremental complexity, is drawn from Downey and McCartin [260] and Kierstead and Trotter [443]. Downey, Fellows, McCartin, and Rosamond [252] introduced the notion of complete FPT inapproximability. The proof that natural $W[P]$-complete problems are multiplicatively and polylogarithmically hard to FPT-approximate is due to Eickmeyer, Grohe, and Grüber [268]. This improved earlier work of Alekhnovich and Razborov [27]. These results were strengthened by Marx [516].

The exercises on *additive* parameterized approximation (e.g. Exercise 31.6.5) are taken from Downey, Fellows, and McCartin [251]. Jansen, Kratsch, Marx, and Schlotter [415] showed that UNARY BIN PACKING parameterized by the number of bins is $W[1]$-hard, *but* there is a FPT approximation algorithm which either gives a certificate that there is no size k solution or gives one of size $k + 1$.

Recent work not reported here uses this methodology on FPT problems to obtain approximation algorithms with faster running times. Brankovic and Fernau [110] gave a reduction rule-based algorithm giving a $\frac{2\ell+1}{\ell+1}$ FPT-approximation for VERTEX COVER running in time depending on a recurrence relation which is fast. For

$\ell = 1$ and hence a $\frac{3}{2}$ approximation FPT-approximation algorithm, the running time is $O^*(1.09^k)$. This work seems of great potential.

31.8   Summary

In this chapter we met the idea of parameterized approximation. Examples were given of applications of parameterized approximation. We showed methods of demonstrating complete inapproximability, meaning that there is no approximate solution for any computable function $g(k)$.

Parameterized Counting and Randomization

32

32.1 Introduction

As we saw in Chap. 1, classical complexity is not only concerned with decision problems, but also with search, enumeration, and counting problems. In this chapter, we will address some of the issues related to parametric versions of these notions, and also look at parameterized randomization. Counting is thought to be *harder* than decision, but the hope is that the parameterized version will yield some ingress towards tractability. An early result towards this programme is the following.

Theorem 32.1.1 (Arvind and Raman [48]) *Counting k vertex covers of a graph* $G = (V, E)$ *is FPT.*

Proof The simplest proof will use the method of bounded search trees. That is, use the method of Theorem 3.2.1, hence taking an edge $e = uv$, and observing that at least one of the vertices must be in *any* VERTEX COVER. Hence we can begin a tree with possibilities $\{u, v\}$. Note that if we include u we might have a k-VERTEX COVER also including v. In any case this method will clearly yield a FPT algorithm running in time $O(2^{k^2+k} + 2^k|V|)$. □

In the next section, we will give a framework for demonstrating likely parametric intractability of parameterized problems. For example, even the simple case of VERTEX COVER above immediately presents us with a problem. As we have seen, the best algorithms for this problem employ kernelization. Is there an analog of the kernel methods for #VERTEX COVER? The answer is yes. The first example was due to Thurley: recall that the Melhorn–Buss kernel has the property that we delete all vertices of degree above k and then solve the reduced problem. Clearly no vertex of degree above k can *ever* be part of a size k vertex cover unless there is a size $k - 1$ vertex cover. This observation yields the following result.

Theorem 32.1.2 (Thurley [646]) *The Buss kernel can be used to solve #VERTEX COVER in time* $O((1 + \sqrt{k})^k k^+ k|G|)$.

R.G. Downey, M.R. Fellows, *Fundamentals of Parameterized Complexity*,
Texts in Computer Science, DOI 10.1007/978-1-4471-5559-1_32,
© Springer-Verlag London 2013

Proof This is Exercise 32.8.1.   □

Thurley also explored many other counting and kernelization properties. For example, he showed how to derive FPT results for counting problems using crown reductions and the Nemhauser–Trotter kernelization. Space limitations preclude us from including these, but certainly this is exciting and there is still a lot of work needed here.

Whilst we have seen significant development of techniques for the design and construction of parameterized decision algorithms, systematic development of parameterized counting algorithms is at present fairly rudimentary.

Some metatheorems in this direction have been proven. For example, as we mentioned in Chap. 14, Courcelle, Makowsky, and Rotics (e.g. [180]) and others have considered counting and evaluation problems on graphs where the range of counting is definable in monadic second-order logic. They show that these problems are fixed-parameter tractable, where the parameter is the treewidth of the graph. Andrzejak [39], and Noble [548] have shown that for graphs $G = (V, E)$ of treewidth at most k, the Tutte polynomial of G can be computed in polynomial time and evaluated in time $O(|V|)$, despite the fact that the general problem is #P complete; Makowsky has shown that the same bounds hold for the colored Tutte polynomial on colored graphs $G = (V, E, c)$ of treewidth at most k. As we also mentioned in Chap. 14, Frick and Grohe have introduced the idea of locally tree-decomposable classes of structures, and have shown that with such structures, counting problems definable in first-order logic can be solved in fixed-parameter linear time [332].

32.2   Classical Counting Problems and #P

The classical hardness theory for counting functions was initiated by Valiant [649]. For more on the classical theory we refer the reader to Welsh [662].

Definition 32.2.1 (Witness function) Let $w : \Sigma^* \to \mathcal{P}(\Gamma^*)$, and let $x \in \Sigma^*$. We refer to the elements of $w(x)$ as *witnesses* for x. We associate a decision problem $A_w \subseteq \Sigma^*$ with w:

$$A_w = \{x \in \Sigma^* \mid w(x) \neq \emptyset\}.$$

In other words, A_w is the set of strings that have witnesses.

Definition 32.2.2 (#P-Valiant [649]) The class #P is the class of witness functions w such that:
1. there is a polynomial-time algorithm to determine, for given x and y, whether $y \in w(x)$;
2. there exists a constant $k \in \mathbb{N}$ such that for all $y \in w(x)$, $|y| \leq |x|^k$. (The constant k can depend on w.)

Of course, #P is the counting analog of NP. As with any hardness theory, the ingredients we need will be a hardness core (presumably counting accepting paths of a nondeterministic Turing Machine running in polynomial time), along with an appropriate notion of reduction for problem comparison.

Definition 32.2.3 (Counting reduction) Let

$$w : \Sigma^* \to \mathcal{P}(\Gamma^*),$$

$$v : \Pi^* \to \mathcal{P}(\Delta^*)$$

be counting problems, in the sense of [649]. A *polynomial-time many-one counting reduction* from w to v consists of a pair of polynomial-time computable functions

$$\sigma : \Sigma^* \to \Pi^*,$$

$$\tau : \mathbb{N} \to \mathbb{N}$$

such that

$$|w(x)| = \tau(|v(\sigma(x))|).$$

When such a reduction exists, we say that w *reduces to* v.

Intuitively, if one can easily count the number of witnesses of $v(y)$, then one can easily count the number of witnesses of $w(x)$.

There is a particularly convenient kind of reduction that preserves the number of solutions to a problem exactly. We call such a reduction *parsimonious*.

Definition 32.2.4 (Parsimonious reduction) A counting reduction σ, τ is *parsimonious* if τ is the identity function.

Armed with the concept of a counting reduction, we can define the class of #P-*complete* problems. With not too much effort, the proof of Cook's Theorem can be made parsimonious, and hence, for example, we can define the following conveniently generic counting problem.

#3SAT
Instance: A Boolean formula X in 3CNF form.
Question: How many satisfying assignments does X have?

Theorem 32.2.1 (Valiant [649]) *#3SAT is #P complete.*

One famous result is the following:

Theorem 32.2.2 (Valiant) *The problem of counting the number of perfect matchings in a bipartite graph is #P complete.*

Determining if a bipartite graph has a perfect matching, is achievable in polynomial time, whereas the counting version is hard.

32.3 #$W[1]$—A Parameterized Counting Class

As in the classical case, we use a witness function to formalize the association between parameterized counting problems and their corresponding decision problems. The basic counting ideas were introduced by Flum and Grohe [310] and McCartin [522, 523]. We will follow the account of McCartin.

Definition 32.3.1 (Parameterized witness function) Let $w : \Sigma^* \times \mathbb{N} \to \mathcal{P}(\Gamma^*)$, and let $\langle \sigma, k \rangle \in \Sigma^* \times \mathbb{N}$. The elements of $w(\langle \sigma, k \rangle)$ are *witnesses* for $\langle \sigma, k \rangle$. We associate a parameterized language $L_w \subseteq \Sigma^* \times \mathbb{N}$ with w:

$$L_w = \big\{ \langle \sigma, k \rangle \in \Sigma^* \times \mathbb{N} \mid w(\langle \sigma, k \rangle) \neq \emptyset \big\},$$

L_w is the set of problem instances that have witnesses.

Definition 32.3.2 (Parameterized counting problem) Let $w : \Sigma^* \times \mathbb{N} \to \mathcal{P}(\Gamma^*)$ be a parameterized witness function. The corresponding *parameterized counting problem* can be considered as a function $f_w : \Sigma^* \times \mathbb{N} \to \mathbb{N}$ that, on input $\langle \sigma, k \rangle$, outputs $|w(\langle \sigma, k \rangle)|$.

We note here that "easy" parameterized counting problems can be considered to be those in the class that we might call "FFPT", the class of functions of the form $f : \Sigma^* \times \mathbb{N} \to \mathbb{N}$ where $f(\langle \sigma, k \rangle)$ is computable in time $g(k)|\sigma|^\alpha$, where g is an arbitrary function, and α is a constant not depending on k. For our hardness programme, we will need the following.

Definition 32.3.3 (Parameterized counting reduction) Let

$$w : \Sigma^* \times \mathbb{N} \to \mathcal{P}(\Gamma^*),$$
$$v : \Pi^* \times \mathbb{N} \to \mathcal{P}(\Delta^*)$$

be (witness functions for) parameterized counting problems. A *parameterized counting reduction* from w to v consists of a parameterized transformation

$$\rho : \Sigma^* \times \mathbb{N} \to \Pi^* \times \mathbb{N}$$

and a function

$$\tau : \mathbb{N} \to \mathbb{N}$$

running in time $f(k)n^\alpha$ (where $n = |\sigma|$, α is a constant independent of both f and k, and f is an arbitrary function) such that

$$\big| w(\langle \sigma, k \rangle) \big| = \tau \big(\big| v(\rho(\langle \sigma, k \rangle)) \big| \big).$$

When such a reduction exists, we say that w *reduces to* v.

As in the classical case, if one can easily count the number of witnesses of $v(\langle \sigma', k' \rangle)$, then one can easily count the number of witnesses of $w(\langle \sigma, k \rangle)$.

> **Definition 32.3.4** (Parsimonious parameterized counting reduction) A parameterized counting reduction ρ, τ is *parsimonious* if τ is the identity function.

There is the obvious analog of the W-hierarchy for counting classes, based on parameterized counting reductions. The core parameterized counting problems are defined as follows.

#WEIGHTED WEFT t DEPTH h CIRCUIT SATISFIABILITY (WCS(t, h))

Input: A weft t depth h decision circuit C.
Parameter: A positive integer k.
Output: The number of weight k satisfying assignments for C.

This gives us the analog of the W-hierarchy as the W-*counting hierarchy*. In the same way as for the W-analog of Cook's Theorem, we can prove a counting analog.

32.4 The Counting Analog of Cook's Theorem

In this section we prove that the following fundamental parameterized counting problem is #$W[1]$-complete.

#SHORT TURING MACHINE ACCEPTANCE

Instance: A nondeterministic Turing machine M, and a string x.
Parameter: A positive integer k.
Output: $acc_M(x, k)$, the number of $\leq k$-step accepting computations of machine M on input x.

Theorem 32.4.1 (McCartin, Flum, and Grohe) #SHORT TURING MACHINE ACCEPTANCE *is complete for* #$W[1]$.

Proof The proof makes use of the (suitably modified) series of parameterized transformations described in Chap. 21. Thus, we will only describe in detail the alterations that are required, to ensure that each of these may be considered as a parsimonious parameterized counting reduction. The most important alteration that we make, is the reworking of the final step in the proof of Lemma 32.4.1. Most of the transformations described in Chap. 21 are, in fact, parsimonious; we just need to argue that this is the case.

Recall from Chap. 21, the proof of the $W[1]$-completeness of SHORT TURING MACHINE ACCEPTANCE began with some preliminary *normalization* results about weft 1 circuits.

As in Chap. 21, let $F(s)$ denote the family of s-normalized circuits. Note that an s-normalized circuit is a weft 1, depth 2 decision circuit, with the *or* gates on level 1 having fanin bounded by s.

We show that there is a parsimonious parameterized counting reduction from $w_{F(1,h)}$ (the standard parameterized witness function for $F(1, h)$) to $w_{F(s)}$ (the standard parameterized witness function for $F(s)$), where $s = 2^h + 1$. Thus, any parameterized counting problem $f_v \in \#W[1]$ can, in fact, be reduced to the following problem (where s is fixed in advance and depends on f_v). We may define the problem of #WEIGHTED s-NORMALIZED CIRCUIT SATISFIABILITY as the counting version for s-normalized circuits.

Lemma 32.4.1 (McCartin [522], implicit in Flum–Grohe [310]) $w_{F(1,h)}$ *reduces to* $w_{F(s)}$, *where* $s = 2^h + 1$, *via a parsimonious parameterized counting reduction.*

Proof Let C be a circuit in $F(1, h)$ and let k be a positive integer. We follow the proof of Theorem 21.2.1. Recall that this describes a parameterized m-reduction that, on input $\langle C, k \rangle$, produces a circuit $C' \in F(s)$ and an integer k', such that for every weight k input accepted by C there exists a unique weight k' input accepted by C'. The transformation proceeds in four stages, and follows that given in Chap. 21, with alterations *only* to the last step in order to ensure parsimony. At this stage the reader should go back and review the first three steps of the proof of Theorem 21.2.1.

The first three steps culminate in the production of a tree circuit, C', of depth 4, that corresponds to a Boolean expression, E, in the following form:

$$E = \prod_{i=1}^{m} \sum_{j=1}^{m_i} E_{ij}$$

where:
(1) m is bounded by a function of h,
(2) for all i, m_i is bounded by a function of h,
(3) for all i, j, E_{ij} is either:

$$E_{ij} = \prod_{k=1}^{m_{ij}} \sum_{l=1}^{m_{ijk}} x[i, j, k, l]$$

or

$$E_{ij} = \sum_{k=1}^{m_{ij}} \prod_{l=1}^{m_{ijk}} x[i, j, k, l],$$

where the $x[i, j, k, l]$ are literals (i.e., input Boolean variables or their negations) and for all i, j, k, m_{ijk} is bounded by a function of h. The family of circuits corresponding to these expressions has weft 1, with the large gates corresponding to the E_{ij}. (In particular, the m_{ij} are *not* bounded by a function of h.)

The argumentation of these three steps shows that any weight k input accepted by C will also be accepted by C', any weight k input rejected by C will also be rejected by C'. Thus, the witnesses for C' are *exactly* the witnesses for our original circuit C.

In the fourth step, soon to be described, we employ additional nondeterminism. Let C denote the normalized depth 4 circuit received from the previous step, corresponding to the Boolean expression E described above.

We produce an expression E' in product-of-sums form, with the size of the sums bounded by $2^h + 1$, that has a unique satisfying truth assignment of weight

$$k' = k + (k+1)\left(1 + 2^h\right)2^{2^h} + m + \sum_{i=1}^{m} m_i$$

corresponding to each satisfying truth assignment of weight k for C.

The idea will be to employ extra variables in E' so that τ', a weight k' satisfying truth assignment for E', encodes both τ, a weight k satisfying truth assignment for E and a "proof" that τ satisfies E. Thus, τ' in effect guesses τ and also checks that τ satisfies E.

We build E' so that the *only* weight k' truth assignments that satisfy E' are the τ''s that correspond to τ's satisfying E.

Here is the new step 4.

Step 4'. Employing additional nondeterminism.

Let C denote the normalized depth 4 circuit received from Step 3 of the proof of Theorem 21.2.1, and corresponding to the Boolean expression E described above.

In this step we produce an expression E' in product-of-sums form, with the size of the sums bounded by $2^h + 1$, that has a unique satisfying truth assignment of weight

$$k' = k + (k+1)\left(1 + 2^h\right)2^{2^h} + m + \sum_{i=1}^{m} m_i$$

corresponding to each satisfying truth assignment of weight k for C.

The idea is to employ extra variables in E' so that τ', a weight k' satisfying truth assignment for E', encodes both τ, a weight k satisfying truth assignment for E and a "proof" that τ satisfies E. Thus, τ' in effect guesses τ and also checks that τ satisfies E.

For convenience, assume that the E_{ij} for $j = 1, \ldots, m'_i$ are sums-of-products and the E_{ij} for $j = m'_i + 1, \ldots, m_i$ are products-of-sums. Let $V_0 = \{x_1, \ldots, x_n\}$ denote the variables of E.

The set V of variables of E' is $V = V_0 \cup V_1 \cup V_2 \cup V_3 \cup V_4$, where

$V_0 = \{x_1, \ldots, x_n\}$ the variables of E,

$V_1 = \left\{x[i, j] : 1 \leq i \leq n, 0 \leq j \leq \left(1 + 2^h\right)2^{2^h}\right\}$,

$V_2 = \left\{u[i, j] : 1 \leq i \leq m, 1 \leq j \leq m_i\right\}$,

$$V_3 = \big\{w[i,j,k] : 1 \le i \le m, 1 \le j \le m_i', 0 \le k \le m_{ij}\big\},$$

$$V_4 = \big\{v[i,j,k] : 1 \le i \le m, m_i'+1 \le j \le m_i, 0 \le k \le m_{ij}\big\}.$$

We first intend to extend a weight k truth assignment for V_0 that satisfies E, into a unique weight k' truth assignment for V. Suppose τ is a weight k truth assignment for V_0 that satisfies E; so, we build τ' as follows:

(1) For each i such that $\tau(x_i) = 1$ and for each j, $0 \le j \le (1+2^h)2^{2^h}$ set $\tau'(x[i,j]) = 1$.

(2) For each i, $1 \le i \le m$, choose the lexicographically least index j_i such that E_{i,j_i} evaluates to 1 (this is possible, since τ satisfies E) and set $\tau'(u[i,j_i]) = 1$.

   (a) If, in (2), E_{i,j_i} is a sum-of-products, then choose the lexicographically least index k_i such that

$$\prod_{l=1}^{m_{i,j_i,k_i}} x[i,j_i,k_i,l]$$

evaluates to 1, and set $\tau'(w[i,j_i,k_i]) = 1$.

      Here, we are choosing the first input line to the sum E_{i,j_i} that makes it true.

   (b) If, in (2), E_{i,j_i} is a product-of-sums, then for each product-of-sums $E_{i,j'}$ with $j' < j_i$, choose the lexicographically least index k_i such that

$$\prod_{l=1}^{m_{i,j',k_i}} \neg x[i,j',k_i,l]$$

evaluates to 1, and set $\tau'(v[i,j',k_i]) = 1$.

      Here, for each product $E_{i,j'}$ that precedes E_{i,j_i}, we are choosing the first input line that makes $E_{i,j'}$ false.

(3) For $i = 1, \ldots, m$ and $j = 1, \ldots, m_i'$ such that $j \ne j_i$, set $\tau'(w[i,j,0]) = 1$.

   For $i = 1, \ldots, m$ and $j = m_i'+1, \ldots, m_i$ such that $j \ge j_i$, set $\tau'(v[i,j,0]) = 1$.

(4) For all other variables, v, of V, set $\tau'(v) = 0$.

Note that τ' is a truth assignment to the variables of V having weight

$$k' = k + (k+1)\big(1+2^h\big)2^{2^h} + m + \sum_{i=1}^{m} m_i$$

and that there is exactly one τ' for V corresponding to each τ for V_0.

We now build E' so that the *only* weight k' truth assignments that satisfy E' are the τ's that correspond to τs satisfying E. The expression E' is a product of subexpressions: $E' = E_1 \wedge \cdots \wedge E_{9c}$.

We first want to associate with each x_i, a variable from V_0, the block of variables $x[i,j] : 0 \le j \le (1+2^h)2^{2^h}$ from V_1. We want to ensure that if x_i is set to 1 then all of the associated $x[i,j]$ are also set to 1, and that if x_i is set to 0 then all of the associated $x[i, j]$ are set to 0. The following two subexpressions accomplish this task.

$$E_1 = \prod_{i=1}^{n}(\neg x_i + x[i, 0])(\neg x[i, 0] + x_i),$$

$$E_2 = \prod_{i=1}^{n} \prod_{j=0}^{2^{2^h}+1}(\neg x[i, j] + x[i, j+1 \ (\mathrm{mod}\ r)]), \quad r = 1 + (1 + 2^h)2^{2^h}.$$

For each i, $1 \le i \le m$, j, $1 \le j \le m_i$, if we set $u[i, j] = 1$ then this represents the fact that E_{ij} is satisfied. We need to ensure that the collection of $u[i, j]$s set to 1 represents the satisfaction of E, our original expression over V_0.

$$E_3 = \prod_{i=1}^{m} \sum_{j=1}^{m_i} u[i, j].$$

For each i, $1 \le i \le m$, we must ensure that we set exactly one variable $u[i, j_i]$ to 1, and that j_i is the lexicographically least index, such that E_{i, j_i} evaluates to 1. Exactly one j_i for each i:

$$E_4 = \prod_{i=1}^{m} \prod_{j=1}^{m_i-1} \prod_{j'=j+1}^{m_i}(\neg u[i, j] + \neg u[i, j']).$$

If E_{ij} is a sum-of-products and $u[i, j] = 1$ then we do not set the default, $w[i, j, 0]$. Thus, it must be the case that some $w[i, j, k], k \ne 0$ is able to be chosen, and E_{ij} must be satisfied by τ:

$$E_5 = \prod_{i=1}^{m} \prod_{j=1}^{m'_i}(\neg u[i, j] + \neg w[i, j, 0]).$$

To make the reduction parsimonious, we must also ensure that if E_{ij} is a sum-of-products and $u[i, j] = 0$ then we *do* set the default, $w[i, j, 0]$:

$$E_{5a} = \prod_{i=1}^{m} \prod_{j=1}^{m'_i}(u[i, j] + w[i, j, 0]).$$

If E_{ij} is a product-of-sums and $u[i, j] = 1$ then E_{ij} is satisfied by τ:

$$E_6 = \prod_{i=1}^{m} \prod_{j=m'_i+1}^{m_i} \prod_{k=1}^{m_{ij}}\left(\neg u[i, j] + \sum_{l=1}^{m_{ijk}} x[i, j, k, l]\right).$$

Again, to make this reduction parsimonious, for each i and j where E_{ij} is a product-of-sums, we must restrict the $v[i, j, k]$ chosen. If we set $u[i, j] = 1$, then for all $E_{ij'}$, $j' \geq j$, that are product-of-sums we set the default $v[i, j, 0]$, for all $E_{ij'}$, $j' < j$, that are product-of-sums we do not set the default:

$$E_{6a} = \prod_{i=1}^{m} \prod_{j=1}^{m'_i} \prod_{j'=m'_i+1}^{m_i} \left(\neg u[i, j] + v[i, j', 0]\right),$$

$$E_{6b} = \prod_{i=1}^{m} \prod_{j=m'_i+1}^{m_i} \prod_{j'=j}^{m_i} \left(\neg u[i, j] + v[i, j', 0]\right).$$

If $u[i, j] = 1$ then all other choices $u[i, j']$ with $j' < j$ must have $E_{ij'}$ not satisfied by τ. In the case of sum-of-products, we ensure directly that none of the input lines can evaluate to 1; in the case of product-of-sums, we ensure that the default variable $v[i, j', 0] \neq 1$, therefore forcing some $v[i, j', k]$ with $k \neq 0$ to be set to 1:

$$E_{7a} = \prod_{i=1}^{m} \prod_{j=1}^{m'_i} \prod_{j'=1}^{j-1} \prod_{k=1}^{m_{ij'}} \left(\neg u[i, j] + \sum_{l=1}^{m_i j' k} \neg x[i, j', k, l]\right),$$

$$E_{7b} = \prod_{i=1}^{m} \prod_{j=m'_i+1}^{m_i} \prod_{j'=1}^{m'_i} \left(\neg u[i, j] + \sum_{l=1}^{m_i j' k} \neg x[i, j', k, l]\right),$$

$$E_{7c} = \prod_{i=1}^{m} \prod_{j=m'_i+1}^{m_i} \prod_{j'=m'_i+1}^{j-1} \left(\neg u[i, j] + \neg v[i, j', 0]\right).$$

For each $E_{i,j}$ that is a sum-of-products, we must set exactly one variable $w[i, j, k]$ to 1. If we set $w[i, j, k_i] = 1$, where $k_i \neq 0$, then we need to ensure that k_i is the lexicographically least index such that

$$\prod_{l=1}^{m_{i,j,k_i}} x[i, j, k, l]$$

evaluates to 1. We have

$$E_{8a} = \prod_{i=1}^{m} \prod_{j=1}^{m'_i} \prod_{k=0}^{m_{ij}-1} \prod_{k'=k+1}^{m_{ij}} \left(\neg w[i, j, k] + \neg w[i, j, k']\right),$$

$$E_{8b} = \prod_{i=1}^{m} \prod_{j=1}^{m'_i} \prod_{k=1}^{m_{ij}} \prod_{l=1}^{m_{ijk}} \left(\neg w[i, j, k] + x[i, j, k, l]\right),$$

$$E_{8c} = \prod_{i=1}^{m} \prod_{j=1}^{m_i'} \prod_{k=1}^{m_{ij}} \prod_{k'=1}^{k-1} \left(\neg w[i,j,k] + \sum_{l=1}^{m_{ijk'}} \neg x[i,j,k',l] \right).$$

For each $E_{i,j}$ that is a product-of-sums, we must set exactly one variable $v[i,j,k]$ to 1. If we set $v[i,j,k_i] = 1$, where $k_i \neq 0$, then we need to ensure that k_i is the lexicographically least index such that

$$\prod_{l=1}^{m_{i,j,k_i}} \neg x[i,j,k,l]$$

evaluates to 1. We have

$$E_{9a} = \prod_{i=1}^{m} \prod_{j=m_i'+1}^{m_i} \prod_{k=0}^{m_{ij}-1} \prod_{k'=k+1}^{m_{ij}} \left(\neg v[i,j,k] + \neg v[i,j,k'] \right),$$

$$E_{9b} = \prod_{i=1}^{m} \prod_{j=m_i'+1}^{m_i} \prod_{k=1}^{m_{ij}} \prod_{l=1}^{m_{ijk}} \left(\neg v[i,j,k] + \neg x[i,j,k,l] \right),$$

$$E_{9c} = \prod_{i=1}^{m} \prod_{j=m_i'+1}^{m_i} \prod_{k=1}^{m_{ij}} \prod_{k'=1}^{k-1} \left(\neg v[i,j,k] + \sum_{l=1}^{m_{ijk'}} x[i,j,k',l] \right).$$

Let C be the weft 1, depth 4, circuit corresponding to E. Let C' be the s-normalized circuit, with $s = 2^h + 1$, corresponding to E'. We now argue that the transformation described here, $\langle C, k \rangle \to \langle C', k' \rangle$, is, in fact, a parsimonious parameterized counting reduction.

If τ is a witness for $\langle C, k \rangle$ (that is, τ is a weight k truth assignment for V_0 that satisfies E), then τ' is a witness for $\langle C', k' \rangle$ (that is, τ' is a weight k' truth assignment for V that satisfies E'). This is evident from the description of E' and τ'.

In addition, we argue that τ' is the only extension of τ that is a witness for $\langle C', k' \rangle$. Suppose τ' is an extension of τ, and that τ is a witness for $\langle C, k \rangle$:

1. The clauses of E_1 and E_2 ensure that for each i such that $\tau(x_i) = 1$ and for each $j, 0 \leq j \leq (1+2^h)2^{2^h}$ we must have $\tau'(x[i,j]) = 1$.
2. The clause E_4 ensures that, for each i, $1 \leq i \leq m$, we must choose exactly one index j_i such that E_{i,j_i} evaluates to 1 and set $\tau'(u[i,j_i]) = 1$, and we must set $\tau'(u[i,j']) = 0$ for all $j' \neq j_i$. The clauses E_{7a}, E_{7b}, and E_{7c} ensure that j_i is the lexicographically least possible index.
   (a) If, in (2), E_{i,j_i} is a sum-of-products, then the clauses E_5 and $E_{8a}, \ldots, E_{8c}$ ensure that we must choose the lexicographically least index $k_i \neq 0$ such that

$$\prod_{l=1}^{m_{i,j_i,k_i}} x[i,j_i,k_i,l]$$

evaluates to 1, and set $\tau'(w[i, j_i, k_i]) = 1$, and we must set $\tau'(w[i, j_i, k']) = 0$ for all $k' \neq k_i$.

Clauses E_{5a} and E_{8a} ensure that for $j = 1, \ldots, m_i'$ such that $j \neq j_i$, we must set $\tau'(w[i, j, 0]) = 1$, and $\tau'(w[i, j, k]) = 0$ for all $k \neq 0$.

Clauses E_{6a} and E_{9a} ensure that for $j = m_i' + 1, \ldots, m_i$ we must set $\tau'(v[i, j, 0]) = 1$, and $\tau'(v[i, j, k]) = 0$ for all $k \neq 0$.

(b) If, in (2), E_{i, j_i} is a product-of-sums, then clauses E_{7c} and $E_{9a}, \ldots, E_{9c}$ ensure that for each product-of-sums $E_{i, j'}$ with $j' < j_i$, we must choose the lexicographically least index $k_i \neq 0$ such that

$$\prod_{l=1}^{m_{i, j', k_i}} \neg x[i, j', k_i, l]$$

evaluates to 1, and set $\tau'(v[i, j', k_i]) = 1$, and that for all $k' \neq k_i$ we must set $\tau'(v[i, j', k']) = 0$.

Clauses E_{6b} and E_{9a} ensure that for all $j = m_i' + 1, \ldots, m_i$ such that $j \geq j_i$, we must set $\tau'(v[i, j, 0]) = 1$, and $\tau'(v[i, j, k]) = 0$ for all $k \neq 0$.

Clauses E_{5a} and E_{8a} ensure that for $j = 1, \ldots, m_i'$ we must set $\tau'(w[i, j, 0]) = 1$, and $\tau'(w[i, j, k]) = 0$ for all $k \neq 0$.

Now suppose υ' is witness for $\langle C', k' \rangle$. We argue that the restriction υ of υ' to V_0 is a witness for $\langle C, k \rangle$.

Claim 1 υ *sets at most k variables of V_0 to 1.*

If this were not so, then the clauses in E_1 and E_2 would together force at least $(k + 1) + (k + 2)(1 + 2^h)2^{2^h}$ variables to be 1 in order for υ' to satisfy E', a contradiction as this is more than k'.

Claim 2 υ *sets at least k variables of V_0 to 1.*

The clauses of E_{7a} ensure that υ' sets at most m variables of V_2 to 1. The clauses of E_{8a} ensure that υ' sets at most $\sum_{i=1}^{m} m_i'$ variables of V_3 to 1. The clauses of E_{9a} ensure that υ' sets at most $\sum_{i=1}^{m} m_i - m_i'$ variables of V_4 to 1.

If Claim 2 were false then for υ' to have weight k' there must be more than k indices j for which some variable $x[i, j]$ of V_1 has the value 1, a contradiction in consideration of E_1 and E_2.

The clauses of E_3 and the arguments above show that υ' necessarily has the following restricted form:

(1) Exactly k variables of V_0 are set to 1.
(2) For each of the k in (1), the corresponding $(1 + 2^h)2^{2^h} + 1$ variables of V_1 are set to 1.
(3) For each $i = 1, \ldots, m$ there is exactly one j_i for which $u[i, j_i] \in V_2$ is set to 1.
(4) For each $i = 1, \ldots, m$ and $j = 1, \ldots, m_i'$ there is exactly one k_i for which $w[i, j, k_i] \in V_3$ is set to 1.

(5) For each $i = 1, \ldots, m$ and $j = m'_i + 1, \ldots, m_i$ there is exactly one k_i for which $v[i, j, k_i] \in V_4$ is set to 1.

To argue that υ satisfies E, it suffices to argue that υ satisfies every E_{i,j_i} for $i = 1, \ldots, m$.

If E_{i,j_i} is a sum-of-products then the clauses of E_5 ensure that if $\upsilon'(u[i, j]) = 1$ then $k_i \neq 0$. This being the case, the clauses of E_{8b} force the literals in the k_ith subexpression of E_{i,j_i} to evaluate in a way that shows E_{i,j_i} to evaluate to 1.

If E_{i,j_i} is a product-of-sums then the clauses of E_6 ensure that if $\upsilon'(u[i, j]) = 1$ then E_{i,j_i} evaluates to 1. $\qquad\square$

Lemma 32.4.1 allows us to now state the following parameterized counting analog of the analogous statement of Chap. 21.

Theorem 32.4.2

$$\#W[1] = \bigcup_{s=1}^{\infty} \#W[1, s]$$

where $\#W[1, s]$ is the class of parameterized counting problems whose associated witness functions reduce to $w_{\mathcal{F}(s)}$, the standard parameterized witness function for $\mathcal{F}(s)$, the family of s-normalized decision circuits.

As in Chap. 21, we now want to show that $\#W[1]$ collapses to $\#W[1, 2]$. The road-map is the same.

Theorem 32.4.3 $\#W[1, s] = \#\text{Antimonotone } W[1, s]$ *for all $s \geq 2$.*

Using Theorem 32.4.3 we can prove the following:

Theorem 32.4.4 $\#W[1] = \#W[1, 2]$.

Theorem 32.4.5 *The following are complete for $\#W[1]$:*
1. #INDEPENDENT SET;
2. #CLIQUE;
3. #SHORT TURING MACHINE ACCEPTANCE.

Proof All of the relevant reductions from Chap. 21 are parsimonious. $\qquad\square$

32.5 The #W-Hierarchy

As proven by Flum and Grohe [310], and McCartin [522], the reasoning above can be extended to the whole W-hierarchy. We state the following (appropriate analogs exists also for $W[\text{SAT}]$ and $W[\text{P}]$). Its proof is left as an exercise to the reader (Exercise 32.8.5).

Theorem 32.5.1 (Flum and Grohe, McCartin) #WEIGHTED t-NORMALIZED SAT-
ISFIABILITY *is complete for* #W[t].

The general "theorem" is that all of the hardness results proving $W[t]$ and other
completeness results (*excluding* ones about $M[t]$) have parsimonious versions.

32.6 An Analog of Valiant's Theorem

Valiant's Theorem, Theorem 32.2.2, asserts that counting bipartite matchings is #P
complete. In the present section, we prove an attractive result of Flum and Grohe,
which can be thought of as a parameterized analog. We consider the following four
problems.

#k-(DIRECTED) CYCLE (PATH)

Input: A (directed) graph G.
Parameter: A positive integer k.
Output: The number of (directed) k-cycles (k-paths) in G.

Theorem 32.6.1 (Flum and Grohe [310]) #k-(DIRECTED) CYCLE (PATH) *are all*
#W[1]-*complete under randomized parameterized Turing reductions.*

Proof We follow the proof of Flum and Grohe [310]. No other proof is known. We
first show that #k-DIRECTED CYCLE reduces to #k-CYCLE. Given a digraph D,
replace each vertex v of D, by a path of length p, and each directed path in D by an
undirected path of length q, and call the result G. Notice that each cycle in G has
length $np + mq$ for some $n, m \in \mathbb{N}$, and each directed cycle in D becomes a cycle
of size $k(p + q)$ in G. The observation is that we can set the parameters p, q so that
we must choose $n = m$. First we must choose them so that $k(p + q) \neq np + mq$
if $n < m$, and for $m > 2k$, this holds if we choose $0 < p \leq q$. Then it is easy to
compute parameters p, and q which also satisfy the inequality for $0 \leq n < m \leq 2k$,
by avoiding $\binom{2k+1}{2}$ equalities.

The next step is to prove that #k-CYCLE reduces to #k-PATH. Again this is combinatorial. We assume $k \geq 3$ (loops are easy), and, given G, construct a new graph
G' as follows. For each edge $e = xy$ of G, let $G_e(\ell, m)$ by adding new vertices
$v_1, \ldots, v_\ell, w_1, \ldots, w_m$, and edges $v_i v$, $w_j w$ for $i \in \{1, \ldots, \ell\}$ and $j \in \{1, \ldots, m\}$.
The main observation is that the number x_e of paths of length $k + 1$ from v_1 to w_1
in $G_e(\ell, m)$ is precisely the number of cycles of length k in G containing e. We
arrange a parsimonious Turing reduction from #k-CYCLE to #k-PATH. v_i and w_j
can only be endpoints of paths in $G_e(\ell, m)$, and each path can have at most one
endpoint from amongst $\{v_1, \ldots, v_\ell\}$, and $\{w_1, \ldots, w_m\}$. We introduce the following
notation.

1. $x = x_e$ the number of paths of length $k + 1$ from v_1 and w_1 in $G_e(1, 1)$.
2. y the number of paths of length $k + 1$ in $G_e(1, 1)$ containing v_1 and not w_1.
3. z the number of paths of length $k + 1$ in $G_e(1, 1)$ containing w_1 and not v_1.

4. d the number of paths of length $k + 1$ containing neither v_1 nor w_1.

Then if $p = p_{\ell,m}$ denotes the number of paths of length $(k + 1)$ in $G_e(\ell, m)$, we have

$$p = d + \ell m x + \ell m + m z.$$

For $0 \le \ell, m \le 1$, this is a system of 4 linear equations in the variables d, x, y, z and the matrix is nonsingular; meaning that it has a unique solution giving $x = x_e$.

Replacing each edge of an undirected graph by two directed edges allows us to deduce that #k-PATH parsimoniously reduced to #k-DIRECTED PATH.

The next part of the hardness proof is to show that #k-CLIQUE (parsimoniously) reduces to #k-DIRECTED CYCLE. This is nontrivial. The proof detours through another problem. Let $h : H \to G$ be a homomorphism, and for $i \ge 1$, let k_i denote the number of vertices v of G with $|h^{-1}(v)| \ge i$. Then the *type* is the polynomial

$$t_h(X) = \prod_{i \ge 1}(X - i + 1)^{k_i} = \prod_{v \in G}(X)_{|h^{-1}(v)|}.$$

Here $(X)_j$ is defined as $(X)_0 = 1$ and $(X)_{j+1} = (X)_j(X - j)$. Let D_k denote the directed cycle with k vertices $1, \ldots, k$ in cyclic order. The problem considered by Flum and Grohe is

#TDC

Input: A directed graph G and a polynomial $t(X)$.
Parameter: A positive integer k.
Output: The number of homomorphisms $h : D_k \to G$ of type $t(X)$.

Lemma 32.6.1 (Flum and Grohe [310]) #TDC *reduces to* #k-DIRECTED CYCLE *via a counting Turing reduction.*

Proof Given G and $\ell, m \ge 1$, let $G_{\ell,m}$ be the graph defined as follows.

$$V(G_{\ell,m}) = V(G) \times [\ell] \times [m],$$

$$E(G_{\ell,m}) = \{(a, i, j)(a', i', j') \mid [i = \ell \wedge i' = 1 \wedge aa' \in E(G)]$$
$$\vee [a = a' \wedge i' = i + 1]\}.$$

Given an embedding $\iota : D_{k \cdot \ell} \to G_{\ell,m}$, Flum and Grohe define the *projection* of this embedding as a homomorphism π_ι, which maps $j \in D_k$ to the first component of $\iota((j - 1)\ell + 1)$; namely $\pi_\iota(j) = b$ iff $\iota((j - 1)\ell + 1) = (b, s, t)$ some $s \in [\ell]$ and $t \in [m]$. For each homomorphism $h : D_k \to G$ there are $t_h(\ell)^m$ many embeddings $\iota : D_{k\ell} \to G_{\ell,m}$ with projection $\pi_\iota = h$.

Let T denote the set of all types of homomorphisms from D_k into some G. Then for each type $t \in T$, if we let x_t be the number of homomorphisms $h : D_k \to G$ with $t_h = t$, we see that

$$b_\ell = \sum_{t \in T} x_t \ell t(m)^\ell,$$

is the number of embeddings $\iota : D_{k\ell} \to G_{\ell,m}$.

Since the types of homomorphisms in T are degree k polynomials, for $t \neq t' \in T$ there are at most k x with $t(x) = t'(x)$, and hence there is an $m \leq k \cdot |T|^2$ such that for all distinct t, t', $t(m) \neq t'(m)$. Fixing such an m, we can let $\bar{b} = (b_1, \ldots, b_{|T|})$, $\bar{x} = (x_t)_{t \in T}$, and $A = (a_{\ell t})_{1 \leq \ell \leq |T|}$ with $a_{\ell t} = \ell t(m)^\ell$. Note that $A\bar{x} = \bar{b}$. The matrix $(\frac{1}{\ell} a_{\ell t})_{\{1 \leq \ell \leq |T| \wedge t \in T\}}$ is a Vandermonde matrix[1] and consequently nonsingular. Thus, A is nonsingular. We conclude that $\bar{x} = A^{-1}\bar{b}$.

Our reduction is the following.

1. Compute T and the suitable m.
2. For $\ell \in [|T|]$, compute $G_{\ell,m}$.
3. Using an oracle for #k-DIRECTED CYCLE, and noting that b_ℓ is $k\ell$ times the number of cycles of length $k\ell$ in $G_{\ell,m}$, for $\ell \in [|T|]$, calculate b_ℓ the number of embeddings $\iota : D_{\ell k} \to G_{\ell,m}$.
4. Compute A and calculate $\bar{x} = A^{-1}\bar{b}$.
5. Return x_t, where $t(X)$ is the input polynomial; return 0 if $t \notin T$.

Note that T, A, and m only depend upon k. □

The next step is to establish certain asymptotic behavior. Fix $\ell, k \geq 1$, and let $\Omega(k, \ell)$ denote the space of mappings $f : [k\ell] \to [k]$ such that $f^{-1}(i) = \ell$ for $i \in [k]$.

Lemma 32.6.2 (Flum and Grohe [310]) *Let $k \geq 1$ and $\mathcal{H} = (H, E)$ be a directed graph with $H = V(\mathcal{H}) = [k]$ and $E \neq H^2$. Then[2]*

$$\lim_{\ell \to \infty} \Pr_{f \in \Omega(k,\ell)}(f \text{ is a homomorphism from } D_{k\ell} \text{ to } \mathcal{H}) = 0.$$

Proof This is basically a calculation. Let $xy \in H^2 - E$, and $m \leq k\ell$. Say that $(i_1, \ldots, i_m) \in H^m$ is *good* if $i_j i_{j+1} \neq xy$ for all $j \in [m-1]$ and *bad* otherwise. For random $(i_1, \ldots, i_m)$ we have

$$\Pr\big((i_1, \ldots, i_m) \text{ good}\big) \leq \Pr\left(\left(\forall j \in \left[\frac{m}{2}\right]\right)[i_{2j-1}i_{2j} \neq xy]\right) = \left(1 - \frac{1}{k^2}\right)^{\lfloor \frac{m}{2} \rfloor}.$$

For all $(i_1, \ldots, i_m)$,

$$\Pr_{f \in \Omega(k,\ell)}\big((\forall j \in [m]) f(j) = i_j\big) \leq \left(\frac{\ell}{k\ell - m}\right)^m.$$

Therefore,

$$\Pr_{f \in \Omega(k,\ell)}(f \text{ is a homomorphism from } D_{k\ell} \text{ to } \mathcal{H})$$

---

[1] That is, the terms of each row form a geometric progression. These are well studied and known to always be nonsingular, with a standard expression for the determinant.

[2] Here we are choosing f uniformly at random.

$$\leq \sum_{(i_1,\dots,i_m)\in H^m \text{ good}} \Pr\big(f(j)=i_j \text{ for } j \in [m]\big)$$

$$\leq \sum_{(i_1,\dots,i_m)\in H^m \text{ good}} \left(\frac{\ell}{k\ell-m}\right)^m$$

$$\leq \left(\frac{\ell}{k\ell-m}\right)^m k^m \Pr\big((i_1,\dots,i_m)\in H^m \text{ good}\big)$$

$$\leq \left(\frac{k\ell}{k\ell-m}\right)^m \left(1-\frac{1}{k^2}\right)^{\lfloor \frac{m}{2}\rfloor}.$$

Now for $\varepsilon > 0$, there does exist a $m = m(\varepsilon, k)$ where

$$\left(1-\frac{1}{k^2}\right)^{\lfloor \frac{m}{2}\rfloor} \leq \frac{\varepsilon}{2}.$$

Also for each m there exists an $\ell(m)$ such that for all $\ell \geq \ell(m)$,

$$\left(\frac{k\ell}{k\ell-m}\right)^m = \left(\frac{1}{1-\frac{m}{k\ell}}\right)^m \leq \frac{1}{(1-\frac{m}{\ell})^m} \leq 2.$$

It follows that for all $\ell \geq \ell(m(\varepsilon, k))$ that

$$\Pr_{f\in\Omega(k,\ell)}(f \text{ is a homomorphism from } D_{k\ell} \text{ to } \mathcal{H}) \leq \varepsilon. \qquad \square$$

Lemma 32.6.3 (Flum and Grohe [310]) #CLIQUE *reduces to* #TDC *via Turing reduction.*

Proof Given a graph H, let H' denote the digraph with the $V(H') = V(H)$, and

$$E(H') = \{vv \mid v \in H\} \cup \{vw \mid vw \in vw \in E^H\}.$$

If H has k vertices and $\ell \geq 1$, let $a_{H\ell}$ denote the number of homomorphisms of type $(X)_\ell^k$ from $D_{k\ell}$ into H' so that each point has exactly ℓ pre-images. Let $a_H^{\mathbb{N}} = (a_{Hi} \mid i \geq 1)$ and $a_H^\ell = (a_{H1},\dots,a_{H\ell})$, and consider them as vectors in the vector spaces $\mathbb{Q}^{\mathbb{N}}$ and $\mathbb{Q}^\ell$ respectively.

Now let $K = K_k$ have the vertex set $\{1,\dots,k\}$, and let $\mathbb{K}$ be the set of all graphs (one per isomorphism type) on these vertices with $\mathbb{K}^- = \mathbb{K} - i\{K_k\}$. We will use $\langle S \rangle$ to denote the linear span of S.

The first claim is that $a_K^{\mathbb{N}} \notin \langle\{a_H^{\mathbb{N}}\}\rangle$. By definition, $\Omega(k,\ell)$ is the set of all mappings $h : [k\ell] \to [k]$ with $h^{-1}(i) = \ell$ for all $i \in [k]$. Note that $a_{K\ell} = |\Omega(k,\ell)|$. By Lemma 32.6.2, for all $H \in \mathbb{K}^-$,

$$\lim_{\ell\to\infty} \frac{a_{H\ell}}{|\Omega(k,\ell)|} = 0.$$

Suppose that $a_K^{\mathbb{N}} \in \langle\{a_H^{\mathbb{N}}\}\rangle$. Then $a_K^{\mathbb{N}} = \sum_{i=1}^n \lambda_i a_{G_i}^{\mathbb{N}}$ with $G_i \in \mathbb{K}^-$. If we take ℓ large enough that for all $i \in [n]$,

$$\frac{a_{G_i\ell}}{a_{K\ell}} = \frac{a_{G_i\ell}}{|\Omega(k, \ell)|} < \frac{1}{\sum_{i\in[n]} |\lambda_i|},$$

then

$$a_{K\ell} = \sum_{i\in[n]} \lambda_i a_{G_i\ell} \le a_{K\ell} \sum_{i\in[n]} |\lambda_i| \frac{a_{G_i\ell}}{a_{K\ell}} < a_{K\ell} \sum_{i\in[n]} |\lambda_i| \frac{1}{\sum_{i\in[n]} |\lambda_i|}.$$

This contradiction shows that $a_K^{\mathbb{N}} \notin \langle\{a_H^{\mathbb{N}}\}\rangle$.

The second claim is that we can compute $\ell = \ell(k)$ such that

$$a_K^\ell \notin \langle\{a_H^\ell \mid H \in \mathbb{K}^-\}\rangle.$$

Following Flum and Grohe, we establish this by defining for $p \in \mathbb{N} \cup \{\mathbb{N}\}$, $V_p = \langle\{a_H^p \mid H \in \mathbb{K}^-\}\rangle$, and can view V_i as a subspace of V_j for all $j \ge i$ and $j = \mathbb{N}$. We can compute a sequence $\mathbb{B}_1 \subseteq \mathbb{B}_2 \subseteq \cdots \subseteq \mathbb{K}^-$, such that for all $i \ge 1$, $\mathbb{B}_i = \{a_H^i \mid H \in \mathbb{K}^-\}$ is a basis of V_i. $V_{\mathbb{N}}$ is finite dimensional, and hence there is an n with $\mathbb{B}_i = \mathbb{B}_n$ for all $i \ge n$.

If $a_K^i \in V_i$ for all $i \ge 1$, we could express a_K^i as a unique linear combination of $\{a_H^i \mid H \in \mathbb{B}_i\}$. But then for all $i \ge n$, these combinations would be the same, and this would prove that $a_K^{\mathbb{N}} \in V_{\mathbb{N}}$. This contradicts the first claim. Therefore there is some ℓ with $a_{K\ell} \notin V_\ell$, and this can be computed from k, and completes the proof of the claim.

To finish the proof of the lemma, fix $k \ge 1$, and define the parameters $K, \mathbb{K}, \mathbb{K}^-$, and compute a_H^p as above. Compute $\ell = \ell(k)$.

Let G be given. For each $H \in \mathbb{K}$, let x_H denote the number of subsets $A \subseteq V(G)$ such that the subgraph induced by G on A is isomorphic to H. The goal is to determine the number $x_{K_k} = x_K$. For $1 \le i \le \ell$, let b_i denote the number of homomorphisms from D_{ki} into G' of type $(X)_i^k$ and let $b^\ell = (b_1, \ldots, b_\ell)$. Note that the b_i can be determined by queries to an oracle for #TDC. For $i \in [\ell]$,

$$b_i = \sum_{H\in\mathbb{K}} x_H a_{Hi},$$

and hence

$$b^\ell = \sum_{H\in\mathbb{K}} x_H a_H^\ell.$$

Since $a_K^\ell \notin \langle\{a_H^\ell \mid H \in \mathbb{K}^-\}\rangle$, the coefficient for x_K can be computed by solving a system of linear equations. This completes the proof of the lemma. □

We note that #TDC is a generalization of #k-DIRECTED CYCLE, and hence we have the theorem since now all are hard for $W[1]$ under counting Turing reductions, since #CLIQUE is hard. □

32.7   Parameterized Randomization

In this section we will look at parameterized analogs of the randomized computation we mentioned in Definition 1.3.1.

Of course we have already met parameterized randomization in Theorem 8.4.1 when we met Koutis' method using multilinear detection for randomized versions of color coding. As with counting, there are few—perhaps even fewer—systematic techniques for the development of parameterized randomized algorithms, though the area has been advancing hand-in-hand with parameterized complexity since the invention of color coding. There still seems to be no systematic development, for example, of randomized kernelization.

In the present section, we will look at the hardness theory for parameterized randomization. In particular, we will look at an analog of the Valiant–Vazirani Theorem that UNIQUE SAT is as hard as SAT we met in Chap. 1.

Towards precise definitions, recall that in Chap. 28, we defined the class $G[t]$ (Uniform $G[t]$) is the class of parameterized languages $L \subseteq \Sigma^* \times N$ for which there is a parameterized (uniform) family of weft t circuits $\mathcal{F} = \{C_{n,k}\}$ such that for all x and k, with $n = |x|$, $(x, k) \in L \Leftrightarrow C_{n,k}(x) = 1$. In that chapter, we observed that Uniform $G[P] = \text{FPT}$. Recall that in Definition 28.1.1, we made the following definitions:

(a) For any class $\mathcal{C}$ of parameterized languages, $\exists \cdot \mathcal{C}$ stands for the class of parameterized languages A such that for some $B \in \mathcal{C}$ there are nice parametric connections (n, k, n', k', n'', k'') giving for all (x, k), $(x, k) \in A \Leftrightarrow (\exists y \in \Sigma^{n'})[\, \text{wt}(y) = k' \wedge (xy, k'') \in B]$. (Here $n = |x|$, $n' = |y|$, and $n'' = n + n'$ and $\text{wt}(x)$ denotes the weight of x.)

(b) For all $t \geq 1$, $N[t]$ stands for $\exists \cdot \text{Uniform-}G[t]$, and $N[P]$ stands for $\exists \cdot \text{Uniform-}G[P]$.

In a corresponding way, we can define "bounded weight" versions of the other familiar class operators $\forall$, $\oplus$, and BP. Combining the latter two formally, we find that a language A belongs to $\text{BP} \cdot \oplus \cdot \mathcal{C}$ if there exists $B \in \mathcal{C}$ and nice connections giving for all (x, k),

$$(x, k) \in A \implies \Pr_{y \in \{0,1\}^{n'}, \text{wt}(y) = k'} \left[\left| \{ z \in \{0, 1\}^{n''} : \text{wt}(z) = k'' \right. \right.$$
$$\left. \left. \wedge \, (xyz, k''') \in B \} \right| \text{ is odd} \right] > \frac{3}{4},$$

while $(x, k) \notin A \Rightarrow \Pr[\ldots] < 1/4$. If the latter probability is zero (i.e., we have one-sided error), then we write $A \in \text{RP} \cdot \oplus \cdot G[t]$.

As in Chap. 28, if $\mathcal{C}$ is any class of parameterized languages, then by $\langle \mathcal{C} \rangle$ we denote the parameterized languages that are reducible to a language in $\mathcal{C}$, and refer to this as the *FPT-closure* of $\mathcal{C}$.

We will now pursue an analog of randomized reduction from a parameterized language.

Definition 32.7.1 (Downey, Fellows, and Regan [256]) A *randomized (FPT, many-one) reduction* from a parameterized language L to a parameterized language L' is a randomized procedure that transforms (x, k) into (x', k'), subject to the following conditions:

(1) The running time of the procedure is bounded by $f(k)|x|^c$ for some constant c and arbitrary function f (i.e., the procedure is fixed-parameter tractable).

(2) There is a function f' and a constant c' such that for all (x, k),

$$(x, k) \in L \quad \Longrightarrow \quad \text{Prob}\big[(x', k') \in L'\big] \geq 1/f'(k)|x|^{c'},$$

$$(x, k) \notin L \quad \Longrightarrow \quad \text{Prob}\big[(x', k') \in L'\big] = 0.$$

In Chap. 28, we gave the usual definition of the $W[t]$ hierarchy in terms of the WEIGHTED CIRCUIT SATISFIABILITY problem. We consider here the following unique-solution variant.

UNIQUE WCS(t, h)

Instance: A circuit C of weft t and overall depth $t + h$.
Parameter: A positive integer k.
Question: Is there a unique input of Hamming weight k that is accepted by C?

Definition 32.7.2 For all $t \geq 1$, UNIQUE $W[t]$ is the class of parameterized languages L such that for some h, L is FPT and many-one reducible to UNIQUE WCS(t, h).

Our proof will make use of a technical but generally useful lemma showing that a restricted form of WEIGHTED t-NORMALIZED SATISFIABILITY is complete for $W[t]$. This lemma is essentially implicit in earlier work. The variant is defined as follows.

SEPARATED t-NORMALIZED SATISFIABILITY

Instance: A t-normalized Boolean expression E over a set of variables V that is partitioned into k disjoint sets $V_1, \ldots, V_k$ of equal size, $V_i = \{v_{i,1}, \ldots, v_{i,n}\}$ for $i = 1, \ldots, k$.
Parameter: A positive integer k.
Question: Is there a truth assignment of weight k making exactly one variable in each of the V_i *true* and all others *false*, and that furthermore satisfies the condition that if $v_{i,j}$ is *true*, then for all $i' > i$ and $j' \leq j$, $v_{i',j'}$ is *false*.

Lemma 32.7.1 (Downey, Fellows, and Regan [256]) SEPARATED t-NORMALIZED SATISFIABILITY *is complete for* $W[t]$ *for all* $t \geq 1$.

Proof We give separate arguments for t even and t odd. For t even we reduce from MONOTONE t-NORMALIZED SATISFIABILITY and use the construction described in Chap. 23. Suppose the parameter is k and that F is the monotone expression. The reduction is to a normalized expression F' and the parameter $k' = 2k$. The key point is that the variables for F' consist of $2k$ disjoint blocks, and that any weight $2k$ truth assignment for F' must make exactly one variable *true* in each block. The blocks can be padded so that they are of equal size. Including additional enforcement for the condition in the definition of SEPARATED t-NORMALIZED SATISFIABILITY is straightforward. It is possible for this to be done in such a way that monotonicity is preserved.

For t odd we similarly employ the reduction described in Chap. 23, starting from ANTI-MONOTONE t-NORMALIZED SATISFIABILITY. In this case, anti-monotonicity can be preserved. $\square$

Theorem 32.7.1 (Downey, Fellows, and Regan [256]) *For all $t \geq 1$ there is an FPT many-one randomized reduction of $W[t]$ to* UNIQUE $W[t]$.

At this stage we invite the reader to recall the proof of the classical Valiant–Vazirani Theorem. It relied on Lemma 32.7, which stated that if we let S be a nonempty subset of V. Let $\{b_1, \ldots, b_n\}$ be a randomly chosen basis of V and let $V_j = \{b_1, \ldots, b_{n-j}\}^\perp$, so that $\{0\} = V_0 \subset V_1 \subset \cdots \subset V_n = V$ is a randomly chosen tower of subspaces of V, and V_j has dimension j. Then the probability

$$\Pr\left(\exists j \left(|V_j \cap S| = 1\right)\right) \geq \frac{3}{4}.$$

The reader should recall that we applied this to the case that S was the collection of satisfying assignments of a given 3SAT formula X. With high probability we could use this to get a unique satisfying assignment for a "hashed" formula, $X' = X'_j$. Of course we could use the same methodology applied to S as the *weight k* satisfying assignments.

This argument indeed works and could easily be seen to show that #$W[P]$ is hard for $W[P]$ under randomized parameterized reductions (Exercise 32.8.7). The issue in the proof below is to show that we can achieve the same result while still controlling weft.

Proof of Theorem 32.7.1 We reduce from SEPARATED t-NORMALIZED SATISFIABILITY. Let E be the relevant t-normalized Boolean expression over the k blocks of n variables:

$$X_i = \left\{x[i, 1], \ldots, x[i, n]\right\} \quad \text{for } i = 1, \ldots, k.$$

Let X denote the union of the X_i, and assume for convenience (with no loss of generality) that n is a power of 2, $n = 2^s$, and that $k - 1$ divides s.

We describe how to produce (by a randomized procedure) a weft t expression E' of bounded depth, and an integer k' so that the conditions defining a randomized reduction are met.

The reduction procedure consists of the following steps:

(1) Randomly choose $j \in \{1, \ldots, k \log n\}$.
(2) Randomly choose j length n 0–1 vectors

$$y_i = \big(y[i, 1], \ldots, y[i, n]\big), \quad 1 \leq i \leq j.$$

(3) Randomly choose $m \in \{1, \ldots, 12\}$.
(4) Output

$$E' = E_1 \wedge E_2 \wedge \cdots \wedge E_9 \quad \text{and} \quad k'$$

where the constituent subexpressions E_i and the weight parameter k' are as described below.

The set X' of variables for E' is

$$X' = X'_1 \cup X'_2 \cup X'_3$$

where

$$X'_1 = \big\{u[a, b, c] : 1 \leq a \leq m, 1 \leq b \leq k, 1 \leq c \leq n\big\},$$
$$X'_2 = \big\{v[a, b] : 1 \leq a \leq k(k - 1), 1 \leq b \leq n\big\},$$
$$X'_3 = \big\{w[a, b] : 1 \leq a \leq m - 1, 1 \leq b \leq k\big\}.$$

We next describe the various constituent subexpressions of E'.

The subexpression E_1. Write $X'_1(i)$ to denote the variables of X'_1 that have first index i, for $i = 1, \ldots, m$. That is,

$$X'_1(i) = \big\{u[i, b, c] : 1 \leq b \leq k, \ 1 \leq c \leq n\big\}.$$

Note that the set $X'_1(i)$ can be paired in a natural way with the set of variables X of the expression E by the correspondence:

$$x[b, c] \leftrightarrow u[i, b, c].$$

Let $E_1(i)$ denote the expression obtained from E (essentially, a copy of E) by substituting the variables of $X'_1(i)$ for the variables of X according to this correspondence:

$$E_1 = \prod_{i=1}^{m} E_1(i).$$

The role of E_1 is to hold each of the m copies of the variables of E accountable for satisfying a copy of E.

The subexpression E_2.

$$E_2 = \prod_{a=1}^{m} \prod_{b=1}^{k} \prod_{1 \leq c < c' \leq n} \big(\neg u[a, b, c] \vee \neg u[a, b, c']\big).$$

The role of E_2 is to enforce that at most one variable is set true in each "block" of the variables of X_1' (there are km blocks, corresponding the m copies X, each copy consisting in a natural way of k blocks).

The subexpression E_3.

$$E_3 = \prod_{a=1}^{m} \prod_{b=1}^{k} \prod_{c=1}^{n} \prod_{b'=b+1}^{k} \prod_{c'=1}^{c} \left(u[a, b, c] \rightarrow \neg u[a, b', c'] \right).$$

The role of E_3 is to enforce the ascending-order condition on truth assignments (with respect to the k blocks of variables); this condition occurs in the definition of SEPARATED t-NORMALIZED SATISFIABILITY. This condition is enforced for each of the m copies of the variables of E.

The subexpression E_4. We view X_3' as consisting of $m - 1$ blocks:

$$X_3'(a) = \left\{ w[a, b] : 1 \leq b \leq k \right\},$$

$$E_4 = \prod_{a=1}^{m-1} \prod_{1 \leq b < b' \leq k} \left(\neg w[a, b] \vee \neg w[a, b'] \right).$$

The role of this subexpression is to enforce that at most one variable is set true in each of the blocks of X_3' in any satisfying truth assignment for E'.

The subexpression E_5 is

$$E_5 = \prod_{a=1}^{k(k-1)} \prod_{1 \leq b < b' \leq n} \left(\neg v[a, b] \vee \neg v[a, b'] \right).$$

The role of this subexpression is to enforce that at most one variable is set true in each of the $k(k - 1)$ blocks of X_2'.

The subexpressions E_6 and E_7 are

$$E_6 = \prod_{a=1}^{m-1} \prod_{b=1}^{k} \prod_{b'=1}^{b-1} \prod_{c \neq c': 1 \leq c, c' \leq n} \left(\neg w[a, b] \vee \neg u[a, b', c] \vee \neg u[a+1, b', c'] \right),$$

$$E_7 = \prod_{a=1}^{m-1} \prod_{b=1}^{k} \prod_{1 \leq c' \leq c \leq n} \left(\neg w[a, b] \vee \neg u[a, b, c] \vee \neg u[a+1, b, c'] \right).$$

The $m - 1$ variables that are set true in the blocks of X_3' in a satisfying assignment for E' provide evidence that the m "solutions" for E recorded in the m blocks of X_1' are distinct and recorded in the m blocks in increasing lexicographic order. The nature of this evidence is an indication of the first of the k choice blocks in which two consecutive solutions differ. The subexpressions E_6 and E_7 enforce the increasing lexicographic ordering based on this evidence.

The subexpressions E_8 and E_9. In order to describe the subexpressions E_8 and E_9, we first must construct an *interpretation* of the variables of X_2'. This consists of the following information:

(1) Each $a \in \{1, \ldots, k(k-1)\}$ is assigned a subset $J_a \subseteq \{1, \ldots, j\}$ so that $|J_a| = \log n/(k-1)$ and $\bigcup_{1 \leq a \leq k(k-1)} = \{1, \ldots, j\}$.

(2) Each even-cardinality subset $S_\alpha \subseteq \{1, \ldots, k\}$ is assigned a unique 0–1 vector α of length $k-1$. (Note that this is possible, since there are 2^{k-1} such even-cardinality subsets.)

(3) Each variable $v[a, b] \in X_2'$ is interpreted as assigning an even-cardinality subset $S(j', a, b)$ to each $j' \in J_a$. This assignment is made in the following way. The index b can be regarded as a 0–1 vector of length $\log n$. This index vector can be read as a sequence of $|J_a|$ blocks of size $k-1$. If the rth block is α then the rth element of J_a is assigned the even-cardinality subset S_α.

$$E_8 = \prod_{p=1}^{m} \prod_{1 \leq a \leq k(k-1)} \prod_{1 \leq b \leq n} \prod_{j' \in J_a} \prod_{r \in S(j',a,b)} \prod_{q:y[j',q]=0} \left(\neg v[a, b] \vee \neg u[p, j', q]\right),$$

$$E_9 = \prod_{p=1}^{m} \prod_{1 \leq a \leq k(k-1)} \prod_{1 \leq b \leq n} \prod_{j' \in J_a} \prod_{r \notin S(j',a,b)} \prod_{q:y[j',q]=1} \left(\neg v[a, b] \vee \neg u[p, j', q]\right).$$

The variables that are set true in X_2' in a satisfying truth assignment for E' are intended to indicate a proof (that can be checked by a weft-1 circuit) that each of the m weight k truth assignments that are solutions for E recorded in the m blocks of X_1' are orthogonal to the randomly chosen length n 0–1 vectors y_i. The proof that is indicated consists of showing that an even subset of the k positions set to true in X_1' have corresponding positions that are 1 in the y_i. A variable $v[a, b]$ indicates part of such a proof, according to the interpretation mechanism described above. The subexpressions E_8 and E_9 provide an enforcement for the interpretation.

The parameter. The description of the reduction is completed by specifying the parameter that accompanies E'.

$$k' = mk + k(k-1) + (m-1).$$

We now argue for the correctness of the reduction. Half of this is easy. If E is not satisfiable by a weight k truth assignment, then because of E_2 and E_1 there is no weight k' truth assignment that satisfies E' (never mind whether it is unique).

For the other half we must argue that if E has a weight k truth assignment, then with the required probability bound, E' has a unique weight k' truth assignment. Let $X_0 = \{x[1], \ldots, x[n]\}$. The weight k truth assignments to X that satisfy the additional conditions that define SEPARATED t-NORMALIZED SATISFIABILITY can be put in a natural $1:1$ correspondence weight k truth assignments to X_0. The correspondence is that, if the rth variable assigned the value 1 in X_0 is $x[s]$, then $x[r, s]$ is assigned 1 in the truth assignment for X. Because of this correspondence we can speak of a weight k truth assignment to X_0 that satisfies E.

It follows from the arguments of the Valiant–Vazirani section of Chap. 1 that if there is any weight k truth assignment to X_0 that satisfies E (and noting that there are no more than n^k such assignments), then with probability at least $\frac{1}{24k \log n}$ there are exactly m distinct weight k truth assignments that satisfy E and that are hashed by the function

$$h\big(x[1], \ldots, x[n]\big)[s] = \bigoplus_{i=1}^{n}\big(x[i] \wedge y[s, i]\big)$$

to 0^j.

We argue that in this case, E' is uniquely satisfied by a weight k' truth assignment to X'. The subexpressions E_1, E_2, E_3, E_4, E_6, and E_7 can be satisfied if the m distinct truth assignments are represented in lexicographically increasing ascending order in the blocks of X'_1, and if the evidence for the lexicographic ordering is represented in X'_3. It is easy to check that if there are exactly m distinct weight k truth assignments that satisfy E, then there is a unique truth assignment to $X'_1 \cup X'_3$ that satisfies these subexpressions, and it must have weight $mk + (m - 1)$. The key point for this assertion is that the subexpressions E_6 and E_7 are sufficiently restrictive, so that not only is increasing lexicographic ordering enforced, but also the evidence for this is uniquely determined.

In the above situation, the subexpressions E_5, E_8, and E_9 can be satisfied by a weight $k(k - 1)$ assignment to X'_2 that represents the hash function condition. Because this is also uniquely determined, there is a unique weight k' truth assignment for E'.

The subexpressions E_2 through E_9 have weft 1, and therefore the weft of E' is the same as the weft of E. $\square$

The reader should note that we can now conclude that virtually all $W[t]$-complete problems are hard under randomized reductions either in the case of the "unique version" or in the case of a version with a "uniqueness promise". For an example, we consider the following.

UNIQUE k-INDEPENDENT SET

Instance: A graph G.
Parameter: A positive integer k.
Question: Does G have a unique size k independent set?

k-INDEPENDENT SET WITH A UNIQUENESS PROMISE

Instance: A graph G.
Parameter: A positive integer k.
Promise: If G has a size k independent set, it has a unique one.
Question: Does G have a size k independent set?

Theorem 32.7.2 (Müller [538])

1. UNIQUE k-INDEPENDENT SET *is hard for* $W[1]$ *under randomized polynomial reductions.*
2. k-INDEPENDENT SET WITH A UNIQUENESS PROMISE *is hard for* $W[1]$ *under randomized polynomial-time reductions.*

The point here is that we can take the hardness result of Theorem 32.7.1, and apply the parsimonious reductions for the hard instance of UNIQUE $W[t]$, to other complete members of the class. In the case of $W[1]$ we can parsimoniously reduce the constructed instance of UNIQUE 3SAT (which is also an instance of k-3SAT WITH A UNIQUENESS PROMISE by the construction) to UNIQUE k-INDEPENDENT SET and to k-INDEPENDENT SET WITH A UNIQUENESS PROMISE.

Moritz Müller was the first to show that various combinatorial problems were hard in unique form, and he did so using a different technique, based on a different hashing technique.

For more on this, and refinements concerning one-sided and two-sided errors, and the difficult problem of probability amplification, we refer the reader to Montoya and Müller [536].

32.8   Exercises

Exercise 32.8.1 Prove Theorem 32.1.2.

Exercise 32.8.2 (Thurley [646]) Formulate a counting version of HITTING SET and show that it is FPT.

Exercise 32.8.3 (Cygan [185]) Using iterative compression, or otherwise, prove that the following is FPT.

#CONNECTED VERTEX COVER
Input: A graph G.
Parameter: A positive integer k.
Output: The number of size k connected vertex covers of G.

Exercise 32.8.4 Prove Theorem 32.2.1.

Exercise 32.8.5 Formulate precise definitions for the statement of, and prove, Theorem 32.5.1. Part of this will be to demonstrate that #k-DOMINATING SET is #$W[2]$-complete.

Exercise 32.8.6 (McCartin [522]) Prove that #PERFECT CODE is #$W[1]$-complete.

Exercise 32.8.7 Use the method described to show that #$W[P]$ is hard for $W[P]$ under randomized parameterized reductions.

Exercise 32.8.8 Prove that the following re-parameterization of counting k-cycles is FPT.

Bounded #k-(DIRECTED) CYCLE (PATH)
Input: A (directed) graph G.
Parameter: Positive integers k, ℓ.
Question: Does G have at least ℓ k-cycles?

Exercise 32.8.9 (Lin and Yijia Chen [492])   Consider the following problem.

EDGE INDUCED SUBGRAPH
Input: A graph G.
Parameter: A positive integers k.
Question: Does G have at k edge induced subgraph?
 (i) Prove that EDGE INDUCED SUBGRAPH is FPT.
      (Hint: This is not easy. The only known algorithm uses the Frick–Grohe Metatheorem on
      first-order logic on bounded degree subgraphs and Ramsey-type theorems.)
 (ii) Prove the counting version #EDGE INDUCED SUBGRAPH is #$W[1]$ hard, by a
      reduction from #INDEPENDENT SET.

32.9   Historical Notes

The first explicit work on the hardness of parameterized counting was in [310], and
in McCartin's Ph.D. thesis [522] eventually published in [523]. The hardness of
counting cycles was proven in [310]. It is unknown if this reduction can be made
into an m-reduction or even a parsimonious m-reduction. There are several other
notable #P-complete problems for which the existence problem is easy, which have
no parameterized analysis, as the techniques seem very hard. The Lin–Chen result
in the Exercises is one of the few other hardness results in this area. There is some
systematic positive work in this area, particularly on some kernelizations, and in
counting in graphs of bounded widths. The work of Courcelle, Frick, Grohe, Flum
and Makowsky and others here is discussed in Chap. 14. There is still a lot of
work being done in this respect. For example, Cygan [185] showed that counting
connected vertex covers is FPT. As our exercise says, this is most easily done using
iterative compression. Damaschke [194] has demonstrated that bounding degree can
accelerate some FPT results for decision and for counting. Arvind and Raman [48]
and later Thurley [646] and others showed how some elementary techniques can be
modified for counting problems. There are too many results to state them all here,
but it suffices to say that there are good research programmes here. Randomization
as a positive technique certainly pre-dated explicit parameterized complexity, and
has been a mainstay of the area, as witnessed by color-coding and multilinear detection. The development of positive methodology for elementary techniques has been
a bit slower. One notable example is the parameterized problem

DENSE k-VERTEX SUBGRAPH
Input: A graph G integers k, m.
Parameter: A positive integer k.
Problem: Find a set of k vertices inducing at least m edges.

This problem is FPT by general metatheorems on graphs of bounded degree d (as it
is first order), but Leizhen Cai, Chan, and Chan [119] gave an elementary randomized algorithm running in time $O^*(2^{d(k+1)})$. They used a method called the *random
separation method*. This method has also been exploited by Cai and Yang [129]

on certain hereditary problems. Chitnis, Cygan, Hajiaghayi, Marcin Pilipczuk and Michał Pilipczuk [158] gave an elementary randomized $O^*(2^{O(q+k)})$ algorithm for the problem below (using the "randomized contraction method").

BALANCED SEPARATION

Input: A graph G integers k, q.
Parameter: Positive integer k, q.
Problem: Find a set S of $\leq k$ vertices such that $G \setminus S$ has two components of size at least q.

Lokshtanov and Marx [517] demonstrated that the problem (p, q)-CLUSTERING below, can be solved using elementary randomized methods in time $O^*(2^{O(q)})$.

(p, q)-CLUSTERING

Input: A graph G integers p, q.
Parameter: A positive integer q.
Problem: Partition $V(G)$ into $V_1, \ldots, V_m$ such that for each i, $|V_i| \leq p$ and at most q edges leave V_i.

There are many other examples, and this is certainly an area of great current interest, particularly relating to things like k-MULTIWAY CUT.

The first systematic treatment of the negative theory was Downey, Fellows, and Regan [256]. At the time they were interested in not only an analog of Valiant–Vazirani but of Toda's Theorem, which states that the polynomial-time hierarchy is a subset of the problems reducible to #P. It is unclear exactly what such an analog would look like. Two possibilities discussed in [256] were

(1) $N[t] \subseteq BP \cdot \oplus \cdot G[t]$ and
(2) $H[t] \subseteq BP \cdot \oplus \cdot G[t]$.

Another possibility might be that $AW[*] \subseteq FPT^{\#W[P]}$. The problem is that *every* known proof of Toda's Theorem uses randomization in an essential way, and then amplification. In [256], the authors stated that here are several obstacles to a proof of, for example, the statement (1) above. Among these is the matter that the proof of Theorem 32.7.1 uses $kn \log n$ random bits, while the definition of the BP· operator provides only $k \log n$ random bits. Furthermore, a method of probability amplification would be needed (also employing only $k \log n$ random bits). Downey, Fellows, and Regan did not know how to achieve this with weft-1 circuits. At the time, they mentioned that the question (of whether (1) and (2) hold) is quite interesting, since together with Theorem 32.7.1 the answers could show that UNIQUE CLIQUE is as hard any parameterized problem in the $W[t]$ hierarchy. The challenge was taken up by Montoya and Müller. Müller showed that several "unique" combinatorial problems were hard under randomized reductions. We have chosen a different approach here based on parsimonious reductions plus the basic unique $W[t]$ hardness results. Montoya and Müller systematically developed and clarified randomized parameterization. They explored the question of amplification. For example, Montoya [535] showed that for any computable function g, a parameterized L is in $BP \cdot \oplus \cdot FPT$ iff $(x, k) \in L$ can be decided with a $W[P]$-randomized algorithm with (two sided) error $|x|^{-g(k)}$. The best proof of this can be found in Müller and Montoya [536],

Sect. 4, where expander graphs are used. Nevertheless, the analog of Toda's Theorem remains elusive. Müller and Montoya [536] also prove that if $W[P] = \text{FPT}$ then $\text{BP} \cdot \oplus \cdot \text{FPT} = \text{FPT}$, the analog of the Sipser–Gács Theorem that if $P = NP$ then $\text{BPP} = P$. Space considerations preclude us from including this material, and we refer the reader to [535] for the latest piece of this puzzle.

32.10 Summary

We introduced parameterized counting and randomized parameterized reductions. Most $W[t]$-complete problems were observed to have #-analogs. For example, #CLIQUE was proven to be #$W[1]$-complete. Using this, Flum and Grohe proved that counting k-cycles is also #$W[1]$-complete using parameterized counting Turing reductions. Randomized reductions allowed for an analog of Valiant–Vazirani's Theorem; namely that UNIQUE $W[t]$ is hard for $W[t]$ under randomized reductions.

Part VIII
Research Horizons

Research Horizons

33

In this final chapter, we describe what we think are some of the most important open problems of the field. In the first book, Downey and Fellows [247], we offered two lists of problems, 18 altogether, that we then thought significant and especially challenging. As this book goes to press, 12 of these have been resolved! Many of the solutions to these problems involved significant new ideas and advances in the field.

Interestingly, ten of the 12 solutions yielded FPT results (even for some problems that we originally conjectured to be $W[1]$-hard). The fate (and current status) of all these problems has been nicely reviewed by Fedor Fomin and Daniel Marx, who generously assisted with the making of these new lists [327].

We offer here four lists of problems. For the sake of continuity with the original effort, any problems on the lists in the first book that have not yet been settled, are included somewhere here.

33.1 The Most Infamous

We list here a few open problems that have absorbed massive unsuccessful efforts by various researchers, and apparently require fresh ideas. All of these truly meet the bar for *infamous open problems* of the field.

$K_{t,t}$ SUBGRAPH

Instance: A graph G and positive integer t.
Parameter: t.
Question: Does G contain the complete bipartite graph $K_{t,t}$ as a subgraph?

Almost everyone considers that this problem should obviously be $W[1]$-hard, and almost everyone quickly finds the same easy (but erroneous) proof. It is rather an embarrassment to the field that the question remains open after all these years!

R.G. Downey, M.R. Fellows, *Fundamentals of Parameterized Complexity*,
Texts in Computer Science, DOI 10.1007/978-1-4471-5559-1_33,
© Springer-Verlag London 2013

The problem is equivalent to a number of other problems, perhaps the most attractive of which is the following seemingly elemental problem about hypergraphs: given a hypergraph $\mathcal{H} \subseteq 2^X$, can one find r elements $H_1, H_2, \ldots, H_r \in \mathcal{H}$, such that $|\bigcap_{i=1}^{r} H_i| \geq k$? Here the parameter is (k, r).

If, to everyone's amazement, this problem is actually in FPT, then one could try to sneak up on this by considering restrictions on X and the kinds of subsets $H \subseteq X$ that are allowed in $\mathcal{H}$. Nothing much is known.

On the other hand, if $K_{t,t}$ SUBGRAPH is $W[1]$-hard, then there is an excellent possible extension. Note that the family of graphs $\mathcal{F} = \{K_{t,t} \mid t \in \mathbb{N}\}$ has unbounded treewidth. By the Plehn–Voigt Theorem of Chap. 15, we know that if H is a graph of treewidth at most w, it can be determined in time $f(|H|)|G|^{w+1}$ whether H is a subgraph of a graph G. It follows that if $\mathcal{F} = \{H_t \mid t \in \mathbb{N}\}$ is a computable family of graphs of treewidth bounded by a fixed constant w, then the following problem is FPT.

FAMILY $\mathcal{F}$ SUBGRAPH

Given: A computable family $\mathcal{F}$ of graphs $\{H_t \mid t \in \mathbb{N}\}$.
Input: A graph G.
Parameter: t.
Question: Is H_t a subgraph of G?

The big open question is the following. *Is* FAMILY $\mathcal{F}$ SUBGRAPH FPT *if and only if $\mathcal{F}$ is a family of graphs of uniformly bounded treewidth?*

EVEN SET

Instance: An undirected red/blue bipartite graph $G = (\mathcal{R}, \mathcal{B}, E)$.
Parameter: A positive integer k.
Question: Is there a nonempty set of at most k vertices $R \subseteq \mathcal{R}$, such that each member of $\mathcal{B}$ has an even number of neighbors in R?

The exact version of the problem, where $|R| = k$, is $W[1]$-hard [244]. Vardy [657] proved the NP-completeness of the problem in 1997, settling a longstanding open problem. There are other equivalent ways of stating the problem, showing that this problem appears naturally in many contexts:

- Given a hypergraph H, is there a nonempty set S of at most k vertices, such that $|e \cap S|$ is even for every hyperedge e?
- Given a matrix A over the two-element field GF[2], is there a nonzero vector $\mathbf{x}$ having at most k nonzero coordinates and satisfying $A\mathbf{x} = 0$?
- Given a binary linear code defined by a matrix A over GF[2], are there two codewords with Hamming-distance at most k?
- Given a binary matroid represented by a matrix A over GF[2], does it have a cycle of length at most k?

Related material on linear codes can be found in Downey, Fellows, Vardy, and Whittle [259].

FPT APPROXIMATION FOR DOMINATING SET

Input: A graph G and a positive integer k.
Parameter: k.
Output: Either the correct information that G has no dominating set of size at most k, or the correct information (perhaps with a witness) that G does have a dominating set of size $\leq g(k)$.

The research question is whether there is any function g such that the above specification admits an FPT algorithm. One can ask similarly about an FPT APPROXIMATION FOR INDEPENDENT SET. One might suspect, but it is unknown, whether either of these problems reduces to the other.

CLIQUEWIDTH

Input: A graph G and a positive integer k.
Parameter: k.
Question: Is the cliquewidth of G bounded by k?

As noted in Sect. 16.5, there is an FPT approximation algorithm for this problem. Is it FPT?

POLYNOMIAL KERNEL FOR DIRECTED FEEDBACK VERTEX SET

Input: A directed graph G and a positive integer k.
Parameter: k.
Question: Does G have a directed feedback vertex set of size at most k?

The problem was shown to be in FPT by Chen, Liu, Lu, O'Sullivan, and Razgon in 2008 [150]. The running time of the algorithm is $4^k k! n^{O(1)}$. It remains open if there exist a single exponential algorithm for DIRECTED FEEDBACK VERTEX SET even on planar graphs.

The research question we highlight here is whether the problem admits a polynomial (many : 1) kernel. If not, one could further ask if it admits polynomial Turing kernelization. This is arguably the most famous concrete open problem concerning polynomial kernelization.

33.2   Think Positive!

At various points in this book, we have tried to stress what we see as a key revelation of modern mathematical algorithmics: that fixed-parameter tractability and deep combinatorial structure theory seem to be close fellow-travelers, with graph minor theory being the primordial example. This we have termed *Langston's Thesis*, as Langston was the first to have stressed this. By now, there are wonderful further examples, such as the recent FPT solution to the open problem posed in the first book concerning PLANAR DIRECTED DISJOINT PATHS by Cygan, Marx, Pilipczuk, Pilipczuk [189] and the truly spectacular (recently announced, but yet unpublished) result of Geelen, Gerards, and Whittle [343] that MATROID MINOR TESTING is in FPT for representable matroids. Another nice example is the recent

FPT result on permutation pattern matching by Guillemot and Marx (where the parameter is the size of the pattern), based on a new "width measure" structure theory of permutations [364].

Here we list a few open problems for which we conjecture positive outcomes that may similarly depend upon the development of new and deep combinatorial structure theory, seriously augmenting the algorithmic toolkit of the field.

POLYNOMIAL KERNELS FOR WITHIN-k-VERTICES OF MINOR IDEAL $\mathcal{F}$

Input: A graph G and a positive integer k.
Parameter: k.
Question: Is it possible to delete at most k vertices of G to obtain a graph $G' \in \mathcal{F}$?

The research question here is whether this problem always admits a polynomial kernel, for any minor-closed class (ideal) $\mathcal{F}$. Note that for any minor-closed class (ideal) $\mathcal{F}$, the problem is in nonuniform FPT by the Graph Minor Theorem. The conjecture that the problem admits a polynomial kernel for every minor ideal $\mathcal{F}$ essentially rests on the intuition that a deepening of graph minor theory around polynomial kernelization may be possible.

BOUNDED DEGREE GRAPH ISOMORPHISM

Input: Graphs G and H of maximum degree d.
Parameter: d.
Question: Are G and H isomorphic?

The problem was shown to be in XP by Luks [502]. We conjecture that this problem may be in FPT, but requiring substantial new investigations of the parameterized complexity of problems about permutation groups, perhaps a whole new structure theory in this area. A similar question can be asked for GRAPH ISOMORPHISM PARAMETERIZED BY TREEWIDTH.

BRANCHWIDTH OF REPRESENTABLE MATROIDS

Input: A matroid M and a positive integer k.
Parameter: k.
Question: Is the branchwidth of M bounded by k?

For the definition of matroid branchwidth, see Geelen, Gerards, and Whittle [342], or Hliněný and Whittle [397]. Oum and Seymour [551] described an FPT approximation algorithm for this problem. Is this problem FPT? We conjecture so. We think a positive outcome will depend on a deeper algorithmic structure theory of representable matroids.

MINIMUM LINEAR ARRANGEMENT PARAMETERIZED BY VERTEX COVER NUMBER

Input: A graph G, of vertex cover number at most k, and a positive integer m.
Parameter: k.
Question: Does G have a layout of *minimum linear arrangement* cost at most m?

Our enthusiasm for this concrete problem is based on its connection to INTEGER LINEAR PROGRAMMING. The problem above is easily reducible to a restricted form of QUADRATIC INTEGER PROGRAMMING that may well be FPT. Conceivably, it might also be approached by the recent (and deep) FPT results of Hemmecke, Onn and Romanchuk [387] concerning nonlinear optimization. It is known that this problem admits an EPTAS where the parameter is $(\frac{1}{\varepsilon}, k)$. We think that the problem is FPT, but that the positive resolution is likely to require new general tools.

CHAIN MINOR ORDERING

Input: A finite poset Q.
Parameter: A finite poset P.
Question: Is P a chain minor of Q?

Recall from Chap. 17 that this is defined as follows. Let $P = (V, <)$ be a poset. A chain is a sequence of elements $x_1 < x_2 < \cdots < x_n$. We say that $P = (V, <)$ is a chain minor $P' = (V', <)$ if there is a partial mapping $\rho : V' \to V$ with the following property: for every chain C of P, there is a chain C' of P' such that ρ restricted to C' is an isomorphism of chains from C' to C. Gustedt [369] showed that the problem is in XP and that the chain minor relation is a well-quasi-ordering.

A recent paper by Blasiok and Kaminski [68] has shown this order-testing problem is in FPT. However, their algorithm requires time $O^*(2^{O(2^{O(k)})})$. Is there an order-testing algorithm that runs in time $O^*(2^{O(k)})$? As Blasiok and Kaminski remark, such an improvement may depend on a deeper structure theory of chain minors of posets.

33.3 More Tough Customers

Here we suspect negative results. A goodly number of these are passed along from the lists in the first book, and we add a few more. As pointed out by Fomin and Marx in their review of the problem lists in the first book—while developments in FPT techniques have zoomed ahead, methods for showing likely non-FPT-ness (e.g. $W[1]$-hardness) have remained largely stationary, except for some slightly more efficient patterns of gadgeteering.

But there is a new light on the horizon in regards hardness demonstrations—the emergence of (forms of) the Exponential Time Hypothesis as a workable and more powerful intractability hypothesis. It might seem heretical, but for working purposes in investigating likely non-FPT problem complexity, it seems to us that ETH (and its variants) are the new main tool.

Given the above as a methodological preface, we expect negative outcomes for the following parameterized problems, many inherited from the list in [247].

But we like surprises!

POLYNOMIAL PRODUCT IDENTITY

Input: Two sets of k multivariate polynomials p_i and q_i for $i = 1, \ldots, k$.
Parameter: k.
Question: Does the following identity hold?

$$\prod_{i=1}^{k} p_i = \prod_{i=1}^{k} q_i ?$$

The polynomials in the input are given by listing the monomials with nonzero coefficients. Note that there is no bound on the number of variables or on the degree of the polynomials. By multiplying out each product, we get at most n^k monomials and we can compare the two sides to test for equality. Therefore the problem is in XP.

As discussed in Sect. 8.4, the Schwartz–Zippel Lemma provides a way of solving the problem in randomized polynomial time and therefore it is in randomized FPT. Is the problem in deterministic FPT? We think that there is an insuperable barrier.

BOUNDED HAMMING WEIGHT DISCRETE LOGARITHM

Input: An n-bit prime p, a generator g of F_p^*, an element $a \in F_p^*$.
Parameter: A positive integer k.
Question: Is there a positive integer x whose binary representation has at most k
                   1's (that is, x has a Hamming weight of k) such that $a = g^x$?

Here F_p^* is the multiplicative group of nonzero integers modulo p. Element $g \in F_p^*$ is a generator of group F_p^* if for every element a, there exists an integer x with $a = g^x$. Note that there is a unique $1 \leq x \leq p - 1$ with $a = g^x$, but the problem definition does not insist that x should be less than p. To show that the problem is in XP, we need to argue that the representation of x is at most kn bits long (hence there are at most $(kn)^k$ different possibilities for x to try). See Fellows and Koblitz [294] for discussion and related problems.

The famous DISCRETE LOGARITHM problem is to find the unique $1 \leq x \leq p-1$ with $a = g^x$; the hardness of some cryptosystems are based on the assumed hardness of this problem. Because the problem definition does not require $x \leq p$, it is not completely obvious how the two problems relate to each other.

POLYMATROID RECOGNITION

Input: A k-polymatroid M.
Parameter: k.
Question: Is M hypergraphic?

Let E be a finite set. A polymatroid is a function $\rho : 2^E \to \mathbb{Z}$ with the following properties:
1. $\rho(\emptyset) = 0$,
2. $\rho(A) \leq \rho(B)$ for every $A \subseteq B \subseteq E$, and
3. $\rho(A) + \rho(B) \geq \rho(A \cap B) + \rho(A \cup B)$.

A k-polymatroid is a polymatroid with $\rho(e) \leq k$ for every $e \in E$. Given a hypergraph H with vertex set V and edge set E, the hypergraphic polymatroid of H is a function $\chi_H : 2^E \to \mathbb{Z}$ defined by

$$\chi_H(A) = |\overline{A}| - \kappa(H|A),$$

where $\overline{A}$ is the set of vertices contained in the edge set A, and $\kappa(H|A)$ is the number of components of the hypergraph H restricted to A (see Vertigan and Whittle [658] for more details). A polymatroid is hypergraphic, if it is the hypergraphic polymatroid of a hypergraph.

A word of caution should be said on how the polymatroid is given in the input. One possibility is that it is given by an oracle, but then the problem does not fit the framework of complexity theory defined by problems as languages (but it is still an interesting question if $f(k) \cdot n^{O(1)}$ oracle calls are sufficient for the problem).

SHORTEST VECTOR

Input: A basis $X = \{x_1, x_2, \dots, x_n\} \subset \mathbb{Z}^n$ for a lattice $\mathcal{L}$.
Parameter: A positive integer k.
Question: Is there a nonzero vector $x \in \mathcal{L}$, such that $\|x\|^2 \leq k$?

Here $\|x\|$ denotes the Euclidean (ℓ_2) norm of $x = (a_1, \dots, a_b)$, defined as $\sqrt{\sum_{i=1}^n a_i^2}$. The problem was shown to be NP-hard under randomized reduction by Ajtai in 1998 [20], settling a longstanding open problem. The problem is in XP: every vector x with $\|x\|^2 \leq k$ contains at most k nonzero coordinates. It could be interesting to investigate the problem for other ℓ_p norms as well.

SHORT GENERALIZED HEX

Input: An undirected graph G with two distinguished vertices v_1 and v_2.
Parameter: A positive integer k.
Question: Does player one have a winning strategy of at most k moves in Generalized Hex?

In Generalized Hex two players play on a graph with white and black pebbles. Player one plays with white and player two with black pebbles. Player one starts by placing a white pebble on a vertex of G. Then alternately players make moves, at each move a pebble is placed on an occupied vertex. Player one wins if he can construct a path of white vertices from v_1 to v_2.

To the best of our knowledge, the problem remains open. Downey and Fellows [247] proposed that the problem is a good candidate for AW[*]-completeness. Towards this goal, Allan [607] showed that the problem is in AW[*].

The following problem was mentioned in Exercise 26.5.7. It's complexity is a longstanding problem. The classical problem is known to be in both NP and co-NP.

PARAMETERIZED PARITY GAME

Instance: A parity game with k colors.
Parameter: k.
Question: Does 0 have a winning strategy?

The reader should recall from Exercise 26.5.7 that this problem concerns winning strategies for infinite games played on finite graphs. A *game graph* is a tuple $G - (V = V_0 \sqcup V_1, E, v_0)$, with the directed bipartite graph sink-free. v_0 is the initial vertex. Players 0 and 1, construct an infinite path $p = v_0, v_1, \dots$ called a *play*, where $v_i v_{i+1}$ must be an edge. The *game* is a pair (G, W) where $W \subseteq 2^V$ is a relation called the *winning condition*. Let $\inf(p)$ denote the set of vertices visited infinitely often in the play p. We say 0 *wins* iff $W(\inf(p))$ holds. A *parity game* is one where W is a pair $(c, [k])$ where c is a k-coloring of the vertices so $c(v) \in [k]$. 0 wins if the maximum coloring occurring infinitely often is even.

The point is that Björklund, Sandberg, Vorobyov [67] proved that this problem has the same parameterized complexity as a related game called the PARAMETER-IZED RABIN GAME. The miniature (i.e., $k \log n$) version, of the PARAMETERIZED PARITY GAME, MINI-PARITY GAME, is FPT, but Björklund, et al. have shown that the miniature version of the PARAMETERIZED RABIN GAME is $M[1]$-hard (see Exercise 26.5.7). Thus if a parameterized reduction could be found from MINI-RABIN GAME to PARAMETERIZED RABIN GAME, this would demonstrate the hardness of the PARAMETERIZED PARITY GAME, in spite of the fact that it is probably not NP-complete.

Another problem of some importance comes out of the work of Eiter and Gottlob [269] about hypergraph transversals. We will state this as a logic problem.

MONET

Input: Two monotone formulas X and Y, with X in CNF form and Y in DNF form with term size bounded by k.

Parameter: k.

Question: Is X equivalent to Y?

An equivalent form is: given two hypergraphs with hyperedges of size bounded by k, decide if one consists of the minimal transversals (i.e. a minimal hitting set) of the other. There are many papers demonstrating that restricted versions of this problem are in P. Some results concerning the parameterized complexity of the problem can be found in Elbassioni, Hagen, and Rauf [271].

33.4   Exemplars of Programs

These problems are chosen as representatives of programmatic research directions we think important.

LOCAL SEARCH SPEEDUP FOR THE EUCLIDEAN TRAVELLING SALESMAN PROBLEM

Input: A set S of points in the plane, a tour T through S, and a positive integer k.

Parameter: k.

Question: Is there a tour T' in the k-change neighborhood of T such that $\text{cost}(T') < \text{cost}(T)$?

The investigation of "k-local search" problems was initiated by Marx, who showed that the usual network version of TSP is hard for $W[1]$, even for networks that obey the triangle inequality; see Marx [514].

A number of FPT results are known for similar problems about *speeding up local search*, as well as a number of hardness results. (For example, see Marx [515], Szeider [634] or Fomin, Lokshtanov, Raman, and Saurabh [321].) Jiong Guo observed that most of the hardness proofs depend upon instances where the current solution S is a ridiculously bad solution, and the hardness proof "works" because the problem legislation forces the improved solution S' to be close to S. Marx proposed to mitigate this by exploring a slightly different parameterized problem definition paradigm termed *Permissive Local Search*. Here we are given an instance I and a current solution S. The quest is for an FPT algorithm that either:

- finds a better solution S' (possibly at *arbitrary* distance from S in the solution space); or
- determines that there is no better solution within k classical local search steps from S.

Little is currently known about the parameterized complexity of permissive local search problems.

GRAPH GENUS PARAMATERIZED BY TREEWIDTH
Input: A graph G of treewidth bounded by k, and a positive integer g.
Parameter: k.
Question: Is the genus of G bounded by g?

We suspect that this problem is FPT. It would also be interesting to know if it is FPT when the parameter is the vertex cover number of G. The program here is sometimes called *parameter ecology*. There seem to be many interesting concrete parameterized problems similar to the above to investigate.

IMPROVING THE DIAMETER OF PLANAR GRAPH
Input: A planar graph G and a positive integer k.
Parameter: k.
Question: Can we add edges to G to obtain a planar graph G' such that the diameter of G' is bounded by k?

We do know that the above problem is non-uniformly FPT, by using the fact that for each fixed k the YES instances are a minor order ideal, and hence characterized by a finite obstruction set. (See also the discussion about this problem in Chap. 19.) We also know that for every fixed k, the YES instances have bounded treewidth. The challenge is to find a "reasonable" FPT algorithm for this problem. Programmatically, it represents the interesting research direction of investigating *operators* on ideals in well-quasi-orders, as well as a direction that resembles *contraction bidimensionality* as per Fomin, Golovach, and Thilikos [315].

APPROXIMATING TSP FOR GRAPHS EXCLUDING H AS A MINOR
Input: An edge-weighted graph G, where G excludes H as a minor.
Parameter: $(H, 1/\varepsilon)$.
Question: A traveling salesman tour of cost within a multiplicative factor of $(1 + \varepsilon)$ of optimal.

APPROXIMATE CORRELATION CLUSTERING FOR A TARGETED NUMBER p OF CLUSTERS
Input: A graph G.
Parameter: $(p, 1/\varepsilon)$.
Question: A p-cluster edit involving a number of edge edit operations (either *add* or *delete*) within a multiplicative factor of $(1 + \varepsilon)$ of optimal.

Both of the above important problems are in XP, because they have PTASs by Demaine, Hajiaghayi, Kwarabayashi [215] and Giotis and Guruswami [346]. The question is whether the above problem specifications admit FPT algorithms. In other words, do the corresponding optimization problems admit EPTASs (*efficient PTASs*). Programmatically, there are many PTASs where this question has yet to be explored.

INCREMENTAL DIRECTED FEEDBACK VERTEX SET

Input: Digraphs D and D' where D is obtained from D' by deleting a single arc a, a positive integer k, and a feedback vertex set S for D.

Parameter: k.

Question: Is there a feedback vertex set S' for D' with $|S'| = |S|$ and $d_v(S, S') \leq k$? (Here d_v denotes the *vertex edit distance*.)

We conjecture that this problem is FPT. The reader should note the similarity to the *compression problem*, for an iterative compression approach to the (parameterized) DFVS problem. The significance of the problem is that it could be useful as a subroutine in a greedy heuristic for DFVS. This works as follows:

(1) Order the arcs of the digraph D. (There may be various strategic choices about how to do this, but this is not important for this sketch.)

(2) We begin with the empty digraph, for which the empty set of vertices is a directed feedback vertex set.

(3) Add arcs one at a time until we recover D (we call this the *inductive route*, and it is formally just the same as what is done in iterative compression FPT). If adding an arc creates a cycle, then a simple greedy heuristic would add one more vertex to the current solution.

The problem above captures the strategy, intuitively described: "Maybe by slightly modifying the current solution S, we can avoid increasing the size of the solution."

This problem-definitional paradigm can be deployed pretty much universally, giving the possibility of what we call *FPT-turbocharged* greedy heuristics. A successful example is described by Hartung and Niedermeier [382].

In programmatic terms, we think the investigation of parameterized problems for which FPT algorithms might provide useful subroutines of heuristics is a *very* promising direction for research.

Part IX
Appendices

Appendix 1: Network Flows and Matchings   **34**

In this appendix, we develop some basic machinery which allows us to begin work in *topological graph theory*, an area where the emphasis is on connectivity and "shape".

34.1 Network Flows

We begin with flows in networks. The material is of interest in itself, and might be familiar with students who have done basic operations research, or some basic algorithms course. We will give a brief reminder without proofs.

34.1.1 Basic Results

First we give a typical problem. We want to send oil from s to t, via a sequence of pipes of varying capacity. We wish to maximize this flow.

To model this problem, we associate a directed graph with the pipes. Let the vertices be s and t together with the places where pipes intersect. We give weights to the arcs (edges) which are the capacities of the pipes. Figure 34.1 gives an example of such a weighted-digraph. For the rest of this section, let s be *the source* and t be *the sink*. The graphs should have no loops or multiple edges. The vertex s to vertex t should be strongly connected. All weights should be positive.

Next we develop some notation. Let N be a network, $V(N)$ be its vertex set and $E(N)$ be the edge set of N. Let $x \in V(N)$. The set $\{v \in V(N) : (x, v) \in E(N)\}$, of vertices *after* x will be denoted by $A(x)$. The set $\{u \in V(N) : (u, x) \in E(N)\}$, of vertices *before* x will be denoted by $B(x)$.

In Fig. 34.2, $A(x) = \{p, b\}$ and $B(x) = \{b, a\}$.

Convention is to use *lower case* letters for individual vertices, e.g. p, s, t, a, b. Also, we use *upper case* letters for sets of vertices, e.g. A, B, C. Now we describe the collection of edges between to sets of vertices. Let A and B both be subsets

R.G. Downey, M.R. Fellows, *Fundamentals of Parameterized Complexity*,
Texts in Computer Science, DOI 10.1007/978-1-4471-5559-1_34,
© Springer-Verlag London 2013

Fig. 34.1 A graph of an oil
pipe system

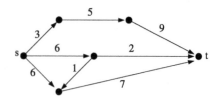

Fig. 34.2 A network

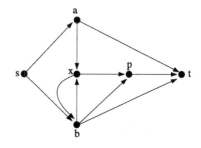

of $V(N)$. Then

$$(A, B) = \big\{(x, y) : x \in A \text{ and } (x, y) \in E(N) \text{ and } y \in B\big\}.$$

An example follows, using the network in Fig. 34.2. Suppose $X = \{s, a, p\}$ and
$Y = \{x, b\}$. Then

$$(X, Y) = \big\{(s, b), (a, x)\big\}$$

and

$$(Y, X) = \big\{(x, p), (b, p)\big\}.$$

Suppose that h is any function, from vertices to vertices, e.g. flow or capacity.
Then for sets A and B, both subsets of $V(N)$, we define

$$h(A, B) = \sum_{(x,y)\in(A,B)} h(x, y).$$

So in the example above, letting $h(a, x) = 3$ and $h(s, b) = 7$, we find that

$$h(X, Y) = h(s, b) + h(a, x) = 10.$$

Some trivial consequences follow. We abuse notation and write a for $\{a\}$ when
the context makes it clear which choice was intended.
1. $(V(N), a) = (B(a), a)$.
2. $(a, V(N)) = (a, A(a))$.
3. $h(X, Y \cup Z) = h(X, Y) + h(X, Z) - h(X, Y \cap Z)$.
   Now we come to a key definition.

Definition 34.1.1 Let N be a network, where every edge (x, y) has a weight $C(x, y)$, called the *capacity* of that edge. A *flow* f in a network N is a function $f : V(N) \times V(N) \to \mathbb{R}$ which satisfies the following three conditions
1. $f(x, y) \geq 0$ for all $(x, y) \in E(N)$. So all flows are non-negative.
2. $f(x, y) \leq C(x, y)$, for all $(x, y) \in E(N)$. So the flow cannot exceed the capacity.
3. There is a fixed positive value q such that, for every $x \in V(N)$,

$$f\big(x, A(x)\big) - f\big(B(x), x\big) = \begin{cases} q & \text{if } x = s, \\ -q & \text{if } x = t, \\ 0 & \text{otherwise.} \end{cases}$$

So, apart from the source and the sink, the flow into any vertex should equal the flow out. We call the value in condition (3) the *value* of the flow.

We continue to build up notation. For any set of vertices X, we write $\overline{X}$ for $V(N) - X$. If $s \in X$ and $t \in \overline{X}$ then we call $(X, \overline{X})$ a *cut*. Next we extend the idea of edge capacity to the capacity of a cut. This will calculate how much you can send from one part of a network to rest. Let $C(X, \overline{X})$ be the *capacity of the cut* $(X, \overline{X})$, and we calculate it as follows:

$$C\big(X, \overline{X}\big) = \sum_{(x,y)\in(X,\overline{X})} C(x, y).$$

Similarly, for a flow f, we denote the *flow of the cut* $(X, \overline{X})$ by $f(X, \overline{X})$ and calculate it as follows:

$$f\big(X, \overline{X}\big) = \sum_{(x,y)\in(X,\overline{X})} f(x, y).$$

The following is straightforward (Exercise 34.1.1).

Theorem 34.1.1 *Suppose f is a flow with value q. Then if $(X, \overline{X})$ is any cut, the following two statements hold*
1. $q = f(X, \overline{X}) - f(\overline{X}, X)$. *So the amount out of X less the amount into X is the amount leaving the source and entering the sink.*
2. $f(X, \overline{X}) \geq 0$ and $q \leq f(X, \overline{X})$.

Corollary 34.1.1 *The maximum value a flow of a network can take, is no larger than the minimum value the capacity of a cut can take.*

Proof Exercise 34.1.2. □

34.1.2 Exercises

Exercise 34.1.1 Prove Theorem 34.1.1.

Exercise 34.1.2 Hence prove Corollary 34.1.1.

Fig. 34.3 Before the path

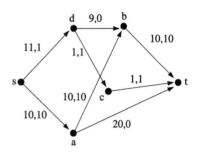

34.1.3 The Ford–Fulkerson Algorithm

The summary of the algorithm below is

Accentuate the positive and eliminate the negative!

This will be done across a minimum cut.

One of the key idea used for this algorithm help maximize a given flow, is how to extend a given flow. Let $s = a_1$ and $t = a_n$.

Definition 34.1.2 An undirected path of vertices $a_1 a_2 \ldots a_n$, with no repetitions, from s to t, is called an *augmenting path* if
1. One of (a_i, a_{i+1}) or (a_{i+1}, a_i) is a directed edge in the network.
2. If $(a_i, a_{i+1}) \in E$ then $f(a_i, a_{i+1}) < C(a_i, a_{i+1})$.
3. If $(a_{i+1}, a_i) \in E$ then $f(a_{i+1}, a_i) > 0$.

Below is an example of an augmenting path in a network flow. Remember that the first number is the capacity of the edge, and the second number is the flow through the edge. An augmenting path is s, d, b, a, t. Note that, even though $f(a, b) = C(a, b)$, in the path the orientation is (b, a), so we only require that $f(a, b) > 0$. The table below explains how we can use the augmenting path to increase the value of the flow.

	$f(X, Y)$	$C(X, Y)$	Possible net change
(s, d)	1	11	10
(d, b)	0	9	9
(b, a)	-10	0	10
(a, t)	0	20	20

As the smallest possible net change is 9, we can augment the flow in Fig. 34.3 by 9 units along the augmenting path. The new, augmented flow, is in Fig. 34.4. Note that we lowered the arc (b, a) by 9, as the path goes the other way.

The Ford–Fulkerson algorithm is one that simply looks for augmenting paths as long as it can, and augments the paths to their capacity. The details can be found in any standard textbook. (There are other methods of maximum flow, but we will not deal with them here.) The key lemma that makes it work is the following, which is proven by running the Ford–Fulkerson algorithm.

Fig. 34.4 After the path

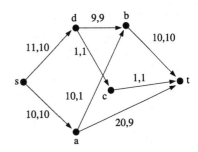

Fig. 34.5 A maximum flow
with a with a min cut

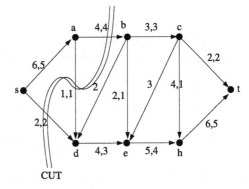

Lemma 34.1.1 *Suppose that a transportation network has a flow from s to t of value q. Then either there is an augmenting path, or there is a cut whose capacity is q.*

Corollary 34.1.2 (Max Flow Min Cut theorem) *In a network, the maximum flow value is equal to the minimum cut capacity.*

Consider the following example. There is no augmenting path. The minimum cut

$$\big(\{s, a\}, \{b, c, d, e, h, t\}\big)$$

is described in Fig. 34.5.

34.1.4 Matching Theory

Next we look at a problem associated with bipartite graphs. Recall that a graph G is bipartite if the vertices of G can be partitioned into two cells X and Y such that all edges are incident with a vertex from each cell. Equivalently, a graph is bipartite if all its circuits contain an even number of vertices. Also, a graph is bipartite if it is two colorable.

Fig. 34.6 People and jobs

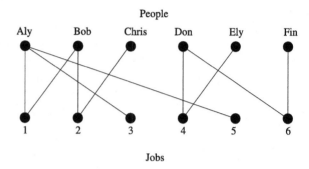

A typical matching problem is the following. There are 6 people and 6 jobs. Some people can only do some jobs. Each job needs just one person. Is it possible to match every person with a job.

A *matching* in a bipartite graph is a collection of edges for which no two edges are incident to a common vertex. A matching in the example above (Fig. 34.6) is

$$\{(Aly, 1), (Bob, 2), (Don, 4), (Fin, 6)\}.$$

A matching is *maximal* if it cannot be extended. So M is a maximal matching if there is no larger matching N that contains M as a subset. A matching is *maximum sized* if there is no larger matching. So no matching in the graph has more edges.

A matching is *complete* if every vertex is incident with an edge in the matching.

We can use network flows to construct maximum matchings. The method is as follows. Let G be a bipartite graph with parts A and B. Orient the edges of G, so that all edges go from A to B. Add two vertices, s and t, to the digraph and add directed edges (s, A) all of value 1, and similarly (B, t). Then it is clear if you run network flow on this graph, all of A can be matched iff the value of the maximum flow is $|A|$.

We also present another algorithm, which is quicker, but does not necessarily find a maximum-sized matching. It will find a maximal matching.

We wish to match $X = \{x_1, \ldots, x_n\}$ to $Y = \{y_1, \ldots, y_m\}$. The algorithm runs in $|X|$ steps. At each step k, match x_k to some y_k not already matched. If you can find the match, then extend your matching. If you cannot do match x_k to anything new, then rematch by undoing the least possible. Figure 34.7 might help.

So the matching found here is

$$(1, e), (2, b), (3, a), (4, f), (5, d), (6, c).$$

The algorithm will find a matching or it may collapse. Remember it does not guarantee a maximum-sized matching, just a maximal matching. By chance, the example in Fig. 34.7, gives a complete matching.

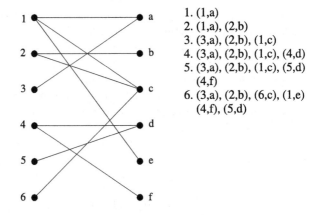

Fig. 34.7 An example of first-fit back-track

1. (1,a)
2. (1,a), (2,b)
3. (3,a), (2,b), (1,c)
4. (3,a), (2,b), (1,c), (4,d)
5. (3,a), (2,b), (1,c), (5,d)
   (4,f)
6. (3,a), (2,b), (6,c), (1,e)
   (4,f), (5,d)

34.2 Berge's Criterion

Some of our results will need matching in general graphs, rather than ones in bipartite graphs. A *matching* of size m in a graph G is a set M of m edges no two of which share a common vertex. A vertex is said to be *matched* by M if it lies in an edge of M.

The basic problem is: given a graph find a matching that is as big as possible.

We will develop techniques for finding a matching that is as big as possible. We remind the reader that there is a difference between a *maximum-sized* matching and a *maximal* matching. If M is a maximal matching, then it is impossible to add edges to M and still have a matching. Remember it is quite possible for a matching to be *maximal but not maximum sized*.

The following is immediate.

Proposition 34.2.1 *In any graph, the greedy algorithm find a maximal matching in linear time.*

Finding maximum-sized matchings in general graphs is *much* more difficult. Before we do this we will need to develop some machinery, useful in its own right.

Recall that if A and B are sets, then the *symmetric difference* of A and B, denoted $A \bigtriangleup B$ is defined by $A \bigtriangleup B = (A \cup B) - (A \cap B)$. Note that $(A \cup B) - (A \cap B) = (A - B) \cup (B - A)$.

Lemma 34.2.1 *Let M_1 and M_2 be matchings of the graph $G = (V, E)$, and let $E' = M_1 \bigtriangleup M_2$. Then the connected components of $G = (V, E')$ are of the following types*:
(1) *a single vertex*;
(2) *a cycle with an even number of edges whose edges are alternately in M_1 and M_2*;
(3) *a path whose edges are alternately in M_1 and M_2 and whose end vertices are matched by one of V_1 or V_2 but not both.*

Proof Consider a component. Assume that it is not of type (1). A vertex in the component has degree at most two since it is incident with at most one edge of M_1 and at most one edge of M_2. Thus the sum of the degrees is at most $2n$. Since the component is connected, it has at least $n - 1$ edges so, by the handshaking lemma, the sum of the degrees is either $2(n - 1)$ or $2n$.

If the sum of the degrees is $2n$, then each vertex has degree 2, so the component is a cycle. Since no vertex can be in two edges of M_1 or M_2, the edges must alternate.

If the sum of the degrees is $2n - 2$, there are $n - 1$ edges and the component is a tree. Since each vertex has degree two or less, it must be a path. Again edges must alternate. End vertices have degree 1, so are matched by M_1 or M_2 but not both. $\square$

Definition 34.2.1 (Alternating path) If M is a matching, then a path $v_0 e_1 v_1 e_2$ $\ldots e_n v_n$ is an *alternating path* for M whenever e_i is in M, e_{i+1} is not, and whenever e_i is not in M, e_{i+1} is in M.

Alternating paths can be used to find a maximum-sized matching. The next two theorems show how.

Lemma 34.2.2 *Let M be a matching in a graph G, and let P be an alternating path with edge set E' starting and ending at unmatched vertices. Then*

$$M \,\Delta\, E'$$

is a matching with one more edge than M.

Proof Every other edge of P is in M. Since P begins and ends at unmatched vertices, for some number k, P has k edges in M and $k + 1$ edges not in M. Let $M' = M \cap P$. The first and last vertices of P are unmatched and all other vertices in P are matched by M', so no edge in $M - M'$ contains any vertex in P. Hence the edges of $M - M'$ have no vertices in common with the edges of $E' - M'$. Also the edges of $E' - M'$ have no vertices in common. Thus

$$\left(M - M'\right) \cup \left(E' - M'\right) = M \,\Delta\, P$$

is a matching. But this matching has $m - k + k + 1 = m + 1$ edges. $\square$

Thus in general graphs we have a notion analogous to that in bipartite graphs and network flows. That is an *augmenting* path for M is an M-alternating paths with first and last vertex not matched. It would seem that all we would need to do is to find augmenting paths to build maximum-sized matchings. Alas, things are not so easy.

We will henceforth always assume that the alternating path has *first* edge and vertex *not* in M. Lemma 34.2.2 tells us that if we can find an alternating path, then we can increase the size of a matching. What if we cannot find an alternating path?

Theorem 34.2.1 (Berge's criterion—Berge [60]) *Let M be a matching of the graph G. Then M is a maximum-sized matching if and only if there is no alternating path connecting unmatched vertices.*

Proof If there is such an alternating path, then M is not maximum sized by Lemma 34.2.2.

Assume that there is no such alternating path. We need to show that M has maximum size. The technique is: let N be a maximum-sized matching. We will show that M and N have the same size. To show that M and N have the same size we show that $M - N$ and $N - M$ have the same size.

Recall that $M \bigtriangleup N = (M - N) \cup (N - M)$. Let G' be the graph on V whose edge set is $M \bigtriangleup N$. We will show that every component of G' has as many edges in $M - N$ as in $N - M$. This will show that $|M - N| = |N - M|$.

By Lemma 34.2.1, components of G' are of three types. Type 1 components consist of a single vertex. No problem. Type 2 components consist of a cycle whose edges alternate between edges in M and N. Clearly such a component has as many edges in M as N. Type 3 components are paths whose edges alternate between M and N. If both end vertices of such a path is matched by N then it is an alternating path for M, contradicting the assumption that no such path exists. Say both end vertices are matched N. Then by Theorem 9.2, we can find a matching with more edges that N, contradicting the fact that N is maximum sized. Hence one end is matched by M and the other by N. It follows that there are as many edges of M in the path as there are of N. $\qquad\square$

34.3 Edmonds' Algorithm

Berge's criterion is technically extremely useful in proving results about matchings. It tells us that the problem of finding a maximum-sized matching reduces to that of finding alternating paths. It is straightforward to modify breadth-first search to find such paths in bipartite graphs, and this is what the back-tracking algorithm does.

For non-bipartite graphs the problem is much harder. Edmonds [267] shows that such paths can be found in polynomial time, so there is a good algorithm in this case too. We now look at Edmonds' algorithm.

Definition 34.3.1 (Edmonds [267]) Let M be a matching in a graph $G = (V, E)$. A *flower* is composed of a *stem*, which is a M-alternating path of even length from an unmatched vertex r called the *root*, to a vertex b, and a *blossom* B from the *base b*. B is a cycle consisting of two edges from b are not in M, and then every second edge in the cycle is in M.

Edmonds' idea is to *shrink* the blossom. Thus, if we find a blossom in a matching of G, then we will shrink the blossom by contracting B to the single point b. Let G_B denote the graph obtained by this contraction, and $M_B = M - B$ the induced matching on G_B.

Lemma 34.3.1 *If G_B contains an M_B-augmenting path, then G contains an M-augmenting path*

Proof Take an augmenting path A in G_P and note that lemma is immediate if it does not involve b. Thus we suppose $b \in A$.

Case 1: b is the first or last element of A. These are symmetric so let bc be the first edge of A. Then $bc \notin M_B$. There must be a vertex $d \in B$ with $dc \in E - M$. By definition of B, there is an even M-alternating path C from b to d, and also by definition this path ends with an edge of M. Then the path $C, dc, \hat{A}$ is an M-alternating path of G, where $\hat{A}$ is the path of A after c.

Case 2: b is internal in A. Then $ab, bc \in A$, say. If $ab \in M_B$ and $bc \notin M_B$, let A_a be the portion of A up to a, and A_c the portion after c. $ab \in E(G)$ and again we can find $d \in B$ and an edge dc such that $dc \in E - M$, and hence even M-alternating path C from b to c. The required path is A_a, ab, C, dc, A_c. The other case is similar. □

Lemma 34.3.2 *Suppose that G contains an M-augmenting path. Then it also has an M_B augmenting path.*

Proof Let the flower be P, b, B. Let A be an M-augmenting path. If A does not meet B, we are done. By reversing A if necessary, we may assume A does not start in the flower. Since A enters the flower, it must enter through and edge not in the matching meeting an edge in the flower in the matching. If this is not in the blossom, then we can use P to finish a derived M_B-augmenting path. If the path M enter the flower in the blossom B, we can route through b and then P to make an augmenting path. This concludes this proof. □

Corollary 34.3.1 *M is a maximum-sized matching in G iff M_B is a maximum-sized matching in G_B.*

This all combined to give the algorithm below. First we need a definition. For a matching M, define a M-alternating walk to be a (perhaps not simple) path with edges alternating in M and out of M.

Algorithm 34.3.1 (Edmonds' matching algorithm)
Input: G and a matching M.
Output: An M-augmenting path or a certificate that M is a maximum-sized matching.
Part 1 Find an M-augmenting simple path or an M-flower.
1. Let X be the set of unmatched vertices. and Y be the set of vertices with an edge not in M to a vertex in X.
2. Build digraph $D = (V, E)$ with $E = \{uv \mid \exists x(ux \in E(G) - M \wedge x, v \in M)\}$.
3. Use breadth-first search in D to find a shortest walk P from a vertex in X to a vertex in Y of length at least 1. Now let P' be P, followed by an edge to a vertex in X. If no such alternating walk exists, then stop, we have a maximum-sized matching. Else:

4. P' is an M-alternating walk between two unmatched vertices. If P' is a simple path it is an M-augmenting path and we augment. If not P is an M-flower and go to Part 2.

Part 2

1. Find M-blossom B on P.
2. Shrink B, and use recursion on G_B and M_B. If G_B has no M_B augmenting path, then M is maximum matching. Otherwise, expand M_B-augmenting path to an M-augmenting path.

With reasonably careful implementation this algorithm yields a $O(|V||E|^2)$ algorithm as it takes about $O(|V||E|)$ time per augmentation. The most efficient algorithm is to be found in Micali and Vazirani [531], and runs in time $O(\sqrt{|V|}|E|)$ time.

34.4 Bipartite Graphs and VERTEX COVER

Let G is a bipartite graph with parts V_1 and V_2. We are concerned with the relationship between VERTEX COVER and matchings in G. The key observation is the following.

- If M is a matching and C is a vertex cover, then $|M| \leq |C|$ because each edge of M contains at least one distinct vertex of C.

Theorem 34.4.1 (König's minimax theorem) *In a bipartite graph the size of a minimum-sized vertex cover and the size of a maximum-sized matching are equal.*

Proof We already know that $|M| \leq |C|$ for any matching M and vertex cover C. We now show that given a maximum-sized matching M we can construct a vertex cover of the same size as M. The theorem will follow from this.

Assume then that M is a maximum-sized matching in the bipartite graph G with parts V_1 and V_2. Let U_1 be the set of *unmatched* vertices in V_1. If U_1 is empty, then $|M| = |V_1|$ and we may let $C = V_1$.

Assume U_1 is non-empty. Let A denote the set of vertices connected by alternating paths to vertices in U_1. Let $A_2 = V_2 \cap A$ and let $A_1 = V_1 \cap A$. In other words,

$$A_2 = \{v \in V_2 | \text{an alternating path connects } v \text{ to a vertex in } U_1\}$$

and

$$A_1 = \{v \in V_1 | \text{an alternating path connects } v \text{ to a vertex in } U_1\}.$$

The set

$$C = A_2 \cup (V_1 - A_1)$$

is a vertex cover since, if an edge contains a member of A_1, it is covered by a member of A_2, while any other edge is covered by a vertex in $V - A_1$.

Also each member of C is matched. This is because members of $V_1 - U_1$ are matched, and $(V_1 - A_1) \subseteq (V_1 - U_1)$ so members of $V_1 - U_1$ are matched. What about members of A_2? If a member of A_2 were not matched, then there would exist an alternating path connecting unmatched vertices in G. By Berge's criterion we would be able to increase the size of M contradicting the fact that M is maximum. So members of A_2 are indeed matched.

Furthermore, each member of A_2 is matched to something in A_1; again because otherwise we could use Berge's criterion to increase the size of the matching. Hence no two members of C lie in the same edge of M. Thus $|C| \le |M|$, so $|C| = |M|$. $\square$

We remark that König's Duality Theorem does *not* work for non-bipartite graphs. It is a is a minimax result. Such results occur often in optimization. Generally minimax results go something like this. You know things of type A have size less than or equal to things of type B. This means that whenever you find a thing of type A with size equal to that of type B, the type A thing is maximum sized and the type B thing is minimum sized.

34.5   Matching and Co-graphic Matroids

In the Sect. 34.1 we solved a basic problem "Find a maximum-sized matching in a bipartite graph", and in this section we will look at an important generalization and meet our old friend the greedy algorithm which you no doubt will have met in a course on finding minimum weight spanning trees.

Now we have a bipartite graph $G = (V_1 \sqcup V_2, E)$ V_1 and the *vertices* in V_1 have weights on them. The goal is to achieve a matching of minimum total weight. For example, V_2 represents jobs needed to be done in a building project. V_1 represents contractors. Each contractor charges an hourly rate, and is qualified to do some jobs but not others. The task is to match contractors to jobs so that the total cost is minimized.

To model these types of problems we assume that in the bipartite graph G, the vertices in V_1 have numbers associated with them, called *costs*. We call a subset X of V_1 *matchable* if it can be matched, that is, if some matching in G matches all the vertices of X.

Algorithm 34.5.1 (Vertex weighted matching algorithm)
1. Let $I = \emptyset$, $X = V_1$.
   Repeat Steps 2 through 4 until X becomes empty.
2. From the set X, pick an element x with minimum cost.
3. If $I \cup \{x\}$ is matchable, replace I by $I \cup \{x\}$.
4. Replace X by $X - \{x\}$.

Note that, in Step 3, we can use the algorithm of the previous section as a subroutine to decide if $I \cup \{x\}$ is feasible. So the algorithm is efficient. Does it work and if so why?

Here is another algorithm that you are probably familiar with. The problem is to find a *minimum weight spanning tree* in an edge weighted connected graph $G = (V, E)$. We recall the following.

Algorithm 34.5.2 (Kruskal's algorithm)
1. Let $I = \emptyset$, $X = E$.
   Repeat Steps 2 through 4 until X becomes empty.
2. From the set X, pick an element x with minimum cost.
3. If $I \cup \{x\}$ is a forest, replace I by $I \cup \{x\}$.
4. Replace X by $X - \{x\}$.

We have totally different problems, but extremely similar algorithms for solving the problems. In fact both algorithms are examples of the *greedy algorithm*. We greedily choose the cheapest possible option, and, if possible, we grab it.

To place these problems in a general setting, let S be a set, and assume that we have a collection of subsets of S that we call *independent*.

- For example in the weighted matching problem, S is the set V_1 and the independent sets are the matchable sets, while in the minimum weight spanning tree problem, S is the set E of edges of the graph and the independent sets are the edge sets of forests.

Each member of S has a cost, and the problem is to find a maximum-sized independent set of minimum cost.

Algorithm 34.5.3 (The greedy algorithm)
1. Let $I = \emptyset$, $X = S$.
   Repeat Steps 2 through 4 until X becomes empty.
2. From the set X, pick an element x with minimum cost.
3. If $I \cup \{x\}$ is independent, replace I by $I \cup \{x\}$.
4. Replace X by $X - \{x\}$.

The greedy algorithm is great because it is very fast—there is no back-tracking. Unfortunately, the greedy algorithm does not always work.

Example Let $S = \{a, b, c, d\}$, where $c(a) = 1$, $c(b) = 100$, $c(c) = 2$, and $c(d) = 3$. Here $c(x)$ is the cost of x. Assume that the independent sets are

$$\{\emptyset, \{a\}, \{b\}, \{c\}, \{d\}, \{a, b\}, \{c, d\}\}.$$

The greedy algorithm applied to this problem chooses $\{a, b\}$ at a cost of 101, whereas the correct solution is $\{c, d\}$ at a cost of 5.

One general setting where is does work is the setting of *matroids*.

Definition 34.5.1 (Matroid)

1. As set S with a collection of subsets of S are *independent sets* if they obey the *subset rule* below.

   Whenever J is independent and $I \subseteq J$, then I is independent. Also $\emptyset$ is independent.

2. A set S with a collection of independent subsets is called a *matroid* it additionally satisfies the *expansion rule* below.

   Whenever I and J are independent, and $|I| < |J|$, then there is an element $x \in J - I$ such that $I \cup \{x\}$ is independent.

A *basis* of a matroid is a maximal independent set. The following is an easy exercise.

Lemma 34.5.1 *If B_1 and B_2 are bases of a matroid, then $|B_1| = |B_2|$.*

Theorem 34.5.1 *Let S be a set with a collection of independent subsets that forms a matroid. Assume the elements of S have costs. Then the greedy algorithm selects a minimum-cost basis.*

Proof It follows from the expansion rule that the greedy algorithm continues to add elements to I until it reaches a maximal independent set, that is, it selects a basis.

Suppose it selects the basis B, and A is another basis. List the elements of B as $(b_1, b_2, \ldots, b_k)$ and A as $(a_1, a_2, \ldots, a_k)$ in order of increasing cost. By Step 2 of the greedy algorithm, $\text{cost}(b_1) \leq \text{cost}(a_1)$. If $\text{cost}(b_i) \leq \text{cost}(a_i)$ for all i, then the cost of B is no more than that of A, so assume that for some i, but for no previous i, $\text{cost}(b_i) > \text{cost}(a_i)$. Then the sets

$$\{a_1, a_2, \ldots, a_i\}$$

and

$$\{b_1, b_2, \ldots, b_{i-1}\}$$

are both independent. But by the expansion property, for some a_j,

$$\{b_1, b_2, \ldots, b_{i-1}, a_j\}$$

is independent. Since $\text{cost}(a_j) \leq \text{cost}(a_i) < \text{cost}(b_i)$, a_j would have been selected by the greedy algorithm; a contradiction. Hence the cost of b_i is no more than the cost of a_i and B is a minimum-cost basis. $\square$

The following shows that the theorem above solves the vertex weighted matching problem.

Lemma 34.5.2 *Let G be a bipartite graph with parts V_1 and V_2. Then the subsets of V_1 that are matchable form the independent sets of a matroid.*

Proof Note that a set is independent if and only if it can be matched. Say $I \subseteq J$, and J is independent. Then, since J can be matched it is clear that I can be matched, so that I is also independent. This shows that the Subset Rule holds.

Now for the Expansion Rule. Suppose that I and J are independent and $|I| < |J|$. Then there is a matching M_1 of I into V_2 and a matching M_2 of J into V_2. Let G' be the graph on $V_1 \cup V_2$ with edge set $M_1 \, \Delta \, M_2$.

We know that each component of G' is one of three types. Since $|M_2| > |M_1|$, at least one of these components must be an alternating path with more edges from M_2 than M_1. Each vertex of this path touched by an edge of M_1 is also touched by an edge of M_2. One vertex x touched by an edge of M_2 is not touched by an edge of M_1. let E' be the set of edges of this path. Now, $M_1 \, \Delta \, E'$ is a matching with one more edge that M_1.

Clearly $M_1 \, \Delta \, E'$ is a matching of $I \cup \{x\}$ into V_2. Hence $I \cup \{x\}$ is independent. By our choice of x we know that x is in J. Thus the Expansion Rule holds and it follows that we have a matroid. $\square$

A *graphic* matroid is one formed by taking the *edges* of a graph and regarding forming a cycle a becoming a dependent set. Thus we can extract Kruskal's Theorem from the following observation.

Lemma 34.5.3 *The forests of a graph $G = (V, E)$ form the independent sets of a matroid.*

Proof Recall that a forest on v vertices with c connected components consists of c trees and hence has $v - c$ edges. Clearly any subset of a forest is a forest, so the subset rule holds. Assume that R and S are forests with r and s edges, respectively, and suppose that $r < s$. If no edge of S can be added to R to give an independent set, then adding any edge of S to R gives a cycle.

This means that each edge of S connects two points in the same connected component of (V, R). Hence each component of (V, S) is a subset of a connected component of (V, R). Now, if R has k connected components, then $r = v - k$, where $v = |V|$. (Note that many components can be isolated vertices.) But S has at least as many components as R, hence $s \leq v - k$, that is, $s \leq r$ a contradiction. We deduce that the expansion rule holds. $\square$

Corollary 34.5.1 *Kruskal's algorithm finds a minimum-cost spanning tree.*

Proof Note that a maximum sized forest of a connected graph is a spanning tree. $\square$

We will also be concerned with a class of matroids formed under duality called *co-graphic matroids*. These have sets S of edges of the graph and are regarded as independent if $G - S$ is connected. The reader is invited to verify that this is indeed a matroid.

34.6   The MATROID PARITY PROBLEM

The MATROID PARITY PROBLEM is a generalization of the a special spanning tree problem for graphs. A *perfect pairing* of the groundset E of a matroid is a partition of E into blocks of size 2. For such a pair $e\overline{e}$, we sometimes call $\overline{e}$ the parity *mate* of e.

MATROID PARITY PROBLEM

Instance: A matroid $(E, \mathcal{F})$ and a perfect pairing $\{a_1\overline{a_1}, \dots, a_n\overline{a_n}\}$ of elements of E.

Task: Find the largest subset $P \in \mathcal{F}$, such that, for all i, either both $a_i \in P$ and $\overline{a_i} \in P$ or neither $a_i \in P$ nor $\overline{a_i} \in P$.

This a problem studied long ago by, for example, Lovász [500, 501]. An important example of this problem is the *spanning tree parity problem* which takes an undirected graph with edges partitioned into pairs, find a spanning forest with as many pairs as possible.

Space considerations preclude us from including a proof of the following important result.

Theorem 34.6.1 (Gabow and Stallman [334]) *The MATROID PARITY PROBLEM for co-graphic matroids is solvable in polynomial time. More precisely, it is solvable in time $O(mn \log^6 n)$ where m is the size of $\mathcal{F}$.*

The proof is not that difficult but somewhat technical, and relies basically on adapting Edmond's paths, trees, and flowers to matroids, and reducing to the linear case. Similar methods have been used by others, such as Orlin [550]. We will use the Gabow and Stallman result in Chap. 5.

Appendix 2: Menger's Theorems

35

35.1 Connectivity

Recall that a graph G is connected iff for all vertices x and y or G, there is a path joining x to y. Similarly if G is a directed graph then G is connected if there is a directed edge from x to y in G. We say that G is n-connected if for all such x and y there are n disjoint paths connecting x to y. We can also specialize this to two particular vertices v and w of G. Namely v and w are called n-connected iff there are at least n (edge) disjoint paths connecting v to w in G.

Suppose that G is a nontrivial graph with an isthmus. Then G cannot be 2-connected! Menger's Theorems tells us that such *separations* determine connectivity.

A set of edges E is called an *n-edge separation* or *n-edge cutset* iff the removal of these n edges disconnects the graph. Similarly for a fixed v and w we say that E is an *n*-cutset or *n*-separation of v and w iff the removal of these n edges disconnects v and w.

Theorem 35.1.1 (Menger [529]) *If the minimum sized v, w cutset in a directed (multi-)graph G has size n, then v is n-edge connected to w in G.*

Proof Let $G = (V, D)$ be a directed graph and v, w in G as above, satisfying the hypotheses of the theorem.

Now regard v as the source and w as the sink, and give each edge capacity 1. Let E be a minimal v, w edge cutset in G. Thus if all but one of the edges is removed then is a path from v to w and yet if we remove them all then there is no such path. Suppose that U is the set of all vertices x such that there is a path from v to x in $(V, D - E)$, and let C be the partition $(U, V - U)$. Then C is a cut. Moreover the only edges from U to $V - U$ are in E and hence the capacity of C is $|E| = n$.

R.G. Downey, M.R. Fellows, *Fundamentals of Parameterized Complexity*,
Texts in Computer Science, DOI 10.1007/978-1-4471-5559-1_35,
© Springer-Verlag London 2013

Summarizing, if the minimum v, w cutset has size n then the minimum cut capacity is n. Now by the Ford–Fulkerson Theorem[1] there is a flow of size n from v to w.

We need to prove that this means that the flow constitutes n *disjoint* paths from v to w. This is an easy induction as we now see. Let S be the set of all k such that there are k disjoint paths from v to w in G. Clearly 1 is in S. Assume that m is in S. Then there these m paths constitute a flow of size m. Remove the edges of these m paths. Call the result $\hat{G}$. Then there is still a flow of $n - m$ in $\hat{G}$. Hence if $m < n$, there is a path from v to w in $\hat{G}$, and this is disjoint from S. □

Menger's Theorem has a more standard vertex separation version. We say that a set of vertices S is an n-vertex cutset or n separation in G if the removal of the n vertices (and of course the edges from them) of S disconnects G.

Theorem 35.1.2 (Menger [529]) *If the minimum sized v, w vertex separation in a directed graph G has size n, then v is n-(vertex-)connected to w in G.*

Proof Take each vertex x of G and replace it by an edge $x_1 x_2$ and then for any edges xy of G replace it by $x_2 y$, and any edge px replace it by px_1. Thus we "split" the vertices. This forms G'. A vertex cutset in G is an edge cutset in G'. □

Menger's Theorem is also true in the undirected case.

Theorem 35.1.3 (Menger [529]) *If the minimum sized v, w cutset in undirected graph G has size n, then v is n-edge connected to w in G.*

Proof Take G and for each edge $e = xy$ of G replace it by a pair of directed edges xy and yx. call the result G'. Notice that then an n-edge cutset in G will be a v, w-cutset of size $2n$ in G'. Now, again note that the reasoning of the first theorem shows that the maximum capacity of a cut in G' is $\leq 2n$ and at least n. Thus by Ford–Fulkerson there is a flow of size n from v to w in G', and for each such cut $C = (U, G' - U)$, this flow has all edges from U to $G' - U$ saturated, and all from $G' - U$ to U zero. Below we will show that in any such flow we can only use one of xy and yx, and hence the value of the flow will be n and we can lift the flow back to the undirected case.

Specifically, for any other edge pq which has been replaced by two pq and qp, suppose that some path of the flow uses both pq and another uses qp. Let these be

$$v = x_0 \ldots p(pq)q \ldots w, \quad \text{and}$$
$$v = x_0 \ldots q(qp)p \ldots w.$$

---

[1] Of course, the original proof of this result predated the Ford–Fulkerson Max-Flow/Min-Cut Theorem. In fact, Menger's original ideas were part of the proof of the Ford–Fulkerson Theorem.

Then we can replace these paths by two that use neither pq nor qp. Namely, start at v, travel to p using the first path, and then to w using the second, and for the other one, travel from v to q using the second, then from q to w using the first. Thus we can eliminate all simultaneous occurrences of both pq and qp, meaning that the paths will correspond to edge disjoint paths of G.   □

The reader should prove for themselves the vertex form of Menger's Theorem for undirected graphs (Exercise 35.2.1).

35.2   Exercises

Exercise 35.2.1 Prove the vertex version of Menger's Theorem for undirected graphs.

Exercise 35.2.2
(i) Use the Ford–Fulkerson algorithm to find the largest n such that a is n edge connected to b in the directed graph below.

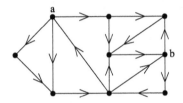

(ii) Find the maximum sized vertex disjoint paths in from a to b.

References

1. J. Abello, M. Fellows, J. Stillwell, On the complexity and combinatorics of covering finite complexes. Australas. J. Comb. **4**, 103–112 (1991) (p. 354)
2. K. Abrahamson, R. Downey, M. Fellows, Fixed-parameter intractability II, in *Proceedings of 10th Annual Symposium on Theoretical Aspects on Computer Science, STACS 93*, Würzburg, Germany, February 25–27, 1993, ed. by P. Enjalbert, A. Finkel, K. Wagner. LNCS, vol. 665 (Springer, Berlin, 1993), pp. 374–385 (pp. 481, 485, 492, 496, 536)
3. K. Abrahamson, R. Downey, M. Fellows, Fixed-parameter tractability and completeness. IV. On completeness for $W[P]$ and PSPACE analogues. Ann. Pure Appl. Log. **73**(3), 235–276 (1995) (pp. 474, 475, 481, 485, 487, 488, 492, 495, 496, 501, 502, 503, 506, 536, 538, 542, 553, 565, 569, 641, 745)
4. K. Abrahamson, M. Fellows, Cutset regularity beats well-quasi-ordering for bounded treewidth, Preprint (1989) (pp. 250, 263, 277, 333, 334, 338)
5. K. Abrahamson, M. Fellows, Finite automata, bounded treewidth, and well-quasiordering, in *Proceedings of the AMS Summer Workshop on Graph Minors, Graph Structure Theory*, ed. by N. Robertson, P. Seymour. Contemporary Mathematics, vol. 147 (Am. Math. Soc., Providence, 1993), pp. 539–564 (pp. 234, 248, 254, 258, 260, 263, 277, 333, 334, 335, 338)
6. K. Abrahamson, M. Fellows, J. Ellis, M. Mata, On the complexity of fixed parameter problems, in *Proceedings of 30th Annual Symposium on Foundations of Computer Science, FOCS 1989*, Research Triangle Park, North Carolina, USA, 30 October–1 November 1989 (IEEE Comput. Soc., Los Alamitos, 1989), pp. 210–215 (pp. 21, 405, 406, 474, 485, 488, 567, 568)
7. K. Abrahamson, M. Fellows, M. Langston, B. Moret, Constructive complexity. Discrete Appl. Math. **34**(1–3), 3–16 (1991) (p. 358)
8. F. Abu-Khzam, A quadratic kernel for 3-set packing, in *Theory and Applications of Models of Computation, Proceedings of 6th Annual Conference, TAMC 2009*, Changsha, China, May 18–22, 2009, ed. by J. Chen, B. Cooper. LNCS, vol. 5532 (Springer, Berlin, 2009), pp. 81–87 (p. 141)
9. F. Abu-Khzam, R. Collins, M. Fellows, M. Langston, H. Suters, C. Symons, Kernelization algorithms for the vertex cover problem: theory and experiments, in *ALENEX/ANALC, Proceedings of the Sixth Workshop on Algorithm Engineering and Experiments and the First Workshop on Analytic Algorithmics and Combinatorics*, New Orleans, LA, USA, January 10, 2004, ed. by L. Arge, G. Italiano, R. Sedgewick (SIAM, Philadelphia, 2004), pp. 62–69 (p. 100)
10. F. Abu-Khzam, C. Markarian, A degree-based heuristic for strongly connected dominating-absorbent sets in wireless ad-hoc networks, in *8th International Conference on Innovations in Information Technology*, March 2012 (p. 63)
11. F. Abu-Khzam, A. Mouawad, A decentralized load balancing approach for parallel search-tree optimization, in *Proceedings of the 13th International Conference on Parallel and Distributed Computing, Applications and Technologies (PDCAT-2012)*, Beijing, China, December 2012 (p. 63)

12. A. Adachi, S. Iwata, T. Kasai, Classes of pebble games and complete problems. SIAM J. Comput. **8**(4), 576–586 (1979) (pp. 510, 511, 517, 518)

13. A. Adachi, S. Iwata, T. Kasai, Some combinatorial game problems require $\Omega(n^k)$ time. J. ACM **31**(2), 361–376 (1984) (pp. 510, 511, 517, 518)

14. L. Addario-Berry, M. Hallett, J. Lagergren, Towards identifying lateral gene transfer events, in *Proceedings of Pacific Symposium on Biocomputing 2003*, ed. by R. Altman, K. Dunker, L. Hunter, T. Jung, T. Klein (World Scientific, Singapore, 2003), pp. 279–290 (p. 74)

15. L.M. Adleman, Two theorems on random polynomial time, in DBLP [566], pp. 75–83. http://doi.ieeecomputersociety.org/10.1109/SFCS.1978.37 (p. 586)

16. I. Adler, Directed tree-width examples. J. Comb. Theory, Ser. B **97**(5), 718–725 (2007) (p. 312)

17. I. Adler, Open problems related to computing obstruction sets. Manuscript, Humboldt Universität, September 2008 (p. 372)

18. I. Adler, M. Grohe, S. Kreutzer, Computing excluded minors, in *Proceedings of the Nineteenth Annual ACM–SIAM Symposium on Discrete Algorithms, SODA 2008*, San Francisco, California, USA, January 20–22, 2008, ed. by S.-H. Teng (SIAM, Philadelphia, 2008), pp. 641–650 (p. 372)

19. I. Adler, S. Kolliopoulos, P. Krause, D. Lokshtanov, S. Saurabh, D. Thilikos, Tight bounds for linkages in planar graphs, in *Proceedings of 38th International Colloquium on Automata, Languages and Programming (ICALP 2011)*, Zürich, Switzerland, July 4–8, 2011, ed. by L. Aceto, M. Henzinger, J. Sgall. LNCS, vol. 6755 (Springer, Berlin, 2011), pp. 110–121 (p. 365)

20. M. Ajtai, The shortest vector problem in ℓ_2 is NP-hard for randomized reductions, in *Proceedings of 30th ACM Symposium on Theory of Computing (STOC '98)*, Dallas, TX, USA, May 23–26, 1998, ed. by J. Vitter (ACM, New York, 1998), pp. 10–19 (p. 683)

21. J. Alber, H. Bodlaender, H. Fernau, T. Kloks, R. Niedermeier, Fixed parameter algorithms for dominating set and related problems on planar graphs. Algorithmica **33**(4), 461–493 (2002) (pp. 46, 561)

22. J. Alber, H. Bodlaender, H. Fernau, R. Niedermeier, Fixed parameter algorithms for planar dominating set and related problems, in *Algorithm Theory—SWAT 2000, Proceedings of 7th Scandinavian Workshop on Algorithm Theory*, Bergen, Norway, July 2000, ed. by M. Halldórsson. LNCS, vol. 1851 (Springer, London, 2000), pp. 97–110 (pp. 46, 561)

23. J. Alber, F. Dorn, R. Niedermeier, Experimental evaluation of a tree decomposition-based algorithm for vertex cover on planar graphs. Discrete Appl. Math. **145**(2), 219–231 (2005) (pp. 185, 305)

24. J. Alber, H. Fan, M.R. Fellows, H. Fernau, R. Niedermeier, F.A. Rosamond, U. Stege, A refined search tree technique for DOMINATING SET on planar graphs. J. Comput. Syst. Sci. **71**(4), 385–405 (2005) (pp. 30, 31, 46, 314)

25. J. Alber, M. Fellows, R. Niedermeier, Polynomial-time data reduction for dominating set. J. ACM **51**(3), 363–384 (2004) (pp. 46, 365)

26. J. Alber, R. Niedermeier, Improved tree decomposition based algorithms for domination-like problems, in *LATIN 2002: Theoretical Informatics. Proceedings of 5th Latin American Symposium*, Cancun, Mexico, April 2002, ed. by S. Rajsbaum. LNCS, vol. 2286 (Springer, Berlin, 2002), pp. 613–627 (p. 305)

27. M. Alekhnovich, A. Razborov, Resolution is not automatizable unless W[P] is tractable, in *Proceedings of 42nd Annual Symposium on Foundations of Computer Science, FOCS 2001*, Las Vegas, Nevada, USA, 14–17 October 2001 (IEEE Comput. Soc., Los Alamitos, 2001), pp. 210–219 (pp. 623, 643)

28. B. Allen, M. Steel, Subtree transfer operations and their induced metrics on evolutionary trees. Ann. Comb. **5**, 1–13 (2000) (pp. 70, 71, 81, 89, 626)

29. N. Alon, Ranking tournaments. SIAM J. Discrete Math. **20**, 137–142 (2006) (p. 170)

30. N. Alon, Y. Azar, G. Woeginger, T. Yadid, Approximation schemes for scheduling on parallel machines. J. Sched. **1**, 55–66 (1998) (p. 141)

31. N. Alon, L. Babai, A. Itai, A fast and simple randomized parallel algorithm for the maximal independent set problem. J. Algorithms **7**, 567–583 (1986) (p. 160)

32. N. Alon, F. Fomin, G. Gutin, M. Krivelevich, S. Saurabh, Spanning directed trees with many leaves. SIAM J. Discrete Math. **23**(1), 466–476 (2009) (pp. 91, 106)

33. N. Alon, G. Gutin, E. Kim, S. Szeider, A. Yeo, Solving MAX-r-SAT above a tight lower bound, in *Proceedings of the Twenty-First Annual ACM–SIAM Symposium on Discrete Algorithms, SODA 2010*, Austin, Texas, USA, January 17–19, 2010, ed. by M. Charikar (SIAM, Philadelphia, 2010), pp. 511–517 (pp. 63, 162, 163, 164, 169)

34. N. Alon, S. Gutner, Linear time algorithms for finding a dominating set of fixed size in degenerated graphs. Algorithmica **54**(4), 544–556 (2009) (pp. 31, 314, 316)

35. N. Alon, D. Lokshtanov, S. Saurabh, Fast FAST, in *Proceedings of 36th International Colloquium on Automata, Languages and Programming (ICALP 2009), Part I*, Rhodes, Greece, July 5–12, 2009, ed. by S. Albers, A. Marchetti-Spaccamela, Y. Matias, S. Nikoletseas, W. Thomas. LNCS, vol. 5555 (Springer, Berlin, 2009), pp. 49–58 (pp. 155, 157, 158, 159, 160, 161, 162, 169, 565)

36. N. Alon, R. Yuster, U. Zwick, Color-coding. J. Assoc. Comput. Mach. **42**(4), 844–856 (1995) (pp. 143, 145, 148, 168)

37. N. Alon, R. Yuster, U. Zwick, Color coding, in *Encyclopedia of Algorithms*, ed. by M.-Y. Kao (Springer, Berlin, 2008), pp. 1–99 (p. 169)

38. S. Alstrup, P. Lauridsen, M. Thorup, Generalized dominators for structured programs, in *Proceedings of the Third International Symposium on Static Analysis (SAS '96)*, Aachen, Germany, September 24–26, 1996, ed. by R. Cousot, D. Schmidt. LNCS, vol. 1145 (Springer, Berlin, 1996), pp. 42–51 (p. 567)

39. A. Andrzejak, An algorithm for the Tutte polynomials of graphs of bounded treewidth. Discrete Math. **190**(1–3), 39–54 (1998) (pp. 282, 283, 646)

40. D. Angluin, Local and global properties in networks of processors, in *Proceedings of 12th ACM Symposium on Theory of Computing (STOC '80)*, Los Angeles, California, USA, April 28–April 30, 1980, ed. by R. Miller, S. Ginsburg, W. Burkhard, R. Lipton (ACM, New York, 1980), pp. 82–93 (p. 354)

41. D. Archdeacon, A Kuratowski Theorem for the projective plane, PhD thesis, Ohio State University, 1980 (p. 350)

42. E. Arkin, Complexity of cycle and path problems in graphs, PhD thesis, Stanford University, 1986 (p. 369)

43. S. Arnborg, Efficient algorithms for combinatorial problems on graphs with bounded decomposability—a survey. BIT Numer. Math. **25**(1), 2–23 (1985) (p. 266)

44. S. Arnborg, D. Corneil, A. Proskurowski, Complexity of finding embeddings in a k-tree. SIAM J. Algebr. Discrete Methods **8**(2), 277–284 (1987) (p. 202)

45. S. Arnborg, B. Courcelle, A. Proskurowski, D. Seese, An algebraic theory of graph reduction. J. ACM **40**(5), 1134–1164 (1993) (p. 371)

46. S. Arnborg, J. Lagergren, D. Seese, Easy problems for tree-decomposable graphs. J. Algorithms **12**(2), 308–340 (1991) (pp. 266, 276, 277)

47. S. Arora, C. Lund, R. Motwani, M. Sudan, M. Szegedy, Proof verification and hardness of approximation problems, in *Proceedings of 33rd Annual Symposium on Foundations of Computer Science, FOCS 1992*, Pittsburgh, Pennsylvania, USA, 24–27 October 1992 (IEEE Comput. Soc., Los Alamitos, 1992), pp. 14–23 (pp. 58, 173, 177)

48. V. Arvind, V. Raman, Approximation algorithms for some parameterized counting problems, in *Algorithms and Computation: Proceedings of 13th International Symposium, ISAAC 2002*, Vancouver, BC, Canada, November 21–23, 2002, ed. by P. Bose, P. Morin. LNCS, vol. 2518 (Springer, Berlin, 2002), pp. 453–464 (pp. 645, 671)

49. G. Bader, C. Hogue, An automated method for finding molecular complexes in large protein interaction networks. BCM Bioinformatics **4** (2003) (p. 313)

50. T. Baker, J. Gill, R. Solovay, Relativizations of the $P = NP$? question. SIAM J. Comput. **4**(4), 431–442 (1975) (p. 567)

51. R. Balasubramanian, M. Fellows, V. Raman, An improved fixed-parameter algorithm for vertex cover. Inf. Process. Lett. **65**(3), 163–168 (1998) (p. 53)

52. J. Balcázar, J. Díaz, J. Gabarró, *Structural Complexity*. EATCS Monographs on Theoretical Computer Science. Book 11, vols. 1 and 2 (Springer, Berlin, 1987/1989) (p. 13)

53. Y. Bar-Hillel, M. Perles, E. Shamir, On formal properties of simple phrase structure grammars. Z. Phon. Sprachwiss. Kommun.forsch. **14**(2), 143–172 (1961) (pp. 230, 234)

54. R. Bar-Yehuda, S. Even, A local-ratio theorem for approximating the weighted vertex cover problem, in *Analysis and Design of Algorithms for Combinatorial Problems*, ed. by G. Ausiello, M. Lucertini. North-Holland Mathematics Studies, vol. 109 (Elsevier Science, Amsterdam, 1985), pp. 27–46. Annals of Discrete Mathematics, vol. 25 (pp. 58, 88)

55. R. Bar-Yehuda, M. Halldórsson, J. Naor, H. Shachnai, I. Shapira, Scheduling split intervals. SIAM J. Comput. **36**(1), 1–15 (2006) (p. 401)

56. C. Barefoot, L. Clark, J. Douthett, R. Entringer, M. Fellows, Cycles of length 0 modulo 3 in graphs, in *Graph Theory, Combinatorics, and Applications*, vol. 1 (Wiley, New York, 1991), pp. 87–101 (p. 369)

57. C. Bazgan, Schémas d'approximation et complexité paramétrée, Rapport de stage de DEA d'Informatique á Orsay, Université Paris-Sud, 1995 (pp. 174, 175, 182)

58. A. Becker, R. Bar-Yehuda, D. Geiger, Randomized algorithms for the loop cutset problem. J. Artif. Intell. Res. **12**, 219–234 (2000) (p. 166)

59. P. Bellenbaum, R. Diestel, Two short proofs concerning tree decompositions. Comb. Probab. Comput. **11**(6), 541–547 (2002) (p. 344)

60. C. Berge, Two theorems in graph theory. Proc. Natl. Acad. Sci. USA **43**(9), 842–844 (1957) (p. 697)

61. D. Berwanger, A. Dawar, P. Hunter, S. Kreutzer, DAG-width and parity games, in *Proceedings of 23rd Annual Symposium on Theoretical Aspects of Computer Science, STACS 2006*, Marseille, France, February 23–25, 2006, ed. by B. Durand, W. Thomas. LNCS, vol. 3884 (Springer, Berlin, 2006), pp. 524–536 (p. 312)

62. S. Bessy, F. Fomin, S. Gaspers, C. Paul, A. Perez, S. Saurabh, S. Thomassé, Kernels for feedback arc set in tournaments. J. Comput. Syst. Sci. **77**(6), 1071–1078 (2011) (p. 167)

63. N. Betzler, M. Fellows, C. Komusiewicz, R. Niedermeier, Parameterized algorithms and hardness results for some graph motif problems, in *Combinatorial Pattern Matching. Proceedings of 19th Annual Symposium, CPM 2008*, Pisa, Italy, June 2008, ed. by P. Ferragina, G. Landau. LNCS, vol. 5029 (Springer, Berlin, 2008), pp. 31–43 (p. 168)

64. N. Betzler, R. Niedermeier, J. Uhlmann, Tree decompositions of graphs: saving memory in dynamic programming. Discrete Optim. **3**(3), 220–229 (2006) (p. 305)

65. D. Bienstock, N. Robertson, P. Seymour, R. Thomas, Quickly excluding a forest. J. Comb. Theory, Ser. B **52**, 274–283 (1991) (pp. 340, 350, 632, 633)

66. A. Björklund, T. Husfeldt, P. Kaski, M. Koivisto, Fourier meets Möbius: fast subset convolution, in *Proceedings of 39th ACM Symposium on Theory of Computing (STOC '07)*, San Diego, California, June 11–June 13, 2007, ed. by D. Johnson, U. Feige (ACM, New York, 2007), pp. 67–74 (p. 305)

67. H. Björklund, S. Sandberg, S. Vorobyov, On fixed-parameter complexity of infinite games, in *The Nordic Workshop on Programming Theory (NWPT'03)*, Åbo Akademi University, Turku, Finland, October 29–31, 2003, ed. by K. Sere, M. Waldén, A. Karlsson (Åbo Akademi, Department of Computer Science, Turku, 2003), pp. 62–64 (pp. 505, 684)

68. J. Blasiok, M. Kaminski, Chain minors are FPT, in *Proceedings of IPEC 2013* (2013) (pp. 333, 681)

69. H. Bodlaender, Classes of graphs with bounded tree-width, Technical Report RUU-CS-86-22, Department of Information and Computing Sciences, Utrecht University, The Netherlands, 1986 (p. 67)

70. H. Bodlaender, Dynamic programming on graphs with bounded tree-width, in *Proceedings of 15th International Colloquium on Automata, Languages and Programming (ICALP 1988)*, Tampere, Finland, July 11–15, 1988, ed. by T. Lepistö, A. Salomaa. LNCS, vol. 317 (Springer, Berlin, 1988), pp. 103–118 (p. 191)

71. H. Bodlaender, On linear time minor tests and depth-first search, in *Algorithms and Data Structures, Proceedings of Workshop WADS 1989*, Ottawa, Canada, August 17–19, 1989, ed. by F. Dehne, J.-R. Sack, N. Santoro. LNCS, vol. 382 (Springer, Berlin, 1989), pp. 577–590 (pp. 293, 294, 300)

72. H. Bodlaender, On disjoint cycles, Technical Report RUU-CS-90-29, Department of Information and Computing Sciences, Utrecht University, The Netherlands, August 1990 (pp. 36, 293)

73. H. Bodlaender, A tourist's guide through treewidth, Technical Report RUU-CS-92-12, Department of Information and Computing Sciences, Utrecht University, The Netherlands, March 1992 (pp. 191, 193)

74. H. Bodlaender, A linear-time algorithm for finding tree-decompositions of small treewidth. SIAM J. Comput. **25**(6), 1305–1317 (1996) (pp. 195, 197, 198, 202)

75. H. Bodlaender, Discovering treewidth, Technical Report RUU-CS-2005-18, Department of Information and Computing Sciences, Utrecht University, The Netherlands, 2005 (p. 191)

76. H. Bodlaender, A cubic kernel for feedback vertex set, in *Proceedings of 24th Annual Symposium on Theoretical Aspects of Computer Science, STACS 2007*, Aachen, Germany, February 22–24, 2007, ed. by W. Thomas, P. Weil. LNCS, vol. 4393 (Springer, Berlin, 2007), pp. 320–331 (p. 130)

77. H. Bodlaender, A tutorial on kernelization, in *Parameterized and Exact Computation. 6th International Symposium, IPEC '11, Revised Selected Papers*, Saarbrücken, Germany, September 6–8, 2011, ed. by D. Marx, P. Rossmanith. LNCS, vol. 7112 (Springer, Berlin, 2011) (p. 82)

78. H. Bodlaender, Fixed-parameter tractability of treewidth and pathwidth, in *The Multivariate Algorithmic Revolution and Beyond: Essays Dedicated to Michael R. Fellows on the Occasion of His 60th Birthday*, ed. by H. Bodlaender, R. Downey, F. Fomin, D. Marx. LNCS, vol. 7370 (Springer, Berlin, 2012), pp. 196–227 (p. 211)

79. H. Bodlaender, B. de Fluiter, Intervalizing *k*-colored graphs, in *Proceedings of 22nd International Colloquium on Automata, Languages and Programming (ICALP 1995)*, Szeged, Hungary, July 10–14, 1995, ed. by Z. Fülöp, F. Gécseg. LNCS, vol. 944 (Springer, Berlin, 1995), pp. 87–98 (p. 457)

80. H. Bodlaender, B. de Fluiter, Reduction algorithms for constructing solutions in graphs with small treewidth, in *Computing and Combinatorics, Proceedings of Second Annual International Conference, COCOON'96*, Hong Kong, June 1996, ed. by J. Cai, C.K. Wong. LNCS, vol. 1090 (Springer, Heidelberg, 1996), pp. 199–298 (p. 371)

81. H. Bodlaender, B. de Fluiter, Reduction algorithms for graphs of small treewidth. Inf. Comput. **167**(2), 86–119 (2001) (pp. 367, 371)

82. H. Bodlaender, R. Downey, M. Fellows, M. Hallett, T. Wareham, Parameterized complexity analysis in computational biology. Comput. Appl. Biosci. **11**(1), 49–57 (1994) (p. 36)

83. H. Bodlaender, R. Downey, M. Fellows, D. Hermelin, On problems without polynomial kernels. J. Comput. Syst. Sci. **75**(8), 423–434 (2009) (pp. 572, 573, 576, 577, 578, 580, 581, 588, 619)

84. H. Bodlaender, R. Downey, M. Fellows, T. Wareham, The parameterized complexity of sequence alignment and consensus. Theor. Comput. Sci. **147**(1–2), 31–54 (1994) (pp. 36, 522)

85. H. Bodlaender, R. Downey, F. Fomin, D. Marx (eds.), *The Multivariate Algorithmic Revolution and Beyond: Essays Dedicated to Michael R. Fellows on the Occasion of His 60th Birthday*. LNCS, vol. 7370 (Springer, Berlin, 2012) (p. 21)

86. H. Bodlaender, P. Evans, M. Fellows, Finite string complexity of annotations of strings and trees, in *Combinatorial Pattern Matching. Proceedings of 7 Annual Symposium, CPM '96*, Laguna Beach, California, USA, June 1996, ed. by D. Hirschberg, G. Myers. LNCS, vol. 1075 (Springer, Berlin, 1996), pp. 384–391 (p. 371)

87. H. Bodlaender, M. Fellows, M. Hallett, Beyond NP-completeness for problems of bounded width (extended abstract): hardness for the w hierarchy, in *Proceedings of 26th ACM Symposium on Theory of Computing (STOC '94)*, Montréal, Québec, Canada, May 23–

May 25, 1994, ed. by F. Leighton, M. Goodrich (ACM, New York, 1994), pp. 449–458. http://dl.acm.org/citation.cfm?id=195229 (pp. 38, 277, 412, 642)

88. H. Bodlaender, F. Fomin, D. Lokshtanov, E. Penninkx, S. Saurabh, D. Thilikos, (Meta) kernelization, in *Proceedings of 50th Annual IEEE Symposium on Foundations of Computer Science, FOCS 2009*, Atlanta, Georgia, USA, October 25–27, 2009 (IEEE Comput. Soc., Los Alamitos, 2009), pp. 629–638 (pp. 365, 366, 367, 368, 369, 371)

89. H. Bodlaender, T. Hagerup, Parallel algorithms with optimal speedup for bounded treewidth. SIAM J. Comput. **27**, 1725–1746 (1998) (p. 371)

90. H. Bodlaender, P. Heggernes, Y. Villanger, Faster parameterized algorithms for minimum fill-in. Algorithmica **61**(4), 817–838 (2011) (p. 210)

91. H. Bodlaender, B. Jansen, S. Kratsch, Cross-composition: a new technique for kernelization lower bounds, in *Proceedings of 28th International Symposium on Theoretical Aspects of Computer Science, STACS 2011*, Dortmund, Germany, March 10–12, 2011, ed. by T. Schwentick, C. Dürr. Leibniz International Proceedings in Informatics, vol. 9 (Schloss Dagstuhl–Leibniz-Zentrum fuer Informatik, Dagstuhl, 2011), pp. 165–176 (pp. 589, 616, 619)

92. H. Bodlaender, B. Jansen, S. Kratsch, Kernel bounds for path and cycle problems, in *Parameterized and Exact Computation. 6th International Symposium, IPEC '11, Revised Selected Papers*, Saarbrücken, Germany, September 6–8, 2011, ed. by D. Marx, P. Rossmanith. LNCS, vol. 7112 (Springer, Berlin, 2011), pp. 145–158 (pp. 45, 82, 83, 106)

93. H. Bodlaender, B. Jansen, S. Kratsch, Preprocessing for treewidth: a combinatorial analysis through kernelization, in *Proceedings of 38th International Colloquium on Automata, Languages and Programming (ICALP 2011)*, Zürich, Switzerland, July 4–8, 2011, ed. by L. Aceto, M. Henzinger, J. Sgall. LNCS, vol. 6755 (Springer, Berlin, 2011), pp. 437–448 (p. 210)

94. H. Bodlaender, T. Kloks, Efficient and constructive algorithms for the pathwidth and treewidth of graphs. J. Algorithms **21**(2), 358–402 (1996) (pp. 195, 203, 205)

95. H. Bodlaender, A. Koster, Safe separators for treewidth, Technical Report UU-CS-2003-027, Department of Information and Computing Sciences, Utrecht University, 2003 (pp. 206, 207, 208, 210)

96. H. Bodlaender, A. Koster, Combinatorial optimization on graphs of bounded treewidth. Comput. J. **51**(3), 255–269 (2008) (p. 211)

97. H. Bodlaender, A. Koster, Treewidth computations I. Upper bounds, Technical Report UU-CS-2008-032, Department of Information and Computing Sciences, Utrecht University, 2008 (pp. 210, 211)

98. H. Bodlaender, A. Koster, Treewidth computations II. Lower bounds. Inf. Comput. **209**(7), 1103–1119 (2011) (pp. 210, 211)

99. H. Bodlaender, A. Koster, F. van den Eijkhof, L. van der Gaag, Pre-processing for triangulation of probabilistic networks, in *Proceedings of the Seventeenth Conference on Uncertainty in Artificial Intelligence (UAI '01)*, Seattle, WA, USA, August 2–5, 2001, ed. by J. Breese, D. Koller (Morgan Kauffman, San Francisco, 2001), pp. 32–39 (p. 209)

100. H. Bodlaender, D. Thilikos, Constructive linear time algorithms for branchwidth, in *Proceedings of 24th International Colloquium on Automata, Languages and Programming (ICALP 1997)*, Bologna, Italy, July 7–11, 1997, ed. by P. Degano, R. Gorrieri, A. Marchetti-Spaccamela. LNCS, vol. 1256 (Springer, Berlin, 1997), pp. 627–637 (p. 302)

101. H. Bodlaender, S. Thomassé, A. Yeo, Analysis of data reduction: transformations give evidence for non-existence of polynomial kernels, Technical Report UU-CS-2008-030, Department of Information and Computing Sciences, Utrecht University, 2008 (pp. 588, 592, 619)

102. H. Bodlaender, J. van Leeuwen, R. Tan, D. Thilikos, On interval routing schemes and treewidth, Technical Report UU-CS-1996-41, Department of Information and Computing Sciences, Utrecht University, 1996 (pp. 355, 356)

103. B. Bollobiás, *Extremal Graph Theory* (Dover, New York, 2004) (p. 294)

104. A. Bonami, Étude des coefficients de Fourier des fonctions de $L^p(G)$. Ann. Inst. Fourier (Grenoble) **20**(2), 335–402 (1970) (p. 163)

105. K. Booth, Isomorphism testing in graphs, semigroups and finite automata are polynomially equivalent problems. SIAM J. Comput. **7**, 273–279 (1978) (p. 403)

106. R. Boppana, M. Sipser, The complexity of finite functions, in *Handbook of Theoretical Computer Science*, vol. A, ed. by J. van Leeuwen (MIT Press, Cambridge, 1990), pp. 757–804 (p. 527)

107. J. Borda, Mémoire sur les élections au scrutin. Histoire de l'Académie Royale des Sciences (1781) (p. 170)

108. R. Borie, G. Parker, C. Tovey, Automatic generation of linear-time algorithms from predicate calculus descriptions of problems on recursively constructed graph families. Algorithmica **7**, 555–581 (1992) (pp. 265, 277)

109. V. Bouchitte, M. Habib, NP-completeness properties about linear extensions. Order **4**(2), 143–154 (1987) (p. 38)

110. L. Brankovic, H. Fernau, Combining two worlds: parameterized approximation for VERTEX COVER, in *Algorithms and Computation: Proceedings of 21st International Symposium, ISAAC 2010*, Jeju Island, Korea, December 15–17, 2010, ed. by O. Cheong, K.-Y. Chwa, K. Park. LNCS, vol. 6507 (Springer, Berlin, 2010), pp. 390–402 (p. 643)

111. G. Brightwell, P. Winkler, Counting linear extensions. Order **8**(3), 225–242 (1991) (p. 175)

112. G. Brightwell, P. Winkler, Counting linear extensions is #*P*-complete, in *Proceedings of 23rd ACM Symposium on Theory of Computing (STOC '91)*, New Orleans, Louisiana, USA, May 6–May 8, 1991, ed. by C. Koutsougeras, J.S. Vitter (ACM, New York, 1991), pp. 175–181 (p. 175)

113. R. Büchi, Weak second-order arithmetic and finite automata. Z. Math. Log. Grundl. Math. **6**, 66–92 (1960) (pp. 276, 277)

114. H. Buhrman, J. Hitchcock, NP-hard sets are exponentially dense unless coNP $\subseteq$ NP/poly, in *Proceedings of the 23rd Annual IEEE Conference on Computational Complexity, CCC 2008*, College Park, Maryland, USA, June 23–26, 2008 (IEEE Comput. Soc., Los Alamitos, 2008), pp. 1–7 (p. 575)

115. J. Buss, J. Goldsmith, Nondeterminism within p. SIAM J. Comput. **22**(3), 560–572 (1993) (pp. 49, 88, 405, 567)

116. L. Cai, Fixed parameter tractability and approximation problems, Project report, June 1992 (pp. 174, 182)

117. L. Cai, Fixed-parameter tractability of graph modification problems for hereditary properties. Inf. Process. Lett. **58**(4), 171–176 (1996) (pp. 45, 67, 68, 69, 89)

118. L. Cai, Parameterized complexity of cardinality constrained optimization problems. Comput. J. **51**(1), 102–121 (2008) (pp. 404, 405)

119. L. Cai, S.M. Chan, S.O. Chan, Random separation: a new method for solving fixed-cardinality optimization problems, in *Parameterized and Exact Computation. Proceedings of Second International Workshop, IWPEC '06*, Zürich, Switzerland, September 13–15, 2006, ed. by H. Bodlaender, M. Langston. LNCS, vol. 4169 (Springer, Berlin, 2006), pp. 239–250 (p. 671)

120. L. Cai, J. Chen, On the amount of nondeterminism and the power of verifying, in *Mathematical Foundations of Computer Science 1993, Proceedings of 18th International Symposium, MFCS '93*, Gdansk, Poland, August 30–September 3, 1993, ed. by A. Borzyszkowski, S. Sokolowski. LNCS, vol. 711 (Springer, Berlin, 1993), pp. 311–320 (pp. 541, 569)

121. L. Cai, J. Chen, On fixed-parameter tractability and approximability of NP optimization problems. J. Comput. Syst. Sci. **54**(3), 465–474 (1997) (pp. 173, 174, 175, 176, 178, 179, 182, 569)

122. L. Cai, J. Chen, R. Downey, M. Fellows, Advice classes of parameterized tractability. Ann. Pure Appl. Log. **84**, 119–138 (1997) (pp. 20, 88, 491, 494)

123. L. Cai, J. Chen, R. Downey, M. Fellows, On the parameterized complexity of short computation and factorization. Arch. Math. Log. **36**(4–5), 321–337 (1997) (pp. 415, 418, 419, 420, 422, 425, 426)

124. L. Cai, J. Chen, R. Downey, M. Fellows, The parameterized complexity of short computation and factorization. Arch. Math. Log. **36**(4/5), 321–337 (1997) (pp. 406, 457)

125. L. Cai, X. Huang, Fixed-parameter approximation: conceptual framework and approximability results, in *Parameterized and Exact Computation. Proceedings of Second International Workshop, IWPEC '06*, Zürich, Switzerland, September 13–15, 2006, ed. by H. Bodlaender, M. Langston. LNCS, vol. 4169 (Springer, Berlin, 2006), pp. 96–108 (pp. 623, 643)

126. L. Cai, D. Juedes, Subexponential parameterized algorithms collapse the W-hierarchy, in *Proceedings of 28th International Colloquium on Automata, Languages and Programming (ICALP 2001)*, Crete, Greece, July 8–12, 2001, ed. by F. Orejas, P. Spirakis, J. van Leeuwen. LNCS, vol. 2076 (Springer, Berlin, 2001), pp. 273–284 (pp. 541, 542, 543, 547, 566, 570)

127. L. Cai, D. Juedes, On the existence of subexponential parameterized algorithms. J. Comput. Syst. Sci. **67**(4), 789–807 (2003) (pp. 541, 542, 543, 546, 547, 561, 566, 570)

128. L. Cai, E. Verbin, L. Yang, Firefighting on trees, $(1 − 1/e)$-approximation, fixed-parameter tractability and a subexponential algorithm, in *Algorithms and Computation: Proceedings of 19th International Symposium, ISAAC 2008*, Gold Coast, Australia, December 15–17, 2008, ed. by S.-H. Hong, H. Nagamochiand, T. Fukunaga. LNCS, vol. 5369 (Springer, Berlin, 2008), pp. 258–269 (pp. 104, 166)

129. L. Cai, B. Yang, Parameterized complexity of even/odd subgraph problems, in *7th International Conference on Algorithms and Complexity, CIAC 2010*, Rome, Italy, May 26–28, 2010, ed. by T. Calamoneri, J. Diaz. LNCS, vol. 6078 (Springer, Berlin, 2010), pp. 85–96 (p. 671)

130. Y. Cao, J. Chen, Y. Liu, On feedback vertex set: new measure and new structures, in *Proceedings of the 12th Scandinavian Symposium and Workshops on Algorithm Theory*, ed. by H. Kaplan (Springer, Berlin, 2010), pp. 93–104 (pp. 36, 118, 119, 120, 121, 122, 128, 129, 130)

131. K. Cattell, M. Dinneen, A characterization of graphs with vertex cover up to five, in *Proceedings of the International Workshop on Orders, Algorithms, and Applications (ORDAL '94)*. LNCS, vol. 831 (Springer, London, 1994), pp. 86–99 (pp. 328, 364)

132. K. Cattell, M. Dinneen, R. Downey, M. Fellows, M. Langston, On computing graph minor obstruction sets. Theor. Comput. Sci. **233**, 107–127 (2000) (pp. 328, 364)

133. K. Cattell, M. Dinneen, M. Fellows, Obstructions to within a few vertices or edges of acyclic, in *Algorithms and Data Structures, Proceedings of 4th International Workshop, WADS 1995*, Kingston, Ontario, Canada, August 16–18, 1995, ed. by S. Akl, F. Dehne, J.-R. Sack, N. Santoro. LNCS, vol. 955 (Springer, Berlin, 1995), pp. 415–427 (pp. 328, 364)

134. K. Cattell, M. Dinneen, M. Fellows, A simple linear-time algorithm for finding path-decompositions of small width. Inf. Process. Lett. **57**, 197–203 (1996) (pp. 628, 629, 630, 631, 632)

135. M. Cesati, Perfect code is W[1]-complete. Inf. Process. Lett. **81**(3), 163–168 (2002) (pp. 407, 426)

136. M. Cesati, The Turing way to parameterized complexity. J. Comput. Syst. Sci. **67**(4), 654–685 (2003) (pp. 380, 457, 488)

137. M. Cesati, M.D. Ianni, Computation models for parameterized complexity. Math. Log. Q. **43**, 179–202 (1997) (pp. 457, 578)

138. J. Cheetham, F. Dehne, A. Rau-Chaplin, U. Stege, P. Taillon, Solving large FPT problems on coarse-grained parallel machines. J. Comput. Syst. Sci. **67**(4), 691–706 (2003) (p. 88)

139. J. Chen, B. Chor, M. Fellows, X. Huang, D. Juedes, I. Kanj, G. Xia, Tight lower bounds for certain parameterized NP-hard problems. Inf. Comput. **201**, 216–231 (2005) (pp. 556, 557, 561)

140. J. Chen, H. Fernau, I. Kanj, G. Xia, Parametric duality and kernelization: lower bounds and upper bounds on kernel size. SIAM J. Comput. **37**(4), 1077–1106 (2007) (pp. 29, 141, 572)

141. J. Chen, F. Fomin, Y. Liu, S. Lu, Y. Villanger, Improved algorithms for feedback vertex set problems. J. Comput. Syst. Sci. **74**(7), 1188–1198 (2008) (p. 130)

142. J. Chen, D. Friesen, W. Jia, I. Kanj, Using nondeterminism to design efficient deterministic algorithms. Algorithmica **40**(2), 83–97 (2004) (pp. 134, 135, 141)

143. J. Chen, X. Huang, I. Kanj, G. Xia, On the computational hardness based on linear FPT-reductions. J. Comb. Optim. **11**(2), 231–247 (2006) (pp. 506, 557)

144. J. Chen, X. Huang, I. Kanj, G. Xia, Strong computational lower bounds via parameterized complexity. J. Comput. Syst. Sci. **72**(8), 1346–1367 (2006) (p. 557)

145. J. Chen, I. Kanj, Constrained minimum vertex covers of bipartite graphs: improved algorithms. J. Comput. Syst. Sci. **67**, 833–847 (2003) (p. 88)

146. J. Chen, I. Kanj, W. Jia, Vertex cover: further observations and further improvements. J. Algorithms **41**(2), 280–301 (2001) (pp. 27, 50, 57, 88)

147. J. Chen, I. Kanj, L. Perkovic, E. Sedgwick, G. Xia, Genus characterizes the complexity of graph problems: some tight results, in *Proceedings of 30th International Colloquium on Automata, Languages and Programming (ICALP 2003)*, Eindhoven, The Netherlands, June 30–July 4, 2003, ed. by J. Baeten, J.K. Lenstra, J. Parrow, G. Woeginger. LNCS, vol. 2719 (Springer, Berlin, 2003), pp. 845–856 (p. 561)

148. J. Chen, I. Kanj, G. Xia, Improved upper bounds for vertex cover. Theor. Comput. Sci. **411**(40–42), 3736–3756 (2010) (pp. 27, 88)

149. J. Chen, J. Kneis, S. Lu, D. Mölle, S. Richter, P. Rossmanith, S.-H. Sze, F. Zhang, Randomized divide-and-conquer: improved path, matching, and packing algorithms. SIAM J. Comput. **38**(6), 2526–2547 (2009) (pp. 153, 155, 169)

150. J. Chen, Y. Liu, S. Lu, B. O'Sullivan, I. Razgon, A fixed-parameter algorithm for the directed feedback vertex set problem. J. ACM **55**(5), 1–19 (2008) (pp. 36, 129, 135, 679)

151. J. Chen, S. Lu, S.-H. Sze, F. Zhang, Improved algorithms for path, matching, and packing problems, in *Proceedings of the Eighteenth Annual ACM–SIAM Symposium on Discrete Algorithms, SODA 2007*, New Orleans, Louisiana, USA, January 7–9, 2007, ed. by N. Bansal, K. Pruhs, C. Stein (SIAM, Philadelphia, 2007), pp. 298–307 (pp. 146, 148)

152. J. Chen, J. Meng, On parameterized intractability: hardness and completeness. Comput. J. **51**(1), 39–59 (2008) (p. 458)

153. Y. Chen, J. Flum, M. Grohe, Bounded nondeterminism and alternation in parameterized complexity theory, in *Proceedings of the 18th Annual IEEE Conference on Computational Complexity, CCC 2003*, Aarhus, Denmark, July 7–10, 2003 (IEEE Comput. Soc., Los Alamitos, 2003), pp. 13–29 (pp. 504, 505, 506)

154. Y. Chen, J. Flum, M. Grohe, An analysis of the W^*-hierarchy. J. Symb. Log. **72**, 513–534 (2007) (pp. 182, 455, 459)

155. Y. Chen, M. Grohe, An isomorphism between subexponential and parameterized complexity theory. SIAM J. Comput. **37**, 1228–1258 (2007) (p. 548)

156. Y. Chen, M. Grohe, M. Grüber, On parameterized approximability, in *Parameterized and Exact Computation, Proceedings of Second International Workshop, IWPEC '06*, Zürich, Switzerland, September 13–15, 2006, ed. by H. Bodlaender, M. Langston. LNCS, vol. 4169 (Springer, Berlin, 2006), pp. 109–120 (pp. 623, 624, 626, 627, 641, 643)

157. J. Cheriyan, J. Reif, Directed s–t numberings, rubber bands, and testing digraph k-vertex connectivity. Combinatorica **14**(4), 435–451 (1994) (p. 92)

158. R. Chitnis, M. Cygan, M. Hajiaghayi, M. Pilipczuk, M. Pilipczuk, Designing FPT algorithms for cut problems using randomized contractions, in *IEEE 53rd Annual Symposium on Foundations of Computer Science, FOCS 2012*, New Brunswick, New Jersey, USA, October 20–23, 2012 (IEEE Comput. Soc., Los Alamitos, 2012), pp. 460–469 (p. 672)

159. B. Chor, M. Fellows, D. Juedes, Linear kernels in linear time, or how to save k colors in $O(k^2)$ steps, in *Graph-Theoretic Concepts in Computer Science: 30th International Workshop, WG 2004, Revised Papers*, Bad Honnef, Germany, June 2004, ed. by J. Hromkovič,

M. Nagl, B. Westfechtel. LNCS, vol. 3353 (Springer, Berlin, 2004), pp. 257–269 (pp. 100, 101, 102, 104, 105, 106)

160. M. Chrobak, M. Ślusarek, On some packing problems related to dynamic storage allocation. RAIRO Theor. Inform. Appl. **22**, 487–499 (1988) (p. 638)

161. W. Cohen, R. Schapire, Y. Singer, Learning to order things, in *Advances in Neural Information Processing Systems 10, NIPS Conference*, Denver, Colorado, USA, 1997, ed. by M. Jordan, M.K. Sara Solla (MIT Press, Cambridge, 1998), pp. 451–457 (p. 170)

162. J. Conway, C. Gordon, Knots and links in spatial graphs. J. Graph Theory **7**(4), 445–453 (1983) (p. 17)

163. S. Cook, The complexity of theorem proving procedures, in *Proceedings of 3rd Annual ACM Symposium on Theory of Computing (STOC '71)*, Shaker Heights, Ohio, USA, May 3–May 5, 1971, ed. by M. Harrison, R. Banerji, J. Ullman (ACM, New York, 1971), pp. 151–158 (pp. 4, 12, 382)

164. D. Coppersmith, Rectangular matrix multiplication revisited. J. Complex. **13**, 42–49 (1997) (p. 561)

165. D. Coppersmith, S. Winograd, Matrix multiplication via arithmetic progressions. J. Symb. Comput. **9**(3), 251–280 (1990) (pp. 307, 560)

166. B. Courcelle, On context-free sets of graphs and their monadic second-order theory, in *Graph-Grammars and Their Application to Computer Science 3rd International Workshop*, Warrenton, Virginia, USA, December 2–6, 1986, ed. by H. Ehrig, M. Nagl, G. Rozenberg, A. Rosenfeld. LNCS, vol. 291 (Springer, Berlin, 1987), pp. 133–146 (p. 276)

167. B. Courcelle, The monadic second-order logic of graphs. I. Recognizable sets of finite graphs. Inf. Comput. **85**(1), 12–75 (1990) (p. 276)

168. B. Courcelle, The monadic second-order logic of graphs. III. Tree-decompositions, minors and complexity issues. RAIRO Theor. Inform. Appl. **26**(3), 257–286 (1992) (p. 277)

169. B. Courcelle, The monadic second-order theory of graphs V: On closing the gap between definability and recognizability. Theor. Comput. Sci. **80**, 153–202 (1991) (p. 277)

170. B. Courcelle, The monadic second-order theory of graphs VI: On several representations of graphs by relational structures. Discrete Appl. Math. **54**, 117–149 (1994) (p. 275)

171. B. Courcelle, The monadic second-order theory of graphs VII: Graphs as relational structures. Theor. Comput. Sci. **101**, 3–33 (1992) (pp. 276, 309, 311)

172. B. Courcelle, The monadic second-order theory of graphs VIII: Orientations. Ann. Pure Appl. Log. **72**, 103–143 (1995) (p. 275)

173. B. Courcelle, The monadic second-order theory of graphs X: Linear orderings. Theor. Comput. Sci. **160**, 87–144 (1996) (p. 275)

174. B. Courcelle, Basic notions of universal algebra for language theory and graph grammars. Theor. Comput. Sci. **163**(1–2), 1–54 (1996) (p. 275)

175. B. Courcelle, The expression of graph properties and graph transformations in monadic second-order logic, in *Handbook of Graph Grammars and Computing by Graph Transformation*, vol. 1, ed. by G. Rozenberg (World Scientific, Singapore, 1997), pp. 313–400 (p. 277)

176. B. Courcelle, R. Downey, M. Fellows, A note on the computability of graph minor obstruction sets for monadic second-order ideals. J. Univers. Comput. Sci. **3**, 1194–1198 (1997) (pp. 328, 338)

177. B. Courcelle, I. Durant, Automata for the verification of monadic second-order graph properties. J. Appl. Log. **10**(4), 368–409 (2012) (p. 272)

178. B. Courcelle, J. Engelfriet, *Graph Structure and Monadic Second Order Logic: A Language-Theoretic Approach*. Encyclopedia of Mathematics and Its Applications, vol. 138 (Cambridge University Press, Cambridge, 2012) (pp. 237, 246, 277, 311)

179. B. Courcelle, P. Franchi-Zannettacci, Attribute grammars and recursive program schemes. Theor. Comput. Sci. **17**, 163–191, 235–257 (1982) (p. 276)

180. B. Courcelle, J. Makowsky, U. Rotics, Linear time solvable optimization problems on graphs of bounded clique-width. Theory Comput. Syst. **33**(2), 125–150 (2000) (pp. 310, 311, 646)

181. B. Courcelle, M. Mosbah, Monadic second-order evaluations on tree-decomposable graphs. Theor. Comput. Sci. **109**, 49–82 (1993) (p. 277)

182. B. Courcelle, S. Olariu, Upper bounds to the clique width of graphs. Discrete Appl. Math. **101**(1–3), 77–144 (2000) (p. 309)

183. B. Courcelle, S. Oum, Vertex-minors, monadic second-order logic, and a conjecture of Seese. J. Comb. Theory, Ser. B **97**(1), 91–126 (2007) (p. 275)

184. R. Crowston, M. Fellows, G. Gutin, M. Jones, F. Rosamond, S. Thomassé, A. Yeo, Simultaneously satisfying linear equations over $\mathbb{F}_2$: MaxLin2 and Max-r-Lin2 parameterized above average, in *IARCS Annual Conference on Foundations of Software Technology and Theoretical Computer Science (FSTTCS 2011)*, IIT Guwahati, India, December 12–14, 2011, ed. by S. Chakraborty, A. Kumar. Leibniz International Proceedings in Informatics, vol. 13 (Schloss Dagstuhl–Leibniz-Zentrum fuer Informatik, Dagstuhl, 2011), pp. 229–240 (p. 169)

185. M. Cygan, Deterministic parameterized connected vertex cover, arXiv:1202.6642 (pp. 670, 671)

186. M. Cygan, H. Dell, D. Lokshtanov, D. Marx, J. Nederlof, Y. Okamoto, R. Paturi, S. Saurabh, M. Wahlström, On problems as hard as CNFSAT, arXiv:1112.2275. To appear in *CCC 2012* (p. 559)

187. M. Cygan, F. Fomin, E. van Leeuwin, Parameterized complexity of firefighting revisited, in *Parameterized and Exact Computation, 6th International Symposium, IPEC '11, Revised Selected Papers*, Saarbrücken, Germany, September 6–8, 2011, ed. by D. Marx, P. Rossmanith. LNCS, vol. 7112 (Springer, Berlin, 2011), pp. 13–26 (pp. 275, 288, 401, 616)

188. M. Cygan, S. Kratsch, M. Pilipczuk, M. Pilipczuk, M. Wahlström, Clique cover and graph separation: new incompressibility results, in *ICALP 2012* (2012), pp. 254–265 (pp. 591, 616)

189. M. Cygan, D. Marx, M. Pilipczuk, M. Pilipczuk, The planar directed k-Vertex-Disjoint Paths problem is fixed-parameter tractable, arXiv:1304.4207 (p. 679)

190. M. Cygan, J. Nederlof, M. Pilipczuk, M. Pilipczuk, J. van Rooij, J.O. Wojtaszczyk, Solving connectivity problems parameterized by treewidth in single exponential time, in *IEEE 52nd Annual Symposium on Foundations of Computer Science, FOCS 2011*, Palm Springs, CA, USA, October 22–25, 2011, ed. by R. Ostrovsky (IEEE Comput. Soc., Los Alamitos, 2011) (pp. 169, 305, 558)

191. M. Cygan, M. Pilipczuk, M. Pilipczuk, J.O. Wojtaszczyk, Kernelization hardness of connectivity problems in d-degenerate graphs, in *Graph-Theoretic Concepts in Computer Science: 36th International Workshop, WG 2010, Revised Papers*, Zarós, Crete, Greece, June 2010 (Springer, Berlin, 2010), pp. 147–158 (p. 618)

192. J. Daligault, G. Gutin, E. Kim, A. Yeo, FPT algorithms and kernels for the directed k-leaf problem. J. Comput. Syst. Sci. **76**(2), 144–152 (2010) (p. 106)

193. J. Daligault, S. Thomassé, On finding directed trees with many leaves, in *Parameterized and Exact Computation, Proceedings of 4th International Workshop, IWPEC '09*, Copenhagen, Denmark, September 2009, ed. by J. Chen, F. Fomin. LNCS, vol. 5917 (Springer, Berlin, 2009), pp. 86–97 (pp. 91, 92, 93, 94, 95, 98, 106)

194. P. Damaschke, Degree-bounded techniques accelerate some parameterized graph algorithms, in *Parameterized and Exact Computation, Proceedings of 4th International Workshop, IWPEC '09*, Copenhagen, Denmark, September 2009, ed. by J. Chen, F. Fomin. LNCS, vol. 5917 (Springer, Berlin, 2009), pp. 98–109 (p. 671)

195. P. Damaschke, Parameterizations of hitting set of bundles and inverse scope, Manuscript (p. 43)

196. B. DasGupta, X. He, T. Jiang, M. Li, J. Tromp, L. Zhang, On the linear-cost subtree-transfer distance between phylogenetic trees. Algorithmica **25**(2–3), 176–195 (1999) (p. 70)

197. M. Davis, Unsolvable problems, in *The Handbook of Mathematical Logic*, ed. by J. Barwise (North-Holland, Amsterdam, 1977), pp. 567–594 (p. 420)

198. M. Davis, H. Putnam, A computing procedure for quantification theory. J. ACM **7**, 201–215 (1960) (p. 113)

199. A. Dawar, M. Grohe, S. Kreutzer, Locally excluding a minor, in *Proceedings of the 22nd Annual IEEE Symposium on Logic in Computer Science (LICS '07)* (IEEE Comput. Soc., Los Alamitos, 2007), pp. 270–279 (p. 365)

200. A. Dawar, S. Kreutzer, Domination problems in nowhere-dense classes, in *IARCS Annual Conference on Foundations of Software Technology and Theoretical Computer Science (FSTTCS 2009)*, IIT Kanpur, India, December 15–17, 2009, ed. by R. Kannan, N. Kumar (Schloss Dagstuhl–Leibniz-Zentrum fuer Informatik, Dagstuhl, 2009), pp. 157–168 (p. 365)

201. J.-A.-N. de Caritat le marquis de Condorcet, Essai sur l'application de l'analyse à la probabilité des décisions rendues à la pluralité des voix, 1785 (p. 170)

202. B. de Fluiter, Algorithms for graphs of small treewidth, PhD thesis, Utrecht University, 1997 (pp. 368, 369, 371)

203. F. Dehne, M. Fellows, H. Fernau, E. Prieto, F. Rosamond, Nonblocker: parameterized algorithmics for minimum dominating set, in *Proceedings of the 32nd Conference on Current Trends in Theory and Practice of Computer Science (SOFSEM '06)* (Springer, Berlin, 2006), pp. 237–245 (p. 139)

204. F. Dehne, M. Fellows, F. Rosamond, P. Shaw, Greedy localization, iterative compression, and modeled crown reductions: new FPT techniques, an improved algorithm for set splitting, and a novel $2k$ kernelization for vertex cover, in *Parameterized and Exact Computation, Proceedings of First International Workshop, IWPEC 2004*, Bergen, Norway, September 14–17, 2004, ed. by R. Downey, M. Fellows, F. Dehne. LNCS, vol. 3162 (Springer, Berlin, 2004), pp. 271–281 (pp. 117, 129, 130, 134, 141)

205. H. Dell, D. Marx, Kernelization of packing problems, in *Proceedings of the Twenty-Third Annual ACM–SIAM Symposium on Discrete Algorithms, SODA 2012*, Kyoto, Japan, January 17–19, 2012, ed. by Y. Rabani (SIAM, Philadelphia, 2012), pp. 68–81 (pp. 595, 597)

206. H. Dell, D. van Melkebeek, Satisfiability allows no nontrivial sparsification unless the polynomial-time hierarchy collapses, in *Proceedings of 42nd ACM Symposium on Theory of Computing (STOC 2010)*, Cambridge, MA, June 6–June 8, 2010, ed. by L. Schulman (ACM, New York, 2010), pp. 251–260 (pp. 58, 595, 596, 598, 616, 619)

207. E. Demaine, F. Fomin, M. Hajiaghayi, D. Thilikos, Bidimensional parameters and local treewidth. SIAM J. Discrete Math. **18**(3), 501–511 (2005) (p. 563)

208. E. Demaine, F. Fomin, M. Hajiaghayi, D. Thilikos, Fixed-parameter algorithms for (k, r)-center in planar graphs and map graphs. ACM Trans. Algorithms **1**(1), 33–47 (2005) (p. 561)

209. E. Demaine, F. Fomin, M. Hajiaghayi, D. Thilikos, Subexponential parameterized algorithms on bounded-genus graphs and H-minor-free graphs. J. ACM **52**(6), 866–893 (2005) (pp. 562, 563, 570)

210. E. Demaine, M. Hajiaghayi, Equivalence of local treewidth and linear local treewidth and its algorithmic applications, in *Proceedings of the Fifteenth Annual ACM–SIAM Symposium on Discrete Algorithms, SODA 2004*, New Orleans, Louisiana, USA, January 11–14, 2004, ed. by I. Munro (SIAM, Philadelphia, 2004), pp. 840–849 (pp. 284, 289)

211. E. Demaine, M. Hajiaghayi, Bidimensionality: new connections between FPT algorithms and PTASs, in *Proceedings of the Sixteenth Annual ACM–SIAM Symposium on Discrete Algorithms, SODA 2005*, Vancouver, British Columbia, Canada, January 23–25, 2005 (SIAM, Philadelphia, 2005), pp. 590–601 (p. 565)

212. E. Demaine, M. Hajiaghayi, Bidimensionality, in *Encyclopedia of Algorithms*, ed. by M.-Y. Kao (Springer, Berlin, 2008), pp. 88–90 (pp. 289, 565)

213. E.D. Demaine, M. Hajiaghayi, The bidimensionality theory and its algorithmic applications. Comput. J. **51**(3), 292–302 (2008) (p. 289)

214. E. Demaine, M. Hajiaghayi, Linearity of grid minors in treewidth with applications through bidimensionality. Combinatorica **28**(1), 19–36 (2008) (pp. 274, 563)

215. E. Demaine, M. Hajiaghayi, K.-i. Kawarabayashi, Contraction decomposition in H-minor-free graphs and algorithmic applications, in *Proceedings of 43rd ACM Symposium on The-*

ory of Computing (STOC '11), San Jose, CA, USA, June 6–June 8, 2011, ed. by L. Fortnow, S. Vadhan (ACM, New York, 2011), pp. 441–450 (p. 686)

216. E. Demaine, M. Hajiaghayi, D. Marx, Minimizing movement: fixed-parameter tractability, in *Algorithms—ESA 2009: Proceedings of 17th Annual European Symposium*, Copenhagen, Denmark, September 7–9, 2009, ed. by A. Fiat, P. Sanders. LNCS, vol. 5757 (Springer, Berlin, 2009), pp. 718–729 (p. 168)

217. E. Demaine, M. Hajiaghayi, D. Thilikos, Exponential speedup of fixed-parameter algorithms for classes of graphs excluding single-crossing graphs as minors. Algorithmica **41**, 245–267 (2005) (pp. 561, 564)

218. E. Demaine, M. Hajiaghayi, D. Thilikos, The bidimensional theory of bounded-genus graphs. SIAM J. Discrete Math. **20**(2), 357–371 (2006) (p. 563)

219. L. Dennenberg, Y. Gurevich, S. Shelah, Definability by constant-depth polynomial-size circuits. Inf. Control **70**, 216–240 (1986) (p. 146)

220. R. Diestel, Graph minors 1: A short proof of the path-width theorem. Comb. Probab. Comput. **4**, 27–30 (1995) (pp. 340, 350)

221. G. Ding, B. Oporowski, D. Sanders, D. Vertigan, Partitioning graphs of bounded treewidth. Combinatorica **18**(1), 1–12 (1998) (pp. 346, 347)

222. M. Dinneen, Too many minor order obstructions (for parameterized lower ideals). J. Univers. Comput. Sci. **3**(11), 1199–1206 (1997) (p. 349)

223. I. Dinur, S. Safra, The importance of being biased, in *Proceedings of 34th ACM Symposium on Theory of Computing (STOC '02)*, Montréal, Québec, Canada, May 19–May 21, 2002, ed. by J. Reif (ACM, New York, 2002), pp. 33–42 (pp. 58, 619)

224. G. Dirac, S. Schuster, A theorem of Kuratowski. Indag. Math. **16**(3), 343–348 (1954) (p. 323)

225. M. Dom, J. Guo, F. Hüffner, R. Niedermeier, A. Truß, Fixed-parameter tractability results for feedback set problems in tournaments, in *3rd International Conference on Algorithms and Complexity, CIAC 2006*, Rome, Italy, May 29–31, ed. by T. Calamoneri, J. Díaz (2006), pp. 320–331 (p. 156)

226. M. Dom, D. Lokshtanov, S. Saurabh, Incompressibility through colors and IDS, in *Proceedings of 36th International Colloquium on Automata, Languages and Programming (ICALP 2009), Part I*, Rhodes, Greece, July 5–12, 2009, ed. by S. Albers, A. Marchetti-Spaccamela, Y. Matias, S. Nikoletseas, W. Thomas. LNCS, vol. 5555 (Springer, Berlin, 2009), pp. 378–389 (pp. 592, 593, 594, 617, 618)

227. M. Dom, D. Lokshtanov, S. Saurabh, Kernelization lower bounds through colors and IDs, to appear (pp. 592, 593, 594, 617)

228. J. Doner, Tree acceptors and some of their applications. J. Comput. Syst. Sci. **4**, 406–451 (1970) (p. 246)

229. F. Dorn, Dynamic programming and fast matrix multiplication, in *Algorithms—ESA 2006: Proceedings of 14th Annual European Symposium*, Zurich, Switzerland, September 11–13, 2006, ed. by Y. Azar, T. Erlebach. LNCS, vol. 4168 (Springer, Berlin, 2006), pp. 280–291 (pp. 29, 305, 306, 308)

230. F. Dorn, Designing subexponential algorithms: problems, techniques and structures, PhD thesis, University of Bergen, July 2007 (p. 563)

231. F. Dorn, F. Fomin, D. Lokshtanov, V. Raman, S. Saurabh, Beyond bidimensionality: parameterized subexponential algorithms on directed graphs, in *Proceedings of 27th International Symposium on Theoretical Aspects of Computer Science, STACS 2010*, Nancy, France, March 4–6, 2010, ed. by J.-Y. Marion, T. Schwentick. Leibniz International Proceedings in Informatics, vol. 5 (Schloss Dagstuhl–Leibniz-Zentrum fuer Informatik, Dagstuhl, 2010), pp. 251–262 (p. 565)

232. F. Dorn, F. Fomin, D. Thilikos, Fast subexponential algorithm for non-local problems on graphs of bounded genus, in *Algorithm Theory—SWAT 2006, Proceedings of 10th Scandinavian Workshop on Algorithm Theory*, Riga, Latvia, July 6–8, 2006, ed. by L. Arge, R. Freivalds. LNCS, vol. 4059 (Springer, Berlin, 2006), pp. 172–183 (p. 306)

233. F. Dorn, F. Fomin, D. Thilikos, Catalan structures and dynamic programming in H-minor-free graphs, in *Proceedings of the Nineteenth Annual ACM–SIAM Symposium on Discrete Algorithms, SODA 2008*, San Francisco, California, USA, January 20–22, 2008, ed. by S.-H. Teng (SIAM, Philadelphia, 2008), pp. 631–640 (pp. 306, 564)

234. F. Dorn, F.V. Fomin, D.M. Thilikos, Subexponential parameterized algorithms. Comput. Sci. Rev. **2**(1), 29–39 (2008) (p. 565)

235. F. Dorn, E. Penninkx, H. Bodlaender, F. Fomin, Efficient exact algorithms on planar graphs: exploiting sphere cut decompositions. Algorithmica **58**(3), 790–810 (2010) (pp. 305, 563)

236. R. Downey, Parameterized complexity for the skeptic, in *Proceedings of the 18th Annual IEEE Conference on Computational Complexity, CCC 2003*, Aarhus, Denmark, July 7–10, 2003 (IEEE Comput. Soc., Los Alamitos, 2003), pp. 147–168 (p. 70)

237. R. Downey, The birth and early years of parameterized complexity, in *The Multivariate Algorithmic Revolution and Beyond: Essays Dedicated to Michael R. Fellows on the Occasion of His 60th Birthday*, ed. by H. Bodlaender, R. Downey, F. Fomin, D. Marx. LNCS, vol. 7370 (Springer, Berlin, 2012), pp. 17–38 (pp. 21, 88, 300, 406, 535)

238. R. Downey, V. Estivill-Castro, M. Fellows, E. Prieto, F. Rosamond, Cutting up is hard to do: the parameterised complexity of k-cut and related problems, in *Computing: the Australasian Theory Symposium, CATS 2003*. Electronic Notes in Theoretical Computer Science, vol. 78 (Elsevier, Amsterdam, 2003), pp. 209–222 (pp. 404, 541, 542, 544, 545, 546, 570)

239. R. Downey, P. Evans, M. Fellows, Parameterized learning complexity, in *Proceedings of the Sixth Annual Conference on Computational Learning Theory, COLT'93*, ed. by P. Clote, J. Remmel (ACM, New York, 1993), pp. 51–57 (p. 413)

240. R. Downey, M. Fellows, Fixed parameter intractability (extended abstract), in *Proceedings. Seventh Annual Structure in Complexity Theory Conference*, Victoria University, British Columbia, Canada, June 22–25, 1992, ed. by K. Abrahamson, R. Downey, M. Fellows (IEEE Comput. Soc., Los Alamitos, 1992), pp. 36–50 (pp. 88, 382, 426)

241. R. Downey, M. Fellows, Fixed-parameter tractability and completeness. Congr. Numer. **87**, 161–178 (1992) (pp. 21, 25, 36, 44, 46, 294, 382, 426, 429, 444)

242. R. Downey, M. Fellows, Fixed-parameter tractability and completeness. III. Some structural aspects of the W hierarchy, in *Complexity Theory: Current Research, Dagstuhl Workshop*, February 2–8, 1992, ed. by K. Ambos-Spies, S. Homer, U. Schöning (Cambridge University Press, Cambridge, 1993), pp. 191–225 (p. 18)

243. R. Downey, M. Fellows, Fixed-parameter tractability and completeness. I. Basic results. SIAM J. Comput. **24**(4), 873–921 (1995) (pp. 21, 88, 106, 382, 426, 429, 444, 459, 497, 530)

244. R. Downey, M. Fellows, Fixed-parameter tractability and completeness II: On completeness for $W[1]$. Theor. Comput. Sci. **141**(1–2), 109–131 (1995) (pp. 21, 406, 407, 424, 425, 426, 429, 613, 678)

245. R. Downey, M. Fellows, Parameterized computational feasibility, in *Feasible Mathematics II*, ed. by P. Clote, J. Remmel. Progress in Computer Science and Applied Logic, vol. 13 (Birkhäuser, Boston, 1995), pp. 219–244 (pp. 44, 46, 65, 88, 413, 456)

246. R. Downey, M. Fellows, Threshold dominating sets and an improved characterization of $W[2]$. Theor. Comput. Sci. **209**(1), 124–140 (1998) (pp. 455, 456, 459)

247. R. Downey, M. Fellows, *Parameterized Complexity*. Monographs in Computer Science (Springer, Berlin, 1999) (pp. vii, xi, xv, xviii, xix, 18, 36, 46, 89, 130, 147, 165, 166, 263, 300, 324, 343, 347, 348, 461, 468, 471, 475, 488, 495, 518, 535, 538, 565, 569, 641, 643, 677, 681, 683)

248. R. Downey, M. Fellows, B. Kapron, M. Hallett, T. Wareham, The parameterized complexity of some problems in logic and linguistics, in *Logical Foundations of Computer Science, Proceedings of Third International Symposium, LFCS'94*, St. Petersburg, Russia, July 11–14, 1994, ed. by A. Nerode, Y. Matiyasevich. LNCS, vol. 813 (Springer, Berlin, 1994), pp. 89–101 (pp. 475, 488)

249. R. Downey, M. Fellows, N. Koblitz, Techniques for exponential parameterized reductions in vertex set problems, Unpublished manuscript, 1996 (p. 147)

250. R. Downey, M. Fellows, M. Langston, The computer journal special issue on parameterized complexity: foreword by the guest editors. Comput. J. **51**(1), 1–6 (2008) (pp. viii, xi)

251. R. Downey, M. Fellows, C. McCartin, Parameterized approximation problems, in *Parameterized and Exact Computation, Proceedings of Second International Workshop, IWPEC '06*, Zürich, Switzerland, September 13–15, 2006, ed. by H. Bodlaender, M. Langston. LNCS, vol. 4169 (Springer, Berlin, 2006), pp. 121–129 (pp. 623, 643)

252. R. Downey, M. Fellows, C. McCartin, F. Rosamond, Parameterized approximation of dominating set problems. Inf. Process. Lett. **109**(1), 68–70 (2008) (pp. 639, 641, 643)

253. R. Downey, M. Fellows, U. Stege, Computational tractability: the view from Mars. Bull. Eur. Assoc. Theor. Comput. Sci. **69**, 73–97 (1999) (p. 88)

254. R. Downey, M. Fellows, V. Raman, The complexity of irredundant sets parameterized by size, Unpublished (p. 403)

255. R. Downey, M. Fellows, K. Regan, Descriptive complexity and the W-hierarchy, in *DIMACS Workshop on Proof Complexity and Feasible Arithmetics*, April 1996, ed. by P. Beame, S. Buss. DIMACS Series in Discrete Mathematics and Theoretical Computer Science, vol. 39 (Am. Math. Soc., Providence, 1997), pp. 119–134 (p. 182)

256. R. Downey, M. Fellows, K. Regan, Parameterized circuit complexity and the W hierarchy. Theor. Comput. Sci. **191**(1–2), 97–115 (1998) (pp. 497, 525, 527, 528, 530, 531, 613, 664, 665, 672)

257. R. Downey, M. Fellows, U. Stege, Parameterized complexity: a framework for systematically confronting computational intractability, in *Contemporary Trends in Discrete Mathematics: from DIMACS and DIMATIA to the Future, DIMATIA-DIMACS Conference*, Štiřín Castle, May 19–25, 1997, ed. by R. Grahm, J. Kratochvíl, J. Nesetril, F. Roberts. DIMACS Series in Discrete Mathematics and Theoretical Computer Science, vol. 49 (Am. Math. Soc., Providence, 1999), pp. 49–100 (pp. 50, 535)

258. R. Downey, M. Fellows, U. Taylor, The parameterized complexity of relational database queries and an improved characterization of $W[1]$, in *Combinatorics, Complexity & Logic, Proceedings of DMTCS '96*, Singapore, ed. by D. Bridges, C. Calude, J. Gibbons, S. Reeves, I. Witten (Springer, Berlin, 1996), pp. 194–213 (pp. 400, 446, 447, 448, 459, 500, 506)

259. R. Downey, M. Fellows, A. Vardy, G. Whittle, The parametrized complexity of some fundamental problems in coding theory. SIAM J. Comput. **29**(2), 545–570 (1999) (p. 678)

260. R. Downey, C. McCartin, Online problems, pathwidth, and persistence, in *Parameterized and Exact Computation, Proceedings of First International Workshop, IWPEC 2004*, Bergen, Norway, September 14–17, 2004, ed. by R. Downey, M. Fellows, F. Dehne. LNCS, vol. 3162 (Springer, Berlin, 2004), pp. 13–24 (pp. 636, 642, 643)

261. R. Downey, D. Thilikos, Confronting intractability via parameters. Comput. Sci. Rev. **5**(4), 279–317 (2011) (p. 561)

262. M. Drescher, A. Vetta, An approximation algorithm for the max leaf spanning arborescence problem. ACM Trans. Algorithms **6**(3), 46 (2010) (p. 589)

263. A. Drucker, New limits to classical and quantum instance compression, in *FOCS 2012* (2012), pp. 609–618 (pp. 580, 619)

264. C. Dwork, R. Kumar, M. Naor, D. Sivakumar, Rank aggregation methods for the web, in *WWW '01, Proceedings of the 10th International Conference on World Wide Web*, Hong Kong, China, May 1–5, 2001 (2001), pp. 613–622 (p. 170)

265. C. Dwork, R. Kumar, M. Naor, D. Sivakumar, Rank aggregation revisited, Manuscript, 2001 (p. 170)

266. H.-D. Ebbinghaus, J. Flum, *Finite Model Theory* (Springer, Berlin, 1995) (pp. 182, 286)

267. J. Edmonds, Paths, trees and flowers. Can. J. Math. **17**, 449–467 (1965) (pp. 12, 101, 697)

268. K. Eickmeyer, M. Grohe, M. Grüber, Approximation of natural $W[P]$-complete minimisation problems is hard, in *Proceedings of the 23rd Annual IEEE Conference on Compu-*

tational Complexity, CCC 2008, College Park, Maryland, USA, June 23–26, 2008 (IEEE Comput. Soc., Los Alamitos, 2008), pp. 8–18 (pp. 641, 643)

269. T. Eiter, G. Gottlob, Identifying the minimal transversals of a hypergraph and related problems. SIAM J. Comput. **24**, 1278–1304 (1995) (p. 684)

270. M. El-Zahar, J. Schmerl, On the size of jump-critical ordered sets. Order **1**, 3–5 (1984) (p. 39)

271. K. Elbassioni, M. Hagen, I. Rauf, Some fixed-parameter tractable classes of hypergraph duality and related problems, in *Parameterized and Exact Computation, 3rd International Workshop, IWPEC '08, Revised Selected Papers*, Victoria, Canada, May 14–16, 2008, ed. by M. Grohe, R. Niedermeier. LNCS, vol. 5018 (Springer, Berlin, 2008), pp. 91–102 (p. 684)

272. J.A. Ellis, I.H. Sudborough, J.S. Turner, The vertex separation and search number of a graph. Inform. Comput. **113**(1), 50–79 (1994) (p. 629)

273. D. Eppstein, Subgraph isomorphism in planar graphs and related problems, in *Proceedings of the Sixth Annual ACM–SIAM Symposium on Discrete Algorithms*, San Francisco, California, 22–24 January 1995, ed. by K. Clarkson (SIAM, Philadelphia, 1995), pp. 632–640 (p. 578)

274. P. Erdös, Some remarks on the theory of graphs. Bull. Am. Math. Soc. **53**, 292–294 (1947) (p. 600)

275. P. Erdös, A. Hajnal, On chromatic number of graphs and set-systems. Acta Math. Hung. **17**(1–2), 61–99 (1966) (p. 313)

276. P. Erdös, J. Moon, On sets on consistent arcs in tournaments. Can. Math. Bull. **8**, 269–271 (1965) (p. 170)

277. P. Erdős, L. Pósa, On independent circuits contained in a graph. Can. J. Math. **17**, 347–352 (1965) (p. 294)

278. V. Estivill-Castro, M. Fellows, M. Langston, F. Rosamond, FPT is P-time extremal structure I, in *Algorithms and Complexity in Durham 2005, Proceedings of the First ACiD Workshop*. Texts in Algorithmics, vol. 4 (King's College, Dusham, 2005), pp. 1–41 (pp. 39, 131, 133)

279. L. Euler, Solutio problematis ad geometriam situs pertinentis. Comment. Acad. Sci. Petropolitanae **8**, 128–140 (1741) (p. 337)

280. S. Even, R. Tarjan, A combinatorial problem which is complete for polynomial space. J. ACM **23**, 710–719 (1976) (p. 499)

281. R. Fagin, Generalized first order spectra and polynomial-time recognizable sets, in *Complexity of Computation*, ed. by R. Karp. SIAM–AMS Proceedings, vol. 7 (Am. Math. Soc., Providence, 1974), pp. 43–73 (pp. 176, 182)

282. R. Fagin, Finite model theory—a personal perspective. Theor. Comput. Sci. **116**(1), 3–31 (1993) (p. 182)

283. U. Feige, M. Goemans, Approximating the value of two prover proof systems, with applications to MAX 2SAT and MAX DICUT, in *Proceedings of the 3rd Israel Symposium on the Theory of Computing and Systems (ISTCS'95)*, Tel Aviv, Israel, January 4–6, 1995 (IEEE Comput. Soc., Los Alamitos, 1995), pp. 182–189 (p. 58)

284. M. Fellows, A category of graphs for computer science, PhD thesis, University of California, San Diego, 1985 (p. 354)

285. M. Fellows, The lost continent of polynomial time: preprocessing and kernelization, in *Parameterized and Exact Computation, Proceedings of Second International Workshop, IWPEC '06*, Zürich, Switzerland, September 13–15, 2006, ed. by H. Bodlaender, M. Langston. LNCS, vol. 4169 (Springer, Berlin, 2006), pp. 276–277 (p. xix)

286. M. Fellows, F. Fomin, D. Lokshtanov, F. Rosamond, S. Saurabh, S. Szeider, C. Thomassen, On the complexity of some colorful problems parameterized by treewidth. Inf. Comput. **209**(2), 143–153 (2011) (pp. 277, 558)

287. M.R. Fellows, S. Gaspers, F.A. Rosamond, Parameterizing by the number of numbers. Theory Comput. Syst. **50**(4), 675–693 (2012) (pp. 137, 140, 141)

288. M. Fellows, P. Heggernes, F. Rosamond, C. Sloper, J. Telle, Finding k disjoint triangles in an arbitrary graph, in *Graph-Theoretic Concepts in Computer Science*. LNCS, vol. 3353 (Springer, Berlin, 2004), pp. 235–244 (p. 135)

289. M. Fellows, P. Heggernes, F. Rosamond, C. Sloper, J.A. Telle, Exact algorithms for finding k disjoint triangles in an arbitrary graph, in *Graph-Theoretic Concepts in Computer Science: 30th International Workshop, WG 2004, Revised Papers*, Bad Honnef, Germany, June 2004, ed. by J. Hromkovič, M. Nagl, B. Westfechtel. LNCS, vol. 3353 (Springer, Berlin, 2004), pp. 235–244 (p. 139)

290. M. Fellows, D. Hermelin, M. Müller, F. Rosamond, A purely democratic characterization of $W[1]$, in *Parameterized and Exact Computation, Proceedings of Third International Workshop, IWPEC '08*, Victoria, Canada, May 2008, ed. by M. Grohe, R. Niedermeier. LNCS, vol. 5018 (Springer, Berlin, 2008), pp. 103–114 (pp. 402, 457)

291. M. Fellows, D. Hermelin, F. Rosamond, S. Vialette, On the parameterized complexity of multiple-interval graph problems. Theor. Comput. Sci. **410**, 53–61 (2009) (pp. 401, 578)

292. M. Fellows, C. Knauer, N. Nishimura, P. Ragde, F. Rosamond, U. Stege, D. Thilikos, S. Whitesides, Faster fixed-parameter tractable algorithms for matching and packing problems. Algorithmica **52**(2), 167–176 (2008) (pp. 136, 141, 168)

293. M. Fellows, N. Koblitz, Kid krypto, in *Advances in Cryptology—CRYPTO '92, Proceedings of 12th Annual International Cryptology Conference*, Santa Barbara, California, USA, August 16–20, 1992. LNCS, vol. 740 (Springer, Berlin, 1992), pp. 371–389 (p. 84)

294. M. Fellows, N. Koblitz, Fixed-parameter complexity and cryptography, in *Proceedings of the 10th International Symposium on Applied Algebra, Algebraic Algorithms and Error-Correcting Codes (AAECC 10)*, Puerto Rico, May 1993, ed. by G. Cohen, T. Mora, O. Moreno. LNCS, vol. 673 (Springer, Berlin, 1993), pp. 121–131 (p. 682)

295. M. Fellows, N. Koblitz, *Combinatorial Cryptosystems Galore!* Contemporary Mathematics, vol. 168 (Am. Math. Soc., Providence, 1994), pp. 51–61 (p. 84)

296. M. Fellows, J. Kratochvíl, M. Middendorf, F. Pfeiffer, The complexity of induced minors and related problems. Algorithmica **13**, 266–282 (1995) (p. 369)

297. M. Fellows, M. Langston, Nonconstructive proofs of polynomial-time complexity. Inf. Process. Lett. **26**, 157–162 (1987/88) (pp. 18, 203, 327, 355, 356)

298. M. Fellows, M. Langston, Layout permutation problems and well-partially ordered sets, in *Advanced Research in VLSI, Proceedings of the Fifth MIT Conference*, March 1988, ed. by J. Allen, F. Leighton (MIT Press, Cambridge, 1988), pp. 315–327 (p. 348)

299. M. Fellows, M. Langston, Nonconstructive tools for proving polynomial-time complexity. J. ACM **35**, 727–739 (1988) (pp. 18, 327, 350, 371)

300. M. Fellows, M. Langston, An analogue of the Myhill–Nerode theorem and its use in computing finite-basis characterizations, in *Proceedings of 30th Annual Symposium on Foundations of Computer Science, FOCS 1989*, Research Triangle Park, North Carolina, USA, 30 October–1 November 1989 (IEEE Comput. Soc., Los Alamitos, 1989), pp. 520–525 (pp. 350, 361, 371)

301. M. Fellows, M. Langston, On search, decision and the efficiency of polynomial time algorithms, in *Proceedings of 21st ACM Symposium on Theory of Computing (STOC '89)*, Seattle, Washington, USA, May 15–May 17, 1989, ed. by D. Johnson (ACM, New York, 1989), pp. 501–512. http://dl.acm.org/citation.cfm?id=73055 (pp. 291, 300, 328, 360)

302. M. Fellows, M. Langston, On search, decision, and the efficiency of polynomial-time algorithms. J. Comput. Syst. Sci. **49**(3), 769–779 (1994) (p. 359)

303. M. Fellows, F. Rosamond, U. Rotics, S. Szeider, Clique-width is NP-complete. SIAM J. Discrete Math. **23**(2), 909–939 (2009) (p. 311)

304. W. Fernandez de la Vega, On the maximal cardinality of a consistent set of arcs in a random tournament. J. Comb. Theory, Ser. B **35**, 328–332 (1983) (p. 170)

305. H. Fernau, Graph separator algorithms: a refined analysis, in *Graph-Theoretic Concepts in Computer Science: 28th International Workshop, WG 2002, Revised Papers*, Ceský Krumlov, Czech Republic, June 2002, ed. by L. Kučera. LNCS, vol. 2573 (Springer, Berlin, 2002), pp. 186–197 (pp. 207, 561)

306. H. Fernau, F. Fomin, D. Lokshtanov, D. Raible, S. Saurabh, Y. Villanger, Kernel(s) for problems with no kernel: on out-trees with many leaves, in *Proceedings of 26th International Symposium on Theoretical Aspects of Computer Science, STACS 2009*, Freiburg, Germany, February 26–28, 2009, ed. by S. Albers, J.-Y. Marion, T. Schwentik. Leibniz International Proceedings in Informatics, vol. 3 (Schloss Dagstuhl–Leibniz-Zentrum fuer Informatik, Dagstuhl, 2009), pp. 421–432 (pp. 91, 99, 106, 579, 590)

307. H. Fernau, D. Juedes, A geometric approach to parameterized algorithms for domination problems on planar graphs, in *Mathematical Foundations of Computer Science 2004, Proceedings of 29th International Symposium, MFCS 2004*, Prague, Czech Republic, August 22–27, 2004, ed. by J. Fiala, V. Koubek, J. Kratochvíl. LNCS, vol. 3153 (Springer, Berlin, 2004), pp. 488–499 (p. 561)

308. P. Fishburn, Intransitive indifference in preference theory: a survey. Oper. Res. **18**, 207–228 (1970) (p. 637)

309. J. Flum, M. Grohe, Fixed-parameter tractability, definability, and model checking. SIAM J. Comput. **31**(1), 113–145 (2001) (pp. 182, 281, 399, 503, 504, 506)

310. J. Flum, M. Grohe, The parameterized complexity of counting problems, in *Proceedings of 43rd Symposium on Foundations of Computer Science, FOCS 2002*, Vancouver, BC, Canada, 16–19 November 2002 (IEEE Comput. Soc., Los Alamitos, 2002), pp. 538–547 (pp. 182, 648, 650, 657, 658, 659, 660, 661, 671)

311. J. Flum, M. Grohe, Parameterized complexity and subexponential time. Bull. Eur. Assoc. Theor. Comput. Sci. **84**, 71–100 (2004) (pp. 541, 542, 545, 565, 570)

312. J. Flum, M. Grohe, *Parameterized Complexity Theory*. Texts in Theoretical Computer Science. An EATCS Series (Springer, Berlin, 2006) (pp. viii, xi, 15, 16, 182, 277, 285, 289, 399, 458, 475, 488, 504, 506, 613)

313. J. Flum, M. Grohe, M. Weyer, Bounded fixed-parameter tractability and $\log^2 n$ nondeterministic bits. J. Comput. Syst. Sci. **72**(1), 34–71 (2006) (pp. 412, 506, 561, 612)

314. F. Fomin, P. Golovach, D. Lokshtanov, S. Saurabh, Clique-width: on the price of generality, in *Proceedings of the Twentieth Annual ACM–SIAM Symposium on Discrete Algorithms, SODA 2009*, New York, NY, USA, January 4–6, 2009, ed. by C. Mathieu (SIAM, Philadelphia, 2009), pp. 825–834 (p. 311)

315. F. Fomin, P. Golovach, D. Thilikos, Contraction bidimensionality: the accurate picture, in *Algorithms—ESA 2009: Proceedings of 17th Annual European Symposium*, Copenhagen, Denmark, September 7–9, 2009, ed. by A. Fiat, P. Sanders. LNCS, vol. 5757 (Springer, Berlin, 2009), pp. 706–717 (pp. 563, 564, 570, 685)

316. F. Fomin, F. Grandoni, D. Kratsch, Measure and conquer: domination—a case study, in *Proceedings of 32nd International Colloquium on Automata, Languages and Programming (ICALP 2005)*, Lisbon, Portugal, July 11–15, 2005, ed. by L. Caires, G. Italiano, L. Monteiro, C. Palamidessi, M. Yung. LNCS, vol. 3580 (Springer, Berlin, 2005), pp. 191–203 (pp. 113, 115, 130, 593)

317. F. Fomin, F. Grandoni, D. Kratsch, Measure and conquer: a simple $O(2^{0.288n})$ independent set algorithm, in *Proceedings of the Seventeenth Annual ACM–SIAM Symposium on Discrete Algorithms, SODA 2006*, Miami, Florida, USA, January 22–26, 2006 (ACM, New York, 2006), pp. 18–25 (pp. 113, 130)

318. F. Fomin, F. Grandoni, D. Kratsch, Solving connected dominating set faster than 2^n, in *FSTTCS 2006: Foundations of Software Technology and Theoretical Computer Science, Proceedings of 26th International Conference*, Kolkata, India, December 13–15, 2006, ed. by S. Arun-Kumar, N. Garg. LNCS, vol. 4337 (Springer, Berlin, 2006), pp. 152–163 (p. 130)

319. F. Fomin, F. Grandoni, A. Pyatkin, A. Stepanov, Bounding the number of minimal dominating sets: a measure and conquer approach, in *Algorithms and Computation: Proceedings of 16th International Symposium, ISAAC 2005*, Sanya, Hainan, China, December 19–21, 2005, ed. by X. Deng, D. Du. LNCS, vol. 3827 (Springer, Berlin, 2005), pp. 573–582 (p. 130)

320. F. Fomin, D. Kratsch, G. Woeginger, Exact (exponential) algorithms for the dominating set problem, in *Graph-Theoretic Concepts in Computer Science: 30th International Workshop, WG 2004, Revised Papers*, Bad Honnef, Germany, June 2004, ed. by J. Hromkovič, M. Nagl, B. Westfechtel. LNCS, vol. 3353 (Springer, Berlin, 2004), pp. 245–256 (p. 593)

321. F. Fomin, D. Lokshtanov, V. Raman, S. Saurabh, Fast local search algorithm for weighted feedback arc set in tournaments, in *Proceedings of the Twenty-Fourth AAAI Conference on Artificial Intelligence, AAAI 2010*, Atlanta, Georgia, USA, July 11–15, 2010 (AAAI Press, Menlo Park, 2010) (pp. 169, 684)

322. F. Fomin, D. Lokshtanov, V. Raman, S. Saurabh, Bidimensionality and EPTAS, in *Proceedings of the Twenty-Second Annual ACM–SIAM Symposium on Discrete Algorithms, SODA 2011*, San Francisco, California, USA, January 23–25, 2011, ed. by D. Randall (SIAM, Philadelphia, 2011), pp. 748–759 (p. 565)

323. F. Fomin, D. Lokshtanov, S. Saurabh, *Kernelization: Theory of Parameterized Compressibility* (Cambridge University Press, Cambridge, to appear) (pp. 372, 565)

324. F. Fomin, D. Lokshtanov, S. Saurabh, D. Thilikos, Bidimensionality and kernels, in *Proceedings of the Twenty-First Annual ACM–SIAM Symposium on Discrete Algorithms, SODA 2010*, Austin, Texas, USA, January 17–19, 2010, ed. by M. Charikar (SIAM, Philadelphia, 2010), pp. 503–510 (pp. 369, 565)

325. F. Fomin, D. Thilikos, Dominating sets in planar graphs: branch-width and exponential speed-up. SIAM J. Comput. **36**(2), 281–309 (2006) (p. 561)

326. F. Fomin, Y. Villanger, Subexponential parameterized algorithm for minimum fill-in, in *Proceedings of the Twenty-Third Annual ACM–SIAM Symposium on Discrete Algorithms, SODA 2012*, Kyoto, Japan, January 17–19, 2012, ed. by Y. Rabani (SIAM, Philadelphia, 2012), pp. 1737–1746 (pp. 46, 565)

327. F. Fomina, D. Marx, FPT suspects and tough customers: open problems of Downey and Fellows, in *The Multivariate Algorithmic Revolution and Beyond: Essays Dedicated to Michael R. Fellows on the Occasion of His 60th Birthday*, ed. by H. Bodlaender, R. Downey, F. Fomin, D. Marx. LNCS, vol. 7370 (Springer, Berlin, 2012), pp. 457–468 (p. 677)

328. L. Fortnow, R. Santhanam, Infeasibility of instance compression and succinct PCPs for NP. J. Comput. Syst. Sci. **77**(1), 91–106 (2011). Special issue Celebrating Karp's Kyoto Prize (pp. 572, 573, 584, 588, 619)

329. J. Fouhy, Computational experiments on graph width metrics, Master's thesis, Victoria University of Wellington, 2002 (pp. 193, 207, 313)

330. G. Frederickson, R. Janardan, Designing networks with compact routing tables. Algorithmica **3**, 171–190 (1988) (p. 355)

331. M. Fredman, J. Komlós, E. Szemerédi, Storing a sparse table with $O(1)$ worst-case access time, in *Proceedings of 23rd Annual Symposium on Foundations of Computer Science, FOCS 1982*, Chicago, Illinois, USA, 3–5 November 1982 (IEEE Comput. Soc., Los Alamitos, 1982), pp. 165–169 (p. 145)

332. M. Frick, M. Grohe, Deciding first-order properties of locally tree-decomposable graphs, in *Proceedings of 26th International Colloquium on Automata, Languages and Programming (ICALP 1999)*, Prague, Czech Republic, July 11–15, 1999, ed. by J. Wiedermann, P. van Emde Boas, M. Nielsen. LNCS, vol. 1644 (Springer, Berlin, 1999), pp. 331–340 (pp. 284, 286, 289, 646)

333. H. Friedman, N. Robertson, P. Seymour, The metamathematics of the graph minor theorem, in *Logic and Combinatorics: Proceedings of the AMS–IMS–SIAM Joint Summer Research Conference*, Arcata, California, 1985, ed. by S. Simpson. Contemporary Mathematics, vol. 65 (Am. Math. Soc., Providence, 1987), pp. 229–261 (p. 350)

334. H. Gabow, M. Stallmann, An augmenting path algorithm for linear matroid parity. Combinatorica **6**(2), 123–150 (1986) (pp. 123, 704)

335. H. Gaifman, On local and non-local properties, in *Proceedings of the Herbrand Symposium: Logic Colloquium '81*, Marseilles, France, July, 1981, ed. by J. Stern. Studies in

Logic and the Foundations of Mathematics, vol. 107 (North-Holland, Amsterdam, 1982), pp. 105–125 (pp. 286, 289)

336. R. Ganian, Twin cover: beyond vertex cover in parameterized algorithmics, in *Parameterized and Exact Computation, 6th International Symposium, IPEC '11, Revised Selected Papers*, Saarbrücken, Germany, September 6–8, 2011, ed. by D. Marx, P. Rossmanith. LNCS, vol. 7112 (Springer, Berlin, 2011), pp. 259–271 (pp. 83, 288)

337. M. Garey, D. Johnson, *Computers and Intractability. A Guide to the Theory of NP-Completeness* (Freeman, San Francisco, 1979) (pp. xvii, xviii, 7, 20, 33, 37, 82, 83, 84, 85, 87, 171, 173, 175, 176, 412, 421, 487, 492, 592, 625)

338. M. Garey, D. Johnson, L. Stockmeyer, Some simplified NP-complete graph problems. Theor. Comput. Sci. **1**(3), 237–267 (1976) (p. 162)

339. M. Garey, D. Johnson, R. Tarjan, The planar Hamiltonian circuit problem is NP-complete. SIAM J. Comput. **5**, 704–714 (1976) (p. 566)

340. S. Gaspers, S. Szeider, Backdoors to satisfaction, in *The Multivariate Algorithmic Revolution and Beyond: Essays Dedicated to Michael R. Fellows on the Occasion of His 60th Birthday*, ed. by H. Bodlaender, R. Downey, F. Fomin, D. Marx. LNCS, vol. 7370 (Springer, Berlin, 2012), pp. 287–317 (p. vii)

341. F. Gavril, The intersection graphs of subtrees in trees are exactly the chordal graphs. J. Comb. Theory, Ser. B **16**, 47–56 (1974) (pp. 188, 208)

342. J. Geelen, B. Gerards, G. Whittle, Towards a structure theory for matrices and matroids, in *International Congress of Mathematicians*, Madrid, August 22–30, 2006, vol. III, ed. by M. Sanz-Solé, J. Soria, J. Varona, J. Verdera (Eur. Math. Soc., Zürich, 2006), pp. 827–842 (pp. 281, 680)

343. J. Geelen, B. Gerards, G. Whittle, Rota's conjecture, to appear. Result announced at the British Combinatorial Conference, 2013 (p. 679)

344. J. Geske, On the structure of intractable sets (binary relation), PhD thesis, Iowa State University, Ames, Iowa, USA, 1987 (pp. 567, 568)

345. F. Giécseg, M. Steinby, *Tree Automata* (Akad. Kiadó, Budapest, 1984) (pp. 237, 246)

346. I. Giotis, V. Guruswami, Correlation clustering with a fixed number of clusters. Theory Comput. **2**, 249–266 (2006) (p. 686)

347. S. Goldwasser, M. Sipser, Private coins versus public coins in interactive proof systems, in *Proceedings of 18th ACM Symposium on Theory of Computing (STOC '86)*, Berkeley, California, USA, May 28–May 30, 1986, ed. by J. Hartmanis (1986), pp. 59–68. http://doi.acm.org/10.1145/12130.12137 (p. 586)

348. P. Golovach, M. Kamiński, D. Paulusma, D. Thilikos, Induced packing of odd cycles in a planar graph, in *Algorithms and Computation: Proceedings of 20th International Symposium, ISAAC 2009*, Honolulu, Hawaii, December 16–18, 2009, ed. by Y. Dong, D.-Z. Du, O. Ibarra. LNCS, vol. 5878 (Springer, Berlin, 2009), pp. 514–523 (p. 365)

349. P. Golovach, D. Thilikos, Paths of bounded length and their cuts: parameterized complexity and algorithms, in *Parameterized and Exact Computation, Proceedings of 4th International Workshop, IWPEC '09*, Copenhagen, Denmark, September 2009, ed. by J. Chen, F. Fomin. LNCS, vol. 5917 (Springer, Berlin, 2009), pp. 210–221 (p. 277)

350. P. Golovach, D. Thilikos, Paths of bounded length and their cuts: parameterized complexity and algorithms. Discrete Optim. **8**(1), 72–86 (2011) (p. 135)

351. M. Golumbic, U. Rotics, On the clique-width of some perfect graph classes. Int. J. Found. Comput. Sci. **11**(3), 423–443 (2000) (p. 315)

352. R. Govindan, M. Langston, B. Plaut, Graph partitioning and cutting, Technical report, Computer Science Department, University of Tennessee, 1997 (p. 349)

353. J. Gramm, J. Guo, F. Hüffner, R. Niedermeier, Automated generation of search tree algorithms for hard graph modification problems. Algorithmica **39**(4), 321–347 (2004) (pp. 591, 615)

354. J. Gramm, J. Guo, F. Hüffner, R. Niedermeier, Graph modeled data clustering: exact algorithms for clique generation. Theory Comput. Syst. **38**(4), 373–392 (2005) (pp. 45, 46)

355. J. Gramm, J. Guo, F. Hüffner, R. Niedermeier, Data reduction and exact algorithms for clique cover. ACM J. Exp. Algorithmics **13**, 2 (2008) (pp. 591, 615)

356. J. Gramm, A. Nickelsen, T. Tantau, Fixed-parameter algorithms in phylogenetics. Comput. J. **51**(1), 79–101 (2008) (p. 70)

357. J. Gramm, R. Niedermeier, P. Rossmanith, Fixed-parameter algorithms for closest string and related problems. Algorithmica **37**, 25–42 (2003) (pp. 36, 37, 81, 89, 137, 138, 139, 141, 549)

358. M. Grohe, M. Grüber, Parameterized approximability of the disjoint cycle problem, in *Proceedings of 34th International Colloquium on Automata, Languages and Programming (ICALP 2007)*, Wrocław, Poland, July 9–13, 2007, ed. by L. Arge, C. Cachin, T. Jurdzinski, A. Tarlecki. LNCS, vol. 4596 (Springer, Berlin, 2007), pp. 363–374 (pp. 628, 643)

359. M. Grohe, K. Kawarabayashi, D. Marx, P. Wollan, Finding topological subgraphs is fixed-parameter tractable, in *Proceedings of 43rd ACM Symposium on Theory of Computing (STOC '11)*, San Jose, California, USA, June 6–June 8, 2011, ed. by L. Fortnow, S. Vadhan (ACM, New York, 2011), pp. 479–488 (pp. 325, 348, 365)

360. M. Grohe, D. Marx, Descriptive complexity, canonisation, and definable graph structure theory, to appear (pp. 312, 339)

361. Q.-P. Gu, H. Tamaki, Optimal branch decomposition of planar graphs in $O(n^3)$ time. ACM Trans. Algorithms **4**(3), 1–13 (2008) (p. 562)

362. Q.-P. Gu, H. Tamaki, Improved bounds on the planar branchwidth with respect to the largest grid minor size, in *Algorithms and Computation: Proceedings of 21st International Symposium, ISAAC 2010*, Jeju Island, Korea, December 15–17, 2010, ed. by O. Cheong, K.-Y. Chwa, K. Park. LNCS, vol. 6507 (Springer, Berlin, 2010), pp. 85–96 (p. 562)

363. D.J. Guan, Generalized Gray codes with applications. Proc. Natl. Sci. Counc., Repub. China **22**, 841–848 (1998) (p. 112)

364. S. Guillemot, D. Marx, Finding small patterns in permutations in linear time, arXiv: 1307.3073 (p. 680)

365. J. Guo, J. Gramm, F. Hüffner, R. Niedermeier, S. Wernicke, Compression-based fixed-parameter algorithms for feedback vertex set and edge bipartization. J. Comput. Syst. Sci. **72**(8), 1386–1396 (2006) (pp. 109, 111, 112, 129)

366. J. Guo, H. Moser, R. Niedermeier, Iterative compression for exactly solving NP-hard minimization problems, in *Algorithmics of Large and Complex Networks*, ed. by J. Lerner, D. Wagner, K. Zweig. LNCS, vol. 5515 (Springer, Berlin, 2009), pp. 65–80 (pp. 129, 130, 170)

367. Y. Gurevich, L. Harrington, Automata, trees and games, in *Proceedings of 14th Annual ACM Symposium on Theory of Computing (STOC '82)*, San Francisco, California, USA, May 5–May 7, 1982, ed. by H. Lewis, B. Simons, W. Burkhard, L. Landweber (ACM, New York, 1982), pp. 60–65 (p. 276)

368. Y. Gurevich, S. Shelah, Nearly linear time, in *Logic at Botik '89: Symposium on Logical Foundations of Computer Science*, ed. by A. Meyer, M. Taitslin. LNCS, vol. 363 (Springer, Berlin, 1989), pp. 108–118 (pp. 405, 567)

369. J. Gustedt, Well quasi ordering finite posets and formal languages. J. Comb. Theory, Ser. B **65**(1), 111–124 (1995) (pp. 332, 337, 681)

370. G. Gutin, E. Kim, M. Lampis, V. Mitsou, Vertex cover problem parameterized above and below tight bounds. Theory Comput. Syst. **48**(2), 402–410 (2011) (p. 129)

371. G. Gutin, E. Kim, S. Szeider, A. Yeo, A probabilistic approach to problems parameterized above or below tight bounds. J. Comput. Syst. Sci. **77**(2), 422–429 (2011). Preliminary version in *IWPEC'09*, Lecture Notes in Computer Science, vol. 5917 (2009), pp. 234–245 (p. 168)

372. G. Gutin, L. van Iersel, M. Mnich, A. Yeo, Every ternary permutation constraint satisfaction problem parameterized above average has a kernel with a quadratic number of variables. J. Comput. Syst. Sci. **78**(1), 151–163 (2012) (p. 162)

373. M. Gyssens, J. Paredaens, A decomposition methodology for cyclic databases, in *Advances in Database Theory, Vol. 2, Based on the Proceedings of the Workshop on Logical*

Data Bases ADBT 1982, Toulouse, France (Centre d'études et de recherches de Toulouse, Toulouse, 1982), pp. 85–122 (p. 316)

374. T. Hagerup, A strengthened analysis of an algorithm for DOMINATING SET in planar graphs. Discrete Appl. Math. **160**(6), 793–798 (2012) (pp. 31, 33, 44, 46)

375. G. Haggard, D. Pearce, G. Royle, Computing Tutte polynomials. ACM Trans. Algorithms **37**, 1–17 (2010) (p. 59)

376. M. Hallett, G. Gonnet, U. Stege, Vertex Cover revisited: a hybrid algorithm of theory and heuristic, Manuscript, 1998 (pp. 54, 88)

377. M. Hallett, C. McCartin, A faster FPT algorithm for the maximum agreement forest problem. Theory Comput. Syst. **41**(3), 539–550 (2007) (pp. 74, 81, 89, 626)

378. F. Harary, *Graph Theory* (Addison-Wesley, Reading, 1969) (p. 323)

379. D. Harnik, M. Naor, On the compressibility of NP instances and cryptographic applications, in *Proceedings of 47th Annual IEEE Symposium on Foundations of Computer Science, FOCS 2006*, Berkeley, California, USA, 21–24 October 2006 (IEEE Comput. Soc., Los Alamitos, 2006), pp. 719–728 (pp. 573, 619)

380. J. Hartmanis, Gödel, von Neumann and the P =? NP problem. Bull. Eur. Assoc. Theor. Comput. Sci. **38**, 101–107 (1989) (pp. 12, 382)

381. J. Hartmanis, R. Stearns, On the computational complexity of algorithms. Trans. Am. Math. Soc. **117**, 285–306 (1965) (p. 12)

382. S. Hartung, R. Niedermeier, Incremental list coloring of graphs, parameterized by conservation. Theor. Comput. Sci. **494**, 86–98 (2013) (p. 686)

383. N. Hasan, C.L. Liu, Minimum fault coverage in reconfigurable arrays, in *Proceedings of the Eighteenth International Symposium on Fault-Tolerant Computing, FTCS 1988*, Tokyo, Japan, 27–30 June 1988 (IEEE Comput. Soc., Los Alamitos, 1988), pp. 348–353 (p. 88)

384. J. Håstad, Some optimal inapproximability results. J. ACM **48**(4), 798–859 (2001) (p. 162)

385. P. Heggernes, P. van't Hof, B. Lévêque, D. Lokshtanov, C. Paul, Contracting graphs to paths and trees, in *Parameterized and Exact Computation, 6th International Symposium, IPEC '11, Revised Selected Papers*, Saarbrücken, Germany, September 6–8, 2011, ed. by D. Marx, P. Rossmanith. LNCS, vol. 7112 (Springer, Berlin, 2011), pp. 55–66 (p. 615)

386. J. Hein, T. Jiang, L. Wang, K. Zhang, On the complexity of comparing evolutionary trees. Discrete Appl. Math. **71**(1–3), 153–169 (1996) (p. 71)

387. R. Hemmecke, S. Onn, L. Romanchuk, N-fold integer programming in cubic time (hence multiway tables are fixed-parameter tractable) (with Raymond Hemmecke and Lyubov Romanchuk). Math. Program. **137**, 325–341 (2013) (p. 681)

388. D. Hermelin, S. Kratsch, K. Soltys, M. Wahlström, X. Wu, Hierarchies of inefficient kernelizability, arXiv:1110.0976v1 (pp. 398, 579, 612, 614, 618, 619)

389. D. Hermelin, X. Wu, Weak compositions and their applications to polynomial lower bounds for kernelization. Electron. Colloq. Comput. Complex. **18**, 72 (2011) (pp. 596, 606, 607, 608, 611, 619)

390. I. Hicks, Planar branch decompositions I: The ratcatcher. INFORMS J. Comput. **17**(4), 402–412 (2005) (p. 302)

391. I. Hicks, Planar branch decompositions II: The cycle method. INFORMS J. Comput. **17**(4), 413–421 (2005) (p. 302)

392. I. Hicks, N. McMurray, The branchwidth of graphs and their cycle matroids. J. Comb. Theory, Ser. B **97**, 681–692 (2007) (p. 282)

393. G. Higman, Ordering by divisibility in abstract algebras. Proc. Lond. Math. Soc. **2**, 326–336 (1952) (pp. 320, 332, 337)

394. P. Hliněný, Branch-width, parse trees, and monadic second-order logic for matroids. J. Comb. Theory, Ser. B **96**(3), 325–351 (2006) (pp. 282, 289)

395. P. Hliněný, S.-I. Oum, Finding branch-decompositions and rank-decompositions. SIAM J. Comput. **38**(3), 1012–1032 (2008) (p. 311)

396. P. Hliněný, S.-I. Oum, D. Seese, G. Gottlob, Width parameters beyond treewidth and their applications. Comput. J. **51**(3), 326–362 (2008) (pp. 302, 316)

397. P. Hliněný, G. Whittle, Matroid tree-width. Eur. J. Comb. **27**, 1117–1128 (2006) (pp. 281, 289, 680)
398. P. Hliněný, G. Whittle, Addendum to "Matroid treewidth". Eur. J. Comb. **30**(4), 1036–1044 (2009) (pp. 281, 289)
399. S. Homer, A. Selman, *Computability and Complexity Theory*, 2nd edn. (Springer, Berlin, 2011) (p. 13)
400. J. Hopcroft, R. Tarjan, Efficient planarity testing. J. ACM **21**(4), 549–568 (1974) (p. 291)
401. J. Hopcroft, J. Ullmann, *Introduction to Automata Theory, Languages and Computation* (Addison-Wesley, Reading, 1979) (pp. 216, 240)
402. J. Howie, *An Introduction to Semigroup Theory*. London Mathematical Society Monographs (Academic Press, San Diego, 1976) (p. 403)
403. D. Huffmann, The synthesis of sequential switching circuits. J. Franklin Inst. **257**(3–4), 161–190, 275–303 (1954) (pp. 216, 234)
404. F. Hüffner, Algorithm engineering for optimal graph bipartization, in *WEA 2005: 4th International Workshop on Efficient and Experimental Algorithms*, Santorini Island, Greece, May 10–13, 2005, ed. by S. Nikoletseas. LNCS, vol. 3503 (Springer, Berlin, 2005), pp. 240–252 (pp. 109, 112, 129, 306)
405. F. Hüffner, C. Komusiewicz, H. Moser, R. Niedermeier, Fixed-parameter algorithms for cluster vertex deletion. Theory Comput. Syst. **47**, 196–217 (2010) (p. 130)
406. F. Hüffner, S. Wernicke, T. Zichner, Algorithm engineering for color-coding with applications to signaling pathway detection. Algorithmica **52**, 114–132 (2008) (p. 169)
407. P. Hunter, S. Kreutzer, Digraph measures: Kelly decompositions, games and orderings, in *Proceedings of the Eighteenth Annual ACM–SIAM Symposium on Discrete Algorithms, SODA 2007*, New Orleans, Louisiana, USA, January 7–9, 2007, ed. by N. Bansal, K. Pruhs, C. Stein (SIAM, Philadelphia, 2007), pp. 637–644 (p. 312)
408. O. Ibarra, C. Kim, Fast approximation algorithms for the knapsack and sum of subset problems. J. ACM **22**(4), 463–468 (1975) (p. 174)
409. N. Immermann, Expressibility as a complexity measure: results and directions, in *Proceedings of Second Annual Structure in Complexity Theory Conference*, Cornell University, Ithaca, NY, June 16–19, 1987 (IEEE Comput. Soc., Los Alamitos, 1987), pp. 194–202 (p. 182)
410. R. Impagliazzo, R. Paturi, The complexity of k-sat, in *Proceedings of the 14th Annual IEEE Conference on Computational Complexity, CCC 1999*, Atlanta, GA, USA, May 4–6, 1999 (IEEE Comput. Soc., Los Alamitos, 1999), pp. 237–240 (pp. 305, 559, 570)
411. R. Impagliazzo, R. Paturi, F. Zane, Which problems have strongly exponential complexity? J. Comput. Syst. Sci. **63**(4), 512–530 (2001) (pp. 543, 546, 547, 552, 570)
412. S. Irani, Coloring inductive graphs on-line, in *Proceedings of 31st Annual Symposium on Foundations of Computer Science, FOCS 1990*, vol. II, St. Louis, Missouri, USA, October 22–24, 1990 (IEEE Comput. Soc., Los Alamitos, 1990), pp. 470–479 (p. 636)
413. A. Itai, M. Rodeh, Finding a minimum circuit in a graph. SIAM J. Comput. **7**(4), 413–423 (1978) (pp. 33, 35, 36)
414. Y. Iwata, A faster algorithm for dominating set analyzed by the potential method, in *Parameterized and Exact Computation, 6th International Symposium, IPEC '11, Revised Selected Papers*, Saarbrücken, Germany, September 6–8, 2011, ed. by D. Marx, P. Rossmanith. LNCS, vol. 7112 (Springer, Berlin, 2011), pp. 41–54 (p. 130)
415. K. Jansen, S. Kratsch, D. Marx, I. Schlotter, Bin packing with fixed number of bins revisited, in *Algorithm Theory—SWAT 2010, Proceedings of 12th Scandinavian Workshop on Algorithm Theory*, Bergen, Norway, June 21–23, 2010, ed. by H. Kaplan. LNCS, vol. 6139 (Springer, Berlin, 2010), pp. 260–272 (p. 643)
416. W. Jia, C. Zhang, J. Chen, An efficient parameterized algorithm for m-set packing. J. Algorithms **50**(1), 106–117 (2004) (pp. 134, 135, 141)
417. M. Jiang, On the parameterized complexity of some optimization problems related to multiple-interval graphs. Theor. Comput. Sci. **411**, 4253–4262 (2010) (p. 402)

418. M. Jiang, Y. Zhang, Parameterized complexity in multiple-interval graphs: domination, in *Parameterized and Exact Computation, 6th International Symposium, IPEC '11, Revised Selected Papers*, Saarbrücken, Germany, September 6–8, 2011, ed. by D. Marx, P. Rossmanith. LNCS, vol. 7112 (Springer, Berlin, 2011), pp. 27–40 (p. 402)

419. D. Johnson, Approximation algorithms for combinatorial problems, in *Proceedings of 5th ACM Symposium on Theory of Computing (STOC '73)*, Austin, Texas, USA, April 30–May 2, 1973, ed. by A. Aho, et al. (ACM, New York, 1973), pp. 38–49 (pp. 162, 167)

420. D. Johnson, A. Demers, J. Ullman, M. Garey, R. Graham, Worst case performance bounds for simple one dimensional packing algorithms. SIAM J. Comput. **3**(4), 299–325 (1974) (p. 181)

421. T. Johnson, N. Robertson, P. Seymour, R. Thomas, Addendum to Directed tree-width. http://www.math.gatech.edu/~thomas/PAP/diradd.pdf (p. 312)

422. T. Johnson, N. Robertson, P. Seymour, R. Thomas, Directed tree-width. J. Comb. Theory, Ser. B **82**(1), 138–154 (2001) (p. 312)

423. D. Joseph, R. Pruim, P. Young, Collapsing degrees in subexponential time, in *Proceedings of Ninth Annual Structure in Complexity Theory Conference*, University of Wisconsin, Madison, Wisconsin, June 28–July 1, 1994 (IEEE Comput. Soc., Los Alamitos, 1994), pp. 367–382 (p. 567)

424. H.A. Jung, On subgraphs without cycles in tournaments, in *Combinatorial Theory and Its Applications*, vol. II (1970), pp. 675–677 (p. 170)

425. V. Kabanets, Recognizability equals definability for partial *k*-paths, Master's thesis, Simon Fraser University, June 1996 (p. 277)

426. V. Kabanets, Recognizability equals definability for partial *k*-paths, in *Proceedings of 24th International Colloquium on Automata, Languages and Programming (ICALP 1997)*, Bologna, Italy, July 7–11, 1997, ed. by P. Degano, R. Gorrieri, A. Marchetti-Spaccamela. LNCS, vol. 1256 (Springer, Berlin, 1997), pp. 805–815 (p. 277)

427. D. Kaller, Definability equals recognizability for partial 3-trees, in *Graph-Theoretic Concepts in Computer Science 22nd International Workshop, WG 1996, Revised Papers*, Cadenabbia (Como), Italy, June 12–14, 1996, ed. by F. d'Amore, P.G. Franciosa, A. Marchetti-Spaccamela. LNCS, vol. 1197 (Springer, Berlin, 1997), pp. 239–253 (p. 277)

428. I. Kanj, L. Perković, Improved parameterized algorithms for planar dominating set, in *Mathematical Foundations of Computer Science 2002, Proceedings of 27th International Symposium, MFCS 2002*, Warsaw, Poland, August 26–30, 2002. LNCS, vol. 2420 (Springer, Berlin, 2002), pp. 399–410 (p. 561)

429. R. Kannan, Minkowski's convex body theorem and integer programming. Math. Oper. Res. **12**, 415–440 (1987) (pp. 137, 141)

430. H. Kaplan, R. Shamir, R. Tarjan, Tractability of parameterized completion problems on chordal and interval graphs: minimum fill-in and physical mapping (extended abstract), in *Proceedings of 35th Annual Symposium on Foundations of Computer Science, FOCS 1994*, Santa Fe, New Mexico, USA, 20–22 November 1994 (IEEE Comput. Soc., Los Alamitos, 1994), pp. 780–891 (pp. 45, 46, 69)

431. R. Karp, Reducibility among combinatorial problems, in *Complexity of Computer Computations*, ed. by R. Miller, J. Thatcher (Plenum, New York, 1972), pp. 45–68 (pp. 5, 13, 113, 376, 382, 553)

432. R. Karp, R. Lipton, Some connections between uniform and non-uniform complexity classes, in *Proceedings of 12th ACM Symposium on Theory of Computing (STOC '80)*, Los Angeles, California, USA, April 28–April 30, 1980, ed. by R. Miller, S. Ginsburg, W. Burkhard, R. Lipton (ACM, New York, 1980), pp. 302–309 (pp. 7, 13)

433. K. Kawarabayashi, The disjoint paths problem: algorithm and structure, in *WALCOM: Algorithms and Computation, Proceedings of 5th International Workshop, WALCOM 2011*, New Delhi, India, February 2011, ed. by N. Katoh, A. Kumar. LNCS, vol. 6552 (Springer, Berlin, 2011), pp. 2–7 (p. 364)

434. K. Kawarabayashi, Y. Kobayashi, The induced disjoint path problem, in *Integer Programming and Combinatorial Optimization, Proceedings of 13th International Conference,*

IPCO 2008, Bertinoro, Italy, May 26–28, 2008, ed. by A. Lodi, A. Panconesi, G. Rinaldis. LNCS, vol. 5035 (Springer, Berlin, 2008), pp. 47–61 (p. 365)

435. K. Kawarabayashi, B. Reed, Odd cycle packing, in *Proceedings of 42nd ACM Symposium on Theory of Computing (STOC 2010)*, Cambridge, MA, June 6–June 8, 2010, ed. by L. Schulman (ACM, New York, 2010), pp. 695–704 (p. 365)

436. K. Kawarabayashi, P. Wollan, A shorter proof of the graph minor algorithm: the unique linkage theorem, in *Proceedings of 42nd ACM Symposium on Theory of Computing (STOC 2010)*, Cambridge, MA, June 6–June 8, 2010, ed. by L. Schulman (ACM, New York, 2010), pp. 687–694 (pp. 348, 365)

437. J. Kennedy, L. Quintas, M. Sysło, The theorem on planar graphs. Hist. Math. **12**(4), 356–368 (1985) (p. 322)

438. C. Kenyon-Mathieu, W. Schudy, How to rank with few errors, in *Proceedings of 39th ACM Symposium on Theory of Computing (STOC '07)*, San Diego, California, June 11–June 13, 2007, ed. by D. Johnson, U. Feige (ACM, New York, 2007), pp. 95–103 (p. 170)

439. S. Khot, V. Raman, Parameterized complexity of finding subgraphs with hereditary properties. Theor. Comput. Sci. **289**(2), 997–1008 (2002) (pp. 67, 603)

440. S. Khot, O. Regev, Vertex cover might be hard to approximate to within $2 - \epsilon$, in *Proceedings of the 18th Annual IEEE Conference on Computational Complexity, CCC 2003*, Aarhus, Denmark, July 7–10, 2003 (IEEE Comput. Soc., Los Alamitos, 2003) (pp. 57, 89)

441. H. Kierstead, The linearity of first-fit colouring of interval graphs. SIAM J. Discrete Math. **1**, 526–530 (1988) (p. 638)

442. H. Kierstead, J. Qin, Colouring interval graphs with first-fit. Discrete Math. **144**, 47–57 (1995) (p. 638)

443. H. Kierstead, W. Trotter, An extremal problem in recursive combinatorics. Congr. Numer. **33**, 143–153 (1981) (pp. 178, 179, 636, 637, 638, 643)

444. C. Kintala, P. Fischer, Refining nondeterminism in relativised polynomial time bounded computations. SIAM J. Comput. **9**, 46–53 (1980) (pp. 405, 536, 567)

445. D. Kirkpatrick, P. Hell, On the completeness of the generalized matching problem, in *Proceedings of 10th ACM Symposium on Theory of Computing (STOC '78)*, San Diego, California, USA, May 1–May 3, 1978, ed. by R. Lipton, W. Burkhard, W. Savitch, E. Friedman, A. Aho (ACM, New York, 1978), pp. 240–245 (p. 618)

446. L. Kirousis, D. Thilikos, The linkage of a graph. SIAM J. Comput. **25**(3), 626–647 (1996) (p. 313)

447. H. Klauck, A. Nayak, A. Ta-Shma, D. Zuckerman, Interaction in quantum communication. IEEE Trans. Inf. Theory **53**(6), 1970–1982 (2007) (p. 584)

448. S. Kleene, Representation of events in nerve nets and finite automata, in *Automata Studies*, ed. by C. Shannon, J. McCarthy. Annals of Mathematical Studies (Princeton University Press, Princeton, 1956), pp. 3–42 (pp. 216, 224, 226)

449. D. Kleitman, D. West, Spanning trees with many leaves. SIAM J. Discrete Math. **4**(1), 99–106 (1991) (p. 134)

450. T. Kloks, C.-M. Lee, J. Liu, New algorithms for k-face cover, k-feedback vertex set, and k-disjoint cycles on plane and planar graphs, in *Graph-Theoretic Concepts in Computer Science: 28th International Workshop, WG 2002, Revised Papers*, Ceský Krumlov, Czech Republic, June 2002, ed. by L. Kućera. LNCS, vol. 2573 (Springer, Berlin, 2002), pp. 282–295 (p. 561)

451. J. Kneis, A. Langer, *A Practical Approach to Courcelle's Theorem*. Electronic Notes in Theoretical Computer Science, vol. 251 (Elsevier, Amsterdam, 2009), pp. 65–81 (pp. 272, 277)

452. J. Kneis, A. Langer, P. Rossmanith, Improved upper bounds for partial vertex cover, in *Graph-Theoretic Concepts in Computer Science: 34th International Workshop, WG 2008, Revised Papers*, Durham, United Kingdom, June 30–July 2, 2008. LNCS, vol. 5344 (Springer, Berlin, 2008), pp. 240–251 (p. 105)

453. J. Kneis, A. Langer, P. Rossmanith, Courcelle's theorem—a game theoretic approach. Discrete Optim. **8**(4), 568–594 (2011) (pp. 272, 277)

454. Y. Kobayashi, K. Kawarabayashi, Algorithms for finding an induced cycle in planar graphs and bounded genus graphs, in *Proceedings of the Twentieth Annual ACM–SIAM Symposium on Discrete Algorithms, SODA 2009*, New York, NY, USA, January 4–6, 2009, ed. by C. Mathieu (SIAM, Philadelphia, 2009), pp. 1146–1155 (p. 365)

455. N. Koblitz, E-mail communication, 1995 (p. 146)

456. P. Kolaitis, M. Vardi, Conjunctive query containment and constraint satisfaction. J. Comput. Syst. Sci. **61**(2), 302–332 (2000) (p. 316)

457. A. Koster, H. Bodlaender, S. van Hoesel, *Treewidth: Computational Experiments*. Electronic Notes in Discrete Mathematics, vol. 8 (Elsevier, Amsterdam, 2001) (p. 207)

458. A. Koster, S. van Hoesel, A. Kolen, Solving partial constraint satisfaction problems with tree decompositions. Networks **40**(3), 170–180 (2002) (p. 185)

459. I. Koutis, Perfect hashing in three algebraic problems, Manuscript, 2004 (p. 167)

460. I. Koutis, Faster algebraic algorithms for path and packing problems, in *Proceedings of 35th International Colloquium on Automata, Languages and Programming (ICALP 2008), Part I*, Reykjavik, Iceland, July 7–11, 2008, ed. by L. Aceto, I. Damgård, L. Goldberg, M. Halldórsson, A. Ingólfsdóttir, I. Walukiewicz. LNCS, vol. 5125 (Springer, Berlin, 2008), pp. 575–586 (pp. 148, 149, 150, 151, 153, 169)

461. I. Koutis, R. Williams, Limits and applications of group algebras for parameterized problems, in *Proceedings of 36th International Colloquium on Automata, Languages and Programming (ICALP 2009), Part I*, Rhodes, Greece, July 5–12, 2009, ed. by S. Albers, A. Marchetti-Spaccamela, Y. Matias, S. Nikoletseas, W. Thomas. LNCS, vol. 5555 (Springer, Berlin, 2009), pp. 653–664 (pp. 149, 153, 167, 169)

462. A. Koutsonas, D. Thilikos, Planar feedback vertex set and face cover: combinatorial bounds and subexponential algorithms. Algorithmica **60**(4), 987–1003 (2011) (p. 561)

463. D. Kozen, On the Myhill–Nerode theorem for trees. Bull. Eur. Assoc. Theor. Comput. Sci. **47**, 170–173 (1992) (pp. 237, 246)

464. D. Kozen, On regularity preserving functions. Bull. Eur. Assoc. Theor. Comput. Sci. **58**, 131–138 (1996) (p. 236)

465. D. Kozen, *Theory of Computation: Classical and Contemporary Approaches* (Springer, London, 2006) (p. 7)

466. D. Kratsch, S. Kratsch, The jump number problem: exact and parameterized, in *Proceedings of IPEC 2013* (2013) (p. 46)

467. D. Kratsch, M. Liedloff, An exact algorithm for the minimum dominating clique problem, in *Parameterized and Exact Computation, Proceedings of Second International Workshop, IWPEC '06*, Zürich, Switzerland, September 13–15, 2006, ed. by H. Bodlaender, M. Langston. LNCS, vol. 4169 (Springer, Berlin, 2006), pp. 78–89 (p. 113)

468. S. Kratsch, Co-nondeterminism in compositions: a kernelization lower bound for a Ramsey-type problem, in *Proceedings of the Twenty-Third Annual ACM–SIAM Symposium on Discrete Algorithms, SODA 2012*, Kyoto, Japan, January 17–19, 2012, ed. by Y. Rabani (SIAM, Philadelphia, 2012), pp. 114–122 (pp. 597, 598, 600, 602, 619)

469. S. Kratsch, M. Philipczuk, A. Rai, V. Raman, Kernel lower bounds using co-nondeterminism: finding induced hereditary subgraphs, in *Algorithm Theory—SWAT 2012* (Springer, Berlin, 2012), pp. 364–375 (pp. 602, 603, 604, 618)

470. I. Kriz, R. Thomas, The Menger-like property of the tree-width of infinite graphs. J. Comb. Theory, Ser. B **52**(1), 86–91 (1991) (p. 343)

471. J. Kruskal, Well-quasi-ordering, the tree theorem, and Vazsonyi's conjecture. Trans. Am. Math. Soc. **95**(2), 210–225 (1960) (pp. 330, 331, 338)

472. J. Kruskal, The theory of well-quasi-ordering: a frequently rediscovered concept. J. Comb. Theory, Ser. A **13**, 297–305 (1972) (p. 337)

473. C. Kuratowski, Sur le probleme des courbes gauches en topologie. Fundam. Math. **15**, 271–283 (1930) (pp. 322, 326, 338)

474. V. Lacroix, C. Fernandes, M.-F. Sagot, Motif search in graphs: application to metabolic networks. IEEE/ACM Trans. Comput. Biol. Bioinform. **3**(4), 360–368 (2006) (p. 83)

475. J. Lagergen, Algorithms and minimal forbidden minors for tree-decomposable graphs, PhD thesis, Department of Numerical Analysis and Computing Sciences, Royal Institute of Technology, Stockholm, Sweden, March 1991 (pp. 203, 364)

476. M. Lampis, Algorithmic meta-theorems for restrictions of treewidth, in *Algorithms—ESA 2010: Proceedings of 18th Annual European Symposium*, Liverpool, United Kingdom, September 6–8, 2010, ed. by M. de Berg, U. Meyer. LNCS, vol. 6347 (Springer, Berlin, 2010), pp. 549–560 (p. 288)

477. A. Langer, F. Reidl, P. Rossmanith, S. Sikdar, Evaluation of an MSO-solver, in *Proceedings of the Fourteenth Workshop on Algorithm Engineering and Experiments, ALENEX 2012*, The Westin Miyako, Kyoto, Japan, January 16, 2012, ed. by D. Bader, P. Mutzel (SIAM/Omnipress, Philadelphia, 2012), pp. 55–63 (p. 272)

478. A. Langer, P. Rossmanith, S. Sikdar, Linear-time algorithms for graphs of bounded rankwidth: a fresh look using game theory, in *CoRR* (Computing Research Repository, Berlin, submitted). arXiv:1102.0908 (pp. 272, 310)

479. M. Langston, Fixed parameter tractability, a prehistory. A Festschrift contribution devoted to Michael R. Fellows on the occasion of his 60th birthday, in *The Multivariate Algorithmic Revolution and Beyond: Essays Dedicated to Michael R. Fellows on the Occasion of His 60th Birthday*, ed. by H. Bodlaender, R. Downey, F. Fomin, D. Marx. LNCS, vol. 7370 (Springer, Berlin, 2012), pp. 3–16 (pp. xvi, 21, 300, 371)

480. M. Langston, A. Perkins, A. Saxton, J. Scharff, B. Voy, Innovative computational methods for transcriptomic data analysis: a case study in the use of FPT for practical algorithm design and implementation. Comput. J. **51**(1), 26–38 (2008) (pp. 27, 61, 88)

481. D. Lapoire, Recognizability equals monadic second-order definability, for sets of graphs of bounded treewidth, in *Proceedings of 15th Annual Symposium on Theoretical Aspects on Computer Science, STACS 98*, Paris, France, February 1998, ed. by M. Morvan, C. Meinel, D. Krob. LNCS, vol. 1373 (Springer, Berlin, 1998), pp. 618–628 (p. 277)

482. S. Lauritzen, D. Spiegelhalter, Local computations with probabilities on graphical structures and their applications to expert systems. J. R. Stat. Soc., Ser. B, Stat. Methodol. **50**(2), 157–224 (1988) (pp. 185, 189)

483. E. Leggett, D. Moore, Optimization problems and the polynomial time hierarchy. Theor. Comput. Sci. **15**, 279–289 (1981) (p. 173)

484. E. Leiss, The complexity of restricted regular expressions and the synthesis problem for finite automata. J. Comput. Syst. Sci. **23**(3), 348–354 (1981) (p. 226)

485. P. Lemke, The maximum leaf spanning tree problem for cubic graphs is NP-complete, IMA Preprint Series 428, Institute for Mathematics and its Applications, 1988 (p. 65)

486. A. Lempel, S. Even, I. Cederbaum, An algorithm for planarity testing of graphs, in *Theory of Graphs, International Symposium*, Rome, 1966, ed. by P. Rosenstiehl (Gordon and Breach, New York, 1966), pp. 215–232 (p. 92)

487. H. Lenstra, Integer programming with a fixed number of variables. Math. Oper. Res. **8**, 538–548 (1983) (pp. 137, 141)

488. L. Levin, Universal sorting problems. Probl. Inf. Transm. **9**, 265–266 (1973). English translation (pp. 4, 13, 358, 382)

489. L. Libkin, *Elements of Finite Model Theory*. Texts in Theoretical Computer Science. An EATCS Series (Springer, Berlin, 2004) (pp. 182, 286, 289)

490. D. Lichtenstein, Planar formulae and their uses. SIAM J. Comput. **11**(2), 329–343 (1982) (p. 487)

491. O. Lichtenstein, A. Pnueli, Checking that finite state concurrent programs satisfy their linear specification, in *Conference Record of the Twelfth Annual ACM Symposium on Principles of Programming Languages*, New Orleans, Louisiana, USA, January 1985, ed. by M.V. Deusen, Z. Galil, B. Reid (ACM, New York, 1985), pp. 97–107 (pp. 21, 281, 289)

492. B. Lin, Y. Chen, The parameterized complexity of k-edge induced subgraphs, in *Proceedings of 39th International Colloquium on Automata, Languages and Programming (ICALP 2012), Part I*, Warick, UK, July 9–13, 2012, ed. by A. Czumaj, K. Mehlhorn, A. Pitts, R. Wattenhofer. LNCS, vol. 7391 (Springer, Berlin, 2012), pp. 641–652 (p. 671)

493. R. Lipton, R. Tarjan, Applications of a planar separator theorem, in *Proceedings of 18th Annual Symposium on Foundations of Computer Science, FOCS 1977*, Providence, Rhode Island, USA, 31 October–1 November 1977 (IEEE Comput. Soc., Los Alamitos, 1977), pp. 162–170 (p. 175)

494. D. Lokshtanov, D. Marx, S. Saurabh, Known algorithms on graphs of bounded treewidth are probably optimal, in *Proceedings of the Twenty-Second Annual ACM–SIAM Symposium on Discrete Algorithms, SODA 2011*, San Francisco, California, USA, January 23–25, 2011, ed. by D. Randall (SIAM, Philadelphia, 2011), pp. 777–789 (pp. 305, 559)

495. D. Lokshtanov, D. Marx, S. Saurabh, Lower bounds based on the exponential time hypothesis. Bull. Eur. Assoc. Theor. Comput. Sci. **84**, 41–71 (2011) (p. 566)

496. D. Lokshtanov, D. Marx, S. Saurabh, Slightly superexponential parameterized problems, in *Proceedings of the Twenty-Second Annual ACM–SIAM Symposium on Discrete Algorithms, SODA 2011*, San Francisco, California, USA, January 23–25, 2011, ed. by D. Randall (SIAM, Philadelphia, 2011), pp. 760–776 (pp. 305, 548, 549, 550, 551, 558, 566, 570)

497. D. Lokshtanov, N.S. Narayanaswamy, V. Raman, M.S. Ramanujan, S. Saurabh, Faster parameterized algorithms using linear programming, arXiv:1203.0833 [cs.DS] (2012) (pp. 85, 87, 107, 109, 130)

498. A. Lopez, H.-F. Law, A dense gate matrix layout method for MOS VLSI. IEEE Trans. Electron Devices **ED-27**, 1671–1675 (1980) (p. 356)

499. L. Lovász, *Combinatorial Problems and Exercises* (North-Holland, Amsterdam, 1979) (p. 35)

500. L. Lovász, Matroid matching and some applications. J. Comb. Theory, Ser. B **28**, 208–236 (1980) (p. 704)

501. L. Lovász, The matroid matching problem, in *Algebraic Methods in Graph Theory, Vol. II (Colloquium Szeged, 1978)*, ed. by L. Lovász, V. Sós. Colloquia Mathematica Societatis János Bolyai, vol. 25 (North-Holland, Amsterdam, 1981), pp. 495–517 (p. 704)

502. E. Luks, Isomorphism of graphs of bounded valence can be tested in polynomial time. J. Comput. Syst. Sci. **25**, 42–65 (1982) (pp. 38, 680)

503. W. Mader, Existenz *n*-Fach zusammenhängender Teilgraphen in Graphen genügend großer Kantendichte. Abh. Math. Semin. Univ. Hamb. **37**, 86–97 (1972) (p. 291)

504. M. Mahajan, V. Raman, Parameterizing above guaranteed values: MaxSat and MaxCut. J. Algorithms **31**(2), 335–354 (1999) (p. 85)

505. M. Mahajan, V. Raman, S. Sikdar, Parameterizing above or below guaranteed values. J. Comput. Syst. Sci. **75**(2), 137–153 (2009) (p. 162)

506. F. Makedon, I. Sudborough, On minimizing width in linear layouts, in *Proceedings of 10th International Colloquium on Automata, Languages and Programming (ICALP 1983)*, Barcelona, Spain, July 18–22, 1983, ed. by J. Díaz. LNCS, vol. 154 (Springer, Berlin, 1983), pp. 478–490 (p. 352)

507. J. Makowsky, Algorithmic uses of the Feferman–Vaught theorem. Ann. Pure Appl. Log. **126**, 159–213 (2004) (p. 276)

508. J. Makowsky, Coloured Tutte polynomials and Kauffman brackets for graphs of bounded tree width. Discrete Appl. Math. **145**(2), 276–290 (2005) (p. 283)

509. J. Makowsky, From a zoo to a zoology: towards a general theory of graph polynomials. Theory Comput. Syst. **43**, 542–562 (2008) (p. 283)

510. J. Makowsky, J. Mariño, Farrell polynomials on graphs of bounded tree width. Adv. Appl. Math. **30**(1–2), 160–176 (2003) (p. 283)

511. J. Makowsky, J. Mariño, The parametrized complexity of knot polynomials. J. Comput. Syst. Sci. **67**(4), 742–756 (2003) (p. 283)

512. J. Makowsky, U. Rotics, I. Averbouch, B. Godlin, Computing graph polynomials on graphs of bounded clique-width, in *Graph-Theoretic Concepts in Computer Science: 32nd International Workshop, WG 2006, Revised Papers*, Bergen, Norway, June 22–24, 2006, ed. by H. Bodlaender, et al. LNCS, vol. 4271 (Springer, Berlin, 2006), pp. 191–204 (p. 283)

513. D. Marx, Parameterized graph separation problems, in *Parameterized and Exact Computation, Proceedings of First International Workshop, IWPEC '04*, Bergen, Norway, Septem-

ber 2004, ed. by R. Downey, M. Fellows, F. Dehne. LNCS, vol. 3162 (Springer, Berlin, 2004), pp. 71–82 (p. 404)

514. D. Marx, Parameterized complexity and approximation algorithms. Comput. J. **51**(1), 60–78 (2008) (pp. 623, 684)

515. D. Marx, Searching the k-change neighborhood for TSP is W[1]-hard. Oper. Res. Lett. **36**(1), 31–36 (2008) (p. 684)

516. D. Marx, Completely inapproximable monotone and antimonotone parameterized problems, in *Proceedings of the 25th Annual IEEE Conference on Computational Complexity, CCC 2010*, Cambridge, Massachusetts, June 9–12, 2010 (IEEE Comput. Soc., Los Alamitos, 2010), pp. 181–187 (pp. 143, 640, 641, 643)

517. D. Marx, D. Lokstanov, Clustering with local restrictions. Inf. Comput. **222**, 278–292 (2013) (p. 672)

518. L. Mathieson, E. Prieto, P. Shaw, Packing edge disjoint triangles: a parameterized view, in *Parameterized and Exact Computation, Proceedings of First International Workshop, IWPEC '04*, Bergen, Norway, September 2004, ed. by R. Downey, M. Fellows, F. Dehne. LNCS, vol. 3162 (Springer, Berlin, 2004), pp. 127–137 (pp. 131, 139, 141)

519. J. Matoušek, R. Thomas, Algorithms finding tree-decompositions of graphs. J. Algorithms **12**, 1–22 (1991) (p. 203)

520. F. Mazoit, S. Thomassé, Branchwidth of graphic matroids, in *Surveys in Combinatorics*, ed. by A. Hilton, J. Talbot. London Mathematical Society Lecture Note Series, vol. 346 (Cambridge University Press, Cambridge, 2007), pp. 275–286 (p. 282)

521. C. McCartin, An improved algorithm for the jump number problem. Inf. Process. Lett. **79**(2), 87–92 (2001) (pp. 39, 46)

522. C. McCartin, Contributions to parameterized complexity, PhD thesis, Victoria University, Wellington, 2002 (pp. 648, 650, 657, 670, 671)

523. C. McCartin, Parameterized counting problems. Ann. Pure Appl. Log. **138**(1–3), 147–182 (2006) (pp. 411, 648, 671)

524. W. McCulloch, W. Pitts, A logical calculus of the ideas immanent in nervous activity. Bull. Math. Biophys. **5**, 115–133 (1943) (p. 216)

525. R. McNaughton, H. Yamada, Regular expressions and state graphs for automata. IEEE Trans. Electron. Comput. **ED-9**(1), 39–47 (1960) (pp. 220, 226)

526. G. Mealy, A method for synthesising sequential circuits. Bell Syst. Tech. J. **34**(5), 1045–1079 (1955) (p. 216)

527. K. Mehlhorn, Data structures and efficient algorithms 2, in *Graph Algorithms and NP-Completeness*, ed. by W. Brauer, G. Rozenberg, A. Salomaa. EATCS Monographs on Theoretical Computer Science (Springer, Berlin, 1984) (p. 25)

528. A. Meier, J. Schmidt, M. Thomas, H. Vollmer, On the parameterized complexity of default logic and autoepistemic logic, in *Language and Automata Theory and Applications, Proceedings of 6th International Conference, LATA 2012*, A Coruña, Spain, March 5–9, 2012, ed. by A.-H. Dediu, C. Martín-Vide. LNCS, vol. 7183 (Springer, Berlin, 2012), pp. 389–400 (p. 277)

529. K. Menger, Zur allgemeinen Kurventheorie. Fundam. Math. **10**, 96–115 (1927) (pp. 705, 706)

530. J. Mezei, J. Wright, Algebraic automata and context free sets. Inf. Control **11**, 3–29 (1967) (pp. 237, 246)

531. S. Micali, V. Vazirani, An $O(\sqrt{|v|}|E|)$ algorithm for finding maximum matching in general graphs, in *Proceedings of 21st Annual Symposium on Foundations of Computer Science, FOCS 1980*, Syracuse, New York, USA, 13–15 October 1980 (IEEE Comput. Soc., Los Alamitos, 1980), pp. 17–27 (pp. 101, 699)

532. N. Misra, V. Raman, S. Saurabh, S. Sikdar, The budgeted unique coverage problem and color-coding, in *Computer Science—Theory and Applications: Proceedings of 4th International Computer Science Symposium, CSR 2009*, Novosibirsk, Russia, August, 2009, ed. by A. Frid, A. Morozov, A. Rybalchenko, K. Wagner. LNCS, vol. 5675 (Springer, Berlin, 2009), pp. 310–321 (p. 168)

533. B. Mohar, Embedding graphs in an arbitrary surface in linear time, in *Proceedings of 28th ACM Symposium on Theory of Computing (STOC '96)*, Philadelphia, Pennsylvania, USA, May 22–May 24, 1996, ed. by G. Miller (ACM, New York, 1996), pp. 392–397 (pp. 18, 330)

534. B. Monien, How to find long paths efficiently. Ann. Discrete Math. **25**, 239–254 (1985) (p. 296)

535. A. Montoya, The parameterized complexity of probability amplification. Inf. Process. Lett. **109**(1), 46–53 (2008) (pp. 672, 673)

536. J.A. Montoya, M. Müller, Parameterized random complexity. Theory Comput. Syst. **52**(2), 221–270 (2013) (pp. 670, 672, 673)

537. E. Moore, Gedanken experiments on sequential machines, in *Automata Studies*, ed. by C. Shannon, J. McCarthy. Annals of Mathematics Studies, vol. 34 (Princeton University Press, Princeton, 1956), pp. 129–153 (pp. 216, 234)

538. M. Müller, Valiant–Vazirani lemmata for various logics, Technical Report 63, Electronic Colloquium on Computational Complexity, 2008 (p. 669)

539. J. Myhill, Finite automata and representation of events, WADD TR-57-624, Wright-Patterson AFB, Ohio, 1957, pp. 112–137 (pp. 227, 234)

540. A. Naik, K. Regan, D. Sivakumar, Quasilinear time complexity theory, in *Proceedings of 11th Annual Symposium on Theoretical Aspects on Computer Science, STACS 94*, Caen, France, February 1994, ed. by P. Enjalbert, E. Mayr, K. Wagner. LNCS, vol. 775 (Springer, Berlin, 1994), pp. 97–108 (p. 405)

541. M. Naor, L. Schulman, A. Srinivasan, Splitters and near-optimal derandomization, in *Proceedings of 36th Annual Symposium on Foundations of Computer Science, FOCS 1995*, Milwaukee, Wisconsin, 23–25 October 1995 (IEEE Comput. Soc., Los Alamitos, 1995), pp. 182–191 (p. 145)

542. C. Nash-Williams, On well-quasi-ordering finite trees. Math. Proc. Camb. Philos. Soc. **59**(4), 833–835 (1963) (pp. 337, 338, 347)

543. C. Nash-Williams, On well-quasi-ordering infinite trees. Math. Proc. Camb. Philos. Soc. **61**(3), 697–720 (1965) (p. 338)

544. G. Nemhauser, L. Trotter, Vertex packings: structural properties and algorithms. Math. Program. **8**, 232–248 (1975) (pp. 55, 88)

545. A. Nerode, Linear automaton transformations. Proc. Am. Math. Soc. **9**, 541–544 (1958) (pp. 227, 234)

546. R. Niedermeier, *Invitation to Fixed-Parameter Algorithms*. Oxford Lecture Series in Mathematics and Its Applications, vol. 31 (Oxford University Press, Oxford, 2006) (pp. viii, xi, 46, 60, 83, 140, 615)

547. R. Niedermeier, P. Rossmanith, A general method to speed up fixed-parameter-tractable algorithms. Inf. Process. Lett. **73**(3–4), 125–129 (2000) (pp. 60, 113)

548. S. Noble, Evaluation of the Tutte polynomial for graphs of bounded tree-width. Comb. Probab. Comput. **7**, 307–321 (1998) (pp. 282, 646)

549. J. Obdržálek, Algorithmic analysis of parity games, PhD thesis, University of Edinburgh, 2007 (p. 312)

550. J. Orlin, A fast, simpler algorithm for the matroid parity problem, in *IPCO'08, Proceedings of the 13th International Conference on Integer Programming and Combinatorial Optimization* (Springer, Berlin, 2008), pp. 240–258 (p. 704)

551. S. Oum, P. Seymour, Approximating clique-width and branch-width. J. Comb. Theory, Ser. B **96**, 514–528 (2006) (pp. 310, 680)

552. C. Papadimitriou, M. Yannakakis, Optimization, approximation, and complexity classes. J. Comput. Syst. Sci. **43**, 425–440 (1991) (pp. 176, 177, 178, 179, 181, 182)

553. C. Papadimitriou, M. Yannakakis, On limited nondeterminism and the complexity of the VC-dimension. J. Comput. Syst. Sci. **53**, 161–170 (1996) (pp. 26, 144, 412, 446)

554. C. Papadimitriou, M. Yannakakis, On the complexity of database queries, in *PODS '97, Proceedings of the Sixteenth ACM SIGACT–SIGMOD–SIGART Symposium on Princi-*

ples of Database Systems, ed. by A. Mendelzon, M. Özsoyoglu (ACM, New York, 1997), pp. 12–19 (p. 400)

555. M. Pătraşcu, R. Williams, On the possibility of faster SAT algorithms, in *Proceedings of the Twenty-First Annual ACM–SIAM Symposium on Discrete Algorithms, SODA 2010*, Austin, Texas, USA, January 17–19, 2010, ed. by M. Charikar (SIAM, Philadelphia, 2010), pp. 1065–1075 (pp. 559, 560, 561)

556. A. Pavan, A. Selman, S. Sengupta, V. Variyam, Polylogarithmic-round interactive proofs for coNP collapse the exponential hierarchy. Theor. Comput. Sci. **385**(1–3), 167–178 (2007) (p. 596)

557. A. Paz, S. Moran, Nondeterministic polynomial optimization problems and their approximations. Theor. Comput. Sci. **15**, 251–277 (1981) (pp. 444, 522)

558. L. Perković, B. Reed, An improved algorithm for finding tree decompositions of small width. Int. J. Found. Comput. Sci. **11**(3), 365–371 (2000) (pp. 203, 205)

559. G. Philip, V. Raman, S. Sikdar, Solving dominating set in larger classes of graphs: FPT algorithms and polynomial kernels, in *Algorithms—ESA 2009: Proceedings of 17th Annual European Symposium*, Copenhagen, Denmark, September 7–9, 2009, ed. by A. Fiat, P. Sanders. LNCS, vol. 5757 (Springer, Berlin, 2009), pp. 694–705 (pp. 314, 315)

560. G. Philip, V. Raman, S. Sikdar, Polynomial kernels for dominating sets in graphs of bounded degeneracy and beyond. ACM Trans. Algorithms **9**(1), 11 (2012) (pp. 314, 315)

561. N. Pippenger, On simultaneous resource bounds, in *Proceedings of 20th Annual Symposium on Foundations of Computer Science, FOCS 1979*, San Juan, Puerto Rico, 29–31 October 1979 (IEEE Comput. Soc., Los Alamitos, 1979), pp. 307–311 (p. 8)

562. J. Plehn, B. Voigt, Finding minimally weighted subgraphs, in *Graph-Theoretic Concepts in Computer Science: 16th International Workshop, WG 1990, Revised Papers*, Berlin, Germany, June 1990, ed. by R. Möhring. LNCS, vol. 484 (Springer, Berlin, 1990), pp. 18–29 (pp. 295, 299, 300)

563. J. Power, Four NP-complete embedding problems, Logic Paper 29, Monash University, January 1981 (p. 403)

564. E. Prieto, Systematic kernelization in FPT algorithm design, PhD thesis, School of Electrical Engineering and Computer Science, University of Newcastle, Australia, 2004 (pp. 131, 139, 141)

565. E. Prieto, C. Sloper, Looking at the stars, in *Parameterized and Exact Computation, Proceedings of First International Workshop, IWPEC '04*, Bergen, Norway, September 2004, ed. by R. Downey, M. Fellows, F. Dehne. LNCS, vol. 3162 (Springer, Berlin, 2004), pp. 138–148 (pp. 131, 132, 139, 141)

566. Proceedings of 19th Annual Symposium on Foundations of Computer Science, SFCS 1978, Ann Arbor, Michigan, USA, 16–18 October 1978 (IEEE Comput. Soc., Los Alamitos, 1978) (p. 710)

567. W. Pulleyblank, On minimizing setups in precedence constrained scheduling, Technical Report 81185-OR, Institut für Ökonometrie und Operations Research, Universität Bonn, West Germany, 1981. To appear in Discrete Appl. Math. (p. 38)

568. M. Rabin, Decidability of second-order theories, and automata on infinite trees. Trans. Am. Math. Soc. **141**, 1–35 (1969) (p. 276)

569. M. Rabin, D. Scott, Finite automata and their decision problems. IBM J. Res. Dev. **3**, 114–125 (1959) (pp. 217, 220, 221)

570. R. Rado, Partial well-ordering of sets of vectors. Mathematika **1**, 88–95 (1954) (p. 337)

571. V. Raman, S. Saurabh, Parameterized algorithms for feedback set problems and their duals in tournaments. Theor. Comput. Sci. **351**(3), 446–458 (2006) (p. 170)

572. F.P. Ramsey, On a problem of formal logic. Proc. Lond. Math. Soc. **30**, 264–286 (1930) (p. 320)

573. R. Raz, A parallel repetition theorem. SIAM J. Comput. **27**(3), 763–803 (1998). Earlier version in STOC '95 (p. 584)

574. I. Razgon, B. O'Sullivan, Almost 2-SAT is fixed-parameter tractable. J. Comput. Syst. Sci. **75**(8), 435–450 (2009) (p. 130)

575. B. Reed, Finding approximate separators and computing tree width quickly, in *Proceedings of 24th ACM Symposium on Theory of Computing (STOC '92)*, Victoria, British Columbia, Canada, May 4–May 6, 1992, ed. by R. Kosaraju, M. Fellows, A. Wigderson, J. Ellis (ACM, New York, 1992), pp. 221–228 (pp. 203, 634, 643)

576. B. Reed, K. Smith, A. Vetta, Finding odd cycle transversals. Oper. Res. Lett. **32**(4), 299–301 (2004) (p. 129)

577. K. Regan, Finite substructure languages, in *Proceedings of Fourth Annual Structure in Complexity Conference*, University of Oregon, Eugene, Oregon, June 19–22, 1989 (IEEE Comput. Soc., Los Alamitos, 1989), pp. 87–96 (pp. 20, 405, 567, 568)

578. K.B. Reid, On sets of arcs containing no cycles in tournaments. Can. Math. Bull. **12**, 261–264 (1969) (p. 170)

579. H. Rice, Classes of recursively enumerable sets and their decision problems. Trans. Am. Math. Soc. **74**, 358–366 (1953) (p. 327)

580. N. Robertson, D. Sanders, P. Seymour, R. Thomas, Efficiently four coloring planar graphs, in *Proceedings of 28th ACM Symposium on Theory of Computing (STOC '96)*, Philadelphia, Pennsylvania, USA, May 22–May 24, 1996 (ACM, New York, 1996), pp. 571–575 (p. 29)

581. N. Robertson, P. Seymour, Graph minors. I. Excluding a forest. J. Comb. Theory, Ser. B **35**, 39–61 (1983) (pp. 350, 632)

582. N. Robertson, P. Seymour, Graph minors—a survey, in *Surveys in Combinatorics 1985 (Glasgow, 1985)*. London Mathematical Society Lecture Note Series, vol. 103 (Cambridge University Press, Cambridge, 1985), pp. 153–171 (p. 185)

583. N. Robertson, P. Seymour, Graph minors. II. Algorithmic aspects of tree-width. J. Algorithms **7**, 309–322 (1986) (p. 347)

584. N. Robertson, P. Seymour, Graph minors. V. Excluding a planar graph. J. Comb. Theory, Ser. B **41**, 92–114 (1986) (pp. 274, 350)

585. N. Robertson, P. Seymour, Graph minors. IV. Tree-width and well quasi-ordering. J. Comb. Theory, Ser. B **48**(2), 227–254 (1990) (pp. 342, 343)

586. N. Robertson, P. Seymour, Graph minors. X. Obstructions to tree-decomposition. J. Comb. Theory, Ser. B **52**(2), 153–190 (1991) (pp. 302, 315)

587. N. Robertson, P. Seymour, Graph minors. XXII. Irrelevant vertices in linkage problems, Preprint, 1992 (p. 365)

588. N. Robertson, P. Seymour, Graph minors. XIII. The disjoint paths problem. J. Comb. Theory, Ser. B **63**(1), 65–110 (1995) (pp. 18, 185, 202, 327, 348, 364, 365, 371)

589. N. Robertson, P. Seymour, Graph minors. XX. Wagner's conjecture. J. Comb. Theory, Ser. B **92**(2), 325–357 (2004) (pp. 18, 185, 195, 327, 346)

590. N. Robertson, P. Seymour, Graph minors. XXI. Graphs with unique linkages. J. Comb. Theory, Ser. B **99**(3), 583–616 (2009) (p. 365)

591. N. Robertson, P. Seymour, R. Thomas, Quickly excluding a planar graph. J. Comb. Theory, Ser. B **62**(2), 323–348 (1994) (p. 562)

592. E. Rodrigues, M.-F. Sagot, Y. Wakabayashi, Some approximation results for the maximum agreement forest problem, in *Approximation, Randomization, and Combinatorial Optimization: Algorithms and Techniques, Proceedings of 4th International Workshop on Approximation, Algorithms for Combinatorial Optimization Problems, APPROX 2001 and 5th International Workshop on Randomization and Approximation Techniques in Computer Science, RANDOM 2001*, Berkeley, California, USA, August 2001, ed. by M. Goemans, K. Jansen, J. Rolim, L. Trevison. LNCS, vol. 2129 (Springer, Berlin, 2001), pp. 159–169 (p. 71)

593. D. Rose, R. Tarjan, Algorithmic aspects of vertex elimination on directed graphs. SIAM J. Appl. Math. **34**(1), 176–197 (1978) (p. 312)

594. D. Rose, R. Tarjan, G. Lueker, Algorithmic aspects of vertex elimination on graphs. SIAM J. Comput. **5**, 266–283 (1976) (pp. 188, 208)

595. B. Rosser, L. Schoenfeld, Sharper bounds for the Chebyshev functions $\theta(x)$ and $\psi(x)$. Math. Comput. **29**(129), 243–269 (1975) (p. 146)

596. J. Rué, I. Sau, D. Thilikos, Dynamic programming for graphs on surfaces, in *Proceedings of 37th International Colloquium on Automata, Languages and Programming (ICALP 2010)*, Bordeaux, France, July 6–10, 2010, ed. by S. Abramsky, C. Gavoille, C. Kirchner, F.M. auf der Heide, P. Spirakis. LNCS, vol. 6198 (Springer, Berlin, 2010), pp. 372–383 (p. 306)

597. H. Sachs, On spatial representation of finite graphs, in *Finite and Infinite Sets*, ed. by A. Hajnal, L. Loász, V. Sós. Colloquia Mathematica Societatis János Bolyai, vol. 37 (North-Holland, Amsterdam, 1984), pp. 649–662 (p. 17)

598. A. Sahai, S.P. Vadhan, A complete problem for statistical zero knowledge. J. ACM **50**(2), 196–249 (2003) (p. 583)

599. S. Santagata, T. Boggon, C. Baird, C. Gomez, J. Zhao, W. Shan, D. Mysczka, L. Shapiro, G-protein signalling through tubby proteins. Science **292**, 2041–2050 (2001) (p. 62)

600. N. Santoro, R. Khatib, Labeling and implicit routing in networks. Comput. J. **28**(1), 5–8 (1985) (p. 355)

601. I. Sau, D. Thilikos, Subexponential parameterized algorithms for degree-constrained subgraph problems on planar graphs. J. Discrete Algorithms **8**(3), 330–338 (2010) (p. 565)

602. T. Schaefer, Complexity of some two-person perfect information games. J. Comput. Syst. Sci. **16**(2), 185–225 (1978) (pp. 492, 497)

603. J.P. Schmidt, A. Siegel, The spatial complexity of oblivious k-probe hash functions. SIAM J. Comput. **19**(5), 775–786 (1990) (p. 145)

604. C.P. Schnorr, Optimal algorithms for self-reducible problems, in *Proceedings of 3rd International Colloquium on Automata, Languages and Programming (ICALP 1976)*, Edinburgh, UK, July 20–23, 1976, ed. by S. Michaelson, R. Milner. Lecture Notes in Computer Science (Edinburgh University Press, Edinburgh, 1976), pp. 322–337 (p. 358)

605. U. Schöning, Graph isomorphism is in the low hierarchy, in *Proceedings of 4th Annual Symposium on Theoretical Aspects on Computer Science, STACS 87*, Passau, Germany, February 19–21, 1987, ed. by F. Brandenburg, G. Vidal-Naquet, M. Wirsing. LNCS, vol. 247 (Springer, Berlin, 1987), pp. 114–124 (p. 38)

606. J. Schwartz, Fast probabilistic algorithms for verification of polynomial identities. J. ACM **27**(4), 701–717 (1980) (p. 152)

607. A. Scott, On the parameterized complexity of finding short winning strategies in combinatorial games, PhD thesis, University of Victoria, 2009 (p. 683)

608. A. Scott, U. Stege, Parameterized chess, in *Parameterized and Exact Computation, Proceedings of Third International Workshop, IWPEC '08*, Victoria, Canada, May, 2008, ed. by M. Grohe, R. Niedermeier. LNCS, vol. 5018 (Springer, Berlin, 2008), pp. 172–189 (p. 505)

609. A. Scott, U. Stege, Parameterized pursuit-evasion games. Theor. Comput. Sci. **411**(43), 3845–3858 (2010) (p. 505)

610. J. Scott, T. Ideker, R. Karp, R. Sharan, Efficient algorithms for detecting signaling pathways in protein interaction networks. J. Comput. Biol. **13**(2), 133–144 (2006) (p. 169)

611. D. Seese, Entscheidbarkeits- und Interpretierbarkeitsfragen monadischer Theorien zweiter Stufe gewisser Klassen von Graphen, PhD thesis, Humboldt-Universitat, Berlin, 1976 (pp. 274, 275, 276)

612. D. Seese, The structure of models of decidable monadic theories of graphs. Ann. Pure Appl. Log. **53**(2), 169–195 (1991) (pp. 266, 272, 273, 275, 277, 329)

613. C. Semple, M. Steel, *Phylogenetics* (Oxford University Press, Oxford, 2003) (p. 70)

614. S. Seshu, M.B. Reed, *Linear Graphs and Electrical Networks* (Addison-Wesley, Reading, 1961) (p. 170)

615. P. Seymour, R. Thomas, Call routing and the ratcatcher. Combinatorica **14**(2), 217–241 (1994) (pp. 302, 305, 562)

616. S. Shelah, The monadic theory of order. Ann. Math. (2) **102**(3), 379–419 (1975) (p. 276)

617. T. Shlomi, D. Segal, E. Ruppin, R. Sharan, Qpath: a method for querying pathways in a protein–protein interaction network. BMC Bioinform. **7**, 199 (2006) (p. 169)

618. A. Silberschatz, P. Galvin, *Operating System Concepts* (Addison-Wesley, Reading, 1994) (p. 33)

619. S. Simpson, *Subsystems of Second Order Arithmetic*, 2nd edn. Perspectives in Logic (Cambridge University Press, Cambridge, 2009) (p. 350)

620. M. Sipser, A complexity theoretic approach to randomness, in *Proceedings of 15th ACM Symposium on Theory of Computing (STOC '83)*, Boston, Massachusetts, USA, May 25–May 27, 1983, ed. by D. Johnson, et al. (ACM, New York, 1983), pp. 330–335 (pp. 10, 527)

621. D. Sleator, R. Tarjan, Amortized efficiency of list update and paging rules. Commun. ACM **28**, 202–208 (1985) (p. 635)

622. D. Sleator, R. Tarjan, W. Thurston, Rotation distance, triangulations and hyperbolic geometry. J. Am. Math. Soc. **1**(3), 647–681 (1988) (p. 46)

623. A. Slivkins, Parameterized tractability of edge-disjoint paths on directed acyclic graphs, in *Algorithms—ESA 2003: Proceedings of 11th Annual European Symposium*, Budapest, Hungary, September 2003, ed. by G.D. Battista, U. Zwick. LNCS, vol. 2832 (Springer, Berlin, 2003), pp. 483–493 (pp. 628, 642)

624. C. Sloper, J.A. Telle, Techniques for designing parameterized algorithms. Comput. J. **51**(1), 122–136 (2008) (pp. 102, 140)

625. C. Slot, P. van Emde Boas, On tape versus core; an application of space efficient perfect hash functions to the invariance of space. Elektron. Inf.verarb. Kybern. **21**(4/5), 246–253 (1985) (p. 145)

626. J. Spencer, Optimal ranking of tournaments. Networks **1**, 135–138 (1971) (p. 170)

627. J. Spencer, Optimal ranking of unrankable tournaments. Period. Math. Hung. **11**(2), 131–144 (1980) (p. 170)

628. U. Stege, The impact of parameterized complexity to interdisciplinary problem solving, in *The Multivariate Algorithmic Revolution and Beyond: Essays Dedicated to Michael R. Fellows on the Occasion of His 60th Birthday*, ed. by H. Bodlaender, R. Downey, F. Fomin, D. Marx. LNCS, vol. 7370 (Springer, Berlin, 2012), pp. 56–68 (p. 88)

629. L. Stockmeyer, A. Chandra, Provably difficult combinatorial games. SIAM J. Comput. **8**(2), 151–174 (1979) (p. 518)

630. L. Stockmeyer, D. Kozen, A. Chandra, Alternation. J. ACM **28**, 114–133 (1981) (pp. 506, 518)

631. L. Stockmeyer, A. Meyer, Word problems requiring exponential time, in *Proceedings of 5th ACM Symposium on Theory of Computing (STOC '73)*, Austin, Texas, USA, April 30–May 2, 1973, ed. by A. Aho, et al. (ACM, New York, 1973), pp. 1–9 (pp. 6, 13)

632. N. Stojanovi'c, L. Florea, C. Riemer, D. Gumucio, J. Slightom, M. Goodman, W. Miller, R. Hardison, Comparison of five methods for finding conserved sequences in multiple alignments of gene regulatory regions. Nucleic Acids Res. **27**(19), 3899–3910 (1999) (p. 36)

633. M. Sysło, Minimizing the jump number for partially-ordered sets: a graph-theoretic approach, II. Discrete Math. **63**(2–3), 279–295 (1987) (p. 40)

634. S. Szeider, The parameterized complexity of k-flip local search for SAT and MAX SAT. Discrete Optim. **8**(1), 139–145 (2011) (pp. 288, 458, 684)

635. G. Szekeres, H. Wilf, An inequality for the chromatic number of a graph. J. Comb. Theory, Ser. B **4**, 1–3 (1968) (p. 313)

636. R. Tarjan, Depth-first search and linear graph algorithms. SIAM J. Comput. **1**, 146–160 (1972) (p. 291)

637. R. Tarjan, Decomposition by clique separators. Discrete Math. **55**, 221–232 (1985) (p. 207)

638. R. Tarjan, M. Yannakakis, Simple linear time algorithms to test chordality of graphs, test acyclicity of hypergraphs, and selectively reduce acyclic hypergraphs. SIAM J. Comput. **13**(3), 566–579 (1984) (p. 46)

639. S. Tazari, Faster approximation schemes and parameterized algorithms on H-minor-free and odd-minor-free graphs, in *Mathematical Foundations of Computer Science 2010, Proceedings of 35th International Symposium, MFCS 2010*, Brno, Czech Republic, August 23–27, 2010, ed. by P. Hlinený, A. Kucera (Springer, Berlin, 2010), pp. 641–652 (p. 565)

640. J. Thatcher, J. Wright, Generalized finite automata. Not. Am. Math. Soc. **12**, 820 (1965) (p. 246)

641. D. Thilikos, Fast sub-exponential algorithms and compactness in planar graphs, in *Algorithms—ESA 2011: Proceedings of 19th Annual European Symposium*, Saarbrücken, Germany, September 5–9, 2011. LNCS (Springer, Berlin, 2011), pp. 358–369 (p. 565)

642. R. Thomas, Well-quasi-ordering infinite graphs with forbidden finite planar minor. Trans. Am. Math. Soc. **312**(1), 279–313 (1989) (pp. 332, 343, 344)

643. R. Thomas, A Menger-like property of tree-width, the finite case. J. Comb. Theory, Ser. B **48**(1), 67–76 (1990) (pp. 343, 344)

644. K. Thompson, Regular expression search algorithm. Commun. ACM **11**(6), 419–422 (1968) (pp. 218, 221)

645. M. Thorup, All structured programs have small tree-width and good register allocation. Inf. Comput. **142**(2), 159–181 (1998) (p. 185)

646. M. Thurley, Kernelizations for parameterized counting problems, in *Theory and Applications of Models of Computation, Proceedings of 4th International Conference, TAMC 2007*, Shanghai, China, May 22–25, 2007, ed. by J. Cai, B. Cooper, H. Zhu. LNCS, vol. 4484 (Springer, Berlin, 2007), pp. 703–714 (pp. 645, 670, 671)

647. S. Toda, PP is as hard as the polynomial-time hierarchy. SIAM J. Comput. **20**(5), 865–877 (1991) (p. 11)

648. B. Trakhtenbrot, Impossibility of an algorithm for the decision problem on finite classes. Dokl. Akad. Nauk SSSR **70**, 569–572 (1950) (p. 329)

649. L. Valiant, The complexity of computing the permanent. Theor. Comput. Sci. **8**(2), 189–201 (1979) (pp. 646, 647)

650. L. Valiant, V. Vazirani, NP is as easy as detecting unique solutions. Theor. Comput. Sci. **47**, 85–93 (1986) (pp. 11, 13)

651. F. van den Eijkhof, H. Bodlaender, A. Koster, Safe reduction rules for weighted treewidth, Technical Report ZIB-report 02-49, Konrad-Zuse-Zentrum für Inforationstechnik, Berlin, Germany, 2002 (p. 209)

652. J. van Leeuwen, Graph algorithms, in *Handbook of Theoretical Computer Science, Volume A: Algorithms and Complexity*, ed. by J. van Leeuwen (Elsevier/MIT Press, Amsterdam, 1990), pp. 525–631 (p. 190)

653. J. van Rooij, H. Bodlaender, Design by measure and conquer: a faster algorithm for dominating set, in *Proceedings of 25th Annual Symposium on Theoretical Aspects on Computer Science, STACS 2008*, Bordeau, France, February 21–23, 2008, ed. by S. Albers, P. Weil. Leibniz International Proceedings in Informatics, vol. 1 (Schloss Dagstuhl–Leibniz-Zentrum fuer Informatik, Dagstuhl, 2008), pp. 657–668 (pp. 113, 117, 130)

654. J. van Rooij, H. Bodlaender, P. Rossmanith, Dynamic programming on tree decompositions using generalised fast subset convolution, in *Algorithms—ESA 2009: Proceedings of 17th Annual European Symposium*, Copenhagen, Denmark, September 7–9, 2009, ed. by A. Fiat, P. Sanders. LNCS, vol. 5757 (Springer, Berlin, 2009), pp. 566–577 (p. 305)

655. M. Vardi, The complexity of relational query languages (extended abstract), in *Proceedings of 14th ACM Symposium on Theory of Computing (STOC '82)*, San Francisco, California, USA, May 5–May 7, 1982, ed. by H. Lewis, B. Simons, W. Burkhard, L. Landweber (ACM, New York, 1982), pp. 137–146. http://dl.acm.org/citation.cfm?id=802186 (pp. 20, 281, 288)

656. M. Vardi, P. Wolper, An automata-theoretic approach to automatic program verification (Preliminary report), in *Proceedings of the Symposium on Logic in Computer Science (LICS '86)*, Cambridge, Massachusetts, USA, June 16–18, 1986 (IEEE Comput. Soc., Los Alamitos, 1986), pp. 332–344 (p. 21)

657. A. Vardy, Algorithmic complexity in coding theory and the minimum distance problem, in *Proceedings of 29th ACM Symposium on Theory of Computing (STOC '97)*, Baltimore, MD, USA, May 4–6, 1997, ed. by F. Leighton, P. Shor (ACM, New York, 1997), pp. 92–109 (p. 678)

658. D. Vertigan, G. Whittle, Recognising polymatroids associated with hypergraphs. Comb. Probab. Comput. **2**, 519–530 (1993) (p. 682)

659. B. Voy, J. Scharff, A. Perkins, A. Saxton, B. Borate, E. Chesler, L. Branstetter, M. Langston, Extracting gene networks for low dose radiation using graph theoretical algorithms. PLoS Comput. Biol. **2**, 757–768 (2006) (p. 61)

660. K. Wagner, Über einer Eigenschaft der ebener Complexe. Math. Ann. **14**, 570–590 (1937) (pp. 326, 338)

661. K. Weihe, Covering trains by stations or the power of data reduction, in *Proceedings of Algorithms and Experiments (ALEX '98)*, Trento, Italy, February 9–11, 1998, ed. by R. Battiti, A. Bertossi (Elsevier, Amsterdam, 1998), pp. 1–8 (pp. 62, 88)

662. D. Welsh, A. Gale, The complexity of counting problems, in *Aspects of Complexity*, ed. by R. Downey, D. Hirschfeldt. De Gruyter Series in Logic and Its Applications (de Gruyter, Berlin, 2001), pp. 115–154 (p. 646)

663. S. Wernicke, On the algorithmic tractability of single nucleotide polymorphism (SNP) analysis and related problems, PhD thesis, Wilhelm-Schickard-Institute für Informatik, Universität Tübingen, 2003 (p. 112)

664. R. Whitty, Vertex-disjoint paths and edge-disjoint branchings in directed graphs. J. Graph Theory **11**, 349–358 (1987) (p. 104)

665. V. William, Breaking the Coppersmith–Winograd barrier, to appear (pp. 307, 560)

666. R. Williams, Finding paths of length k in $O^*(2^k)$ time. Inf. Process. Lett. **109**(6), 315–318 (2009) (pp. 149, 150, 151, 169)

667. A. Yamaguchi, K. Aoki, H. Mamitsuka, Graph complexity of chemical compounds in biological pathways. Genome Inform. **14**, 376–377 (2003) (p. 185)

668. M. Yannakakis, Computing the minimum fill-in is NP-complete. SIAM J. Algebr. Discrete Methods **2**, 77–79 (1981) (pp. 45, 69)

669. A.C.-C. Yao, Separating the polynomial-time hierarchy by oracles, in *Proceedings of 26th Annual Symposium on Foundations of Computer Science, FOCS 1985*, Portland, Oregon, USA, 21–23 October 1985 (IEEE Comput. Soc., Los Alamitos, 1985), pp. 1–10 (p. 569)

670. C.-K. Yap, Some consequences of non-uniform conditions on uniform classes. Theor. Comput. Sci. **26**, 287–300 (1983) (pp. 8, 9, 13)

671. R. Yuster, Combinatorial and computational aspects of graph packing and graph decomposition. Comput. Sci. Rev. **1**, 12–26 (2007) (p. 597)

672. G. Yuval, An algorithm for finding shortest paths using $N^{(2.81)}$ infinite precision multiplications. Inf. Process. Lett. **4**(6), 155–156 (1976) (p. 307)

Index

Printed in the United States
By Bookmasters